Native Crops in India

With 20 agro-climatic regions, India has the second-largest agricultural land mass. This book presents the latest scientific and technical information on indigenous and domesticated crops grown, consumed, and traded in India, forming part of its agro-economy. It covers the uses of the crops in Indian food products and/or future food developments, highlighting product developments through traditional and innovative processing and engineering and/or microbial technologies to produce nutraceutical and functional food ingredients. The health benefits of these crops, many used in Ayurvedic medicine, are also covered, particularly in regard to alleviating prevalent non-communicable diseases.

Key Features:

- Discusses increasing crop biodiversity, promoting sustainable, climate- and eco-friendly, circular agriculture
- Reviews industry standards/regulations, by-product/s recovery and use, and other value-added processes
- Presents future prospects, particularly in reference to sustainability, responsible water use, crop/food waste reduction, and benefits to the agro-economy

Food Biotechnology and Engineering

Series Editor: Octavio Paredes-López

Volatile Compounds Formation in Specialty Beverages
Edited by Felipe Richter Reis and Caroline Mongruel Eleutério dos Santos

Native Crops in Latin America
Biochemical, Processing, and Nutraceutical Aspects
Edited by Ritva Repo-Carrasco-Valencia and Mabel C. Tomás

Starch and Starchy Food Products
Improving Human Health
Edited by Luis Arturo Bello-Perez, Jose Alvarez-Ramirez, and Sushil Dhital

Bioenhancement and Fortification of Foods for a Healthy Diet
Edited by Octavio Paredes-López, Oleksandr Shevchenko, Viktor Stabnikov, and Volodymyr Ivanov

Latin-American Seeds
Agronomic, Processing and Health Aspects
Edited by Claudia Monika Haros, María Reguera, Norma Sammán, and Octavio Paredes-López

Advances in Plant Biotechnology
In Vitro Production of Secondary Metabolites of Industrial Interest
Edited by Alma Angélica Del Villar-Martínez, Juan Arturo Ragazzo-Sánchez, and Pablo Emilio Vanegas-Espinoza

Bioconversion of Wastes to Value-added Products
Edited by Olena Stabnikova, Oleksandr Shevchenko, Viktor Stabnikov, and Octavio Paredes-López

Health-Promoting Food Ingredients in Food Processing
Edited by Rosalva Mora-Escobedo, Gloria Dávila-Ortiz, Gustavo F. Gutiérrez López, and Jose De J. Berrios

Wild Edible Plants
Improving Foods Nutritional Value and Human Health through Biotechnology
Edited by Sergey Gubsky, Olena Stabnikova, Viktor Stabnikov, and Octavio Paredes-López

Native Crops in India: Biochemical, Processing, and Nutraceutical Aspects
Edited by Devina Lobine and B. Dave Oomah

For more information about this series, please visit: www.routledge.com/Food-Biotechnology-and-Engineering/book-series/CRCFOOBIOENG

Native Crops in India

Biochemical, Processing, and Nutraceutical Aspects

Edited by
Devina Lobine and B. Dave Oomah

CRC Press
Taylor & Francis Group
Boca Raton London New York

CRC Press is an imprint of the
Taylor & Francis Group, an **informa** business

Designed cover image: Shutterstock

First edition published 2026
by CRC Press
2385 NW Executive Center Drive, Suite 320, Boca Raton FL 33431

and by CRC Press
4 Park Square, Milton Park, Abingdon, Oxon, OX14 4RN

CRC Press is an imprint of Taylor & Francis Group, LLC

ISBN: 978-1-032-40285-7 (hbk)
ISBN: 978-1-032-40288-8 (pbk)
ISBN: 978-1-003-35234-1 (ebk)

DOI: 10.1201/9781003352341

Typeset in Times
by Apex CoVantage, LLC

Contents

Preface

India is expected to be the most populous country in the world, with 1.51 billion people in 2031, surpassing the People's Republic of China. Its GDP will continue to grow (average ~4% annually), with increasing per capita income by 5.3% per annum and reduced inflation (4.9 to 3.9% annually). India has a booming middle-class population (~432 million, or one in three people), with an annual household income between €6.1k and €36.8k. It is one of the youngest countries in the world, with 42.7% of the population under 25 and a median age of only 28, compared to 38 in the US and 39 in China.

Over the coming decade, India is projected to be the second-largest contributor to global crop production after China (17% and 30%, respectively). Currently, over 60% of the Indian population consists of farmers, although fewer than 4% of these farmers have adopted sustainable agricultural practices. India is a megadiverse country, housing 11.2% of the globally recorded plant species and over 2.4% of the world's surface area. The country shares 4 (Western Ghats and Sri Lanka, Himalayas, Indo-Burma, and Sundaland) of the 34 global biodiversity hotspots, and one-third of the higher plant species present in these areas are endemic. India is also one of the 12 mega biodiversity centers and ranks seventh in global agricultural biodiversity status. About 117 fruit and nut crop species are cultivated, with 175 wild relatives, of which only 25 species have been domesticated. The Western Ghats and Northeast India are centers of diversity for several important native fruits, including mango, jackfruit, and citrus. India is also home to many underutilized fruit crops.

From a diet perspective, India has a higher Alternative Healthy Eating Index (AHEI), a diet quality metric, than the global mean AHEI score (48.2 vs 40.3). This indicates a higher risk of noncommunicable diseases, particularly diabetes mellitus and cardiovascular diseases (CVD). In fact, the age-adjusted comparative prevalence of diabetes in India was 9.6% in 2021 (74.2 million people with diabetes) and is projected to increase to 10.4% by 2030 and 10.8% by 2045, according to the International Diabetes Federation. Research has indicated the inherent biological differences in lipid and glucose metabolism, inflammatory states, and genetic predisposition for the increased CVD risk among Indians. The first chapter in this book explores the antidiabetic potential of herbal (Ayurvedic) medicine to address this diabetic epidemic, with the history of native Indian plants dating back to the pre-insulin era.

About 80% of the world's population depends on plants or plant-based products for food and health care needs, as plants are rich sources of nutrition and safe medicine. We devoted a chapter to address pain management benefits of native Indian plants. Natural pain management product sales continue to grow, with the global pain management market worth \$79 million in 2021 and projected to grow to \$121 million by 2027. Approximately 50.2 million adults in the United States felt pain on most or every day, according to a 2019 National Health Interview Survey. Conventional treatments or pharmaceuticals do not always work for those experiencing pain because of their side effects. Hence, many individuals desire natural pain management. A good example is the proprietary botanical formula *Rhuleave-K*, for instance, composed of turmeric and boswellia combined with black sesame oil, that clinically provides safe, efficient pain relief comparable to popular OTC drugs. Knee pain accounts for approximately 5% of all primary care visits for adults. The most common causes of knee pain are osteoarthritis, patellofemoral pain (lifetime prevalence ~25%), and meniscal tears (620 million adults or 12% of the general population). Knee osteoarthritis, most common in older patients (≥45 years), affects an estimated 654 million people (or 23% of adults over 40 years old) worldwide. Rheumatoid arthritis (RA) is a debilitating inflammatory disease that targets the joints and erodes quality of life. About 20 million people are globally affected by this disease. Tumor necrosis factor-α inhibitors (TNFIs), although effective and safe for RA treatment, are expensive, thereby limiting access. Although biosimilars are cheaper (\$12,500 per year vs \$7000 for 4 weeks for *Humira*), they only represent 2–3% of the market, with low uptake.

The northeast hilly region of India is rich in fruits, vegetables, and flowers, especially orchids. We dedicated a chapter to orchids, their importance for this region, and the progressive research of ICAR–National Research Center for Orchids, Sikkim, to provide research support to orchid growers in India, preserve orchid germplasm, and develop sustainable use of orchid biodiversity. Valuable alternative uses of orchids are elaborated in this chapter, along with value addition to 75 Indian orchid species. Citrus fruits are also endemic to this region. Sixteen species of citrus (52 varieties and 7 natural hybrids) of Assam have been reported. *Citrus indica* is considered the most primitive citrus species and probably the progenitor of cultivated species. Diverse forms of sour orange, rough lime, and sweet lime are found in this region. Lemons are grown commercially only on a limited scale. Eureka lemon in some regions and Assam lemon in northeastern India are popular varieties under cultivation. Two chapters are dedicated to enhancing the use of citrus waste and adding value to lemon peel. The market is moving toward a greater holistic understanding of health, well-being, and ecosystem preservation. This can be achieved by environmentally sustainable and circular production and waste reduction by adding value to existing industrial by-products or harnessing new technologies.

The global functional food market is currently estimated at $189.5 billion and expected to grow to $285 billion by 2030. The food industry has been focused on health, wellness, and sustainability since the pandemic. Therefore, new start-ups are using the most innovative technologies to disrupt the food industry and create "future food." The nutraceutical industry in India is recent, established about a decade ago; so, growth opportunity is phenomenal. The sector is growing at an average rate of 12% per annum and is expected to reach US$100 billion by 2030. Natural and herbal products are the driving category in the Indian nutraceutical industry. Moringa (*Moringa oleifera*) human health benefits are extensively described in a chapter related to functional foods. Another facet of functional food is explored in another chapter dedicated to various native underutilized fruits of the Himalayas. Some of these fruits yield juice with excellent flavor and can be converted into nutritious blended beverages of high therapeutic value. The underutilized fruit species play a crucial role in mitigating nutritional insecurity and have multipurpose uses, such as ornamental, forest restoration, and social and economic plants. The underutilized fruit crops are hardy and locally adapted; they enhance sustainable farm income under the harsh arid and semi-arid environments and offer diversity in wide-ranging agroecological environments. For example, mulberry (*Morus alba*), one of the fruits discussed in the chapter, is rich in phytonutrients and used as a functional food in the forms of masala, herbal tea, marmalades, juices, yogurt, biscuits, smoothies, capsules, cosmetic oil, and dietary food products. Various pharmacological uses have been suggested for mulberry, including metabolic and cardiac diseases, oxidative stress, and neuro- and hepato-protection. A recent clinical study demonstrated that a modest dose (0.75 g) of mulberry fruit extract consumed with a digestible carbohydrate source can improve glycemic control in individuals with type 2 diabetes.

Custard apples and their by-products are presented in a chapter divulging their nutritional and therapeutic health benefits as well as functional food product development. Custard apple is one of the drought-hardy fruit plants commercially cultivated in a limited area of the Indian Deccan plateau. The demand for custard apple fruit is increasing in domestic and international markets thanks to their sensory, therapeutic, nutritional properties, and pleasant flavor. They are antimalarial, antifeedant, immunosuppressive, cytotoxic, and contain diterpenes and are used to treat HIV. Many cosmetic products using custard apple are available in the market, such as perfumery, soaps, essential oils, cold balms, anti-stress massage oils, pain massage oils, and foot care creams.

Holy basil and native Indian crops, considered anti-COVID-19 agents, are each described in separate chapters. Many crops covered in this book have been and continue to be used in Indian cuisine. A wide array of spices (i.e., turmeric, cumin, black cumin, mustard seeds, nigella seeds) is used in Indian cooking. Moreover, these plants and their parts form Ayurvedic formulations to treat several diseases. Clinical efficacies of the plants and/or their parts or phytocompounds have been included in the chapters. Products and formulations from some of these crops are already available globally and find their way into dietary supplements, functional foods, nutraceuticals, and cosmetics. Products from turmeric have proliferated in the global market. For example, Sabinsa's C3 Reduct® curcumin, a highly bioavailable and easily assimilated form by the body, offers multiple health benefits, delivering outstanding antioxidant and anti-inflammatory properties valuable for product innovation for the global antioxidant and sports nutrition markets. These properties are also beneficial for healthy aging to reduce the risk of osteoarthritis and neurodegeneration. These two markets are valued to reach USD 67.8 billion by 2030 for sports nutrition and USD 7.4 billion by 2031 for the antioxidant market.

We express our sincere thanks and appreciation to all the contributors who, by freely and willingly sharing their knowledge and expertise, have made this book possible. Our gratitude is also extended to the series editor, who reviewed the chapters, and the editorial staff and publishers for their contribution in bringing this work to publication. We hope that this book will serve to further stimulate the development of functional foods and nutraceuticals from native and endogenous Indian crops and provide global consumers with products that prevent and reduce the risk of diseases and help them maintain a healthier life.

Devina Lobine
B. Dave Oomah

Editor Biographies

Dr. Devina Lobine is a passionate biotechnologist and dedicated scholar with a strong commitment to innovation and global impact. She holds a Bachelor's degree in Agricultural Biotechnology and a PhD in Natural Products from the University of Mauritius, completed in close collaboration with Durham University (UK) through a prestigious Commonwealth Scholarship. She further pursued a postdoctoral research fellowship in pharmacology, broadening her expertise across various fields of life sciences. Devina has authored or co-authored over 64 scientific publications in high-impact journals and has been actively involved in teaching and mentoring at several higher education institutions in Mauritius. In her current role at the Mauritius Institute of Biotechnology Ltd (MIBL), she plays a pivotal role in shaping national biotechnology strategies and advancing research and innovation.

An emerging science diplomat, Devina is actively engaged with several esteemed national and international organizations, including the National Youth Environment Council (under the aegis of the Prime Minister's Office), the Commonwealth Scholarship Commission, the Global Young Academy, the International Network for Governmental Science Advice – Africa (INGSA-Africa), and the Academy of Medical Sciences. A strong advocate of transdisciplinary collaboration and evidence-based policy, she also leads/co-leads global initiatives such as the Global Indigenous Youth Summit on Climate Change, empowering Indigenous youth in climate action.

Dr. B. Dave Oomah is a retired (2013) research scientist formerly with Summerland Research and Development Centre, Agriculture and Agri-Food Canada (AAFC), Summerland, British Columbia. He is also an adjunct professor in the Department of Food, Nutrition and Health at the University of British Columbia, Vancouver. He received his Ph.D. in plant science (Cereal Chemistry) from the University of Manitoba, Winnipeg, Manitoba. Dr. Oomah has been active in crop utilization research since 1980 and has authored/co-authored 124 research papers and 43 book chapters. He is co-editor of six books (2022, 2020, 2018, 2017, 2014, and 2000), co-editor of the Special Food Research International Issue on Molecular, Functional and Processing Characteristics of Whole Pulses and Pulse and Nutraceutical Application, co-inventor of a US patent 6261629 on functional, water-soluble protein–fiber products from grains, and a trade secret on fiber-enriched products from flaxseed hulls. He contributed to the Organization for Economic Co-operation and Development (OECD) consensus document on common bean constituents (Safety of Novel Foods and Feeds, No. 27). His research interests have ranged from the processing of cereals, legumes, and oilseeds to the physicochemical characterization of major components from traditional and alternative crops, genetic and seasonal variations of secondary metabolites in oilseeds, antinutritional factors in oilseeds and legumes, and modern methodology assessing functional foods and nutraceuticals. He has special expertise in the area of phytochemicals from flaxseed, pulses, botanicals, fruits, and high-value alternative crops. He has continued his collaboration on dry bean and other research investigations with Universidad Autónoma de Querétaro, Mexico, since 2000. He has mentored and supervised 75 students (1998–2012), including 5 postdoctoral fellows/visiting scientists from Canada, Chile, China, Denmark, and Mexico; 14 Ph.D. students (from Algeria, Mexico, and Canada); 1 M.Sc. student; 27 graduate students, mainly from France; and 4 co-op students. He was a professional member of the American Chemical Society, the American Oil Chemists' Society, and the Institute of Food Technologists. He was an external reviewer on functional foods and nutraceuticals for grant applications to Action Concertée Fonds Nature et Technologies–AEE–MAPAQ–MIC, Québec. He was also a Natural Sciences and Engineering Research Council [NSERC] (GSC-03–Plant Biology and Food Science) committee member (2007–2009) and a member of the NSERC John C. Polanyi Award selection committee in 2008. He served as a member of the AAFC RES 03 Promotion Review Committee for 3 years.

Contributors

Vandita Anand
Department of Biotechnology
Motilal Nehru National Institute of Technology (MNNIT)
Allahabad, Prayagraj, India
ORCID: https://orcid.org/0000-0002-1839-2853

Runu Chakraborty
Food Technology and Biochemical Engineering Department
Jadavpur University
Kolkata, West Bengal, India
ORCID: https://orcid.org/0000-0003-4026-2542

Igguda Achaiah Chethan
JSS College of Pharmacy
JSS Academy of Higher Education and Research
Mysore, India

Lakshman Chandra De
ICAR-NRC for Orchids
Pakyong, Sikkim
ORCID: https://orchid.org/0000-0003-0691-2986

Mani Divya
Advanced Laboratory of Bio-nanomaterials
BioMe Live Analytical Centre
Karaikudi, Tamil Nadu, India
ORCID: https://orcid.org/0000-0003-4541-5827

Veronique François-Newton
School of Health Sciences
University of Technology Mauritius
La Tour Koenig, Pointe aux Sables, Mauritius;
ORCID: https://orcid.org/0009-0003-5469-6226

Rachna Gupta
Department of Food Engineering
National Institute of Food Technology Entrepreneurship and Management
Sonipat, Haryana, India

Seyashree Hazra
Food Technology and Biochemical Engineering Department
Jadavpur University,
Kolkata, West Bengal, India
ORCID: https://orcid.org/0000-0001-5145-9848

Annaelle Hip Kam
Biopharmaceutical Unit
Centre of Biomedical and Biomaterials Research
University of Mauritius Réduit, Mauritius

Shreynish Joy Mawooa
Department of Health Sciences, Faculty of Medicine and Health Sciences
University of Mauritius
Réduit 80837, Mauritius

Najmun Nahar
Department of Food Science
Maulana Abul Kalam Azad University of Technology
Nadia, West Bengal, India
ORCID: https://orcid.org/0000-0002-4037-0140

Vidushi S. Neergheen
Biopharmaceutical Unit
Centre of Biomedical and Biomaterials Research
University of Mauritius, Réduit, Mauritius
ORCID: https://orcid.org/0000-0001-5311-8676

Prabhat K. Nema
Department of Food Engineering
National Institute of Food Technology Entrepreneurship and Management
Sonipat, Haryana, India
ORCID: https://orcid.org/0000-0001-6419-4973

B. Dave Oomah
Formerly with Summerland Research and Development Centre
Agriculture and Agri-Food Canada
Summerland, British Columbia, Canada
ORCID: https://orcid.org/0000-0002-7035-3333

Anjana Pandey
Department of Biotechnology
Motilal Nehru National Institute of Technology (MNNIT)
Allahabad, Prayagraj, India

Manish Putteeraj
School of Health Sciences,
University of Technology Mauritius
La Tour Koenig, Pointe aux Sables, Mauritius

Tanmay Sarkar
Department of Food Processing Technology Malda Polytechnic
West Bengal State Council of Technical Education
Government of West Bengal, Malda, India
ORCID: https://orcid.org/0000-0003-3869-1604

Anupama Singh
Department of Food Engineering
National Institute of Food Technology Entrepreneurship and Management
Sonipat, Haryana, India

Divya Singh
Department of Biotechnology
Motilal Nehru National Institute of Technology (MNNIT)
Allahabad, Prayagraj, India

Jhoti Somanah
School of Health Sciences, University of Technology Mauritius
La Tour Koenig,
Pointe aux Sables, Mauritius
ORCID: https://orcid.org/0009-0006-5418-2999

Shweta Suri
Amity Institute of Food Technology (AIFT)
Amity University
Uttar Pradesh, India

Jaishree Vaijanathappa
Faculty of Life Sciences
JSS Academy of Higher Education and Research
Vacoas-Phoenix, Mauritius

M. Sai Varshini
Department of Pharmacology, JSS College of Pharmacy
JSS Academy of Higher Education & Research,
Ooty, Tamil Nadu, India
ORCID: https://orcid.org/0009-0001-0046-7079

Sekar Vijayakumar
College of Material Science and Engineering
Huaqiao University
Engineering Research Center of Environment-Friendly Functional Materials
Ministry of Education
Xiamen, P.R. China
ORCID: https://orcid.org/0000-0001-8211-069X

Ashish Wadhwani
Faculty of Health Sciences, School of Pharmacy
JSS Academy of Higher Education & Research
Vacoas-Phoenix, Mauritius
ORCID: https://orcid.org/0000-0001-8102-9486

Antidiabetic Potential of Indian Medicinal Plants/Crops

An Overview

1

M. Sai Varshini and Ashish Wadhwani

1.1 INTRODUCTION

India is the largest producer of medicinal herbs and is called the botanical garden of the world. In the last few years, there has been an exponential growth in the field of herbal medicine, and these drugs are gaining popularity both in developing and developed countries because of their natural origin and low side effects. Many traditional medicines in use are derived from medicinal plants, minerals, and organic matter. Numerous medicinal plants, traditionally used for over 1000 years, named rasayana, are present in herbal preparations of Indian traditional health care systems. According to the World Health Organization (WHO), up to 90% of the population in developing countries uses plants and their products as traditional medicine for primary health care. The WHO has listed 21,000 plants that are used for medicinal purposes around the world. Among these, 2500 species are in India. About 800 plants are reported to have antidiabetic potential. Many plant-derived active principles, representing numerous bioactive compounds, have established their role for possible use in the treatment of diabetes (Marles and Farnsworth, 1995).

Indians show a significantly higher age-related prevalence of diabetes compared with several other populations. For a given body mass index (BMI), Asian Indians display higher insulin levels, which is an indicator of peripheral insulin resistance. This insulin resistance in Indians is presumably due to their higher body fat percentage. Excess body fat, typical abdominal deposition pattern, low muscle mass, and racial predisposition may explain the prevalence of hyperinsulinemia and increased development of type 2 diabetes in Asian Indians. Gujarat leads all other states with the highest number of diabetes cases (161,578 diabetics and hypertension cases at 20.5% of the total screened population), according to the *National Health Profile 2015.*

Diabetes is characterized by metabolic dysregulation, primarily of carbohydrate metabolism, manifested by hyperglycemia resulting from defects in insulin secretion, impaired insulin action, or both. Uncontrolled diabetes leads to many complications affecting the vascular system, eyes, nerves, and kidneys, leading to peripheral vascular disease, nephropathy, neuropathy, retinopathy, morbidity, and/or mortality (Sharma et al., 2024).

DOI: 10.1201/9781003352341-1

The most common and effective antidiabetic medicinal plants of Indian origin are *Ficus, Salacia, Trigonella foenum-graecum, Brassica juncea, Eugenia jambolana, Azadirachta indica, Pterocarpus marsupium, Tinospora cordifolia, Ocimum sanctum, Momordica charantia, Coccinia indica, Mangifera indica, Benincasa hispida, Aegle marmelose,* and *Gymnema sylvestre* (Grover et al., 2002; Narwade et al., 2023). This chapter discusses their phytochemistry and antidiabetic potential primarily in preclinical and clinical studies.

1.2 MEDICINAL PLANTS WITH ANTIDIABETIC POTENTIAL

1.2.1 *Ficus*

The genus *Ficus* comprises about 850 species of trees, shrubs, vines, and epiphytes, widely distributed in tropical South America, subtropical Southeast Asia, and Australia (Deepa et al., 2018). *Ficus* species is one of the largest genera of the plant kingdom, with promising phytoconstituents from various classes of compounds, including phenols, flavonoids, sterols, alkaloids, tannins, saponin, and terpenoids. India is home to several species used in Ayurvedic medicine for herbal preparations to treat diabetes. These species and their parts used to treat diabetes include *Ficus amplissima* (bark), *F. benghalensis* (bark, stem, and leaves), *F. glomerata* (leaves), *F. hispida* (raw and mature fruit), *F. microcarpa* (bark, fruits, and leaves), *F. palmata* (raw fruit), *F. racemosa* (root), *F. religiosa* (bark, stem, leaves), *F. semicordata* (bark and ethanol leaf extract), and *F. carica* (leaves, fruit–fig, and root). Among these, *F. carica* and *F. racemosa* were studied in clinical trials.

Ficus carica L. (Figure 1.1) is an important member of the genus *Ficus*. It is ordinarily deciduous and commonly referred to as "fig." The common fig is a tree native to southwest Asia and the eastern

FIGURE 1.1 *Ficus Carica*

FIGURE 1.2 *Ficus racemosa*

Mediterranean, and it is one of the first plants that were cultivated by humans. The fig is an important harvest worldwide for its dry and fresh consumption. Its common edible part is the fruit, which is fleshy, hollow, and a receptacle. The dried fruits of *F. carica* are an important source of vitamins, minerals, carbohydrates, sugars, organic acids, and phenolic compounds. Fresh and dried figs also contain high amounts of fiber and polyphenols. Figs are an excellent source of phenolic compounds, such as proanthocyanidins, whereas red wine and tea, which are two good sources of phenolic compounds, contain phenols in amounts lower than those in figs. Its fruit, root, and leaves are used in traditional medicine to treat various ailments, such as gastrointestinal (colic, indigestion, loss of appetite, and diarrhea), respiratory (sore throats, coughs, and bronchial problems), and cardiovascular disorders and as anti-inflammatory and antispasmodic remedy (Mawa et al., 2013).

Ficus racemosa (Figure 1.2) is an important medicinal plant, found in India, Australia, and Southeast Asia. It is popularly known as "*gular*." The active constituent β-sitosterol, isolated from the leaves and stem bark, has good antidiabetic potential. Many active constituents isolated from various parts of this plant possess useful pharmacological activities. The literature survey proposed that it has multiple pharmacological actions that include antidiabetic, antioxidant, antidiarrheal, anti-inflammatory, antipyretic, antifungal, antibacterial, hypolipidemic, antifilarial, and hepatoprotective effects (Yadav et al., 2015).

In Ayurveda, powdered *F. religiosa* bark (10–20 g) decoction is used to treat diabetes. Aqueous *F. religiosa* bark extract reduced blood glucose, serum triglyceride, and total cholesterol levels, and increased serum insulin, body weight, and collagen content in the liver and skeletal muscle in streptozotocin-induced diabetic rats. The aqueous leaf extract reduced blood glucose, serum cholesterol, and triglyceride levels in alloxan-induced diabetic rats. The ethanol leaf and fruit extracts reduced blood glucose levels by stimulating β-cell insulin release in alloxan-induced diabetic rats (Deepa et al., 2018). The antidiabetic activity of *F. religiosa* is attributed to its anti-inflammatory and immunomodulatory properties in attenuating cytokine TNF-α and elevated glucose (by aqueous bark extract [100 and 200 mg/kg for 4 weeks]) in streptozotocin-induced neonatal rats (Kirana et al., 2011).

1.2.2 Clinical Trials

Many studies explore the antidiabetic effects of *Ficus* and its extracts in animal models, with limited (three) human clinical trials (Khan et al., 2011). Serraclara et al. (1998) investigated the effect of fig leaves (*F. carica*) decoction as a breakfast supplement on insulin-dependent type 1 diabetes mellitus (IDDM) patients (6M/4F, 29 years median age, 21 kg/m^2 average BMI, 7.2% HbA1c, 9 years chronic diabetes). In this randomized crossover trial (n = 4), patients orally supplemented *Ficus* (once daily filtered decoction [13 g leaf sachet boiled for 15 min]) during the first month, followed by non-sweet commercial tea the month after. *Ficus* supplementation significantly ($P < 0.001$) lowered postprandial glycemia (47%) without preprandial differences, medium average capillary profiles (~32%, $P < 0.05$), and average insulin dose (12%) or 15.5% total daily insulin dose. The study demonstrated the short-term hypoglycemic action of *Ficus* supplementation in humans by reducing insulin administration and lowering mean capillary glycemia. This beneficial effect of *Ficus* leaves is attributed to their bioactive components (presumably ficusin) that facilitate glucose uptake. Ficusin, isolated from *F. carica* leaves, reduced blood glucose levels and improved plasma insulin, liver enzymes, and activated glucose transport in diabetic animal models (Deepa et al., 2018). Thus, *Ficus* addition to diet in IDDM can help control postprandial glycemia.

Mazhin et al. (2016) evaluated the hypoglycemic effects of *F. carica* (common leaf fig) decoction in type 2 diabetes mellitus patients (53 years mean age, 7%–8.5% HbA1c) under oral hypoglycemic treatment in a double-blind crossover clinical trial. The decoction was prepared by boiling each pack (13 g *F. carica* dried leaf powder) in 500 mL distilled water for 15 min, followed by filtration after cooling. During the first phase, patients (5M/8F) received fig leaf decoction once daily for 21 days, while green tea was provided as a placebo to another group (4M/9F). The groups exchanged after a one-week washout period for another 21 days. *Ficus carica* treatment significantly ($P < 0.001$) lowered postprandial (2 h) blood sugar levels (16%) compared to placebo in type 2 diabetic patients without changes in other parameters (fasting blood sugar, HemoglobinA1c [HbA1c], and C-peptide). The hypoglycemic effect was presumably not mediated by enhanced insulin secretion, since C-peptide (insulin marker) did not change.

Gul-E-Rana et al. (2013) evaluated the hypoglycemic effect of *F. racemosa* bark in obese diabetic patients (25M/25F, 45–60 years, >30 kg/m^2 average BMI, 4–5 years chronic diabetes) taking oral hypoglycemic drugs (sulfonylurea). The bark contains glycosides, β-sitosterol, and lupeol. *Ficus racemosa* bark (5 mL, ≡100 mg twice daily for 15 days), along with oral hypoglycemic drugs, lowered pre- and postprandial blood glucose levels, and only significantly ($P < 0.05$) in males after 1.5 h post-breakfast (probably due to the sedentary lifestyle of women in this study). This human trial confirms previous *F. racemosa* (bark) hypoglycemic effects to potentiate insulin secretion in experimental animal models.

Another double-blind, randomized, placebo-controlled clinical trial evaluated the antihyperglycemic activity of *F. racemosa* bark on human volunteers (18M/12F). A human equivalent dose (HED) of 1.2 g daily (400 mg, three capsules) was established based on *in vivo* animal studies. One month of *F. racemosa* bark treatment significantly lowered blood glucose levels without altering cholesterol and triglyceride levels. Moreover, the above-normal body mass index of the subjects from both groups did not change significantly during the study period (Ahmed et al., 2011).

1.2.3 *Salacia*

The global *Salacia* market is being aided by the growth of the herbal medicine market, which is expected to grow at a 6% compound annual growth rate [CAGR] in the forecast period (2024–2032). Dietary supplements occupy the largest global *Salacia* market, followed by pharmaceuticals. *Salacia* species, with biological properties concentrated in its roots and leaves, belongs to the *Hippocastanaceae* family. India is home to 18 *Salacia* species, but only five species are traceable, namely, *Salacia chinensis* (Figure 1.3), *S. grandiflora*, *S. oblonga*, *S. reticulata*, and *S. roxburghii* (Vyas et al., 2016). The *Salacia* plant has been used in Ayurvedic medicine to treat diabetes, and *Salacia* extracts are currently consumed in commercial

FIGURE 1.3 *Salacia chinensis*

foods and food supplements (functional foods) in Japan for diabetes and obesity treatment. Salacinol and kotalanol, potent α-glucosidase inhibitors isolated from *Salacia* extracts, inhibit the small intestinal enzymes sucrose, maltase, and isomaltase; salacinol is presumably responsible for *Salacia*'s hypoglycemic activity. Mangiferin is an important component of various *Salacia* species and has therefore been suggested as suitable quality control of *Salacia* species and products. *Salacia oblonga* hot water extract (6.5% yield from dried root) contained 1.4% or 0.74% of mangiferin.

Salacia and its extracts have been investigated for their beneficial therapeutic effects to treat and/or prevent diabetes. We hereby present pertinent clinical trials demonstrating the efficacy of this important antidiabetic plant without delving into *in vivo* and animal studies. In one of the earliest studies, aqueous *S. reticulata* stem extract [SRE] (200 mg) significantly suppressed postprandial hyperglycemia after sucrose (50 g) loading in seven male human volunteers. SRE inhibited α-amylase activity (35 µg/mL IC_{50}), but not that of β-glucosidase, since it dose-dependently suppressed elevated serum glucose level induced by sucrose, maltose, and starch, but not by glucose or lactose (Kawamori and Kawahara, 1998). The same aqueous SRE-containing diet (240 mg daily) for 6 weeks significantly reduced fasting plasma glucose (FPG) levels, glycated hemoglobin (HbA1c), and BMI compared to the control diet group in mild type 2 diabetic patients (10M/10F, 58 years average age, 24 kg/m^2 median BMI) in this placebo-controlled crossover trial (Kajimoto et al., 2000).

Several studies relate to *Salacia*'s benefits in healthy humans. *Salacia oblonga* extract (1000 mg) reduced postprandial glycemia and insulinemia in healthy adults (20M/23F, 24 years mean age, 24 kg/m^2 median BMI) in a randomized double-masked crossover trial (Collene et al., 2005). The same dose of *S. oblonga* extract (1000 mg) significantly reduced postprandial insulin response (29%) and insignificantly lowered postprandial glycemia (23%) compared to control in 39 healthy adults (15M/24F, 26 years mean age, 24 kg/m^2 median BMI) in a double-masked randomized crossover trial (Heacock et al., 2005).

Single *S. chinensis* hydroalcoholic roots and stems extract [1000 mg dose, three capsules] (NR-Salacia developed by Natural Remedies, Bangalore, India) before carbohydrate-rich diet (~600 kcal) significantly lowered plasma glucose level (13%) at 90 min and postprandial (area under the glucose curve (AUC) 0–180 min) glycemia (34%) in healthy adults (12M/2F) compared to placebo (Koteshwar et al., 2013). NR-Salacia attenuated postprandial plasma glucose levels (34% AUC 0–180 min), favorably compared to 20% for acarbose (100 mg after 75 g glucose load), compared to placebo in healthy adults (Hotta et al., 1993). The postprandial glycemia reduction (1000 mg *Salacia* extract) in this randomized double-blind placebo-controlled crossover (19-day washout period) study is similar to the previously reported 23% decline (AUC 0–120 min) in non-diabetic adults and 24% reduction (AUC 0–180 min) in healthy

(20M/23F) subjects (Collene et al., 2005). The hypoglycemic effect of *Salacia* extract (240 and 480 mg) was lower (14% serum glucose AUC 0–180 min) in diabetic patients (53M/13F) (Williams et al., 2007).

Jeykodi et al. (2016) compared the effects of three doses (200, 300, or 500 mg) of ethanol extract from *S. chinensis* roots (SCE) on the glycemic and insulinemic response in normal healthy South Asian Indians (35M, 34 years mean age, 26 kg/m^2 average BMI). Randomized SCE (200, 300, or 500 mg) was given orally with sucrose (75 g) in water (250 mL) to each subject in each study period with a 7-day washout between dosing. SCE (containing 12% polyphenols, 2% mangiferin, and 1% 25,26-oxidofriedelane-1,3-dione by weight) dose-dependently lowered insulin during the sucrose load, indicating carbohydrate malabsorption and interference with insulin-releasing mechanisms. SCE (300 and 500 mg) reduced blood glucose and insulin (decelerating the postprandial hyperglycemic process) in this randomized placebo-controlled crossover study. SCE's insulinemic response was attributed primarily to the α-glucosidase inhibitory activities of its constituents.

Ozaki et al. (2008) evaluated the aqueous extract of Kothala himbutu (*S. reticulata* Wight) in a double-blind placebo-controlled study on healthy or mild type 2 diabetic patients. Drinks containing powdered Kothala himbutu extract (450 mg twice daily) significantly ($P < 0.05$) lowered HbA1c levels in mild type 2 diabetic individuals (11M/3F, 54 years average age, 25 kg/m^2 median BMI), but without significant changes in healthy adults (7M/7F, 38 years mean age, 22 kg/m^2 median BMI) compared to the placebo group. Williams et al. (2007) studied the effect of *S. oblonga* ethanol and water extract [SOE] (IC_{50}-46 μg/mL α-glucosidase inhibition, Tanabe Seiyaku Co. Ltd., Osaka, Japan) on postprandial glycemia and insulinemia in type 2 diabetic patients after high-carbohydrate (≥150 g/d) meal ingestion. *Salacia oblonga* supplementation to a standard liquid control meal significantly lowered acute glycemia (postprandial AUC (14 and 22% for the 240 mg and 480 mg extracts; adjusted peak glucose response, 19 and 27%) and insulinemia (adjusted peak insulin response, 19% and 12% for the 240 mg and 480 mg extracts in type 2 diabetic patients (53M/13F, 61 years average age, 29 kg/m^2 median BMI) in this randomized double-blind 3-period 3-treatment crossover study.

Powdered *S. reticulata* bark (2 g, 2 × 1 g capsules daily) for 60 days significantly ($P < 0.01$) lowered mean fasting (162 vs 153 mg/dL) and postprandial (282 vs 273 mg/dL) blood glucose, HbA1c (10.3% vs 9.2%), and total cholesterol (262 vs 255 mg/dL) levels in 20 type 2 diabetic patients (45–60 years) (Radha and Amrithaveni, 2009). The same dose of *S. chinensis* (1000 mg twice daily) for 6 months stabilized renal function (serum creatinine and creatinine clearance) and significantly reduced cardiovascular risk (serum homocysteine, triglyceride, and IL-6 levels) in 15 diabetic chronic kidney disease (CKD) patients compared to placebo (Singh et al., 2010). Similar treatment with *S. oblonga* (1000 mg twice daily) for 6 months reduced triglyceride levels by 24% in non-diabetic CKD patients (11M/4F, 45 years mean age) and by 18% in diabetic CKD patients (11M/4F, 54 years mean age) (Singh et al., 2015). Moreover, the treatment modulated endothelial function (significant IL-6 reduction) and inflammation (CRP reduction) in diabetic CKD patients. These investigators concluded that *S. oblonga* has significant hypolipidemic action and beneficial effects on multiple cardiovascular risk factors in both diabetic and non-diabetic kidney disease patients.

Salacia reticulata root bark extract (500 mg daily for 6 weeks) significantly ($P < 0.01$) reduced low-density lipoprotein cholesterol and fasting blood sugar levels at 3 and 6 weeks in prediabetes and mild to moderate hyperlipidemia patients ($n = 9$, 57 years mean age) compared to placebo ($n = 9$, 49 years mean age) (Shivaprasad et al., 2013). The leaf extract at the same dose significantly ($P < 0.05$) lowered fasting blood sugar levels in patients ($n = 11$, 43 years mean age) only at 6 weeks compared to the placebo group. *Salacia reticulata*'s improvement in lipid profile and blood glucose levels was attributed to the α-glucosidase inhibitory activity of its phytonutrients in this randomized double-blind placebo-controlled trial. The investigators proposed *Salacia* extract's use as a dietary supplement to manage patients with prediabetes and mild to moderate hyperlipidemia.

Kobayashi et al. (2021) evaluated the efficacy of *S. chinensis* neokotalanol-standardized extract in two randomized double-blind placebo-controlled clinical trials in subjects with borderline and mild diabetic stages (fasting blood glucose ≥100 to <140 mg/dL). The powdered aqueous *S. chinensis* extract contained 0.22% neokotalanol—a potent α-glucosidase inhibitor. The multiple-dose study (all males, 51 years average age) in this placebo-controlled double-blind crossover (6-day washout between two dosing days) was randomized into four groups: low, intermediate, and high *Salacia* extract doses and placebo. *Salacia* administration (at all doses; 150–600 mg *Salacia* extract containing 0.33–1.33 mg neokotalanol) significantly and dose-dependently suppressed blood glucose and lowered insulin levels compared to placebo. *Salacia* administration (600 mg *Salacia* extract) for 12 weeks in the single-dose study (6F/7M, 55 years average age, 30 kg/m^2 median BMI) significantly lowered HbA1c, suppressed postprandial hyperglycemia, and improved insulin sensitivity and other blood glucose parameters compared with placebo (7F/6M, 59 years average age, 29 kg/m^2 median BMI). *Salacia chinensis* extract standardized by neokotalanol exerted beneficial effects in early diabetic patients.

Biscuits containing dried aqueous *S. reticulata* extract (SRE), taken as a snack for 3 months with 1-month washout, reduced HbA1c by 0.25% compared to placebo biscuits in type 2 diabetic patients (Siribaddana et al., 2023). The biscuit (cracker) containing dried aqueous *S. reticulata* extract was a commercially available biscuit, "*Munchee Kothala Himbutu Biscuit,*" in the Sri Lanka market. This triple-blind randomized placebo-controlled two-sequence crossover clinical trial of type 2 diabetic adults (50 years mean age) consisted of two groups: placebo-KH biscuit (42F/29M) and KH biscuits-placebo (51F/11M). The modest HbA1c reduction of the SRE extract was attributed to α-glucosidase enzyme inhibition by *Salacia* phytochemicals (kotalanol, ponkoranol, salaprinol, and mangiferin) at the intestinal brush border without distressing gastrointestinal effects, unlike commercially available α-glucosidase inhibitors.

Clinical trials have also evaluated *Salacia*'s benefits in combination with other products. For example, Nakata et al. (2011) investigated *Salacia*'s beneficial effect on plasma glucose levels in type 2 diabetic (KK-Ay/TaJcl) mice and in humans with premetabolic syndrome in a double-blind randomized study. The human clinical trial consisted of high plasma glucose [HG] group (FPG 100–125 mg/dL) and hyperlipidemia [HL] group (>150 mg/dL triglycerides, or >120 mg/dL LDL, or <40 mg/dL HDL). This study was randomized in each group (48–50 years average age, 23–26 kg/m^2 median BMI; HG group—control, 6 subjects; SI tea, 7 subjects; HL group—control and SI tea treatment, 14 subjects each). SI tea consumption (twice daily for 60 days) significantly lowered FPG and HbA1c of the HL group more than the control (*Salacia* without IP-PAI) and decreased LDL levels. SI tea (combination of *Salacia* tea, fermented flour containing 1% IP-PAI bacteria, and vitamins) significantly reduced LDL levels in plasma lipids of both HG and HL groups. *Salacia* tea reduces high blood glucose levels because it contains the most effective α-glucosidase inhibitors, salacinol and kotalanol. *Salacia* and SI tea had similar effects on FPG level and HbA1c in the HG group.

Jayawardena et al. (2005) investigated the effects of an herbal tea containing *S. reticulata* (Kothala Himbutu tea, Siddhalepa Ayurveda Hospitals, Ratmalana, Sri Lanka) in type 2 diabetic patients in a randomized single-center double-blind crossover clinical trial. The first group (16M/12F, 53 years mean age, 22 kg/m^2 median BMI) received a standard Kothala Himbutu tea for three months, followed by placebo in similar tea bags for a further 3 months. The order was reversed for the second group (12M/11F), receiving placebo first, followed by Kothala Himbutu tea for 3 months each. *Salacia reticulata* tea significantly lowered HbA1c and medication (glibenclamide) dose without carryover effect compared to the placebo group. The tea contains dried 35% *S. reticulata* (barks, roots, and leaves), 35% *Artocarpus heterophyllus* (plant; i.e., roots, bark, stalk, leaves, and flowers), 20% *Tinospora cordifolia* stems, and 10% *Cinnamomum zeylanicum* (plant). The hypoglycemic effect of the *S. reticulata* tea is ascribed to the presence of kotalanol with potent α-glucosidase inhibitor action, similar to that of acarbose.

Faizal et al. (2009) evaluated the hypoglycemic and hypolipidemic effects of an Ayurvedic medicine, "Rajanayamalakadi," in type 2 diabetic patients. Each "Rajanyamalakadi" tablet contains *Curcuma longa* (125 mg), *Emblica officinalis* (125 mg), and *S. oblonga* (250 mg). Patients were divided into six groups based on age and mean fasting blood (FBS) values: normal control subjects (n = 10); 35 to 45 years (n = 15); 46 to 55 years (n = 13); >55 years (n = 15); <145.9 mg/dL FBS (n = 21); >145.9 mg/dL FBS (n = 22).

The treatment, "Rajanyamalakadi" tablets (1–2 tablets, 500 mg each, Bipha Drug Laboratories, Kottayam, Kerala), based on patients' responsiveness and clinical conditions for 3 months, had significant ($P < 0.05$) hypoglycemic and hypolipidemic effects in diabetic patients. It increased HDL cholesterol and insulin levels and glutathione reductase activity, and lowered triglycerides, LDL, free fatty acids, atherogenic index, aspartate aminotransferase (AST), alanine aminotransferase (ALT), FBS, and HbA1c (except insignificant for the group with FBS <146 mg/dL). The bioactive principles of "Rajanyamalakadi" components are responsible for its beneficial hypoglycemic and hypolipidemic effects. *Curcuma longa* curcuminoids and *E. officinalis* polyphenols, with their hypolipidemic and antioxidant properties, can protect pancreatic β-cells from oxidative stress, modulate lipid profile, and ensure proper function. *Salacia oblonga* can minimize long-term diabetic complications in addition to its antidiabetic property. Rajanyamalakadi may be considered a nutraceutical with multiple mechanisms of action (i.e., insulin activation/stimulation, antioxidant, hypolipidemic, antidiabetic, hepatoprotective, and others).

Another product containing *S. oblonga* extract, Glucaffect™ (48 g daily for 8 weeks), significantly lowered fasting blood glucose (30%) and HbA1c in prediabetic overweight subjects (14M/10F, 42 years average age) with metabolic syndrome (≥25 kg/m^2 BMI; >126 mg/dL fasting blood sugar) compared to the control group (15M/11F, 43 years average age) (Belcaro et al., 2009). Thus, Glucaffect™ consumption normalized healthy blood sugar levels in subjects with prediabetic glucose levels, significantly reduced weight (~7.2 kg), and reduced BMI by over 2 kg/m^2 (<25 kg/m^2). The treatment apparently delayed dietary carbohydrate absorption due to α-glucosidase inhibition, thereby decreasing insulin secretion and prolonging satiety. Glucaffect™ fasting blood glucose and HbA1c reduction in this study exceeds those reported for acarbose in diabetic patients. Glucaffect™ (Reliv Inc., Chesterfield, MO, USA) formulation per 12 g serving consists of non-genetically modified origin (non-GMO) low-fat soy flour (8.78 g), soy lecithin (150 mg), ground cinnamon (44 mg), xanthan and guar gum (120 mg each), inulin (400 mg), acesulfame potassium (90 mg), omega-3 fish oil (750 mg), CoQ10 (20 mg), L-glutathione (12.5 mg), alpha-lipoic acid (100 mg), with blood glucose-lowering active components: French maritime pine bark extract Pycnogenol® (15 mg), *Syzygium cumini* extract (500 mg), *S. oblonga* extract (120 mg), *Pterocarpus marsupium* extract Silbinol® (250 mg), and *Lagerstroemia speciosa* extract (4 mg).

Kurian et al. (2014) evaluated the efficacy of the polyherbal combination G-400 (1000 mg daily for 8 weeks) to manage type 2 diabetic patients (50M/39F, 53 and 47 years mean age for men and women, 25 kg/m^2 median BMI, FBG > 140 mg/dL) under medical treatment in a randomized case-controlled study. Polyherbal G-400 treatment attenuated hyperglycemia and hypolipidemia by significantly ($P < 0.001$) lowering postprandial blood glucose and HbA1c, and improving lipid profiles (total cholesterol, lipoprotein cholesterol, and triglycerides) in type 2 diabetic patients without affecting kidney and liver functions. Moreover, G-400 positively modulates the activities of 3-hydroxy-3-methyl glutaryl coenzyme A reductase, malic enzyme, glucose-6-phosphate dehydrogenase, and glucose-6-phosphatase in hepatic tissue. The polyherbal drug G-400 (Kerala Ayurvedic Pharmaceutical Company) consists of *S. oblonga* leaves (30%), *T. cordifolia* stem (10%), *E. officinalis* (10%), *C. longa* (10%), and *G. sylvestre* (40% dry weight). Phytoconstituents of these herbs can inhibit α-glucosidase, stimulate insulin secretion from β-cells, regulate glucose transporter four (GLUT4), reduce hyperlipidemia, and promote β-cell regeneration.

1.2.4 *Trigonella foenum-graecum*

Species of *Trigonella* are widely distributed throughout the world. The plant has been mainly found on the continents of Asia (India and China), parts of Europe, Africa, Australia, and North and South America. It bears two cotyledons and belongs to the *Papilionaceae* subfamily and *Leguminosae* (*Fabaceae*) family. Synonyms include *Methi* or *Mutti* (Hindi) and fenugreek (English). It is found as a wild plant and also cultivated in northern India. The presence of a complex array of important phytochemicals renders fenugreek one of the important medicinal plants. The leaves and seeds of *methi* have been used extensively to prepare extracts and powders for medicinal uses.

FIGURE 1.4

Scientists have reported several medicinal uses of fenugreek seeds, such as remedies for diabetes and hypercholesterolemia, protection against liver free radicals, and breast and colon cancer. These protective roles emanate from secondary metabolites, also known as phytochemicals. The major constituents present in fenugreek seeds are carbohydrates, proteins, lipids, alkaloids, flavonoids, fibers, saponins, steroidal saponins, vitamins, minerals, and nitrogen compounds, which can be categorized under nonvolatile and volatile constituents (Goyal et al., 2016). *Trigonella foenum*-graecum seeds (Figure 1.4) have a wide range of phytoconstituents like trigonelline, protodioscin, trigoneoside, diosgenin, yamogenin, 4-hydroxyisoleucine, and galactomannans. The major compound trigonelline can be used as an antidiabetic and CNS stimulant and can be beneficial in Alzheimer's disease. Hereby, we review *in vivo* animal and clinical studies of fenugreek.

Isolated fibers, saponins, and other proteins from fenugreek seeds, given with meals for 21 days to alloxan-diabetic dogs, showed significant antihyperglycemic and antiglycosuric effects along with a reduction in high plasma glucagon and somatostatin levels (Ribes et al., 1986). Fenugreek is a commonly used herb for the management of diabetes mellitus (Cortez-Navarrete et al., 2023). The antidiabetic effect of fenugreek is associated with different mechanisms of action. In an *in vitro* study, fenugreek seed extract inhibited alpha-amylase activity in a dose-dependent manner. Suppression of starch digestion and absorption induced by fenugreek seed extract was confirmed in normal rats in the same study (Gad et al., 2006). The hypoglycemic effect of fenugreek is also associated with glucose uptake inhibition, since the seed extract inhibited intestinal sodium-dependent glucose uptake. Fenugreek can also regulate plasma glucose levels by delaying gastric emptying and decelerating the postprandial glucose absorption rate due to its high soluble fiber content. It can also modulate the activity of key regulatory gluconeogenic enzymes.

Scientific evidence suggests insulinotropic activity by 4-hydroxyisoleucine, a free amino acid present in fenugreek seeds. *In vitro* studies showed that 4-hydroxyisoleucine increased insulin secretion through a direct effect on both isolated rat and human islets of Langerhans (Sauvaire et al., 1998). Regulation of other pancreatic hormones is also a possible mechanism of the antidiabetic effects of fenugreek. The administration of fenugreek seed extract to alloxan-induced diabetic dogs reduced glucagon and somatostatin levels (Ribes et al., 1986). Furthermore, fenugreek may play a role in regulating glucagon-like peptide-1 (GLP-1) activity. It has been shown that an active compound (N55), isolated from fenugreek seeds, binds to GLP-1 and enhances its potency in stimulating GLP-1 receptor signaling (King et al., 2015).

Several animal experimental models have also demonstrated the hypoglycemic effect of fenugreek. However, we review clinical studies that have evaluated the antidiabetic effect of fenugreek in type 2 diabetes mellitus (T2DM) patients. Rafraf et al. (2014) conducted a randomized, triple-blind, placebo-controlled clinical trial in 88 T2DM patients. Subjects were randomly allocated to receive powdered whole fenugreek seeds (10 g daily) or a placebo for 8 weeks. Fenugreek treatment significantly lowered

FPG levels (P = 0.007), glycated hemoglobin A1c (A1C) levels (P = 0.0001), and the homeostasis model assessment of insulin resistance (HOMA-IR) (P < 0.005) compared to the placebo group.

Verma et al. (2016) carried out a randomized, double-blind, placebo-controlled clinical trial in T2DM patients (108M/46F, 25–60 years). Subjects were randomly assigned to fenugreek seed extract (1 g daily) or a placebo for 90 days in addition to standard antidiabetic therapy (metformin). A patented fenugreek seed extract (Fenfuro™), enriched in approximately 40% furostanolic saponins, was used in the study. Fenfuro treatment significantly lowered FPG levels (7%, 10%, and 22% at 30, 60, and 90 days, compared to 3.2%, 1.2%, and 7.6% reduction for the placebo group) and postprandial plasma glucose (PPG) levels (14%, 21%, and 30% vs 7.6, 9.5, and 17% for placebo at 30, 60, and 90 days). However, A1C reduction was not statistically significant, although almost half (49% vs 18% for placebo) of the patients in the Fenfuro-treated group reduced their antidiabetic therapy (metformin). In summary, Fenfuro proved safe and efficacious in ameliorating the symptoms of type 2 diabetes in humans.

Ranade and Mudgalkar (2017) evaluated the effect of fenugreek in a randomized, single-blind, parallel-group clinical study. Sixty T2DM patients treated with any antidiabetic medication (oral hypoglycemic agent, insulin) were included in the study. Patients were randomized to receive fenugreek seeds soaked in hot water (10 g daily) for 6 months, while patients in the other group only received antidiabetic medication. Fenugreek intervention (21M/9F, 48 years mean age, 25 kg/m^2 median BMI) significantly lowered FPG levels (fenugreek group: 154–122 mg/dL; control group: 160–154 mg/dL; P = 0.0351) and A1C levels (fenugreek group: 7.2%–5.78%; control group: 7.9%–6.96%; P = 0.0201) compared to the control group (24M/6F, 46 years mean age, 26 kg/m^2 median BMI).

Gupta et al. (2001) evaluated the effects of fenugreek seeds on glycemic control and insulin resistance in mild to moderate T2DM patients in a double-blind, placebo-controlled clinical trial. Patients (11M/1F, 49 years mean age, 27 kg/m^2 median BMI) were assigned to receive a hydroalcoholic extract of fenugreek seeds (1 g daily) or a placebo (8M/5F, 53 years mean age, 27 kg/m^2 median BMI) for 2 months. Oral hypoglycemic agents were permitted if blood glucose control was considered inadequate (a combination of sulfonylurea and biguanides). Fenugreek treatment significantly reduced FPG levels (148–120 vs 138–113 mg/dL for placebo, P < 0.05) and PPG (211–181 vs 220–242 mg/dL for placebo, P < 0.5). Insulin sensitivity, determined by the HOMA model, increased significantly (113% vs 92% for placebo, P < 0.05) among patients who received fenugreek. They concluded that adjunct use of fenugreek seeds improves glycemic control and decreases insulin resistance in mild type 2 diabetic patients. This improvement in the glycemic status of T2DM patients is mediated by enhancing insulin sensitivity and reducing insulin resistance, apart from the known mechanisms of reduced glucose absorption.

Najdi et al. (2019) compared the hypoglycemic effect of fenugreek to glibenclamide in a randomized, open-label clinical study in T2DM patients. The study included 12 patients treated with conventional therapy (metformin) who received fenugreek seed (2 g daily, GNC, Pittsburgh, PA, USA) or glibenclamide (5 mg daily) for 12 weeks. Fenugreek administration (n = 5, 51 years mean age, 28 kg/m^2 median BMI) significantly increased fasting insulin levels (7 vs 9.6 U/mL, P = 0.04) and reduced insulin resistance (HOMA-IR, 2.2 vs 3.5, P < 0.01) without significant changes in FBG and A1C. They concluded that fenugreek could be used as adjuvant therapy to antidiabetic drugs to control blood glucose.

Lu et al. (2008) investigated the effect of *T. foenum-graecum* L. total saponins (TFGs) in combination with sulfonylureas in 69 T2DM patients. In this randomized, double-blind, placebo-controlled clinical study, patients (25M/21F, 55 years mean age) received either fenugreek saponin extract (0.35 g/pill; each gram of powder equals 16 g of crude drug, six capsules each time, thrice a day) or Chinese yam placebo (13M/10F, 54 years mean age) for 12 weeks. Significant decreases from baseline to endpoint were reported in FPG levels (8.63 vs 6.79 mmol/L, P < 0.05), 2 h PPG levels (13.34 vs 9.46 mmol/L, P < 0.01), and A1C (8.02% vs 6.56%, P < 0.05) among patients who received fenugreek compared to insignificant changes in the placebo group. Moreover, the total effective rate in the treated group was significantly superior to control (80% vs 44%, P = 0.002). The study showed that the combination of TFGs therapy with sulfonylureas was relatively safe and can lower the blood glucose level and ameliorate clinical symptoms in T2DM treatment.

Sharma and Raghuram (1990) conducted a randomized, crossover-design clinical study in 15 non-insulin-dependent T2DM subjects treated with antidiabetic drugs (glibenclamide, glipizide, or metformin).

Patients (10M/5F, 46 years mean age) were randomized to receive defatted (debittered) fenugreek seed powder (100 g daily) incorporated into chapati (unleavened bread) for 10 days, while the rest received the same in the second period. The effect of the same intervention for 20 days was assessed in a subset of the study. Fenugreek intervention significantly lowered FPG levels (179–137 mg/dL, P < 0.05), PPG (P < 0.01), and serum insulin levels (P < 0.05) compared to the control diet. Ingestion of fenugreek seeds for 10 to 20 days reduced (64%–74%) urinary glucose excretion and serum lipid profile (serum total cholesterol, LDL and VLDL cholesterol, and triglycerides) without changes in HDL cholesterol. Moreover, clinical symptoms (polydipsia and polyuria) improved with fenugreek administration. The investigators proposed that debittered fenugreek seed (25–100 g) incorporation in the daily diet of diabetes patients can effectively serve as supportive therapy to prevent and manage long-term diabetes complications.

Gaddam et al. (2015) designed a prospective 3-year randomized, single-blinded, controlled, parallel clinical trial to study the efficacy of fenugreek in T2DM progression, monitored at baseline and every three months. Prediabetes patients were divided into two groups: one group (n = 74) received fenugreek seed powder [debittered, defatted, and deodorized fenugreek fiber supplemented with vitamins, minerals, and amino acids, SMS Pharmaceuticals Ltd., Jeedimetla, Hyderabad, India] (10 g daily), and the other group (n = 66) served as the control. Fenugreek treatment significantly lowered FPG (104–100 mg/dL, P < 0.05) and PPG levels (143–129 mg/dL, P < 0.01), and increased serum insulin levels (10.2 vs 12 mU/L, P < 0.05). Supplementation with fenugreek was associated with lower diabetes progression, as the conversion rate from impaired fasting glucose and impaired glucose tolerance (IGT) at the end of a 3-year period significantly decreased in the fenugreek group compared to controls.

1.2.5 *Brassica juncea*

Brassica juncea, commonly known as mustard, belongs to the Brassicaceae or Cruciferae family. Synonyms include brown mustard, Indian mustard (Figure 1.5), leaf mustard, oriental mustard, and vegetable mustard. Leaves, stems, and seeds of mustard are edible. Its origin is in Central Asia (Northwest India), along with secondary centers in western China, eastern India, and Burma. This plant appears in various cuisines in India, Africa, Bangladesh, China, Italy, Japan, Nepal, Pakistan, Korea, and America. Mostly, this plant is used for its seeds and its oil, known as mustard oil. The mustard condiment made from the seeds is called brown mustard and is considered to be spicier than yellow mustard. The mustard oil is commercially used in cooking and also for cosmetic purposes, such as for hair growth control (Kumar et al., 2015).

FIGURE 1.5 *Brassica juncea*

One of the crucial chemical constituents of mustard oil is allyl isothiocyanate, which is formed during oil processing from seeds (Yu et al., 2020). Recently, this constituent has been considered crucial for cancer and also for several other therapeutic benefits, such as antimicrobial activity. Glucosinolates are abundant in *B. juncea* leaves (McNaughton and Marks, 2003). Together with glucosinolates, numerous other *B. juncea* secondary metabolites are presumed to be responsible for its crucial health benefits. These bioactive components include glucosinolates [sinigrin (allyl glucosinolate)], isothiocyanates (allyl isothiocyanate, phenyl isothiocyanate), phenolic compounds (sinapic acid, sinapine), fatty acids (α-linolenic acid), kaempferol glycosides, and other flavonoid compounds, such as isorhamnetin 3,7-di-O-β-D-glucopyranoside (isorhamnetin diglucoside), and proteins such as napins and juncin.

The literature reports several preclinical studies, but only a few clinical trials are available. Hereby, we discuss some of the preclinical and clinical studies. Administration of aqueous *B. juncea* extract at varying doses (250, 350, and 450 mg/kg/day) to STZ-induced diabetic rats had dose-dependent effects against hyperglycemia, insulin deficiency, and reduction in blood glucose levels. This antihyperglycemic effect of *B. juncea* seed extract presumably depends on the presence of functional β-cells for insulin release (Thirumalai et al., 2011). Apparently, *B. juncea* seeds noticeably prevented pancreatic β-cell destruction, exerting their antioxidant effect along with antidiabetic activity. This study suggests an interesting antioxidant potential simultaneously with antidiabetic potential of *B. juncea* seeds.

Another study reported that *B. juncea* increased glucose utilization (increased glycogenesis) and decreased gluconeogenic enzymes (Khan et al., 1995). Later, in another study, *B. juncea* significantly attenuated the development of insulin resistance in fructose-enriched-diet-fed rats. Further, it resulted in decreased fasting glucose, cholesterol, and insulin levels. These studies suggest that *B. juncea* plays an important role in the management of diabetes and insulin resistance (Yadav et al., 2004). Grover et al. (2003) studied the synergistic effect of *B. juncea* and *Murraya koenigii* in diabetic nephropathy and analyzed the levels of urine volume, urinary albumin, and serum creatinine. Unfortunately, both failed in significant polyuria reduction. However, *B. juncea* significantly reduced creatinine levels, indicating that it can be used to delay the development of diabetic nephropathy. This research group has also investigated the antihyperlipidemic, antihyperglycemic, and some other beneficial effects of the seeds against different metabolic disorders and associated pathologies.

One clinical study determined the effectiveness of mustard green leaf decoction as an adjunct to drug treatment in controlling blood glucose among Filipinos with T2DM. Participants were randomly assigned to receive mustard green decoction or a placebo solution for eight weeks on top of their oral antihyperglycemic medication. Fasting blood sugar and complete blood counts were determined at baseline, week 4, and week 8, and compared within and across the two groups. Mustard green treatment lowered blood sugar level, while the opposite was noted in the placebo group. The mean FBS levels of the mustard green group were significantly lower (6.10 vs 8.69 mmol/L, $P = 0.004$) than that of the placebo group and compared with the baseline level ($P = 0.008$) in 8 weeks. This study demonstrated that the intake of *B. juncea* decoction can significantly decrease blood sugar levels among type 2 diabetics compared to metformin alone (Chio et al., 2018).

Chaiyasut et al. (2018) evaluated the cholesterol-lowering property of lactic acid bacteria-fermented *B. juncea* (mustard greens) pickle. Twenty-seven different pickle formulas were prepared. Lactic acid content increased gradually, whereas acetic acid level decreased in the pickles. The acceptable level of ethanol was observed in pickles after 30 days, with no methanol content. Pickles showed antimicrobial activity against tested microbes. Cholesterol level in pickle liquid was significantly reduced. The stability and sensory analysis suggested that pickle formula-2 was accepted by the healthy volunteers. LAB5-fermented MG pickle could be a functional food to control cholesterol.

Lett et al. (2013) compared the glycemic response and satiety ratings of potato and leek soup with and without the addition of 5 g of yellow mustard bran in a randomized, repeated-measures design. Ten healthy, non-smoking, moderately active male subjects (21 years average age and 23 kg/m^2 median BMI) were recruited for the study. Yellow mustard bran supplementation significantly lowered mean postprandial blood glucose levels ($P < 0.0001$, $P < 0.0001$, and $P = 0.0059$ at 15, 30, and 90 min, respectively). This study demonstrated the attenuation of postprandial glycemic response following the addition of 5 g of yellow mustard bran to a soup.

Repin et al. (2017) examined the effects of soluble dietary fiber (SDF), including (ethanol-treated) yellow mustard mucilage consumption and fenugreek (CANAFEN®, Emerald Seed Products Ltd., Avonlea, Canada) gum on postprandial glycemic and insulinemic responses and gastric emptying in a human clinical trial. Consumption of chocolate-flavored puddings supplemented with SDF (yellow mustard mucilage 15.5 g fiber [1.15% w/v]; fenugreek gum 5.9 g [0.43% w/v] fibers per 500 mL) significantly lowered glucose (~13%) and insulin (~40%) and attenuated gastric emptying compared to control in patients at risk of T2DM (10M/5F, 55 years average age, 29.5 kg/m^2 median BMI). The investigators proposed that delay of gastric emptying by SDF may be the main mechanism involved in attenuating glycemic and insulinemic responses.

1.2.6 *Eugenia jambolana*

Eugenia jambolana, commonly known as Java plum, belongs to the Myrtaceae family. This tree is famous for its fruits, and the fruit synonyms include Malabar plum, Portuguese plum, Indian blackberry, black plum, jambu, jaman, jambul, and jambool. It is an evergreen tree up to 25 m tall, with young grayish-white stems and coarse and discolored lower bark (Baliga et al., 2011). It is widely distributed in India, Sri Lanka, and Australia. All parts of this tree have been traditionally used in Ayurveda, Unani, Homeopathy, and Siddha medicine in various countries, especially India, Sri Lanka, and Tibet. During ancient times, before the discovery of insulin, Java plum was used for the treatment of diabetes either alone or in combination with other hypoglycemic plants.

In the Indian subcontinent, *E. jambolana* (jamun) is divided into two major types, namely Kaatha jamun and Ras jamun, based on morphological features. Fruits are formed in clusters (Figure 1.6) and do not ripen immediately, because fruit development takes about two months, during which the fruit turns purple with compositional changes of phytoconstituents. Fruits are of the highest importance for their dietary composition, in which even seeds are dried and marketed in several countries. Ripe fruits are used to manufacture health drinks, juices, jellies, wine, and other products.

Barks are acrid, digestive, and astringent. From ancient times, they have been useful in treating sore throat, thirst, asthma, bronchitis, dysentery, biliousness, and ulcers. The ash of the leaves is effective at strengthening the gums and teeth and is used as a dentifrice. The bark is known to have wound-healing properties. In the Siddha system of medicine, jamun is considered to be hematinic, semen-promoting, and capable of decreasing excessive body heat. As per the Unani system of medicine, they are used as a liver

FIGURE 1.6 *Eugenia jambolana*

tonic, to enrich blood, and to strengthen teeth and gums. Multiple studies have suggested that *E. jambolana* possesses a wide range of pharmacological activities, such as treatment of chronic diarrhea, sore throat, and dysentery, and also has antifungal, antibacterial, antiviral, anti-genotoxic, anti-ulcerogenic, anti-inflammatory, anti-allergic, cardioprotective, radioprotective, anticancer, chemopreventive, antioxidant, free radical scavenging, antidiarrheal, hepatoprotective, hypoglycemic, and antidiabetic effects (Baliga et al., 2011).

Multiple studies have reported that the fruit is extremely nutritive, containing minerals (sodium, iron, potassium, phosphorus, calcium, and zinc), water-soluble vitamins (ascorbic acid, niacin, and thiamine), carbohydrates (mannose, maltose, sucrose, glucose, galactose, and fructose), and amino acids (asparagine, alanine, glutamine, tyrosine, and cysteine). Other phytochemicals are also present, such as β-sitosterol, mycaminose, betulinic acid, n-heptacosane, crategolic (maslinic) acid, n-hentriacontane, n-nonacosane, n-triacontanol, noctacosanol, n-dotricontanol, myricetin, quercetin, myricitrin, and flavanol glycosides (acylated flavonol glycosides and myricetin 3-O-(4″-acetyl)-α-L-rhamnopyranosides). Leaves of the tree contain phytochemicals such as α-terpeneol, eucarvone, pinocarveol, muurolol, myrtenol, α-myrtenal, geranyl acetone, cineole, and α-cadinol. The stem contains betulinic acid, friedelan-3-α-ol, friedelin, β-sitosterol-D-glucoside, kaempferol, gallic acid, gallotannin, ellagic acid, ellagitannin, and myricetin. Flowers contain ellagic acids, oleanolic acid, isoquercetin, kampferol, quercetin, and myricetin. Seeds contain gallic acid, jambosine, ellagic acid, 3,6-hexahydroxy diphenoylglucose, corilagin, 4,6-hexahydroxydiphenoylglucose, quercetin, 3-galloylglucose, β-sitosterol, and 1-galloylglucose.

Jamun has been rigorously investigated because of its antidiabetic effect since ancient times. Many parts, such as fruit pulp, seed, and bark, possess antidiabetic effects, while the leaf was ineffective. Based on many preclinical studies, it is concluded that powdered seed was effective in reducing diabetes and its complications. Administration of *E. jambolana* extract alone or in combination with *Musa paradisiaca* significantly reversed serum glucose level, glucose tolerance, lipid profile, muscle glycogen contents, and recovered glucose-6-phosphate and hexokinase in STZ-induced diabetic male rats (Panda et al., 2009). Another study reported that administration of *E. jambolana* seed kernel extract significantly reversed the effects of STZ on levels of triglycerides, cholesterol, phospholipids, and various fatty acids in STZ-induced diabetic rats (Ravi et al., 2005).

Fruit pulp extracts of *E. jambolana* significantly reduced blood glucose levels and improved glucose tolerance in alloxan-induced diabetic rats (Sharma et al., 2006). Furthermore, *E. jambolana* ethanolic extract showed significant hypolipidemic and hypoglycemic effects in alloxan-induced rabbits; it improved fasting glucose levels, glycosylated hemoglobin levels, serum insulin levels, lipid profile, liver, and muscle glycogen content (Sharma et al., 2003). In another study, the combination of *E. jambolana* and *Momordica charantia* exerted significant hyperglycemic and hyperinsulinemic effects in fructose-fed rats by reducing elevated glucose levels, insulin levels, and triglycerides (Vikrant et al., 2001). In addition to the above studies, several animal studies report preventions of diabetes-induced secondary complications, such as damage to kidneys, diabetic cataracts, gastropathy, and peptic ulcers.

Syzygium cumini (L.) Skeels (syn. *S. jambolanum DC*, *E. jambolana* Lam., Jambul, Java plum) played an outstanding role among many medicinal plants known to decrease glucose levels, particularly in Western Europe in the three decades prior to insulin discovery. Jambul seeds were already recognized as diabetic therapeutic and phytotherapeutic in the standard literature at the dawn of the 20th century. Helmstädter (2008) provides the historical backdrop of studies on *S. cumini* use in reducing type 2 diabetes dating back to the pre-insulin era (Helmstädter, 2008). High doses of *S. cumini* seeds (325 mg seed powder, 3–4 times/day [common dose]), seed kernels, or fruits (30 g fruit powder) from India are recommended for blood reduction levels (~30%) based on these studies. Most successful studies used *S. cumini* seed powder or frequently its aqueous or alcoholic extract (typically 100–200 mg/kg used in animals). This bibliographical study proposes the design of a clinical study for type 2 diabetic patients with high doses of a carbohydrate-free aqueous/ethanolic extract of fresh *S. cumini* fruits, seeds, or seed kernels compared to a placebo or standard oral antidiabetic therapy. The actual bioactive compound or group of compounds responsible for the antidiabetic activity of *S.*

cumini has not been reported to date. Several mechanisms of action have been proposed for the blood glucose reduction and other antidiabetic activities of *S. cumini*. These include insulin stimulation from beta cells, oxidative stress reduction, glycogen increase in liver and muscle cells, effects comparable to those of sulfonylurea and biguanide derivatives [increasing blood glucose metabolism and insulin secretion], inhibition of carbohydrate-degrading enzymes, inhibition of the human peroxisome proliferator-activated receptor gamma, up-regulation of Glut-4 glucose transporter, cathepsin-B activity increase, inhibition of insulinase in liver and kidney, and development of insulin-positive cells from pancreatic duct epithelial cells.

Jambul was known as an Indian drug derived from *E. jambolana* that was recommended in the *Medical Record* in 1883. The report cites several cases where jambul or the powdered seed (2–30 g) reduced sugar in the urine of diabetic patients, and in some cases, sugar in urine completely disappeared. Two earlier clinical studies are noteworthy. Srivastava et al. (1983) treated 28 severe diabetic patients with 4 to 24 g *S. cumini* seed powder TDS in gelatin capsules and reported significant mean fasting and postprandial blood sugar levels reduction (–18% and –32%, respectively). Five patients showed adverse reactions (nausea, diarrhea, and epigastric pain). The second study involved 30 "uncomplicated" NIDDM patients receiving 12 g (4 g × 3) *S. cumini* seed powder for three months, with oral glucose tolerance tests performed monthly. Glucose tolerance decreased (up to 30%) after two months of treatment, and diabetic relief symptoms (polyuria, polyphagia, weakness, weight loss) improved progressively (Kohli and Singh, 1993). Furthermore, postprandial glucose levels decreased (–18%) compared to baseline in six type 2 NIDDM patients given chlorpropamide (250 mg/day) for 1 month, prompting the investigators to conclude that the moderate glucose-lowering effect of *S. cumini* was comparable to that of chlorpropamide (Derosa et al., 2022).

Sahana et al. (2010) conducted a 6-month parallel, open-label, randomized controlled trial in 30 middle age (mean age 55–60 years old) newly diagnosed type 2 diabetic patients. The test group (8M/7F) received *E. jambolana* seed-based drug Madhuhara churna [AVA Trust Regd] (5 g twice daily before meals), another group (2M/3F) received metformin (500 mg tablet), and the third group (6M/4F) were on diet restriction. The *E. jambolana* drug significantly lowered fasting blood glucose (7.4% vs 20.2% and 11.8% vs 25.3% for metformin group at the 3rd and 6th month), homeostatic model assessment for insulin resistance (HOMA-IR 10.3–6.5), and elevated high-density lipoprotein value (40–47 mg/dL) at the third month compared to baseline values. The average HbA_{1C} reduction for the 2nd and 3rd groups of patients decreased significantly (9%–7.8% and 7.8%–7.3% at the 3rd month), while the reduction in HbA_{1C} (~8%) was insignificant in patients receiving the *E. jambolana* drug. Nevertheless, *E. jambolana* treatment maintained postprandial blood sugar levels and HbA_{1C} during the study period. The *E. jambolana* drug acts as a weak secretagogue and may have extrapancreatic action due to attenuation of glucose release by enhancing glycogen storage, making it a potential agent to reduce the risk of diabetic complications. The investigators proposed that "Madhuhara churna" can be a potential agent in reducing diabetic complications such as atherosclerosis and cardiovascular disease since it reduced HDL-C levels. Therefore, Madhuhara churna can be an effective complementary therapy in patients with newly diagnosed type 2 diabetes with mildly elevated plasma glucose. It reduces the glycemic state, insulin resistance, and elevates HDL-C levels in type 2 diabetic patients.

According to the website (https://ayurvedaone.co.in/product/madhuhar-churna/), Madhuhara churna is an efficacious combination of Ayurvedic (Premehahara) herbs that effectively manage diabetic complications (nephropathy, neuropathy, and carbuncles) by naturally helping to balance blood glucose levels. The main ingredients of Madhuhara churna are Jambu (*S. cumini*), Karavellaka (*Momordica charantia*), Methika (*T. foenum-graecum*), Saptaparna (*Alstonia scholaris*), Nimba (*Azadirachta indica*), Meshasrungi (*G. sylvestre*), Bilva (*Aegle marmelos*), Guduchi (*T. cordifolia*), Chirayata (*Swertia chirata*), Katuki (*Picrorrhiza kurroa*), Lata karanja (*Caesalpinia crista*), and Shuddha Shilajit (*Asphaltum punjabianum*).

Sidana et al. (2017) described the largest sample size, single-center, double-blind, randomized controlled parallel-designed trial involving 99 type 2 diabetes mellitus (DM) patients with poorly controlled blood sugar levels (>126 mg/dL fasting plasma glucose; >180 mg/dL postprandial blood glucose during continued oral hypoglycemic treatment). *S. cumini* seed powder (5 g twice daily) supplementation reduced fasting plasma glucose of the intervention group (n = 50) by 9.2%, 18.2%, and 30% after 30, 60, and 90 days. The treatment significantly ($P < 0.001$) reduced postprandial blood glucose levels (by 14.7% and 21.6%) after 60 and 90 days compared to minimal increases (by 1.3% and 2.2%) for the placebo group. Moreover, the treatment also significantly reduced (0.7%) HbA_{1C} after 90 days compared to an insignificant increase in the placebo group. Sidana et al. (2017) confirm the beneficial effects of *S. cumini* on diabetic patients (fasting plasma glucose reduction) from earlier studies with smaller sample sizes and longer durations. Hemoglobin $(Hb)A_{1C}$ decrease results in a reduction in diabetes-related complications, including myocardial infarction (MI) and all-cause mortality.

Ahmed et al. (2023) reported the latest clinical evaluation of *E. jambolana*'s hypoglycemic effects in type 2 (NIDDM) patients (fasting blood glucose >140 mg/dL) over a six-month period. However, the details and design of the study were sketchy, following an earlier study for 14 days without providing the actual number of participants for each intervention (Waheed et al., 2006). In their study, type 2 NIDDM patients (n = 20; 10 patients with no medication and 10 patients taking oral hypoglycemic agents; six healthy subjects were used as controls) received *E. jambolana* ground powdered low or high (2 or 4 g thrice daily) doses of aqueous or alcoholic seed extract for 14 days. *E. jambolana* significantly reduced blood glucose, bringing the mean plasma glucose value within the normal diabetic range with the high seed dose. Moreover, glycosuria disappeared after two weeks of treatment with seeds and aqueous extract at both low and high doses for patients with no previous medication.

Altogether, these clinical studies demonstrate the human health antidiabetic effects of *E. jambolana* seed initially described over a century ago. Other parts of the plant have also been evaluated; however, antidiabetic effects were mostly absent, particularly from the leaves. In a randomized, parallel, double-blind trial, non-diabetic volunteers (15M/15F, mean age 22 years, fasting blood glucose >110 mg/dL) received *S. cumini* dry leaf tea (2 g dry leaves in 250 mL water or the same volume of placebo tea) with 75 g of oral glucose after fasting for 10 h. Blood glucose levels were measured after 30, 60, 90, and 120 min of the glucose load. The tea had no antihyperglycemic effect (blood glucose levels 845 ± 256 vs 818 ± 187 area under the curve for placebo; 73–123 vs 70–124 mg% for placebo) (Teixeira et al., 2000).

Aqueous *S. cumini* leaf extract (ASC) (60–1000 μg/mL) dose-dependently inhibits total adenosine deaminase (ADA) activity of serum from hyperglycemic subjects (≥150 mg/dL) and reduces blood glucose levels. ADA modulates insulin bioactivity, and a reduction in ADA activity can potentially contribute to the antioxidant defense system of red cells related to the ADA/DPP-IV-CD26 complex, thereby regulating blood glucose. Serum for the study was obtained from the control group (5F/12M; 35 years mean age, 7.76% HbA1c); intervention [hyperglycemic] group (8F/14M; 43.5 years mean age; 11.94% HbA1c) (Bopp et al., 2009).

The actual bioactive compound or group of compounds responsible for the antidiabetic activity of *S. cumini* has not been reported to date. Several investigators attribute the effects to the myriads of bioactive compounds present in the seeds, particularly the polyphenols and sterols. However, Ratsimamanga et al. (2001) isolated polyphenol and sterol complex-free components from *E. jambolana* seeds with hypoglycemic properties. Fine-ground *E. jambolana* seed powder (0.5 μm) is alcohol-extracted twice (40–70 °C), filtered, the residue solubilized in alkaline (28% ammonium hydroxide) solution, then re-extracted with 80% aqueous ethanol, filtered, and dried. Two products are obtained: mixture 1, containing sodium oxamate, reduces glycaemia in diabetic subjects and is therefore useful to treat diabetes and its complications with no risk of hypoglycemia in monotherapy. This antidiabetic agent does not stimulate insulin secretion, reduces hyperglycaemia, prevents or decreases cataracts, and restores fertility in spontaneous diabetic rats (*Psammomys obesus*). Mixture II contains 2-(tetrahydroxybutyl)-5–(2′,3′,4′-trihydroxybutyl)-pyrazine, 2-(tetrahydroxybutyl)-6–(2′,3′,4′-trihydroxybutyl)-pyrazine, and 2,5-di(tetrahydroxybutyl) pyrazine. Sharma et al. (2011) identified and purified a fraction (LH II) from ethanol *E. jambolana* seed extract that possessed antidiabetic activity in severely diabetic rabbits. However, the chemical constituents or composition of the fraction were not revealed. Therefore, the antidiabetic bioactive compound(s) of *E. jambolana* seed still remain unsolved.

1.2.7 *Azadirachta indica*

Azadirachta indica (Figure 1.7), commonly known as neem, belongs to the *Meliaceae* family. Synonyms include bead tree, margosa, pride of China, holy tree, neem or nim tree, indiar, and lilac tree. Neem is one of the most versatile medicinal plants, widely distributed in the Indian subcontinent. It is typically grown in tropical and semi-tropical regions. Neem trees also grow on islands in southern Iran. Its fruits and seeds are the source of neem oil. Neem is a fast-growing tree that can reach 15 to 20 m in height, and rarely 35 to 40 m. Bark extracts of the neem tree have been used to treat peptic ulcers. Neem oil is larvicidal against mosquitoes, and neem seed extracts are pesticidal and have been used to treat head lice (Abbas et al., 2020).

In India, neem has been in use since ancient times as a traditional medicine against various human ailments, and about 700 herbal preparations based on neem are found in Ayurveda, Siddha, Unani, Amchi, and other local health prescriptions. Neem has also received worldwide attention for its potential as an herbal pesticide and in other healthcare formulations in countries such as China, the USA, France, Germany, and Italy. The medicinal value of neem has been recognized by the US National Academy of Sciences, which published a report in 1992 entitled "Neem–A Tree for Solving Global Problems."

Azadirachta indica L. (neem) exerts a therapeutic role in health management due to its rich source of active ingredients (over 300 different compounds) found in all parts of the tree, including the seed, bark, leaves, and roots. It is a rich source of limonoids that are endowed with potent medicinal properties such as antioxidants, anti-inflammatory, and anticancer activities. The most important limonoids are azadirachtin, salannin, nimbin, nimbidin, nimbolinin, nimbidol, gedunin, and nimbolide. The neem extract has been widely used in herbal pesticide formulations because of its pest repellent properties. Neem contains various steroids, sesquiterpenoids, ester terpenoids, tetranortriterpenoids, and triterpenoids. Other active constituents include sodium nimbinate and quercetin. Leaves contain ingredients such as nimbin, nimbanene, 6-desacetylnimbinene, nimbandiol, nimbolide, ascorbic acid, n-hexacosanol, amino acid, 7-desacetyl-7-benzoylazadiradione, 7-desacetyl-7-benzoylgedunin, 17-hydroxyazadiradione, and nimbiol (Alzohairy, 2016 and references therein). Quercetin and ß-sitosterol, polyphenolic flavonoids, were purified from neem's fresh leaves and were known to have antibacterial and antifungal properties. The seeds hold valuable constituents, including gedunin and azadirachtin.

In March 2020, during the COVID-19 pandemic, in various Southeast Asian countries and Africa, the use of neem leaves was claimed to effectively inhibit the disease. The Malaysian Ministry of Health summarized myths related to the use of leaves to treat COVID-19 and warned of health risks from overconsumption of the leaves against the disease. Neem bark is currently being studied for its potential

FIGURE 1.7 *Azadirachta indica*

to treat COVID-19 (Alzohairy, 2016). Numerous biological and pharmacological activities have been reported, including antibacterial, antifungal, anti-inflammatory, antiarthritic, antipyretic, hypoglycemic, antigastric ulcer, antifungal, antibacterial, and antitumour activities (Amini et al., 2012; Bandyopadhyay et al., 2004; Sultana et al., 2007).

1.2.7.1 Clinical Trials

Neem root bark (70% alcohol) extract showed statistically significant results at an 800 mg/kg dose in diabetes (Patil et al., 2013). Neem extract (250 mg/kg) for 15 days significantly lowered glucose levels compared to the control group (Dholi et al., 2011). In a diabetic murine model, *A. indica* chloroform extract showed good oral glucose tolerance and significantly reduced intestinal glucosidase activity (Bhat et al., 2011). Another important study suggested that leaf extracts of *A. indica* and *Andrographis paniculata* have significant antidiabetic activity and could be a potential source for the treatment of diabetes mellitus (Akter, 2013).

The neem (*A. indica*) genome, sequenced since 2015, has led to a better understanding of phytomedical/phytochemical pathways. The neem tree adapts to the hot and dry climates of arid and semi-arid regions and is therefore economically and environmentally sustainable. Neem leaf is traditionally used to effectively control sugar levels in human blood. It is a natural product known for its potent low sugar and anti-cholesterol effects used to treat different types of diabetes, in addition to its antibacterial, antiviral, antifungal, biopesticidal, and other physiological/biological properties. Neem has been used in folk medicine in India for about 4,000 years.

The neem tree is indigenous to India, although it has spread to almost 80 countries around the world, with an estimated 91 million trees globally (Ezin and Chabi, 2023 and references therein). Many bioactive components have been isolated from neem leaves. For example, the pest control compound AzaA, and medicinal compounds such as nimbin, salanin, azadirachtin, Nimbocinone, nimolinone, kulactone, nimocinolides, isonimocinolide, meldinidiol, vilasinin, margosinolide, and others. Neem extracts are used to treat diabetes and help reduce the risks associated with diabetes (e.g., acidosis, eye disorders, infections, and wounds that do not heal) (Ezin and Chabi, 2023 and references therein).

The antidiabetic effects of *A. indica* (neem) have mostly been investigated in diabetic-induced animal models, although human studies are pertinent. An earlier study showed that oral administration (5 g of aqueous *A. indica* extract or a dry leaf equivalent in capsules) reduced the insulin dose of patients by over 30% to 50% (Shukla et al., 1973). Pingali et al. (2020) evaluated the safety and efficacy of a standardized aqueous *A. indica* leaves and twigs extract (NEEM) on glycemic control, endothelial dysfunction, and systemic inflammation in type 2 diabetic patients. Diabetic patients (n = 78, ~55 years old) on standard metformin therapy (1500–2500 mg/day) received 125, 250, or 500 mg NEEM (125 and 250 mg capsules PhytoBGS®, Natreon Inc., New Brunswick, NJ, USA) or placebo twice daily for 12 weeks in this randomized, double-blind, placebo-controlled clinical trial (India clinical trial registry—CTRI/2018/12/016666). The extract (intervention) significantly reduced postprandial blood glucose (PPBS), FBS, glycosylated hemoglobin (HbA1c) levels, and reflection index (IR) compared to placebo. Moreover, the treatment significantly improved diabetic outcomes (endothelial function, reduced oxidative stress, and systemic inflammation [IL-6 and high-sensitivity C-reactive protein (hsCRP)]) compared to placebo. The treatment (500 mg NEEM) at 12 weeks lowered postprandial and fasting blood sugar levels, insulin resistance (HOMA-IR), and HbA1c values by 22.6%, 19%, 57.4%, and 19.6%, respectively. The blood sugar-lowering activity of *A. indica* leaf extract is attributed to its insulin release effect due to serotonin inhibition, although the actual mechanism of action is still under investigation.

Previous studies suggested meliacinolin, a bioactive component of *A. indica* leaves, for its antihyperglycemic effect due to carbohydrate enzyme (α-glucosidase and α-amylase) inhibition (Pingali et al., 2020). However, PhytoBGS® contains total flavonoids (2.3–3.7% w/w [3% w/w average]) (quercetin-3-O-glucoside, quercetin-3-rutinoside, apigenin rutinoside, rutin derivatives), myo-inositol phosphate (7.3–11.3% w/w [9.3% average]), total polyphenols (5.2–7.7% w/w [6.6% average]), and amino acids (including arginine and methionine (Veeraragavan et al., 2021).

Azadirachta indica leaf powder (4 × 500 mg capsules/day) for 3 months significantly lowered blood pressure and reduced diabetic symptoms (polydipsia, polyphagia, and headache; 33%, 35%, and 38%, respectively) in overweight male diabetic patients (n=30, 40–60 years old, 28 kg/m^2 average BMI) (Kochhar et al., 2009). The blood pressure reduction was attributed to nimbidin, known to dilate blood vessels. Aqueous *A. indica* leaf extract (5 mL/day or 10 g/day for 2 months) significantly ($P < 0.03$) lowered fasting blood sugar levels (120 vs 125 mg/dL for control) in 20 type 2 diabetic patients (≥140 mg/dL fasting plasma glucose, 51 years median age, 27.5 kg/m^2 average BMI). The hypoglycemic effect was attributed to a reduction in carbohydrate absorption from the gut (Dineshkumar et al., 2010).

Freshly prepared *A. indica* leaves extract irrigation (twice a week for four weeks), followed by topical antibiotics (Metrogyl P and Metrogyl gel) application, significantly reduced wound (mean and area scores, 82% vs 56% for normal saline) in diabetic foot ulcer patients (n = 74) (Jayalakshmi et al., 2021). *Azadirachta indica* treatment effectively reduced wound area in more diabetic patients than saline (64% vs 35% of ≥50% wound area reduction). The extract effectively healed ulcers, reduced wound-healing scores, and improved other wound variables (wound area, tissue type, and wound exudate improvement) compared to traditional normal saline. Wound healing in diabetic foot ulcers was attributed to control of sugar levels by *A. indica* leaves. Moreover, *A. indica* leaves extract during dressing was effective in controlling clinical infection, presumably due to its antimicrobial activity against human pathogenic bacteria.

The hypoglycemic effect of *Azadirachta indica* seeds was clinically investigated in type-2 diabetes mellitus patients (Waheed et al., 2006). Powdered, aqueous, and alcoholic extract of *A. indica* (2 g thrice daily) were given to 10 type-2 diabetic patients with no previous medication, 10 diabetic patients taking oral hypoglycemic agents with a history of inadequate control, and six control subjects for 14 days. *Azadirachta indica* treatment had significant hypoglycemic activity and can be successfully combined with oral hypoglycemic agents in type-2 diabetic patients. *Azadirachta indica* seed powder (6 g/day in three divided doses for 40 days) significantly reduced fasting and postprandial blood glucose (149–135 mg/dL and 252–229 mg/dL) levels in non-insulin-dependent diabetic patients (32M/14F, 30–60 years old). Serum triglycerides and LDL cholesterol also decreased significantly (136–122 mg/dL and 157–134 mg/dL), demonstrating the hypocholesterolemic effect of *A. indica* seed powder (Jalaja kumari, 2010).

Patil, Shirahatti and Ramu (2022) reviewed 63 pharmacological investigations, predominantly (76%) on *A. indica* leaf effects on blood glucose and cholesterol levels (27%), oxidative stress (17%), diabetic complications (21%), carbohydrate digestive enzymes (17%), glucose tolerance (10%), and glucose uptake (8%). The antidiabetic activity of *A. indica* is due to the presence of diverse phytochemicals, particularly the flavonoids, surmised to lower blood glucose levels by enhancing insulin secretion. The terpenoids are presumably responsible for reducing blood sugar and improving glucose tolerance. Most *A. indica* animal studies have a common probable mechanism of antihyperglycemic activity. Administration of *A. indica* leaf extract enhances insulin by serotonin inhibition, improves pancreatic islet morphology by suppressing or scavenging reactive oxygen species (ROS), which in turn increases insulin production and reduces blood sugar levels. Another proposed antidiabetic mechanism is increased glucose utilization due to efficient glucose absorption in the gut by *A. indica* leaf extract, thereby increasing insulin receptor sensitivity.

Satyanarayana et al. (2015) studied the effects of *A. indica* aqueous leaf extract on the expression of insulin signaling molecules and glucose oxidation in high-fat and fructose (25%) induced type 2 diabetic male rats. The extract (400 mg/kg bw) for 30 days normalized the altered blood glucose and serum insulin levels, lipid profiles (antihyperlipidemic activity), insulin signaling molecules, and glucose transporter 4 (GLUT4) proteins (ascribed to the lipid-lowering effects of the extract). The antidiabetic (blood glucose reduction, normalized glucose tolerance, and normalized elevated glucose oxidation) effects of *A. indica* extract were superior and more effective than metformin (50 mg/kg)-treated rats. At the molecular level, *A. indica* extracts normalize impaired insulin-signaling molecules (insulin receptor substrate-1, insulin receptor, phosphor-Akt^{Ser473}, phosphor-IRS-1^{Ser636}, phosphor-IRS-1^{Tyr632}, and glucose transporter 4 [GLUT4]) (Satyanarayana et al., 2015). Recently, moisture-resistant film-coated tablets have been developed containing neem leaf ethanol (90%) extract (42.7% of uncoated tablets) with antidiabetic properties. The tablets (30 mg/mL leaf extract concentration) inhibited α-glucosidase enzyme (43 vs 55% for the commercial drug Glucobay) and effectively managed the post-diet fast sugar in mice (Nguyen et al., 2023).

Azadirachta indica leaf phytoconstituents (azadirachtin, nimbin, rutin, quercetin, campesterol, and others) have known pharmacological effects. They lower blood glucose, cholesterol, and triglyceride levels, improve β-cell function, increase insulin secretion, enhance glucose uptake, and inhibit α-amylase and α-glucosidase activities (Ansari et al., 2022). Molecular docking studies suggest other *A. indica* compounds as potential antidiabetic agents. For example, 63 phytochemicals were analyzed by molecular docking to the drug nateglinide and their stability with the insulin receptor ectodomain (Abdullah et al., 2023). 7-deacetyl-oxogedunin had the highest negative docking and can be a supplemental agent to treat type 2 diabetes. Other compounds (vilasinin and nimbidinin) have similar high affinity to nateglinide for the insulin receptor (thereby blocking the insulin receptor) and can therefore delay diabetes mellitus development.

Jalil et al. (2013) identified 3-deacetyl-3-cinnamoyl-azadiractin as the antidiabetic compound from *A. indica* leaf extracts with the best binding properties with phosphoenol pyruvate carboxykinase (PEPCK). The study used molecular docking simulation and screening techniques to search for antidiabetic compounds with the best PEPCK binding properties. PEPCK catalyzes the gluconeogenesis pathway, and its inhibition is associated with insulin sensitivity. *Azadirachta indica* fruit basic limonoids, azadiradione, and gedunin are potential antidiabetic drug candidate leads to control postprandial hyperglycemia by binding and inactivating human pancreatic α-amylase (Ponnusamy et al., 2015). These limonoids inhibit α-amylase (42% and 53%) from secretory cells (AR42J) by binding with aromatic amino acids tryptophan and tyrosine. The tetranortriterpenoid meliacinolin and azadirachtolide from *A. indica* leaves also inhibit porcine pancreatic α-amylase (IC_{50} ~93 μM) and α-amylase in streptozotocin-induced diabetic mice (Perez-Gutierrez and Damian-Guzman, 2012).

1.2.8 *Pterocarpus marsupium*

Pterocarpus marsupium (Figure 1.8) belongs to the family Roxb-Fabaceae and is commonly known as Indian Kino Tree or Malabar Tree in English; Vijayasar or Bija in Hindi; and Asana in Sanskrit. It is indigenous to India, Nepal, and Sri Lanka. It is a moderate to large tree commonly found in the hilly regions

FIGURE 1.8 *Pterocarpus marsupium*

throughout India, specifically in the areas of the Western Ghats, the Karnataka-Kerala region, and the states of Gujarat, Madhya Pradesh, Bihar, and Orissa. It has a place in the Rasayans group of Ayurveda and is mentioned in the Red Data Book because of its decreasing wild population due to the exploitation of the tree for its timber and medicinal bark. Different plant parts have been used for various diseases, such as leaves for boils, sores, skin diseases, and stomach pain; flowers for fever; gum-kino for diarrhea, dysentery, leucorrhea; and bark as an astringent and for toothache. Decoctions of bark and resin have been used traditionally to treat tumors of the gland, urethral discharges, and as an abortifacient. The heartwood possesses astringent, anti-inflammatory, antidiabetic, and anodyne properties. The major phytoconstituents of *P. marsupium* are pterosupin, pterostilbene, liquiritigenin, isoliquiritigenin, epicatechin, kinoin, kinotannic acid, kino-red, beta-eudesmol, carsupin, marsupol, and marsupinol (Katiyar et al., 2016).

This plant has been used as a highly potent antidiabetic agent since ancient times. It possesses blood glucose-lowering, beta-cell protective, and regenerative properties. The hypoglycemic effect of *P. marsupium* has been investigated in numerous experimental studies on various animal species, viz., rats, dogs, and rabbits. The results have shown that *P. marsupium* restored normal insulin secretion by reversing the damage to the beta cells and by repopulating the islets. Numerous studies have shown the hypoglycemic activity of the wood extract in different animal models (Gupta, 1963; Shah, 1967).

Pterostilbene (a constituent derived from *Pterocarpus marsupium* wood) caused hypoglycemia in dogs (at 10 mg/kg IV). Higher doses (20, 30, and 50 mg/kg) caused initial hyperglycemia followed by hypoglycemia lasting for nearly 5 h (Haranath et al., 1958). Oral administration of bark decoction (1 g/100 g body weight for 10 days) showed a hypoglycemic action in alloxanized diabetic rats (Dhanabal et al., 2006). The flavonoid fraction of *P. marsupium* causes pancreatic beta-cell regranulation and may explain the antidiabetic mechanism of the plant (Chakravarthy et al., 1980). Epicatechin, a pure flavonoid isolated from *P. marsupium* ethanol bark extract, also possesses a significant antidiabetic effect (Chakravarthy et al., 1982). Phenolic constituents such as marsupin and pterostilbene significantly lowered blood glucose levels in STZ-diabetic rats, and the effect was comparable to metformin (Manickam et al., 1997). All these studies support the antidiabetic potential of *P. marsupium*.

An Indian open multicenter study assessing the Ayurvedic drug Vijayasar (*P. marsupium*) in the treatment of newly diagnosed or untreated NIDDM showed that the extract controlled fasting and postprandial blood glucose levels in 67 out of 97 patients (69%) by 12 weeks at 2, 3, and 4 g in 73%, 16%, and 10% of patients, respectively. Four patients withdrew from treatment due to excessively high postprandial blood glucose levels. Fasting and postprandial glucose levels fell significantly ($P < 0.001$) by 32 and 45 mg/dL from baseline 151 and 216 mg/dL, respectively, and mean HbA1c decreased significantly ($P < 0.001$) to 9.4% among 93 patients who completed the 12-week treatment. No significant change was observed in the mean levels of lipids. Other laboratory parameters remained stable during the designated treatment period of 12 weeks (Philip, 1998). Hence, it is concluded that Vijayasar is useful in the treatment of newly diagnosed or untreated mild NIDDM patients.

1.2.9 *Tinospora cordifolia*

Tinospora cordifolia, commonly known as giloy, belongs to the *Menispermaceae* family. Synonyms include *gurjo*, *guduchi*, and heart-leaved moonseed. It is a common climbing shrub or vine (Figure 1.9) that grows on other trees and is indigenous to tropical regions of the Indian subcontinent. Its roots, stems, and leaves are used in Ayurvedic medicine due to its medicinal properties, based on centuries of use in Ayurveda. It is a large, deciduous, extensively spreading, climbing vine with several elongated, twining branches. Leaves are simple, alternate, and estipulate with long, round, and pulvinate petioles up to 15 cm long. Its name, heart-leaved moonseed, is derived from its heart-shaped leaves and reddish fruit.

A recent study has shown that 29 endophytes belonging to different taxa were present in the samples collected from *T. cordifolia*. Extracts of the endophytic fungus *Nigrospora sphaerica* obtained from *T. cordifolia* have insecticidal properties against the oriental leafworm moth (*Spodoptera litura*), a polyphagous pest. *Tinospora cordifolia* contains diverse phytochemicals, including alkaloids, phytosterols,

FIGURE 1.9 *Tinospora cordifolia*

glycosides, tinosporide, and other mixed chemical compounds. During the 2020–2022 COVID-19 outbreak in India, the Ministry of AYUSH recommended *T. cordifolia* use as a home remedy for immune support, but such a practice appeared to be associated with hepatitis cases among sick people who used boiled or capsule preparations of the plant.

Many active components derived from the plant, like alkaloids, tannins, cardiac glycosides, flavonoids, saponins, steroids, diterpenoid lactones, aliphatics, and glycosides, have been isolated from different parts of the plant, including root, stem, and whole plant (Upadhyay et al., 2010). The plant has been of great interest to researchers across the globe because of its reported medicinal properties, like antidiabetic, antiperiodic, anti-spasmodic, anti-inflammatory, antiarthritic, antioxidant, anti-allergic, anti-stress, anti-leprotic, antimalarial, hepatoprotective, immunomodulatory, and anti-neoplastic activities (Saha and Ghosh, 2012).

The stem of *T. cordifolia* is widely used in diabetes therapy by regulating blood glucose in traditional Indian folk medicine. Its antidiabetic potential is mediated through mitigating oxidative stress (OS), promoting insulin secretion, and also by inhibiting gluconeogenesis and glycogenolysis, thereby regulating blood glucose (Sangeetha et al., 2011). *Tinospora cordifolia*'s antidiabetic effects have been ascribed to its major phytoconstituents (alkaloids, tannins, cardiac glycosides, flavonoids, saponins, and steroids).

The isoquinoline alkaloid (palmatine, jatrorrhizine, and magnoflorine)-rich fraction from the stem has insulin-mimicking and insulin-releasing effects both in *in vitro* and *in vivo* (Patel and Mishra, 2011). Oral treatments of root extracts regulate blood glucose levels, enhance insulin secretion, and suppress OS markers. *In vitro* studies report initiation and restoration of cellular defense antioxidant markers, including superoxide dismutase (SOD), glutathione peroxidase (GPx), and glutathione (GSH); inhibition of glucose-6-phosphatase and fructose-1,6-diphosphatase; and restoration of glycogen content in the liver. *Tinospora cordifolia* crude stem ethyl acetate, dichloromethane (DCM), chloroform, and hexane extracts inhibited the enzymes salivary and pancreatic amylase and glucosidase, thus increasing the postprandial glucose level with potential application in the treatment of diabetes mellitus (Chougale et al., 2009).

The root extract decreases the levels of glycosylated hemoglobin, plasma thiobarbituric acid reactive substances, hydroperoxides, ceruloplasmin, and vitamin E in diabetic rats (Umamaheswari and Mainzen Prince, 2007). Oral administration of *T. cordifolia* extract in "Ilogen-Excel" formulation (Ayurvedic herbal formulation) composed of eight medicinal plants including *C. longa*, *Strychnos potatorum*, *S. oblonga*, *T. cordifolia*, *Vetivelia zizanioides*, *Coscinium fenestratum*, *Andrographis paniculata*, and *Mimosa pudica* is reported to reduce GSH and vitamin C in blood and urine, glucose and lipids in the serum and tissues in alloxan-diabetic rats, with a subsequent decrease in body weight (Prince et al., 2004). Antioxidant enzymes (GSH, GPx, and SOD levels) and catalase activity decrease in the heart and brain of diabetic rats. *T. cordifolia* root extract (TCE) has hypoglycemic and hypolipidemic effects by increasing body weight, total hemoglobin, and hepatic hexokinase and lowering hepatic glucose-6-phosphatase, serum acid phosphatase (ACP), alkaline phosphatase (ALP), and lactate dehydrogenase (LDH) in diabetic rats (Stanely et al., 2000). The protective effects of TCE were reported in the presence of higher levels of antioxidant molecules and enzymes (Shivananjappa and Muralidhara, 2012). TCE significantly counterbalances the diabetes-associated OS in the maternal liver by lowering the levels of malondialdehyde and ROS and increasing GSH and total thiol levels.

Tinospora cordifolia stem is approved for medicinal use due to its high alkaloid (berberine, palmatine, tembetarine, magnoflorine, tinosporin, tinocardifolin, and others) content. Sharma et al. (2015) presented an in-depth review of *T. cordifolia*'s antidiabetic activities in animal models and some clinical trials (Sharma et al., 2015). Aqueous *T. cordifolia* leaf extract/digest (10 g/200 mL water) significantly reduced blood sugar levels in type 2 diabetic patients (Sai and Srividya, 2002). A similar hypoglycemic effect was demonstrated in another study with two *T. cordifolia* extracts (*Guduchi Ghana*—solidified aqueous extract and *Guduchi Satva*—sedimented starchy aqueous extract). These extracts showed significant hypoglycemic activity, resulting in T2DM symptoms relief. *Guduchi Ghana* had superior glycemic control compared to *Guduchi Satva* (Sharma et al., 2013). However, a previous study reported no significant changes in fasting blood sugar levels, hypoglycemia, or hypolipidemia in T2DM patients (51 years mean age, 27.5 kg/m^2 median BMI) after aqueous *T. cordifolia* leaf extract (5 mL daily) administration for 2 months (Dineshkumar et al., 2010). Similarly, aqueous *T. cordifolia* extract (500 mg daily for 21 days) had no significantly different hematological and biological effects in healthy volunteers (22M/8F, 22.5 years average age) compared to placebo (450 mg lactose + 50 mg starch) in a double-blind, randomized, placebo-controlled trial (Rao and Bairy, 2007).

Roy et al. (2015) studied the effects of *T. cordifolia* stem supplementation on the glycemic and lipidemic profile in diabetic dyslipidemia subjects (n = 29, 50–60 years). Encapsulated mature *Tinospora cordifolia* stem (250 mg twice daily), along with prescribed hypoglycemic agents and statins for 60 days, significantly ($P < 0.005$) reduced waist and hip circumference, waist-stature ratio, and systolic blood pressure ($P = 0.0013$) compared to control. The intervention also significantly ($P= 0.0007$) reduced inflammation (high-sensitivity C-reactive protein [hs-CRP]), dyslipidemic features (by 29% vs 19% for control), and metabolic syndrome (by 13.7% vs 6.7% for the control group). Although the HbA1c reduction was insignificant, it was greater in the experimental arm (*T. cordifolia* stem supplementation), indicating its antidiabetic function due to the presence of alkaloids and other phytonutrients. The investigators speculated that *T. cordifolia* may be acting as an antihyperglycemic drug via peripheral mechanisms, since it increased glycogen synthase and decreased glycogen phosphorylase activity without serum insulin increase or pancreatic β-cell regeneration. The reduction in systolic blood pressure and metabolic syndrome indicates *T. cordifolia*'s anti-hypertensive activity and signifies reduced risk of adverse cardiac events.

Mishra et al. (2015) evaluated *T. cordifolia*'s effect as an add-on therapy to lower blood glucose levels of 100 type 2 diabetic patients (50 years average age, 26 kg/m^2 median BMI) randomized into treatment and control groups. *Tinospora cordifolia* (500 mg thrice daily) for 6 months along with antidiabetic medications significantly ($P \leq 0.005$) reduced fasting blood glucose and HbA1c (15 vs 8% for control) compared to the control group. Powdered *T. cordifolia* stem (50 mg/kg body weight, p.o., daily) for 15 days significantly lowered fasting blood sugar, total cholesterol, β-lipoproteins, and triglycerides in type 2 diabetic patients (30M/30F, 47–57 years old, 24 kg/m^2 average BMI) compared to healthy controls (15M/15F, 18–26 kg/m^2 BMI) (Kumar et al., 2016). The diabetic group was separated into control and treatment (30 subjects each). *Tinospora cordifolia* treatment reversed the high levels of fasting blood sugar (9%), HbA1c (14%), total cholesterol (13%), and triglycerides (18%) compared to healthy controls. The treatment also improved the lipoprotein profile and increased endogenous antioxidant enzymes (GSH, SOD, catalase, GPx, and GR, and attenuated LPO) in type 2 diabetic patients. However, HbA1c and LDL cholesterol showed a non-significant declining trend after *T. cordifolia* therapy compared to the diabetic control group.

Aqueous *T. cordifolia* stem extract (200 or 400 mg/kg body weight) significantly ($P < 0.05$) lowered blood sugar level during oral glucose (50 g) tolerance test in non-insulin-dependent diabetic patients (16M/8F, 48–56 years old) compared to untreated control (Puranik et al., 2014). The extract (2 g/kg body weight) also reduced blood sugar level without changing serum insulin in streptozotocin-induced diabetic albino rats. The study indicated that *T. cordifolia* improvement in glucose tolerance was probably influenced by hepatic or peripheral glucose disposal because diabetic rats were unresponsive to serum insulin or pancreatic β-cell regeneration.

1.2.10 *Ocimum sanctum*

Ocimum sanctum (Figure 1.10), commonly known as tulsi, belongs to the basil *Lamiaceae* family (*Ocimeae* tribe), thought to have originated in north-central India and now grows native throughout the eastern world tropics (Bast et al., 2014). Within Ayurveda, tulsi is known as "The Incomparable One," "Mother Medicine of Nature," and "The Queen of Herbs," and is revered as an "elixir of life" that is without equal for both its medicinal and spiritual properties. Tulsi has been adopted into spiritual rituals and lifestyle practices in India, providing many health benefits that are just beginning to be confirmed by modern science. This emerging science on tulsi, which reinforces ancient Ayurvedic wisdom, suggests that tulsi is a tonic for the body, mind, and spirit that offers solutions to many modern-day health problems (Cohen, 2014). It is present in the Himalayas up to 1800 meters above sea level. It is also grown all over the country and islands. It grows abundantly in Malaysia, Australia, West Africa, and some Arab countries. It is also found in Gir Wildlife Sanctuary and Sasangir National Park. Parts used are tulsi leaves, roots, and seeds.

Daily tulsi consumption is said to prevent disease, promote general health, well-being, and longevity, and assist in dealing with the stresses of daily life. Tulsi is also credited with giving luster to the complexion, sweetness to the voice, and fostering beauty, intelligence, stamina, and calm emotional disposition. In addition to these health-promoting properties, tulsi is recommended as a treatment for a range of conditions including anxiety, cough, asthma, diarrhea, fever, dysentery, arthritis, eye diseases, otalgia, indigestion, hiccups, vomiting, gastric, cardiac and genitourinary disorders, back pain, skin diseases, ringworm, insect, snake and scorpion bites, and malaria (Mohan et al., 2011).

The medicinal properties of tulsi have been studied in numerous scientific studies, including *in vitro*, animal, and human experiments. These studies reveal that tulsi has a unique combination of actions that include: antimicrobial (including antiviral, antibacterial, antiprotozoal, antifungal, antimalarial, and anthelmintic), mosquito repellent, antioxidant, antidiarrheal, anti-inflammatory, anti-cataract, chemopreventive, hepato-protective, radioprotective, cardioprotective, neuro-protective, antidiabetic, anti-hypertensive, anti-hypercholesterolemia, analgesic, anti-carcinogenic, anti-pyretic, immunomodulatory, anti-allergic, memory enhancement, central nervous system depressant, anti-asthmatic, diaphoretic, anti-tussive, anti-fertility, anti-thyroid, antiemetic, anti-ulcer, anti-spasmodic, adaptogenic, anti-arthritic,

FIGURE 1.10 *Ocimum sanctum*

anti-cataract, anti-leukodermal, anti-stress, and anti-coagulant activities. These pharmacological actions help the body and mind cope with a wide range of chemical, physical, infectious, and emotional stresses and restore physiological and psychological function.

Tulsi is enriched with various phytochemicals. The leaves of the plant contain volatile oil, 71% eugenol and 20% methyl eugenol. The fresh plant parts contain phenolic compounds like apigenin, isothymusin, cirsilineol, circimaritin, and rosameric acid. These also consist of two flavonoids: orientin and vicenin, ursolic acid, apigenin-7-O-glucuronide, luteolin-7-O-glucuronide, luteolin, apigenin, molludistin, orientin, sesquiterpenes, and monoterpenes such as elemene, neral, alpha- and beta-pinenes, campesterol, camphene, stigmasterol, cholesterol, and sitosterol. It also contains vitamins like vitamin C, carotene, and minerals such as calcium, phosphorus, copper, chromium, zinc, and iron.

Insulin secretion from the pancreas and clonal pancreatic β cells is enhanced due to the alcoholic and various organic solvent extracts. Tulsi extracts are known to reduce blood sugar levels, and it also has aldose reductase activity, which further helps in alleviating diabetes-related complications. It enhances insulin secretion by stimulating adenylate cyclase (cAMP) and also exerts a direct effect, which leads to the mobilization of calcium ions present within the cell, as well as promoting the entry of calcium ions.

1.2.10.1 Clinical Trials

In one of the oldest clinical studies, aqueous whole dried *Ocimum sanctum* plant extract (20 g/L) after boiling (3 min) and filtration (20 oz daily for 15 days) stabilized glycemia and glucosuria in seven out of 11 diabetic patients on regular insulin and tolbutamides regimens (Luthy and Martinez-Fortun, 1964). The beneficial effect persisted one to two months post-treatment. The extract gradually, but consistently, decreased blood sugar levels in normal non-diabetic patients. The beneficial effect was attributed to hypoglycemic compounds present in the extract. *Ocimum sanctum* (holy basil leaves tea, 5-day run-in period) for 4 weeks, followed by placebo leaves for 4 weeks or reverse, significantly reduced fasting and postprandial blood sugar levels (17.6% and 7.3%) and cholesterol (moderately) in type 2 diabetic patients (n = 40) in a randomized, placebo-controlled crossover trial (Agrawal et al., 1996). Aqueous *O. sanctum* leaves extract (5 mL daily) for 2 months lowered total cholesterol (142 vs 137 mg/dL) and LDL (91 vs 85 mg/dL) level and increased HDL (25 vs 27 mg/dL) level in type 2 diabetic patients (51 years mean age, 27.5 kg/m^2 median BMI) compared to control (Dineshkumar et al., 2010). The extract exhibits hypolipidemic effects that can potentially manage and prevent atherosclerosis formation and coronary heart disease in type 2 diabetes. This was a high-quality antidiabetic clinical trial (3 Jadad score) based on methodology quality (study design, blinding, randomization, and participant dropouts) according to Jamshidi et al. (2018).

Rai et al. (1997) evaluated the effect of *O. sanctum* leaf powder supplementation on glycemic and lipidemic control in non-insulin-dependent diabetes mellitus patients (17M/10F, 55 years mean age, 26 kg/m^2 average BMI) on hypoglycemic drugs (Rai et al., 1997). *Ocimum sanctum* supplementation (1 g fine powdered leaves daily for one month) significantly lowered blood glucose (21%), glycated proteins (11%), total amino acids (13.5%), and uronic acid (14%). The treatment also improved lipid profile by reducing total cholesterol (11%), LDL cholesterol (14%), VLDL cholesterol (16%), and triglycerides (16%). FBS and lipid levels increased slightly (<1%) in the control group (5M/5F). The beneficial effects of *O. sanctum* supplementation were attributed to its bioactive phytochemicals, speculated to influence β cells, enhance insulin secretion, and/or lower peripheral resistance. *Ocimum sanctum* dry leaf powder (3 g) administered on an empty stomach, followed by breakfast after 30 min for 45 days, significantly lowered postprandial blood sugar after 2 h (18% and 7% in males and females, respectively) and blood glucose levels (5%) only in male type 2 diabetic patients (20M/20F) (Gandhi et al., 2016).

Somasundaram et al. (2012) evaluated the antidiabetic effect of *O. sanctum* capsules (powdered aerial part) in addition to glibenclamide in type 2 diabetic patients (39M/21F, 30–65 years) in a randomized control trial. Patients were randomized into two groups; the first group received glibenclamide (5 mg daily), and the other, glibenclamide with *O. sanctum* (500 mg [2 capsules, Himalayan Drug Company, Bangalore] daily for 90 days). *Ocimum sanctum* treatment for 30 days lowered fasting and postprandial blood glucose more effectively than the hypoglycemic drug glibenclamide alone (22% vs 16% and 28%

vs 20% for glibenclamide group), and this drop was sustained for 90 days. Mean fasting and postprandial blood glucose and HbA1c levels decreased significantly in the glibenclamide group (34%, 39%, and 30%, respectively), whereas this reduction was higher in the *Ocimum* treatment group (40%, 44%, and 36%, respectively) on day 90. Moreover, over 85% of patients in the *Ocimum* group did not have any hypoglycemic episodes compared to over 80% of subjects on glibenclamide. *Ocimum sanctum* with glibenclamide was more effective than glibenclamide alone as a hypoglycemic agent. The higher mean reduction in HbA1c (36%) indicates overall *O. sanctum* glycemic control. The study demonstrates the therapeutic use of *O. sanctum* in type 2 diabetes as an adjunct hypoglycemic drug attributed to its bioactive components. The same *Ocimum sanctum* capsule (500 mg daily for 8 weeks) significantly improved body weight, serum lipid profile (increased HDL-C [22%]), plasma insulin, and insulin resistance without significant changes in fasting glucose levels in 16 young overweight/obese healthy subjects (21 years mean age, 25 kg/m^2 median BMI) (Satapathy et al., 2017).

Kochhar et al. (2009) studied the effects of powdered dry *O. sanctum* and *A. indica* leaves or their blend (1:1 ratio) (500 mg capsules) supplementation on the signs, symptoms, anthropometric parameters, and blood pressure of male diabetic subjects. T2DM diabetic male patients (40–60 years, 28 kg/m^2 mean BMI) were divided into three groups (30 patients each) after a one-month control period to receive (2 g powder daily for 3 months) *O. sanctum* (group I), *A. indica* (group II), or a mixture of both (1:1 ratio) powders (group III). All treatments significantly reduced diabetic symptoms (polydipsia, polyuria, polyphagia, and tiredness), but maximum reduction occurred in group III. *Ocimum sanctum*, *A. indica*, and their mixture significantly reduced polydipsia (35%, 33%, and 40%), polyphagia (21%, 35%, and 40%), and headache (27%, 38%, and 40%, respectively) and blood pressure (152, 149, and 151 vs 160, 155, and 159 mm Hg systolic; 92, 86, and 80 vs 98, 91, and 88 mm Hg diastolic). Medicinal plant supplementation did not change patients' anthropometric measurements.

A meta-analysis of randomized clinical trials (RCTs) up to August 2017 revealed that holy basil or tulsi consumption at higher doses (≥ 1 g/day) significantly lowered mean fasting blood glucose (by 26 mg/dL) compared to control interventions in metabolically impaired adults because of its adaptogenic properties. Tulsi supplementation (≥ 1 g/daily) was associated with improved lipid profiles (total, LDL, and VLDL-cholesterol reduction) in older (≥ 40 years) patients. Subgroup analyses indicated variation in fasting blood glucose levels with tulsi consumption due to dosage, age, and the presence of metabolic disease. The dosage of tulsi consumption, participants' age, and presence of metabolic disease significantly affected the reduction in fasting blood glucose level. FBG reductions were more pronounced in patients with T2DM or MetS with high baseline FBG (>126 mg/dL). Tulsi's bioactive phytochemicals enhance insulin secretion through elevated glucose uptake by tissues in animal and *in vitro* models. It improves glycemic control, lipid profile, and cardiometabolic parameters in preclinical studies.

1.2.11 *Momordica charantia*

Momordica charantia, commonly known as bitter melon, belongs to the *Cucurbitaceae* family. Synonyms include Goya, bitter apple, bitter gourd, bitter squash, balsam-pear, and Sanskrit: *Karavellam.* It is a tropical and subtropical vine that is widely grown in Asia, Africa, and the Caribbean for its edible fruit. Its many varieties differ substantially in fruit shape and bitterness. Bitter melon originated in Africa, where it was a dry season staple food of Kung hunter-gatherers. Wild or semi-domesticated variants spread across Asia in prehistory, and it was likely fully domesticated in Southeast Asia. It is widely used in the cuisines of East Asia, South Asia, and Southeast Asia. This herbaceous, tendril-bearing vine grows up to 5 m in length. It bears simple, alternate leaves 4 to 12 cm across, with three to seven deeply separated lobes. Each plant bears separate yellow male and female flowers. *Momordica charantia* (Figure 1.11) has a number of alleged uses including cancer prevention, treatment of diabetes, fever, HIV and AIDS, and infections. The plant has one subspecies and four varieties: *Momordica charantia var. abbreviata, Momordica charantia var. charantia, Momordica charantia ssp. macroloba, Momordica charantia L. var. muricata, and Momordica charantia var. pavel.*

FIGURE 1.11 *Momordica charantia*

The specific chemical constituents of *M. charantia* are momordicin, charantin, galacturonic acid, linoleic acid, spinasterol, and nerolidolcitrulin. *M. charantia* shows certain hypoglycemic effects via different pharmacological modes. It contains bioactive substances such as vicine, along with some antioxidants known to have antidiabetic potential. It obstructs glucose absorption by blocking glucosidase. It helps trigger insulin levels and also improves signaling or sensitivity of insulin by repairing β-cell damage. It suppresses glucose-6-phosphatase and also triggers hepatic glucose-6-phosphate dehydrogenase activity. It also inhibits α-glucosidase and α-amylase. Saponins also increase insulin secretion.

Bitter gourd is grown mostly by smallholder farmers, with over 400,000 ha planted annually in Asia, where the current value of the seed market is €16 million annually. Bitter gourd is phytonutrient-rich with many uses in traditional medicine, particularly in helping to treat type 2 diabetes that afflict up to 500 million people globally, mostly in low- and middle-income countries. Clinical studies confirm the presence of antidiabetic compounds in bitter gourd, and diets supplemented with bitter gourd fruits lower elevated fasting glucose levels (Joseph and Jini, 2013). In 2021–2022, India produced 1,132.44 thousand tonnes of bitter gourd, with five states (Madhya Pradesh, Chhattisgarh, Tamil Nadu, Andhra Pradesh, and Orissa [17%, 13%, 9.7%, 9.4%, and 8.8% of production, respectively]) producing over 115,000 tonnes each of bitter gourd (Agriexchange, 2023).

1.2.11.1 Clinical Trials

Dahlquist et al. (2023) evaluated the current clinical evidence (13 trials with a total of 892 participants) for bitter melon (*M. charantia*) as a strategy to control/manage type 2 diabetes mellitus. *M. charantia*'s antidiabetic effects are ascribed to the presence of phytochemicals, particularly cucurbitanes, saponins, charantins, momoridins, karaviloside XI, momordicoside S, polypeptide-p, and vicine. Bitter melon lowered $HbA1_C$, postprandial and fasting plasma glucose levels in several clinical trials. However, $HbA1_C$ reduction was not statistically significant in some studies, presumably due to their short duration or small sample size. Bitter melon reduced BMI, weight, fat percentage, and fructosamine levels in two trials, although the results were inconsistent. Bitter melon was not standardized to its chemical constituents in the clinical trials, which can explain inconsistencies in outcomes of the studies (Dahlquist et al., 2023).

A systematic review (1960–2018) reports ten clinical studies where *M. charantia* monoherbal formulation (2–6 g daily for at least four weeks) significantly reduced FPG, postprandial glucose (PPG), and glycosylated hemoglobin A1c ($HbA1_C$) among adult type 2 diabetes mellitus patients (n = 1045, 4–16 weeks' follow-up). *M. charantia* also lowered FPG in prediabetes (n = 52), with low-quality evidence and unclear bias risk. Phytochemicals (i.e., saponin, charantin, polypeptide-p, momordicoside, and lipids) with hypoglycemic activity in unripe fruit, seeds, and fruit pulp presumably account for *M. charantia*'s hypoglycemic effect. These phytochemicals in *M. charantia* preparations have diverse glucose-lowering actions, pathways, and modes of action, justifying the use of whole extracts rather than isolated active compounds (Peter et al., 2019).

Charantin, insulin-like peptide [plant (p)-insulin], cucurbitanoids, momordicin, and oleanic acid are primary constituents responsible for *M. charantia*'s hypoglycemic properties (Singh et al., 2011). P-insulin, with disulfide-linked two polypeptide chains, is structurally and pharmacologically similar to bovine insulin. *M. charantia*'s blood sugar-lowering effect is due to increased hepatic glucose utilization, decreased gluconeogenesis by inhibiting two key enzymes (glucose-6-phosphatase and fructose-1,6-bisphosphate), and improved glucose oxidation through the shunt pathway by activating glucose-6-phosphate dehydrogenase. Moreover, *M. charantia* extracts enhance cellular glucose uptake, promote insulin release and potentiate its effect, and increase the number of insulin-producing beta cells in the pancreas of diabetic animals.

Subcutaneous crystallized p-insulin injection from *M. charantia* extract significantly lowered blood sugar levels in nine type 1 diabetes patients compared with controls in a clinical trial (Baldwa et al., 1977). Fasting blood sugar was drawn prior to p-insulin administration, and plasma glucose levels were used to determine p-insulin dosage given to each patient. P-insulin onset was 30 to 60 min after administration, with peak effect ranging from 4 to 12 h, like that of NPH (neutral protamine Hagedorn) insulin, a medication used to treat and manage diabetes mellitus. Polypeptide-p exerted a hypoglycemic effect in juvenile and maturity-onset diabetic patients (15M/4F; 11 juvenile and 8 maturity-onset diabetes) with peak effects at 4 to 8 h in the juvenile diabetic group, compared with 2 h for crystalline bovine insulin. Blood sugar levels decreased for juvenile (n = 5, from 304 to 169 mg%) and maturity-onset (n = 6, from 141 to 101 mg%) diabetes at 4 h. Polypeptide-p preparation in saline solution was administered dose-dependently based on the severity of diabetes mellitus (10, 20, and 30 units for patients with <180, 180–250, and ≥250 mg/100 mL blood sugar levels, respectively); the control group consisted of eight out of the 19 diabetic patients. Polypeptide-p-$ZnCl_2$ was administered to three juvenile patients (Khanna et al., 1981).

Yang et al. (2022) investigated the hypoglycemic effects of *M. charantia* aqueous fruit extract containing 0.17% (~2162 g/moL, molecular weight) of the insulin-like peptide mcIRBP-19 (mcIRBP-19-BGE, Insumate®, Greenyn Biotechnology Co., Ltd., Taichung, Taiwan) in type 2 diabetic patients who failed to achieve treatment goals with antidiabetic medication. Insumate® is a patented peptide sequence: mcIRBP™-19 (composed of 19 amino acids, 2162 Da with high insulin receptor affinity), plant insulin extracted from bitter melon. The peptide can activate insulin receptor activity and directly regulate blood glucose. The intervention (300 mg mcIRBP-19-BGE capsule, twice daily for 12 weeks; 14F/6M, 58 years mean age) lowered FBG and $HbA1_C$ with borderline significance (P = 0.06) compared to the placebo group for patients on antidiabetic medications. However, FBG and $HbA1_C$ decreased significantly ($P < 0.05$) for the subset (10F/4M, 61 years mean age) of type 2 diabetic patients compared to placebo when hypoglycemic medications were ineffective. Moreover, oral mcIRBP-19-BGE administration delayed the reduction of arms and thighs circumference, indicating that it can benefit type 2 diabetic patients in the subset group in terms of sarcopenia and regulate their heart rate. This randomized, placebo-controlled trial demonstrated the hypoglycemic efficacy of mcIRBP-19-BGE (600 mg daily for 3 months) by reducing $HbA1_C$ (~0.5% on average) for type 2 diabetic patients when antidiabetic drugs were ineffective.

An earlier randomized clinical study demonstrated the therapeutic benefits of Insumate®, the bitter melon peptide (mcIRBP-19), in dose-dependently regulating diabetic patients' blood glucose (Hsu et al., 2020). Oral administration of the peptide (600 mg Insumate® [i.e., 600 ppm mcIRBP-19] peptide daily) significantly lowered glycated hemoglobin ($HbA1_C$) levels after two months in 64 diabetic patients (63 years mean age) under proper medical treatment to control blood sugar; this reduction continued until the end of the intervention (3 months, 6.3% $HbA1_C$ reduction). Moreover, the intervention regulated fasting blood glucose by improvement at the beginning and stabilization after two months. Animal studies showed that mcIRNP-19 was a putative insulin receptor binding with a β-hairpin structure and activated the insulin kinase receptor, stimulated the IR-downstream signaling transduction pathway, enhanced glucose transporter isoform 4 (GLUT4) expression, and increased glucose uptake in 3T3-L1 cells and glucose clearance in diabetic mice.

A randomized, double-blind, placebo-controlled clinical trial demonstrated that bitter melon intake for 12 weeks lowered fasting glucose levels and insulin resistance index (HOMA-IR) in type 2 diabetic patients (38M/24F, 58 years average age, 25 kg/m^2 median BMI) (Kim et al., 2020). The bitter melon used in the study was *Momordica charantia* ethanol (70%) extract (2380 mg daily for 12 weeks), standardized to contain 0.1% γ-aminobutyric acid (GABA) as a marker compound. $HbA1_C$ levels did not improve or change with *M. charantia* intervention but were exacerbated in the control group (12M/16F, 60 years mean age, 26.6 kg/m^2 median BMI) with significantly lower baseline fasting glucose levels (131 vs 146 mg/dL). The control [placebo] (57% maltodextrin [~1.36 g daily] and 36% microcrystalline cellulose), containing a higher dextrin level than the extract (49% dextrin), may have significantly ($P = 0.006$) increased fasting blood glucose (131–155 mg/dL in 12 weeks) and $HbA1_C$ (6.9%–7.2%, $P = 0.024$).

Commercial *M. charantia* tablets (1.5 g daily for 8 weeks) effectively controlled glycemia, lowered total cholesterol and oxidative stress, improved HDL-C and insulin resistance in 25 uncomplicated type 2 diabetes patients (40–60 years, 28 kg/m^2 average BMI) (Kumari et al., 2018). The treatment significantly reduced fasting blood glucose, postprandial blood sugar, HbA1c, and oxidative stress (reduced MDA levels), and effectively improved lipid metabolism, thereby alleviating the cardiovascular risk in type 2 diabetic patients. A lower *M. charantia* (1 g daily) dose also improved glycemic profile (significant FPG, PPBS, and systolic blood pressure reduction) and lipid profile (significant total cholesterol reduction) without significant changes in serum insulin levels, insulin resistance, and oxidative stress. All patients were supplemented with stable oral antidiabetic agents (metformin and glibenclamide) in this parallel randomized controlled trial.

Dried *M. charantia* fruit pulp powder (2 g daily [i.e., 4 × 500 mg capsules], Shanghai Kangxi Biotechnology Co., China) for 12 weeks significantly reduced glycated hemoglobin ($HbA1_C$), increased glucose tolerance, and total and first phase of insulin secretion (insulinogenic and Stumvoll index) in type 2 diabetic patients (7F/5M, 50 years average age, 29 kg/m^2 median BMI) without pharmacological treatment (Cortez-Navarrete et al., 2018). Moreover, *M. charantia* administration significantly reduced weight, BMI, fat percentage, and weight circumference (cardiovascular mortality risk factors) without modifying insulin sensitivity (Matsuda index), blood pressure, and lipid profiles in this randomized, double-blind, placebo-controlled clinical trial.

Fuangchan et al. (2011) compared the efficacy of three bitter melon doses (0.5, 1, and 2 g) daily for 4 weeks with metformin (1 g daily) in a multicenter randomized, double-blind, active-control trial. Dried seedless *M. charantia* fruit pulps (2 g daily containing 0.04%–0.05% [w/w] charantin [~0.8–1 mg charantin]) significantly reduced fructosamine levels (glycemic control) from baseline in type 2 diabetic patients (20F/11M, 52 years mean age, 25 kg/m^2 average BMI) with a modest and lesser hypoglycemic effect than metformin. However, the intervention did not reduce FPG and 2 h plasma glucose after oral glucose tolerance test (OGTT). Bitter melon groups administered lower doses (0.5 and 1 g daily) showed no significant reduction in fructosamine levels from baseline (Fuangchan et al., 2011). In conclusion, bitter melon had a modest hypoglycemic effect and significantly reduced fructosamine levels from baseline among patients with type 2 diabetes who received 2 g/day. However, the hypoglycemic effect of bitter melon was less than metformin 1 g/day.

Dried *M. charantia* fruit pulp (6 g daily containing 6.3 mg/day of charantin) significantly ($P < 0.05$) lowered $HbA1_C$ at 8 and 16 weeks and reduced antiglycation activities ($P < 0.03$) at 16 weeks in type 2 diabetic patients (16F/3M, 57 years median age, 25 kg/m^2 average BMI), with 15 patients taking antidiabetic medicine (Trakoon-osot et al., 2013). At 16 weeks, the intervention reduced the reversible (HbA1c) and irreversible (serum AGEs) glycation products, indicating long-term glycemic control and reduced risk or deceleration (delay/retard) of diabetic nephropathy in this two-arm, parallel, randomized, placebo-controlled trial.

Low and high (0.5 and 1.5 g/kg/d) doses of dry powdered *M. charantia* unripe fruits and their aqueous and alcohol extracts for 14 days significantly reduced mean plasma glucose level (hypoglycemic activity) in ten unmedicated type 2 diabetic patients and ten others taking oral hypoglycemic agents (Waheed et al., 2008). Glycosuria disappeared two weeks after low and high doses of *M. charantia* fruit administration. Shade-dried powdered *M. charantia* fresh whole fruit (2 g tablets thrice daily) showed no significant change in blood sugars or fructosamine after 2 and 4 weeks of administration in mild to moderate type 2 diabetic patients (19F/7M, 52 years mean age, 140–200 mg/dL fasting blood glucose, 200–300 mg/dL postprandial blood glucose) (John et al., 2003). The insignificant change in short-term glycemic control/no blood sugar-lowering effect of *M. charantia* is ascribed to the low dose used in this randomized, placebo-controlled study. Charantia® (herbal food supplement from *M. charantia* fruits and seeds) capsules (two [500 mg] capsules three times daily after meals) for 3 months slightly and insignificantly improved HbA1C in newly diagnosed diabetic patients (13F/7M, 59 years median age, 26 kg/m^2 average BMI) in a randomized, double-blind, placebo-controlled clinical trial (Dans et al., 2007). Moreover, the treatment had no significant effect on secondary outcomes (mean fasting blood sugar, serum cholesterol and creatinine, weight, hepatic transaminases [alanine aminotransferase—ALT and aspartate aminotransferase—AST], sodium, and potassium).

Carbon tetrachloride and benzene (CCl_4+ C_6H_6) fraction of whole *M. charantia* fruit methanol extract (200 mg) potentiated the hypoglycemic effects of metformin and glibenclamide (oral hypoglycemic agents) by synergy in 15 noninsulin dependent diabetes mellitus patients (52–65 years) (Tongia et al., 2004). Extract administration for one week significantly increased the hypoglycemic effect (increased fasting and postprandial blood sugar) of oral hypoglycemic agents (metformin half dose, 0.25 g; glibenclamide half dose, 2.5 mg; and half doses of both metformin and glibenclamide). The hypoglycemic effect of the extract was ascribed to the presence of phytochemicals (primarily glycosides and aglycones). A standardized ethanol *M. charantia* extract, further fractionated in ethyl acetate, was supplied as a 10% processed powder (Amsar Pvt. Ltd., Indore, India) (Fernandes et al., 2007). The extract (4 mg/kg as suspension in 1% w/v carboxymethyl cellulose) was administered (150 and 300 mg/kg) orally to rats for 30 days. The extract (300 mg/kg) significantly inhibited the rise in blood sugar levels in glucose-loaded rats, the effect equivalent to that of glibenclamide (4 mg/kg). Moreover, the extract enhanced insulin secretion from the beta cells of the islets of Langerhans, like glibenclamide. It enabled the restoration of protein breakdown and enhanced the hepatic glycogenesis process of diabetic rats. The antihyperglycemic action of the extract presumably potentiated pancreatic insulin secretion from the islets' intact β-cells.

The glucose-lowering effect of a single intake of bitter melon juice has been demonstrated in several studies. Single intake of homogenized suspension of bitter melon pulp (2 g/kg body weight) significantly ($P < 0.001$) lowered fasting and postprandial (2 h after 75 g oral glucose) serum glucose in 100 moderate non-insulin-dependent diabetic subjects (Ahmad et al., 1999). This largest and oldest study showed the highest number (86%) of glycemic response. Oral *M. charantia* preparations significantly improved glucose tolerance of type 2 diabetics in a clinical trial (Welihinda et al., 1986). Seedless juice (100 mL) administered prior to glucose load improved glucose tolerance in 13 of 18 (73%) newly diagnosed type 2 diabetic patients (38 years median age). Water was used as a control for glucose tolerance test on all patients prior to glucose load. A similar study reported blood sugar reduction (54%) after 3 weeks in seven diabetic patients using boiled *M. charantia* (100 g fruit per 100 mL water) (Srivastava et al., 1993). Moreover, glycosylated hemoglobin ($HbA1_C$) decreased 17% on average after the three-week trial. Dried fruit powder (5 g thrice daily) lowered blood sugar (25%) in five diabetes subjects after three weeks in the same study. Hypoglycemic effects in diabetic patients were highly significant at the end of the trial, but cumulative and gradual, unlike that produced by insulin. The gradual effect was ascribed to adaptogenic properties, indicated by delay in the appearance of cataracts (secondary diabetic complications) before hypoglycemia occurred in alloxan-induced diabetic rats.

Lee and Ma (2019) investigated the effect of bitter melon intake on postprandial glucose and insulin levels in sedentary, abdominally obese subjects in a randomized, double-blind, placebo-controlled trial. Single bitter melon fruit pulp juice (100 mL) intake prior (30 min) to an oral glucose test significantly ($P < 0.05$) lowered 2 h postprandial blood glucose (99.5 vs 134 mg/dL for control group), resulting in 26% glucose reduction in eight obese females (31 years mean age, 35 kg/m^2 median BMI, 122 cm waist circumference) without changes in insulin levels between the two groups. Oral glucose (50 g) tolerance test was administered in Asian diabetic outpatients (6M/3F) on antidiabetic drugs (3 on chlorpropamide, 3 on tolbutamide, and one each on glymidine and glibenclamide) before and after *M. charantia* (karela) pulp juice (50 mL) consumption and 8–11 weeks after fried karela (0.23 kg daily) consumption (Leatherdale et al., 1981). Karela juice reduced plasma glucose level and increased insulin concentration (31 vs 25 mU/L). Fried karela consumption significantly reduced glycosylated hemoglobin (18% vs 19.6%). Karela improved glucose tolerance in diabetes, with juice providing pronounced effect, and small improvement with fried karela. The reduction in glycosylated hemoglobin suggests an extrapancreatic effect.

Consumption of *M. charantia* beverage (1.25–3 g bitter melon extract) prior to an oral glucose tolerance test attenuated the postprandial glucose response in 50% of prediabetic participants (3F/2M, 65 years mean age, 26 kg/m^2 median BMI) (Boone et al., 2017). Acute ingestion of the beverage reduced total AUC_{gluc} (–13%), mean glucose concentration (–12.2%), and postprandial blood glucose compared to control. Glycemic response improved independently of an augmented insulin response. In this single-blinded, crossover investigation on prediabetic adults (99–126 mg/dL fasting plasma glucose and/or 5.7%–6.4% $HbA1_C$) (6F/4M, 61 years mean age, 26 kg/m^2 average BMI), subjects displayed two distinct metabolic response patterns to oral glucose tolerance test. Responders had slightly impaired glucose tolerance ($Fast_{peak}$) (i.e., those whose highest glucose concentration was observed 30 min after glucose ingestion). Non-responders had extensively impaired glucose tolerance ($Slow_{peak}$). However, neither group ($Fast_{peak}$ and $Slow_{peak}$) showed significant change in postprandial glycemic response during an oral glucose tolerance test when the beverage was consumed 30 min prior to glucose ingestion. The beverage had no significant insulinogenic effect, indicating that the acute glucose tolerance improvement may be attributable to extrapancreatic mechanisms such as increased glucose transporter isoform 4 (GLUT4) activities or enhanced glucose uptake in the skeletal muscle and liver.

Bitter melon pulp juice (55 mL/24 h) treatment did not alter serum sialic acid (elevated in diabetes) and glucose levels in NIDDM (21M/4F, 47 years mean age, 24 kg/m^2 average BMI) compared to normal control subjects (15M/10F) (Inayat-ur-Rahman et al., 2009). However, total cholesterol concentration was higher in bitter melon-treated patients (192 vs 171 mg/dL) compared to control, suggesting significant hypoglycemic, lipid-lowering, and antidiabetic effects of bitter melon in diabetes management. Serum sialic acid, glucose, and total cholesterol levels significantly increased in NIDDM patients (13M/12F) following rosiglitazone (4 mg/24 h) treatment compared to control. The study showed that bitter melon can effectively manage diabetes and its related complications compared to the oral antidiabetic drug rosiglitazone (Avandia, GlaxoSmithKline).

Richter et al. (2023) suggested that *M. charantia* polysaccharides may be responsible for its hypoglycemic effects. *M. charantia* consists of an acidic, branched heteropolysaccharide (MCBP) fraction composed of mannose, galacturonic acid (GalA), rhamnose, glucose, galactose, xylose, and arabinose, whereas pectic polysaccharide (PS) is the main constituent of the remaining polysaccharide fraction. The MCBP fraction possesses α-amylase inhibitory and angiotensin-converting enzyme (ACE) inhibitory activities. Alpha-amylase inhibition reduces blood glucose excursions by delaying starch/carbohydrate digestion, and ACE inhibitors lower blood pressure. Thus, enzyme inhibition by MCBP fraction can help in the control, treatment, and prevention of T2DM. A water-soluble polysaccharide isolated from *M. charantia* demonstrated significant hypoglycemic effects (Richter et al., 2023 and references therein). *Momordica* seeds also possess hypoglycemic activity. The seed (unknown amount) significantly lowered postprandial blood sugar in adult diabetic patients (14 NIDDM and 6 female IDDM, 30–60 years old) after 2 h breakfast (150–180 vs 350–380 without seed consumption) (Grover and Gupta, 1990).

1.2.12 *Coccinia indica*

Coccinia indica (Figure 1.12), commonly known as the ivy gourd belongs to the *Cucurbitaceae* family. Synonyms include *baby watermelon, little gourd, gentleman's toes, tindora*, or *gherkin*. It is a tropical vine. It is also known as *Cephalandra indica* and is indigenous to Bengal and other parts of

FIGURE 1.12 *Coccinia indica*

India. *C. indica* grows abundantly all over India, Australia, tropical Africa, Fiji, and throughout the oriental countries. *Coccinia indica* is grown largely in Bangalore, Kerala, Tamil Nadu, Andhra Pradesh, Maharashtra, Bihar, and Madhya Pradesh. It is used as a vegetable in southern and coastal India. Immature fruits are popular among diabetic patients and used for cooking since they are rich sources of carbohydrates, protein, and vitamins (A and C). The plant has also been used extensively in Ayurvedic and Unani practice in the Indian subcontinent. Seeds or fragments of the vine can be relocated and lead to viable offspring. This can occur when humans transport organic debris or equipment containing *C. grandis*. Once the ivy gourd is established, it is presumably spread by birds, rats, and other mammals.

The plant contains fatty acids, alkaloids, resins, flavonoids, and proteins. *C. indica* methanol fruit extract contains alkaloids, steroids, tannins, saponins, ellagic acid, phenols, glycosides, lignans, and triterpenoids. Roots contain triterpenoid, saponin coccinioside, flavonoid glycoside ombuin 3-O-arabinofuranoside, lupeol, β-amyrin, β-sitosterol, and stigmast-7-en-3-one. Many clinical trials have proven the antidiabetic effectiveness and safety of these plant parts and derived formulations. Anti-inflammatory, analgesic, and antipyretic activity of the fruit and leaves were also found to be significant. Chronic administration of *C. indica* (fruit) extract (200 mg/kg) for 14 days reduces blood glucose levels of diabetes-induced animals compared to the diabetic control group. Dried whole plant extract was utilized. Ingredients present in the extract act like insulin, correcting the elevated enzymes G-6-P(ase), LDH in the glycolytic pathway, and restoring the LPL activity in the lipolytic pathway with the control of hyperglycemia in diabetes.

Coccinia indica is famous as a marvel plant with immense antidiabetic potential. It possesses hypoglycemic, antidiabetic, hypolipidemic, hepatoprotective, larvicidal, anti-inflammatory, analgesic, and antipyretic activities. Various phytoconstituents reported in *C. indica* are cephalandrol, lupeol, tritriacontane, stigma-7-en-3-one, cephalandrine A, cephalandrine B, β-sitosterol, taraxerone, and taraxerol. Terpenoids are responsible for the antidiabetic activity. Many patented formulations derived from *C. indica* are now distributed increasingly all over the world (Kuriyan et al., 2008).

Dharmatti et al. (2010) identified DRC-1 with superior traits except fruit shelf life among 46 *C. indica* genotypes collected from South India and grown in one location under the same conditions in 2005. DRC-1 recorded the highest yield (84.17 tonnes/ha) with better quality parameters (20 g average fruit weight and 29 kg/plant average fruit yield) in 4 years. Bhatt et al. (2019) standardized a micropropagation protocol to commercially field produce *C. indica* crop. A typical plantation of the micropropagated crop produces 1000 to 1500 kg per acre compared to 150 to 250 kg for the conventional crop. This production method of micropropagated plants is now common practice for this crop.

Coccinia indica homogenized freeze-dried leaves (6 tablets daily for six weeks) significantly (P < 0.001) improved glucose tolerance in ten out of 16 uncontrolled, maturity-onset diabetic patients in a double-blind control trial (Khan et al., 1980). The authors concluded that *C. indica*'s active principle is slow-acting, since the maximum effect occurred only after three weeks of treatment. Dried herb pellets of *C. indica* fresh leaves improved glycemic control comparatively similar to an oral hypoglycemic agent (chlorpropamide—a conventional drug) in a three-arm controlled clinical trial (n = 70) (Kamble et al., 1998). The same group administered dried *C. indica* extract (500 mg/kg body weight) orally to 30 diabetic patients for six weeks. The extract, acting like insulin, corrected the elevated glucose-6-phosphatase, LDL in the glycolytic pathway, and restored the lipolytic activity to control hyperglycemia in diabetes. Kuriyan et al. (2008) investigated *C. cordifolia*'s effectiveness on blood glucose levels of incident type 2 diabetic patients (33M/26F, 47 years mean age, 25 kg/cm^2 average BMI) requiring only dietary or lifestyle modifications in a randomized, double-blind, placebo-controlled trial. Alcohol extract (50%) from *C. cordifolia* leaves and fruits (1 g as two 500 mg capsules daily [≡ 15 g dried herb] for 90 days) significantly lowered fasting and postprandial blood glucose (16% and 18%) and HbA1c (6.1% vs 6.7% at baseline) without significant changes in serum lipid levels and anthropometric parameters compared with that of the placebo group.

Powdered (through 1.4 mm) *C. indica* leaves (6 g twice daily) for 60 days significantly (P < 0.001) lowered fasting and postprandial blood sugar (20% and 24.5%) and HbA1c (8.4%) levels in type 2 diabetic patients (10F/5M, 31–60 years old) (Junaid et al., 2020). The treatment also significantly reduced fasting and postprandial urine sugar levels (39% and 44%) and improved other symptomatic parameters (reduced excessive urination, excessive thirst, turbid urine, and weakness; 55%, 51%, 53%, and 73%, respectively). Barley, used as control in this single-blind study, showed no significant improvement in all laboratory and symptomatic parameters. The study indicated that *C. indica* effectively relieved diabetic symptoms and normalized blood sugar metabolism. *Coccinia grandis* leaves (20 g) significantly lowered blood sugar levels and 2 h postprandially in healthy volunteers (15F/15M, 33 years mean age) in a double-blind phase 1 clinical trial (Munasinghe et al., 2011), indicating its improvement of glucose tolerance. However, one study (Quamri et al., 2017) reported insignificant effect of dried *C. indica* leaves (7.5 g daily) for 8 weeks on dyslipidemia (21F/19M, 44 years average age) associated with type 2 diabetic patients. The control group in this single-blind, randomized, standard control clinical trial received metformin (500 mg tablets twice daily) before meals.

The mechanism of *C. indica*'s action is not well understood, but it is apparently an insulin mimetic (Kuriyan et al., 2008). Phytochemicals from *C. indica*'s extract, like insulin, normalize the elevated enzymes glucose-6-phosphatase and lactate dehydrogenase in the glycolytic pathway and restore lipase activity in the lipolytic pathway, thereby controlling hyperglycemia in diabetes. Graidist and Purintrapiban (2009), investigating the mechanism of *C. indica*'s hypoglycemic action, demonstrated that aqueous *C. indica* (CI) stem extract regulated glucose transporter 1 (GLUT1) promoter activation, resulting in elevated GLUT1 protein expression. CI extract acts directly through the stimulation of the GLUT1 promoter at the transcription initiation site, akin to metformin, that directly activates the GLUT1 promoter. Jamwal and Kumar (2019) isolated quercetin by bioactivity-guided fractionation of ethyl acetate fraction obtained from methanol extract of *C. indica*'s aerial parts. Quercetin (5 mg/kg, p.o.) significantly reduced glucose levels, increased insulin and serum HDL-cholesterol levels, and improved β-cell functions in type 2 diabetic rats (Jamwal, 2019). Quercetin exhibited significant antidiabetic activity equivalent to the standard drug metformin, leading these investigators to conclude that quercetin was responsible for the antidiabetic activity of *C. indica* aerial parts.

1.2.13 *Mangifera indica*

Mangifera indica (Figure 1.13), commonly known as mango, is a juicy fruit which belongs to the family Anacardiaceae. It is grown in many parts of the world, particularly in tropical countries, and is also considered to be the national fruit of India and the Philippines and the national tree of Bangladesh. Nearly 1000 varieties of mango are available, out of which a few are traded commercially in 87 countries.

FIGURE 1.13 *Mangifera indica*

Mango is cultivated on an area of approximately 3.7 million ha worldwide. India is the global leader among mango-producing countries, accounting for about 50% of the total world mango production (Kehar and Misra, 2020 and references therein). Mango is considered "king of Asian fruits" and the national fruit of India. India's mango exports (27,330 MT) to 41 countries witnessed a 19% increase at $47.98 bn during the 2023 season (www.business-standard.com, January 18, 2024). In terms of production, mango is the second-most cultivated tropical crop after banana. Mango fruits are an important source of micronutrients, vitamins, and other phytochemicals. Along with these, they provide energy, dietary fiber, carbohydrates, proteins, fats, and phenolic compounds. Stems, bark, leaves, and seeds of mango have an important role in the treatment of diabetes. Herbal medicines are used as alternative agents to help lower blood glucose levels, with minimal side effects and low cost. Among those herbal plants, mango (*M. indica* L.) is one of the effective plants used for the treatment of diabetes in different communities (Samanta et al., 2019).

Mango is a rich source of vitamins, organic acids, carbohydrates, amino acids, phenolic acids (e.g., gallic acid, caffeic acid, and tannic acid), and certain volatile compounds. Many of the pharmacological properties present in mango might be due to the presence of phenolic acids. These phenolic compounds possess potent antioxidant activity, which is helpful for protecting body tissues against oxidative stress, with their antidiabetic, antioxidant, anti-inflammatory, antilithiatic, and anticarcinogenic properties. Apart from the fruit, mango flesh, leaf, and stem bark have also been reported to have antidiabetic, antilithiatic, and free radical-scavenging properties. Even the stem and bark of this plant have antioxidant, anti-inflammatory, and immunomodulator activities, and have been formulated (tablet, capsule, syrup, and cream) in various food supplements such as antioxidant, analgesic, anti-inflammatory, and immunomodulatory agents, and have been extensively used to prevent and cure diseases like HIV/AIDS, cancer, asthma, gastric, and dermatological disorders.

The major chemical constituent present in mango leaf is mangiferin. Other constituents include selected anthocyanidins such as delphinidin, peonidin, and cyanidin; leucoanthocyanins; catechin; and gallic tannins. The mangiferin content of mango leaves ranges from about 2% to 15%, depending on the variety and geographic source, while extracts of mango leaves contain up to approximately 60%

mangiferin. Mango leaf oil is rich in sesquiterpenes and also contains mangiferin, δ-3-carene, α-gurjunene, β-selinene, and β-caryophyllene. Mango leaves also contain alkaloids (0.84 mg/100 g), phenols (0.09 mg/100 g), flavonoids (11.24 mg/100 g), saponins (3.22 mg/100 g), and tannins (0.45 mg/100 g). These also contain high amounts of polyphenolic antioxidants, including xanthonoids, mangiferin, and gallic acid. Other flavonoid and flavonol constituents include quercetin, catechin, and epicatechin.

Roots and bark are used as astringent, acrid, refrigerant, styptic, antisyphilitic, vulnerary, antiemetic, anti-inflammatory, and constipating. They are useful in vitiated conditions of *pitta*, metrorrhagia, calonorrhagia, pneumorrhagia, leucorrhea, syphilis, uteritis, wounds, ulcers, and vomiting. The juice of fresh bark has a marked action on mucous membranes in menorrhea, leucorrhea, bleeding piles, and diarrhea. Leaves are used as astringent, refrigerant, styptic, constipating, and in conditions like cough, hiccup, hyperdipsia, burning sensation, hemoptysis, hemorrhages, wounds, hemorrhoids, ulcers, diarrhea, pharyngopathy, dysentery, and stomachopathy. Leaf ashes are useful in burns and scalds, and smoke from burning leaves is inhaled for relief of throat diseases. Flowers are used as astringent, refrigerant, styptic, vulnerary, constipating, and hematinic.

The dried flowers are useful in vitiated conditions of *pitta*, hemorrhages, hemoptysis, wounds, ulcers, anorexia, dyspepsia, uroedema, gleet, catarrh of the bladder, diarrhea, chronic dysentery, and anemia. The unripe fruits are acidic, acrid, antiscorbutic, refrigerant, digestive, and carminative. The ripe fruits are refrigerant, sweet, emollient, laxative, cardiotonic, hemostatic, aphrodisiac, and tonic. They are also used in vitiated conditions of *vata* and *pitta*, anorexia, dyspepsia, cardiopathy, hemoptysis, hemorrhages from uterus, lungs, and intestine, emaciation, and anemia. The seed kernel is a rich source of protein (8.5%) and gallic acid. It is sweet, acrid, astringent, refrigerant, anthelmintic, constipating, hemostatic, vulnerary, and a uterine tonic. It is useful in vitiated conditions of *pitta* and cough, helminthiasis, chronic diarrhea, dysentery, hemorrhages, hemoptysis, hemorrhoids, ulcers, bruises, leucorrhea, menorrhagia, diabetes, heartburn, and vomiting.

Badmus et al. (2018) studied a significant increase in the fasting blood glucose concentrations in alloxan-induced diabetic rats. *M. indica* ethanol leaf extract significantly lowered fasting blood glucose levels in diabetic rats compared with untreated rats. Luka and Mohammed (2012) reported that *M. indica* aqueous leaf extract (400 mg/kg body weight) significantly reduced blood glucose levels in diabetic rats. However, the mechanism of action of the plant extract was unknown. The extract also significantly decreased serum cholesterol levels in diabetic rats (Sudha Madhuri and Mohanvelu, 2017).

Long-term (21 days) administration of methanol and aqueous extracts of *M. indica* was effective in decreasing blood glucose levels and normalizing other biochemical parameters in diabetic rats. A single-dose study of the extract had no hypoglycemic effect on normal rats. Further studies are required to define the active principles present in these extracts. They also confirmed that oral administration of *M. indica* seed kernel extracts lowered total cholesterol and triglyceride levels in diabetic rats compared to diabetic controls (Irondi et al., 2016). The extracts of *M. indica* leaves and stem bark showed significant antihyperglycemic effects in type 2 diabetic model rats when the extracts were fed simultaneously with glucose. Single oral administration (250 mg/kg body weight) produced a potent and strong hypoglycemic effect in type 2 rats. The leaf extract of *M. indica* can be used for its antidiabetic properties using normoglycemic, glucose-induced hyperglycemia, and STZ-induced diabetic mice. The aqueous extract of *M. indica* leaves possesses hypoglycemic activity.

1.2.13.1 Clinical Studies

Mango has low glycemic value (51 glycemic index) and glycemic load for 120 g mango. Clinical studies on the effects of mangoes' consumption on glycemia and related metabolic factors show that overall mango consumption decreased inflammation, systolic blood pressure, and overall blood glucose levels and improved HbA1c and insulin levels (Kehar and Misra, 2020). However, these clinical trials involved a limited number of subjects for short periods, with only one study in diabetes patients, with all studies (except one) conducted in the USA and hence may not be applicable to Indian varieties of mangoes. Mango consumption (208.5 g edible mango portion containing 25 g digestible carbohydrate) modulated both postprandial glucose and insulin response in 10 NIDDM female patients (8 treated with sulfonylureas and 2 by diet alone) (52 years average age, 26 kg/m^2 median BMI, including 5 obese [> 25 kg/m^2 BMI] subjects) (Roongpisuthipong et al., 1991).

Zarasvand et al. (2023) investigated the antidiabetic properties of different parts of the mango plant in managing type 2 diabetes mellitus in animal models and humans in a systematic review. Twenty-eight animal and human studies evaluated mango's antidiabetic properties from leaf (31%), flesh (38%), seed kernel (7%), peel (14%), stem bark (7%), and by-product (3%), supporting mango's glucose-lowering properties. Several mechanisms of action are proposed for mango's antidiabetic effects including enzyme inhibition (α-amylase and α-glucosidase), improved antioxidant status, insulin sensitivity and glucose uptake, and gene regulation (glucose transporter 4, insulin receptor substrate 1, and phosphoinositide 3-kinase). A short-duration study (<24 h) examined the impact of mango leaf in human subjects. Mango tea had no effect on blood glucose of normal and diabetes patients, although the antidiabetic impact of mango leaf extracts was significant on blood glucose in animal models (Zarasvand et al., 2023 and references therein).

Eight studies evaluated the effects of mango fruit flesh (fresh, frozen, pureed) or dried mango powder in healthy subjects and those with T2DM or obesity. *Langra* mango (250 g) produced higher glycemic and insulin response compared to white bread (63 g as reference meal) in non-obese diabetic subjects (7M/6F, 39 years mean age, 24.6 kg/m^2 median BMI). An equi-carbohydrate amount (50 g) of mango produced a higher glycemic response (122 glycemic index of mango) than bread and absolute C-peptide increment over basal value [ACP] (1.05 vs 2.73 ng/mL for bread) (Fatema et al., 2003).

Mango was considered an intermediate food choice compared with white bread based on glycemic index (59 vs 100 GI for white bread). Guevarra and Panlasigui (2000) determined mango's glucose responses among T2DM (4M/6F, 55 years mean age, 23 kg/m^2 median BMI) patients using white bread as control. Mango (Bulacan variety, 144 g with 25 g available carbohydrate) significantly ($P \leq 0.05$) lowered blood glucose areas compared to bread (50 g), with a lower glycemic index (59) attributed to fruit characteristics (fiber and sugar [3% fructose] contents, presence of antinutrients [0.03% phytic acid], acidity, and physical characteristics) that contribute to slow digestion and absorption. The study proposes that mango, classified as intermediate GI food, can be eaten by diabetics without significantly increasing their blood glucose levels.

Ray et al. (2017) compared the glycemic response of equal quantities of mango with white bread among T2DM individuals and age-matched (45–65 years) normal controls (n = 6). The peak and total IAUC (fasting and postprandial blood sugar) response to mango (50 g *Alphonso* variety) was significantly ($P < 0.001$ and $P < 0.05$ for diabetic and normal, respectively) lower than an equal quantity of bread in both groups. The glycemic bread equivalent for mango is only 14, indicating that 50 g mango produces a similar glycemic response as that of 14 g of white bread. The study indicates that mango does not result in postprandial hyperglycemia in normal or diabetic subjects (because of high moisture content resulting in low glycemic load), and its consumption can have therapeutic value against oxidative stress–related disorders.

Mango supplementation (400 g mango pulp daily for 6 weeks) had no effect on lipid levels, but lowered plasma AUC inflammatory cytokines [IL-8 and MCP-1, by 46% and 33%] in an obese (4M/5F, 28 years mean age, 35 kg/m^2 median BMI) group (Fang et al., 2018). This indicated improved acute inflammatory response in obesity after mango supplementation. Mango supplementation lowered systolic blood pressure (by 4 mm Hg) and significantly ($P < 0.05$) increased gastric inhibitory peptide [GIP] (47%) only in lean participants (9M/3F, 25.6 years average age, 23 kg/m^2 median BMI), and reduced (18%) HbA1c (glycemic control) [indicating improved long-term glucose homeostasis] and plasminogen activator inhibitor-1 [PAI-I] (18% and 20%) in obese participants. The study demonstrated that 6 weeks of mango supplementation exerts beneficial effects in lean and obese individuals by lowering blood pressure in lean individuals and lowering inflammatory cytokines (IL-8 and MCP-1), PAI-I, and HbA1c in obese individuals, thus reducing the risk of developing obesity-related chronic diseases.

Freeze-dried mango (*Tommy Atkins* variety) pulp supplementation (10 g daily for 12 weeks) significantly reduced blood glucose in males (−4.45 mg/dL, P = 0.018) and females (−3.56 mg/dL, P = 0.003), and hip circumference only in males (−3.3 cm, P = 0.048), and increased insulin levels in males (2.2 μU/mL, P = 0.032) (11M/9F, 36.5 years average age, 35 kg/m^2 median BMI) (Evans et al., 2014). Changes were insignificant in anthropometric measurements, body weight or composition, and glycated hemoglobin of obese adults in this pilot study. Mango supplementation was unable to modulate body composition in obese individuals, unlike animal models of diet-induced obesity (i.e., high-fat diet–fed mice) using the same product (1% by weight). The study demonstrates that mango consumption has no effect on body weight or fat in obese individuals but can still benefit blood glucose improvement. The mango pulp (4.4 mg/kg mangiferin) powder contained 89.6% carbohydrate, 4% protein, 1.6% fat, and 13.4% fiber by weight, with each 10 g bag (equivalent to ~½ of mango fruit [100 g]) providing 39 kcal.

Careless mango fruit powder (100 or 300 mg daily for 4 weeks) significantly improved HbA1c concentration and blood flow in healthy individuals with early-stage impaired glucose metabolism (42M/33F, 57 years mean age, 26 kg/m^2 median BMI). The improved glucose tolerance was attributed to mangiferin's ability to activate the cellular master regulators AMPK and Sirt 1, and thereby enhance glucose uptake and utilization in skeletal muscle. The same mango (100%) fruit powder improved glucose utilization and protected against hepatic fat accumulation and insulin resistance in fat-fed mice (Zarasvand et al., 2023 and references therein).

1.2.14 *Benincasa hispida*

Benincasa hispida, commonly known as ash gourd, belongs to the Cucurbitaceae family. It is highly cultivated throughout India, especially in Kerala, although it is native to Japan and Java. *B. hispida* is considered one of the famous crops under the Cucurbitaceae family, grown mainly for its fruits and well-renowned for its nutritional and medicinal properties (Islam et al., 2021). Synonyms, also known as *kundur fruit, chalkumra, wax gourd, winter gourd, ash gourd, winter melon, white gourd, tallow gourd, Chinese preserving melon, ash pumpkin,* and *(alu) puhul,* a creeper grown for its very big-sized fruit, is eaten as a green mature vegetable or greens. Fuzzy, fine hair coats the outer side of the young fruit and have solid, thick, sweet, white flesh. The mature fruit sheds its hairs and forms a waxy, white coating, hence the name "*wax gourd*." The gourd wax coating increases its storage. It can grow up to 80 cm long, has broad leaves, and yellow flowers. The taste is rather bland. *B. hispida* is a native of South and Southeast Asia. However, it is commonly grown all over Asia, including Japan, Burma, Ceylon, Sri Lanka, Java, and Australia.

In India, *B. hispida* (Figure 1.14) is used as a winter season vegetable for various diseases. Its medicinal properties have also been recognized in the Ayurvedic system of medicine, spiritual traditions of

FIGURE 1.14 *Benincasa hispida*

India, and Yoga. In Vietnam, its soup (cooked with pork short ribs) is traditionally used by breastfeeding mothers. In north India and almost all regions in Bangladesh, it is added with pulses like moong, which, usually crushed along with wax gourd, makes a dish locally called *bori*, which, after sun drying, is used in curry dishes and eaten with rice or chapati. The fruit is also used in peptic ulcer, diabetes mellitus, urinary infection, hemorrhages from internal organs, insanity, epilepsy, and other nervous disorders in Ayurveda. The fruit is sweet and traditionally used as a cooling, styptic, antiperiodic, laxative, diuretic, tonic, aphrodisiac, and cardiotonic, and also in jaundice, dyspepsia, urinary calculi, blood disease (e.g., hemorrhages from internal organs), insanity, epilepsy, asthma, diabetes, vitiated conditions of *pitta*, fever, menstrual disorders, and balancing the body heat.

The edible portion of *B. hispida* contains moisture (94–97 g/100 g), proteins (0.30–0.70 g/100 g), carbohydrates (1.10–4 g/100 g), fat (0.02–0.20 g/100 g), fiber (0.50–2.1 g/100 g), and ash (0.27–0.70 g/100 g). Vitamins present in the edible portion (mg per 100 g) of this plant are vitamin C (1.35–68), thiamin (0.02–0.04), riboflavin (0.02–0.31), niacin (0.20–0.46), and vitamin E. Major minerals in the edible portion (mg per 100 g) include sodium (0.14–6), potassium (77–131), calcium (5–23.3), iron (0.2–0.5), and phosphorus (225–235) (Nagarajaiah and Prakash, 2015). The fruit contains water-soluble polysaccharides, such as arabinogalactans (Mazumder et al., 2005). The fruit pulp contains homogalacturonan, β-(1$\rightarrow$4)-D-galactan, acidic arabinan, and natural sugars (e.g., glucose and fructose) (Mazumder et al., 2004). The mature fruit also contains organic acids such as malic and citric acid.

The leaf contains alkaloids, flavonoids, steroids; and the fruit contains amino acids, pectic polysaccharides, hemicellulose polysaccharides, terpenes and terpenoids, flavonoid C–glycosides, sterols, proteins, phenols, alkaloids, glycosides, tannins, saponins, hydroxybenzoic acids, flavonols, hydrocinnamic acids, and triterpenes (Arbotante and Arriola, 2016). The seeds contain proteins, carbohydrates, phenolic compounds, amino acids, flavonoids, sterols, glycosides, alkaloids, fixed oils and fats, phenolic compounds, steroids, and unsaturated fatty acids. The peel contains alkaloids, steroids, saponins, flavonoids, carbohydrates, tannins, oxalates, carotenoids, and phytate. The fruit contains many volatile compounds, including (*E,E*)-2,4-nonadienal, (*E*)-2-hexenal, (*E,E*)-2,4-heptadienal, n-hexyl formate, n-hexanal, (*Z*)-3-hexenal, 2-methyl pyrazine, 1-octen-3-ol, (*E*)-2-heptenal, 2,3,5-trimethylpyrazine, 2,5-dimethylpyrazine, 2-ethyl-5-methyl pyrazine, and 2,6-dimethylpyrazine. *B. hispida* is rich in phenolic compounds. Several other bioactive compounds include isomultiflorenyl acetate, isovitexin, 5-gluten-3-β-ylacetate, multiflorenol, 1-sinapoylglucose, alnusenol, and benzylalcohol-*O*-α-L-arabinopyranosyl-(1–6)-β-D-glucopyranoside. The most representative phytochemicals present in *B. hispida* are linoleic, palmitic, oleic, stearic, and gallic acids, lupeol, β-sitosterol, ascorbic acid, and β-carotene.

The methanol *Benincasa hispida* stem extract significantly and dose-dependently lowered blood glucose level in alloxan-induced diabetic rats (Battu et al., 2007). In another study, chloroform fruit extract dose-dependently ameliorated the derangements in lipid metabolism in alloxan-induced diabetic albino rats after 14 days of treatment (Patil et al., 2011). The study reveals that methanol, ethanol, and aqueous peel extracts showed significant α-amylase inhibitory activity. The ethanol and ethyl ethanoate leaf extracts lowered blood glucose levels of diabetic mice in a dose-dependent manner. Lipids are fatty organic substances that are the largest source of energy for the body. The vast majority of fats are stored in solid form in various organs or skin, and a small part circulates in the blood in liquid form (Tsatsakis et al., 2019). Methanolic fruit extract reduced food intake, suggesting anorectic activity in mice (Kumar and Vimalavathini, 2004). Hexane fraction from the aqueous fruit extract inhibited adipocyte differentiation by blocking leptin gene expression, peroxisome proliferator-activated receptor gamma (PPARγ), and CCAAT enhancer-binding protein alpha (C/EBPα), resulting in the reduction of lipid accumulation and increased release of glycerol and intracellular triglycerides in 3T3-L1 cells.

In the only human study, Amirthaveni and Priya (2011) investigated the hypoglycemic and hypolipidemic effect of ash gourd (*B. hispida*) and curry leaves (*Murraya koenigii*) in 20 type 2 diabetic and hyperlipidemic (40–65 years) patients relative to a similar unsupplemented control group. Fresh salad (100 g ash gourd and 1 g [10 leaves] curry leaves) with skim milk powder (5 g) daily for 90 days significantly ($P < 0.01$) lowered blood glucose (fasting and postprandial) levels and blood pressure, reduced body weight and obesity grades, and improved lipid profile compared to the control group. *Benincasa hispida* supplementation significantly ($P < 0.01$) reduced mean total cholesterol, LDL cholesterol, VLDL cholesterol, and triglycerides, and increased HDL-cholesterol. The intervention normalized BMI (from 30% to 80%, 20–25 kg/m^2 BMI) and reduced grade I obesity [25–30 kg/m^2] from 70% to 20% after 90 days compared to no changes in the control group (Amirthaveni and Priya, 2011).

The study provided an impetus to develop oral antidiabetic tablets containing *B. hispida* and *Murraya koenigii* powders as active ingredients and major excipients (Sreeshma et al., 2019). *Benincasa hispida* fruit pulp and leaves were dried, ground to powder (200 mg), and mixed with *M. koenigii* powder (300 mg) along with binder to produce tablets. The tablet (200 mg/kg body weight) showed no toxicity, restored body weight, lowered plasma glucose level by half, and reduced plasma cholesterol after 21 days of treatment in alloxan-induced diabetic rats, similar to acarbose (25 mg/kg, p.o.).

1.2.15 *Aegle marmelose*

Aegle marmelos, commonly known as bel fruit, also spelled bael, and also called Bengal quince, belongs to the Rutaceae family and is cultivated mainly for its fruit. This plant is native to India, Bangladesh, and throughout Southeast Asia. The unripe fruit is traditionally used as a remedy for dysentery and other digestive ailments. The ripe fruit is sweet, aromatic, and cooling. The tree's wood is yellowish white and hard but not durable (*Bel Fruit | Fruit and Tree | Britannica*, n.d.). The slow-growing trees bear strong spines and alternate compound leaves with three leaflets. The sweet-scented white flowers are borne in panicle clusters and are sometimes used in perfumes. The fruit is pyriform (pear-shaped) to oblong and 5 to 25 cm in diameter. It has a very hard, woody gray or yellow rind and sweet, thick, orange-colored pulp.

Aegle marmelos (Figure 1.15) leaves are among the 10 most highly traded medicinal plants in India and have huge demand in foreign markets. Aegeline is a known constituent of the bael leaf and is consumed as a dietary supplement for weight loss. The plant contains several phytoconstituents belonging to the category of coumarins (marmenol), alkaloids (aegeline), glycosides, triterpenoids, anthraquinones, sterols, and volatile oils. Several modern scientific studies have authenticated its anti-diarrhoeal, antidiabetic, gastroprotective, analgesic, antibacterial, anti-inflammatory, anti-thyroid, antioxidant, anti-proliferative, anti-histaminic, and anti-fertility activities (Niazi et al., 2013).

FIGURE 1.15 *Aegle marmelos*

Methanol *A. marmelos* leaf extract contains coumarin glycosides like marmenol (7-geranyloxy-coumarin), scopoletin (7-hydroxy-6-methoxy coumarin), along with several other known compounds like praealtin D, trans-coumaryl tyramine, N-p-cis, montainine, valencic, trans-cinnamic, 4-methoxy-benzoic and betulinic acids, and rutaretin (Panda and Kar, 2006). (R)-(-) Aegeline, a β-hydroxyamide alkaloid, has been isolated from *A. marmelos* leaves, along with various other alkaloids (marmeline [N-2-hydroxy-2–(4-hydroxyphenyl)ethylcinnamide, N-2-hydroxy-2–[4-(3′,3′dimethylallyloxy)phenyl]ethylcinnamide], N-4-methoxy styrylcinnamide, and N-2–[4-(3′,3′-dimethylallyloxy)phenyl]ethylcinnamide (Govindachari and Premila, 1983). Alloimperatorin (a furocoumarin) and o-isopentenylhalfordinol were present in the leaf extract (Kim et al., 2021). Limonene (82.4%) and (Z)-β-ocimene (5.1%), along with p-menth-1-en-3β, 5β-diol, are the major volatile oils isolated from the *A. marmelos* leaves (Niazi et al., 2013). *Aegle marmelos* stem bark methanol extract contains S-(+)-marmesin (furanocoumarin), while the ethanol extract yields butylated hydroxylanisole, 3-methyl-4-chromanone, butyl-p-tolyl sulfate, methyl linoleate, 5,6-dimethoxy-1-indanone, palmitic acid, and 5-methoxypsoralen on further fractionation with carbon tetrachloride, petroleum ether, ethyl acetate, and methanol.

Methanol fruit extract contains imperatorin (furocoumarin), along with o-(3,3-dimethylallyl)halofordinol, N-2-ethoxy-2–(4-phenyl)ethylcinnamide, marmeline, aegeline, alloimperatorin, xanthotaxol, o-methylhalfordinol, umbelliferone, marmelosin, ascorbic acid and essential oils (Manandhar et al., 1978). Fruit pulp contains neutral polysaccharides containing arabinose, galactose, and glucose in 2:3:14 molar ratios. Bael gum polysaccharide contains D-galactose (71%), L-arabinose (12.5%), L-rhamnose (6.5%), and D-galacturonic acid (7%). Seeds contain carbohydrates (38.5%) consisting of glucose, galactose, rhamnose, and arabinose, and proteins (66.6%) along with a newly reported pyranocoumarin, luvangentin. Roots are known to possess auroptene, umbelliferone, marmine, lupeol, and skimmanine.

Methanol *A. marmelos* extract reduced blood sugar in alloxan-diabetic rats. Blood sugar was reduced from the 6th day after continuous administration of the extract and by 54% on the 12th day. Alloxan-induced oxidative stress was significantly lowered by the administration of *A. marmelos* extract. These results indicate that *A. marmelos* extract effectively reduced the alloxan-induced oxidative stress and reduced blood sugar levels (Sabu and Kuttan, 2004). Another study reported the lipid-lowering property of *A. marmelos* aqueous extract on STZ-induced diabetic rats (Sachdewa et al., 2001). The extract showed significant effects on levels of total cholesterol, low-density lipoprotein, triglycerides, and high-density lipoprotein. This study suggested the anti-hyperlipidemic effect of *A. marmelos*.

The antidiabetic effects of *A. marmelos* have been demonstrated in the following clinical human trials. *Aegle marmelos* leaves decoction (5 g powder daily) for one month significantly lowered postprandial blood glucose levels (159 vs 191 mg/dL baseline) in adult type 2 diabetic patients (35–60 years) and those on standard hypoglycemic therapy (135 vs 201 mg/dL) (Ismail, 2009). *Aegle marmelos* intervention with oral hypoglycemic agents effectively normalized blood glucose levels in diabetic patients probably acting similarly to insulin (i.e., enhancing peripheral glucose utilization, correcting impaired hepatic glycolysis, and limiting gluconeogenic formation). Mature *A. marmelos* leaves (combination of powdered and boiled) (4 g daily) for 8 weeks lowered blood and urine sugar levels (fasting and postprandial) in 38 male diabetic patients (40–60 years) on hypoglycemic drugs (5 or 10 mg sulfonylurea) (Aarti et al., 2009). Moreover, a significant blood sugar-lowering effect occurred only after four weeks of *A. marmelos* intervention compared to placebo and for patients on sulfonylurea. *Aegle marmelos* administration eliminated urinary sugar in 50% and 37.5% of patients in the 5 mg and 10 mg sulfonylurea groups, respectively. The beneficial *A. marmelos* hypoglycemic effect was attributed to the presence of alkaloids.

Aegle marmelos leaf, pulp, and seed powder (2 g daily for 30 days) supplementation significantly ($P \leq 0.01$) lowered blood sugar levels (fasting and postprandial) and blood pressure and ameliorated the lipid profile of non-insulin-dependent diabetics (Singh and Kochhar, 2012). Supplementation of *A. marmelos* leaf, pulp, and seed powder ($n = 30$ each) along with nutrition counseling significantly controlled glycaemia and hyperlipidemia in diabetic male patients (35–65 years). For example, nutrition counseling reduced fasting blood glucose in *A. marmelos* leaf, pulp, and seed powder supplementation more than without counseling (16.1% vs 9.8%, 10.8% vs 7.3%, and 11.4% vs 8.5%, respectively). The leaf, pulp, and seed supplementation had the highest, moderate, and lowest effects on lipid profiles, respectively.

Maximum blood sugar (fasting and postprandial) reduction occurred with leaf and pulp supplementation, respectively.

Sharma and Sharma (2013) evaluated the efficacy of *A. marmelos* supplementation on blood sugar, glycosylated hemoglobin (HbA1c), and blood pressure levels in type 2 diabetes mellitus in a randomized double-blind placebo-controlled trial. Adult type 2 diabetic patients (≥ 30 years) on unmedicated conventional treatment (diet and exercise for 3 months) were randomized into three groups: *A. marmelos* supplementation 250 mg (27F/23M) or 600 mg (32F/18M) daily or placebo (29F/21M) for 3 months. Aqueous extract powder of *A. marmelos* leaves (Anjun Extraction Pvt. Ltd. Chittorgar, India) significantly ($P < 0.05$) lowered systolic blood pressure, fasting blood glucose, and HbA1c at both doses, with more effective glycaemic response and hypertensive control at the higher dose. Apparently, *A. marmelos* regulates blood glucose levels in diabetic conditions by impeding energy release from the metabolic cycle and reducing blood glucose levels by regulating the glycogenolysis pathway. In another trial, *A. marmelos* fruit pulp powder (2 g daily incorporated in khakra) for 60 days significantly lowered fasting blood glucose, HbA1c, and cholesterol (total, LDL, and VLDL) levels in six diabetic subjects (30–60 years, 250 mg/dL fasting blood glucose) in a randomized controlled study (Sharma et al., 2016). The study demonstrated *A. marmelos*'s significant serum lipid reduction with hyperlipidemia in type 2 diabetic patients.

Nigam and Nambiar (2019) investigated the clinical efficacy of *A. marmelos* leaf juice (20 g/100 mL daily) for 60 days on anthropometric, blood glucose levels, lipid profile, liver, and kidney functions of type 2 diabetic patients (19F/11M, 51 years mean age) in a randomized controlled trial. *Aegle marmelos* supplementation significantly lowered blood pressure (6.5% SBP and 4.6% DBP), fasting blood glucose, and HbA1c (20%), postprandial blood glucose (15%), cholesterol (8% total and 15% LDL), triglycerides (11%), and liver function (13%–19%) and increased antioxidant activity (18%). The juice reduced body fat, FBG, HbA1c, and improved lipid profiles and antioxidant activity compared to the control group. The antidiabetic activity of *A. marmelos* leaf juice was ascribed to its bioactive components (i.e., aegeline 2, scopoletin, and sitosterol). The hypoglycemic effect of *A. marmelos* leaves probably enhances peripheral glucose utilization, corrects impaired hepatic glycolysis, and limits its gluconeogenic formation, similar to insulin (Nigam and Nambiar, 2019).

Tiwari et al. (2023) enriched the principal active compounds in *A. marmelos* fruit pulp to develop a phytopharmaceutical ingredient applicable to drug design. *Aegle marmelos* fruit powder (5 g) was extracted by reflux (65 °C, 2 h) with methanol (50 mL), cooled, filtered, and the residue re-extracted (3 x) to obtain a dry filtrate. Coumarins were enriched from the filtrate (~5% yield) containing alkaloids, phenols, and tannins and evaluated for *in vitro* antidiabetic potential (α-amylase inhibition assay). The anti-hyperglycaemic activity of the enriched extract (27 μg IC_{50}) was equivalent to metformin. The *A. marmelos* coumarins-enriched extract contained about 50% of the total coumarins content, represented by marmelosin (30% w/w content), marmesin, psoralen, and scopoletin (9%, 4%, and 2% w/w content, respectively) (Tiwari et al., 2023).

1.2.16 *Gymnema sylvestre*

Over 40 *Gymnema* species are currently being investigated for their use in diabetes, including *G. sylvestre*, credited in the Ayurvedic system, *G. montanum*, *G. latifolium*, and *G. indorum*. These species are widely distributed in India (Assam, Bengal, Bihar, Haryana, Kerala, Madhya Pradesh, Maharashtra, Punjab, Tamil Nadu, and Uttar Pradesh) (Kahksha et al., 2022). India has about 14 of the 37 global species of the *Gymnema* R. Br. (Apocynaceae) genus, which include *G. sylvestre*, commonly known as "Gurmar" or "Madhunashini" (Vaishnav et al., 2023). *G. sylvestre*, known for its antidiabetic properties, is distributed throughout the entire Indian Peninsular region, extending to Uttar Pradesh and Bihar, and is found in central and southern India. It is a perennial, slow-growing woody climber (8–10 m high) with pale yellow flowers. The plant contains various triterpenoids along with other valuable phytoconstituents (alkaloids, coumarins, essential oils, and flavonoids). In Ayurvedic medicine, *G. sylvestre* is recommended for glycosuria control and also prescribed as a diuretic. Its primary application was and continues

FIGURE 1.16 *Gymnema sylvestre*

to be recommended today for adult-onset diabetes (NIDDM) in India. *Gymnema* leaves exert gradual hypoglycemic action, unlike the rapid effects of hypoglycemic drugs. The leaves elevate insulin levels in healthy humans by regenerating β cells in the pancreas, lower serum cholesterol and triglycerides, and reduce intestinal glucose uptake due to the presence of phytoconstituents, primarily gymnemic acid. *G. sylvestre* is one of the most highly traded medicinal plants in India and is an important ingredient in the preparation of many polyherbal formulations for diabetes (D-400 or Diabecon, IME-9, and BGR-34). The primary antidiabetic bioactive phytochemicals in this herb are gymnemic acids, gymnema saponins (oleanane or demmarene triterpenoid saponins in *G. sylvestre*), gymnemasides, and other constituents (stigmasterol, anthraquinones, flavones, phytins, and resins). These compounds from various *G. sylvestre* leaf extracts exhibited comparable α-amylase and α-glucosidase inhibitory activity to acarbose and often superior antidiabetic activity to the standard hypoglycemic drug, metformin, in *ex vivo* or animal models.

The leaves (generally 2–5 cm long and 10–12.5 cm occasionally, 1.2–3 cm broad) when chewed, paralyze the taste glands for a few hours against sweet and bitter tastes and are used to reduce blood glucose levels. Hence, the leaves (Figure 1.16) are widely used to treat diabetes and are commonly found in most herbal drug markets and local shops in India (Krishnamurthy et al., 2015 and references therein). The fresh and dry weights of leaves varied between 104–466 g and 48–186 g/plant per year after field planting of 35 elite lines from *G. sylvestre* accessions. Gymnemic acid, ranging from 2.8% to 5.1%, increased with age and was directly proportional to foliage characteristics.

1.2.16.1 Clinical Trials

Herein, we focus primarily on the extensive human clinical trials of *G. sylvestre*, its extracts, or combinations with other bioactive components on diabetes, although several animal studies illustrate a similar mode of action and function. Human studies demonstrate significant blood glucose reduction during gymnema therapy. These studies suggest regeneration and/or repair of β cells in type 2 diabetics on gymnema supplementation. These claims are supported by increased endogenous serum insulin levels in diabetic patients after gymnema supplementation. Other studies report that gymnemic acids suppress elevated blood glucose levels by inhibiting glucose uptake in the intestine.

In one of the earliest antidiabetic studies, Gharpurey (1926) reported reduced urine sugar levels of diabetics administered *G. sylvestre* leaves. Another study showed that *G. sylvestre* lowered fasting and postprandial blood sugar levels (15%–30%) in normal and moderate diabetic patients (Shanmugasundaram et al., 1981). In nine normal healthy adults (25–40 years), *G. sylvestre* (2 g dry leaf powder daily for 10 days) marginally altered serum insulin response, causing mild blood sugar level reduction. In the single 40-year-old female diabetic patient, the same *G. sylvestre* treatment significantly lowered blood sugar levels and serum insulin response, indicating that it stimulated insulin release with only partial upswing in circulating insulin levels.

Sun-dried powdered *G. sylvestre* leaves (10 g daily) for 21 days improved the glucose tolerance test of seven mild insulin-independent diabetic patients without affecting liver functions. *Gymnema sylvestre* intervention significantly ($P < 0.05$) lowered fasting and postprandial (2 h) blood glucose levels on the 11th day of the trial, with the 2 h blood glucose level returning towards the fasting levels (121 vs 104 mg/dL). This normalization of the blood glucose level within a fortnight is presumably due to increased insulin secretion and/or peripheral glucose utilization. Later, the same group evaluated the hypoglycemic effect of *G. sylvestre* leaf powder (10 g daily for 7 days) in 16 normal subjects and 43 mild diabetics (Balasubramaniam et al., 1992). From the 7th day, 36 mild diabetics were treated with tolbutamide for one week, while the remaining seven diabetics continued with *G. sylvestre* intervention for another two weeks. Seven days of *G. sylvestre* administration significantly lowered mean fasting blood glucose of both normal and diabetic groups. *Gymnema sylvestre* intervention for 3 weeks improved glucose tolerance for the seven diabetic patients, while fasting blood glucose did not differ significantly for tolbutamide-treated diabetics. This demonstrated that *G. sylvestre* leaf powder had a comparable hypoglycemic effect to tolbutamide. However, the difference in fasting blood glucose between the 7th and 11th days of group 1A subjects was an order of magnitude higher (10×) than that of group 1B subjects. This superior hypoglycemic effect of *G. sylvestre* presumably may be due to differences in their mode of action or the concentration of active hypoglycemic agents. Lipid profiles of diabetic patients administered *G. sylvestre* improved (reduced serum triacylglycerol, free fatty acid, and cholesterol levels), while those of normal subjects were unaffected. This significant decrease in fat mobilization in blood indicates that *G. sylvestre* influenced lipid metabolism and can mediate carbohydrate metabolism. It also suggests that *G. sylvestre* acts like insulin in promoting cholesterol catabolism.

Paliwal et al. (2009) investigated the effect of gurmar (*G. sylvestre*) naturally dried leaf powder intervention on the blood glucose level of 20 non-insulin-dependent diabetic Rajasthan women. Gurmar intervention (6 g [3 packets, 2 g each] daily for one month) significantly ($P < 0.01$) lowered fasting and postprandial blood glucose levels in diabetic women (40–60 years, 29.5 kg/m^2 average BMI) not taking hypoglycemic drugs. Powdered *G. sylvestre* leaves (1 g daily for 30 days) significantly ($P < 0.05$) reduced fasting serum glucose (37%) (Paliwal et al., 2009). However, the decrease in lipid profiles (cholesterol, low-density lipoproteins, and triglycerides; 13%, 19%, and 5%, respectively) was insignificant in eight type 2 diabetic patients (48 years mean age, 29 kg/m^2 median BMI) on hyperglycemic drugs (Li et al., 2015).

Various *G. sylvestre* extracts have been used to treat obesity and diabetes in animal and human studies. Aqueous leaf extract potentiates insulin release from pancreatic β-cells and lowers blood glucose in diabetic rats and in human trials (Baskaran et al., 1990). The primary antidiabetic bioactive phytochemical compounds from various *G. sylvestre* leaf extracts exhibited comparable α-amylase and α-glucosidase inhibitory activity to acarbose and often superior antidiabetic activity to the standard hypoglycemic drug, metformin, in *ex vivo* or animal models. Compound 2a (Δ^{15} oleanane glycoside) isolated from aqueous *G. sylvestre* leaf extract had an antihyperglycemic effect (α-glucosidase inhibition in oral sucrose tolerance tests in mice) comparable to acarbose. Hydroalcoholic (70% v/v) *G. sylvestre* dried leaf powder extract had superior α-amylase and α-glucosidase inhibitory activity and a consistent dose-dependent increase in glucose uptake by yeast cells compared to metformin. *Ex vivo* antidiabetic potential on intestinal and skeletal muscle tissues confirmed a marked decline in glucose uptake compared to metformin (Kahksha et al., 2022).

Khare et al. (1983) evaluated the hypoglycemic activity of an aqueous decoction of shade-dried *G. sylvestre* leaves (10 g/100 mL equivalent) in normal and diabetic subjects. *Gymnema sylvestre* treatment (2 g thrice daily for 10 days) significantly ($P < 0.05$) reduced fasting blood sugar (80 vs 69 mg/100 mL) in normal subjects (6M/4F, 19–25 years), but the reduction was insignificant in the postprandial glucose test

(30 and 120 min). *Gymnema sylvestre* administration (2 g thrice daily for 15 days) significantly lowered fasting ($P < 0.02$) and postprandial ($P < 0.05$ and 0.01 for 30 and 120 min, respectively) blood sugar levels in diabetic patients (4M/2F, 35–50 years). The study demonstrated that *Gymnema sylvestre* has hypoglycemic activity in both normal subjects and diabetic patients (Khare et al., 1983).

Gymnema sylvestre leaf extract (500 mg daily for 3 months) supplementation lowered blood glucose (14% fasting and 22% postprandial) and HbA1c, reduced polyphagia and fatigue (56%) in free-living type 2 diabetic patients (18M/21F, 56 years mean age) using oral hypoglycemic agents [OHA] (Kumar et al., 2010). After *Gymnema sylvestre* [GS] supplementation, two diabetic patients (from the experimental group) discontinued their conventional OHA therapy, while the control group increased their drug consumption. GS therapy significantly reduced waist circumference ($P < 0.01$) and blood pressure (systolic [$P < 0.001$] and diastolic [$P = 0.04$]) of the experimental group compared to significant increases in weight, BMI, and systolic blood pressure of the control group (11M/8F, 56 years mean age). GS improved insulin secretion and reduced insulin resistance, evidenced by increased insulin/glucose (I/G) and lowered glucose/insulin (G/I) ratios. GS supplementation improved liver function (reduced liver enzyme levels), kidney function, and lipid profiles (reduced TC, LDL-C, and L/H ratio). This study demonstrated that GS ameliorates clinical symptoms of diabetes (i.e., excess hunger, tiredness, polyuria, and nocturia) and biophysical profiles (marginal reduction in weight, BMI, WC, HA, and fat percentage). This quasi-experimental study demonstrated effective type 2 diabetes management of *G. sylvestre* supplementation by controlling weight, blood pressure, lowering atherogenic lipids, improving metabolic control, and reducing the risk of secondary diabetic complications.

Shanmugasundaram et al. (1990) disclosed the preparation and composition of oral hypoglycemic agents from *G. sylvestre* leaves that effectively treat and prevent type 1 and type 2 diabetes and impaired glucose tolerance. The agent GSHAN is prepared by gently fragmenting dried leaves, steeped (15–20 °C, 8–10 days) in an aqueous propanol/butanol/amyl alcohol solution (15%–20% v/v), concentrated by evaporation, salted (10% concentration), acidified (pH 3), and the precipitate (8% yield) filtered, washed, and dried. GSHAN (200 mg thrice daily) along with insulin for 3 months improved glucose metabolism in type 1 diabetic patients (4M/2F, 8–38 years) by significantly reducing fasting blood glucose and HbA1c levels. Moreover, half of the patients reduced their daily insulin dosage during GSHAN treatment, and one patient diagnosed with type 1 diabetes for only six months eliminated insulin administration altogether. GSHAN intervention for four months significantly reduced cholesterol, triglycerides, free fatty acids, and blood urea nitrogen and increased C-peptide in type 1 diabetic patients, indicating an increase in endogenous insulin production. GSHAN therapy (200 mg thrice daily) along with insulin for 4 months significantly reduced fasting blood glucose and HbA1c within 2 months in type 2 diabetic patients (4M/2F, 48–56 years). Most patients reduced their pharmaceutical hypoglycemic drugs while on GSHAN therapy. The same GSHAN intervention significantly reduced fasting blood glucose in patients (4M/2F, 50–65 years) with impaired glucose tolerance even within 2 months.

Water-soluble acidic fraction of *G. sylvestre* leaves ethanol extract [GS_4] (400 mg [2 capsules] daily for 16–18 months) lowered fasting blood glucose, HbA1c, and glycosylated plasma protein levels and insulin requirements in insulin-dependent diabetic mellitus [IDDM] patients (14M/9F, 10–31 years old and 3M/1F, 44–50 years old) on insulin therapy (Shanmugasundaram et al., 1990). Moreover, GS_4 therapy normalized serum lipids, although HbA1c and glycosylated plasma protein (marker of β-cell regeneration/repair—serum C-peptide assay) remained higher than controls. Reduction was insignificant in serum lipids, HbA1c, and glycosylated plasma proteins after 10–12 months follow-up. However, insulin dosage was reduced to nearly half of the initial amount, while mean blood glucose decreased (232–152 mg/dL). Blood urea remained within the normal range, before, during, and after 18 months of GS_4 therapy. For the first 6 to 8 months, GS_4 therapy significantly ($P < 0.001$) lowered HbA1c, although it remained significantly higher than normal values, almost normalized cholesterol level, and reduced triglycerides, free fatty acids, and serum amylase. Moreover, GS_4 therapy increased C-peptide levels, indicating greater endogenous insulin availability. GS_4-treated patients developed hypoglycemic episodes, and their insulin dose was steadily reduced by 10 units. This study demonstrated that prolonged oral *G. sylvestre* extract administration reduces insulin requirements by improving endogenous insulin availability, enhances glucose homeostasis (reduced HbA1c and glycosylated plasma protein), controls diabetes-associated hyperlipidemia, reduces serum amylase activity, and increases β-cell function (inferred from increased C-peptide levels). These investigators speculated that GS_4 treatment enhanced insulin, probably by regeneration/revitalization of residual β-cells in IDDM. The control group consisted of 37 type 1 diabetic patients on continuous insulin therapy.

GS_4 (400 mg daily) for 18–20 months supplementing oral hyperglycemic drugs significantly ($P < 0.001$) lowered blood glucose, glycosylated hemoglobin, and glycosylated plasma protein, simultaneously reducing conventional drug dosage in type 2 diabetic patients (19M/3F, 40–60 years primarily on glibencamide therapy) (Baskaran et al., 1990). Over one fifth of the patients (5/22) maintained their glucose homeostasis with GS_4 alone and discontinued their conventional drug. Moreover, GS_4 supplementation elevated serum insulin levels, indicating β-cell repair/regeneration in type 2 diabetic patients. Secondary hypoglycemic symptoms developed after several weeks in all GS_4-supplemented patients, necessitating dose reduction or complete withdrawal of conventional drugs (glibenclamide or tolbutamide). GS_4, within 4 weeks of supplementation, alleviated nagging pain in the limbs of female diabetic patients. Plasma lipid (cholesterol, triglycerides, and free fatty acids) levels decreased significantly in GS_4-supplemented patients, indicating reduced hyperlipidemia, while it remained elevated in the control group. The control type 2 diabetic patients (17M/8F, 41–63 years old) on conventional drugs alone were studied over 10 to 12 months. No significant changes occurred in fasting blood glucose, HbA1c, and glycosylated plasma proteins of the control group, and drug requirements increased in contrast with GS4 supplementation. This indicated that GS4 may be enhancing hormone synthesis and/or its release on a 24-h basis. The investigators postulated that GS_4 supplementation apparently regenerates/repairs or removes damaged pancreas tissue in type 2 diabetes mellitus, inducing elevated glucose secretion, enhancing glycemic control, or inhibiting the process of insulin inactivation. This study demonstrated that GS_4 therapy elevates circulating insulin levels, thereby correcting insulin insufficiency in type 2 diabetes.

Adi et al. (2020) compared the efficacy of homeopathic medicine *G. sylvestre* 6 CH and *G. sylvestre* (to its) mother tincture in type 2 diabetes mellitus treatment with three- and six-month follow-up. *Gymnema sylvestre* 6 CH and mother tincture (Belgavi, Karnataka, India) (30 drops in a quarter glass of water, twice daily, 30 min after meals) significantly ($P < 0.0001$) lowered fasting blood sugar (7.2 vs 8.5 and 7.4 vs 8.4 mg/dL for baseline) after 3 months and (6.4 vs 8.5 and 6.6 vs 8.4 mg/dL for baseline) after 6 months in 15 type 2 diabetic outpatients (30–65 years). Similarly, postprandial blood sugar decreased significantly ($P < 0.0001$) in both treated groups after 3 and 6 months.

Gymnema sylvestre homeopathic mother tincture (Sri Ganganagar, Rajasthan, India) was administered (200 μL/kg body weight, twice daily) to 30 type 2 diabetic outpatients (30–65 years) and compared to a similar control group without remedy (Sharma et al., 2020). Homeopathic *G. sylvestre* mother tincture treatment significantly ($P < 0.0001$) lowered fasting and postprandial blood sugar after three, six, and nine months compared to control.

In the US, *G. sylvestre* is sold under several brands (Nature's Way, Natrol, Pro Beta™, and Beta Fast GXR®). These products promote healthy glucose levels and pancreas function. Beta Fast GXR® (Informulab®, Omaha, NE) tablet (400 mg twice daily for 90 days) lowered mean daily preprandial and postprandial blood glucose levels and HbA1c (11%, 13%, and 0.6%, respectively) in diabetic patients (37M/28F) (Joffe and Freed, 2001). The hypoglycemic effects were more profound (15%, 21%, and 0.8% reduction in mean daily preprandial and postprandial blood glucose levels and HbA1c, respectively) in a subset of diabetics with high HbA1c (≥ 9%). Further reductions (18%, 28%, and 1.2% in mean daily preprandial and postprandial blood glucose levels and HbA1c, respectively) occurred in the poorest controlled patients (HbA1c ≥ 10%). Moreover, 11 patients (16%) reduced their prescription drug intake. *Gymnema sylvestre* (Beta Fast GXR®) supplementation improves glycemic control, significantly reduces postprandial blood glucose and HbA1c, thereby reducing diabetic complications in type 2 diabetes. Beta Fast GXR® is a concentrated, extended-release, herbal dietary supplement containing *G. sylvestre* leaf extract (400 mg standardized to 25% per tablet), recommended at two tablets (breakfast and dinner) daily.

Nani et al. (2023) evaluated the effects of *G. sylvestre* in combination with inositol (40:1 [myo: D-chiro]), α-lactalbumin, and zinc on glucose and lipid profiles of type 2 diabetic patients. The intervention group (33M/14F, 67 years mean age, 28 kg/m^2 median BMI) was treated with antidiabetic drugs and two sachets daily of Eudiamet® 40:1 (Lo.Li. Pharma S.r.l., Rome, Italy) for 6 months, each containing myo-Ins (1950 mg), D-chiro-Ins (50 mg), *G. sylvestre* (250 mg), and zinc (7.5 mg). The dietary supplement did not impair glycemic parameters or the background antidiabetic therapy. Differences in lipid parameters (total cholesterol, HDL, LDL, and triglycerides) were not significant between the two arms. This randomized, placebo-controlled, double-blind, two-arm parallel clinical trial demonstrated that dietary combined *G. sylvestre* supplementation improved the lipid profiles of type 2 diabetic outpatients. It improved levels of lipid markers and induced greater weight loss compared with the control group (20M/8F, 69 years mean age, 29 kg/m^2 median BMI).

Preuss et al. (2004) reported that a daily dose of *G. sylvestre* leaf extract (400 mg) combined with water-soluble calcium salt of hydroxycitric acid [HCA-SX] (4,667 mg) and niacin-bound chromium (4 mg) for 8 weeks improved blood lipid profiles in moderately obese subjects. The intervention consisted of standardized *G. sylvestre* extract (Gymnema [GSE, GYM-250 containing 25% gymnemic acid], InterHealth Nutraceuticals, Inc. Benicia, CA) combined with *Garcinia cambogia* 60% HCA extract (Super CitriMax [HCA-SX]) and niacin-bound chromium 10% (ChromeMate). GSE combined with HCA-SX and chromium significantly reduced appetite, body weight, BMI, lipid metabolites (triglycerides, total cholesterol), serum leptin, and increased HDL, serotonin levels, and urinary fat metabolite excretion. This randomized, double-blind, placebo-controlled human study was conducted in 30 obese subjects (21–50 years, BMI > 26 kg/m^2) (Preuss et al., 2004).

Khan et al. (2019) summarized ten clinical study results on *G. sylvestre* (1983–2017) with up to 30 months' duration. *Gymnema sylvestre* administration for 90 days significantly lowered blood glucose levels (fasting and postprandial), along with triglyceride reduction in 58 type 2 diabetic patients (Kumar et al., 2010). *Gymnema sylvestre* leaf extract for 6 to 30 months reduced plasma glucose levels, HbA1c, and external insulin dose in 64 type 1 diabetic patients (Shanmugasunddaram et al., 1990). *Gymnema sylvestre* leaf powder (capsule) for 30 days significantly ($P < 0.05$) lowered blood glucose levels, triglycerides, and LDL cholesterol levels in 32 type 2 diabetic patients (Li et al., 2015).

Devangan et al. (2021) systematically reviewed the effects of GS supplementation on glycemic control in type 2 diabetes mellitus [T2DM]. The meta-analysis of 10 interventional studies (1988–2017) involving 419 participants (40–62 years mean age) showed that GS supplementation is an effective therapy to manage type 2 diabetes mellitus, since it effectively improves glycemic control and enhances lipid profiles of T2DM patients. GS supplementation significantly ($P < 0.0001$) lowers blood glucose (fasting and postprandial), glycated hemoglobin [HbA1c], triglycerides, and total cholesterol compared to baseline. Kahksha et al. (2022), and total cholesterol compared to baseline. Kahksha et al. (2022) reviewed studies (2016–2022) exploring the antidiabetic potential of traditional Gymnema species, most importantly *G. sylvestre*. The antidiabetic action is mostly due to the presence of pharmacologically active phytoconstituents (i.e., gymnemic acids, triterpenoid saponin glycosides, and others).

In a recent comprehensive systematic review (Zamani et al., 2023), involving six randomized placebo-controlled trials *G. sylvestre* supplementation was associated with improvements in lipid profiles of type 2 diabetic patients by decreasing their triglycerides, total cholesterol, and low-density lipoproteins. GS supplementation had favorable effects on fasting blood glucose and diastolic blood pressure, with insignificant changes in BMI and weight circumference. Changes in fasting blood sugar level were linearly related to trial duration, according to the meta-regression analysis. The quality of evidence was moderate for lipid profile (LDL, HDL) and anthropometric parameters (BMI, WC, and DBP), but low for hypoglycemia (fasting blood sugar, HbA1c, and oral glucose tolerance test).

Zuniga et al. (2017) demonstrated the effect of *G. sylvestre* administration on metabolic syndrome, insulin sensitivity, and insulin secretion. A randomized, double-blind, placebo-controlled clinical trial was carried out in 24 patients (without pharmacological treatment), 30 to 60 years old, with a diagnosis of MetS in accordance with the modified International Diabetes Federation criteria. Patients were randomly assigned to receive *G. sylvestre* or placebo twice daily before breakfast and dinner in 300 mg capsules for a total of 600 mg per day for 12 weeks. *Gymnema sylvestre* administration significantly reduced BW (81.3 vs 77.9 kg, $P = 0.02$), BMI (31.2 vs 30.4 kg/m^2, $P = 0.02$), and VLDL levels (0.45 vs

0.35 mmol/dL, P = 0.05) without modifying the components of MetS, insulin secretion, and insulin sensitivity.

Gaytan Martinez et al. (2021) evaluated the effect of *G. sylvestre* administration on glycemic control, insulin secretion, and insulin sensitivity in patients with IGT. A randomized, double-blind, placebo-controlled clinical trial was conducted in 30 patients with IGT. Fifteen patients randomly received 300 mg b.i.d. *G. sylvestre* and the other 15 received the placebo in the same amount. There was a significant reduction in 2-h OGTT (9.1 vs 7.8 mmol/L, P = 0.003), A1C (5.8% vs 5.4%, P = 0.025), body weight, body mass index, and low-density lipoprotein cholesterol levels in the *G. sylvestre* group, with an increment in the Matsuda index (1.8% vs 2.4%, P = 0.008). At the end of the intervention, 46.7% of the patients obtained normal A1C values. In conclusion, *G. sylvestre* administration in IGT patients decreased 2-h OGTT and A1C, increased insulin sensitivity, and also improved anthropometric measures and lipid profile.

1.3 CURRENT AND FUTURE PERSPECTIVES

Almost 439 million people are estimated to be diabetic, primarily adults, including 300 million with type 2 diabetes mellitus by 2030 (Ezin and Chabi, 2023). In India, DM currently affects over 65.1 million people and is expected to reach 80 million people by 2025, making India the diabetes capital of the world. Jodhpur (population over one million) is the second-largest city in Rajasthan with high DM (12.5% hyperglycemic population in 2013). Neem leaves scored the second-highest disease consensus index (0.64) among 21 antidiabetic plants in an ethnomedicinal survey of 50 type 2 diabetic patients conducted in the Sursagar constituency of Jodhpur (Goyal, 2015). Interestingly, Neem (*A. indica*) is also one of the ancient traditional herbs found in many (~700) preparations/formulations in Ayurvedic and other local health prescriptions (Siddha, Unani, and Amchi). Indigenous remedies have been used to treat diabetes in India since the 6th century BC (Grover et al., 2002). Over 1200 plant species (725 different genera) with hypoglycemic activity have been identified in ethnobotanical studies of traditional herbal remedies used globally for diabetes (Shekelle et al., 2005). The Indian pharmacopoeia is particularly rich in herbal treatments for diabetes, with prescriptions containing 80% of the 20 antidiabetic plants most widely used worldwide (Marles and Farnsworth, 1995). Four herbs described in this chapter (*C. indica*, fenugreek, *G. sylvestre*, and *E. jambolana*) were also the most studied Ayurvedic therapies for diabetes mellitus, according to a systematic review (Shekelle et al., 2005). These herbs and others have demonstrated glucose-lowering effects and often HbA1c reduction based on their extensive reported clinical studies. Many antidiabetic Ayurvedic formulations (e.g., Madhuhara churna) contain combinations of several herbs that act synergistically to treat/prevent diabetes. However, despite these latest advances, diabetic and prediabetic rates are 11.4% and 15.3%, respectively, in India. Obesity prevalence (28.6%) is closely related to diabetes, while abdominal obesity stands at 39.5%, according to the recent report (July 2023) of the Indian Council of Medical Research-India Diabetes (Koe, 2023). Lage and Boye (2020) emphasized significant economic healthcare costs benefits associated with HbA1c reduction among type 2 diabetes patients. Their multivariate analysis of T2D patients (n = 77,622) in the US revealed that HbA1c reduction (1%) significantly ($P < 0.0001$) decreased all-cause total healthcare costs (2%) and diabetes-related total healthcare costs (13%), resulting in $429 and $736 annual cost savings.

Novel technologies, such as phytosome or fiber encapsulation, are already being used to increase the bioavailability of the active ingredients present in these antidiabetic herbs. Some commercial antidiabetic formulations have also been standardized based on the bioactive phytochemicals of these herbs. Many of these antidiabetic herbs and (Ayurvedic) formulations have the potential to improve human health and increase their economic benefits, as well as reduce overall escalating healthcare costs.

REFERENCES

Aarti, S., Shashi, S., & Nidhi, S. (2009). Hypoglycemic effect of bael patra (*Aegle marmelos*) in NIDDM patients. *Journal of Dairying, Foods and Home Sciences*, *28*(3/4), 233–236.

Abbas, G., Ali, M., Hamaed, A., Al-Sibani, M., Hussain, H., & Al-Harrasi, A. (2020). *Azadirachta indica*: The medicinal properties of the global problems-solving tree. In: M. Ozturk, M. Pešić, & D. Egamberdieva (eds.), *Biodiversity and Biomedicine* (pp. 305–316). Academic Press. https://doi.org/10.1016/B978-0-12-8-819541-3.00017-7

Abdullah, A., Biswas, P., Sahabuddin, Md., Mubasharah, A., Khan, D. A., Hossain, A., Roy, T., Rafi, N. Md. R., Dey, D., Hasan, Md. N., Bibi, S., Moustafa, M., Shati, A., Hassan, H., & Garg, R. (2023). Molecular dynamics simulation and pharmacoinformatic integrated analysis of bioactive phytochemicals from *Azadirachta indica* (Neem) to treat diabetes mellitus. *Journal of Chemistry*, *2023*, 1–19. https://doi.org/10.1155/2023/4170703

Adi, B. S., Adi, G. B., & Jamadade, A. K. (2020). A comparison of the efficacy of *Gymnema sylvestre* 6 Ch and *Gymnema sylvestre* mother tincture in cases of type 2 diabetes mellitus. *World Journal of Current Medical and Pharmaceutical Research*, *2*(2), 133–138. https://doi.org/10.37022/WJCMPR.2020.2208

Agrawal, P., Rai, V., & Singh, R. B. (1996). Randomized placebo-controlled, single blind trial of holy basil leaves in patients with noninsulin-dependent diabetes mellitus. *International Journal of Clinical Pharmacology and Therapeutics*, *34*(9), 406–409.

Agriexchange. (2023) Indian production of bitter gourd, National Horticulture Board (NHB), 2021–2022. https://agriexchange.apeda.gov.in/Home.aspx; accessed November 22, 2023

Ahmad, N., Hassan, M. R., Halder, H., & Bennoor, K. S. (1999). Effect of *Momordica charantia* (Karolla) extracts on fasting and postprandial serum glucose levels in NIDDM patients. *Bangladesh Medical Research Council Bulletin*, *25*(1), 11–13.

Ahmed, F., Hudeda, S., & Urooj, A. (2011). Antihyperglycemic activity of *Ficus racemosa* bark extract in type 2 diabetic individuals. *Journal of Diabetes*, *3*(4), 318–319. https://doi.org/10.1111/j.1753-0407.2011.00152.x

Ahmed, Z., Ansari, M. A., Ali, I., Soomro, F., Kumar, G., & Joher, I. (2023). Clinical evaluation of *Eugenia Jambolana* hypoglycemic effects in patients diagnosed with type II mon-insulin independent diabetes. *Pakistan Journal of Medical & Health Sciences*, *17*(1), Article 01. https://doi.org/10.53350/pjmhs2023171404

Akter, R. (2013). Comparative studies on antidiabetic effect with phytochemical screening of *Azadirachta indicia* and *Andrographis paniculata*. *IOSR Journal of Pharmacy and Biological Sciences*, *5*(2), 122–128. https://doi.org/10.9790/3008-052122128

Alzohairy, M. A. (2016). Therapeutics role of *Azadirachta indica* (Neem) and their active constituents in diseases prevention and treatment. *Evidence-Based Complementary and Alternative Medicine*, *2016*, 1–11. https://doi.org/10.1155/2016/7382506

Amini, J., Safaie, N., Salmani, M., & Shams-bakhsh, M. (2012). Antifungal activity of three medicinal plant essential oils against some phytopathogenic fungi. *Trakia Journal of Sciences*, *10*, 1–8.

Amirthaveni, D. M., & Priya, V. (2011). Hypoglycemic and hypolipidemic effect of ash gourd (*Benincasa hispida*) and curry leaves (*Murraya koenigii*). *International Journal of Current Research*, *3*.

Ansari, P., Akther, S., Hannan, J. M. A., Seidel, V., Nujat, N. J., & Abdel-Wahab, Y. H. A. (2022). Pharmacologically active phytomolecules isolated from traditional antidiabetic plants and their therapeutic role for the management ofdiabetes mellitus. *Molecules*, *27*(13), 4278. https://doi.org/10.3390/molecules27134278

Arbotante, C., & Arriola, E. (2016). Investigation of the bioactive properties and hypoglycemic effects of ethanol, hexane and ethyl ethanoate extracts from Kondol leaves (*Benincasa hispida* Cogniaux). *American Journal of Clinical Pathology*, *146*(suppl_1), 83. https://doi.org/10.1093/ajcp/aqw163.008

Badmus, J. A., Adedosu, O. T., Jimoh, R. A., & Saraki, M. A. (2018). Ethanol leaves extract of *Mangifera indica* (L.) exhibits protective, antioxidative, and antidiabetic effects in rats. *Asian Pacific Journal of Health Sciences*, *5*(1), Article 1. https://doi.org/10.21276/apjhs.2018.5.1.42

Balasubramaniam, K., Arasaratnam, V., Nageswaran, A., Anushiyanthan, S., & Mugunthan, N. (1992). Studies on the effect of *Gymnema sylvestre* on diabetes. *Journal of the National Science Foundation of Sri Lanka*. *20*(1), Article 1. https://doi.org/10.4038/jnsfsr.v20i1.8061

Baldwa, V. S., Bhandari, C. M., Pangaria, A., & Goyal, R. K. (1977). Clinical trial in patients with diabetes mellitus of an insulin-like compound obtained from plant source. *Upsala Journal of Medical Sciences*, *82*(1), 39–41. https://doi.org/10.3109/03009737709179057

Baliga, M. S., Bhat, H. P., Baliga, B. R. V., Wilson, R., & Palatty, P. L. (2011). Phytochemistry, traditional uses and pharmacology of *Eugenia jambolana* Lam. (black plum): A review. *Food Research International*, *44*(7), 1776–1789. https://doi.org/10.1016/j.foodres.2011.02.007

Bandyopadhyay, U., Biswas, K., Sengupta, A., Moitra, P., Dutta, P., Sarkar, D., Debnath, P., Ganguly, C. K., & Banerjee, R. K. (2004). Clinical studies on the effect of Neem (*Azadirachta indica)* bark extract on gastric secretion and gastroduodenal ulcer. *Life Sciences*, *75*(24), 2867–2878. https://doi.org/10.1016/j.lfs.2004.04.050

Baskaran, K., Kizar Ahamath, B., Radha Shanmugasundaram, K., & Shanmugasundaram, E. R. (1990). Antidiabetic effect of a leaf extract from *Gymnema sylvestre* in non-insulin-dependent diabetes mellitus patients. *Journal of Ethnopharmacology*, *30*(3), 295–300. https://doi.org/10.1016/0378-8741(90)90108-6

Bast, F., Rani, P., & Meena, D. (2014). Chloroplast DNA phylogeography of holy basil (*Ocimum tenuiflorum*) in Indian subcontinent. *The Scientific World Journal*, *2014*, 1–6. https://doi.org/10.1155/2014/847482

Battu, G. R., Mamidipalli, S. N., Parimi, R., Raja Kumar, V., Patchula, R. P., & Mood, L. R. (2007). Hypoglycaemic and antihyperglycemic effect of alcoholic extract of *Benincasa hispida* in normal and in alloxan induced diabetic rats. *Pharmacognosy Magazine*, *3*, 101–105.

Bel fruit | fruit and tree | Britannica. (n.d.). Retrieved February 21, 2023, from https://www.britannica.com/plant/bel-fruit

Belcaro, G., Cesarone, M., Silvia, E., Ledda, A., Stuard, S., G, V., Dougall, M., Cornelli, U., Hastings, C., & Schönlau, F. (2009). Daily consumption of Reliv Glucaffect for 8 weeks significantly lowered blood glucose and body weight in 50 subjects. *Phytotherapy Research: PTR*, *23*(12), 1673–1677. https://doi.org/10.1002/ptr.2793

Bhat, M., Kothiwale, S. K., Tirmale, A. R., Bhargava, S. Y., & Joshi, B. N. (2011). Antidiabetic properties of *Azardiracta indica* and *Bougainvillea spectabilis*: *In Vivo* studies in murinediabetes model. *Evidence-Based Complementary and Alternative Medicine*, *2011*, 1–9. https://doi.org/10.1093/ecam/nep033

Bhatt, D., Bhatt, P., Patel, D., & Jagdale, D. (2019). Commercial production of micropropagated *Coccinia indica* (Tondli)—A success story. *Journal of AgriSearch*, *6*(Special), 59–63.

Boone, C. H., Stout, J. R., Gordon, J. A., Redd, M. J., Church, D. D., Oliveira, L. P., Fukuda, D. H., & Hoffman, J. R. (2017). Acute effects of a beverage containing bitter melon extract (CARELA) on postprandial glycemia among prediabetic adults. *Nutrition & Diabetes*, *7*(1), e241. https://doi.org/10.1038/nutd.2016.51

Bopp, A., De Bona, K. S., Bellé, L. P., Moresco, R. N., & Moretto, M. B. (2009). *Syzygium cumini* inhibits adenosine deaminase activity and reduces glucose levels in hyperglycemic patients. *Fundamental & Clinical Pharmacology*, *23*(4), 501–507. https://doi.org/10.1111/j.1472-8206.2009.00700.x

Chaiyasut, C., Kesika, P., Sirilun, S., Peerajan, S., & Sivamaruthi, B. S. (2018). Formulation and evaluation of lactic acid bacteria fermented *Brassica juncea* (Mustard Greens) pickle with cholesterol lowering property. *Journal of Applied Pharmaceutical Science*, *8*(4), 033–042. https://doi.org/10.7324/JAPS.2018.8405

Chakravarthy, B. K., Gupta, S., & Gode, K. D. (1982). Functional beta cell regeneration in the islets of pancreas in alloxan induced diabetic rats by (-)-epicatechin. *Life Sciences*, *31*(24), 2693–2697. https://doi.org/10.1016/0024-3205(82)90713-5

Chakravarthy, B. K., Saroj, G., Gambhir, S., & Gode, K. (1980). Pancreatic beta-cell regeneration—A novel antidiabetic mechanism of *Pterocarpus marsupium*, Roxb. *Indian Journal of Pharmacology*, *12*, 123–127.

Chio, S. P., Chua, M. E., Coralde, M., Covar, R. C., Dating, M. S., Francisco, J. C., Gabay, K. K., Gatbonton, B. M., Giron, J. J., Go, H. J., Hernandez, C. I., Hou, T., Ignacia, M. A., Ilagan, A. V., & Butacan, R. J.-A. (2018). Effectiveness of *Brassica juncea* (mustard green) leaf decoction as an adjunct in the treatment of type 2 diabetes mellitus among Filipinos: A randomized clinical trial. *Health Sciences Journal*, *7*(2). https://www.herdin.ph/index.php?view=research&cid=70244

Chougale, A. D., Ghadyale, V. A., Panaskar, S. N., & Arvindekar, A. U. (2009). Alpha glucosidase inhibition by stem extract of *Tinospora cordifolia*. *Journal of Enzyme Inhibition and Medicinal Chemistry*, *24*(4), 998–1001. https://doi.org/10.1080/14756360802565346

Cohen, M. M. (2014). Tulsi—*Ocimum sanctum*: A herb for all reasons. *Journal of Ayurveda and Integrative Medicine*, *5*(4), 251–259. https://doi.org/10.4103/0975-9476.146554

Collene, A. L., Hertzler, S. R., Williams, J. A., & Wolf, B. W. (2005). Effects of a nutritional supplement containing *Salacia oblonga* extract and insulinogenic amino acids on postprandial glycemia, insulinemia, and breath hydrogen responses in healthy adults. *Nutrition (Burbank, Los Angeles County, Calif.)*, *21*(7–8), 848–854. https://doi.org/10.1016/j.nut.2004.11.018

Cortez-Navarrete, M., Martínez-Abundis, E., Pérez-Rubio, K. G., González-Ortiz, M., & Méndez-del Villar, M. (2018). *Momordica charantia* administration improves insulin secretion in type 2 diabetes mellitus. *Journal of Medicinal Food*, *21*(7), 672–677. https://doi.org/10.1089/jmf.2017.0114

Cortez-Navarrete, M., Pérez-Rubio, K. G., & Escobedo-Gutiérrez, M. de J. (2023). Role of fenugreek, cinnamon, *Curcuma longa*, berberine and *Momordica charantia* in type 2 diabetes mellitus treatment: A review. *Pharmaceuticals*, *16*(4), Article 4. https://doi.org/10.3390/ph16040515

Dahlquist, A., Jandali, D., Nauman, M. C., & Johnson, J. J. (2023). Clinical application of *Momordica charantia* (Bitter Melon) for reducing blood sugar in type 2 diabetes mellitus. *International Journal of Nutrition*, *7*(4), 8–26. https://doi.org/10.14302/issn.2379-7835.ijn-23-4737

Dans, A. M. L., Villarruz, M. V. C., Jimeno, C. A., Javelosa, M. A. U., Chua, J., Bautista, R., & Velez, G. G. B. (2007). The effect of *Momordica charantia* capsule preparation on glycemic control in type 2 diabetes mellitus needs further studies. *Journal of Clinical Epidemiology*, *60*(6), 554–559. https://doi.org/10.1016/j.jclinepi.2006.07.009

Deepa, P., Sowndhararajan, K., Kim, S., & Park, S. J. (2018). A role of Ficus species in the management of diabetes mellitus: A review. *Journal of Ethnopharmacology*, *215*, 210–232. https://doi.org/10.1016/j.jep.2017.12.045

Derosa, G., D'Angelo, A., & Maffioli, P. (2022). The role of selected nutraceuticals in management of prediabetes and diabetes: An updated review of the literature. *Phytotherapy Research: PTR*, *36*(10), 3709–3765. https://doi.org/10.1002/ptr.7564

Devangan, S., Varghese, B., Johny, E., Gurram, S., & Adela, R. (2021). The effect of *Gymnema sylvestre* supplementation on glycemic control in type 2 diabetes patients: A systematic review and meta-analysis. *Phytotherapy Research*, *35*(12), 6802–6812. https://doi.org/10.1002/ptr.7265

Dhanabal, S. P., Kokate, C. K., Ramanathan, M., Kumar, E. P., & Suresh, B. (2006). Hypoglycaemic activity of *Pterocarpus marsupium* Roxb. *Phytotherapy Research: PTR*, *20*(1), 4–8. https://doi.org/10.1002/ptr.1819

Dharmatti, P. R., Patil, R. V., Patil, S. S., & Athani, S. I. (2010). A new Coccinia (*Coccinia indica*) DRC-1, A variety boon to vegetable growers. *Journal of Farm Sciences*, *21*(1), Article 1. http://14.139.155.167/test5/index.php/kjas/article/view/1125

Dholi, S. K., Raparla, R., Mankala, S. K., & Nagappan, K. (2011). *In vivo* antidiabetic evaluation of Neem leaf extract in alloxan induced rats. *Journal of Applied Pharmaceutical Science*, *1*(4), 100–105.

Dineshkumar, B., Analava, M., & Manjunatha, M. (2010). Antidiabetic and hypolipidaemic effects of few common plants extract in Type 2 diabetic patients at Bengal. *Dubai Diabetes and Endocrinology Journal*, *18*, 59–65. https://doi.org/10.1159/000497694

Evans, S. F., Meister, M., Mahmood, M., Eldoumi, H., Peterson, S., Perkins-Veazie, P., . . . & Lucas, E. A. (2014). Mango supplementation improves blood glucose in obese individuals. *Nutrition and Metabolic Insights*, *7*, 77–83. https://doi.org/10.4137/NMI.S17028

Ezin, V., & Chabi, I. B. (2023). *Azadirachta indica*: Its biological, pharmacological, antidiabetic potential, and omics applications. In: D. Pandita, A. Pantita & C. Banu (eds.), *Antidiabetic Plants for Drug Discovery: Pharmacology, Secondary Metabolite Profiling and Ingredients with Insulin Mimetic Activity* (pp. 1–22). Apple Academic Press, Inc.

Faizal, P., Suresh, S., Satheesh Kumar, R., & Augusti, K. T. (2009). A study on the hypoglycemic and hypolipidemic effects of an ayurvedic drug Rajanyamalakadi in diabetic patients. *Indian Journal of Clinical Biochemistry: IJCB*, *24*(1), 82–87. https://doi.org/10.1007/s12291-009-0014-1

Fang, C., Kim, H., Barnes, R. C., Talcott, S. T., & Mertens-Talcott, S. U. (2018). Obesity-associated diseases biomarkers are differently modulated in lean and obese individuals and inversely correlated to plasma polyphenolic metabolites after 6 weeks of mango (*Mangifera indica* L.) consumption. *Molecular Nutrition & Food Research*, *62*(14), 1800129.

Fatema, K., Ali, L., Rahman, M. H., Parvin, S., & Hassan, Z. (2003). Serum glucose and insulin response to mango and papaya in type 2 diabetic subjects. *Nutrition Research*, *23*(1), 9–14.

Fernandes, N. P., Lagishetty, C. V., Panda, V. S., & Naik, S. R. (2007). An experimental evaluation of the antidiabetic and antilipidemic properties of a standardized *Momordica charantia* fruit extract. *BMC Complementary and Alternative Medicine*, *7*(1), 29. https://doi.org/10.1186/1472-6882-7-29

Fuangchan, A., Sonthisombat, P., Seubnukarn, T., Chanouan, R., Chotchaisuwat, P., Sirigulsatien, V., Ingkaninan, K., Plianbangchang, P., & Haines, S. T. (2011). Hypoglycemic effect of bitter melon compared with metformin in newly diagnosed type 2 diabetes patients. *Journal of Ethnopharmacology*, *134*(2), 422–428. https://doi.org/10.1016/j.jep.2010.12.045

Gad, M. Z., El-Sawalhi, M. M., Ismail, M. F., & El-Tanbouly, N. D. (2006). Biochemical study of the anti-diabetic action of the Egyptian plants fenugreek and balanites. *Molecular and Cellular Biochemistry*, *281*(1–2), 173–183. https://doi.org/10.1007/s11010-006-0996-4

Gaddam, A., Galla, C., Thummisetti, S., Marikanty, R. K., Palanisamy, U. D., & Rao, P. V. (2015). Role of Fenugreek in the prevention of type 2 diabetes mellitus in prediabetes. *Journal of Diabetes and Metabolic Disorders*, *14*, 74. https://doi.org/10.1186/s40200-015-0208-4

Gandhi, R., Chauhan, B., & Jadeja, G. (2016). Diabetes mellitus. *Ocimum sanctum L.* (Tulsi), hypoglycaemia, PP2BS, HBA1C.

Gaytán Martínez, L. A., Sánchez-Ruiz, L. A., Zuñiga, L. Y., González-Ortiz, M., & Martínez-Abundis, E. (2021). Effect of *Gymnema sylvestre* administration on glycemic control, insulin secretion, andinsulin sensitivity inpatients with impaired glucose tolerance. *Journal of Medicinal Food*, *24*(1), 28–32. https://doi.org/10.1089/jmf.2020.0024

Gharpurey, K. G. (1926). *Gymnema sylvestre* in the treatment of diabetes. *The Indian Medical Gazette*, *61*(3), 155.

Govindachari, T. R., & Premila, M. S. (1983). Some alkaloids from *Aegle marmelos*. *Phytochemistry*, *22*(3), 755–757. https://doi.org/10.1016/S0031-9422(00)86977-0

Goyal, M. (2015). Traditional plants used for the treatment of diabetes mellitus in Sursagar constituency, Jodhpur, Rajasthan—An ethnomedicinal survey. *Journal of Ethnopharmacology*, *174*, 364–368. https://doi.org/10.1016/j.jep.2015.08.047

Goyal, S., Gupta, N., & Chatterjee, S. (2016). Investigating therapeutic potential of *Trigonella foenum-graecum* L. as our defense mechanism against several human diseases. *Journal of Toxicology*, *2016*, 1250387. https://doi.org/10.1155/2016/1250387

Graidist, P., & Purintrapiban, J. (2009). The biological activity of *Coccinia indica* on glucose transporter 1 (GLUT1) promoter. *Songklanakarin Journal of Science & Technology*, *31*(3), 247–253.

Grover, J. K., & Gupta, S. R. (1990). Hypoglycemic activity of seeds of *Momordica charantia*. *European Journal of Pharmacology*, *183*(3), 1026–1027.

Grover, J. K., Yadav, S., & Vats, V. (2002). Medicinal plants of India with anti-diabetic potential. *Journal of Ethnopharmacology*, *81*(1), 81–100.

Grover, J. K., Yadav, S. P., & Vats, V. (2003). Effect of feeding *Murraya koeingii* and *Brassica juncea* diet on [correction] kidney functions and glucose levels in streptozotocin diabetic mice. *Journal of Ethnopharmacology*, *85*(1), 1–5. https://doi.org/10.1016/s0378-8741(02)00355-0

Guevarra, M. T. B., & Panlasigui, L. N. (2000). Blood glucose responses of diabetes mellitus type II patients to some local fruits. *Asia Pacific Journal of Clinical Nutrition*, *9*(4), 303–308.

Gul-e-Rana, null, Karim, S., Khurhsid, R., Saeed-ul-Hassan, S., Tariq, I., Sultana, M., Rashid, A. J., Shah, S. H., & Murtaza, G. (2013). Hypoglycemic activity of *Ficus racemosa* bark in combination with oral hypoglycemic drug in diabetic human. *Acta Poloniae Pharmaceutica*, *70*(6), 1045–1049.

Gupta, A., Gupta, R., & Lal, B. (2001). Effect of *Trigonella foenum-graecum* (fenugreek) seeds on glycaemic control and insulin resistance in type 2 diabetes mellitus: A double blind placebo controlled study. *The Journal of the Association of Physicians of India*, *49*, 1057–1061.

Gupta, S. S. (1963). Effect of Gymnema sylvestre and *Pterocarpus marsupium* on glucose tolerance in albino rats. *Indian Journal of Medical Sciences*, *17*, 501–505.

Haranath, P. S., Ranganatharao, K., Anjaneyulu, C. R., & Ramanathan, J. D. (1958). Studies on the hypoglycemic and pharmacological actions of some stilbenes. *Indian Journal of Medical Sciences*, *12*(2), 85–89.

Heacock, P. M., Hertzler, S. R., Williams, J. A., & Wolf, B. W. (2005). Effects of a medical food containing an herbal α-glucosidase inhibitor on postprandial glycemia and insulinemia in healthy adults. *Journal of the American Dietetic Association*, *105*(1), 65–71. https://doi.org/10.1016/j.jada.2004.11.001

Helmstädter, A. (2008). *Syzygium cumini* (L.) SKEELS (Myrtaceae) against diabetes—125 years of research. *Die Pharmazie*, *63*(2), 91–101.

Hotta, N., Kakuta, H., Sano, T., Matsumae, H., Yamada, H., Kitazawa, S., & Sakamoto, N. (1993). Long-term effect of acarbose on glycaemic control in non-insulin-dependent diabetes mellitus: A placebo-controlled double-blind study. *Diabetic Medicine*, *10*(2), 134–138. https://doi.org/10.1111/j.1464-5491.1993.tb00030.x

Hsu, P.-K., Pan, F. F. C., & Hsieh, C.-S. (2020). mcIRBP-19 of bitter melon peptide effectively regulates diabetes mellitus (DM) patients' blood sugar levels. *Nutrients*, *12*(5), 1252. https://doi.org/10.3390/nu12051252

Inayat-ur-Rahman, Malik, S. A., Bashir, M., Khan, R., & Iqbal, M. (2009). Serum sialic acid changes in non-insulin-dependant diabetes mellitus (NIDDM) patients following bitter melon (*Momordica charantia*) and rosiglitazone (Avandia) treatment. *Phytomedicine*, *16*(5), 401–405. https://doi.org/10.1016/j.phymed.2009.01.001

Irondi, E. A., Oboh, G., & Akindahunsi, A. A. (2016). Antidiabetic effects of *Mangifera indica* kernel flour-supplemented diet in streptozotocin-induced type 2 diabetes in rats. *Food Science & Nutrition*, *4*(6), 828–839. https://doi.org/10.1002/fsn3.348

Islam, M. T., Quispe, C., El-Kersh, D. M., Shill, M. C., Bhardwaj, K., Bhardwaj, P., Sharifi-Rad, J., Martorell, M., Hossain, R., Al-Harrasi, A., Al-Rawahi, A., Butnariu, M., Rotariu, L. S., Suleria, H. A. R., Taheri, Y., Docea, A. O., Calina, D., & Cho, W. C. (2021). A literature-based update on *Benincasa hispida* (Thunb.) Cogn.: Traditional uses, nutraceutical, and phytopharmacological profiles. *Oxidative Medicine and Cellular Longevity*, *2021*, e6349041. https://doi.org/10.1155/2021/6349041

Ismail, M. (2009). Clinical evaluation of antidiabetic activity of Bael leaves. *World Applied Sciences Journal*. https://www.semanticscholar.org/paper/Clinical-Evaluation-of-Antidiabetic-Activity-of-Ismail/746093454b75cd6e488aaf620522ea956694a96d

Jalaja kumari, D. (2010). Hypoglycemic effect of *Moringa oleifera* and *Azadirachta indica* in type-2 diabetes. *Bioscan*, *5*(20), 211–214.

Jalil, A., Ashfaq, U. A., Shahzadi, S., Rasul, I., Rehman, S.-U., Shah, M., & Javed, M. R. (2013). Screening and design of anti-diabetic compounds sourced from the leaves of neem (*Azadirachta indica*). *Bioinformation*, *9*(20), 1031–1035. https://doi.org/10.6026/97320630091031

Jamshidi, N., Da Costa, C., & Cohen, M. (2018). Holybasil (tulsi) lowers fasting glucose and improves lipid profile in adults with metabolic disease: A meta-analysis of randomized clinical trials. *Journal of Functional Foods*, *45*, 47–57. https://doi.org/10.1016/j.jff.2018.03.030

Jamwal, A., & Kumar, S. (2019). Antidiabetic activity of isolated compound from *Coccinia indica*. *Indian Journal of Pharmaceutical Education and Research*, *53*(1), 151–159. https://doi.org/10.5530/ijper.53.1.20

Jayalakshmi, M. S., Thenmozhi, P., & Vijayaraghavan, R. (2021). Plant leaves extract irrigation on wound healing in diabetic foot ulcers. *Evidence-Based Complementary and Alternative Medicine*, *2021*, e9924725. https://doi.org/10.1155/2021/9924725

Jayawardena, M. H. S., de Alwis, N. M. W., Hettigoda, V., & Fernando, D. J. S. (2005). A double blind randomised placebo controlled cross over study of a herbal preparation containing *Salacia reticulata* in the treatment of type 2 diabetes. *Journal of Ethnopharmacology*, *97*(2), 215–218. https://doi.org/10.1016/j.jep.2004.10.026

Jeykodi, S., Deshpande, J., & Juturu, V. (2016). *Salacia* extract improves postprandial glucose and insulin response: A randomized double-blind, placebo controlled, crossover study in healthy volunteers. *Journal of Diabetes Research*, *2016*, 1–9. https://doi.org/10.1155/2016/7971831

Joffe, D. J., & Freed, S. H. (2001). Effect of extended release *Gymnema sylvestre* leaf extract (Beta Fast GXR) alone or in combination with oral hypoglycemics or insulin regimens for type 1 and type 2 diabetes. *Diabetes in Control Newsletter*, *76*(1), 30 Oct 2001.

John, A. J., Cherian, R., Subhash, H. S., & Cherian, A. M. (2003). Evaluation of the efficacy of bitter gourd (*Momordica charantia*) as an oral hypoglycemic agent—A randomized controlled clinical trial. *Indian Journal of Physiology and Pharmacology*, *47*(3), 363–365.

Joseph, B., & Jini, D. (2013). Antidiabetic effects of *Momordica charantia* (bitter melon) and its medicinal potency. *Asian Pacific Journal of Tropical Disease*, *3*(2), 93–102. https://doi.org/10.1016/S2222-1808(13)60052-3

Junaid, M., Singh, B., Vats, S., Thapliyal, S., & Thapliyal, S. (2020). Anti-hyperglycemic effect of Kundru (Coccinia indica) on patients of diabetes mellitus type–2. *International Journal of Ayurveda and Pharma Research*, 29–35. https://doi.org/10.47070/ijapr.v8i11.1681

Kahksha, Alam, O., Naaz, S., Sharma, V., Manaithiya, A., Khan, J., & Alam, A. (2022). Recent developments made in the assessment of the antidiabetic potential of gymnema species—From 2016 to 2020. *Journal of Ethnopharmacology*, *286*, 114908. https://doi.org/10.1016/j.jep.2021.114908

Kajimoto, O., Kawamori, S., Shimoda, H., Kawahara, Y., Hirata, H., & Takahashi, T. (2000). Effects of a diet containing *Salacia reticulata* on mild type 2 diabetes in humans. *Nippon Eiyo Shokuryo Gakkaishi*, *53*(5), 199–205. https://doi.org/10.4327/jsnfs.53.199

Kamble, S. M., Kamlakar, P. L., Vaidya, S., & Bambole, V. D. (1998). Influence of *Coccinia indica* on certain enzymes in glycolytic and lipolytic pathway in human diabetes. *Indian Journal of Medical Sciences*, *52*(4), 143–146.

Katiyar, D., Singh, V., & Ali, M. (2016). Phytochemical and pharmacological profile of *Pterocarpus marsupium*: A review. *The Pharma Innovation Journal*, *5*(4), 31–39.

Kawamori, S., & Kawahara, Y. (1998). Effects of an aqueous extract of *Salacia reticulata*, a useful plant in Sri Lanka, on postprandial hyperglycemia in rats and humans. *Journal of Japanese Society of Nutrition and Food Science (Japan)*, *51*(5), 279–287.

Kehar, S., & Misra, A. (2020). Mango: A fruit too far in patients with diabetes? (or is it?). *Diabetes & Metabolic Syndrome*, *14*(2), 135–136. https://doi.org/10.1016/j.dsx.2020.01.014

Khan, A. K., Akhtar, S., & Mahtab, H. (1980). Treatment of diabetes mellitus with *Coccinia indica*. *British Medical Journal*, *280*(6220), 1044.

Khan, B. A., Abraham, A., & Leelamma, S. (1995). Hypoglycemic action of *Murraya koenigii* (curry leaf) and *Brassica juncea* (mustard): Mechanism of action. *Indian Journal of Biochemistry & Biophysics*, *32*(2), 106–108.

Khan, F., Sarker, Md. M. R., Ming, L. C., Mohamed, I. N., Zhao, C., Sheikh, B. Y., Tsong, H. F., & Rashid, M. A. (2019). Comprehensive review on phytochemicals, pharmacological and clinical potentials of *Gymnema sylvestre*. *Frontiers in Pharmacology*, *10*. https://www.frontiersin.org/articles/10.3389/fphar.2019.01223

Khan, K. Y., Khan, M. A., Ahmad, M., Hussain, I., Mazari, P., Fazal, H., Ali, B., & Khan, I. Z. (2011). Hypoglycemic potential of genus Ficus L.: A review of ten years of plant based medicine used to cure diabetes (2000–2010). *Journal of Applied Pharmaceutical Science*, *1*(6), 223–227.

Khanna, P., Jain, S. C., Panagariya, A., & Dixit, V. P. (1981). Hypoglycemic activity of polypeptide-p from a plant source. *Journal of Natural Products*, *44*(6), 648–655. https://doi.org/10.1021/np50018a002

Khare, A. K., Tondon, R. N., & Tewari, J. P. (1983). Hypoglycaemic activity of an indigenous drug (*Gymnema sylvestre*, 'Gurmar') in normal and diabetic persons. *Indian Journal of Physiology and Pharmacology*, *27*(3), 257–258.

Kim, H. J., Seo, Y. J., Htwe, K. M., & Yoon, K. D. (2021). Chemical constituents from *Aegle marmelos* fruits. *Natural Product Sciences*, *27*(4).

Kim, S. K., Jung, J., Jung, J. H., Yoon, N., Kang, S. S., Roh, G. S., & Hahm, J. R. (2020). Hypoglycemic efficacy and safety of Momordica charantia (bitter melon) in patients with type 2 diabetes mellitus. *Complementary Therapies in Medicine*, *52*, 102524. https://doi.org/10.1016/j.ctim.2020.102524

King, K., Lin, N.-P., Cheng, Y.-H., Chen, G.-H., & Chein, R.-J. (2015). Isolation of positive modulator of glucagon-like peptide-1 signaling from *Trigonella foenum-graecum* (Fenugreek) seed. *The Journal of Biological Chemistry*, *290*(43), 26235–26248. https://doi.org/10.1074/jbc.M115.672097

Kirana, H., Jali, M. V., & Srinivasan, B. P. (2011). The study of aqueous extract of *Ficus religiosa* Linn. on cytokine TNF-α in type 2 diabetic rats. *Pharmacognosy Research*, *3*(1), 30–34.

Kobayashi, M., Akaki, J., Ninomiya, K., Yoshikawa, M., Muraoka, O., Morikawa, T., & Odawara, M. (2021). Dose-dependent suppression of postprandial hyperglycemia and improvement of blood glucose parameters by *Salacia chinensis* extract: two randomized, double-blind, placebo-controlled studies. *Journal of Medicinal Food*, *24*(1), 10–17. https://doi.org/10.1089/jmf.2020.4751

Kochhar, A., Sharma, N., & Sachdeva, R. (2009). Effect of supplementation of Tulsi (*Ocimum sanctum*) and Neem (*Azadirachta indica*) leaf powder on diabetic symptoms, anthropometric parameters and blood pressure of non insulin dependent male diabetics. *Studies on Ethno-Medicine*, *3*(1), 5–9. https://doi.org/10.1080/09735070.2009.11886330

Koe, T. (2023, November 7). 'One-stop solution': Dr. Reddy's launches e-commerce website specializing in diabetic-friendly supplements. *NutraIngredients Asia*. Nutraingredients-asia.com

Kohli, K. R., & Singh, R. H. (1993). A clinical trial of jambu (*Eugenia jambolana*) in non insulin dependent diabetes mellitus. *Journal of Research in Ayurveda and Siddha*, *14*(3/4), 89–97.

Koteshwar, P., Raveendra, K. R., Allan, J. J., Goudar, K. S., Venkateshwarlu, K., & Agarwal, A. (2013). Effect of NR-Salacia on post-prandial hyperglycemia: A randomized double blind, placebo-controlled, crossover study in healthy volunteers. *Pharmacognosy Magazine*, *9*(36), 344–349. https://doi.org/10.4103/0973-1296.117831

Krishnamurthy, R., Chandorkar, M. S., Pathak, J. M., Adedayo, A. D., & Gupta, R. (2015). Selection of elite lines from accessions of *Gymnema sylvestre* (Gudmar) based on characterization of foliage and gymnemic acid yield. *International Journal of Medicinal Plants Photon*, *108*, 596–605.

Kumar, A., & Vimalavathini, R. (2004). Possible anorectic effect of methanol extract of *Benincasa hispida* (Thunb). Cogn, fruit. *Indian Journal of Pharmacology*, *36*(6), 348–350.

Kumar, S. N., Mani, U. V., & Mani, I. (2010). An open label study on the supplementation of *Gymnema sylvestre* in type 2 diabetics. *Journal of Dietary Supplements*, *7*(3), 273–282. https://doi.org/10.3109/19390211.2010.505901

Kumar, V., Mahdi, F., Singh, R., Mahdi, A. A., & Singh, R. K. (2016). A clinical trial to assess the antidiabetic, antidyslipidemic and antioxidant activities of *Tinospora cordifolia* in management of type-2 diabetes mellitus. *International Journal of Pharmaceutical Sciences and Research*, *7*(2), 757–764. https://doi.org/10.13040/IJPSR.0975-8232.7(2).757-64

Kumar, V., Thakur, Dr. A., Barothia, N., & Chatterjee, S. (2011). Therapeutic potentials of *Brassica juncea*: An overview. *TANG [Humanitas Medicine]*, *1*, 1–17. https://doi.org/10.5667/tang.2011.0005

Kumari, S., Dash, I., & Behera, K. K. (2018). Therapeutic effect of *Momordica charantia* on blood glucose, lipid profile and oxidative stress in type 2 diabetes mellitus patients: A randomised controlled trial. *Journal of Clinical and Diagnostic Research*. *12*(9), BC21–BC25. https://doi.org/10.7860/JCDR/2018/36354.12036

Kurian, G. A., Manjusha, V., Nair, S. S., Varghese, T., & Padikkala, J. (2014). Short-term effect of G-400, polyherbal formulation in the management of hyperglycemia and hyperlipidemia conditions in patients with type 2 diabetes mellitus. *Nutrition*, *30*(10), 1158–1164. https://doi.org/10.1016/j.nut.2014.02.026

Kuriyan, R., Rajendran, R., Bantwal, G., & Kurpad, A. V. (2008). Effect of supplementation of *Coccinia cordifolia* extract on newly detected diabetic patients. *Diabetes Care*, *31*(2), 216–220. https://doi.org/10.2337/dc07-1591

Lage, M. J., & Boye, K. S. (2020). The relationship between HbA1c reduction and healthcare costs among patients with type 2 diabetes: Evidence from a US claims database. *Current Medical Research and Opinion*, *36*(9), 1441–1447.

Leatherdale, B. A., Panesar, R. K., Singh, G., Atkins, T. W., Bailey, C. J., & Bignell, A. H. (1981). Improvement in glucose tolerance due to *Momordica charantia* (karela). *British Medical Journal (Clinical Research Ed.)*, *282*(6279), 1823–1824.

Lee, C., & Ma, T. (2019). Effect of acute bitter melon intake on postprandial glucose and insulin in sedentary, abdominally obese persons. *Functional Foods in Health and Disease*, *9*(6), 384. https://doi.org/10.31989/ffhd.v9i6.605

Lett, A. M., Thondre, P. S., & Rosenthal, A. J. (2013). Yellow mustard bran attenuates glycaemic response of a semi-solid food in young healthy men. *International Journal of Food Sciences and Nutrition*, *64*(2), 140–146. https://doi.org/10.3109/09637486.2012.728201

Li, Y., Zheng, M., Zhai, X., Huang, Y., Khalid, A., Malik, A., Shah, P., Karim, S., Azhar, S., & Hou, X. (2015). Effect of *Gymnema sylvestre*, *Citrullus colocynthis* and *Artemisia absinthium* on blood glucose and lipid profile in diabetic human. *Acta Poloniae Pharmaceutica*, *72*(5), 981–985.

Lu, F., Shen, L., Qin, Y., Gao, L., Li, H., & Dai, Y. (2008). Clinical observation on *Trigonella foenum-graecum* L. total saponins in combination with sulfonylureas in the treatment of type 2 diabetes mellitus. *Chinese Journal of Integrative Medicine*, *14*(1), 56–60. https://doi.org/10.1007/s11655-007-9005-3

Luka, C. D., & Mohammad, A. (2012). Evaluation of antidiabetic property of aqueous extract of *Mangifera indica* leaf on normal and alloxan-induced diabetic rats. *Journal of Natural Product and Plant Resources*, *2*(2), 239–243.

Luthy, N., & Martinez-Fortun, O. (1964). A study of a possible oral hypoglycemic factor in albahaca morada (*Ocimum sanctum* L.). *Ohio Journal of Science*, *64*, 223–224.

Manandhar, M. D., Shoeb, A., Kapil, R. S., & Popli, S. P. (1978). New alkaloids from *Aegle marmelos*. *Phytochemistry*, *17*(10), 1814–1815. https://doi.org/10.1016/S0031-9422(00)88714-2

Manickam, M., Ramanathan, M., Jahromi, M. A., Chansouria, J. P., & Ray, A. B. (1997). Antihyperglycemic activity of phenolics from *Pterocarpus marsupium*. *Journal of Natural Products*, *60*(6), 609–610. https://doi.org/10.1021/np9607013

Marles, R. J., & Farnsworth, N. R. (1995). Antidiabetic plants and their active constituents. *Phytomedicine*, *2*, 131–189.

Mawa, S., Husain, K., & Jantan, I. (2013). *Ficus carica* L. (Moraceae): Phytochemistry, traditional uses and biological activities. *Evidence-Based Complementary and Alternative Medicine*, *2013*, 974256. https://doi.org/10.1155/2013/974256

Mazhin, S. A., Asadi Zaker, M., Shahbazian, H. B., Azemi, M. E., & Madanchi, N. (2016). *Ficus carica* leaves decoction on glycemic factors of patients with type 2 diabetes mellitus: A double-blind clinical trial. *Jundishapur Journal of Natural Pharmaceutical Products*, *11*(1). https://doi.org/10.17795/jjnpp-25814

Mazumder, S., Lerouge, P., Loutelier-Bourhis, C., Driouich, A., & Ray, B. (2005). Structural characterisation of hemicellulosic polysaccharides from *Benincasa hispida* using specific enzyme hydrolysis, ion exchange chromatography and MALDI-TOF mass spectroscopy. *Carbohydrate Polymers*, *59*(2), 231–238. https://doi.org/10.1016/j.carbpol.2004.09.014

Mazumder, S., Morvan, C., Thakur, S., & Ray, B. (2004). Cell wall polysaccharides from Chalkumra (*Benincasa hispida*) fruit. Part I. Isolation and characterization of pectins. *Journal of Agricultural and Food Chemistry*, *52*(11), 3556–3562. https://doi.org/10.1021/jf0343130

McNaughton, S. A., & Marks, G. C. (2003). Development of a food composition database for the estimation of dietary intakes of glucosinolates, the biologically active constituents of cruciferous vegetables. *The British Journal of Nutrition*, *90*(3), 687–697. https://doi.org/10.1079/bjn2003917

Mishra, S., Verma, N., Bhattacharya, S., Usman, K., Himanshu, D., Singh, P., Anjum, B., & Verma, N. (2015). Effect of *Tinospora cordifolia* as an add—On therapy on the blood glucose levels of patients with Type 2 diabetes. *International Journal of Basic & Clinical Pharmacology*, *4*(3), 537–541. https://doi.org/10.18203/2319-2003.ijbcp20150035

Mohan, L., Amberkar, M. V., & Kumari, M. (2011). *Ocimum sanctum* linn (TULSI)—An overview. *International Journal of Pharmaceutical Sciences Review and Research*, *7*(1), 51–53.

Munasinghe, M. A. A. K., Abeysena, C., Yaddehige, I. S., Vidanapathirana, T., & Piyumal, K. P. B. (2011). Blood sugar lowering effect of *Coccinia grandis* (L.) J. Voigt: Path for a new drug for diabetes mellitus. *Experimental Diabetes Research*, *2011*, 1–4. https://doi.org/10.1155/2011/978762

Nagarajaiah, S. B., & Prakash, J. (2015). Chemical composition and bioactive potential of dehydrated peels of *Benincasa hispida, Luffa acutangula*, and *Sechium edule*. *Journal of Herbs, Spices & Medicinal Plants*. https://www.tandfonline.com/doi/full/10.1080/10496475.2014.940437

Najdi, R. A., Hagras, M. M., Kamel, F. O., & Magadmi, R. M. (2019). A randomized controlled clinical trial evaluating the effect of *Trigonella foenum-graecum* (fenugreek) versus glibenclamide in patients with diabetes. *African Health Sciences*, *19*(1), 1594–1601. https://doi.org/10.4314/ahs.v19i1.34

Nakata, K., Taniguchi, Y., Yoshioka, N., Yoshida, A., Inagawa, H., Nakamoto, T., Yoshimura, H., Miyake, S.-I., Kohchi, C., Kuroki, M., & Soma, G.-I. (2011). A mixture of *Salacia oblonga* extract and IP-PA1 reduces fasting plasma glucose (FPG) and low-density lipoprotein (LDL) cholesterol levels. *Nutrition Research and Practice*, *5*(5), 435–442. https://doi.org/10.4162/nrp.2011.5.5.435

Nani, A., Bertuzzi, F., Meneghini, E., Mion, E., & Pintaudi, B. (2023). Combined inositols, α-lactalbumin, *Gymnema sylvestre* and zinc improve the lipid metabolic profile of patients with type 2 diabetes mellitus: A randomized clinical trial. *Journal of Clinical Medicine*, *12*(24), 7650. https://doi.org/10.3390/jcm12247650

Narwade, S., Bangar, G., & Sudrik, A. (2023). A review on herbal medicinal plants used in diabetic treatment. *Journal of Pharmacognosy and Phytochemistry*, *12*(1), 597–603.

National Horticulture Board (NHB). (2021–2022). *India Production of BITTERGOURD*. Retrieved December 16, 2023, from https://agriexchange.apeda.gov.in/India%20Production/India_Productions.aspx?hscode=1068

Nguyen, N. N. T., Duong, X. C., Nguyen, K. N., Nguyen, T. N. V. M., Nguyen, T. T. D., Le, T. T. Y., Le, T. C. T., Nguyen, T. T. T., & Pham, D. T. (2023). Development and *in-vitro/in-vivo* evaluation of film-coated tablets containing *Azadirachta indica* A. Juss leaf extracts for diabetes treatment. *Journal of Applied Pharmaceutical Science*, *13*(1), 193–200. https://doi.org/10.7324/JAPS.2023.130119

Niazi, J., Bansal, Y., & Kaur, N. (2013). *Aegle Marmelos: A Phytochemical and Phytopharmacological Review. Pharmatutor-Art-1728*. https://www.pharmatutor.org/articles/aegle-marmelos-phytochemical-and-phytopharmacological-review

Nigam, V., & Nambiar, V. S. (2019). *Aegle marmelos* leaf juice as a complementary therapy to control type 2 diabetes—Randomised controlled trial in Gujarat, India. *Advances in Integrative Medicine*, *6*(1), 11–22. https://doi.org/10.1016/j.aimed.2018.03.002

Ozaki, S., Tamura, H., & Kataoka, K. (2008). Effects of repeated overdoses of Kothala-kimbutu extract in drinks. *Journal of The Japanese Society for Food Science and Technology-Nippon Shokuhin Kagaku Kogaku Kaish*, *55*, 481–486. https://doi.org/10.3136/nskkk.55.481

Paliwal, R., Kathori, S., & Upadhyay, B. (2009). Effect of Gurmar (*Gymnema sylvestre*) powder intervention on the blood glucose levels among diabetics. *Studies on Ethno-Medicine*, *3*(2), 133–135. https://doi.org/10.1080/09735070.2009.11886350

Panda, D. K., Ghosh, D., Bhat, B., Talwar, S. K., Jaggi, M., & Mukherjee, R. (2009). Diabetic therapeutic effects of ethyl acetate fraction from the roots of *Musa paradisiaca* and seeds of *Eugenia jambolana* in streptozotocin-induced male diabetic rats. *Methods and Findings in Experimental and Clinical Pharmacology*, *31*(9), 571–584. https://doi.org/10.1358/mf.2009.31.9.1435645

Panda, S., & Kar, A. (2006). Evaluation of the antithyroid, antioxidative and antihyperglycemic activity of scopoletin from *Aegle marmelos* leaves in hyperthyroid rats. *Phytotherapy Research: PTR*, *20*(12), 1103–1105. https://doi.org/10.1002/ptr.2014

Patel, M. B., & Mishra, S. (2011). Hypoglycemic activity of alkaloidal fraction of *Tinospora cordifolia*. *Phytomedicine: International Journal of Phytotherapy and Phytopharmacology*, *18*(12), 1045–1052. https://doi.org/10.1016/j.phymed.2011.05.006

Patil, P., Patil, S., Mane, A., & Verma, S. (2013). Antidiabetic activity of alcoholic extract of Neem (*Azadirachta Indica*) root bark. *National Journal of Physiology, Pharmacy and Pharmacology*, *3*(2), 142. https://doi.org/10.5455/njppp.2013.3.134–138

Patil, R. N., Patil, R. Y., Ahirwar, B., & Ahirwar, D. (2011). Evaluation of antidiabetic and related actions of some Indian medicinal plants in diabetic rats. *Asian Pacific Journal of Tropical Medicine*, *4*(1), 20–23.

Patil, S. M., Shirahatti, P. S., & Ramu, R. (2022). *Azadirachta indica* A. Juss (neem) against diabetes mellitus: A critical review on its phytochemistry, pharmacology, and toxicology. *The Journal of Pharmacy and Pharmacology*, *74*(5), 681–710. https://doi.org/10.1093/jpp/rgab098

Perez-Gutierrez, R. M., & Damian-Guzman, M. (2012). Meliacinolin: A potent α-glucosidase and α-amylase inhibitor isolated from *Azadirachta indica* leaves and *in vivo* antidiabetic property in streptozotocin-nicotinamide-induced type 2 diabetes in mice. *Biological & Pharmaceutical Bulletin*, *35*(9), 1516–1524. https://doi.org/10.1248/bpb.b12-00246

Peter, E. L., Kasali, F. M., Deyno, S., Mtewa, A., Nagendrappa, P. B., Tolo, C. U., Ogwang, P. E., & Sesaazi, D. (2019). *Momordica charantia* L. lowers elevated glycaemia in type 2 diabetes mellitus patients: Systematic review and meta-analysis. *Journal of Ethnopharmacology*, *231*, 311–324. https://doi.org/10.1016/j.jep.2018.10.033

Philip, S. (1998). Flexible dose open trial of Vijayasar in cases of newly-diagnosed non-insulin-dependent diabetes mellitus. Indian Council of Medical Research (ICMR), Collaborating Centres, New Delhi. *The Indian Journal of Medical Research*, *108*, 24–29.

Pingali, U., Ali, M. A., Gundagani, S., & Nutalapati, C. (2020). Evaluation of the effect of an aqueous extract of *Azadirachta indica* (Neem) leaves and twigs on glycemic control, endothelial dysfunction and systemic inflammation in subjects with type 2 diabetes mellitus—A randomized, double-blind, placebo-controlled clinical study. *Diabetes, Metabolic Syndrome and Obesity: Targets and Therapy*, *13*, 4401–4412. https://doi.org/10.2147/DMSO.S274378

Ponnusamy, S., Haldar, S., Mulani, F., Zinjarde, S., Thulasiram, H., & RaviKumar, A. (2015). Gedunin and Azadiradione: Human pancreatic alpha-amylase inhibiting limonoids from Neem (*Azadirachta indica*) as anti-diabetic agents. *PLoS ONE*, *10*(10), e0140113. https://doi.org/10.1371/journal.pone.0140113

Preuss, H. G., Bagchi, D., Bagchi, M., Rao, C. V. S., Satyanarayana, S., & Dey, D. K. (2004). Efficacy of a novel, natural extract of (–)-hydroxycitric acid (HCA-SX) and a combination of HCA-SX, niacin-bound chromium and *Gymnema sylvestre* extract in weight management in human volunteers: A pilot study. *Nutrition Research*, *24*(1), 45–58. https://doi.org/10.1016/j.nutres.2003.09.007

Prince, P. S. M., Kamalakkannan, N., & Menon, V. P. (2004). Restoration of antioxidants by ethanolic *Tinospora cordifolia* in alloxan-induced diabetic Wistar rats. *Acta Poloniae Pharmaceutica*, *61*(4), 283–287.

Puranik, N., Kazi, S. A. K., Chaitra, B., Shivakumar, J., Omprakash, K.V., & Devi, S. (2014). Improvement in glucose tolerance due to *Tinospora Cordifolia* (Willd.). *International Journal of Health Sciences and Research*, *4*(5), 37–41.

Quamri, M. A., Begum, S., Siddiqui, M., & Alam, Md. A. (2017). Efficacy of Kanduri (*Coccinia indica*) in diabetes associated dyslipidemia—A randomized single blind standard controlled study. *International Journal of Unani and Integrative Medicine*, *1*(1), 01–04. https://doi.org/10.33545/2616454X.2017.v1.i1a.1

Radha, R., & Amrithaveni, M. (2009). Role of medicinal plant *Salacia reticulata* in the management of type II diabetic subjects. *Ancient Science of Life*, *29*(1), 14–16.

Rafraf, M., Malekiyan, M., Asghari-Jafarabadi, M., & Aliasgarzadeh, A. (2014). Effect of fenugreek seeds on serum metabolic factors and adiponectin levels in type 2 diabetic patients. *International Journal for Vitamin and Nutrition Research. Internationale Zeitschrift Fur Vitamin-Und Ernahrungsforschung. Journal International De Vitaminologie Et De Nutrition*, *84*(3–4), 196–205. https://doi.org/10.1024/0300-9831/a000206

Rai, V., Iyer, U., & Mani, U. V. (1997). Effect of Tulasi (*Ocimum sanctum*) leaf powder supplementation on blood sugar levels, serum lipids and tissue lipids in diabetic rats. *Plant Foods for Human Nutrition (Dordrecht, Netherlands)*, *50*(1), 9–16. https://doi.org/10.1007/BF02436038

Ranade, M., & Mudgalkar, N. (2017). A simple dietary addition of fenugreek seed leads to the reduction in blood glucose levels: A parallel group, randomized single-blind trial. *Ayu*, *38*(1–2), 24–27. https://doi.org/10.4103/ayu.AYU_209_15

Rao, Y. K., & Bairy, L. K. (2007). Safety of aqueous extract of *Tinospora cordifolia* (Tc) in healthy volunteers: A double blind randomised placebo controlled study. *Iranian Journal of Pharmacology and Therapeutics*, *6*(1), 59–61.

Ratsimamanga, A. R., Ratsimamanga, S. R., Rasoanaivo, P., Leboul, J., Provost, J., & Reisdorf, D. (2001). *Mixtures Derived from Grains of Eugenia Jambolana Lamarck Preparation and Use of Said Mixtures and Some of Their Constituents as Medicaments* (United States Patent US6194412B1). https://patents.google.com/patent/US6194412B1/en

Ravi, K., Rajasekaran, S., & Subramanian, S. (2005). Antihyperlipidemic effect of *Eugenia jambolana* seed kernel on streptozotocin-induced diabetes in rats. *Food and Chemical Toxicology: An International Journal Published for the British Industrial Biological Research Association*, *43*(9), 1433–1439. https://doi.org/10.1016/j.fct.2005.04.004

Ray, K. S., Paharia, N. V., & Singhania, P. S. (2017). Mango: Yes or no for individuals with diabetes. *Endocrinological Diabetes Clinical & Medical Research*, *1*, 103.

Repin, N., Kay, B. A., Cui, S. W., Wright, A. J., Duncan, A. M., & Douglas Goff, H. (2017). Investigation of mechanisms involved in postprandial glycemia and insulinemia attenuation with dietary fibre consumption. *Food & Function*, *8*(6), 2142–2154. https://doi.org/10.1039/c7fo00331e

Ribes, G., Sauvaire, Y., Da Costa, C., Baccou, J. C., & Loubatieres-Mariani, M. M. (1986). Antidiabetic effects of subfractions from fenugreek seeds in diabetic dogs. *Proceedings of the Society for Experimental Biology and Medicine. Society for Experimental Biology and Medicine (New York)*, *182*(2), 159–166. https://doi.org/10.3181/00379727-182-42322

Richter, E., Geetha, T., Burnett, D., Broderick, T. L., & Babu, J. R. (2023). The effects of *Momordica charantia* on type 2 diabetes mellitus and Alzheimer's disease. *International Journal of Molecular Sciences*, *24*(5), 4643. https://doi.org/10.3390/ijms24054643

Roongpisuthipong, C., Banphotkasem, S., Komindr, S., & Tanphaichitr, V. (1991). Postprandial glucose and insulin responses to various tropical fruits of equivalent carbohydrate content in non-insulin-dependent diabetes mellitus. *Diabetes Research and Clinical Practice*, *14*(2), 123–131.

Roy, K., Shah, R., & Iyer, U. (2015). *Tinospora cordifolia* stem supplementation in diabetic dyslipidemia: An open labelled randomized controlled trial. *Functional Foods in Health and Disease*, *5*(8), 265. https://doi.org/10.31989/ffhd.v5i8.208

Sabu, M. C., & Kuttan, R. (2004). Antidiabetic activity of *Aegle marmelos* and its relationship with its antioxidant properties. *Indian Journal of Physiology and Pharmacology*, *48*(1), 81–88.

Sachdewa, A., Raina, D., Srivastava, A. K., & Khemani, L. D. (2001). Effect of *Aegle marmelos* and *Hibiscus rosa sinensis* leaf extract on glucose tolerance in glucose induced hyperglycemic rats (Charles foster). *Journal of Environmental Biology*, *22*(1), 53–57.

Saha, S., & Ghosh, S. (2012). *Tinospora cordifolia*: One plant, many roles. *Ancient Science of Life*, *31*(4), 151–159. https://doi.org/10.4103/0257-7941.107344

Sahana, D. A., Shivaprakash, G., Baliga, R., MR, A. P., Ganesh, J., & Pai, M. R. S. M. (2010). Effect of *Eugenia jambolana* on plasma glucose, insulin sensitivity and HDL-C levels: Preliminary results of a randomized clinical trial. *Journal of Pharmacy Research*, *3*(6), 1268–1270.

Sai, K. S., & Srividya, N. (2002). Blood glucose lowering effect of the leaves of *Tinospora cordifolia* and *Sauropus androgynus* in diabetic subjects. *Journal of Natural Remedies*, 28–32. https://doi.org/10.18311/jnr/2002/341

Samanta, S., Chandra, R., Ganguli, S., Reddy, A. G., & Banerjee, J. (2019). Anti-diabetic activity of mango *(Mangifera indica):* A review. *MOJ Bioequivalence & Bioavailability*, *6*(2), 23–26.

Sangeetha, M. K., Balaji Raghavendran, H. R., Gayathri, V., & Vasanthi, H. R. (2011). Tinospora cordifolia attenuates oxidative stress and distorted carbohydrate metabolism in experimentally induced type 2 diabetes in rats. *Journal of Natural Medicines*, *65*(3–4), 544–550. https://doi.org/10.1007/s11418-011-0538-6

Satapathy, S., Das, N., Bandyopadhyay, D., Mahapatra, S. C., Sahu, D. S., & Meda, M. (2017). Effect of Tulsi (*Ocimum sanctum* Linn.) supplementation on metabolic parameters and liver enzymes in young overweight and obese subjects. *Indian Journal of Clinical Biochemistry: IJCB*, *32*(3), 357–363. https://doi.org/10.1007/s12291-016-0615-4

Satyanarayana, K., Sravanthi, K., Shaker, I. A., & Ponnulakshmi, R. (2015). Molecular approach to identify antidiabetic potential of *Azadirachta indica. Journal of Ayurveda and Integrative Medicine*, *6*(3), 165–174. https://doi.org/10.4103/0975-9476.157950

Sauvaire, Y., Petit, P., Broca, C., Manteghetti, M., Baissac, Y., Fernandez-Alvarez, J., Gross, R., Roye, M., Leconte, A., Gomis, R., & Ribes, G. (1998). 4-Hydroxyisoleucine: A novel amino acid potentiator of insulin secretion. *Diabetes*, *47*(2), 206–210. https://doi.org/10.2337/diab.47.2.206

Serraclara, A., Hawkins, F., Pérez, C., Domínguez, E., Campillo, J. E., & Torres, M. D. (1998). Hypoglycemic action of an oral fig-leaf decoction in type-I diabetic patients. *Diabetes Research and Clinical Practice*, *39*(1), 19–22. https://doi.org/10.1016/s0168-8227(97)00112-5

Shah, D. S. (1967). A preliminary study of the hypoglycemic action of heartwood of *Pterocarpus marsupium* roxb. *The Indian Journal of Medical Research*, *55*(2), 166–168.

Shanmugasundaram, E. R., Rajeswari, G., Baskaran, K., Rajesh Kumar, B. R., Radha Shanmugasundaram, K., & Kizar Ahmath, B. (1990). Use of *Gymnema sylvestre* leaf extract in the control of blood glucose in insulin-dependent diabetes mellitus. *Journal of Ethnopharmacology*, *30*(3), 281–294. https://doi.org/10.1016/0378-8741(90)90107-5

Shanmugasundaram, K. R., Panneerselvam, C., Samudram, P., & Shanmugasundaram, E. R. (1981). The insulinotropic activity of *Gymnema sylvestre*, R. Br. An Indian medical herb used in controlling diabetes mellitus. *Pharmacological Research Communications*, *13*(5), 475–486. https://doi.org/10.1016/s0031-6989(81)80074-4

Sharma, K., Shukla, S., & Chauhan, E. S. (2016). Evaluation of *Aegle marmelos* (Bael) as hyperglycemic and hyperlipidemic diminuting agent in type II diabetes mellitus subjects. *The Pharma Innovation*, *5*(5), 43–46.

Sharma, P., & Sharma, S. (2013). A randomised, double-blind, placebo-controlled trial of "*Aegle marmelos*" supplementation on glycaemic control and blood pressure level in type 2 diabetes mellitus. *Australian Journal of Medical Herbalism*. https://www.semanticscholar.org/paper/A-randomised%2C-double-blind%2C-placebo-controlled-of-2-Sharma-Sharma/96238a37fdcdd356a8241c3572a595ae65060625

Sharma, P., Singh, C., Chakraborty, P., & Reddy e, S. (2020). A verification study on R.N. Chopra findings of *Gymnema sylvestre* to reduce blood sugar levels in type 2 diabetes mellitus cases. *Indian Journal of Scientific Research*, *10*, 32661–32665.

Sharma, R., Amin, H., Galib, & Prajapati, P. K. (2015). Antidiabetic claims of *Tinospora cordifolia* (Willd.) Miers: Critical appraisal and role in therapy. *Asian Pacific Journal of Tropical Biomedicine*, *5*(1), 68–78. https://doi.org/10.1016/S2221-1691(15)30173-8

Sharma, R., Choudhary, A., Kaur, H., Singh, G., Sharma, K., Kumar, A., . . . & Mehta, S. (2024). Introduction to diabetes: An overview. In: A. Husen (ed.), *Antidiabetic Medicinal Plants and Herbal Treatments* (pp. 1–24). CRC Press. https://doi.org/10.1201/b23347-1

Sharma, R., Kumar, V., Ashok, B. K., Galib, R., Prajapati, P. K., & Ravishankar, B. (2013). Evaluation of hypoglycaemic and antihyperglycaemic activities of Guduchi Ghana in Swiss albino mice. *International Journal of Green Pharmacy (IJGP)*, *7*(2), Article 2. https://doi.org/10.22377/ijgp.v7i2.313

Sharma, R. D., & Raghuram, T. C. (1990). Hypoglycaemic effect of fenugreek seeds in non-insulin dependent diabetic subjects. *Nutrition Research*, *10*(7), 731–739. https://doi.org/10.1016/S0271-5317(05)80822-X

Sharma, S. B., Tanwar, R. S., Nasir, A., & Prabhu, K. M. (2011). Antihyperlipidemic effect of active principle isolated from seed of *Eugenia jambolana* on alloxan-induced diabetic rabbits. *Journal of Medicinal Food*, *14*(4), 353–359. https://doi.org/10.1089/jmf.2010.1227

Sharma, S. B., Nasir, A., Prabhu, K. M., & Murthy, P. S. (2006). Antihyperglycemic effect of the fruit-pulp of *Eugenia jambolana* in experimental diabetes mellitus. *Journal of Ethnopharmacology*, *104*(3), 367–373. https://doi.org/10.1016/j.jep.2005.10.033

Sharma, S. B., Nasir, A., Prabhu, K. M., Murthy, P. S., & Dev, G. (2003). Hypoglycaemic and hypolipidemic effect of ethanolic extract of seeds of *Eugenia jambolana* in alloxan-induced diabetic rabbits. *Journal of Ethnopharmacology*, *85*(2–3), 201–206. https://doi.org/10.1016/s0378-8741(02)00366-5

Shekelle, P. G., Hardy, M., Morton, S. C., Coulter, I., Venuturupalli, S., Favreau, J., & Hilton, L. K. (2005). Are Ayurvedic herbs for diabetes effective? *The Journal of Family Practice*, *54*(10), 876–886.

Shivananjappa, M. M., & Muralidhara. (2012). Abrogation of maternal and fetal oxidative stress in the streptozotocin-induced diabetic rat by dietary supplements of *Tinospora cordifolia. Nutrition (Burbank, Los Angeles County, Calif.)*, *28*(5), 581–587. https://doi.org/10.1016/j.nut.2011.09.015

Shivaprasad, H. N., Bhanumathy, M., Sushma, G., Midhun, T., Raveendra, K. R., Sushma, K. R., & Venkateshwarlu, K. (2013). *Salacia reticulata* improves serum lipid profiles and glycemic control in patients with prediabetes and mild to moderate hyperlipidemia: A double-blind, placebo-controlled, randomized trial. *Journal of Medicinal Food*, *16*(6), 564–568. https://doi.org/10.1089/jmf.2013.2751

Shukla, R., Singh, S., & Bhandari, C. R. (1973). Preliminary clinical trial on antidiabetic actions of *Azadirachta indica. Medicine and Surgery*, *13*, 11–12.

Sidana, S., Singh, V., Meena, B., Beniwal, S., Singh, K., Kumar, D., & Singla, R. (2017). Effect of Syzygium cumini (jamun) seed powder on glycemic control: A double-blind randomized controlled trial. *Journal of Medical Society*, *31*(3), 185–189.

Singh, J., Cumming, E., Manoharan, G., Kalasz, H., & Adeghate, E. (2011). Medicinal chemistry of the anti-diabetic effects of *Momordica charantia*: Active constituents and modes of actions. *The Open Medicinal Chemistry Journal*, *5*, 70–77. https://doi.org/10.2174/1874104501105010070

Singh, R. G., Rathore, S. S., Kumar, R., Usha, Agarwal, A., & Dubey, G. P. (2010). Nephroprotective role of *Salacia chinensis* in diabetic CKD patients: A pilot study. *Indian Journal of Medical Sciences*, *64*(8), 378–384.

Singh, R. G., Rathore, S. S., Wani, I. A., Usha, Agrawal, A., & Dubey, G. P. (2015). Effects of *Salacia oblonga* on cardiovascular risk factors in chronic kidney disease patients: A prospective study. *Saudi Journal of Kidney Diseases and Transplantation: An Official Publication of the Saudi Center for Organ Transplantation, Saudi Arabia*, *26*(1), 61–66. https://doi.org/10.4103/1319-2442.148736

Singh, U., & Kochhar, A. (2012). Effect of supplementation of bael (*Aegle Marmelos* L.) and nutrition counseling on blood glucose, lipid profile and blood pressure of non-insulin dependent diabetics. *International Journal of Green and Herbal Chemistry*, *1*(3), 284–295.

Siribaddana, S., Medagama, A., Wickramasinghe, N., Siribaddana, N. M., Agampodi, S., & Fernando, D. (2023). The effect of *Salacia reticulata* extract biscuits on blood sugar control of type 2 diabetes mellitus patients: A two-period, two-sequence, crossover, randomized, triple-blind, placebo-controlled, clinical trial. *Cureus*. https://doi.org/10.7759/cureus.45921

Somasundaram, G., Manimekalai, K., Salwe, K. J., & Pandiamunian, J. (2012). Evaluation of the antidiabetic effect of *Ocimum sanctum* in type 2 diabetic patients. *International Journal of Life Science and Pharma Research*, *5*, 75–81.

Sreeshma, K. K., Chandran, C. S., Vatakeel, B., & Vishnu, A. S. (2019). Development and evaluation of tablets containing powders of *Benincasa Hispida* and *Murraya Koenigii*: A novel anti-diabetic product. *International Journal of Research in Engineering, Science and Management*, *2*(8), 31–35.

Srivastava, Y., Bhatt, H., Gupta, O. P., & Gupta, P. S. (1983). Hypoglycemia induced by *Syzygium cumini* Linn. seeds in diabetes mellitus. *Asian Medical Journal*, *26*(7), 489–492.

Srivastava, Y., Venkatakrishna-Bhatt, H., Verma, Y., Venkaiah, K., & Raval, B. H. (1993). Antidiabetic and adaptogenic properties of *Momordica charantia* extract: An experimental and clinical evaluation. *Phytotherapy Research*, *7*(4), 285–289. https://doi.org/10.1002/ptr.2650070405

Stanely, P., Prince, M., & Menon, V. P. (2000). Hypoglycaemic and other related actions of *Tinospora cordifolia* roots in alloxan-induced diabetic rats. *Journal of Ethnopharmacology*, *70*(1), 9–15. https://doi.org/10.1016/s0378-8741(99)00136-1

Sudha Madhuri, A., & Mohanvelu, R. (2017). Evaluation of antidiabetic activity of aqueous extract of *Mangifera Indica* leaves in alloxan induced diabetic rats. *Biomedical and Pharmacology Journal*, *10*(02), 1029–1035. https://doi.org/10.13005/bpj/1200

Sultana, B., Anwar, F., & Przybylski, R. (2007). Antioxidant activity of phenolic components present in barks of *Azadirachta indica*, *Terminalia arjuna*, *Acacia nilotica*, and *Eugenia jambolana* Lam. Trees. *Food Chemistry*, *104*(3), 1106–1114. https://doi.org/10.1016/j.foodchem.2007.01.019

Teixeira, C. C., Rava, C. A., Mallman da Silva, P., Melchior, R., Argenta, R., Anselmi, F., Almeida, C. R., & Fuchs, F. D. (2000). Absence of antihyperglycemic effect of jambolan in experimental and clinical models. *Journal of Ethnopharmacology*, *71*(1–2), 343–347. https://doi.org/10.1016/s0378-8741(00)00185-9

Thirumalai, T., Therasa, S. V., Elumalai, E., & David, E. (2011). Hypoglycemic effect of *Brassica juncea* (seeds) on streptozotocin induced diabetic male albino rat. *Asian Pacific Journal of Tropical Biomedicine*, *1*(4), 323–325. https://doi.org/10.1016/S2221-1691(11)60052-X

Tiwari, R., Mishra, S., Danaboina, G., Pratap Singh Jadaun, G., Kalaivani, M., Kalaiselvan, V., Dhobi, M., & Raghuvanshi, R. S. (2023). Comprehensive chemo-profiling of coumarins enriched extract derived from *Aegle marmelos* (L.) Correa fruit pulp, as an anti-diabetic and anti-inflammatory agent. *Saudi Pharmaceutical Journal*, *31*(9), 101708. https://doi.org/10.1016/j.jsps.2023.101708

Tongia, A., Tongia, S. K., & Dave, M. (2004). Phytochemical determination and extraction of *Momordica charantia* fruit and its hypoglycemic potentiation of oral hypoglycemic drugs in diabetes mellitus (NIDDM). *Indian Journal of Physiology and Pharmacology*, *48*(2), 241–244.

Trakoon-Osot, W., Sotanaphun, U., Phanachet, P., Porasuphatana, S., Udomsubpayakul, U., & Komindr, S. (2013). Pilot study: Hypoglycemic and antiglycation activities of bitter melon (*Momordica charantia* L.) in type 2 diabetic patients. *Journal of Pharmacy Research*, *6*, 859–864. https://doi.org/10.1016/j.jopr.2013.08.007

Tran, N., Pham, B., & Le, L. (2020). Bioactive compounds in anti-diabetic plants: From herbal medicine to modern drug discovery. *Biology*, *9*(9), 252. https://doi.org/10.3390/biology9090252

Tsatsakis, A., Docea, A. O., Calina, D., Tsarouhas, K., Zamfira, L.-M., Mitrut, R., Sharifi-Rad, J., Kovatsi, L., Siokas, V., Dardiotis, E., Drakoulis, N., Lazopoulos, G., Tsitsimpikou, C., Mitsias, P., & Neagu, M. (2019). A mechanistic and pathophysiological approach for stroke associated with drugs of abuse. *Journal of Clinical Medicine*, *8*(9), 1295. https://doi.org/10.3390/jcm8091295

Umamaheswari, S., & Mainzen Prince, P. S. (2007). Antihyperglycaemic effect of "Ilogen-Excel", an ayurvedic herbal formulation in streptozotocin-induced diabetes mellitus. *Acta Poloniae Pharmaceutica*, *64*(1), 53–61.

Upadhyay, A. K., Kumar, K., Kumar, A., & Mishra, H. S. (2010). *Tinospora cordifolia* (Willd.) Hook. F. and Thoms. (Guduchi)—validation of the Ayurvedic pharmacology through experimental and clinical studies. *International Journal of Ayurveda Research*, *1*(2), 112–121. https://doi.org/10.4103/0974-7788.64405

Vaishnav, K., Tiwari, V., Durgapal, A., Meena, B., & Rana, T. S. (2023). Estimation of genetic diversity and population genetic structure in *Gymnema sylvestre* (Retz.) R. Br. Ex Schult. Populations using DAMD and ISSR markers. *Journal of Genetic Engineering and Biotechnology*, *21*(1), 42. https://doi.org/10.1186/s43141-023-00497-7

Veeraragavan, M. A., Sen, C. K., & Kalidindi, S. R. (2021). *Anti-Diabetic Activity of Neem Extract and Synergistic Combinations of Urolithins A and B* (European Union Patent EP3876968A1). https://patents.google.com/patent/EP3876968A1/en

Verma, N., Usman, K., Patel, N., Jain, A., Dhakre, S., Swaroop, A., Bagchi, M., Kumar, P., Preuss, H. G., & Bagchi, D. (2016). A multicenter clinical study to determine the efficacy of a novel fenugreek seed (*Trigonella foenum-graecum*) extract (Fenfuro™) in patients with type 2 diabetes. *Food & Nutrition Research*, *60*, 32382. https://doi.org/10.3402/fnr.v60.32382

Vikrant, V., Grover, J. K., Tandon, N., Rathi, S. S., & Gupta, N. (2001). Treatment with extracts of *Momordica charantia* and *Eugenia jambolana* prevents hyperglycemia and hyperinsulinemia in fructose fed rats. *Journal of Ethnopharmacology*, *76*(2), 139–143. https://doi.org/10.1016/s0378-8741(01)00218-5

Vyas, N., Mehra, R., & Makhija, R. (2016). Salacia—The new multi-targeted approach in diabetics. *Ayu*, *37*(2), 92–97. https://doi.org/10.4103/ayu.AYU_134_13

Waheed, A., Miana, G. A., & Ahmad, S. I. (2006). Clinical investigation of hypoglycemic effect of seeds of *Azadirachta-inidca* in type-2 (NIDDM) diabetes mellitus. *Pakistan Journal of Pharmaceutical Sciences*, *19*(4), 322–325.

Waheed, A., Miana, G. A., Sharafatullah, T., & Ahmad, S. I. (2008). Clinical investigation of hypoglycemic effect of unripe fruit on *Momordica charantia* in type-2 (NIDDM) diabetes mellitus. *Pakistan Journal of Pharmacology*, *25*(1), 7–12.

Welihinda, J., Karunanayake, E. H., Sheriff, M. H., & Jayasinghe, K. S. (1986). Effect of *Momordica charantia* on the glucose tolerance in maturity onset diabetes. *Journal of Ethnopharmacology*, *17*(3), 277–282. https://doi.org/10.1016/0378-8741(86)90116-9

Williams, J. A., Choe, Y. S., Noss, M. J., Baumgartner, C. J., & Mustad, V. A. (2007). Extract of *Salacia oblonga* lowers acute glycemia in patients with type 2 diabetes. *The American Journal of Clinical Nutrition*, *86*(1), 124–130. https://doi.org/10.1093/ajcn/86.1.124

Yadav, R. K., Nandy, B. C., Maity, S., Sarkar, S., & Saha, S. (2015). Phytochemistry, pharmacology, toxicology, and clinical trial of *Ficus racemosa*. *Pharmacognosy Reviews*, *9*(17), 73–80. https://doi.org/10.4103/0973-7847.156356

Yadav, S. P., Vats, V., Ammini, A. C., & Grover, J. K. (2004). *Brassica juncea* (Rai) significantly prevented the development of insulin resistance in rats fed fructose-enriched diet. *Journal of Ethnopharmacology*, *93*(1), 113–116. https://doi.org/10.1016/j.jep.2004.03.034

Yang, Y.-S., Wu, N.-Y., Kornelius, E., Huang, C.-N., & Yang, N.-C. (2022). A randomized, double-blind, placebo-controlled trial to evaluate the hypoglycemic efficacy of the mcIRBP-19-containing *Momordica charantia* L. fruit extracts in the type 2 diabetic subjects. *Food & Nutrition Research*, *66*. https://doi.org/10.29219/fnr.v66.3685

Yu, J. C., Jiang, Z.-T., Li, R., & Chan, S. M. (2020). Chemical composition of the essential oils of *Brassica juncea* (L.) Coss. Grown in different regions, Hebei, Shaanxi and Shandong, of China. *Journal of Food and Drug Analysis*, *11*(1). https://doi.org/10.38212/2224-6614.2729

Zamani, M., Ashtary-Larky, D., Nosratabadi, S., Bagheri, R., Wong, A., Rafiei, M. M., Asiabar, M. M., Khalili, P., Asbaghi, O., & Davoodi, S. H. (2023). The effects of *Gymnema sylvestre* supplementation on lipid profile, glycemic control, blood pressure, and anthropometric indices in adults: A systematic review and meta-analysis. *Phytotherapy Research*, *37*(3), 949–964. https://doi.org/10.1002/ptr.7585

Zarasvand, S. A., Mullins, A. P., Arjmandi, B., & Haley-Zitlin, V. (2023). Antidiabetic properties of mango in animal models and humans: A systematic review. *Nutrition Research (New York)*, *111*, 73–89.

Zuñiga, L. Y., González-Ortiz, M., & Martínez-Abundis, E. (2017). Effect of *Gymnema sylvestre* administration on metabolic syndrome, insulin sensitivity, and insulin secretion. *Journal of Medicinal Food*, *20*(8), 750–754. https://doi.org/10.1089/jmf.2017.0001

Medicinal and Nutraceutical Fruits from the Himalayas

2

Jaishree Vaijanathappa and Igguda Achaiah Chethan

2.1 INTRODUCTION

Across the world, mountain regions are known for their high levels of biodiversity, and this includes a diverse range of plant species. Many of these plants produce edible fruits, berries, nuts, and other food items. Among the 36 global biodiversity hotspots, the Himalayas represent the 4th position, thus have global significance and receive special attention toward biodiversity conservation and management. Indigenous native fruit crops contribute only 6.56% of the total corresponding Indian area, with high productivity (11.47 tons/ha). Wild edible berries in the Himalayan region serve as valuable bio-resources and are often harvested for both subsistence and commercial purposes due to their good yield and nutritional value. The diversity of high-value plant species, distribution and availability for conservation identification of potential areas of cultivation, traditional use practices, nutraceutical and nutritional properties, demand, supply, and marketing thus become vital for mainstreaming wild bio-resources for sustainable development (Pathak et al., 2023). The Government of India sponsored a Mission on Integrated Development of Horticulture (MIDH, then National Horticulture Mission [NHM]) in 2005/2006 to provide special impetus to establish orchards of underutilized fruit species. The morphological characteristics of these underutilized fruit crops enable them to be drought-responsive and survive in adverse arid and semi-arid environments. These underutilized fruit crops represent the next generation of futuristic crops that can improve the farmer's income through sustainable production systems under a climate-change scenario (Meena et al., 2022).

Himalayan berry contains more than 400 nutrients and is replete with various compounds such as vitamins A, B-1, B-2, C, and E, and 25 minerals and 18 amino acids. The vitamin C concentration is 20 times more compared to amla (Indian gooseberry; *Phyllanthus emblica*). The Himalayan berries are beneficial in the treatment of cold, obesity, cancer, liver, blood pressure, cholesterol, diabetes, heart disorders, breathing disorders, and other ailments. They are also effective in treating mental, physical, and sexual weakness. Consuming Himalayan berry juice keeps the body young and disease-free for a longer period (Painuli et al., 2021).

The leaves of the Himalayan species are more cordate, or heart-shaped, with more finely serrated edges than those of the Pacific variety, which has more grooves. The five petals of the Himalayan blackberry are generally fuller and wider than the Pacific blackberry, and the thorns are more abundant on the non-native (Bertucci, 2017).

DOI: 10.1201/9781003352341-2

The following plants from the Himalayan region are reported as fruits of nutraceutical importance and livelihood sources of the Indian Himalayan Region (Belwal et al., 2019) and are discussed in this chapter.

1. *Morus alba*
2. *Rubus ellipticus*
3. *Myrica esculenta*
4. *Berberis aristata*

2.2 *MORUS ALBA*

Morus alba (mulberry genus) species are believed to have evolved on the low Himalayan slopes bordering India and China. India is the second-largest mulberry producer (280,000 hectares of land cultivation) after China (626,000 hectares) (Ahmed et al., 2023 and references therein). *Morus alba*, known as white mulberry, common mulberry, and silkworm mulberry, is a fast-growing, small- to medium-sized mulberry tree which grows to 10–20 m (33–66 ft) tall. It is generally a short-lived tree with a lifespan comparable to that of humans, although there are some specimens known to be more than 250 years old. In the young, vigorous shoots, the leaves may be up to 30 cm (12 in) long, and deeply and intricately lobed, with the lobes rounded. On older trees, the leaves are generally 5–15 cm (2.0–5.9 in) long, unlobed, cordate at the base and rounded to acuminate at the tip and serrated on the margins. Generally, the trees are deciduous in temperate regions, but trees grown in tropical regions may be evergreen.

Bajpai et al. (2015) studied ten quantitative morphological characters in 56 *Morus alba* trees representing three natural populations from the trans-Himalayan Ladakh region. The altitude of collection sites ranged from 2815 to 3177 m above sea level (asl). Coefficient of variation (CV) showed high phenotypic variation in *M. alba*. Linear regression analysis revealed that leaf and fruit size decrease with an increase in altitude. High CV was observed for leaf length, leaf width, petiole length, leaf area, internodal distance, number of nodes, bud length, fruit length, fruit width, and fruit weight. Similarly, a high phenotypic plasticity index was observed for bud length, leaf length, leaf width, petiole length, leaf area, internodal distance, number of nodes, fruit length, fruit width, and fruit weight. For every 100 m increase in altitude, leaf length, leaf width, and leaf area decreased by 1 cm, 0.8 cm, and 16.6 cm^2, respectively. Analysis of covariance showed a predominant altitudinal effect on the morphological characters in comparison to the population effect. A small change in the altitude caused a significant change in the plant morphological characteristics (Figure 2.1).

The flowers are single-sex catkins, with male catkins 2–3.5 cm (0.8–1.4 in) long and female catkins 1–2 cm (0.4–0.8 in) long. Male and female flowers are usually found on separate trees, although they may occur on the same tree. The fruit is 1–1.5 cm (0.4–0.6 in) long. In the wild it is deep purple, but in many cultivated plants it varies from white to pink. It is sweet but bland, unlike the more intense flavor of the red mulberry and black mulberry. The seeds are widely dispersed by droppings of birds that eat the fruit (Barua et al., 2019; Bajpai et al., 2014). The mulberry belongs to the genus *Morus* of the Moraceae family. There are 24 *Morus* species and one subspecies, with at least 100 known varieties. Mulberry is found from temperate to subtropical regions of the Northern Hemisphere to the tropics of the Southern Hemisphere, and they can grow in a wide range of climatic, topographical, and soil conditions. These are widely spread throughout all regions from the tropics to the subarctic and from sea level to altitudes as high as 4000 m. The white mulberry is scientifically notable for the rapid plant movement involved in pollen release from its catkins. The stamens act as catapults, releasing stored elastic energy in just 25 μs. The resulting movement is approximately 380 miles per hour (610 km/h), about half the speed of sound, making it the fastest known movement in the plant kingdom.

Morus alba represents one of Northeast India's indigenous and underutilized fruit crops used by local inhabitants in rural areas. Several plant parts have therapeutic uses; for example, carminative, antiseptic (bark), asthma, fever, bronchitis, dysentery, and toothache (decoction), worms, jaundice, and dysentery

FIGURE 2.1 Different plant parts of Morus alba A) Shrub, B) Flowers, C) Raw and ripe fruits.

(leaf, bark, root, and fruit). *Morus alba* synonym *Morus indica* contains protein, fat, carbohydrate, and fiber (0.10, 0.48, 8.1, and 0.10 g/100 g), and minerals (Ca, P, and Fe; 70, 30, and 2.3 mg/100 g pulp). Its processed products include jam, juice/beverages/squash, and dehydrated products (Barua et al., 2019).

Mulberry (*Morus alba*) is one of the 19 climate-resistant important underutilized fruit crops of Indian arid and semi-arid regions. *M. alba* fruits are rich in vitamins, minerals, dietary fiber, sugar, amino acids, carotenoids, flavonoids, and phytosterols. They are used as a functional food in masala, herbal tea, marmalades, juices, yogurt, biscuits, smoothies, capsules, and dietary food products (i.e., pekmez, kome, and pestil). Their pharmacological uses are indicated for obesity, cardiac diseases, diabetes, hypercholesterolemia, oxidative stress, and for their antifungal, antiaging and hepatoprotective activities (Table 2.1) (Meena et al., 2022 and references therein).

Four mulberry (*Morus*) species (i.e., *M. alba*, *M. indica*, *M. laevigata*, and *M. serrata*) are found in India throughout the Himalayas (Bindroo et al., 2012). *M. alba* is a comparatively earlier introduction with ~200-year-old trees. The most common local *M. alba* varieties prevalent in Kashmir are Chattatul (Mirgund), Chattatul (Zaingeer), Brantul, Buta-Tul, Zaga-Tul, and Ptsari-Tul. *Morus alba* fruit (approximately 3.5 g weight, ~72% moisture content) contained total soluble carbohydrates, glucose, fructose, reducing sugars, and inulin (3.4, 3.1, 3.0, 1.7, and 0.04 g/100 g fresh weight, respectively) (Batiha et al., 2023). The high sugar content enables its use in vegetable and fruit industries to manufacture cakes, bread, and other food products. Moreover, the fruits are used fresh or frozen and dried to make syrups and wines (Dadhwal and Banerjee, 2023).

Bajpai et al. (2014) studied thirty-seven *Morus alba* genotypes from the trans-Himalaya for their polyphenolic contents and antioxidant activity. The plant source (sampling location) determined the polyphenolic content (total phenolic, total flavonoid, and total anthocyanin contents) and antioxidant (ferric reducing potential). The ranges were (1.8–8.3 mg gallic acid equivalent [GAE]/g dw) for total polyphenol content, for total flavonoid content (0.4–0.9 mg quercetin equivalent [QE]/g dw), (0–1.2 mg cyanidin-3-glucoside equivalent/g dw) for total anthocyanin contents, and (34.2–46.2 μmol Fe (II)/g dw) for ferric reducing potential. The antioxidant activity (expressed as IC_{50}) of *M. alba* fruits varied the most by the DPPH assay (1–25 fold), followed by ABTS, lipid peroxidation inhibition, and nitric oxide free radical.

Singh et al. (2015) evaluated the polyphenolic contents, antioxidant, anti-elastase, anti-collagenase, anti-tyrosinase, and anti-inflammatory activities of 13 wild Himalayan (Uttarakhand) including *Morus alba* edible fruit. The total phenolic and flavonoid contents on a fresh weight basis were 623 mg GAE/100 g and 670 mg CE/100 g for *M. alba* 80% acetone extracts. However, the DPPH and FRAP antioxidant activities were 490 and 395 mg ascorbic acid equivalent [AAE]/100 g fw, respectively. *M. alba* exhibited the lowest anti-inflammatory activity diclofenac equivalent human red blood cell [HRBC] membrane stabilization activity/100 g fw. *M. alba* extract exhibited considerable anti-collagenase activity (~1740–1767 mg epigallocatechin gallate [ECGC] equivalent/100 g fw) while displaying the lowest anti-elastase (122 mg ECGC equivalent/100 g fw) activity.

2.2.1 Phytoconstituents of *Morus alba*

The fruit contains anthocyanins (~22 mg cyanidin-3-glucoside equivalent/100 g fresh weight). *M. alba* contains alkaloids (i.e., calystegins B2, C1, and 1-deoxynorjirimycin), terpenoids (betulinic and usolic acids and uvaol), flavonoids (astragalin, atalantoflavone), anthocyanins, chalcones (Morachalcones B and C), phenolic acids, stillbenoids and coumarins (Ahmed et al., 2023 and references therein). Some of the common polyphenols (rutin, chlorogenic acid, caffeic acid, quercetin, gallic acid, kaempferol, and apigenin are illustrated below (Wang et al., 2013):

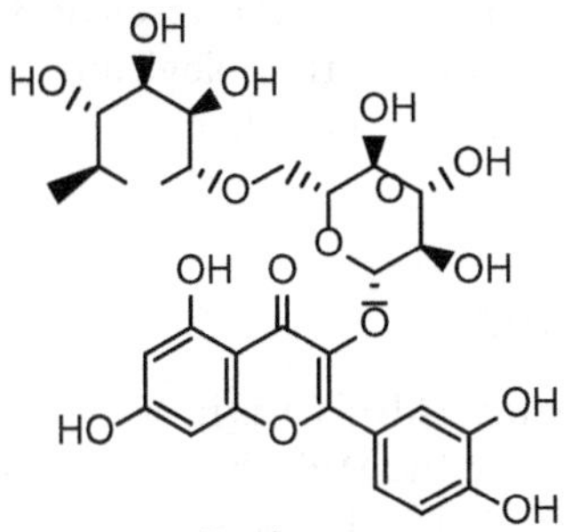

Rutin

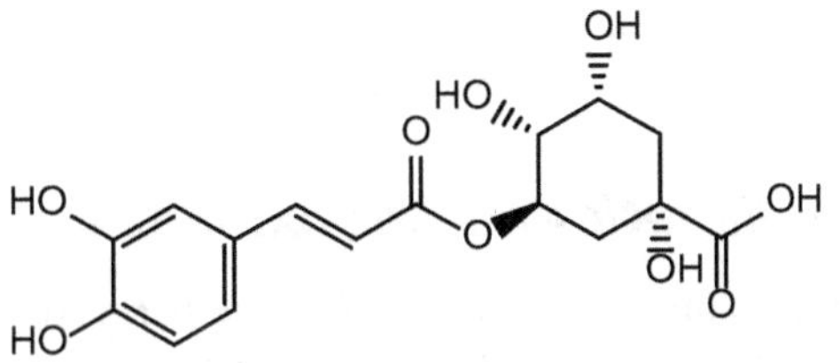

Chlorogenic acid

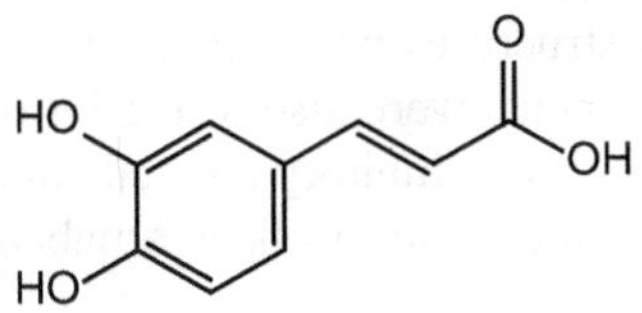

Caffeic acid

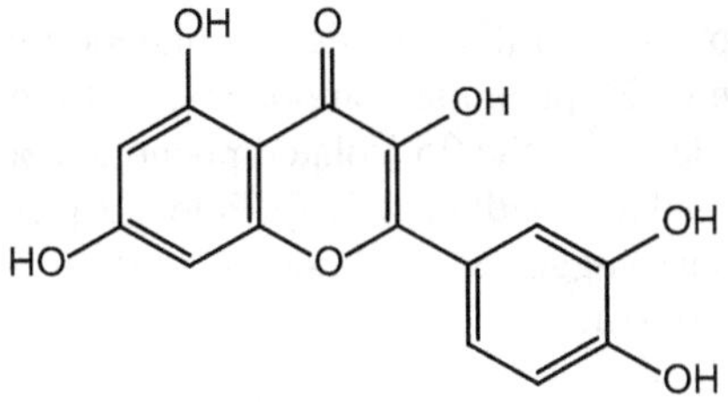

Quercetin

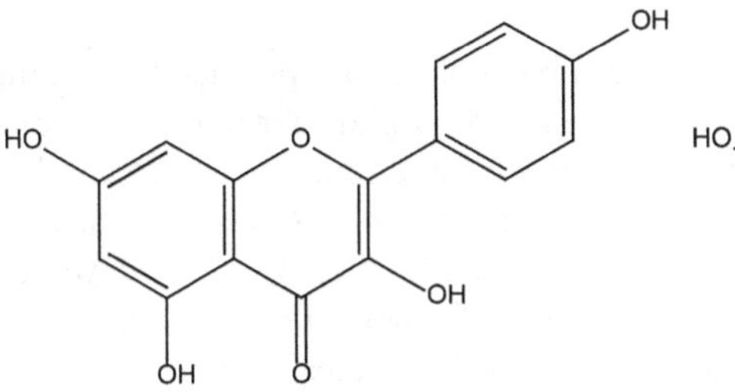

kaempferol

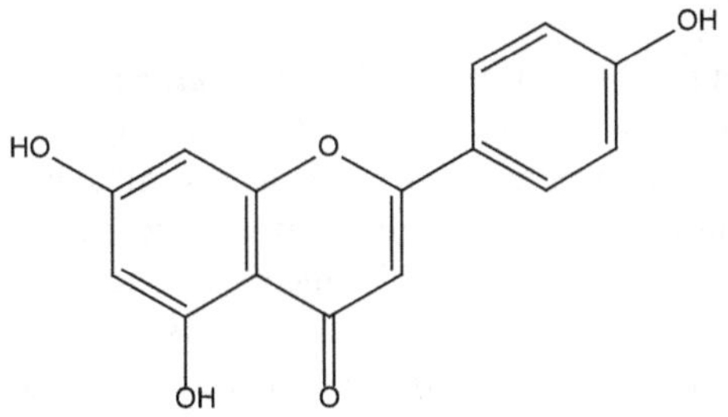

apigenin

2.2.2 Pharmacological Screening

2.2.2.1 Antidiabetic Activity

The active polysaccharides from *M. alba* fruit (MFP) were used to evaluate the antidiabetic activities of different fractions on Type 2 Diabetes Mellitus (T2DM) rats and elucidate the mechanism underlying these activities. The T2DM rats were treated with the fractions of MFP. The disease models induced by a high-fat diet and low-dose streptozotocin injection were compared to normal and metformin-treated diabetic rats. The hypolipidemic and antioxidant effects of the MFP as a dietary supplement were evaluated in rats (fed 4 weeks of either a high-fat or a normal diet supplemented with 5% or 10% MFP). MFP administration to rats on a high-fat diet significantly reduced levels of serum and liver triglyceride, total cholesterol, serum low-density lipoprotein cholesterol, and atherogenic index, and increased serum high-density lipoprotein cholesterol. In addition, the serum and liver content of thiobarbituric acid-related substances, a lipid peroxidation product, significantly decreased, while the superoxide dismutase (SOD) of red blood cells and liver, as well as blood glutathione peroxidase (GSH-Px) activities, significantly increased. No significant changes in lipid profile in the serum and liver were observed in rats on a normal diet supplemented with MFP, but blood and liver antioxidant status improved, as measured by SOD and GSH-Px activity, and lipid peroxidation was reduced. These beneficial effects of MFP on hyperlipidemic rats might be attributed to its dietary fiber, fatty acids, phenolics, flavonoids, anthocyanins, vitamins and trace elements content (Jiao et al., 2017).

A systematic review demonstrated that *Morus alba* leaf extracts improved glucose uptake based on 29 studies in rats (Morales Ramos et al., 2021). Phytoconstituents (chlorogenic acid, isoquercetin and quercitrin and 22 others known to have significant curative effects on diabetes) present in *M. alba*'s leaves have hypoglycemic properties and ameliorate diabetic nephropathy. Pharmacologically, the leaves regulate glucose absorption, produce insulin secretion, exert antioxidant, anti-inflammatory, antihyperglycemic, and hyperlipidemic activities, and manage obesity.

2.2.2.2 Antidiabetic and Antioxidant Activities

The ethyl acetate-soluble extract of *M. alba* (mulberry) fruit extract (MFE) was investigated for the antidiabetic and antioxidant activities. In the *in vitro antioxidant activity*, the MFE showed potent α-glucosidase inhibitory activity and radical-scavenging activities against DPPH and superoxide anion radicals. The fruit extract significantly lowered fasting blood glucose and glycosylated serum protein and increased antioxidant enzymatic activities (superoxide dismutase (SOD), catalase (CAT), glutathione peroxidase (GSH-Px)) in streptozotocin (STZ)-induced diabetic mice. Bioactivity-guided fractionation of the MFE led to the isolation of 25 phenolic compounds, and their structures were reported on the basis of mass spectra and NMR data. All the 25 isolated phenolic compounds were also tested for their α-glucosidase inhibitory activity and antioxidant activity. Potent α-glucosidase inhibitory and radical-scavenging activities of these phenolics suggested that they may be partly responsible for the antidiabetic and antioxidant activities of mulberry fruit.

2.2.3 *Morus alba* Patents

Himalayan plants, particularly *Morus alba*, have been the subject of several patents. For example, *Morus alba* is mixed with other botanicals (plant ingredients) to provide a plant ferment that reduces insulin resistance, body weight, body mass index, whole body fat, and/or hip circumference (Yung-Hsiang et al., 2022). Consumption of the plant ferment containing *M. alba* active ingredients has various beneficial physiological effects (i.e., anti-cancer, anti-aging, anti-inflammatory, increases blood vessel elasticity, protects eyes, helps digestion, and prevents gray hair). The plant ferment (*Lactobacillus plantarum* and *Streptococcus thermophilus*) can be used to manage weight since it elevates the cellular expression level of fat-metabolizing genes and suppresses fat accumulation genes, regulates intracellular energy metabolism, and reduces blood sugar concentration.

In ERα/ERβ+ breast tissue, ethanol or aqueous *M. alba* extract activates estrogen response element gene by repressing the tumor necrosis factor [TNF]-RE-controlled gene TNF-α (Cohen, 2010). The lowest effective *M. alba* extract dose (1.2 μg) represses TNF-α activation of TNF-RE in the presence of ERβ and ERα (110% and 103%). The extracts are claimed to provide treatment for climacteric symptoms, breast and/or uterine cancer, and osteoporosis. Mary Kay Inc. (Addison, TX, USA) recently patented a method to topically treat skin by applying a composition including *Morus alba* fruit extract, *Acmella oleracea* extract, tetrahexadecyl ascorbate, and retinol (Kalahasti et al., 2023). The composition reduces lines and wrinkles, evens and lightens skin tone, increases skin radiance, reduces photo damage, increases elasticity, skin barrier function and firmness, reduces sagging skin, and increases skin antioxidant capacity and laminin expression, and inhibits matrix metalloproteinases [MMPs; MMP-1, -3, and -9], pro-inflammatory cytokines (i.e., lipoxygenase, IL-6, IL-8, TNF-α, or VEGF), and skin elastase expression.

Montaldo et al., (2017) disclosed a formulation for oral administration (i.e., tablets, capsules, or powder) comprising Berberine and *Morus alba* extract to treat excess weight with beneficial effect on the cardiovascular system. The weight reduction is due to the synergistic effect of berberine's clinically demonstrated triglyceride- and cholesterol-lowering activities and *M. alba*'s active ingredient 1-deoxynojirimicin (DNJ) [mostly from the bark or leaves]. DNJ inhibits intestinal glucose absorption and accelerates hepatic glucose metabolism, thereby modulating blood glucose response and beneficially altering lipoprotein profile. DNJ content of *M. alba* fruit varies (13.5–31.1 mg/kg dry weight basis) depending on ripening stages (Lee and Hwang, 2017).

2.3 *MYRICA ESCULENTA*

M. esculenta, commonly known as "Soh Phienarn," and also known as *kaphal* or *kafal*, is found in the hilly parts of northern India and Nepal, particularly in the Uttarakhand regions of Garhwal and Kumaon, southern Bhutan, and western Nepal at heights of 900–1800 m (3,000–6,000 ft). It is among the highly valued wild edible fruits growing between 900 and 2100 m above sea level in the Indian Himalayan region (Rawat et al., 2011). In the Indian Himalayan Region, over 675 wild edibles are known, of which *Myrica esculenta* (family Myricaceae) is one. It is also found in Ziro, Arunachal Pradesh, Himachal Pradesh, and Meghalaya. Its berries are edible and popular in the area. It is the state fruit of Uttarakhand, India.

M. esculenta tree is between 6 and 8 meters tall (20 to 26 ft). Bark is fragile and delicate. Joint leaves are 30–60 cm (1–2 ft) long, with leaflets in pairs that range 6 to 9 cm in size and are 19 mm wide (0.75 in). Flowers are white and found in bunches. The fruit (670 mg average weight) has a firm endocarp and is a globose, succulent drupe (10.3 g), 1.1–1.3 cm diameter (0.4–0.51 in). The astringent-tasting seeds are triangular. *Shweta* (white) and *Rakta* are the two types identified by Ayurveda based on the color of the blossom (red) (Figure 2.2).

M. esculenta has dark green leaves (2–6 in long, 0.5–1.5 in wide) with resinous dots beneath. The unripe fruit is green, changing color with maturity to red upon ripening. The fruit is used locally in Uttarakhand to reduce oxidative stress and prevent certain degenerative diseases. In Meghalaya, the fruits are eaten raw and are also used to make refreshing drinks. The fruit has high commercial value, estimated at Rs. 100,000–890,000 per season. The species is recognized in the Ayurvedic and Unani systems of medicine. The fruit pulp represents up to 75% of the whole fruit, with 40% juice content (3.68% acidity and 12.7% total sugars). The fruit has relatively high iron content (40.4 mg/100 g). The leaves have relatively high protein content (useful as a dietary supplement) and energy (239% total energy) (Sawian et al., 2023). It contains protein, fat, carbohydrate, and fiber (2.65%, 1.36%, 21.55%, and 1.44% g/100 g), minerals (Ca, Fe, and K; 463, 40.4, and 763 mg/100 g pulp), with 109 kcal/100 g pulp energy.

FIGURE 2.2 Different plant parts of Myrica esculenta (A) shrub, (B) flowers, (C) raw fruits, and (D) ripe fruits.

2.3.1 Traditional Uses

M. esculenta is a well-recognized medicinal plant from ancient Ayurveda and the Unani system of medicine. Fruits and roots are used to prepare Ayurvedic formulations such as *Chyawanprash* and *Brahmarasayan* to enhance digestion, memory, intelligence, concentration, and physical strength (Parmar and Kaushal, 1982). Other commercially manufactured Ayurvedic formulations, which contain fruits or bark of *M. esculenta*, include *Katphaladi churna*, *Kaas-har churna*, *Katphala taila*, *Katphala kvatha*, *Khadiradi gutika*, *Maha vatagajankush rasa*, *Brihatphala ghrita*, *Pusyanuga churna*, *Arimedadi taila*, *Bala taila*, and *Mahavisagarbha taila*, used to treat various ailments such as rheumatoid arthritis, diarrhea, dysentery, headache, menorrhagia, and other menstrual disorders (Table 2.1) (Kala, 2007; Pundir et al., 2015).

2.3.2 Phytoconstituents of *Myrica esculenta*

The plant contains a vast amount of phytocompounds, including gallic acid, epigallocatechin 3-O-gallate, epigallocatechin-(4β→8)-epigallocatechin-3-O-gallate, castalagin, catechin, chlorogenic acid, *p*-coumaric acid, ethyl-β-D-glucopyranoside, 3-hydroxybenzaldehyde, isovanillin, ferulic acid, myricetin, myricitrin, lupeol, oleanolic acid, trihydroxytaraxaranoic acid, dihydroxytaraxerane, dihydroxytaraxaranoic acid, prodelphinidin dimer, myricanol, myricanone, and others (Bhatt et al., 2023). Tannin and ascorbic acids were isolated from *M. esculenta* fruits (Kabra et al., 2019a). Caffeic acid, trans-cinnamic acid, and ellagic acid were also isolated from fruits, while leaves contained other compounds ethyl-β-D-glucopyranoside; 3-hydroxybenzaldehyde, isovanillin, 4-(hydroxymethyl) phenol, 4-methoxybenzoic acid (Yang et al., 2011; Patel et al., 2017) (Figure 2.3).

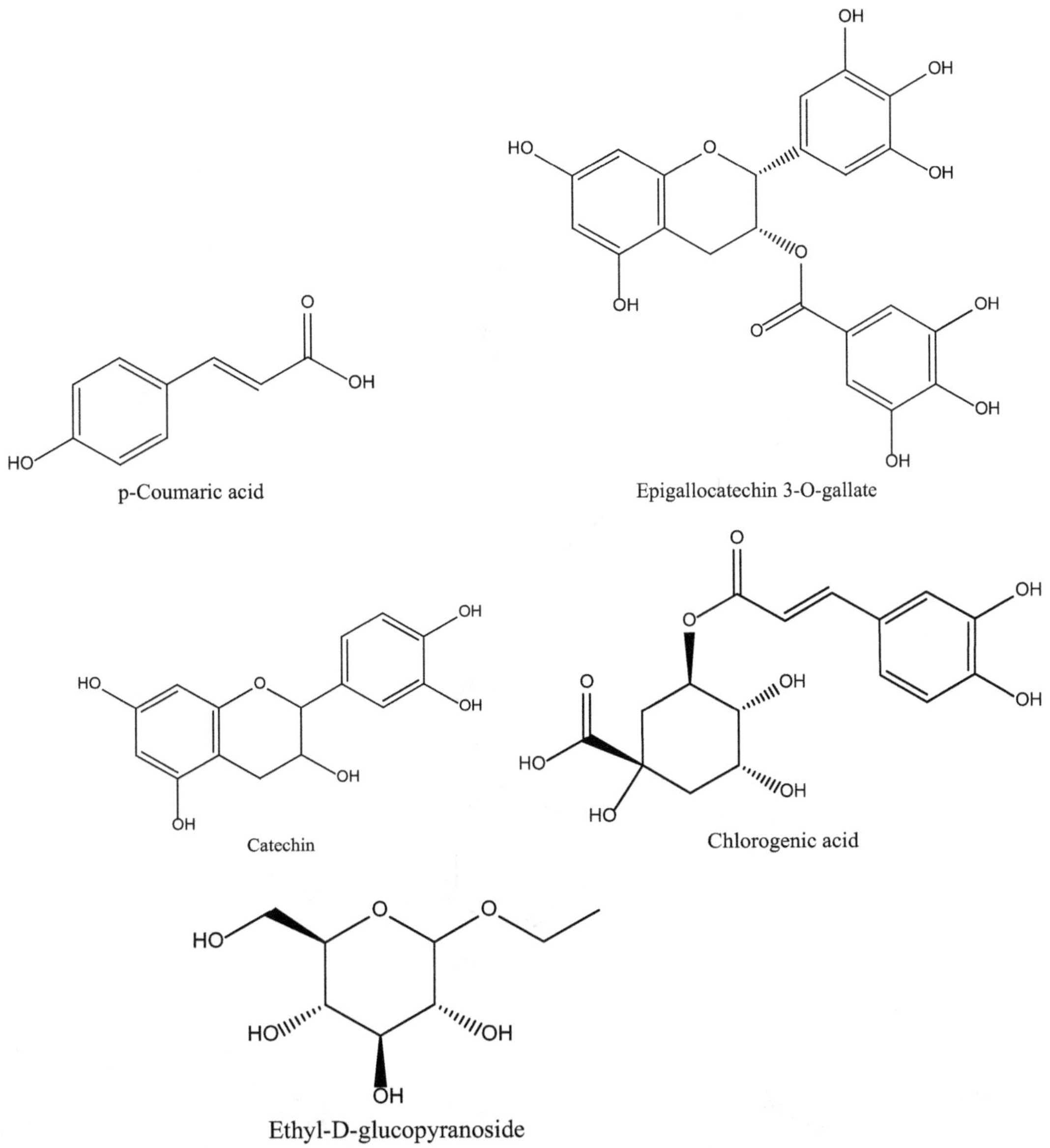

p-Coumaric acid

Epigallocatechin 3-O-gallate

Catechin

Chlorogenic acid

Ethyl-D-glucopyranoside

FIGURE 2.3A Phytoconstituents of *Myrica esculenta*

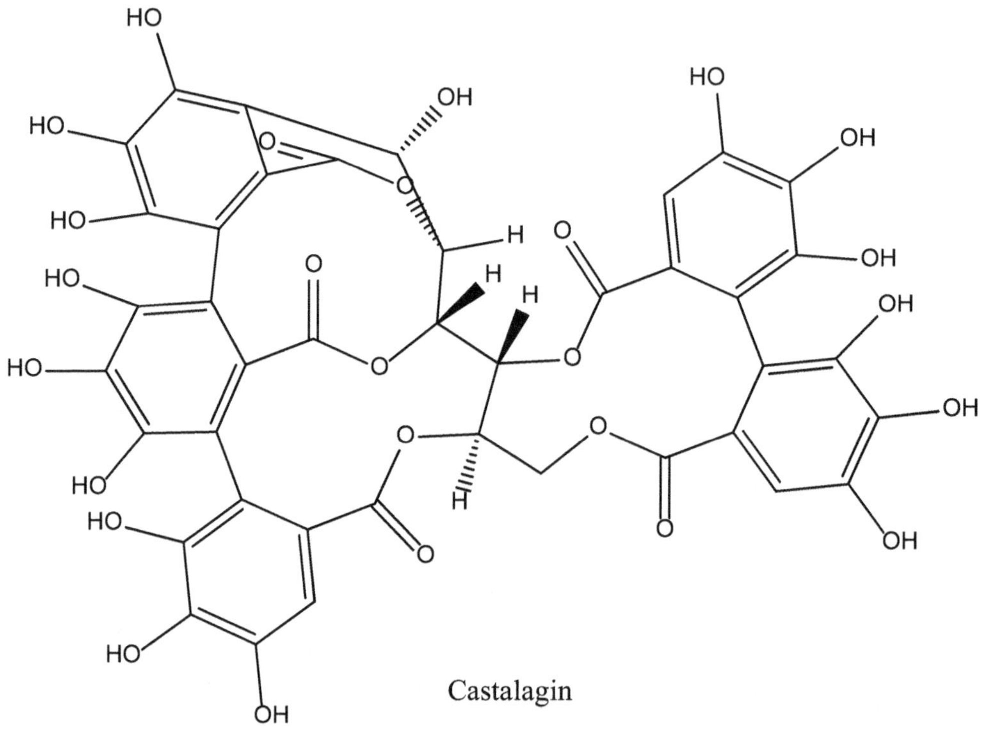

Castalagin

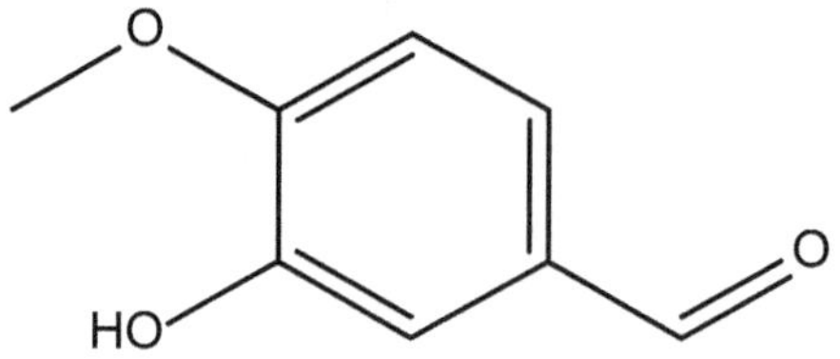

Isovanillin

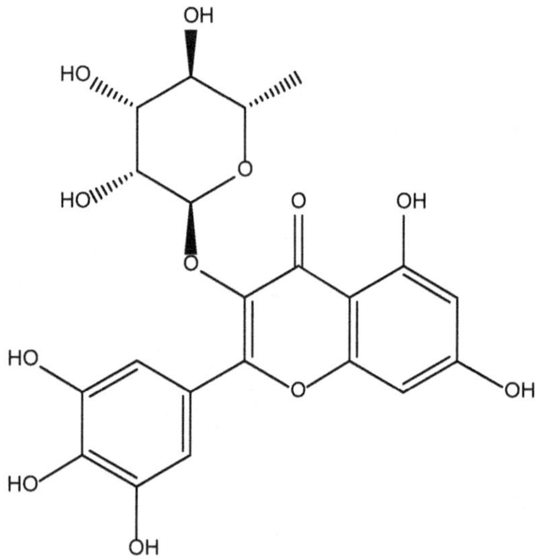

Myricitrin

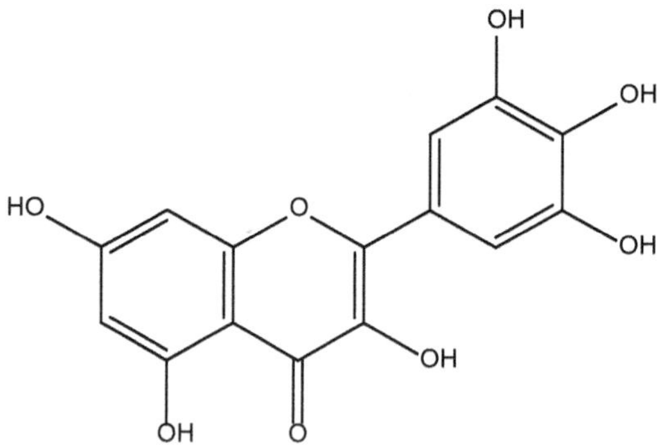

Myricetin

FIGURE 2.3B (Continued)

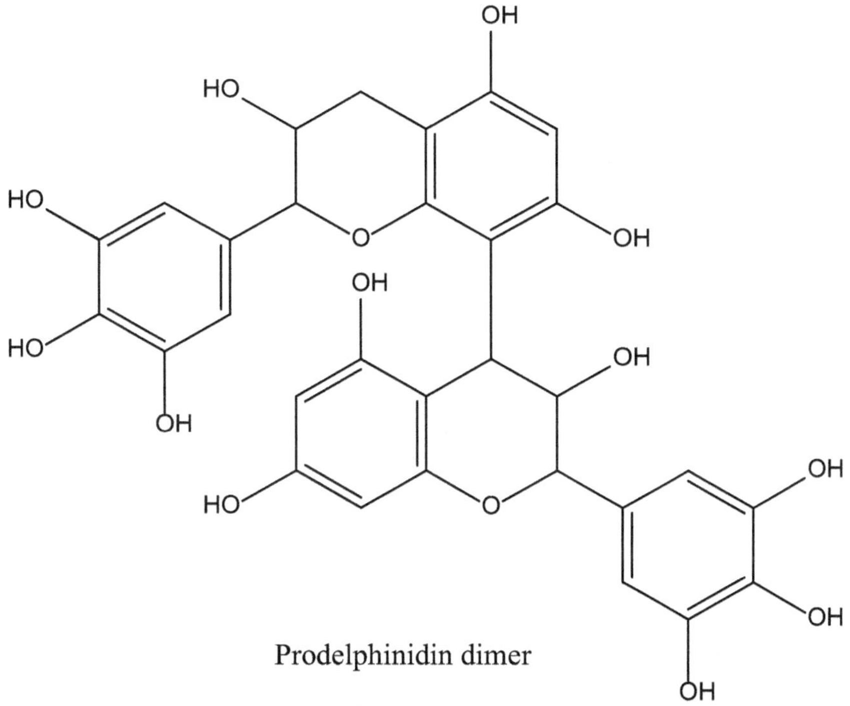

Prodelphinidin dimer

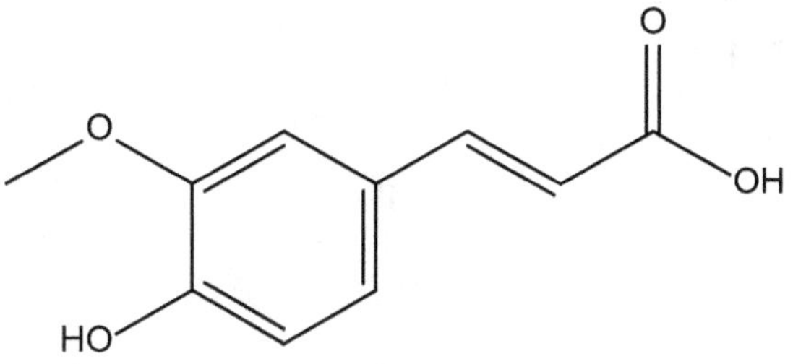

Ferulic acid

FIGURE 2.3C (Continued)

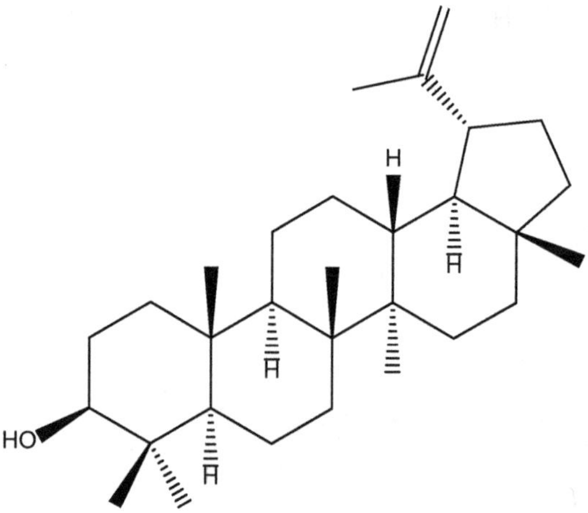

Lupeol

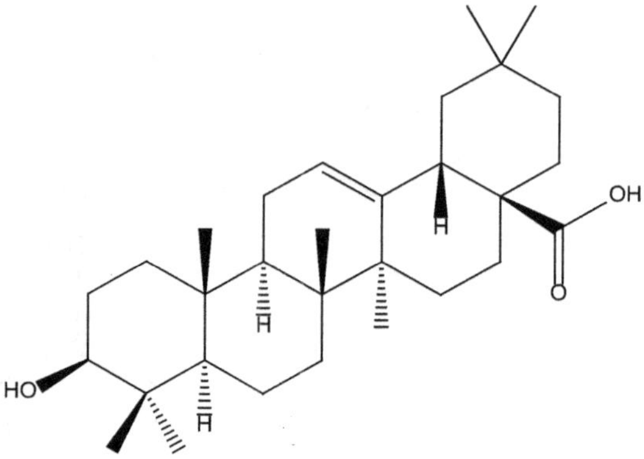

Oleanolic acid

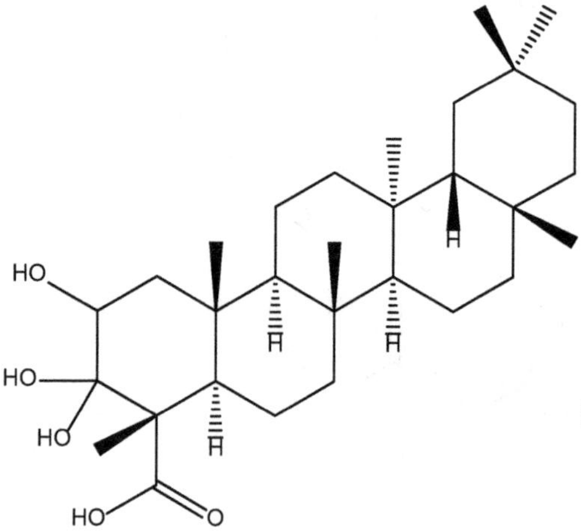

Trihydroxytaraxeranoic acid

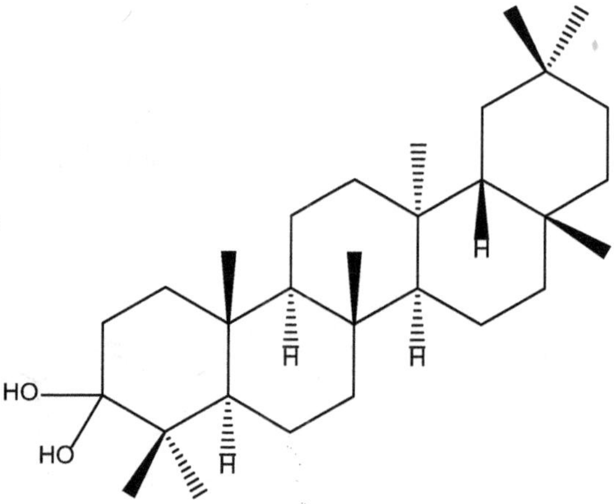

Dihydroxytaraxerane

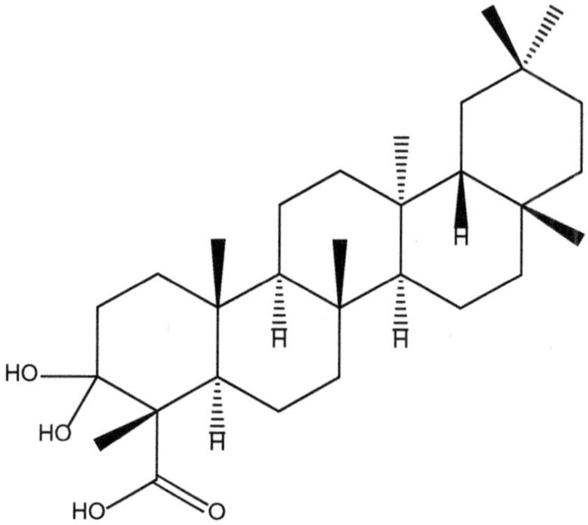

Dihydroxytaraxeranoic acid

FIGURE 2.3D (Continued)

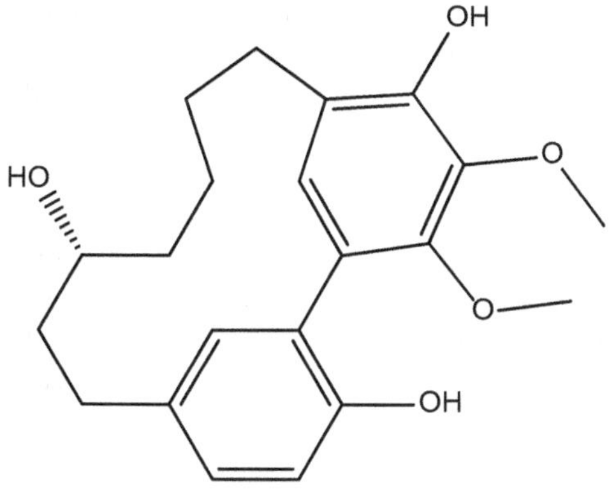

Myricanol

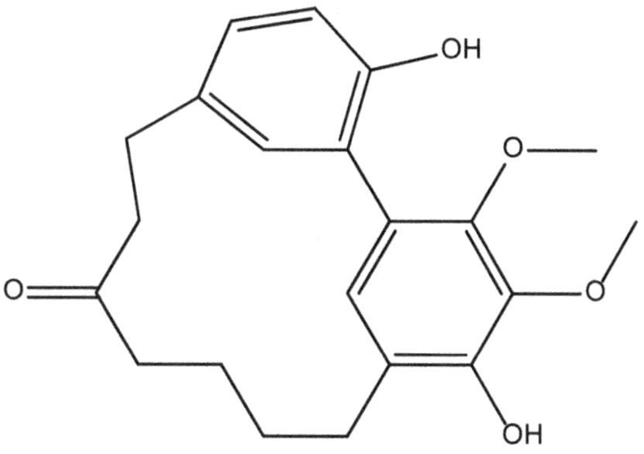

Myricanone

FIGURE 2.3E (Continued)

2.3.3 Pharmacological Screening

M. esculenta fruit extracts and isolated phytoconstituents have been evaluated for different biological activities such as analgesic, anti-asthmatic, anticancer, antioxidant, anti-inflammatory, antidiabetic, anti-ulcer, anxiolytic, hepatoprotective, chemopreventive, hypotensive, and wound healing activities by using *in vitro* methods and *in vivo* animal models, which scientifically prove the traditional utilization of this plant.

2.3.3.1 Analgesic Activity

M. esculenta methanol fruit extract significantly ($P < 0.05$) and dose-dependently increased paw licking and jumping on a hot plate until 50 min, compared to a 60 min maximum in indomethacin-treated mice, indicating analgesic activity (Pant et al., 2014). The ethyl acetate fraction of methanol leaf extract (ME-EtAC) exhibited a significant analgesic effect in acetic acid-induced writhing and tail immersion assays, while methanol leaf extract (200 mg/kg) inhibited writhing (55%) in the acetic acid-induced writhing model (Sood and Shri, 2018 and references therein).

2.3.3.2 Antiasthmatic Activity

The bark of *M. esculenta* was evaluated for the treatment of asthma. Oral bark ethanol extract (75 mg/kg) exerted remarkable antiasthmatic activity through several mechanisms, which include antianaphylactic activity in egg albumin-induced guinea pigs, spasmolytic activity in histamine- and acetylcholine-induced contractions, bronchodilatory activity, histamine release inhibition, and leukocyte inhibition (Sood and Shri, 2018 and references therein). Moreover, the ethanol extract exhibited antiallergic activity by inhibiting eosinophil accumulation. However, the aqueous bark extract displayed more potent antiasthmatic activity at lower doses (27 and 54 mg/kg, p.o.) than the ethanol extract.

2.3.3.3 Anticancer Activity

Acetone and methanol extracts of *M. esculenta* fruits exert anticancer activities by reducing (70%–92%) cancer ($C_{33}A$, SiHa, and HeLa) cell viability without cytotoxicity toward normal transformed cell lines (Saini et al., 2013). The methanol fruit extract exhibited moderate anticancer activity (50%, 48%, and

46% inhibition of Hep G2, HeLa, and MDA-MB-231 cancer cell lines) at 5 mg/mL concentration in the methylthiazolyl tetrazolium (MTT) assay. The anticancer activity increased with increasing extract concentration and may be attributed to the presence of bioactive compounds such as ferulic acid and gallic acid, determined by LC/MS analysis (Mann et al., 2015). The anticancer activity is presumably due to the presence of chemopreventive agents. Alam et al. (2000) reported the chemopreventive effect of polyphenolic-rich *Myrica* ethanol (95%) extract that significantly enhanced the activity of antioxidant enzymes (glutathione reductase, glutathione peroxidase, glucose-6-phosphate dehydrogenase, and catalase) and phase II metabolizing enzymes (glutathione-S-transferase and quinone reductase) in the skin of Swiss albino female mice. The extract suppressed cumene hydroperoxide-mediated cutaneous oxidative stress, probably by inducing glutathione S-transferase and quinone reductase activities, thereby controlling cancer.

2.3.3.4 Antioxidant Activity

Methanol extract of *M. esculenta* fruit and fruit pulp was studied using three different *in vitro* radical scavenging assays, viz., 1,1-diphenyl-2-picrylhydrazyl (DPPH) assay, 2,2-azinobis (3-ethylbenzothiazoline-6-sulphonic acid) (ABTS) assay, and ferric reducing antioxidant power (FRAP) assay (Rawat et al., 2011). Pulp from ripened *M. esculenta* fruits collected from distantly located wild populations (175–2100 m asl) varied significantly in total phenolic and flavonoid contents (1.78–2.51 mg GAE/g fw and 1.31–1.59 mg QE/g fw). Antioxidant activity also varied among the populations (1–1.8 mM for ABTS, 1.5–2.6 mM for DPPH, and 1.5–3 mM AAE/100 g fw for FRAP assay). Total phenolics by HPLC varied from 8.6 to 20.2 mg/100 g fw, with the highest amounts of chlorogenic acid, gallic acid, catechin, and *p*-coumaric acid (5.7, 5, 27, and 0.35 mg/100 g fw, respectively) detected in the fruit extract. Moreover, the antioxidant activity was positively correlated with total phenolic and flavonoid content of the pulp extract.

2.3.3.5 Hepatoprotective Activity

A polyherbal Ayurvedic formulation, Herbitars (50 and 100 mg/kg), containing *M. esculenta* (5 mg/g), exhibited a hepatoprotective effect against carbon tetrachloride (CCl4)-induced hepatotoxicity in Wistar rats. The intervention significantly reduced the levels of thiobarbituric acid-reactive substances and hydroperoxides, and increased antioxidant enzymes (superoxide dismutase, catalase, glutathione peroxidase) activities, as well as reduced glutathione levels in liver and kidney tissues of CCl4-induced rats (Samundeeswari et al., 2011).

2.3.3.6 Antipyretic Effect

M. esculenta methanol fruit extract was screened for antipyretic activity in the Brewer's yeast-induced pyrexia mouse model. Oral administration of the extract (50 and 100 mg/kg) produced significant antipyretic activity comparable to that produced by paracetamol (Pant et al., 2014).

2.4 *RUBUS ELLIPTICUS*

The plant, often referred to as ainselu, golden evergreen raspberry, golden Himalayan raspberry, or yellow Himalayan raspberry, is an Asian species of thorny fruiting shrub in the rose family. The plant, also called Indian raspberry, belongs to the Rosaceae family. Different parts of the plant have been used in ailments like diabetes, diarrhea, gastralgia, dysentery, epilepsy, and as a wound-healing agent, anti-fertility agent, antimicrobial, analgesic, and renal tonic.

As a substantial shrub, the golden Himalayan raspberry can reach 4.5 meters (15 feet) in height and has thick stems. Its trifoliate, elliptic, or obovate leaves have long, serrated bristles. The leaves can reach up to 5 to 10 cm (2 to 4 in) in length. It blooms in the Himalayas from February to April and has five-petaled, short, white flowers that are borne in bunches. Elephants and birds both lust over the fruit because of its taste and ease of pick. Although *Rubus ellipticus* is rarely picked for domestic use, it has a delicious flavor. As soon as the fruit is removed from the thorny bush, it spoils and deteriorates quickly.

In Tibetan villages, this plant's bark is utilized for medical purposes, as an antidiuretic and kidney tonic. Moreover, its juices can be used to treat sore throats, colic, fever, and cough. Another product made from the plant is a bluish-purple dye. The golden Himalayan raspberry is indigenous to China, India, and Pakistan, mainly in its temperate Himalayan area. It grows as a weed in open grasslands and only seldom in forests in India's Himalayan states, such as Himachal Pradesh and Uttarakhand, in their upper altitudes between 1,500 and 2,100 m (4900 to 6900 ft). It can frequently be seen in the local pine forests. The golden Himalayan raspberry can also be found in mesic or wet forests. It has evolved to survive in both complete shade and under intense light from the sun. Birds quickly disperse its seeds, as they do with other *Rubus* species. It can also spread by cutting or reproduce asexually. It can flourish either in open meadows or under the moist forest canopy. Elephants also eat the fruits of the Himalayan raspberry, which can support enormous populations of fruit flies from its rotting berries. All plant parts have been used in traditional medicine to treat respiratory ailments, diabetes, and gastrointestinal disorders, and as an anti-infective agent (Table 2.1, Figure 2.4).

FIGURE 2.4 Different plant parts of Rubus ellipticus (A) shrub, (B) flowers, and (C) ripen fruits.

2.4.1 Phytoconstituents from *R. ellipticus*

Rubus ellipticus fruits are rich sources of natural bioactive compounds (phenolics, flavonoids, anthocyanins, terpenoids, tannins, saponins, steroids, alkaloids, and β-carotene). These and other valuable phytochemicals (glycosides, coumarins, polyphenols, and carotenoids) are present in other plant parts (Kewlani et al., 2023 and references therein). Phytochemical analysis of *R. ellipticus* fruit revealed the presence of flavonoids, carbohydrates, anthocyanins, terpenoids, steroids, tannins, and phenolic compounds. L-ascorbic acid, ellagic acid, quercetin, kaempferol, gallic acid, and catechin were isolated and characterized. Ferulic, trans-cinnamic, and ellagic acids increase, whereas gallic acid, caffeic acid, and phloridzin decrease during *R. ellipticus* fruit ripening (Belwal et al., 2019). Fruits are rich sources of phenolic acids (gallic, chlorogenic, caffeic, ellagic, m-coumaric, hydroxybenzoic, ferulic, vanillic, and trans-cinnamic acids). Fruits are also rich in tannins (628 mg/g dw in methanol and 34 mg/g fw in acidified methanol extract). Various compounds (2,4-bis(1,1-dimethylethyl), benzene propanoic acid, 3,5-bis(1,1-dimethylethyl)-4-hydroxy-methyl ester, and n-hexadecanoic acid) have been isolated in different *R. ellipticus* solvent fractions. Total flavonoid in fruits varied in different solvents; 217 mg/g in dried fruit extracted with methanol (100%), 2.76–4.65 mg/g fw extracted in acidified methanol (80%), and 4.34 mg/g fw in acidified acetone. Kaempferol (17.4 mg/g), phloridzin, and (+)-catechin were the primary flavonoid compounds identified by HPLC. Total anthocyanins of *R. ellipticus* fruits were extracted with different solvents; 3.18 and 12 mg/100 g, and 1.71 mg CGE/100 g in methanol (100%, 80%, and 70%, respectively). Cyanin and delphinidin were the prominent anthocyanins (Kewlani et al., 2023 and references therein).

2.4.2 *In vitro* Activities

Phenolic contents and antioxidant activity have been the focus of many *Rubus ellipticus* studies. Badhani et al. (2015) investigated the phenolic contents and antioxidant activities of ripe fruits from ten *Rubus ellipticus* genotypes collected from the Alora district of Uttarakhand. These genotypes varied significantly in total phenolics (256–331 mg GAE/100 g fw), anthocyanins, and ascorbic acid of ripe fruits (0.04–0.28 and 10.65–40.15 mg/100 g fw for monomeric anthocyanins and ascorbic acid, respectively) (Badhani et al., 2015). A previous study by the same investigators showed that wild Himalayan *Rubus ellipticus* fruits acidified methanol (80% v/v) extract contained total phenolics (3.95 mg GAE/g fw), total flavonoid (5 mg/g fw), total anthocyanins, vitamin C and β-carotene (0.58, 4.46 and 1.81 mg/100 g fw, respectively) (Badhani et al., 2011). HPLC detected only two phenolic compounds (gallic and caffeic acids, 40.5 mg/100 g fw each) that contributed to high antioxidant activities (4.25, 3.10, and 29.22 mM ascorbic acid equivalent [AAE]/100 g fw in the ABTS, FRAP, and DPPH assays, respectively) in the fruit extract.

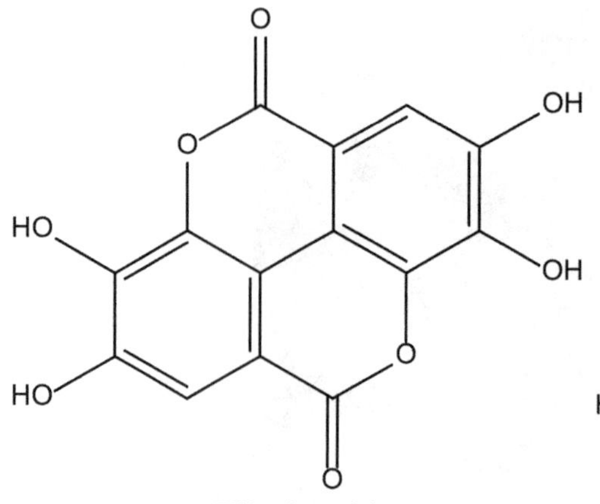

Ellagic acid

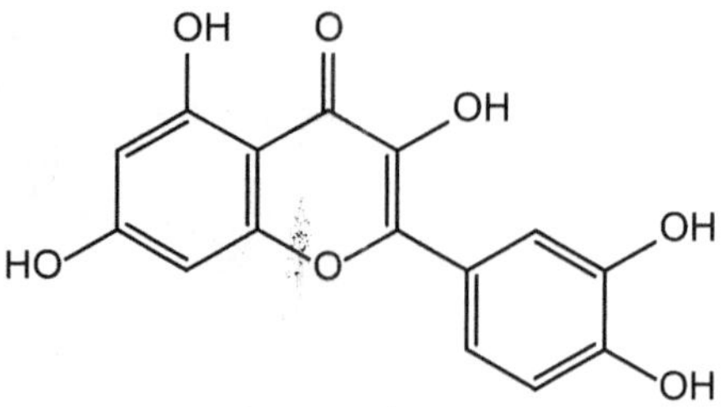

Quercetin

Various solvents have been used to extract *R. ellipticus* phenolics, resulting in variable antioxidant activities. Thus, *Rubus ellipticus* fruits harvested from Uttarakhand contained the highest and lowest amount of total phenolics in acid-acetone and 80% methanol extracts (899 and 550 mg GAE/100 g fresh weight; 690 mg GAE/100 g fw for acidified methanol [pH 2] extract) (Saini et al., 2014). The methanol extract exhibited potent antiproliferative activity (40–60% reduced viability) against cervical cancer (C33A) cells. Aqueous methanol (70%) extract of fresh *Rubus ellipticus* fruit from the Western Ghats rich in total phenolics (72 mg GAE/100 g), total flavonoids (86 mg QE/100 g), anthocyanins (1.7 mg CGE/100 g), and ascorbic acid (44 mg AAE/100 g), exhibited high antioxidant activity (196 μg AAE/mL) on a fresh fruit basis (Karuppusamy et al., 2011). Water extract of fresh *Rubus ellipticus* fruits from the Himalayan region of North Pakistan contained high total phenolics (83 mg GAE/100 g fw) and DPPH scavenging activity (92%). The acetone extract had the highest flavonols (200 mg rutin equivalent [RE]/100 g fw) content, exhibiting the highest DPPH scavenging activity (95%) and maximum FRAP value (137 μM GAE/100 g fw) among 20 plant species evaluated in this study (Shan et al., 2019).

2.4.3 Pharmacological Screening

2.4.3.1 Antidiabetic Activity

Petroleum ether, ethanolic, and aqueous extracts of *R. ellipticus* fruit were evaluated for antidiabetic activity. In an acute toxicity study, no toxic neurological and behavioral symptoms were observed for the plant extracts up to a dose of 2000 mg/kg. Alloxan (2,4,5,6-tetraoxypyrimidine; 2,4,5,6-pyrimidinetetrone) is a toxic glucose analogue, which selectively destroys insulin-producing pancreatic β-cells and causes non-insulin-dependent diabetes mellitus (NIDDM) when administered to animal species. In this study, the protective effects of *R. ellipticus* fruit on glucose tolerance test and alloxan-induced diabetes were evaluated. An effort was also made to investigate the acute toxicity (LD50) of the extracts and the qualitative presence of phytoconstituents in the extracts. Male Wistar albino rats of either sex (weighing 150–200 g) were studied for glucose tolerance test and alloxan-induced diabetes. Blood glucose levels were estimated at different time intervals on the 1st, 4th, 7th, and 15th day of the treatment with *R. ellipticus* extracts and glibenclamide. These levels were compared with the diabetic control group. All *R. ellipticus* fruit extracts exhibited significant antidiabetic effects in both the experimental models of diabetes mellitus. The results justified the traditional use of fruits in the treatment of diabetes (Sharma and Kumar, 2011).

2.4.3.2 Antioxidant and Antiproliferative Activities of R. ellipticus *Phenolics*

R. ellipticus acidic methanol fruit extracts from ten different genotypes varied considerably in antioxidant activities (3.34–4.58 mM ABTS; ~27 mM DPPH and 1.9–3.43 mM ascorbic acid equivalents [AAE]/100 g fw FRAP) (Singh et al., 2023). Antioxidant activity (ABTS assay) of methanol extract from *Rubus ellipticus* was approximately twice that of the standards rutin (1.9×) and butylated hydroxytoluene (2.2×). The ferric reducing antioxidant power of the extract was half that of the standards (57 and 51% for rutin and BHT, respectively). The high polyphenolic contents (phenolics, tannins, flavonoids, and anthocyanins) are responsible for the high antioxidant activity of *Rubus ellipticus* methanol extract. The methanol extracts contained kaempferol (17.4 mg/g extract) that supported its antioxidant and anticancer (50% viable Caco-2 cells at 10 μg/mL) activities (Muniyandi et al., 2019). Sharma and Kumar (2011) evaluated *Rubus ellipticus* fruits (60% petroleum ether, 80% ethanol and aqueous) extracts for antioxidant activities (DPPH radical scavenging and reducing power). The ethanol extract (200 μL) had comparable DPPH radical scavenging activity (61%) to BHA standard (6 μL), whereas the BHA standard (6 μL) had twice (63 vs 31%) the DPPH radical scavenging activity of the aqueous extract (150 μL). The reducing power (Fe^{3+}/Fe^{2+}) of the petroleum ether, ethanol, and aqueous extracts (200 mg/L) was comparable (92% antioxidant activity) to that of the ascorbic acid standard (10 mg/L). However, the ethanol extract showed maximum antioxidant activity, probably due to the high amounts of phytoconstituents (flavonoids, tannins, and others).

George et al. (2014) investigated the *in vitro* antioxidant potential of *Rubus ellipticus* leaf, stem, and root extracts in Soxhlet using petroleum ether, chloroform, acetone, and methanol. The root chloroform and petroleum extracts showed the highest total phenolics (80 and 66 g gallic acid equivalents [GAE]/100 g for *R. ellipticus* and *R. niveus*, respectively) and flavonoids (309 and 265 mg rutin equivalents [RE]/g for *R. ellipticus* and *R. niveus*, respectively) contents. The *R. ellipticus* leaf methanol extract (RELM) showed superior antioxidant capacity relative to that of natural antioxidant ascorbic acid (78 mg ascorbic acid equivalent [AAE]/g extract) and better results for nitric oxide, superoxide, and hydroxyl radical scavenging activities (71%, 69%, and 62%, respectively). RELM was then administered (250 and 100 mg/kg dissolved in 0.1% carboxymethyl cellulose) in Swiss albino male mice to study the *in vivo* antioxidant levels after 30 days of treatment. RELM extract exhibited maximum radical scavenging activity and demonstrated significant cellular antioxidant activities (i.e., catalase, superoxide dismutase, glutathione peroxidase, glutathione S-transferase, glutathione reductase, and reduced glutathione). RELM (250 mg/kg) significantly ($P < 0.01$) increased antioxidant enzymes (catalase, SOD, and glutathione) in blood and glutathione [GSH], glutathione peroxidase [GPX—most important in physiological conditions], glutathione reductase [GR], and catalase ($P < 0.05$) in liver of treated mice. The high GSH and GPX levels in RELM play an important role in lipid peroxidation. This antioxidant effect was ascribed to the phenolic/flavonoid constituents in RELM.

Chauhan and Sharma (2021) extracted *Rubus ellipticus* phytochemicals using hexane, petroleum ether, chloroform, methanol, and water. The phenolic contents of the methanol and water extracts were 8.6 and 32.23 mg/g, whereas their flavonoid contents were comparable (3.30 and 2.86 mg/g). The methanol extract exhibited the highest antioxidant activity (ABTS, DPPH, and FRAP). Three primary phenolic compounds (quercetin, benzoic acid, and phenylacetic acid [447, 103, and 103.3 μg/mL, respectively]) were identified by HPLC in the methanol extract. Benzoic acid and quercetin showed anticancer activity against HeLa cancer cells (77 and 137 μg/mL IC_{50}) by overexpressing HDAC3 and suppressing or downregulating DNMT1, DNMT3A, HDAC1, and HDAC2 genes.

R. ellipticus fruit extracts were analyzed for phenolic contents and antioxidant, antibacterial, and antiproliferative activities. Phenolics were extracted from *R. ellipticus* fruits using 80% aqueous solvents containing methanol, acidic methanol, acetone, and acidic acetone. The estimation revealed that the acidic acetone extracts contained the highest level of total phenolics (899 mg GAE/100 g) and flavonoids (433.5 mg CE/100 g). Free radical scavenging activities of extracts against 1,1′-diphenyl picryl hydrazyl, 2,2′-azino-bis(3-ethylbenzothiazoline-6-sulfonic acid), superoxide, and linoleate hydroperoxide radicals and in ferric reducing activity, the acetone and acidic acetone extracts showed the highest activity. In the metal chelating or antibacterial activity, the extracts did not show any activity. Acetone and methanol extracts showed potent antiproliferative activity against human cervical cancer cells. Therefore, the study concluded that the yellow Himalayan raspberry is a potent source of phytochemicals having super antioxidant and potent antiproliferative activities (Badhani et al., 2015; Saini et al., 2014).

2.4.4 Nutraceutical Profile of *R. ellipticus*

R. ellipticus fruits have high nutritional value due to the presence of carbohydrates, crude fiber, protein, and lipid (86.4%, 3.53%, 4.4%, and 2.7%, respectively), along with high energy value (374 kcal). Lower carbohydrate (73%) and higher crude fiber (7.9%) have also been reported in *R. ellipticus* fruits. The fruits varied in moisture (64.4–80.6%), ash (1.3–4.1%), vitamin C (4.1–44 mg/100 g), reducing sugars (2.2–4.9%), non-reducing sugars (4.2–5.2 mg/100 g), and ascorbic acid content (10.7–40.2 mg/100 g fw) (Kewlani et al., 2023 and references therein). The fruit also contains many essential minerals (Ca, Mg, K, P, Fe, and Zn [450, 119, 680, 1.26, 4.25, and 12.88 mg/100 g, respectively]).

2.4.5 Zinc Oxide Nanoparticles of *R. ellipticus*

The aqueous *Rubus ellipticus* fruit extract was used to prepare ZnO-nanoparticles (ZnO-NPs) through a green synthesis method. The structural, optical, and morphological properties of ZnO-NPs were investigated using X-ray diffraction (XRD), FTIR, UV-vis spectrophotometer, X-ray photoelectron spectroscopy (XPS), field emission scanning electron microscopy, and transmission electron microscope (FESEM). The Rietveld refinement confirmed the phase purity of ZnO-NPs with hexagonal wurtzite crystalline structure and p-63-mc space group with an average 20 nm crystallite size. XPS revealed the presence of an oxygen chemisorbed species on the surface of ZnO-NPs. The nanoparticles exhibited significant *in vitro* antioxidant activity due to attachment of the hydroxyl phenol group on the surface of the nanoparticles. Maximum antibacterial and antifungal activity of ZnO-NPs was observed against *Bacillus subtilis* (31 μg/mL) and *Rosellinia necatrix* (15.6 μg/mL). The anticancer activity revealed 52.41% of A549 cell death (IC_{50}: 158 μg/mL) at 200 μg/mL concentration of nanoparticles, whereas photocatalytic activity showed ~17.5% degradation of the methylene blue within 60 min, with a final 72.7% dye degradation efficiency. These results suggest the medicinal potential of the synthesized ZnO-NPs and, therefore, can be recommended for use in wastewater treatment and medicinal purposes by pharmacological industries (Dhatwalia et al., 2022).

2.5 *BERBERIS ARISTATA*

Berberis aristata is commonly known as Indian barberry and is an Indian medicinal plant from the Berberidaceae family. It is an Ayurvedic herb used since ancient times. It is also known as *Indian berberi, Daruharidra, Daruhaldi, Darvi*, and *Chitra.* The genus contains roughly 450–500 species of deciduous evergreen shrubs found in Asia, Europe, and America's temperate and subtropical climates. *Berberis asiatica* is indigenous to the Himalayas and widely distributed in Himachal Pradesh (600–2700 m altitude) and Assam. It can also be found naturally in the Nilgiri Mountains of southern India and Sri Lanka. *B. aristata* extracts and its formulations are useful in the treatment of diarrhea, hemorrhoids, gynecological disorders, HIV-AIDS, osteoporosis, diabetes, eye and ear infections, wound healing, jaundice, skin diseases, and malarial fever (Choudhary et al., 2021).

Berberis aristata is characterized by an erect spiny shrub (~2–3 m or 6.6–9.8 ft high). It is a woody plant, with bark that appears yellow to brown from the outside and deep yellow from the inside. The bark is covered with three-branched thorns, which are modified leaves, and can be removed by hand in longitudinal strips. The leaves are arranged in tufts of five to eight and are approximately 4.9 cm (1.9 in) long and 1.8 cm (0.71 in) wide. The leaves are deep green on the dorsal surface and light green on the ventral surface. The leaves are simple with pinnate venation, leathery in texture, and toothed, with many small indentations along the leaf margin (Figure 2.5).

The blossoming season begins in mid-March and continues through April. The developing yellow blooms are complete and hermaphroditic. The fully bloomed flowers (average 12.5 mm) are grouped in a racemose inflorescence (11–16 blooms per raceme) along a central stalk. The flower is polysepalous (three big and three little sepals) and polypetalous (six petals altogether). The androecium, the male reproductive component, is polyandrous and has six (5–6 mm long) stamens. The gynoecium (4–5 mm [0.16–0.20 in] long) is the only female reproductive structure, consisting of a short style and a wide stigma. The shrub produces bunches of succulent, acidic, edible berries with therapeutic characteristics that are vivid red. The fruits begin to mature in the second week of May and continue to do so until the end of June. The berries are around 7 mm (0.28 in) long, 4 mm (0.16 in) wide, and weigh 227 mg (0.0080 oz).

FIGURE 2.5 Different plant parts of Berberis aristata (A) Shrub, (B) Flowers, (C) Unripe fruits and (D) Ripe fruits.

2.5.1 Phytoconstituents from *B. aristata*

Several bioactives were reported from *B. aristata*, that is, berberine, palmatine, columbamine, berbamine, oxyberberine, jatrorrhizine, oxycanthine and tetrahydropalmatine (Chander et al., 2017; Potdar et al., 2012). The plant contains 2.4% berberine, the main alkaloid and secondary metabolite known for its biological activity against many chronic diseases; 1.73% tannins; 2.65% ash; 16.44% starch; and 11.8% alcohol-soluble extractive. The fruit contains anthocyanins (24.6 mg cyanidin-3-glucoside equivalent/100 g fresh weight) (Ahmed et al., 2023). *B. asiatica* pulp had the lowest moisture, highest protein, fiber and phytic acid (6 mg/100 g) contents among other Himalayan *Berberis* species, while the pulp and seed had higher carbohydrate than other *Berberis* species (Andola et al., 2011b) (Figure 2.6).

2.5.2 Pharmacological Properties of *B. aristata*

The fruits have antioxidant and anti-inflammatory properties due to the presence of rutin and can presumably treat cancer, diabetes, hypercholesterolemia, and hypertension.

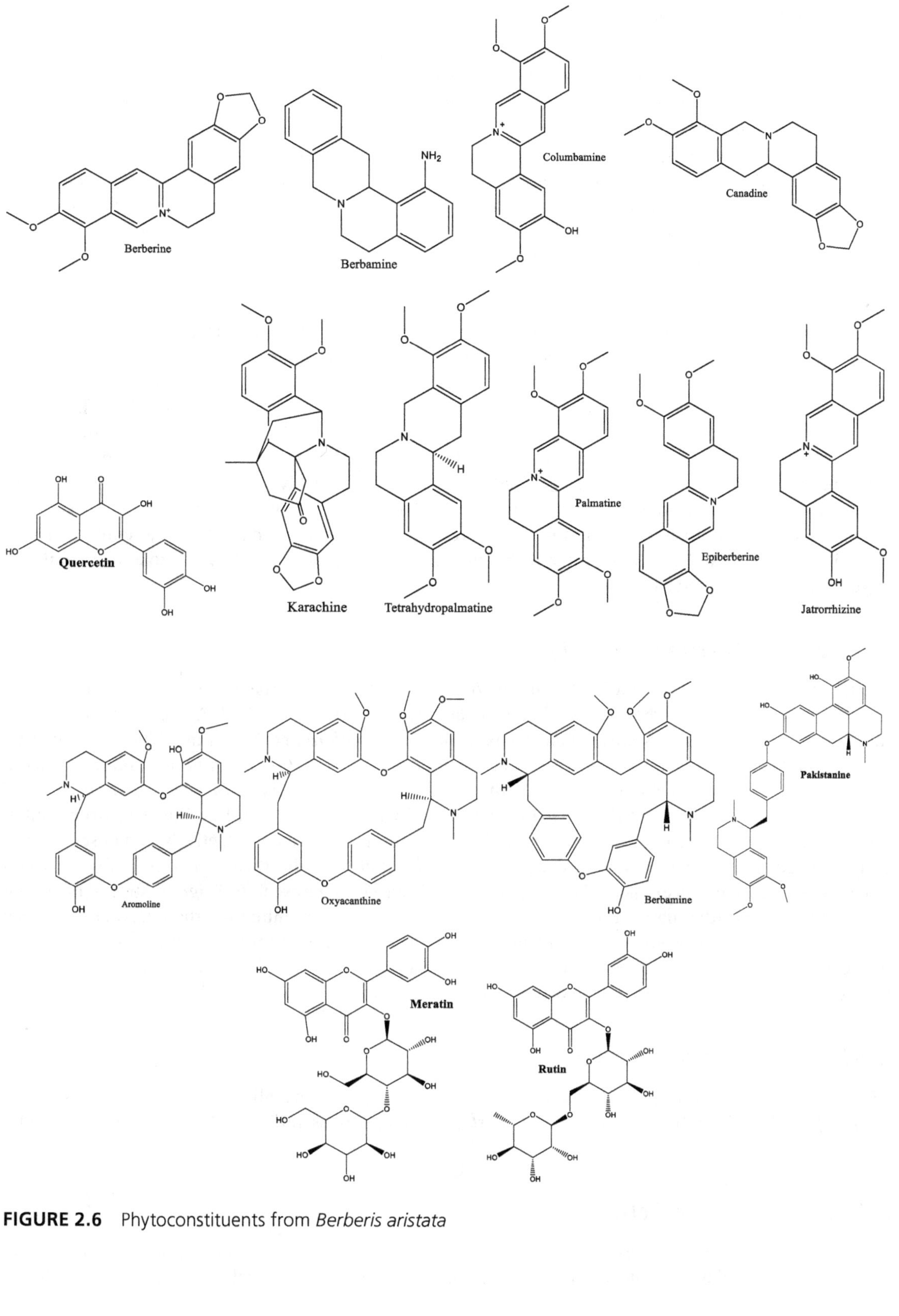

FIGURE 2.6 Phytoconstituents from *Berberis aristata*

2.5.2.1 Antimicrobial Activity

The alkaloid extract of *Berberis aristata* plant showed antibacterial effect against trachoma. Berberine, a phytoconstituent of the plant, showed significant antimicrobial activity against numerous microbes including viruses, bacteria, fungi, protozoans, helminths, and chlamydia (Singh et al., 2007). The herbal gel formulation containing *B. aristata* extract was effective medicine against skin infections in *Staphylococcus aureus*, *Pseudomonas aeruginosa,* and *Corynebacterium* tested bacteria (Padma et al., 2009). Moreover, berberine is known to intercalate with DNA to inhibit *Propionibacterium acnes* and *Staphylococcus epidermis* implicated in the development of inflammatory acne. Its root and hexane extracts exhibit antifungal activity against different fungal pathogens (Sharma et al., 2008). *B. aristata* extract and/or its constituents (primarily berberine) have important clinical significance to prevent or treat several infectious diseases due to their antibacterial, antiviral, and antiprotozoan activities.

Gliardia intestinalis, a parasitic protozoan, is a global cause of intestinal infection resulting in severe and explosive diarrhea. Berberine salts and extracts inhibit *Giardia* trophozoites by inducing morphological damage to trophozoites. In a placebo-controlled clinical trial, 40 subjects received placebo (vitamin B-complex syrup), berberine hydrochloride (5 mg/kg daily), or metronidazole for six days. Berberine treatment markedly reduced gastrointestinal symptoms (superior to that of the standard drug metronidazole) and *Giardia*-positive stools (68% reduction) (Hawrelak, 2003 and references therein). Crude extracts have greater antiprotozoal activity than isolated berberine salts, probably because of the synergistic effect of berberine with other isoquinoline alkaloids present in the plant. In a previous clinical study, berberine (10 mg/kg daily) administered orally for 10 days resulted in satisfactory parasitological cure compared with other antigiardial drugs (quinacrine hydrochloride, furazolidone, and metronidazole) in 32 children (5 years mean age) (Gupte, 1975).

2.5.2.2 Antidiarrheal Activity

B. aristata plant extract was studied for the *in vivo* and *in vitro* antidiarrheal activity. The results of the studies revealed that the berberine constituent isolated from roots and barks of *B. aristata* plant showed inhibition of the secretory response of enterotoxins of *Vibrio cholera* and *E. coli* in rabbit ligated intestinal loop model and infant mouse assay. Also, crude dried formulation of *B. aristata* plant inhibits cholera toxin-induced diarrhea (Joshi et al., 2011). Ethanol and aqueous extracts of *Berberis aristata* bark are effective antidiarrheal agents attributed to the presence of berberine, a standard antidiarrheal drug used in pharmacology. The extracts (31.25–500 mg/kg body weight) pre-treatment effectively and dose-dependently reduced the number exhibiting diarrhea and total number of diarrheal episodes in Swiss albino mice. Berberine and the extracts showed antibacterial activity against four *Shigella* strains including *Shigella dysenteria* (the main causative agent of the disease), thereby inhibiting the intestinal secretory response induced by enterotoxins of *Shigella* species (Joshi et al., 2011). Moreover, the minimum inhibitory concentration and minimum bactericidal concentration values of berberine against *Shigella* strains were comparable to the standard drug ciprofloxacin.

2.5.2.3 Antidiabetic Activity

The ethanolic extract of *B. aristata* showed antidiabetic activity against alloxan induced diabetes in rats (Goli, 2007). The alcoholic stem extract of *B. aristata* plant possesses antihyperglycemic activity (Wasana et al., 2020).

2.5.2.4 Anticancer Activity

The anticancer activity of *B. aristata* methanolic extract was studied against human colon cancer cell line. *B. aristata* methanol extracts dose dependently dependent inhibited HT29 cells (Mazumder et al., 2010).

Moreover, berberine isolated from *B. aristata* significantly inhibited carcinogenesis induced by 20-methylcholanthrene or N-nitrosodiethylamine in a dose-dependent manner (Anis et al., 2001).

2.5.2.5 Cardiotonic Property of B. aristata Fruits

The fruit extract of the plant *B. aristata* was evaluated for inotropic action. The biochemical study was conducted in healthy rabbits to evaluate the cardiovascular property of the plant. The study revealed a significant decrease in serum cholesterol, triglycerides and low-density lipoprotein level and an increase in fibrinogen and thrombin levels (Ahmed et al., 2009). Gilani et al. (1999) revealed that the cardiotonic activity of *B. aristata* fruit extract is concentrated in the n-butanolic fraction (BF) based on activity-directed fractionation of isolated cardiac tissues from reserpinized guinea pigs. BF exhibited dose-dependent positive inotropic action with little effect on heart rate. The selective inotropic effect did not involve β-adrenoceptors stimulation but mediated a cyclic AMP-independent mechanism and cooperatively modulated the calcium–dependent interaction of actin and myosin.

2.5.2.6 Hepatoprotective Effects

B. aristata roots have been used to treat jaundice in Ayurveda because of its hepatoprotective effect. Aqueous and methanol extracts (200 and 300 mg/kg) of dried aerial part of *B. aristata* and berberine had comparable hepatoprotective effect to the standard silymarine (100 mg/kg) against CCI4 induced liver injury (Komal et al., 2011 and references therein). The aqueous and methanol extracts provided significant ($P < 0.001$) hepatoprotective action by reducing serum marker enzymes (glutamate oxaloacetate and glutamate transaminase), elevated serum alkaline phosphatase, serum acid phosphatase, and bilirubin. In another study, aqueous methanol *Berberis aristata* fruits extract (500 mg/kg, orally twice daily for 2 days) pre-treatment in rats significantly ($P < 0.05$) prevented paracetamol (640 mg/kg) as well as CCI4 (1.5 mL/kg)-induced rise in serum transaminases (Gilani and Janbaz, 1995). The demonstrated hepatoprotective action was partly mediated by inhibition of microsomal drug-metabolizing enzymes. Later, the same investigators showed that berberine (4 mg/kg, orally twice daily for 2 days) pre-treatment prevented acetaminophen- or CCL4- induced rise in serum alkaline phosphatase and aminotransferase, indicating antihepatotoxic action in rats (Janbaz and Gilani, 2000). Moreover, the post-treatment response with three successive aqueous methanol *Berberis aristata* fruits extract (500 mg/kg every 6 h) was similar to berberine (4 mg/kg every 6 h) in reducing acetaminophen-induced hepatic damage, indicating selective curative effect mediated by inhibition of microsomal drug-metabolizing enzymes by both the aqueous methanol fruit extract and berberine.

Berberis aristata, *Morus alba*, *Myrica esculenta* and *Rubus ellipticus* from the Himalayan region are reported as fruits (Table 2.1) of nutraceutical importance and livelihood sources of the Indian Himalayan Region (IHR) (Belwal et al., 2019). The Himalayan crops discussed in this chapter are rich in minerals, particularly potassium (in fruits), calcium (except in *M. esculenta* pulp and *B. asiatica* seed), and magnesium (*M. esculenta* fruit and leaves) (Table 2.2). *Berberis asiatica* and *Myrica esculenta* are good iron sources, especially *M. esculenta* leaves. *B. asiatica* pulp had the lowest moisture and phytic acid, and the highest protein and fiber contents among other Himalayan *Berberis* species (Andola et al., 2011). Its pulp and seed had higher carbohydrate than other *Berberis* species (i.e., *aristata*, *lyceum*, *jaeschkeana*, and *pseudumbellata*). *Berberis asiatica* is the most widely distributed species of the *Berberis* genus in the Western Himalayas and Northeast India. The plant contains 2.4% berberine, the main alkaloid and secondary metabolite known for its biological activity against many chronic diseases; 1.73% tannins; 2.65% ash; 16.44% starch; and 11.8% alcohol-soluble extractive. The fruit contains anthocyanins (24.6 mg cyanidin-3-glucoside equivalent/100 g fresh weight) (Ahmed et al., 2023).

Myrica esculenta has dark green leaves (2–6 in long, 0.5–1.5 in wide) with resinous dots beneath. The unripe fruit is green, changing color with maturity to red upon ripening. The fruit is used locally in Uttarakhand to reduce oxidative stress and prevent certain degenerative diseases. In Meghalaya, the

TABLE 2.1 Details of traditional uses, species genera, synonym, and fruits

DETAILS	*MORUS ALBA*	*RUBUS ELLIPTICUS*	*MYRICA ESCULENTA*	*BERBERIS ARISTATA*
Traditional Uses	Popularly, fruits, roots, and leaves of *Morus alba* are used to treat dizziness, insomnia, premature aging, and DM2	The bark is used in Tibetan villages, mainly as a renal tonic and antidiuretic. juices are used to treat coughs, fevers, colic and sore throat. The plant is also used to make a bluish-purple dye.	It is widely used in folk medicine to treat several ailments (asthma, cough, chronic bronchitis, ulcers, inflammation, anemia, fever, diarrhea, ear, nose, and throat disorders).	The plant is used traditionally in Indian system as an antibacterial, antiperiodic, antidiarrheal and anticancer and to treat ophthalmic infections.
Fruits	The fruit is 1–1.5 cm long. In the wild it is deep purple and white to pink in many cultivated plants. It is sweet, bland and the more intense flavor of the red mulberry and black mulberry. The seeds are widely dispersed in the droppings of birds that eat the fruit.	It is sweet, though not commonly harvested for domestic use. The fruit perishes quickly after plucking from the thorny bush.	Fruit is a globose, succulent drupe, with a hard endocarp; diameter 1.1–1.3 cm; average mass 670 mg (10.3 g)	The fruits start ripening from the 2nd week of May and continue throughout June. The berries are approximately 7 mm (0.28 in) long, 4 mm (0.16 in) in diameter, and weigh about 227 mg (0.0080 oz)
Ripening Time	1 to 2 months	2 months	2 months	2 months
Synonym	white mulberry, common mulberry, and silkworm mulberry	ainselu, golden evergreen raspberry, golden or yellow Himalayan raspberry	box myrtle, bayberry and kaphal	Indian barberry, Mara manjal, tree turmeric

Source: Adapted from Elisana Lima Rodrigues et al. (2019), Bhadauria and Rathore (2023), Kabra et al. (2019a) and Rathi et al. (2013)

fruits are eaten raw and also used to make refreshing drinks. The fruit has high commercial value estimated at Rs. 100,000–890,000 per season. The species is recognized in the Ayurvedic and Unani systems of medicines. The fruit pulp represents up to 75% of the whole fruit, with 40% juice content (3.68% acidity and 12.7% total sugars). The fruit has relatively high iron content (40.4 mg/100 g). The leaves have relatively high protein content (useful as diet supplements) and energy (239% total energy) (Sawian et al., 2023).

Four mulberry (*Morus*) species (i.e., *M. alba*, *M. indica*, *M. laevigata*, and *M. serrata*) are found in India throughout the Himalayas (Bindroo et al., 2012). *M. alba* is a comparatively earlier introduction with ~200-year-old trees. The most common local *M. alba* varieties prevalent in Kashmir are Chattatul (Mirgund), Chattatul (Zaingeer), Brantul, Buta-Tul, Zaga-Tul, and Ptsari-Tul. *Morus alba* fruit contained total soluble carbohydrates, glucose, fructose, reducing sugars, and inulin (3.4, 3.1, 3.0, 1.7, and 0.04 g/100 g fresh weight, respectively) (Batiha et al., 2023). The high sugar content enables its use in vegetable and fruit industries to manufacture cakes, bread, and other food products. Moreover, the fruits are used fresh or frozen and dried to make syrups and wines (Dadhwal and Banerjee, 2023 and references therein).

Myrica esculenta is amongst the highly valued wild edible fruits growing between 900 and 2100 m above sea level (asl) in the Indian Himalayan Region (IHR) (Rawat et al., 2011). Ripened *M. esculenta*

TABLE 2.2 Composition of *Berberis asiatica*, *Myrica esculenta*, *Morus alba*, and *Rubus ellipticus*

	BERBERIS ASIATICA		MYRICA ESCULENTA			MORUS ALBA	RUBUS ELLIPTICUS
COMPONENT	*SEED*	*PULP*	*PULP*	*LEAVES*	*FRUIT*	*FRUIT*	*FRUIT*
Moisture (%)	63.3	72				81.7	66.4
Protein (%)	5.9	5.5		1.06	0.96	1.55	3.68
Fiber (%)	4.2	8.1				1.47	
Fat (%)	5.3	3.1				0.48	
Ash (%)	3.8	4.2				0.57	1.3
Carbohydrate (%)	21.6	15.1	0.97	4.62	7.8	14.2	27.12
Minerals (mg/100 g)							
Copper	1.1	2.3		17	0.4	0.42–0.64	
Zinc	1.67	3.17		232	21.6	1.49–1.96	
Iron	12.5	52.3	4.0	171	40.4	2.82–4.67	
Magnesium	1.1	5.1	10	1482	840	120–190	
Sodium	239	47.1		1241	81	10	
Potassium	301	393	190	686	775	1620–2130	680
Calcium	198	396	40	4259	463	190–370	450
Cobalt	8.15	3.5					
Lithium	6.5	6.6					
Phosphorus			10	9.1	24	240–310	
Manganese				345	3.2	1.23–1.94	
Vitamin B (mg/100 g)							
Riboflavin	0.95	1.8				3.1	
Thiamine	15	20.5					
Niacin						0.088	
Anti-nutrients (mg/100 g)							
Tannins	2.4	0.58					
Phytic acid	2.7	0.37					

Source: Adapted from Andola et al. (2011), Sawian et al. (2023), Batiha et al. (2023) and Singh et al. (2023) for *B. asiatica*, *M. esculenta*, *M. alba*, and *R. ellipticus*, respectively

TABLE 2.3 Total phenolics, flavonoids, and anthocyanin composition and antioxidant activity

	PHENOLICS	*FLAVONOIDS*	*ANTHOCYANINS*	*ABTS*	*DPPH*	*FRAP*
PLANT	*(MG GAE/G)*	*(MG QE/G)*	*(MG C3GE/G)*	*MM AAE/100 G FW*		
Berberis asiatica	6.56	5.24	2.7	4	24	2
Myrica esculenta	9.21	4.73	3.2	6	25	3
Morus alba	2.36	4.77	29.53	2	5	1
Rubus ellipticus	2.56–3.31	NS	0.04–0.28	3.3–4.6	26.4–27.7	2.2–3.4
Rubus ellipticus[#]	2.92	4.65	0.12*			
Morus alba	1.87–3.87	0.59	0.19–0.24			

Source: Adapted from Bhatt et al. (2017) for *Berberis asiatica*, *Myrica esculenta* and *Morus alba*; Badhani et al. (2015) for *Rubus ellipticus*; #indicates average values of ten genotypes; Bajpai et al. (2014) for *Morus alba* (all data expressed on dry weight); *data is in mg/100 fresh weight; NS-not specified; GAE—gallic acid equivalent; QE—quercetin equivalent; C3GE—cyanidin-3-glucoside equivalent; AAE—ascorbic acid equivalent. When compared to the standards butylated hydroxytoluene (BHT; 13.18 g/mL; $P < 0.01$), butylated hydroxy anisole.

fruits collected from distantly located wild populations (175–2100 m asl) varied significantly in total phenolic and flavonoid contents (1.75–2.51 mg GAE/g fw and 1.31–1.59 mg QE/g fw). Antioxidant activity also varied among the populations (1–1.8 mM for ABTS; 1.5–2.6 mM for DPPH; and 1.5–3 mM AAE/100 g fw for FRAP assay). Total phenolics by HPLC varied from 8.6 to 20.2 mg/100 g fw, with highest amounts of chlorogenic acid, gallic acid, catechin, and *p*-coumaric acid (5.7, 5, 27, and 0.35 mg/100 g fw, respectively) detected in the fruit extract. Antioxidant compounds varied significantly among the Himalayan fruits, with *Myrica esculenta* and *Morus alba* containing the highest and lowest amounts, respectively (Table 2.3). The antioxidant activities also varied by fruit and types of assays used, with ABTS and DPPH radical scavenging activities and ferric reducing antioxidant power (FRAP) positively correlated ($r^2 \geq 0.928$) with total phenolics (Bhatt et al., 2017).

Chlorogenic acid, gallic acid, and catechin were the most common phenolic acids in the Himalayan fruits, containing the highest amounts of compounds (Table 2.4). Moreover, *Myrica esculenta* had the broadest phenolic acid spectrum relative to the other fruits. The ripe fruits, except *Rubus ellipticus*, were also rich in anthocyanins, cyanin in particular (Belwal et al., 2019). Phenolic acid content and composition of the fruits varied depending on the extraction conditions and quantification methods used by the investigators. Bhatt et al. (2017) investigated the nutraceutical potential of ten wild edible fruits of the Indian Himalayan Region, including the four selected in this chapter. Phenolics, flavonoids, and anthocyanins were quantified in acidic methanol extract (methanol and 1N HCl in 80:20 ratio) of fruits after seed removal. Ripe fruits from ten *Rubus ellipticus* genotypes collected from Alora district of Uttarakhand were investigated for their phenolic contents and antioxidant activities (Badhani et al., 2015). Genotypes varied significantly in gallic acid (22.7–97.6 mg/100 g), catechin (20.9–79.5 mg/100 g), and chlorogenic acid (3.2–10.1 mg/100 g). Caffeic acid (0.82–5.43 mg/100 g) was present in seven genotypes and absent in three genotypes.

TABLE 2.4 Phenolic composition of fruits from Himalayan crops.

PARAMETERS (MG/100 G)	MYRICA ESCULENTA	BERBERIS ARISTATA	RUBUS ELLIPTICUS	MORUS ALBA	MYRICA ESCULENTA	BERBERIS ARISTATA	RUBUS ELLIPTICUS
	A				*B*		*C*
Anthocyanins							
Cyanin	423.5	353	4.6	313.5			
Delphinidin	110.9	39	1.1	81.7			
Polyphenolics							
Chlorogenic acid	2.3	94.2	0.18	9	11.24	11.9	5.34
Gallic acid	9.2	309.9	55	25.2	52.16	18.4	62.2
Catechin	175	466		30.9	6.36	7.99	52.36
Caffeic acid		1.42		0.15		137.28	2.05
m-Coumaric acid	0.18	1.89		0.05			
Rutin	2.6	1.59					
p-Coumaric acid	0.12	0.25				0.38	
4-Hydroxybenzoic acid	1.6	0.64					
3-Hydroxybenzoic acid		32.7		5.5			
Ferulic acid	0.98		0.89				
Ellagic acid		4.63	0.44				
trans Cinnamic acid			0.44				
Vanillic acid				16			

Source: Adapted from Muniyandi et al. (2019), Belwal et al. (2019), Bhatt et al. (2017) and Badhani et al. (2015) for A, B and C, respectively

2.6 READY-TO-SERVE (RTS) BEVERAGES

Beverages (RTS) have been developed from *M. esculenta* and *R. ellipticus* fresh fruits and their blend as functional foods (Krishna et al., 2016). Fresh juice (100 mL/L) and blend (50 mL of each juice) were adjusted to 0.3% acidity (with citric acid) and 13% total soluble solids prior to refrigerated storage (5 °C) for 10 weeks. The beverages contained soluble solids (10 °Brix), total and reducing sugars (9.7% and 4.5%), with 33 soluble solids:acid ratio and other desirable attributes (i.e., antioxidant components). Storage affected vitamin C content the most and carotenoids the least during beverage storage. Monomeric anthocyanins in *M. esculenta* RTS beverage decreased linearly (2.3–29%) during storage. Moreover, anthocyanin degradation was higher in blended beverages than in *M. esculenta* RTS, presumably due to lower ascorbic content. Storage steadily reduced flavonol content, initially increased flavonoids and total phenolics for two weeks, followed by linear and precipitous declines for flavonoids and total phenolics, respectively. Antioxidant activity of the beverage followed the same trend as total phenolics during storage, with significant reduction after 10 weeks of storage (Table 2.5).

Trained sensory panel preferred *M. esculenta* beverage the most, and the blend was preferred the least based on sensory attributes (i.e., overall appearance, color, and flavor ranking). Organoleptic acceptability of fruit beverages decreases significantly during storage, prominently for *M. esculenta* and blended beverages, due to discoloration (reduction/instability) of red color resulting from pigment hydrolysis and potential ascorbic acid bleaching (Krishna et al., 2016). The study recommends consumption of these RTS beverages within 8-week refrigerated storage (4 °C). These RTS beverages can be part of the high-value global botanical beverage market (estimated at $608 million by the end of 2022), driven by consumer preferences toward functional (health halo), flavor, and convenient products (Ataman, 2023).

Krishna et al. (2012) developed antioxidant-rich beverages from mulberry fruits (Central Institute for Arid Horticulture [CIAH] genotypes CIAH Mulberry Selections-1 and -2 or CIAH M-1 and CIAH M-2) grown in Bikaner, Rajasthan. Genotypes CIAH M-1 and M-2 are purple red and greenish white at maturity. Squashes containing 25% mulberry juice (45% total soluble solids, 1% acidity) were diluted three times with water and presented to a sensory panel for quality evaluation. The fruit beverage serving (100 mL) contained ascorbic acid (1.5 and 1.9 mg), total phenolics (29.8 and 23.2 mg), flavonoids (12.6 and 10.4 mg), and flavanol (2.9 and 2.8 mg) for CIAH M-1 and M-2, respectively. Antioxidant activities of CIAH M-1 and M-2 genotypes were 98.5 and 42.8 μM Trolox equivalent [TE]/100 mL (CUPRAC method), 55.7 and 36.3 μM TE/100 mL (FRAP assay), and 78.6 and 69.2% DPPH inhibition. The sensory quality of

TABLE 2.5 Ready-to-serve beverages from Himalayan Berberis fruits

				M. ESCULENTA	R. ELLIPTICUS	*BLEND*
ATTRIBUTES	M. ESCULENTA	R. ELLIPTICUS	BLEND	**CHANGE (%) DURING STORAGE*		
Ascorbic acid (mg/100 mL)	0.2	2.1	0.98	100	61.9	100
Total carotenoids (μg/100 mL)	29.7	516.9	297.5	40.1	19.4	26.6
Flavanol (mg PGE/100 mL)	8.6	5.3	6.2	14	15.1	14.5
Flavonoids (mg QE/100 mL)	25.4	16.2	20.6	7.5	8.6	5.3
Anthocyanins (mg C3GE/L)	1.58		0.77	30.4		35.1
Total phenols (mg GAE/100 mL)	45.2	14.5	27.9	19.9	21.4	29.4
Antioxidant (mM TE/L)	4.95	1.47	3.19	31.3	25.2	37.3
Organoleptic acceptability	85	72	65.2	26.8	16.4	38.5

Source: Adapted from Krishna et al. (2016); * values indicate % change in attribute after 10 weeks storage; PGE—pholoroglucinol equivalent

CIAH M-1 was significantly superior to that of CIAH M-2 for color and appearance, body/texture, flavor, taste, and overall quality, presumably because of the dark-fruited genotype containing higher antioxidant activities.

Kumar and Kumar (2016) evaluated the storage stability of probiotic milk (*Lactobacillus rhamnosus* LBS 2 culture [5% v/v containing 10^7-10^8 CFU/mL]) products containing pasteurized *R. ellipticus* juice (10% v/v). The fruit-fortified milk was analyzed for antioxidant activity during storage (1–15 days, 4 °C) at 5-day intervals. Storage reduced pH (5.7 vs 3.3) and increased acidity (0.23%–0.67%) during storage of fruit-fortified probiotic milk. The *R. ellipticus*-fortified probiotic milk retained a probiotic value of 6.6 log CFU/mL only on the first day of storage, with no growth observed on day 10 of storage. Antioxidant activity (DPPH radical scavenging assay) of the fortified probiotic milk decreased (~33%) during storage, nevertheless retaining a considerable amount of antioxidants.

Other products already in the market include *Rubus ellipticus* root extract used as a moisturizer/humectant in cosmetics and pharmaceuticals to protect skin from UV-induced damage and improve wrinkles. *Myrica esculenta* is sold as a powder named Kaiphal (Yuvika from Amazon). *Morus alba* fruit extract is combined with sodium phytate, glycerin, and alcohol (TEGO® enlight, Evonik personal care product), marketed as a 100% natural effective brightening blend for various types of cosmetic products, suitable for natural cosmetic products and a very effective option to target age spot correction.

2.7 CLINICAL TRIALS

Mulberry (*Morus alba*) leaves are rich sources of iminosugars also known as azasugars (i.e., 1-deoxynojirimycin [DNJ], N-methyl-DNJ, 2-O-α-D-galactopyranosyl-DNJ, and fagomine), flavonoids (rutin, isoquercitrin, astragalin), chlorogenic acid, terpene lactones (loliolide), and phenolic compounds (moracin M, steppogenin-4'-O-β-D-glucoside, and mulberroside) (Alugoju and Tencomnao 2024 and references therein). The pharmacological properties of *M. alba* include antioxidant, anti-microbial, anti-inflammatory, anti-cancer, anti-skin aging, cardioprotective, hepatoprotective, renoprotective, neuroprotective, and antidiabetic activities. Some *Morus alba* phytochemicals (astragalin, isoquercitrin, roseoside II and scopolin) inhibit α-glucosidase activity and glucose absorption (Prabhakar and Doble, 2008). Alugoju and Tencomnao (2024) reported studies on the antidiabetic effect of different *M. alba* parts (leaf, fruit and roots) and extracts or phytochemical constituents.

We present only human clinical studies demonstrating the antidiabetic effects of *M. alba*, its constituents, and/or extracts, although these benefits are also observed and supported in many animal studies. Nakamura et al. (2009) developed three confections containing aqueous ethanol (50%) *Morus alba* leaves extract (~0.77% DNJ) to investigate their antidiabetic efficacy in humans. The (1:10 *M. alba* extract [3 g]: sucrose) ratio had the most suppressive effect on elevated postprandial blood glucose and insulin in ten healthy females (22 years average age, 21 kg/m^2 median BMI). Moreover, *M. alba* extract addition to confections inhibits intestinal sucrase, thus creating a prebiotic effect. The study illustrates that *M. alba*-containing functional foods (i.e., confections) can regulate glucose and contribute to the prevention and quality of life for prediabetic and diabetic patients.

Extracts of mulberry fruit (1.5 g containing 0.5% w/w 1-deoxynojirimycin [DNJ]) and mulberry leaf (1 g containing 0.8% w/w DNJ) added to rice porridge (~50 g available carbohydrate [control]) significantly ($P < 0.05$) reduced post-prandial (2 h) blood glucose and insulin responses in healthy adults (31 years mean age, 22 kg/m^2 median body mass index [BMI]) in a randomized, balanced clinical study (Mela et al., 2020). The same group (Mela et al., 2023) investigated the effects of several doses (0.04–1.5 g) of mulberry fruit extract added to boiled rice on postprandial blood glucose and serum insulin response in healthy Indian adults (~34 years mean age, ~22 kg/m^2 median BMI; 42M/42F and 45M/39F for trials 1 and 2, respectively). These clinical trials demonstrated that a low dose (0.37 g containing ~2 mg DNJ) produced consistent reduction in postprandial blood glucose and serum insulin response to boiled rice with no apparent malabsorption or intolerance. In the latest clinical trial (Mela et al., 2024) commercially available aqueous mulberry fruit extract (0.75 g containing 3.75 mg DNJ) added to boiled rice acutely reduced postprandial (2 h) plasma glucose (22.4%) and serum insulin (17.5%) responses in diabetic subjects. The clinical trial was a small pilot, randomized, double-blind, multicenter, 3-period-balanced order cross-over (within-subject) trial involving diabetic subjects (11M/13F, 51 years mean age, 27.5 kg/m^2 median BMI, 7.2% mean hemoglobin A1C [HbA1c]). A lower dose (0.37 g containing ~1.85 mg DNJ) used in the trial produced less consistent results presumably due to the level being close to the lower efficacy limit.

The iminosugars are potent glycosidase enzyme inhibitors and can therefore exert hypoglycemic effects. Consumption of DNJ-rich mulberry leaf powder significantly reduced postprandial blood glucose in humans, suggesting its beneficial role as a food supplement to regulate glucose levels in diabetic patients (Kimura et al., 2007). Asai et al. (2011) investigated the efficacy of long term (12 weeks) DNJ-enriched mulberry leaf extract (~1.5% DNJ) supplementation or its single injection on postprandial glucose levels in humans with impaired glucose metabolism. Long-term supplementation of the leaf extract (6 mg daily for 12 weeks followed by 4-week post-treatment observation) significantly ($P < 0.001$) improved postprandial glycemic control (increased serum 1,5-anhydroglucitol) in subjects (21M/12F, 54 years median age) with impaired glucose metabolism compared to placebo (22M/10F) group in this randomized double-blind placebo-controlled clinical trial.

Józefczuk et al. (2017) examined the impact of mulberry leaf extract addition to cornflakes on postprandial glycemic effects in a randomized single-blind placebo-controlled cross-over clinical trial. Mulberry leaf extract (36 mg DNJ) supplemented single cornflakes (50 g with 100 mL low-fat milk) lowered participants' starch digestion and absorption, indicating potential to manage diabetic postprandial glucose levels. In a randomized controlled, two-parallel group trial, mulberry leaf extracts (12 mg DNJ, with 120 mL water thrice daily) supplementation for 12 weeks reduced insulin resistance and plasma glucose (fasting and postprandial) in obese prediabetes (19F/9M, 53 years mean age, 31 kg/m^2 median BMI) (Thaipitakwong et al., 2020). In another study, *M. alba* leaf extract (300 mg twice daily for 12 weeks) significantly reduced insulin and lipid peroxidation levels while raising HDL cholesterol levels in type 2 diabetic patients in a randomized double blind placebo controlled trial (Taghizadeh et al., 2022).

Mulberry leaves consistently improve lipid profiles of dyslipidemia patients according to several clinical studies (Thaipitakwong et al., 2020). Mulberry leaves (0.367 mg DNJ/tablet, three tablets daily before meals) administered for 12 weeks significantly ($P < 0.05$) lowered total cholesterol, LDL-C, and triglycerides and increased HDL-C compared to baseline levels in 23 early-stage dyslipidemia patients. Moreover, the lipid reduction of mulberry leaves was superior to lifestyle modifications alone (Aramwit et al., 2011). Daily consumption of mulberry leaf tea (3 g thrice daily before meals) effectively lowered blood lipids (total cholesterol, triglycerides, and LDL-C) after 8 weeks intervention compared to baseline levels in 46 dyslipidemia patients (Banchobphutsa and Jarasphol, 2012). Mulberry leaf capsules (corresponding to 36 mg DNJ daily) for 12 weeks had moderate effect on plasma triglycerides and lowered very small, dense, highly atherogenic LDL particles in ten hypertriglyceridemia (TG ≥ 200 mg/dL) patients (Kojima et al., 2010). The leaves also effectively improved lipid profiles in 24 type 2 diabetic patients with abnormal lipid levels. The lipid (total cholesterol, triglycerides, LDL-C, VLDL-C, and plasma free fatty acids) lowering effect of the mulberry leaves (500 mg capsules daily) after 4 weeks was comparable to glibenclamide (5 mg daily) (Andallu et al., 2001). In Thailand, mulberry tea also known as Mon Tea is a therapeutic remedy for diabetes, whereas the plant is used in Chinese medicine to control blood pressure.

Kaewsedam et al. (2022) produced non-thermally pasteurized microfiltered (0.22 μm) mulberry fruit juice (8° Brix) supplemented with isomalto oligosaccharide (IMO) and evaluated its prebiotic activity after fecal fermentation. Mulberry juice with 2 and 8% IMO significantly increased bifidobacteria [predominantly *B. scardovii* and *B. stercoris*] (5% and 17.5%) in 24 h fermentation and biologically active metabolites of polyphenols and anthocyanin in the gut. Butyric acid concentrations of mulberry juice with 2 and 8% IMO increased at 48 h fermentation, thereby increasing its potential anti-inflammatory activity that can reduce colonic inflammation and risk of colon cancer. The study indicated that IMO-supplemented mulberry juice promoted bacteria beneficial to gut health.

Lown et al. (2017) investigated the glycaemic and insulinaemic response of Reducose®, a proprietary mulberry leaf extract co-administered with maltodextrin (50 g) in a double-blind randomized crossover clinical trial. Mulberry leaf extract (250 and 500 mg) significantly ($P < 0.005$) and dose dependently reduced postprandial blood glucose and suppressed elevated total insulin in normoglycemic healthy adults (25F/17M, 29 years average age, 23 kg/m^2 median BMI) without side effects compared to the placebo group. Moreover, mulberry leaf extract had no effect on average glucose or insulin response until 30 min after ingestion, probably due to delay in the release of the active constituent DNJ because of the use of encapsulation material (hydroxypropyl methylcellulose). Constituents other than DNJ, fagomine, a glycoprotein for example, may be responsible for enhancing insulin sensitivity to glucose metabolism.

The same Reducose® (mulberry leaf extract, 250 mg) consumed during a complete meal (510 kcal) reduced postprandial glucose response in lean adults (11F/19M, 31 years mean age, 23 kg/m^2 median BMI) in a randomized crossover study (Gheldof et al., 2022). Mulberry leaf extract ([5% Reducose®, Phynova, Whitney, UK] containing 12.5 mg DNJ) was reconstituted in water (200 mL) when consumed before the meal and mixed in the standardized rice during the entire meal simultaneously with carbohydrates. Water (200 mL) before the meal was the control for the study. Mulberry leaf extract lowered postprandial glucose response of the complete meal as evidenced by significant 2 h-iAUC reduction (34%), presumably because the DNJ reached the small intestine simultaneously with carbohydrates in the food to compete for binding with α-glucosidase enzymes. The mulberry leaf extract intervention is proposed as a convenient solution for subjects with impaired glucose tolerance or diabetes to control their daily blood glucose. Controlling postprandial glucose response is important in the management and prevention of type 2 diabetes.

Mulberry leaf extract has been consistently reported to lower acute blood glucose response upon carbohydrate ingestion, which was confirmed in a meta-analysis (Phimarn et al., 2017). *Morus alba* leaves significantly reduced postprandial glucose within 90 min, but not at 120 min, without altering other glycemic and lipidemic markers based on a meta-analysis that included 13 RCTs involving 436 participants (223 and 213 receiving *M. alba* and placebo, respectively) (Phimarn et al., 2017). DNJ, the main active ingredient in mulberry leaf extract, is responsible for the postprandial glucose response due to its potent α-glucosidase and glycogen phosphorylase inhibition. Mulberry leaf extract at 6–36 mg DNJ effectively lowered postprandial glucose response in healthy subjects, in people with impaired glucose metabolism or T2D (Gheldof et al., 2022 and references therein).

Alvin et al. (2011) evaluated the efficacy of 75% mulberry extract oil as a skin lightening agent to treat facial melasma, an aesthetically undesirable skin hyperpigmentation condition, in a randomized, single-blind, placebo-controlled open 2-arm clinical trial. Mulberry extract (75% mulberry in a coconut oil base) as a topical treatment for 8 weeks significantly ($P < 0.05$) improved melasma area and severity [MASI] score (1.2 vs 0.06 for placebo [virgin coconut oil]), reduced Mexameter values indicating lightening of the pigmentation, and increased quality of life in 25 healthy Filipino women (47 years mean age). The lightening effect of mulberry is ascribed to inhibition of tyrosinase enzyme (due to skin-whitening bioactives, primarily mulberroside F, Moracin J, steppogenin, mulberrofuran G, albanol B, and kuwanon G) involved in the melanogenesis pathway. Mulberry extract (0.39%) is a potent tyrosinase inhibitor (50% compared to 5.5% for hydroquinone [gold standard for melasma] and 10% for kojic acid) with strong superoxide scavenging activity.

Berberine, a potent hypoglycemic agent, lowers glucose, waist circumference, systolic blood pressure, triglycerides, total insulin secretion, and increases insulin sensitivity in subjects with metabolic syndrome, and improves blood lipid levels of mild hyperlipidemic subjects and metabolic profile in patients with non-alcohol fatty liver disease (Rondanelli et al., 2023 and references therein). Rondanelli et al. (2023) evaluated the potential benefit of oral Berberine Pytosome™ intake in improving glycemic and lipid values in subjects with impaired fasting blood glucose (IFG) in a randomized double-blind placebo-controlled clinical trial. Brevis™ (two 550 mg tablets daily) for eight weeks significantly lowered glycemia, lipid profile (total cholesterol, total cholesterol/HDL, triglycerides), insulin, visceral adipose tissue [VAT], and fat mass in overweight subjects (14F/10M, 59 years mean age and 30 kg/m^2 median BMI). Berberine Pytosome™ supplementation significantly reduced glycemia, improved lipid profile, and lowered cardiovascular disease risk (lower ApoB/ApoA ratio), thereby modulating glycemic profile. Brevis™ (Indena SpA, Milan, Italy) is a dietary ingredient consisting of berberine phospholipids with optimized bioabsorption by formulating *Berberis aristata* extract with a specific Pytosome™ that increases its bio-accessibility and tolerability (Rondanelli et al., 2023). Brevis™ tablet (550 mg of Berberine Phytosome™) corresponds to 188 mg berberine. In an earlier clinical study, Brevis™ (two 550 mg tablets daily) administration for 60 days improved glycemic, lipid, and insulin resistance profiles in 12 women (27 years average age, 25 kg/m^2 median BMI) with polycystic ovary syndrome (PCOS). This one-group pretest-post-test study demonstrated Berberine Phytosome™ to be an effective insulin sensitizer improving metabolic and hormonal disorders in PCOS women.

Fogacci et al. (2023) evaluated the efficacy and safety profile of the new Zeta Colest nutraceutical formulation containing *Berberis aristata* dry extract (400 mg, one pill daily for 8 weeks) in low-risk CVD patients with elevated LDL-C levels. The Zeta Colest pill consisted of *Berberis aristata* dry extract (400 mg with 340 mg berberine), 98 mg red yeast rice extract (2.9 mg monacolins), 87.5 mg milk thistle dry extract (70 mg silymarin), 50 mg guggul dry extract (1.3 mg guggulsterone). The intervention significantly ($P < 0.05$) reduced total cholesterol, non-high-density lipoprotein cholesterol, and high-sensitivity C-reactive protein in overweight free-living Italians (49M/31F, 53 years average age, 26 kg/m^2 median BMI) with suboptimal cholesterolemia. This double-blind placebo-controlled randomized clinical trial demonstrated that the *Berberis aristata* nutraceutical improved lipid metabolism, vascular function, and inflammation in patients with polygenic hypercholesterolemia.

Trimarco et al. (2015) compared the effects of two nutraceuticals, with and without *Morus alba* leaf extract, in a double-blind randomized crossover trial. Combination A (Armolipid Plus, RottapharmSpA) is approved in Italy to control dyslipidemia and contains policosanol, red yeast rice (Monakolin K 3 mg), berberine 500 mg, astaxanthine, folic acid, and coenzyme Q10. Combination B contains berberine (531.25 mg), red yeast rice powder (220 mg, Monacolin K 3.3 mg), and 200 mg *Morus alba* leaf extract and has been approved in Italy (LopiGLIK™, Akademy Pharma). Combination B therapy (*M. alba* leaf extract intervention) significantly ($P < 0.005$) reduced total and LDL cholesterol, improved glucose metabolism (significant reduction in serum fasting glucose, insulin, and HbA1c), and improved insulin sensitivity more than combination A. This clinical trial for 4 weeks treatment in dyslipidemic subjects (11M/12F, 60 years mean age, 29 kg/m^2 median BMI showed that *Morus alba* leaf extract addition to berberine and Monacolin K improved (insulin sensitivity on hepatic gluconeogenesis and potentiated cholesterol-lowering effect of the combination) their effect on plasma cholesterol and glucose metabolism.

2.8 CURRENT AND FUTURE PERSPECTIVES

The native crops (*Berberis asiatica*, *Morus alba*, *Myrica esculenta* and *Rubus ellipticus*) described in this chapter are reported as fruits of nutraceutical importance and livelihood sources of the Indian Himalayan Region (IHR) (Belwal et al., 2019). *Rubus ellipticus* has high economic value and market potential. The fruits are commercially cultivated to collect nectar sugar as honey, and produce value-added edible products (squash, jam, jelly, alcoholic beverages, herbal wines, toothpaste, health beverage, yogurt, and ice cream) (Kewlani et al., 2023 and references therein). *R. ellipticus* fruit jam generates a net return of 117 rupees (Rs) per person, while the fruit costs 50 rupees in rural areas with an estimated Rs 217 per person daily net benefit based on input and output costs (Rs 203 and 420, respectively). Assam State Biodiversity Board fixed *R. ellipticus* fruits market value at Rs 30–40/kg, while the global price is generally around \$1.54/kg. *R. ellipticus* root is also used by pharmaceutical and cosmetic industries for clinically tested products (e.g. BG80) to protect skin from UV damage and improve wrinkles. In West Himalaya, the potential *Myrica esculenta* fruit yield is 2–4.2 tonnes/ha at different sites, with ~3–7% of which is harvested for income generation (Bhatt et al., 2000). *M. esculenta* is popular with local people for its delicious fruit and processed products such as squash, syrup, and jam. The fruit contributes 50,000–450,000 rupees per year (US$1430–12,860) to the economy of nearby villages, accounting for high total harvested fruit value (Rs 100,000–890,000 per season [US$ 2857–25,430]). *M. esculenta* is estimated to have high income generating potential (US$ 714–3036/ha) exceeding the economic potential of other promising trees in the region.

Proprietary mulberry leaf products (e.g. Reducose®) have already been used in clinical trials and other products (e.g. berberine, formulated/standardized extracts) from these Himalayan crops are readily available in powders or other forms. Patents described in this chapter provide evidence that these Himalayan crops and/or their value added products can be used in the food, nutraceutical, pharmaceutical and cosmetic industries.

REFERENCES

Ahmed, K. M., Khan, M., & Shivananda, B. (2009). Cardiovascular diseases and role of medicinal plants as a re-emerging health aid. *Pharmacognosy Reviews*, *3*(5), 8–21

Ahmed, M., Bose, I., Goksen, G., & Roy, S. (2023). Himalayan sources of anthocyanins and its multifunctional applications: A review. *Foods*, *12*(11), 2203. https://doi.org/10.3390/foods12112203

Alam, A., Iqbal, M., Saleem, M., Ahmed, S. U., & Sultana, S. (2000). *Myrica nagi* attenuates cumene hydroperoxide-induced cutaneous oxidative stress and toxicity in Swiss albino mice. *Pharmacology & Toxicology*, *86*(5), 209–214. https://doi.org/10.1034/j.1600-0773.2000.d01-37.x

Alugoju, P., & Tencomnao, T. (2024). Phytochemical constituents, antidiabetic potential and other pharmacological activities of mulberry (*Morus alba* L.). In A. Husen (ed.), *Antidiabetic Medicinal Plants and Herbal Treatments* (pp. 345–362). CRC Press.

Alvin, G., Catambay, N., Vergara, A., & Jamora, M. J. (2011). A comparative study of the safety and efficacy of 75% mulberry (*Morus alba*) extract oil versus placebo as a topical treatment for melasma: A randomized, single-blind, placebo-controlled trial. *Journal of Drugs in Dermatology: JDD*, *10*(9), 1025–1031.

Andallu, B., Suryakantham, V., Srikanthi, B. L., & Reddy, G. K. (2001). Effect of mulberry (Morus indica L.) therapy on plasma and erythrocyte membrane lipids in patients with type 2 diabetes. *Clinica Chimica Acta*, *314*(1–2), 47–53. https://doi.org/10.1016/s0009-8981(01)00632-5

Andola, H. C., Gaira, K. S., Rawal, R. S., Rawat, M. S., & Bhatt, I. D. (2011a). Influence of environmental factors on production of berberine content in *Berberis asiatica* Roxb. Ex DC in Kumaun West Himalaya, India. *Journal of Herbs, Spices & Medicinal Plants*, *17*(4), 329–338. https://doi.org/10.1080/10496475.2011.626892

Andola, H. C., Rawal, R. S., & Bhatt, I. D. (2011b). Comparative studies on the nutritive and anti-nutritive properties of fruits in selected Berberis species of West Himalaya, India. *Food Research International*, *44*(7), 2352–2356. https://doi.org/10.1016/j.foodres.2010.07.017

Anis, K. V., Rajeshkumar, N. V., & Kuttan, R. (2001). Inhibition of chemical carcinogenesis by berberine in rats and mice. *Journal of Pharmacy and Pharmacology*, *53*(5), 763–768. https://doi.org/10.1211/0022357011775901

Aramwit, P., Petcharat, K., & Supasyndh, O. (2011). Efficacy of mulberry leaf tablets in patients with mild dyslipidemia. *Phytotherapy Research*, *25*(3), 365–369. https://doi.org/10.1002/ptr.3270

Asai, A., Nakagawa, K., Higuchi, O., Kimura, T., Kojima, Y., Kariya, J., . . . & Oikawa, S. (2011). Effect of mulberry leaf extract with enriched 1-deoxynojirimycin content on postprandial glycemic control in subjects with impaired glucose metabolism. *Journal of Diabetes Investigation*, *2*(4), 318–323. https://doi.org/10.1111/j.2040-1124.2011.00101.x

Ataman, D. (2023). From functional to ethical, identifying the health halo in botanicals. *Food Navigator-USA*, 26-Aug-2023. https://www.foodnavigator-usa.com/Article/2023/08/25

Badhani, A., Rawat, S., Bhatt, I. D., & Rawal, R. S. (2015). Variation in chemical constituents and antioxidant activity in yellow Himalayan (*Rubus ellipticus* Smith) and hill raspberry (*Rubus niveus* Thunb.). *Journal of Food Biochemistry*, *39*(6), 663–672. https://doi.org/10.1111/jfbc.12172

Badhani, A., Sakalani, S., & Mishra, A. (2011). Variation in biochemical's and antioxidant activity of some wild edible fruits of Uttarakhand. *Report and Opinion*, *3*(3), 1–10.

Bajpai, P. K., Warghat, A. R., Dhar, P., Kant, A., Srivastava, R. B., & Stobdan, T. (2014). Variability and relationship of fruit color and sampling location with antioxidant capacities and bioactive content in *Morus alba* L. fruit from trans-Himalaya, India. *LWT-Food Science and Technology*, *59*(2), 981–988. https://doi.org/10.1016/j.lwt.2014.07.055

Bajpai, P. K., Warghat, A. R., Yadav, A., Kant, A., Srivastava, R. B., & Stobdan, T. (2015). High phenotypic variation in Morus alba L. along an altitudinal gradient in the Indian trans-Himalaya. *Journal of Mountain Science*, *12*(2), 446–455. https://doi.org/10.1007/s11629-013-2875-2

Banchobphutsa, Y., & Jarasphol, R. (2012). The efficacy of Morus alba leaf tea in patients with dyslipidemia. In *Proceeding of the 6th Mae Fah Luang Annual Research Conference in Dermatology and Aesthetic Dermatology*, pp. 206–225, Mae Fah Luang University, Chiang Rai. https://en/mfu.ac.th/antiaging/File_PDF/Research_PDF55/Proceeding_7.pdf

Barua, U., Das, R. P., Gogoi, B., & Baruah, S. R. (2019). Underutilized fruits of Assam for livelihood and nutritional security. *Agricultural Reviews*, *40*(3), 175–184.

Batiha, G. E., Al-Snafi, A. E., Thuwaini, M. M., Teibo, J. O., Shaheen, H. M., Akomolafe, A. P., . . . & Papadakis, M. (2023). *Morus alba*: A comprehensive phytochemical and pharmacological review. *Naunyn-Schmiedeberg's Archives of Pharmacology*, *396*(7), 1399–1413. https://doi.org/10.1007/s00210-023-02434

Belwal, T., Pandey, A., Bhatt, I. D., Rawal, R. S., & Luo, Z. (2019). Trends of polyphenolics and anthocyanins accumulation along ripening stages of wild edible fruits of Indian Himalayan region. *Scientific Reports*, *9*(1), 5894. https://doi.org/10.1038/s41598-019-42270-2

Bertucci, A. K. (2017). *Native Plants: Blackberries: The good, bad and thorny*. Shasta Chapter of the California Native Plant Society. http://redding.com/story/life/2017/01/05

Bhadauria, P., & Rathore, K. S. (2023). Isolation and identification of phenolic profiles in selected Himalayan wild berries and determination of their antimicrobial activity. *International Journal of Pharmaceutical Quality Assurance*, *14*(2), 310–315.

Bhatt, I. D., Rawal, R. S., & Dhar, U. (2000). The availability, fruit yield, and harvest of *Myrica esculenta* in Kumaun (west Himalaya), India. *Mountain Research and Development*, *20*(2), 146–153.

Bhatt, I. D., Rawat, S., Badhani, A., & Rawal, R. S. (2017). Nutraceutical potential of selected wild edible fruits of the Indian Himalayan region. *Food Chemistry*, *215*, 84–91. https://doi.org/10.1016/j.foodchem.2016.07.143

Bhatt, S. C., Kumar, V., Gupta, A. K., Mishra, S., Naik, B., Rustagi, S., & Preet, M. S. (2023). Insights on bio-functional properties of *Myrica esculenta* plant for nutritional and livelihood security. *Food Chemistry Advances*, *3*, 100434. https://doi.org/10.1016/j.focha.2023.100434

Bindroo, B. B., Chowdhuri, S., & Ghosh, M. K. (2012). Diversity~ distribution and conservation of mulberry [Morus sp] in Himalayas. *Journal of Crop and Weed*, *8*(1), 26–30.

Chander, V., Aswal, J. S., Dobhal, R., & Uniyal, D. P. (2017). A review on Pharmacological potential of Berberine; an active component of Himalayan Berberis aristata. *Journal of Phytopharmacology*, *6*(1), 53–58.

Chauhan, O., & Sharma, S. B. (2021). Gene expression analysis of therapeutically active antiproliferative polyphenolic fractions extracted from *Rubus ellipticus* on human cervical cancer cells (HeLa). *Uttar Pradesh Journal of Zoology*, *42*(22), 178–194. https://www.mbimph.com/index.php/UPJOZ/article/view/2584

Choudhary, S., Kaurav, H., Madhusudan, S., & Chaudhary, G. (2021). Daruharidra (*Berberis aristata*): Review based upon its Ayurvedic properties. *International Journal for Research in Applied Sciences and Biotechnology*, *8*(2), 98–106. https://doi.org/10.31033/ijrasb.8.2.12

Cohen, I. (2010). Estrogenic extracts of *Morus alba* and uses thereof. *US Patent 7,815,949 B2*, issued October 19, 2010. http://patents.google.com/patent/US7815949B2/en

Dadhwal, R., & Banerjee, R. (2023). Ethnopharmacology, pharmacotherapeutics, biomedicinal and toxicological profile of *Morus alba* L.: A comprehensive review. *South African Journal of Botany*, *158*, 98–117. https://doi.org/10.1016/j.sajb.2023.05.006

Dhatwalia, J., Kumari, A., Chauhan, A., Mansi, K., Thakur, S., Saini, R. V., . . . & Kumar, R. (2022). *Rubus ellipticus* sm. Fruit extract mediated zinc oxide nanoparticles: A green approach for dye degradation and biomedical applications. *Materials*, *15*(10), 3470. https://doi.org/10.3390/ma15103471

Fogacci, F., Giovannini, M., D'Addato, S., Grandi, E., & Cicero, A. F. (2023). Effect of dietary supplementation with a new nutraceutical formulation on cardiometabolic risk factors: A double-blind, placebo-controlled, randomized clinical study. *Archives of Medical Science–Atherosclerotic Diseases*, *8*(1), 53–59. https://doi.org/10.5114/amsad/166571

George, B. P., Thangaraj, P., Chandran, R., & Saravanan, S. (2014). A comparative study on *in vitro* and *in vivo* antioxidant properties of *Rubus ellipticus* and *Rubus niveus*. *Pharmacologia*, *5*, 247–255. https://doi.org/10.5567/pharmacologia.2014.247.255

Gheldof, N., Francey, C., Rytz, A., Egli, L., Delodder, F., Bovetto, L., Piccardi, N., & Darimont, C. (2022). Effect of different nutritional supplements on glucose response of complete meals in two crossover studies. *Nutrients*, *14*(13), 2674. https://doi.org/10.3390/nu14132674

Gilani, A. H., Janbaz, K. H., Aziz, N., Herzig, M. J., Kazmi, M. M., Choudhary, M. I., & Herzig, J. W. (1999). Possible mechanism of selective inotropic activity of the n-butanolic fraction from *Berberis aristata* fruit. *General Pharmacology: The Vascular System*, *33*(5), 407–414. https://doi.org/10.1016/s0306-3623(99)00035-x

Gilani, A. U. H., & Janbaz, K. H. (1995). Preventive and curative effects of Berberis aristata Fruit extract on paracetamol-and CCl4-induced hepatotoxicity. *Phytotherapy Research*, *9*(7), 489–494.

Goli, D. (2007). *Anti-Diabetic Activity of Stem Bark of Berberis Aristata DC in Alloxan Induced Diabetic Rats* (Doctoral dissertation, RGUHS). http://13.232.72.61:8080/jspui/handle/123456789/4054

Gupte, S. (1975). Use of berberine in treatment of giardiasis. *American Journal of Diseases of Children*, *129*(7), 866–866.

Hawrelak, J. (2003). Giardiasis: Pathophysiology and management. *Alternative Medicine Review: A Journal of Clinical Therapeutic*, *8*(2), 129–142.

Janbaz, K. H., & Gilani, A. H. (2000). Studies on preventive and curative effects of berberine on chemical-induced hepatotoxicity in rodents. *Fitoterapia*, *71*(1), 25–33. https://doi.org/10.1016/s0367-326x(99)00098-2

Jiao, Y., Wang, X., Jiang, X., Kong, F., Wang, S., & Yan, C. (2017). Antidiabetic effects of *Morus alba* fruit polysaccharides on high-fat diet-and streptozotocin-induced type 2 diabetes in rats. *Journal of Ethnopharmacology*, *199*, 119–127. https://doi.org/10.1016/j.jep.2017.02.003

Joshi, P. V., Shirkhedkar, A. A., Prakash, K., & Maheshwari, V. L. (2011). Antidiarrheal activity, chemical and toxicity profile of *Berberis aristata*. *Pharmaceutical Biology*, *49*(1), 94–100.

Józefczuk, J., Malikowska, K., Glapa, A., Stawińska-Witoszyńska, B., Nowak, J. K., Bajerska, J., . . . & Walkowiak, J. (2017). Mulberry leaf extract decreases digestion and absorption of starch in healthy subjects—A randomized, placebo-controlled, crossover study. *Advances in Medical Sciences*, *62*(2), 302–306. https://doi.org/10.1016/j.advms.2017.03.002

Kabra, A., Martins, N., Sharma, R., Kabra, R., & Baghel, U. S. (2019a). *Myrica esculenta* Buch.-Ham. ex D. Don: A natural source for health promotion and disease prevention. *Plants (Basel, Switzerland)*, *8*(6), 149. https://doi.org/10.3390/plants8060149

Kabra, A., Sharma, R., Hano, C., Kabra, R., Martins, N., & Baghel, U. S. (2019b). Phytochemical composition, antioxidant, and antimicrobial attributes of different solvent extracts from *Myrica esculenta* Buch.-Ham. ex. D. Don leaves. *Biomolecules*, *9*(8), 357. https://doi.org/10.3390/biom9080357

Kaewsedam, T., Youravong, W., Li, Z., & Wichienchot, S. (2022). Modulation of gut microbiota and their metabolites by functional mulberry juice non-thermally pasteurized using microfiltration. *Functional Foods in Health and Disease*, *12*(9), 547–563. https://doi.org/10.31989/ffhd.v12i9.980

Kala, C. P. (2007). Prioritization of cultivated and wild edibles by local people in the Uttaranchal hills of Indian Himalaya. *Indian Journal of Traditional Knowledge*, *6*(1), 239–244.

Kalahasti, G., Jacoby, P., Sanchez, M., Burkes-Henderson, S., & Gan, D. (2023). Cosmetic composition. *US Patent 11,684,566 B2* issued June 27, 2023. http://patents.google.com/patent/US11684566B2/en

Karuppusamy, S., Muthuraja, G., & Rajasekaran, K. M. (2011). Antioxidant activity of selected lesser known edible fruits from Western Ghats of India. *Indian Journal of Natural Products and Resources*, *2*(2), 174–178.

Kewlani, P., Tiwari, D., Rawat, S., & Bhatt, I. D. (2023). Pharmacological and phytochemical potential of *Rubus ellipticus*: A wild edible with multiple health benefits. *The Journal of Pharmacy and Pharmacology*, *75*(2), 143–161. https://doi.org/10.1093/jpp/rgac053

Kimura, T., Nakagawa, K., Kubota, H., Kojima, Y., Goto, Y., Yamagishi, K., . . . & Miyazawa, T. (2007). Food-grade mulberry powder enriched with 1-deoxynojirimycin suppresses the elevation of postprandial blood glucose preparation in humans. *Journal of Agricultural and Food Chemistry*, *55*(14), 5869–5874. https://doi.org/10.1021/jf062680g

Kojima, Y., Kimura, T., Nakagawa, K., Asai, A., Hasumi, K., Oikawa, S., & Miyazawa, T. (2010). Effects of mulberry leaf extract rich in 1-deoxynojirimycin on blood lipid profiles in humans. *Journal of Clinical Biochemistry and Nutrition*, *47*(2), 155–161. https://doi.org/10.3164/jcbn.10-53

Komal, S., Ranjan, B., Neelam, C., Birendra, S., & Kumar, S. N. (2011). *Berberis aristata*: A review. *International Journal of Research in Ayurveda & Pharmacy*, *2*(2), 383–388.

Krishna, H., Attri, B. L., Kumar, A., & Ahmed, N. (2016). Changes in phenolic contents and antioxidant capacity of bayberry (*Myrica esculenta* Buch. Ham. ex D. Don) and yellow Himalayan raspberry (*Rubus ellipticus* Smith) based health beverages. *Indian Journal of Traditional Knowledge*, *15*(3), 417–424.

Krishna, H., Singh, D., Singh, R. S., Sharma, S. K., & Parashar, A. (2012). Bioactive compounds rich squash from mulberry (*Morus* sp.) for nutritional security. *Crop Improvement PAU*, *39*(Special Issue), 1103–1104.

Kumar, A., & Kumar, D. (2016). Development of wild fruits fortified probiotic milk using *Lactobacillus rhamnosus* culture. *Biological Forum—An International Journal*, *8*(2), 22–28.

Kumar, G. S., Jayaveera, K. N., Kumar, C. K., Sanjay, U. P., Swamy, B. M., & Kumar, D. V. (2007). Antimicrobial effects of Indian medicinal plants against acne-inducing bacteria. *Tropical Journal of Pharmaceutical Research*, *6*(2), 717–723.

Lee, Y., & Hwang, K. T. (2017). Changes in physicochemical properties of mulberry fruits (*Morus alba* L.) during ripening. *Scientia Horticulturae*, *217*, 189–196. https://doi.org/10.1016/j.scienta.2017.01.042

Lown, M., Fuller, R., Lightowler, H., Fraser, A., Gallagher, A., Stuart, B., . . . & Lewith, G. (2017). Mulberry-extract improves glucose tolerance and decreases insulin concentrations in normoglycaemic adults: Results of a randomised double-blind placebo-controlled study. *PLoS ONE*, *12*(2), e0172239. https://doi.org/10.1371/journal.pone.0.172239

Mann, S., Satpathy, G., & Gupta, R. K. (2015). *In vitro* evaluation of bio-protective properties of underutilized *Myrica esculenta* Buch.–Ham. ex D. Don fruit of Meghalaya. *Indian Journal of Natural Products and Resources*, *6*, 183–188.

Mazumder, P. M., Das, S., Das, S., & Das, M. K. (2010). Cytotoxic activity of methanolic extracts of *Berberis aristata* DC and Hemidesmus indicus R. Br. *MCF7 cell line. Journal of Current Pharmaceutical Research*, *1*, 12–15.

Meena, V. S., Gora, J. S., Singh, A., Ram, C., Meena, N. K., Rouphael, Y., . . . & Kumar, P. (2022). Underutilized fruit crops of Indian arid and semi-arid regions: Importance, conservation and utilization strategies. *Horticulturae*, *8*(2), 171. https://doi.org/10.3390/horticulturae8020171

Mela, D. J., Alssema, M., Hiemstra, H., Hoogenraad, A. R., & Kadam, T. (2024). Effect of low-dose mulberry fruit extract on postprandial glucose and insulin responses: A randomized pilot trial in individuals with type 2 diabetes. *Nutrients*, *16*(14), 2177. https://doi.org/10.3390/nu16142177

Mela, D. J., Cao, X. Z., Dobriyal, R., Fowler, M. I., Lin, L., Joshi, M., . . . & Zhang, Z. (2020). The effect of 8 plant extracts and combinations on post-prandial blood glucose and insulin responses in healthy adults: A randomized controlled trial. *Nutrition & Metabolism*, *17*, 1–12. https://doi.org/10.1186/s12986-020-00471-x

Mela, D. J., Cao, X. Z., Govindaiah, S., Hiemstra, H., Kalathil, R., Lin, L., . . . & Verhoeven, C. (2023). Dose–response efficacy of mulberry fruit extract for reducing post-prandial blood glucose and insulin responses: Randomised trial evidence in healthy adults. *British Journal of Nutrition*, *129*(5), 771–778. https://doi.org/10.1017/S0007114522000824

Montaldo, I., Zanardi, A., & Maus, J. (2017). Oral formulation comprising berberine and *Morus alba* extract. *World Intellectual Property Organization Patent WO 2017/008909 A1* issued January 19, 2017. http://patents.google.com/patent/WO2017008909A1/en

Morales Ramos, J. G., Esteves Pairazamán, A. T., Mocarro Willis, M. E. S., Collantes Santisteban, S., & Caldas Herrera, E. (2021). Medicinal properties of *Morus alba* for the control of type 2 diabetes mellitus: A systematic review. *F1000Research*, *10*, 1022. https://doi.org/10.12688/100research.55573.1

Muniyandi, K., George, E., Sathyanarayanan, S., George, B. P., Abrahamse, H., Thamburaj, S., & Thangaraj, P. (2019). Phenolics, tannins, flavonoids and anthocyanins contents influenced antioxidant and anticancer activities of Rubus fruits from Western Ghats, India. *Food Science and Human Wellness*, *8*(1), 73–81. https://doi.org/10.1016/j.fshw.2019.03.005

Nakamura, M., Nakamura, S., & Oku, T. (2009). Suppressive response of confections containing the extractive from leaves of *Morus alba* on postprandial blood glucose and insulin in healthy human subjects. *Nutrition & Metabolism*, *6*(1), 29. https://doi.org/10.1186/1743-7075-6-29

Nakamura, S. (2011). Hypoglycemic effect of *Morus alba* leaf extract on postprandial blood glucose and insulin levels in patients with type 2 diabetes treated with sulfonylurea hypoglycemic agents. *Journal of Diabetes & Metabolism*, *9*. https://cir.nii.ac.jp/crid/1571417124241918080

Padma, D., Pooja, G., & Ravi, S. (2009). *In vitro* and *in vivo* efficacy of a herbal formulation against skin infections. *Journal of Pure and Applied Microbiology*, *3*(1), 199–204.

Painuli, S., Semwal, P., Cruz-Martins, N., & Bachheti, R. K. (2021). Medicinal plants of Himalayan forests. In A. Husen, R. K. Bachheti & A. Bachheti (eds.), *Non-Timber Forest Products*. Springer. https://doi.org/10.1007/978-3-030-73077-2_8

Pant, G., Prakash, O., Chandra, M., Sethi, S., Punetha, H., Dixit, S., & Pant, A. K. (2014). Biochemical analysis, pharmacological activity, antifungal activity and mineral analysis in methanolic extracts of *Myrica esculenta* and *Syzygium cumini*: The Indian traditional fruits growing in Uttarakhand Himalaya. *Indian Journal of Pharmaceutical and Biological Research*, *2*(1), 26–34.

Parmar, C., & Kaushal, M. K. (1982). Wild fruits of the Sub-Himalayan region. In *Wild Fruits of the Sub-Himalayan Region* (136 p.). Kalyani Publishers.

Patel, V. G., Patel, K. G., Patel, K. V., & Gandhi, T. R. (2017). Development of Standardisation parameters and Isolation of Phytomarker Myricetin from stem bark of *Myrica esculenta* Buch. Ham. Ex d. Don. *Journal of Pharmacognosy and Phytochemistry*, *6*(2), 29–34.

Pathak, R., Pant, V., Negi, V. S., Bhatt, I. D., & Belwal, T. (2023). Introduction to Himalayan region and wild edible diversity. In *Himalayan Fruits and Berries* (pp. 1–12). Academic Press.

Phimarn, W., Wichaiyo, K., Silpsavikul, K., Sungthong, B., & Saramunee, K. (2017). A meta-analysis of efficacy of *Morus alba* Linn. to improve blood glucose and lipid profile. *European Journal of Nutrition*, *56*, 1509–1521. https://doi.org/10.1007/s00394-016-1197-x

Potdar, D., Hirwani, R. R., & Dhulap, S. (2012). Phyto-chemical and pharmacological applications of Berberis aristata. *Fitoterapia*, *83*(5), 817–830. https://doi.org/10.1016/j.fitote.2012.04.012

Prabhakar, P. K., & Doble, M. (2008). A target based therapeutic approach towards diabetes mellitus using medicinal plants. *Current Diabetes Reviews*, *4*(4), 291–308. https://doi.org/10.2174/157339908786241124

Pundir, S., Tomar, S., Upadhyay, N., & Sharma, V. (2015). Antioxidant, anti-inflammatory and analgesic activity of bioactive fraction of leaves of *Myrica esculenta* Buch.-Ham along with its pharmacognostic and chromatographic evaluation. *International Journal of Biology, Pharmacy and Allied Sciences*, *4*, 6509–6524.

Rathi, B., Sahu, J., Koul, S., & Kosha, R. L. (2013). Detailed pharmacognostical studies on *Berberis aristata* DC plant. *Ancient Science of Life*, *32*(4), 234–240.

Rawat, S., Jugran, A., Giri, L., Bhatt, I. D., & Rawal, R. S. (2011). Assessment of antioxidant properties in fruits of *Myrica esculenta*: A popular wild edible species in Indian Himalayan Region. *Evidence-based Complementary and Alternative Medicine: eCAM*, *2011*, 512787. https://doi.org/10.1093/ecam/neq055

Rodrigues, E. L., Marcelino, G., Silva, G. T., Figueiredo, P. S., Garcez, W. S., Corsino, J., . . . & Freitas, K. D. C. (2019). Nutraceutical and medicinal potential of the Morus species in metabolic dysfunctions. *International Journal of Molecular Sciences*, *20*(2), 301. https://doi.org/10.3390/ijms20020301

Rondanelli, M., Gasparri, C., Petrangolini, G., Allegrini, P., Avenoso, D., Fazia, T., . . . & Riva, A. (2023). Berberine phospholipid exerts a positive effect on the glycemic profile of overweight subjects with impaired fasting blood glucose (IFG): A randomized double-blind placebo-controlled clinical trial. *European Review for Medical and Pharmacological Sciences*, *27*, 6718–6727. https://doi.org/10.26355/eurrev_202307_33142

Saini, R., Dangwal, K., Singh, H., & Garg, V. (2014). Antioxidant and antiproliferative activities of phenolics isolated from fruits of Himalayan yellow raspberry (*Rubus ellipticus*). *Journal of Food Science and Technology*, *51*(11), 3369–3375. https://doi.org/10.1007/s13197-012-0836-3

Saini, R., Garg, V., & Dangwal, K. (2013). Effect of extraction solvents on polyphenolic composition and antioxidant, antiproliferative activities of Himalyan bayberry (*Myrica esculenta*). *Food Science and Biotechnology*, *22*(4), 887–894. https://doi.org/10.1007/s10068-013-0160-3

Samundeeswari, C., Rajadurai, M., Periasami, R., & Kanchana, G. (2011). Hepatoprotective effect of herbitars, a polyherbal against CCl4 induced hepatotoxicity in rats. *Journal of Pharmaceutical Research*, *4*, 676–679.

Sawian, C. E., Susngi, A. M., Manners, B., & Sawian, J. T. (2023). Myrica esculenta. In T. Belwal, I. Bhatt & H. Devkota (eds.), *Himalayan Fruits and Berries: Bioactive Compounds, Uses and Nutraceutical Potential* (pp. 287–303). Academic Press.

Shan, S., Huang, X., Shah, M. H., & Abbasi, A. M. (2019). Evaluation of polyphenolics content and antioxidant activity in edible wild fruits. *BioMed Research International*, *2019*, 1381989. https://doi.org/10.1155/2019/1381989

Sharma, R. S., Mishra, V., Singh, R., Seth, N., & Babu, C. R. (2008). Antifungal activity of some Himalayan medicinal plants and cultivated ornamental species. *Fitoterapia*, *79*(7–8), 589–591. https://doi.org/10.1016/j.fitote.2008.06.004

Sharma, U. S., & Kumar, A. (2011) *In vitro* antioxidant activity of *Rubus ellipticus* fruits. *Journal of Advanced Pharmaceutical Technology & Research*, 2, 47–50. https://doi.org/10.4103/2231-4040.79805

Singh, B., Singh, L., Kewlani, P., Joshi, V. C., & Bhatt, I. D. (2023). Rubus spp. (*Rubus armeniacus*, *Rubus ellipticus*, *Rubus fruticosus*, *Rubus nepalensis*, *Rubus niveus*, *Rubus occidentalis*). In T. Belwal, I. Bhatt & H. Devkota (eds.), *Himalayan Fruits and Berries: Bioactive Compounds, Uses and Nutraceutical Potential* (pp. 381–394). Academic Press.

Singh, H., Lily, M. K., & Dangwal, K. (2015). Evaluation and comparison of polyphenols and bioactivities of wild edible fruits of North-West Himalaya, India. *Asian Pacific Journal of Tropical Disease*, *5*(11), 888–893. https://doi.org/10.1016/S2222-1808(15)60951-3

Singh, M., Srivastava, S., & Rawat, A. K. (2007). Antimicrobial activities of Indian Berberis species. *Fitoterapia*, *78*(7–8), 574–576. https://doi.org/10.1016/j.fitote.2007.03.021

Sood, P., & Shri, R. (2018). A Review on ethnomedicinal, phytochemical and pharmacological aspects of *Myrica esculenta*. *Indian Journal of Pharmaceutical Sciences*, *80*(1).

Taghizadeh, M., Mohammad Zadeh, A., Asemi, Z., Farrokhnezhad, A. H., Memarzadeh, M. R., Banikazemi, Z., Shariat, M., & Shafabakhsh, R. (2022). *Morus alba* leaf extract affects metabolic profiles, biomarkers inflammation and oxidative stress in patients with type 2 diabetes mellitus: A double-blind clinical trial. *Clinical Nutrition ESPEN*, *49*, 68–73. https://doi.org/10.1016/j.clnesp.2022.03.027

Thaipitakwong, T. (2020). Efficacy and safety of mulberry leaves on glycemic control in patients with obesity and patients with type 2 diabetes. *Chulanongkorn University Theses and Dissertations (Chule ETD)*, *297*. https://digital.car.chula.ac.th/chulaetd/297

Thaipitakwong, T., Supasyndh, O., Rasmi, Y., & Aramwit, P. (2020). A randomized controlled study of dose-finding, efficacy, and safety of mulberry leaves on glycemic profiles in obese persons with borderline diabetes. *Complementary Therapies in Medicine*, *49*, 102292. https://doi.org/10.1016/j.ctim.2019.102292

Trimarco, V., Izzo, R., Stabile, E., Rozza, F., Santoro, M., Manzi, M. V., . . . & Trimarco, B. (2015). Effects of a new combination of nutraceuticals with *Morus alba* on lipid profile, insulin sensitivity and endotelial function in dyslipidemic subjects. A cross-over, randomized, double-blind trial. *High Blood Pressure & Cardiovascular Prevention: The Official Journal of the Italian Society of Hypertension*, *22*(2), 149–154. https://doi.org/10.1007/s40292-015-0087-2

Wang, Y., Xiang, L., Wang, C., Tang, C., & He, X. (2013). Antidiabetic and antioxidant effects and phytochemicals of mulberry fruit (*Morus alba* L.) polyphenol enhanced extract. *PLoS ONE*, *8*(7), e71144. https://doi.org/10.1371/journal.pone.0071144

Wasana, K. G. P., Attanayake, A. P., Jayatilaka, K. A. P. W., & Weerarathna, T. P. (2020). Natural drug leads as novel dipeptidyl peptidase-IV inhibitors targeting the management of type 2 diabetes mellitus. *Journal of Complementary Medicine Research*, *11*(1), 43–53. https://doi.org/10.5455/jcmr.2020.11.01.06

Yang, W., Tang, C. M., Li, X., Zhou, Y., Wang, L., & Li, L. (2011). Study on the chemical constituents of *Myrica esculenta*. *Journal of Yunnan University (Natural Sciences)*, *33*, 453–457.

Yung-Hsiang, L., Pei-Yi, W., & Huan-You, L. (2022). Plant ferment and weight loss method for using the same. *US Patent 20220273748 A1* issued September 1, 2022. http://patents-review.com/a/20220273748-plant-ferment-weight-loss-method.html

Analgesic Plants in India

3

B. Dave Oomah

3.1 INTRODUCTION

Most US population (11–40%) experience daily acute or chronic pain. Pain can be classified into three main categories: nociceptive, inflammatory, and pathological (neuropathic and dysfunctional). Neuropathic pain has diverse pathogenetic origins and is actually the result of a final common pathway attained by many different, variable, very specific molecular and cellular routes. Neuropathic pain has 12 different origins with different molecular interactions including diabetic peripheral neuropathy, post-mastectomy pain syndrome, post-stroke neuropathy, chemotherapy-induced peripheral neuropathy, and migraine (Smith and Trần 2022). Pain management/relief medications such as opioids and other pain killers result in opioid use disorder or addiction to non-steroidal anti-inflammatory drugs (NSAIDs) causing gastrointestinal disorders. These side effects are minimal or not encountered when medicinal plants are used to alleviate or treat pain. Hence, 80% of the global population uses herbal analgesic bioactive compounds to manage and relieve pain in traditional therapy (Uritu et al. 2018). Plant phytochemicals relieve pain by inhibiting calcium from entering the cell, affecting prostaglandins synthetic pathways or inhibiting COX and LOX enzymes (Aleebrahim-Dehkordy et al. 2017). Many nonopioid therapeutics treat pain by nociception, particularly activating pain-relieving effects in the central nervous system via a G protein-coupled receptor (GPCR), the α_{2A}-adenergic receptor (α_{2A}AR) (Fink et al. 2022). However, the analgesic agents, clonidine and dexmedetomide known therapeutics targeting the α_{2A}AR are strongly sedative, thereby drastically restricting their use. Fink et al. (2022) computationally docked over 301 million molecules into highly similar α_{2A}AR binding site to identify two new agonists that overcome sedation.

Analgesics often described as 'mild' and 'strong' had no internationally agreed classification according to Twycross (1984). However, oversimplified basic definition generally described non-steroidal anti-inflammatory drugs (NSAIDs) (e.g., aspirin) to act peripherally, whereas narcotics (opioid) operate within the central nervous system (centrally). Medically, analgesics are simply a way to elevate a patient's pain threshold. Analgesics should be administered regularly and prophylactically since persistent pain requires preventive therapy. The World Health Organization advocates the selection of appropriate analgesic therapy based on pain intensity of cancer patients according to the three step analgesic ladder. Moreover, the Pain Management Index (PMI) was developed as a simple and objective tool to evaluate the quality of analgesic prescriptions, with breakthrough pain recognized as a separate entity (Enting et al. 2007 and references therein). The PMI is the single most often outcome measure for quality of pain treatment. India among 14 other countries had the highest (79%) inadequate treatment rate (Saxena et al. 1999).

DOI: 10.1201/9781003352341-3

Eighty three families of medicinal plants (out of 242 plants from 184 genera) are used to relieve 25 types of pain in 19 indigenous tribes in four states of Western Himalayan region (UT Jammu and Kashmir, Himachal Pradesh, Ladakh, and Uttarakhand) (Balkrishna et al. 2022). *Artemisia* genus and *Saussurea costus* were the primary analgesic plant species used in powder, decoction, and paste forms by tribal people, in spite of the high analgesic potential of leaves and roots of these medicinal species. Ayurvedic preparations that rely heavily on native plants/crops can reduce pain and increase function in osteoarthritic people and help manage symptoms in type-2 diabetics according to NCCIH's overview (Platkin et al. 2022). India is also one of the largest producers of opium alkaloids in the world with cultivation confined to Uttar Pradesh, Madhya Pradesh and Rajasthan. The opium is considered the best known pain killer from times immemorial. The major *Opium poppy* alkaloids are: morphine (7–17%), codeine (2.1–4.4%), thebaine (1–3%), nacrotics (3–10%) and papverine (0.5–3%) (Nalina et al. 2022). From a pharmaceutical perspective, pain killing drugs (morphine, oxycodone and illegal street drugs [heroin, fentanyl] activate mu opioid receptors or nerve cells relieving pain, but also cause euphoria contributing to addiction.

3.2 TURMERIC—*CURCUMA LONGA*

India is the world's top turmeric producer with an annual turn-over of nearly 400 tons, or 6,500 kg/hectare in 2022 (Shankar et al. 2022 and references therein). It is also the largest global turmeric exporter accounting for over 50% of the international trade fulfilling 90% of the world's demand [APEDA, 2018]. Dry turmeric production in India is estimated at 0.513 million MT compared to last year's 0.467 million MT according to Agriwatch (2023). The global turmeric production is ~1.1 million tonnes per annum with India dominating world production (80%), export (~0.18 l million T), and consumption. India devoted 349, 642 hectares under turmeric cultivation with an estimated total production of 13, 30,932 tonnes (dry turmeric) during 2021–22 (Spice Board 2023). Turmeric (including standardize turmeric extracts with high curcumin levels) from *Curcuma longa* ranks 5th among the top selling herbal supplements in 2021 in the United States with total sales of 111,658,256 USD (15% increase from 2020) (Smith et al. 2022). Turmeric occupies ~6% of the total area under spices and condiments cultivation in India (Choudhary and Rahi 2018). Turmeric global market is ~37,000 tons valued at US$40,160 million.

India has about 40 species of the *Curcuma* genus; *Curcuma longa* being the most important species of the genus. About 50 named cultivars of turmeric have been reported in India with limited number of elite cultivars grown commercially (Sahoo et al. 2019 and references therein). Eighteen improved *Curcuma longa* varieties were evaluated for their genetic variation in turmeric morphological characteristics (Aswathi et al. 2023). These varieties developed by various research institutions in India were field grown under the same environmental conditions in Kozhikode, Kerala, India. Varieties differed in fresh rhizome yield per plant (RY, mean 0.44 kg, range 0.16–0.73 kg), dry rhizome yield per plant (DY, mean 0.07 kg, range 0.02–0.10 kg), percentage of dry rhizome yield per plant (DYP, mean 15.6%, range10.1–24.4%) and curcumin content (CUR, mean 4.33%, range 1.65–7.14%). Turmeric genotypes grown in the southern and eastern part of India (Prayagraj agro-climatic zone) varied in growth and yield parameters. Rhizomes yield of 15 genotypes ranged from 13.2 to 27.7 t/ha, with (19.8–61.9 g) dry rhizomes per plant variation (Annie et al. 2022). Genotype 'Pragati' with the highest yield parameters had the highest cost benefit ratio (4.78), gross return (11,07,600 Rs/ha) and net profit (9,16,233 Rs/ha).

The fat soluble polyphenolic pigments, curcuminoids are responsible for the bright yellow color and medicinal therapeutic properties of turmeric. Curcuminoids (2–5% of the rhizome mass) accounts for turmeric's characteristic yellow color; turmeric redness index (TRI = a*/b*), 0.16–0.50 (Pal et al. 2020). The curcuminoids include the major bioactive constituent, curcumin, demethoxycurcumin, bis-demethoxycurcumin and cyclocurcumin (Lepcha et al. 2022 and references therein). Curcumin content varied among

11 cultivars from 10 environments with genotype accounting for 17.7%, whereas environment and G x E interaction explained 42.9 and 39.3% of the total variation (Anandaraj et al. 2014). Curing methods affect chemical composition of turmeric (cv Suranjana) with variation in curcumin (4.2–5.2%), essential oil (3.3–4.4%), and total phenolic (457–647 mg GAE/100 g in ethanol extracts of cured and dried samples) contents, dry recovery (15.8–24.2%) and drying time (20.8–29 h) (Lepcha et al. 2022).

3.2.1 Composition

Neelam et al. (2015) profiled the elemental composition of rhizomes from turmeric plants cultivated in 50 different locations in Uttarakhand, the sub-tropical to temperate and alpine regions of the Himalayas. Mineral content varied widely with concentration [mg/kg] ranges: (Fe, 21 to 3463; Cu, 1.1–198; Zn, 3.6–212 and Mn, 5.6–111). Ranges of other elements (%) were: (P, 0.04–0.99; Ca, 0.19 –1.42; Mg, 0.09–0.85; K, 0.57–2.99 and N, 9.52–35.3). Chemical composition (mg/g) of rhizomes also varied in ascorbic acid (0.12–2.62), total sugar (13.8–56), reducing sugar (1.2–11.2), non-reducing sugars (8.6 –49.7) and total protein (4.9–24) (Neelam et al. 2015). Rhizome essential oil was evaluated for eight commercial elite cultivars grown under the same environmental conditions (Sahoo et al. 2019). Cultivars varied in fresh rhizome yield (300–436 g/plant), essential oil yield (1.6–3.7 mL/plant), and essential oil content (0.5–0.9%). Sesquiterpenes (hydrocarbons and oxygenated) were the predominant class accounting for 58 to 71 % of the total identified essential oil in all cultivars. The major identified essential oils were Ar-turmerone (39.5–45.5%), curlone (9.8–11.7%), α-phellandrene (5.5–7.7%), eucalyptol (3.2–5.5%), β-Himachalene (1.6–5.5%) and α-copen-11-ol (2.3–5.4%). These quantitative and qualitative differences in rhizome yield and essential oils were due mainly to genetics since all cultivars were grown under the same environment (Sahoo et al. 2019).

Pal et al. (2020) evaluated 45 turmeric genotypes differing in rhizome color from the sub-Himalayan region. These genotypes varied in curcuminoid (0.14–5.12%), total phenol (2.86–9.96 mg GAE/g), total flavonoid (1.05–11.05 mg QE/g dry rhizome), iron (47.15–55.09 mg/100 g) contents. Turmeric varieties with 5–6% curcuminoid content are considered good for commercial use; however, Lakadong, a Meghalaya (Indian) turmeric variety is reported to have 7.5% curcuminoid content. Curcuminoid isomers (mg/kg) also varied significantly among the genotypes; curcumin (12.6–25900), demethoxycurcumin [DMC] (142– 19800) and bisdemethoxycurcumin [BDC] (9.7–10000) and total (1400– 55700) (Table 3.1). One genotype (Turmeric Collection Pundibari 2 [TCP 2]) had the highest curcuminoids content and superior bioactive potential (antioxidant activity 78 and 89 µg/mL DPPH and ABTS IC_{50}). TCP 2 also contained polyphenolics (p-coumaric, catechin, sinapic acid and quercetin [162, 108, 417 and 2746 mg/kg], respectively) and 33 volatile compounds including ar-turmeronm, curlone, eucalyptol and chemigran (28.6, 10, 10 and 13%, respectively (Pal et al. 2020).

Rhizomes are generally dehydrated for preservation and because dried rhizomes are also the major and most desirable form for international trade. Post-harvest drying methods affect turmeric composition

TABLE 3.1 Curcuminoid contents (mg/kg) from selected genotypes

GENOTYPE	*BDMC*	*DMC*	*CURCUMIN*	*TOTAL CURCUMIN*	*RATIO*
TCP 154	9.72	142.9	1247	1400	128/15/1
TCP 244	32439	5580	1183	39202	1/4.7/27.4
TCP 67	12229	30532	6139	48900	1.0/5.0/2.0
TCP 2	10000	19800	25900	55700	2.6/2/1
LSD (P=0.05)	10.17	987.51	8.73	20.98	
CV	0.08	9.1	0.14	0.07	

Source: Adapted from Pal et al. (2020); BDMC—bisdemethoxycurcumin;

DMC—demethoxycurcumin; Ratio –curcumin/DMC/BDC

particularly curcumin reduction by sun, hot air and freeze drying methods (72, 61 and 55%, respectively) (Chumroenphat et al. 2021). Drying reduced total curcuminoids content (1165–1742 µg/g vs 3795 µg/g db for fresh rhizome) due to elevated polyphenol oxidase activity. However, drying increases total phenolic contents (6 folds), apparently due to curcumin degradation and transformation to other phenolic compounds, particularly vanillin and ferulic acid. Antioxidant activity decreased upon drying (145 vs 919 mg TE/100 g db [fresh rhizome] for DPPH and 27 vs 56 mg FeSO4/g db for FRAP scavenging activity). However, ABTS scavenging activity increased over 3 folds upon drying (32–60 vs 17 mg TE/100 g db for fresh rhizome).

Extraction affects curcuminoid composition and isomer ratios (Table 3.2). Soxhlet extraction with acetone provided the highest curcuminoids yield (43.5%; 3.4 g total extract) with curcumin, demethoxycurcumin and bisdemethoxycurcumin comprising 22.8, 14.2 and 6.5%, respectively (Revathy et al. 2011). Solvents other than hexane and methanol (chloroform, ethylacetate and hexane/methanol) extracted comparable amounts of curcuminoids (~35%) consisting of curcumin, demethoxycurcumin and bisdemethoxycurcumin (~19, 12, and 5%, respectively). Curcumin, demethoxycurcumin and bisdemethoxycurcumin comprised ~52, 33 and 15% of the total curcuminoids, respectively.

Higher yield (4.12%) is obtained by novel technologies such as ethanol-modified subcritical water extraction. However, ultrasound-assisted extraction (UAE) and conventional solvent extraction (CSE), both with 25% ethanol-water mixture produced the lowest yield (0.84%). Nevertheless, these technologies have similar ratios of curcumin, demethoxycurcumin and bisdemethoxycurcumin (55.8, 22.7 and 21.5%, respectively) (Rezaei et al. 2023). Hexane (Soxhlet) extracted the most curcumin 83% at 2.71% yield. Soxhlet (except hexane as solvent) generally extracts curcuminoid isomers curcumin/demethoxycurcumin/bisdemethoxycurcumin in 3.4/2.1/1 ratio. This ratio is modified to 2.6/1.1/1 when turmeric is extracted by green technology.

3.2.2 Clinical Studies

3.2.2.1 Analgesic Effects in Orofacial Region

Pain in the orofacial region generally originates from musculoskeletal, neurovascular, or neuropathic disorders. Almost ⅕ (~22%) of US individuals (predominantly females and younger 15–45 years old individuals) experience orofacial pain during any given 6-month period. Sterniczuk et al. (2022) systematically

TABLE 3.2 Extraction yield (%) of curcumin and its isomers

EXTRACTION	*CURCUMIN*	*DMC*	*BDMC*	*YIELD (G)*	*RATIO*
Revathy et al. (2011)					
Acetone	22.8	14.2	6.5	3.49	3.5/2.2/1
Methanol	15.68	9.9	4.73	4.31	3.3/2.1/1
Ethylacetate	18.76	11.6	5.2	3.2	3.6/2.2/1
Hexane/MeOH	18.1	11.2	6.1	3.62	3/1.8/1
Chloroform	19.7	12.15	5.05	3.09	3.9/2.4/1
Hexane	6.5	1.02	0.04	0.9	162.5/25.5/1
Rezaei et al. (2023)					
SSE	55.4	22.46	22.14	4.12	2.1/1/1
UAE	55.47	23.04	21.49	0.85	2.6/1.1/1
CSE	56.6	22.56	20.84	0.84	2.7/1.1/1
Hexane (Soxhlet)	83.27	16.29	0.44	2.71	189.3/37/1

DMC—demethoxycurcumin; BDMC—bisdemethoxycurcumin; Ratio –curcumin/DMC/BDMC;

SSE—subcritical solvent extraction; UAE—ultrasound-assisted extraction; CSE—conventional solvent extraction

reviewed randomized controlled trials (RCTs) evaluating curcumin's effectiveness in attenuating self-rated pain in the orofacial region (ORF). Curcumin exerted analgesic effects and effectively lowered self-rated ORF pain in most trials (15/19 [15 out of 19] RCTs).

Three RCTs had low and eight had moderate bias risk (Table 3.3). Curcumin (80 mg Nano Curcumin SinaCurcumin™ capsules) lowered VAS pain score and lesion size in oral lichen planus (OLP) patients (25F/4M, 52 years mean age) in a phase 3 parallel clinical trial without significant difference with the prednisone (10 mg/d) group (Kia et al. 2020a). The same curcumin nanomicelle (SinaCurcumin 80 mg capsules twice daily for 7 weeks) significantly lowered oral mucositis (OM—the most prevalent chemotherapy and radiotherapy complications) severity and pain grade (only in the seventh week) in OM patients (10F/15M, 55 years mean age). Moreover, curcumin effectively prevented and treated head and neck radiotherapy and OM induced chemotherapy (Kia et al. 2021). However, high curcumin dose (2 g [4 x 500 mg tablets twice daily] for four weeks had no effect in OLP patients (10F/2M, 49 years average age) compared to placebo (3F/5M, 53 years mean age). All patients in the study received routine OLP treatment (0.5 mg Dexamethasone mouthwash and 100,000 Units Nystatin suspension) that may have masked curcumin's positive effect (Amirchaghmaghi et al. 2016).

Meghana et al. (2020) compared curcumin gel (Curenext—10 mg *Curcuma longa* rhizome extract in 50 g tube) effects with noneugenol periodontal dressing (COE pack) on pain of post periodontal flap surgery patients in a cross over split mouth study. The chronic generalized moderate periodontitis patients (13F/7M, 38 years mean age) were randomized into periodontal dressing group COE pack or Curenext oral gel applied gently on the operated site 24 h post-surgery twice daily for 1 week. All patients were prescribed 600 mg ibuprofen preoperatively followed by 1 tablet every 8 h for the first 24 h and analgesic (Si Opus Sit—SOS) for 1 week. The curcumin group required significantly less analgesic than the

TABLE 3.3 Analgesic effects of curcumin in orofacial region based on selected randomized controlled studies

REFERENCE	*TREATMENT*	*CURCUMIN'S OUTCOME/BENEFITS*
Kia et al. (2020a)	4 Weeks -80 mg nano-curcumin capsules daily	Effective as prednisone in oral lichen planus (OLP) treatment
Kia et al. (2021)	7 Weeks-160 mg nanomicelle cucumin capsules daily	Effectively lowered pain in chemotherapy patients
Amirchaghmaghi et al. (2016)	4 Weeks-2 g curcumin daily in corticosteroid presence	No analgesic effect in OLP treatment
Meghana et al. (2020)	Curnext oral gel (10 mg *C. longa* rhizome extract)	Positive effect on post-surgery pain as periodontal dressing
Mugilan et al. (2020)	Curnext oral gel placed into the extracted socket	Greater pain reduction than control
Deshmukh and Bagewadi (2014)	4 Days-10 mg curcumin oral gel (3 x daily)	Efficiency similar to 0.1% triamcinolone in minor recurrent aphthous stomatitis (RAS) treatment
Kia et al. (2020b)	4 Weeks-5% curcumin oral gel (thrice daily)	Effect similar to 0.1% triamcinolone in alleviating aphthous ulcers
Kia et al. (2015)	4 Weeks-5% curcumin oral gel (thrice daily)	Outcome similar to 0.1% triamcinolone in oral lichen planus treatment
Mansourian et al. (2017)	28 Days-topical curcumin orabase	Analgesic efficacy similar to triamcinolone in corticosteroid presence
Anil et al. (2019)	7 Days–0.5% curcumin micro adhesive film	Lowered numerical pain score and discomfort post periodontal surgery
Srivastava et al. (2021)	3 Months-100 mg curcumin lozenge (thrice daily)	Pain outcome similar to corticosteroids in oral submucous fibrosis treatment

Source: Adapted from Sterniczuk et al. (2022); OLP—oral lichen planus

periodontal dressing group and had the same positive effects on post-surgery pain as periodontal dressing. Curnext oral gel in another study (11 patients [6 test/5 control], 57.6 years mean age) significantly reduced central and distal pain after one week post extraction in the presence of standard analgesic (Hifenac) and antibiotic (500 mg Novamox). Curcumin had greater pain reduction in Hifenac and Novamox presence (Mugilan et al. 2020). Pain reduction was attributed to curcumin's analgesic property in inhibiting pro inflammatory cytokines (IL-6, IL-8 and PGE2). Moreover, curcumin promoted wound healing evidenced by significant reduction in width of extraction socket compared to control.

RAS—recurrent aphthous stomatitis is a commonly occurring oral cavity disease generally treated with steroids (e.g. Triamcinolone acetonide gel). Minor RAS are small size (< 1 cm) painful oral cavity ulcers that heal within 10–14 days without scarring. Curenext oral gel (Abbot Pharmaceuticals; 3 x daily for four days) was as efficient as 0.1% Triamcinolone in treating minor RAS patients (29F/31M, 32 years mean age) (Deshmukh and Bagewadi 2014). However, in an earlier study (Manifar et al. 2012), 2% curcumin fluid gel (Orphanidis Pharma Research GmbH and Orphan Teb Pars) significantly relieved pain (Perceived Pain Rating Scale) at day 4 (2.99–1.17 vs 2.24–1.51 for placebo pain score) and reduced ulcer diameter (4.71–2.64 vs 5–3.2 mm for undescribed placebo gel) in minor RAS patients (13F/15M, 36 years mean age). Curcumin oral gel (5%, 3 x daily for 4 weeks) was also as effective as 0.1% Triamcinolone in alleviating aphthous ulcers (Kia et al. 2020b). Oral Lichen Planus (OLP) is a chronic inflammatory mucocutaneous disease often treated with topical and systemic corticosteroids. Remission was complete in nine OLP patients (36%) treated with curcumin (3 x 5% curcumin oral gel daily for 4 weeks; 15F/10M, 49 years mean age) and eight patients (32%) receiving 0.1% Triamcinolone (21F/4M, 52 years mean age) (Kia et al. 2015).

Graft-versus-host disease (GVHD) is one of the most frequent complications of allogenic hematopoietic stem cell transplantation. GVHD causes oral sensitivity and pain generally treated with good oral hygiene and local and topical steroids use or corticosteroids for chronic oral GVHD. VAS pain score at treatment onset and at 14 and 28 days were not significantly different between the curcumin (5F/8M, 35 years mean age) and control (FDA approved corticosteroid drug—Triamcinolone) groups (6F/7M, 39 years mean age). All patients had systemic GVHD and were under corticosteroid (prednisone) and immunosuppressant (cyclosporine) treatment (Mansourian et al. 2017). In a split mouth placebo controlled study, 0.5% curcumin micro adhesive film (4–5 mm rectangular strip) was used in healthy individuals (8F/7M, 42 years mean age) after periodontal flap surgery (Anil et al. 2019). Antibiotics (500 mg amoxicillin) were prescribed 8 hourly for 5 days and analgesic (100 mg Aceclofenac) taken as rescue drug as required. Curcumin lowered numerical pain scores and discomfort significantly at 2 and 3 h after periodontal surgery accounting for curcumin's anti-inflammatory properties. Moreover, analgesic uptake was significantly lower in the curcumin group on the first two postoperative days confirming curcumin's analgesic effect (Anil et al. 2019). Oral submucous fibrosis (OSMF) is considered an Indian disease. Curcumin (100 mg curcumin and 10 mg clove oil lozenges; TurmNova®, 3 x daily for 3 months) was highly bioavailable in OSMF patients (71M/9F, 33.5 years mean age) evidenced by high plasma curcumin level. Curcumin's pain outcome was similar to corticosteroids (8 mg Dexamethasone + 2% lignocaine +1500 IU hyaluronidase, twice per week for 3 months) in OSMF treatment (Srivastava et al. 2021). Curcumin controls pain by three proposed mechanisms: it inhibits COX pathway and thus suppress pain similar to NSAIDs; it has an antagonizing effect on transient receptor potential vanilloid type 1 ion channel and it activates and hence stimulates potassium ATP channels producing analgesic action (Anil et al. 2019 and references therein).

Oral *Curcuma longa* administration (200 mg) was ineffective for pain and discomfort management after surgical periodontal therapy compared to mefenamic acid (500 mg) of stage 3/grade C periodontitis (Al-Askar et al. 2022). In this parallel-arm designed trial, periodontitis post-operative patients received curcumin capsules (200 mg, Terry Natural Products, CuraMed, Green Bay, WI, USA; 26M/12F, 58.4 years mean age) or the control (one 500 mg mefenamic acid tablet; 28M/10F, 57.2 years mean age). Antibiotics (500 mg amoxicillin and 400 mg metronidazole, 3x daily [every 8 h]) were prescribed for all patients for 7 days. Patients in the treatment group were advised to take 2 curcumin capsules (or 1 mefanic acid capsule for the control group) immediately after surgery and then every 8 h for 3 days after which all patients took analgesics as needed for pain. Differences were not significant in the mean pain and discomfort numerical rating and verbal rating scales among patients in the test and control groups evaluated after

24, 48 and 72 h. The investigators posited that the results may have been biased due to lack of operator blinding and low dose curcumin used in the study (Al-Askar et al. 2022).

Cleft surgery patients experience acute pain regardless of analgesic administration. Maulina et al. (2022) evaluated the efficacy of a curcumin patch (100 mg) as an adjuvant analgesic agent to manage postoperative pain in postbioplasty (corrective cleft lip surgery) and/or postpalatoplasty (corrective cleft palate surgery) patients. Application of the curcumin patch (6 x 10 cm) significantly reduced pain score after 8 h in post-surgery infants (33M, 0–18 months/22F 19–36 months). Curcumin's analgesic effect was attributed to its high bioavailability due to delivery and release through the transdermal route. Moreover, curcumin presumably alleviated inflammation by lowering the release of inflammation mediators (particularly curcumin's well-known prostaglandin-E2 downregulation) after surgery.

3.2.2.2 Analgesic Effects in Osteoarthritis

Arthritis is a degenerative joint disorder in several forms (osteoarthritis, rheumatoid arthritis, psoriatic arthritis) that often results in disabling pain. Knee osteoarthritis generally affects older people (over 60 years old, 9.6% men and 18% women). It is the most common osteoarthritis (~80%) disease burden impacting over 300 million people globally and ~30% of Indians (Pal et al. 2016). Pain reduction is the major goal in arthritis treatment to alleviate joint inflammation, daily wear VAS and tear of joints and muscle strains.

Musculoskeletal disorders are major focus for curcumins beneficial effects on clinical outcomes representing 17% of 389 citations in the latest review of clinical trials (Panknin et al. 2023). Turmeric dietary supplements (containing 98% curcuminoid-enriched extract) are used by one-third of documented epidemiological studies related to rheumatoid arthritis and ~¼ of women diagnosed with breast cancer. A large share (55%) of the US turmeric dietary supplement market consists of products formulated to enhance curcumin bioavailability (Panknin et al. 2023). Musculoskeletal system (MSK) disorders have the most (79%) reported double blind randomized controlled trial (D-RCT) using enhanced bioavailability products (61%) for an average 2.2 month duration. Osteoarthritis and muscle related (soreness) outcomes were the most common (46 and 36%) studied MSK disorders. Curcumin showed positive effects or equivalency

TABLE 3.4 Analgesic effects of formulated curcumin extracts on selected randomized controlled knee osteoarthritis studies

REFERENCE	*TREATMENT*	*CURCUMIN'S OUTCOME/BENEFITS*
Panda et al. (2018)	7–60 days-500 mg Curene® capsule	Curene®reduced WOMAC pain linearly
Nakagawa et al. (2014)	8 weeks-180 mg/day curcumin	Lowered knee pain VAS score and rescue medication
Madhu et al. (2013)	42 days-500 mg Turmacin™ capsule	Alleviated pain and improved physical function
Panahi et al. (2014)	6 weeks-1500 mg C3 Complex®	Alleviated knee OA pain and reduced rescue medication
Henrotin et al. (2019)	1 month–186.6 mg Flexofytol	Analgesic knee OA pain reduction over paracetamol
Kuptniratsaikul et al. (2009)	6 weeks–1 g curcuminoids daily	Reduced OA pain as effectively as ibuprofen
Kuptniratsaikul et al. (2014)	4 weeks-1500 mg *C. domestica* extract	Reduced pain and improved physical function as ibuprofen
Srivastava et al. (2016)	4 months-500 mg Haridra capsules	Reduced pain and physical function scores

Source: Adapted from Paultre et al. (2021); OA—osteoarthritis; WOMAC—Western Ontario and McMaster Universities Osteoarthris Index; VAS—visual analogue scale

to non-steroidal anti-inflammatory drugs (NSAIDs) in most clinical outcomes such as pain (92%), function (86%) and inflammation, oxidative stress and cartilage degradation biomarkers (80%). Moreover, the need for rescue medications was often reduced in curcumin treatments. Curcumin had beneficial effects on muscle/exercise related outcomes (82% pain, 75% functional outcomes, 58% creatinine kinase reduction and 56% muscle damage biomarkers) (Panknin et al. 2023).

Daily et al. (2016) systematically evaluated all randomized clinical trials (RCTs) of turmeric extracts and curcumin efficacy for treating arthritis symptoms. Three among 8 selected RCTs reported significant reduction of the major arthritis outcome (PVAS—pain visual analogue score), and 4 other studies showed lowered WOMAC (Western Ontario and McMaster Universities Osteoarthritis Index). PVAS was not significantly different between turmeric/curcumin and pain medicine in five RCTs. These RCTs demonstrate the efficacy of turmeric extract (1 g curcumin/d for 8–12 weeks) in arthritis treatment (Daily et al. 2016). Another study (Rolfe et al. 2020) found moderate evidence to support turmeric/curcumin use for pain relief and improvement in physical function in arthritis based on 65 systemic reviews. Eight of these reviews explored turmeric/curcumin effects on arthritis-related disorders, with pain, symptom and function scores and safety as outcome measures. Turmeric/curcumin was favored as 'moderate quality grade' to reduce pain in patients with knee osteoarthritis, knee stiffness and improve physical function compared to either a placebo or standard care (ibuprofen, diclofenac and glucosamine). Pain management was generally 'low' for turmeric/curcumin use in patients with various painful conditions (arthritis, post-surgical).

Several patented bioavailable curcuminoids formulations (e.g., Meriva®, BCM-95®, Curcumin C3 Complex®) demonstrated superior pain reduction parameters and knee OA symptoms (clinical and biochemical endpoints) over prescription drugs. Paultre et al. (2021) selected eight high methodological quality (low to moderate bias risk) studies of randomized trials demonstrating formulated turmeric's benefits to alleviate pain and improve physical function in knee osteoarthritis patients (Table 3.4). Curene® is a bioavailable (sustained release) turmeric extract containing naturally derived curcuminoids formulated with proprietary Aquasome® technology. Curene® (500 mg daily) reduced total Western Ontario and McMaster Universities Osteoarthritis Index (WOMAC) and Visual Analogue Scale (VAS) scores compared to placebo (Panda et al. 2018). Curene® significantly and clinically reduced pain, stiffness and improved physical function in knee osteoarthritis subjects (n = 25, 55 years mean age, 25 kg/m^2 average BMI) safely without adverse reaction common with NSAIDs. WOMAC pain reduction increased linearly ($r^2 = 0.985$) with time (10.7, 19.9, 29.6 and 48%, respectively) at 7, 14, 30 and 60 days Curene® treatment. Similarly, WOMAC stiffness and physical function decreased linearly ($r^2 = 0.934$ and 0.986) with Curene® treatment time. Moreover, Curene® reduced rescue medication (2 g/day Paracetamol) use compared to placebo group (75 vs 19 tablets) (Panda et al. 2018).

Theracurmin® (6 capsules/d, 180 mg/d curcumin) administered for 8 weeks significantly lowered knee pain VAS score and celecocib dependence in knee osteoarthritis patients (14F/4M, 72 years mean age, 25 kg/m^2 average BMI, Kellgren/Lawrence grade 2–3) with initial >0.15 VAS scores (Nakagawa et al. 2014). The benefits of Theracurmin® were attributed to the anti-inflammatory and condroprotective effects of curcumin in this randomized, double blind, placebo controlled, prospective clinical study. Theracurmin® is clinically effective in improving endothelial function in postmenopausal women. Components (primarily polysaccharides) other than curcumin in turmeric can have potent anti-inflammatory and/or analgesic benefits (PVAS and WOMAC scores) to alleviate osteoarthritis pain (Madhu et al. 2013). Primary knee osteoarthritis patients (83F/37M) received either placebo (400 mg twice daily) or NR-INF-02 (500 mg twice daily) or glucosamine sulfate (750 mg twice daily) alone or in combination with NR-INF-02 for 42 days. NR-INF-02 significantly reduced VAS, WOMAC and CGIC (Clinical Global Impression Change) scores in osteoarthritis patients (17F/13M, 56.6 years mean age, 27 kg/m^2 average BMI, 23 with moderate OA grade) compared to placebo. The treatment also reduced rescue medication use along with clinical and subjective improvement compared to placebo. NR-INF-02 is a standardized polysaccharide rich *Curcuma longa* extract (Turmacin™, Natural Remedies Pvt. Ltd, Bangalore, India supplied as oral 500 mg gelatin capsule containing 12.6% w/w polysaccharide). NR-INF-02 significantly alleviated pain as the primary outcome over 42 days based on VAS (symptomatic OA improvement), WOMAC (functional pain), and CGIC. The beneficial effects of NF-INF-02 are attributed to the polysaccharides (such as ukonans) of the turmeric extract (Madhu et al. 2013).

Curcuminoid treatment effectively alleviated knee osteoarthritis pain severity in a pilot randomized double blind placebo controlled parallel group clinical trial (Panahi et al. 2014). Curcuminoids (1.5 g/d, 3 x 500 mg C3 Complex® capsules containing 5 mg Bioperine® to enhance oral curcuminoids bioavailability, Sami Labs, LTD, Bangalore, India; 14F/5M, average age 57 years, 29 kg/m^2 average BMI) treatment for 6 weeks significantly reduced pain severity (Western Ontario and McMaster Universities Osteoarthritis Index [WOMAC] and Visual analogue scale [VAS] scores and improved physical function (Lequesne's pain functional index [LPFI]) compared to placebo group. Curcuminoids treatment also enabled patients to reduce their naproxen (250–500 mg) use (84 vs 19% for placebo patients receiving 500–750 mg naproxen). The protective effects of curcuminoids against osteoarthritis are attributed to its anti-inflammatory and antioxidant properties primarily due to NF-κB inhibition (Panahi et al. 2014). Henrotin et al. (2019) compared two doses of bio-optimized turmeric extract (Flexofytol™, Tilman SA, Baillonville, Belgium) to manage symptomatic knee osteoarthritis. The low dose (4 capsules daily, 186.6 mg bio-optimized *Curcuma longa* extract) for 3 months reduced VAS pain and patient global assessment of disease activity in knee OA patients (40F/7M, 61 years mean age, 30 kg/m^2 average BMI). Flexofytol reached a steady state after 1 month and its analgesic effect was additional to that of paracetamol and NSAIDs (Henrotin et al. 2019).

Curcuma domestica ethanol extract (2 g/d [or 1 g curcuminoids/d] for 6 weeks) had similar efficacy as ibuprofen (400 mg twice daily) in knee osteoarthritis patients (mostly overweight elderly women) (n=45, 61 years mean age, 26.4 kg/m^2 average BMI). Curcuma treatment was as effective as ibuprofen in alleviating pain and improving knee functions with a trend toward greater effect in patients receiving *Curcuma domestica* extract (Kuptniratsaikul et al. 2009). The same group repeated a similar, but shorter (4 weeks) study evaluating the efficacy of *Curcuma domestica* ethanol extracts compared to ibuprofen in pain reduction and functional improvement of knee osteoarthritis patients (Kuptniratsaikul et al. 2014). In that multicenter study, 367 primary knee osteoarthritis (≥5 pain score) patients received ibuprofen (1.2 g/d) or *C. domestica* extract (1.5 g/day containing 75–85% curcuminoids; 1/0.59/0.23 curcumin/demethoxycurcumin/bismethoxycurcumin ratio). The curcumin treatment (157F/14M, 60 years mean age, 26.5 kg/m^2 average BMI) was as effective as ibuprofen in pain reduction and functional improvement based on WOMAC scores (Kuptniratsaikul et al. 2014). Curcumin treatment (Haridra®) significantly reduced VAS and WOMAC scores (pain and physical function) than placebo at 60 and 120 days in knee osteoarthritis patients (53F/25M, 50 years mean age, 28 kg/m^2 average BMI, 76% Kellgren/Lawrence grade 3–4) (Srivastava et al. 2016). Furthermore, the intervention significantly reduced IL-1β, ROS and MDA compared to baseline values. Curcumin extract clinically improved all WOMAC parameters (pain, stiffness and physical function) and VAS scores and reduced disease-related oxidative stress biomarkers (IL-1β, ROS, and MDA) in knee osteoarthritis patients. Hence, curcumin reduced knee osteoarthritis related inflammatory process. Turmeric rhizome extract Haridra® (The Himalaya Drug Company, Bangalore, India) contained total ≥95% curcuminoids (0.39/0.2/0.06 curcumin/demethoxycurcumin/bisdemethoxycurcumin ratio) (Srivastava et al. 2016).

Several other studies, undefined on their risk bias demonstrate the efficacy, safety and benefit of turmeric extracts on knee osteoarthritis patients. Singhal et al. (2021) compared the efficacy and safety of bioavailable turmeric extract (BCM-95®, 1 g/d) to paracetamol (650 mg/d) for 6 weeks in knee osteoarthritis patients in a randomized, controlled clinical trial. BCM-95® (also known as Curcugreen™) is a patented turmeric (*Curcuma longa*) rhizome extract combined with turmeric essential oil and standardized curcuminoids levels (Barrett 2019). BCM-95®, formulated by Arjuna Natural Extracts Ltd., (Alwaye, Kerala, India) contains 95% curcuminoid complex (65% curcumin, demethoxycurcumin, and bisdemethoxycurcumin in their natural ratio) together with turmeric essential oil (~45% ar-turmerone). The turmeric treatment improved physical functions and alleviated pain and stiffness of middle age knee osteoarthritis patients (53F/20M, 53 years mean average age, 64% grade II and 9% grade III knee osteoarthritis [Lawrence classification]) as effectively as paracetamol. Moreover, patients in the turmeric extract group attained lower WOMAC pain score and stiffness function score (18% attained ≥50% reduction and 3% attained ≥70% reduction) indicating superior outcome than paracetamol. Turmeric extract significantly ($P < 0.01$) lowered blood inflammation marker TNF-α levels compared to paracetamol (Singhal et al. 2021).

Belcaro et al. (2014) compared the efficacy of a curcumin phospholipid complex (500 mg Meriva® [Indena SpA, Milan] and 500 mg Regenasure® [Glucosamine HCl, Cargill]) tablet/d with chondroitin sulphate and glucosamine HCl (800 mg Chondroitin and 830 mg Glucosamin HCl)/d for 4 month in knee osteoarthritis (Grade 1–2) patients. In this observational, 'real life' scenario study, Meriva® curcumin improved functional performance evidenced by significantly higher Karnofsky index than chondroitin+glucosamine group (31 vs 11% after 4 months) and WOMAC scores (social functions 59 vs 23%; emotional functions 55 vs 17%). Moreover, curcumin in this bioavailable lecithin delivery system significantly improved mobility by increasing the walking distance (over 9 folds) and physical function (~4 folds), decreasing WOMAC stiffness (2 folds), WOMAC pain (1.7 folds) and the need for concomitant drugs (by ~½) compared to (chondroitin+glucosamine), the widely used osteoarthritis treatment at lower than half the medical costs (Belcaro et al. 2014). Bioavailable curcumin (500 and 1000 mg Acumin/day) reduced pain, joint inflammation, swelling and tenderness in 36 rheumatoid arthritis patients. Administration of another bioavailable curcumin (BCM-95, 1 g/d) alone or in combination with diclofenac sodium for 8 weeks alleviated pain and disease conditions in 45 rheumatoid arthritis patients (Kunnumakkara et al. 2023 and references therein).

Many studies demonstrate that turmeric extract formulations (i.e., B-Turmactive, TamaFlex, and Instaflex) relieve joint pain, improve knee function and mobility. These formulations improved Western Ontario and McMaster Universities osteoarthritis index (WOMAC) scores, walking time and distance. WOMAC and VAS scores also decreased in arthritic patients treated with turmeric extract alone and as formulations or combinations (i.e., BCM-95, L173014F2, and AINAT) due to improved levels of critical disease regulators (e.g., resistin, adinopectin, leptin, CRP, TNA-α, TNFR2, IL-1β, reactive oxygen species [ROS], and MDA). Turmeric extract alone and in combination with *Boswellia* reduced PGE2 levels, and improved patients pain scores, function and mobility. Curcumin (200 mg thrice daily for 7 days pre-molar surgery) alleviated acute inflammation associated pain compared to patients receiving a standard non-steroidal anti-inflammatory drug [500 mg mefenic acid]. Nanocurcumin (80 mg/kg daily for 4 months) reduced anti-inflammatory cytokines (IL-6) and associated gene regulators (miR-17 and miR-27) levels in ankylosing spondylitis patients effectively increasing regulatory T-cells populations (Kunnumakkara et al. 2023 and references therein).

The analgesic benefits of turmeric/curcumin in relieving pain extend beyond arthritis and other disorders. The general analgesic mechanism of action of turmeric and curcumin is based on their modification/modulation of NF-κB signaling, pro-inflammatory cytokines (interleukin production and phospholipase A2), COX-2 and 5-LOX activities. Curcumin also modulates expressions of various transcription factors. Turmeric presumably down-regulates pain receptor expression similar to the opioid-related nociception receptor 1 gene (OPRL1) or may be associated with its anti-inflammatory actions by suppressing IL-6 and NF-κB pathways. Generally, turmeric is favored than curcuminoids or curcumin extracts since other constituents can have synergistic anti-inflammatory effects.

A recent review (Kunnumakkara et al. 2023) aptly describes the role of turmeric and curcumin in chronic disease prevention and treatment. The following summarizes studies related to pain relief discussed in that review. For example, several studies report the efficacy of turmeric tablets in relieving abdominal pain and discomfort with potential anti-ulcer activity. Curcumin (550 mg twice daily for the first month followed by 550 mg thrice daily for another month) reduced abdominal pain in patients with inflammatory bowel syndrome. Abdominal pain and other severe IBD symptoms were effectively reduced in 116 patients treated with curcumin in combination with fennel oil (84 mg CU-FEO daily for 30 days). Curcumin (350 mg) in combination with calendula extract (80 mg) for one month reduced pain and inflammatory cytokines in chronic prostatitis or pelvic pain syndrome III patients. Curcumin with LipiSperse (500 mg HydroCurc) and 500 mg maltodextrin after exercise reduced overall muscle pain in young healthy males with strength training experience (Kunnumakkara et al. 2023 and references therein).

PureVida capsules (50 mg curcumin extract, 460 mg fish oil, and 125 mg Hytolive powder/capsule) three times daily for 30 days reduced pain score in 45 postmenopausal breast cancer patients in a multicenter prospective, single arm clinical trial. Curcumin gel application markedly reduced skin reactions and pain in 171 patients with breast cancer and radiation-associated dermatitis in a multicenter, double blind study. Turmeric in combination with mefanic acid for 5 days effectively relieved menstrual pain associated with primary dysmenorrhea. Curcumin (500 mg/d) and piperine (5 mg/d) for 10 days also reduced dysmenorrhea-associated pain in a randomized, triple blind and placebo-controlled clinical trial. Pelvic pain also decreased with soft gel tablets containing curcumin (2 pills daily for 12 weeks in 20 endometriosis patients. In a clinical trial, turmeric gel application (1 g turmeric powder/d) for 37 days relieved pain and reduced pain score in oral cancer patients. Curcuminoids (1 g/day) reduced VAS score and pain in patients after laparoscopic hysterectomy. Combined curcumin (1 g/day) and piperine (10 mg/day) treatment improved muscular pain after 2 weeks in 23 COVID-19 patients (Kunnumakkara et al. 2023 and references therein).

Curcuminoids significantly reduced pain (muscle pain, osteoarthritis and postoperative pain) in the first meta-analysis (search up to September 2014) evaluating its analgesic effect (Sahebkar and Henrotin 2016). This pain relieving effect was independent of administered dose and treatment duration with curcuminoids. The meta-analysis included eight randomized controlled trials [RCTs] (selected out of 48 studies) involving 606 randomized patients with 300 receiving curcuminoids. Curcuminoid supplementation reduced pain (by −19.3 points; $P < 0.0005$) in studies using 100 point VAS for pain severity assessment. The eight studies were rated on the Jadad scale (0–5) for their quality. In the highest rated study (5 Jadad score) oral curcumin (6 g/d) during radiotherapy reduced severity of radiation dermatitis [RDS] (mean RDS 2.6 vs 3.4 for placebo; $P = 0.002$) in middle age breast cancer patients (n=30, [n=14 curcumin] mean age 58 years) (Ryan et al. 2013). In this randomized, double blind, placebo controlled trial, all patients received standard radiation (≥42 Gy total radiation dose) prior to orally taking 12 daily capsules of Curcumin C3 Complex® (Sabinsa Corp., Payson, UT; 500 mg capsules containing ~390 mg curcumin, 75 mg demethoxycurcumin, 12.5 mg bisdemethoxycurcumin) or placebo. Daily intake of 6 g curcuminoids (4680 mg curcumin, 900 mg demethoxycurcumin and 150 mg bisdemethoxycurcumin) significantly increased sensory pain descriptors (gnawing, aching and splitting). However, curcumin and placebo treatments showed no significant difference in total McGill Pain Questionnaire or sensory inventory (Ryan et al. 2013).

Three studies (Kuptniratsaikul et al. 2014; Kuptniratsaikul et al. 2009; Panahi et al. 2014) discussed above were also rated on the Jadad scale (5, 3 and 3 score, respectively) indicating consensus among investigators. Chandran and Goel (2012) evaluated the effectiveness of curcumin alone and in combination with diclofenac sodium in active rheumatoid arthritis patients (2 Jadad score). Curcumin (500 mg BCM-95®, Arjuna Natural Extracts, Kerala, India for 8 weeks, 2M/13F; 48 years mean age; 24.5 kg/m^2 average BMI) improved overall disease activity and American College of Rheumatology (ACR) scores. Pain scale disability was evaluated by the visual analogue scale (VAS; 0, no pain and 100, severe pain). Curcumin treatment reduced pain the most evidenced by the highest VAS score reduction (60%) from baseline. Curcumin treatment also showed the highest percentage of ACR20, ACR50 and ACR70 (93, 73 and 33%, respectively). Curcumin alleviated pain more effectively than diclofenac. BCM-95® has enhanced bioavailability (6-to 8-folds of regular curcumin). Curcumin has been reported to effectively alleviate chronic pain in various experimental models, even the most difficult neuropathic pain (Chandran and Goel 2012 and references therein). ACR20 is defined as tender and swollen joint counts reduction by 20% from baseline.

Curcumin (200 mg twice/d) supplemented 48 h prior to eccentric continuous exercise (downhill running) and continued for 4 days attenuated pain in the lower limb, the right and left anterior thighs of healthy active males (n=9, 33 years average age, 24 kg/m^2 average BMI) (Drobnic et al. 2014). Curcumin treatment also reduced pain intensity and posterior or medial muscle injury, prevented delayed onset muscle soreness and lowered interlukin-8 levels 2 h after exercise (2 Jadad score). The favorable effects of curcumin on exercise induced muscle injury were attributed to its antioxidant properties that boost the body's endogenous antioxidant response and Nrf2-mediated stimulation of the cellular antioxidant system (Drobnic et al. 2014). This randomized placebo controlled trial confirms curcumin's analgesic effects of a previous study where Meriva® (2 g corresponding to 400 mg curcumin) relieved pain in middle aged patients (8F/7M, 50 years mean age) with acute algesic episodes better than 500 mg acetaminophen (Francesco et al. 2013).

Agarwal et al. (2011) evaluated curcumin's effect on pain and postoperative fatigue in laparoscopic cholecystectomy (LC) patients. Curcumin (4 x 500 mg capsules, Indsaff Inc., India) reduced mean pain scores (on the Visual Analog Pain Scale) at the first and second week after laparoscopic surgery (15 vs 30 in controls) with significantly lower analgesic tablet use (7 vs 39 in control) (3 Jadad score). All patients were pain free at week 3. Moreover, curcumin treatment (4M/21F, 37 years mean age) reduced analgesic tablet usage (650 mg Paracetamol; 7 vs 39 in controls)

Merolla et al. (2015) evaluated pain improvement (20 mm on a 100 mm visual analog scale [VAS} after administration of a commercial dietary supplement Tendisulfur® containing *Boswellia serrata* and *Curcuma longa* in subjects with full-thickness rotator cuff tears in the supraspinatus (SSP) treated by arthroscopy. In this prospective randomized trial, 50 patients (27M/23F, mean age 53 years, 27 kg/m^2 average BMI) received Tendisulfur® (2 sachets/d for 15 days and 1 sachet for the next 45 days) starting from the first postoperative day. The placebo group consisted of 50 patients with similar demographics. Conventional analgesic therapy (tramadol and/or paracetamol [maximum 400 mg and 4 g/d, respectively) were prescribed for all patients. Two sachets of Tendisulfur® contains 200 mg each of *B. serrata* titrated to 30% inacetyl-11-keto-β-boswellic acid [AKBA] and dry *C. longa* extract titrated to 95% in curcuminoids. The treatment significantly lowered the overall pain score only in the short term (week 1) without significant changes in pain during activity and two clinical scores (constant Murley score [CMS] and simple shoulder test [SST]). These findings have clinical implications since pain is severe in the immediate postoperative period after arthroscopic SSP repair. Moreover, Tendisulfur® enhanced rehabilitation and minimized complications. Curcumin in the formulation presumably exerts its effect by modulating important targets including transcription factors (e.g., NF-κB, AP-1, β-catenin, pro-inflammatory cytokines [TNF-α, IL-1B and IL-6], enzymes [COX-2, 5-LOX and inducible NOS).

Later, another study confirmed Tendisulfur® Forte benefits in combination with Extracorporeal Shock Wave Therapy (ESWT) in the treatment of shoulder tendinopathy, lateral epicondylitis, and Achilles tendinopathy (Vitali et al. 2019). Patients received either ESWT (6M/9F, mean age 52 years) or ESWT supplemented with Tendisulfur® Forte (2 sachets/d for 1 month, then one sachet/d for an additional month) (4M/11F, mean age 55 years). Tendisulfur® Forte improved overall function and VAS scale scores for Achilles tendinopathy at 30 days and shoulder tendinopathy and elbow epicondylitis at 60 days. Moreover, Tendisulfur® Forte greatly reduced NSAIDs consumption and totally stopped its use in most cases resulting in faster recovery and better outcomes (Vitali et al. 2019).

Curcumin C3 complex® (1 g/d) for 12 weeks supplementation improved physical function and lower-limb muscle strength of sedentary healthy older adults (4F/5M, 79.4 years median age, 31 kg/m^2 average BMI) (Mankowski et al. 2023). The treatment also lowered inflammation biomarkers (lower galectin-3 and higher IL-6 levels) compared to placebo (nitrocellulose). Another study showed that a similar supplementation (1 g/d, but with 10 mg piperine [enhanced bioavailability]) for 8 weeks reduced systemic inflammation biomarkers (TNF-α, IL-6, transforming growth factor beta [TGF-β] and monocyte chemoattractant protein-1 [MCP-1]) in young and middle aged adults (23F/27M, 45 years median age, 25.5 kg/m^2 average BMI) (Panahi et al. 2016). These beneficial effects are attributed to the antioxidant and anti-inflammatory properties of curcumin. Curcumin C3 complex® and chlorogenic acid (103 and 0.32 mg) consumption in bioactive yogurt attenuated inflammatory cytokines (TNFα) in postmenopausal (n=16, 58 years median age, 27 kg/m^2 average BMI) healthy women in a randomized double blind placebo-controlled, crossover study (1 week washout period) (Ahmed Nasef et al. 2022).

Neuropathic pain results from primary lesion or diseases of the somatosensory nervous system prevalent at 9.8% clinically based and 12.4% self-reported in the United States. Damage to the peripheral nerves causes peripheral neuropathic pain (Basu et al. 2021). Many drugs with considerable side effects are used to treat neuropathic pain; first-line drugs (serotonin-norepinephrine reuptake inhibitors); second-line drugs (lidocaine, capsaicin and tramadol) and third-line drugs ([opioids, morphine, oxycodone] and botulinum toxin A). The latest review (Basu et al. 2021) revealed the analgesic effects of curcumin and its formulations in neuropathic and postoperative pain conditions. Nano curcumin alleviated diabetic sensorimotor polyneuropathy (DSPN) that affects 25% of type 2 diabetic patients based on total Toronto Clinical Neuropathy Score. The treatment also decreased fasting blood glucose serum levels and HbA1C indicating hyperglycemia improvement in T2DM individuals. The lecithinized curcumin Meriva® significantly reduced chemotherapy induced side effects and plasma free radicals levels compared to control.

A multi-ingredient formula (800 mg each of dexibuprofen, Lipicur, lipoic acid and curcumin phytosome with 8 mg piperine) reduced neuropathic pain in patients with lumbar sciatica and lumbar tunnel syndrome. Curcumin also reduced medicine use (dexibuprofen by 40%) (Basu et al. 2021 and references therein).

Di Pierro et al. (2013) posited that curcumin can attenuate thermal hyperalgesia associated with diabetic neuropathic pain by inhibiting TNF-α and nitric oxide release. Moreover, curcumin behaves as a combined transient receptor potential cation channel 1(TRPA1) inhibitor and transient receptor potential cation channel subfamily V member 1(TRPV1) desensitizer complemented by inflammatory mediators. Di Pierro et al. (2013) reached these conclusions after comparing the acute pain relieving properties of lecithinized Meriva® curcumin with acetaminophen and the non-steroidal anti-inflammatory agent nimesulide in a pilot study. The analgesic activity of Meriva® (2 g corresponding to 400 mg curcumin) lasted 4 h and was comparable to standard acetaminophen (1 g), but lower than the therapeutic dose (100 mg of nimesulide). Merive® analgesic effect occurred 2 h after administration, similar to acetaminophen, whereas numesulide acted rapidly (1 h post administration) with strongest pain relieving properties. The similar action of curcumin and acetaminophen indicates that they share the same mechanism of action, while curcumin modestly inhibits cyclooxygenases compared to nimesulide potent action on these enzymes. Rapid analgesic effect has been reported within 1–2 h administration of Minerva® in many clinical studies for various chronic diseases (Di Pierro et al. 2013 and references therein). Meriva® contains 20% curcumin, 40% phosphatidylcholine, and 40% microcrystalline cellulose.

3.3 ASHWAGANDHA—*WITHANIA SOMNIFERA*

3.3.1 Composition

India is the largest Ashwagandha producer (worth 3000 million USD in international export market) with over 30 genetic varieties from the arid and semi-arid regions. National production is estimated at 1500 t (~21% of the annual demand) with a wide production gap for an annual demand ~7000 t (Chauhan et al. 2022). The medicinal value of *W. somnifera* is attributed primarily to withanolides, naturally occurring C28 steroidal lactones. The important Ashwagandha compounds are withanolide A (WL-A), withanolide B (WL-B), withanolide D (WL-D), withaferin A (WF-A) and withanoside (WS) 1 to VII. The genetic diversity of withanolides and other breeding traits was evaluated in 25 Indian *W. somnifera* germplasms (Table 3.5) grown in triplicates for two consecutive years in Udaipur, Rajasthan. The average total alkaloid content (TAC) was 0.48% (range 0.08–0.96%) (Chauhan et al. 2022).

Kumar et al. (2018) developed a simple rapid HPLC method to identify and quantify withaferin A (WA), 12-deoxywithastromonolide (12WD) and withanolide A (WDA) in *W. somnifera* roots. Aqueous alcohol (50, 70 or 100%, v/v) produced maximum extract yield, total withanolide and phenolics (21.4, 0.065 and 1.94%), respectively. Cultivated *W. somnifera* varieties contained withanolides (0.0094%WA, 0.0005%12 WD and 0.0433% WDA) (Joshi et al. 2010). However, these withanolides varied in 15 selected accessions (0.001–0.567 % WA, 0.002–0.117 % WDA dry weight basis) (Dhar et al. 2006). Higher withanolides contents have been reported in *W. somnifera* roots depending on extraction solvent and condition (0.092% WA, 0.19% 12WD and 0.388% WDA) (Chatterjee et al. 2010).

Extract yield, chemical composition of the extracts and antioxidant activity of the extracts varied with the extraction process as well as the solvent composition. Green technology (ultrasound and microwave assisted solvent extraction) is a viable alternative to traditional extraction methods (Soxhlet) (Table 3.6). Ethanol extracts had the highest antioxidant and antiradical activities because of their high phenolics and withanolides contents. The water extract had the lowest phenolic and withanolides content although it had maximum yield (mass of extract/mass of dry matter) (Dhanani et al. 2017).

TABLE 3.5 Withanolide contents (ng/μL) of *W. somnifera* roots

	WDL	*WIV*	*WV*	*WB*	*WFA*
Mean	93.17	268.48	37.19	22.25	144.51
Min	12.11	9.75	6.36	2.22	3.27
Max	1134.96	654	73.33	73.26	361.59

Source: Adapted from Chauhan et al. 2022; WDL—Wedeolactone;

WIV—Withanoside IV; WV—Withanoside V; WB—Withanolide B; WFA—Withaferin A

TABLE 3.6 Extraction effects on withanolide yield and contents of *W. somnifera* roots

EXTRACTION		*YIELD*	*TOTAL PHENOLICS*	*WITHANOLIDES (MG/G)*		
METHOD	*SOLVENT*	*(%)*	*(MG GAE/G)*	*WA*	*12 DW*	*TWS*
Soxhlet	ETOH	9.08	35.93	3.57	1.22	4.79
	Aq-ETOH	9.43	21.15	1.25	0.4	1.65
	Water	9.51	17.63	1.14	0.36	1.5
UASE	ETOH	2.99	26.38	6.23	2.02	8.26
	Aq-ETOH	9.26	21.55	1.51	0.5	2.01
	Water	10.67	16.30	0.80	0.23	1.02
MASE	ETOH	10.01	40.96	4.35	1.39	5.73
	Aq-ETOH	12.65	21.85	1.02	0.29	1.31
	Water	12.31	17.95	0.65	0.21	0.86

Source: Adapted from Dhanani et al. 2017; UASE—ultrasound assisted solvent extraction, Values are means of extractions at 5, 10 and 20 min; MASE—microwave assisted solvent extraction, values for aqueous ethanol (Aq-ETOH) and water are averages at 5, 10 and 20 min extractions; ETOH—ethanol; Aq-ETOH—aqueous ethanol (9:1): WA—withanolide A; 12DW—12-deoxy withastramonolide; TWs—total withanolides

Polumackanycz et al. (2023) evaluated the polyphenolic content and composition of individual phenolics in several commercially available *Withania somnifera* products including root powders of Indian origin (Table 3.7). The hydromethanolic extracts were prepared by sonicating root sample (1 g) with methanol water (80:20, v/v; 7 mL) for 20 min in an ultrasonic bath. The aqueous extract was obtained by adding boiling water (100 mL) to the root sample (1 g), resting for 10 min prior to filtration. Polyphenols were more abundant in the aqueous than the hydromethanolic extract. However, caffeic and ferulic acids and rutin were present only in the hydromethanolic extract.

The total phenolic contents of ethanol and water *W. somnifera* root extract were 79.8 and 58.2 mg GAE/g extract, whereas the total flavonoid contents were 4.8 and 1.9 mg QE/g extract (Nile et al. 2022). Methanolic *W. somnifera* roots extract contained phenolic (66 mg GAE/g total phenolic), tannins (5.89 mg GAE/g), flavonoid (650 mg QE/g), and alkaloid (102 mg/g) (Mahto et al. 2022).

3.3.2 Clinical Human Trials

Osteoarthritis (OA) is the most common cause of chronic pain, disability, and poor quality of life. Symptomatic relief is generally provided with analgesic, including nonsteroidal anti-inflammatory drugs (NSAIDs). The prevalence of osteoarthritis (OA) [including knee, hip and hand OA, and other OA] was estimated through a systematic analysis (1990–2019) of prevalence, incidence, and years lived with

TABLE 3.7 Polyphenolic composition of commercially available *Withania somnifera* products

SAMPLES/ EXTRACT	*MG/G DW*			*(μG/G)*						
METHANOL (80%)	*TPC (GAE)*	*TF (QE)*	*TPA (CAE)*	*CAT*	*Q*	*CA*	*FA*	*RUT*	*GA*	*NAR*
13	0.65	0.18	1.00	950	1320	310	255	381	118	
15	1.34	0.16	0.26	810	1320	300	245	743	112	444
16	1.29	0.18	0.30	1060	1400	298	267	392	123	
18	1.57	0.19	0.42	840	1260	298	ND	360	109	
Aqueous										
13	2.57	2.71	1.43	6420	1020					
15	3.48	1.93	1.57	6340	820				870	2.49
16	2.24	2.21	0.99	6530	1060					
18	1.73	2.43	1.44	5980	990					

Source: Adapted from Polumackanycz et al. 2023; TPC—total phenolic compounds; TF—total flavonoids; TPA—total phenolic acids; Individual phenolic compounds: GA—gallic acid; CAT—catechin; CA—caffeic acid; FA—ferulic acid; RUT—rutin; Q—quercetin and NAR—naringenin

disability (YLDs) modelled data from global burden of disease (GBD) 2019 study. In 1990, about 23.46 million individuals in India had OA and that number increased to 62.35 million in 2019. Knee OA was most common OA form, followed by hand OA, with knee OA consistently higher in females than males (Singh et al. 2022). Knee osteoarthritis is likely to become the fourth most important global cause of disability in women and eight most important in men (Ramakanth et al. 2016). Rheumatoid arthritis (RA) is the prototype of a severely painful chronic disease affecting multiple joints, causing swelling and crippling deformities in most patients. RA prevalence is 0.3–0.7% in Indian population, lower than worldwide prevalence (~1%). Analgesic provides immediate pain relief in RA; these include paracetamol/tramadol, non-steroidal anti-inflammatory drugs (NSAIDs) such as ibuprofen, diclofenac and celecoxib (Chopra et al. 2010).

Peripheral sensory neuropathy manifesting as pain affects over 50% (>½) of all diabetic patients. Currently available drugs are ineffective in relieving pain in all patients, in addition to other drawbacks. *Withania somnifera* can benefit poorly controlled diabetic patients with peripheral neuropathy (i.e., complications arising immediately after sudden glycaemic improvement). Ashwagandha leaves are bitter and are used for fever, anthelmintic and for painful swellings; the roots are also used for painful swellings. Many compounds isolated from *W. somnifera* leaf and roots are used in treating rheumatic pain and diabetes. US consumers spent overwhelmingly more (226%, sales totaling 92,326,926 USD) on ashwagandha supplements in 2021 compared to previous year (~64 million USD increase from 2020) (Smith et al. 2022). It ranks seventh on the US top selling supplement list predominantly (96% of all mainstream ashwagandha supplements) marketed as mood support (adaptogens).

Ayurvedic medicine considers *Withania somnifera* as an experimentally proven analgesic anti-inflammatory drug. Sensoril® is a proprietary aqueous extract from a unique blend of Indian grown (Madhya Pradesh) certified organic *Withania somnifera* leaves and roots used in several studies. Sensoril® is a clinically proven multipatented, standardized *W. somnifera* roots and leaves extract marketed by NutraGenesis under exclusive license from Natreon, Inc., for stress relief due to glycowithanolide bioactives that mimic the body's own stress-reducing hormone. It is GRAS-affirmed, can be used 365 days a year, and has an extensive safety record. It has a patented composition with specific bioactive levels of withanolide (≥10%), glycosides (≥32%) oligosaccharides, and withaferin A (≤0.5%). Murthy et al. (2019) evaluated the analgesic activity (centrally mediated antinociceptive responses) of a single *W. somnifera* dose (2 x 500 mg Sensoril® capsules) in healthy human using mechanical pain model. The Randall Selitto test using Ugo Basile analgesymeter measured mechanical pain before and three hours after drug administration in this randomized, double blind placebo controlled cross over study. Healthy male adults (n =

12, 18–40 y, 19.5–25.9 kg/m^2 normal BMI) were randomized to receive either *W. somnifera* or placebo capsules with 10–14 days washout period for cross over between the two drugs. *Withania somnifera* significantly ($P < 0.005$) increased pain threshold force and time (~19%), and pain tolerance force and time (~8.8%) compared to placebo.

The analgesic effects of *W. somnifera* and/or its bioactive components were clearly demonstrated in humans in a randomized, double-blind, placebo-controlled, cross over study with 10–14 days washout period (Nalini et al. 2013). In this study, single oral dose (1 g) of standardized *W. somnifera* aqueous extract (2 x 500 mg Sensoril®) significantly ($P < 0.05$) increased pain threshold time in 12 healthy male (35 years meadian age; 21 kg/m^2 average BMI) participants without any gastric side effects. Pain threshold was evaluated at baseline and 3 h post treatment in a hot air analgesiometer delivering reproducible heat-induced thermal pain on the volar surface of subjects forearm. Each Sensoril® capsule contained 15.7% withanolide glycoside (active ingredient), 0.24% withaferin-A, and 40.2% oligosaccharides (bio-carrier) (Nalini et al. 2013).

The standardized *W. somnifera* aqueous roots and leaves extract Sensoril® was also used to demonstrate analgesic activity in experimentally induced mechanical pain in healthy human volunteers (Murthy et al. 2019). *Withania somnifera* (2 × 500 mg Sensoril® capsules) significantly increased pain threshold and pain tolerance force and time (~20% compared to baseline) 3 h post administration in 12 healthy male (34 years median age, 21 kg/m^2 average BMI) adults. Analgesic activity (mechanical pain model by Randall Selitto tests) was monitored by Ugo Basile Analgesymeter that elucidates centrally mediated anti nociceptive responses, primarily changes above the spinal cord level. This randomized, double-blind, placebo controlled, crossover (10–14 days period) study confirms a similar previous study (Nalini et al. 2013) using hot air pain model.

In knee joint patients (43M/17 F; 58 years mean age), *Withaina somnifera* (250 mg or 125 mg Sensoril® capsule twice daily for 12 weeks) significantly reduced MWOMAC (Modified Western Ontario and McMaster University Osteoarthritis Index), knee swelling index (KSI), and visual analogue scale (VAS) scores for pain stiffness and disability without adverse gastrointestinal effects (Ramakanth et al. 2016). However, the 250 mg (twice daily) Sensoril® treatment was superior due to fast and better symptomatic relief increasing patient compliance and satisfaction in this randomized, double-blind, placebo-controlled study. The MWOMAC score reduction by *W. somnifera* was attributed to its withaferin A-mediated (and presumably serotonin) analgesic and chondroprotective effects. Withaferin A reduced KSI due to its anti-inflammatory activity. Moreover, *W. somnifera* flavonoids and phenolic acids (gallic and vanillic acids, quercetin and kaempferol) can block signal transduction for NF-κB activation thereby inhibiting transcription factors [(NF-κB, activating protein-1 (AP-1) and nuclear factor-erythroid 2-related factor 2 (Nrf2)].

Sensoril® (2 × 250 mg capsule/d containing $\not<$10% total glycowithanolides for 14 days) also significantly improved cognitive and psychomotor performance in twenty young (20–35 years) healthy men in a perspective, double blind, placebo controlled cross over (14 days washout period) study (Pingali et al. 2014). The study indicates that the glycowithanolides can preferentially affect events in the cortical cholinergic-signal transduction cascade partly explaining the cognition-enhancing and memory improving effects of *W. somnifera* (Pingali et al. 2014). *Withaina somnifera* (500 mg or 250 mg Sensoril® capsule twice daily for 12 weeks) significantly improved endothelial function (3–4 folds reflection index reduction), oxidative stress and systemic inflammation biomarkers, lipid parameters and HbA1C compared to baseline and placebo in diabetic patients (Usharani et al. 2014). In this randomized, double blind, placebo controlled, parallel design study, diabetic patients (n= 60, ~55 years old, ~25 kg/m2 BMI) on standard medication (Metformin 1.5–2.5 g/d) received Sensoril® treatment [250 mg (14M/6F); 500 mg (13M/7F)] or placebo (12M/8F) twice daily. Both Sensoril® treatments benefited endothelial function with significant improvement in antioxidant status (nitric oxide, glutathione, malondialdehyde) and systemic inflammation (high sensitive C reactive protein levels). Moreover, *W. somnifera* treatment (500 mg Sensoril®) improved lipid profile (reduced total and LDL cholesterol, and triglyceride levels), generally elevated in diabetic patients. Withanolide glycosides and withaferin A, the active principles of Sensoril® are presumably responsible for its physiologically beneficial (antidiabetic, antioxidant, anti-inflammatory and other) effects (Usharani et al. 2014).

Intake of a standardized ashwagandha extract (Shoden) significantly relieved stress in mildly anxious healthy adults in a randomized, double blind, placebo controlled trial (Lopresti et al. 2019). The treatment [one capsule containing 240 mg *Withania somnifera* ethanol (70%) extract (with 84 mg withanolide glycosides) daily for 60 days] reduced anxiety (42 vs 24% for placebo) in ashwagandha participants (19M/11F, 42 years old, 24.7 kg/m^2 average BMI). The antioxidant and anti-inflammatory activities of standardized *W. somnifera* extract (≮35% total glycowithanolides) are presumably due to its antioxidant and anti-inflammatory effects. Moreover, the study suggests that *W. somnifera*'s anxiolytic effects may occur via its moderating effect on the hypothalamus-pituitary-adrenal axis ((Lopresti et al. 2019).

W. somnifera analgesic effects are attributed to its major metabolites, primarily withanolides, alkaloids and other constituents. This analgesic activity is used in ayurvedic preparations containing *W. somnifera* for various musculoskeletal disorders (including arthritis and rheumatism) (Bhat et al. 2022). *Withania somnifera* root powder (3 g thrice daily with milk for 6 weeks) administered to arthritis patients (68F/9M, 12–60 years old) improved pathological symptoms (reduced morning stiffness, pain swelling tenderness, functional status, fever, and erythrocyte sedimentation rate [inflammatory marker]), particularly in chronic rheumatic arthritis patients in a single blind study (Bikshapathi and Kumari 1999). Primary bioactive constituents of the *W. somnifera* root powder, withanolides and iron were presumed to be responsible for its ant-arthritic effects.

Ghawte et al. (2014) evaluated *Withania somnifera* analgesic and anti-inflammatory effects to manage rheumatoid arthritis in a randomized single blind, placebo controlled study. *W. somnifera* powder (7 g twice daily) for 90 days increased grip strength, reduced pain, morning stiffness, and swelling in rheumatic arthritis patients (29F/31M) compared to placebo. Prostaglandin inhibition was proposed as the mechanism of action to protect against inflammation and cartilage damage associated with osteoarthritis. Ayurvedic treatment (Ashwagandha powder [in water or milk] 5 g twice daily for 3 weeks and *Sidh Makardhwani* 100 mg/day [with honey] for 4 weeks) reduced rheumatoid factor, disease activity score, inflammation and disease remission in rheumatoid arthritis patients (45F/33M, 45–50 years old) in a prospective, open-label, non-randomized, single-centered pilot study (Kumar et al. 2015d).

Ayurvedic formulations (500 mg capsules) containing aqueous ashwagandha dried root extracts (600 mg daily dose) showed no significant pain reduction [active-pain VAS and Western Ontario and McMaster University's OA version LK3 (WOMAC) index] in patients with symptomatic osteoarthritis knees (n = 34) from baseline and placebo. Moreover, no meaningful change occurred in serum cytokines (IL-6, IL-1, and TNF-α) in patients administered ashwagandha, glucosamine, and placebo at baseline and completion of the 4 months randomized, double blind, placebo and oral glucosamine controlled, parallel efficacy, multicentre drug trial study (Chopra et al. 2011). However, only the (daily dose) formulation containing Amalaki (500 mg) [along with *Shunthi* (100 mg) + *Guduchi* (220 mg)] reduced pain (mean pain VAS reduction from 6.3 [baseline] to 4.4 cm [on completion), improved patients' knee status, and requirement of rescue analgesic.

Withania somnifera is a common component of polyherbal anti-rheumatic medications. Seven out of 33 studies (1976-March 2003) that met the inclusion criteria according to Park and Ernst (2005) repeatedly used *W. somnifera*, *Curcuma longa*, *Boswellia serrata*, *Tinospora cordifolia*, and *Zingiber officinale*. These medicinal plants are potential for RA treatment and a formula comprising these plants have been patented for treating RA and osteoarthritis (Patwardhan 1996).

The efficacy of a standardized Ayurvedic formula (RA-1) was evaluated for rheumatoid arthritis treatment in a placebo-controlled RCT (Chopra et al. 2011). RA patients (152F/30M, aged ≥ 18 years with over 6 months RA) were randomized to receive either RA-1 (n = 89) or placebo (n = 93) thrice daily (444 mg total daily dose) immediately after meals for 16 weeks. Patient's assessment included the duration of morning stiffness (min) on the day before the visit, pain severity on a 10 cm VAS, 5-grade patient global assessment and modified Stanford HAQ. Differences were not significant in primary outcome measures between the experimental and placebo group, although this study was highly rated (5 Jadad score). However, the blood hemoglobin level significantly increased and mean RF titer decreased after 16 weeks treatment.

Kulkarni et al. (1991) evaluated the Ayurvedic formula "Articulin-F" (450 mg *W. somnifera*, 100 mg *Boswellia serrata*, 50 mg *Cucurma longa* and 50 mg zinc) on 20 RA patients (14F/6M, 30 years median age with 2 years of RA) in a double-blind, placebo controlled, crossover RCT with 2 weeks wash out period. Patients received two Articulin-F or placebo capsules 3 times daily after food. Articulin-F significantly improved all clinical (pain severity score, morning stiffness, Ritchie articular index, and joint score) and biochemical (hematological ESR and RF) parameters at three months compared to placebo. RA-1 was also used as a monotherapy in a 3 year open label phase (OLP) prospective observational study following a protocol driven 16 week randomized placebo controlled trial involving 165 RA patients (152 F, 45 years mean age) (Chopra et al. 2018). RA-1 contained standardized aqueous extracts from *Withania somnifera* root [90 mg] (Ashwagandha), *Boswellia serrata* [aqueous-alcohol gum extract 90 mg] (Salai Guggul), *Zingiber officinale* [24 mg rhizome] (ginger), and *Curcuma longa* [18 mg rhizome] (curcumin). RA-1 (investigational drug) was administered twice daily—2 tablets (222 mg actives per tablet) following meals. Symptomatic and persistent rheumatoid arthritis patients (n = 57) were fixed on low dose prednisone (7.5 mg/d) and continued RA-1 (Artrex ™, 2 tablets twice daily) throughout the study. RA-1 significantly improved disease outcome (painful joints, swollen joints, pain and HAQ, hand grip strength and walking time) during the first year and sustained till three years. Some patients (10–15%) reported absence of pain/tenderness and swelling in joints with pain VAS ≤ 2 cm and total modified Stanford Health Assessment Questionnaire (HAQ) ≤ 3, and minimal disease even after 24 weeks of RA-1 treatment. Moreover, 40% of RA-1 patients controlled their disease without steroids or disease modifying anti-rheumatic drugs (DMRD). RA-1 effectively treated RA as a monotherapy in about 40% of the cohort (Chopra et al. 2018).

Chopra et al. (2004) evaluated RA-11 (Artrex, Mendar—standardized multiplant Ayurvedic drug [*W. somnifera*, *B. serrata*, *Z. officinale*, and *C. longa*]) currently used to treat arthritis in a randomized, double-blind, placebo-controlled, parallel efficacy, single-center, 32-week drug trial. RA-11(two capsules twice daily after meals) significantly ($P < 0.05$) reduced pain VAS (2.7 vs 1.3) at 16 weeks and improved WOMAC scores ($P < 0.01$) compared to placebo in patients [(mean age 59 years; n = 78 (16 weeks), 62 (32 weeks); 35 F] with symptomatic knee osteoarthritis. RA-11 is a more potent and pure standardized RA-1 variant. In an earlier 16 weeks randomized, placebo controlled, parallel trial, RA-1 (444 mg, 6 tablets daily) showed modest but insignificant efficacy (higher patients with 20% ACR improvement [39 vs. 30%]) in rheumatic arthritis patients (n = 182) compared to placebo (Chopra et al. 2000).

Fatima et al. (2012) evaluated the effect of another polyherbal preparation CardiPro on endothelial function and oxidative stress biomarkers in type 2 diabetic patients. One CardiPro capsule contained *Withania somnifera*, *Terminalia arjuna*, *Embilica officinalis* (100 mg each), *Boerhaavia diffusa* and *Ocimum* sanctum (50 mg each). CardiPro (one capsule thrice daily for 12 weeks) significantly improved endothelial function (reduced reflexion index) and antioxidant status of diabetic patients (16M/10F, 54 years mean age, 25.7 kg/m^2 average BMI) on medication (Metformin 1.5–3 g/d). This randomized, double blind, placebo controlled study showed that CardiPro alleviated endothelial dysfunction, one of the early atherosclerosis prognostic markers and diabetic vascular complications and significantly improved lipid profile (Fatima et al. 2012).

Kanjilal et al. (2021) reviewed the literature on anti-arthritic effects of *Withania somnifera* including clinical trials registered in India from April 2008 to March 2020. Analysis of 77 registered clinical trials revealed that the most common was interventional, single-centered, randomized, double-blind, two arm studies compared with placebo. Moreover, most Phase 2 trials were registered in Maharastra with preference for solid dosage. *Withania somnifera* may be useful in managing arthritis symptoms in patients when administered (6 g) in powder form or extracts in tablets, or capsule (500–1000 mg) for 8–12 weeks.

Withania somnifera roots contain abundant Withanolides (> 2.5%, w/w) and withaferin A (0.66%, w/w) known to interact with cholinergic mechanisms and Nuclear factor-κB. *W. somnifera* modulates morphine analgesic effect *in vivo* and can be useful in improving (enhancing) the opioid-sparing strategies in pain therapy (Caputi et al. 2018). *Withania somnifera* root extract (WSE) administration prolongs morphine-elicited analgesia and reduces the development of morphine analgesic tolerance. In neuroblastoma SH-SYSY cells, WSE influences μ (MOP) and nociception (NOP) opioid receptors gene expression by down regulating (suppressing) morphine-elicited opioid receptors (Caputi et al. 2018).

Withaferin A inhibits activation of the transcription factor NF-κB involved in inflammatory responses. Several inflammatory stimuli, including proinflammatory cytokines such as tumor necrosis factor (TNF-α) and interlukin-1 (IL-1), rapidly activate NF-κB inducing IKKβ kinase activity. Withaferin A exerts its anti-inflammatory effects by binding with IKKβ at Cys179 residue located at catalytic site thereby directly interfering with the kinase activity of IKKβ protein (Heyninck et al. 2014).

Glycowithanolides (Ws, 100 mg/kg, i.p., o.d., for 10 days) containing sitoindosides VII-X and withaferin A from *W. somnifera* roots significantly inhibited the development of tolerance to morphine-induced analgesia in Swiss mice (Ramarao et al. 1995). However, glycowithanolides did not produce any perceptible analgesic effect indicating their potential in alleviating morphine's adverse effects. Ws tolerance to morphine induced analgesia may be associated with changes in central receptors, particularly the dopaminergic receptors in corpus striatum. Furthermore, the sitoinosides apparently facilitate CNS interaction with the immune system of recipients, thereby eliciting immunostimulant effects. Apparently, glycowithanolides (Ws) does not elicit analgesic effects in mice; however, it potentiates morphine induced analgesia by modulating the dopaminergic receptors in corpus striatum (Ramarao et al. 1995).

3.4 BOSWELLIA—*BOSWELLIA SERRATA*

3.4.1 Characteristics

Two *Boswellia* species, *B. ovalifoliolata* and *B. serrata* originate from India, the latter commonly used in Ayurvedic medicine (Pullaiah 2023). *Boswellia serrata* is a large deciduous tree (up to 15 m high), common in rocky areas of Central and Peninsular India. Average gum resin yield is estimated at 1 kg/tree/year with mature healthy tree yielding 2–2.5 kg gum resin/year. India is the main *B. serrata* gum resin distributor with the highest gum resin amount (530 tons, 2008–2013) collected from Sheopur district, Madhya Pradesh. The predominant gum resin component is pentacyclic boswellic acids reported as having potent anti-arthritic, anti-inflammatory and anti-atherosclerotic properties. In the tradition Ayurveda medicinal system, the gum is used to cure arthritis since it is more effective than drugs (e.g., Phenylbutazone) to reduce inflammatory conditions (Pullaiah 2023).

Boswellia serrata oleo gum resin contains 30–60% resin, 5–10% essential oils (organic solvents soluble, primarily mixture of mono-, di-, and sesquiterpenes), and water soluble polysaccharides (~65% arabinose, galactose, xylose). The Imperial Institute (London) provides an exact composition; moisture (10–11%), volatile oil (8.9%), resin (55–57%), gum (20–23%) and 4–5% insoluble matter. Resin, the most important *B. serrata* fraction consists mainly of pentacyclic triterpenic acids, namely boswellic acids (BAs). The major volatile oil constituents are α- and β-pinene (73 and 2%) obtained by *B. serrata* stem bark hydro distillation. The commercial *B. serrata* triterpenoids extract includes αBAs, βBAs, 11-keto-β-boswellic acids (KBAs), acetyl-11-keto-β-boswellic acids (AKBAs) and their esters. Three *B. serrata* gum of Indian origin contained varying amounts of boswellic acids (Table 3.8) that differed from a previous study (Asteggiano et al. 2023; Schmiech et al. 2019). Different amounts of various terpenes have been reported in gum oleoresin. For example, 12% α-thujene and 11.6% methyl chavicol from stem bark, 20% borneol acetate, 13.6% α-terpinol as major components; 50% α-thujene and 14% para-cymene, 32% α-thujene, 20% borneol acetate and 13.6% α-terpineol from leaves (Bhakshu and Ratnam 2023 and references therein).

The frankincense market is estimated to reach US$406.1 million by 2028 from US$193.4 million in 2018 through turnover of raw *Boswellia* and its essential oils (Pullaiah 2023). The aromatic resin, olibanum is used in various products such as health and home care products, aromatherapy, cosmetics, toiletries and dietary supplements. India is the main supplier of *B. serrata* derived frankincense with Madhya Pradesh harvesting 530 tons of resin annually. *B. serrata* is included in the Ayurvedic Pharmacopoeia of India as well as the Pharmacopoeia of Europe and United states.

The air dried *B. serrata* gum resin exudate should contain a minimum of 1% KBA and AKBA each according to the European Pharmacopoiea 6.0. The gum resin of Indian frankincense (*B. serrata*)

TABLE 3.8 Boswellic acids (%) in *Boswellia serrata* resins of Indian origin

SAMPLE	*AKBA*	*KBA*	A-*BA*	B-*BA*	*A-ABA*	*A-BBA*
8	2.76	2.82	3.14	10.31	1.3	3.3
9	2.45	2.41	2.9	7.63	1.2	3.1
10	3.62	1.1	1.74	9.07	1.1	2.3
Mean	2.94	2.11	2.59	9.00	1.20	2.90
Data from Schmiech et al. 2019						
	1.13	0.53	1.64	4.97	1.17	3.65
	1.87	2.23	2.53	4.97	1.17	3.06
	1.36	0.31	1.56	4.79	1.62	4.19
	0.86	0.58	1.32	5.4	0.74	2.31
	1.33	0.98	1.53	4.27	1.01	2.94
	0.93	0.63	0.89	2.83	0.62	1.98
	0.81	0.40	1.22	3.83	0.84	2.36
Mean	1.18	0.81	1.53	4.44	1.02	2.93
Range	0.81–1.87	0.31–2.23	0.89–2.53	2.83–5.40	0.62–1.62	1.98–4.19

Source: Adapted from Asteggiano et al. 2023; AKBA—3-O-Acetyl-11-keto-*β*-boswellic acid;

KBA—11-keto-β-boswellic acid; α-BA—α-boswellic acid; β-BA—β-boswellic acid; A-ABA—3-O-Acetyl-α-boswellic acid; A-BBA—3-O-Acetyl-β-boswellic acid

contains similar KBA and AKBA amounts (3–4.7% and 2.2–2.9%) (Abdel-Tawab et al. 2011). India, the sole producer of *B. serrata* resin exported an average of 87.5 tons/year of *B. serrata* resin (from 1987–1993). The U.S. sales of Boswellia containing dietary supplements increased from US$1.5 to 2.1 million (2013–2016) (McCutcheon 2018).

The boswellic acids yield depends on extraction conditions. For example, three phases partitioning a one-step green bioseparation process yields a maximum of 35–40% AKBA and 25–30% BA with butanol dichloromethane/ethanol/petroleum ether/water (3–4 h). Ultrasonic assisted petroleum ether/ethanol/methanol gives 40–60% AKBA (maximum yield), whereas supercritical CO_2 fluid extraction generates 45–50% maximum boswellic acids. Steam distillation provides the lowest boswellic acid yield (3–10%) (Mishra et al. 2020). Methanol extract of *B. serrata* gum resin contained BBA (16.7–19.7%, w/w), AKBA (3.33–4.41 %, w/w), TCA [3-oxo-tirucallic acid] (13.3–14.4%) and serratol [SRT] (1.42–6.56%) (Mukadam et al. 2023).

The first marketed medicinal *B. serrata* product in India, Sallaki™ was imported into Switzerland under the trade name H15 Gufic™. Sallaki™ or H15 Gufic™ contains 400 mg *B. serrata* dry extract (2.6% KBA and 2.8% AKBA), although other sources report 3.5 or 2.8% KBA and 1.9 or 2.2% AKBA and 8% acetyl-β-boswellic acid (AβBA). Another formulation, BSE-018 (262 mg *B. serrata* dry extract/capsule) contains 3.7% AKBA, 10.5% AβBA, 6.1% KBA, 18.2% βBA, 13.2% αBA and 3.3% AαBA (Abdel-Tawab et al. 2011). Branded *B. serrata* formulations include Boswellin® (Sabinsa 1991) available in pills or tablets (150–250 mg BA per capsule or tablet taken orally 2–3 times daily); Shallaki ® ([Himalayan Drug Co. Makali, Bangaluru] contains 125 mg *B. serrata* per capsule taken twice daily to relieve joint pain; www.himalayahealthcare.com). Rheumatix-X® (Dr. Reddy's Laboratories Ltd., Hyderabad) includes 20 mg "Shallaki" in addition to various other ingredients intended to relieve rheumatic, gouty, osteoarthritis, and sciatic pains (2 capsules twice daily). Colox® (Vegan Society UK) 901 mg natural extracts is a herbal capsule that blocks COX-2 and 5-LOX thereby reducing inflammation, joint pain and stiffness (Mishra et al. 2020). Other brand products are Nutrilite—Glucosamine HCl with Boswellia; Cart Fit (Streamline Pharma Private Ltd.); *Boswellia serrata* extract (Ayurleaf Herbals); Boswellia (Phyto Drug Private Ltd.); Salai Guggul—Wilson Drugs & Pharmaceuticals Private Ltd.) (Rawat et al. 2023).

3.4.2 Human Clinical Trials

One of the earliest report (Menon and Kar 1971) showed that *B. serrata* non-phenolic extracts (300 mg/kg) produced significant analgesia (hot-wire technique) in rats with long lasting effect (> 2 h). This effect was comparable to that of morphine hydrochloride (4.5 mg/kg). Significant analgesia (mechanical pressure method) occurred in 70% of animals at 150 mg/kg extract with maximum effect at 30 min similar to morphine hydrochloride (7.5 mg/kg) and lasting a shorter period (2–3 vs 3–4 h for morphine hydrochloride). The extract generated powerful morphine-like analgesic effect in rats with intrinsic activity intermediate to that of morphine and nalorphine. It reduced spontaneous motor activity causing ptosis and potentiated secobarbitone-induced hypnosis in rats (Menon and Kar 1971).

The prevalence of knee pain osteoarthritis causing functional disability ranges from 22% to 39% in India with females more affected than men with increasing age (Pal et al. 2016). Seven randomized controlled trials from meta-analysis showed that Boswellia and its extract relieved pain and stiffness, improved joint function and were effective and safe option to treat osteoarthritis patients (Yu et al. 2020). Four (Sengupta et al. 2008, 2010; Vishal 2011; Haroyan 2018) out of the seven studies had low bias risk, while the remaining three trials had unclear bias risk due to ill-defined randomization procedures. Boswellic acids significantly reduce glycosaminoglycan degradation in contrast to non-steroidal anti-inflammatory drugs (NSAIDs) that disrupt glycosaminoglycan synthesis. Clinical trials of Boswellia gum-resin alone have shown symptom improvements in osteoarthritis, and rheumatoid arthritis patients (Siddiqui 2011 and references therein).

3-O-acetyl-11-keto-β-boswellic acid enriched (30% AKBA) *Boswellia serrata* extract (5-Loxin®) exerts anti-inflammatory activities due to 5-lipoxgenase enzyme inhibition. Mild to moderate osteoarthritis patients (52 years mean age) received low dose [100 mg/d 5-Loxin® (17F/7M, 26 kg/m^2 average BMI); high dose, 250 mg/d 5-Loxin® (15F/8M, 23 kg/m^2 average BMI)] treatment or placebo (18F/5M, 26 kg/m^2 average BMI) for 90 days (Sengupta et al. 2008). The treatment (low and high dose) clinically and significantly improved pain scores (by 49, 24 and 40% in VAS, LFI and WOMAC in the low dose), and physical ability scores (by 43 and 29% in WOMAC stiffness and functional ability) in OA patients. The high dose group improved VAS, LFI, WOMAC pain, stiffness and functional ability scores by 66, 31, 52, 62 and 49%, respectively. Moreover, the treatment relieved pain and improved physical function after the first week and continued thereafter during the course of the trial. The treatment also reduced (29 and 46% for low and high 5-Loxin® dose) MMP-3 concentration in synovial fluid (the cartilage degrading enzyme generally elevated in OA molecular pathogenesis). The treatment also reduced rescue medication (ibuprofen) use (17 and 72 % higher in the placebo group than in the low and high 5-Loxin® dose groups). 5-Loxin® (250 mg/d) promptly and significantly relieved pain and improved physical ability in OA patients faster than those reported for glucosamine (6 weeks). These clinical benefits in fast pain relief, improvement in physical ability and joint health most likely occur through downregulation of the cartilage degrading enzyme, such as MMP-3 (Sengupta et al. 2008).

The same group later revealed a more effective *Boswellia serrata* gum resin formulation Aflapin® than 5-Loxin® for knee osteoarthritis treatment (Sengupta et al. 2010). Aflapin®, an effective anti-inflammatory agent was compared to the efficacy and tolerability of 5-Loxin® in a double-blind, placebo-controlled, randomized study to treat knee osteoarthritis. Tibio-femoral middle aged (52 years mean age, 25 kg/m^2 average BMI) OA patients received either daily treatment (100 mg 5-Loxin® [16F/3M]; Aflapin® [12F/7M]) or placebo [10F/9M]) for 90 days. Both 5-Loxin® and Aflapin® conferred clinically and statistically significant pain relief and improved physical function in knee osteoarthritis patients. Fast acting Aflapin® improved pain score and functional ability as early as one week after study initiation by inhibiting cartilage degrading enzyme MMP-3 and ICAM-1 that regulates inflammatory response. For example, Aflapin® lowered pain score (VAS, and WOMAC) and WOMAC functional ability (13 vs 8%, 12 vs 9% and 10 vs 8% for 5-Loxin®, respectively) after one week treatment. The higher efficacy of Aflapin® may be attributed to its higher AKBA availability, and enzyme (5-lipoxygenase and MMP-3) inhibition compared to 5-Loxin®. Aflapin® contains at least 20% AKBA (vs 30% for 5-Loxin®) in addition to *B. serrata* non-volatile oil. Pain score (VAS, LFI and WOMAC) and WOMAC stiffness and functional ability scores improved by 47, 36, 62, 60 and 49%, respectively for patients in the Aflapin® group.

A 30-day double-blind, placebo controlled, randomized study assessed the validity of Aflapin® efficacy to manage clinical knee osteoarthritis symptoms (Vishal et al. 2011). Aflapin® (100 mg daily) improved pain score (~15% reduction in VAS and LFI) and functional ability in middle aged (53 years median age, 26 kg/m^2 average BMI) mild to moderate OA patients (19F/11M) as early as 5 days treatment compared to placebo (18F/11M). All pain scores (VAS, LFI and WOMAC), WOMAC stiffness and WOMAC function scores decreased significantly (49 vs 18%, 34 vs 12%, 50 vs 9% for placebo, 48 and 45%, respectively) after 30 days treatment. Aflapin® confers quick and significant pain relief and improves physical ability and quality of life of OA patients. Its therapeutic efficacy and fast action was attributed to mediation of the cellular and molecular mechanisms associated with AO pathological processes (Vishal et al. 2011). The most recent randomized, double blind, placebo controlled study (Karlapudi et al. 2023) confirmed that 100 mg Aflapin® daily ameliorated knee osteoarthritis symptoms as early as five days treatment. The 30 days treatment reduced pain scores (VAS, LFI and WOMAC by 45, 41 and 44%, respectively) and WOMAC stiffness, function and total scores (66, 44 and 48%, respectively). Aflapin® intervention also attenuated inflammatory and cartilage biomarkers (i.e., matrix metalloproteinase-3[MMP-3], tumor necrosis factor-α [TNFα], high-sensitive C-reactive protein [hsCRP] and collagen type II cleavage [C2C]). Aflapin® (Laila Nutraceuticals Research and Development Center, India) is a patented synergistic composition of AKBA enriched *Boswellia serrata* extract and its non-volatile oil that selectively inhibits 5-LOX enzyme. It has anti-inflammatory, cartilage-protective and anti-osteoarthritic effect with higher bioavailability than other commercially available *B. serrata* extracts.

Boswellin® tablet contains *Boswellia serrata* extract (BSE; 169.33 mg) with an average 87.3 mg of total β-boswellic acids (53.27 mg 3-O-acetyl-11-keto-β-boswellic acid [AKBA], 20.83 mg β-boswellic acid [BBA] and 6.06 mg α-boswellic acid [ABA]). Boswellin® significantly improved physical function of knee OA (mild to moderate severity) patients (31F/17M, 59 years median age, 23 kg/m^2 average BMI) by lowering pain and stiffness compared with placebo (Majeed et al. 2019). After 120 days, Boswellin® treatment (one tablet twice daily), reduced pain scores (42 vs 9%VAS for placebo, 52 vs 7% physician's global assessment for pain), WOMAC stiffness (23 vs 7%) and improved physical function (36 vs 5% for 6 min walk test score) and quality of life (42 vs 10% European Q5D quality of life). The treatment also reduced osteophytes (spur) and inflammatory marker (serum high-sensitive C-reactive protein) and improved knee joint gap. These Boswellin® benefits were attributed to the synergistic anti-inflammatory activity of BBA and AKBA in inhibiting 5-LOX and leukotriene biosynthesis resulting in reduced pain stiffness and physical discomfort (Majeed et al. 2019).

Topical *Boswellia serrata* (1 g dry extract) applied three times daily on affected knee for 4 weeks reduced pain severity and knee stiffness and improved knee OA (Kellgren/Lawrence grade II or III) patients (29F/4M, 60 years mean age) daily activity compared to control group (33F/4M, 57 years median age) (Mohsenzadeh et al. 2023). The control group had the same topical formulation (20 mL black seed oil and 80 mL olive oil) without *Boswellia serrata*. All evaluated outcomes (total WOMAC and its subscales, VAS, and PGA (patient global assessment) scores) decreased significantly in both groups, although the drug was more effective than the placebo. The beneficial effects of boswellia were attributed to the anti-inflammatory properties of boswellic acid (AKBA, particularly), even in the presence of standard knee OA treatment (500 mg acetaminophen, three times daily).

Casperome® (CSP, Indena S.p.A.) is one of several formulations developed to increase *Boswellia serrata* bioavailability by combining *B. serrata* extract (BE) with soy lecithin in a 1:1 ratio. Casperome® at equivalent weight dosage had significantly higher AUC for AKBA (4-fold), and AβBA, βBA, αBA and AαBA (2-fold) indicating enhanced absorption than BE in 12 healthy non-smoking subjects (6F/6M, 33 years mean age). This open-label, randomized, crossover study with two treatments, two sequences and two periods had 21 days washout period. Total content of the six major BAs was higher in the BE than the CSP capsule (177 vs 67 mg) due to the dilution factor associated with lecithin and microcrystalline cellulose in CSP (Riva et al. 2016). Casperome® also known as Monoselect AKBA™ (4.5 g daily, i.e., 9 x 500 mg tablets/day reduced and/or stabilized cerebral edema, influenced/reduced steroid consumption and increased quality of life and psychological well-being of primary glioblastoma multiforme (GBM) patients (11M/7F, 38 years mean age) (Di Pierro et al. 2019). GBM is a challenging neurological disease. Oral daily Casperome® (*Boswellia serrata* extract [≥25% boswellic acids] and soy lecithin; 1:1 ratio) supplementation for 4 weeks reduced intestinal pain (by ½; 66 vs 33% for control group 15M/6F) in ulcerative colitis patients (16M/6F, 52 years average age) during minimally symptomatic remission phase. Moreover, Casperome® attenuated the symptoms associated with symptomatic ulcerative colitis in remission and reduced the need for rescue drugs and medical consultations (Pellegrini et al. 2016).

Another formulation solid lipid *Boswellia serrata* particles (SLBSP) (WokVida™ capsules [333 mg equivalent to 100 mg of 40% boswellic acid [by HPLC]) is a solid lipidized (stearic acid) boswellic acid [BA] particle manufactured (Pharmanza Herbal Pvt. Ltd.) to address low BA availability. SLBSP improved knee osteoarthritis symptoms of patients (17F/5M, 58 years median age, mean baseline [6.4 WOMAC, 6 VAS and 10.3 rescue medications]) markedly without modifying the disease (radiographic OA features) and particularly reduced dependency on NSAIDs (Gota et al. 2015). The treatment (3 times daily) significantly improved WOMAC score (18.2 and 14.6% at one and two months, respectively) 30% in two months compared to baseline. It similarly reduced VAS score (19 and 26% at one and two months, respectively) 40% in two months compared to baseline. Dependency on rescue medication decreased (by 67% in the first month and a further 76% in the second month) 92% (overall reduction at 2 months) compared to baseline in this prospective, interventional study.

The efficacy of SLBSP was later compared to a standardized uncomplexed *Boswellia serrata* extract [BSE] (WokVel™, Pharmanza Herbal Pvt. Ltd.) with the same amount of BA in a prospective, interventional double blind, double dummy, placebo controlled clinical trial (Kulkarni et al. 2020). The SLBSP arm of the trial was the same as in the previous study (Gota et al. 2015) with the exact same results, whereas fewer patients participated in the BSE arm (13F/5M, 54 years median age). Both extracts significantly improved symptomatic relief in knee osteoarthritis over the two months study period without difference between the groups. However, SLBSP showed superior efficacy to BSE due to its reduced NSAIDs (100 mg Acelofenac) dependency and better control over proinflammatory cytokines (IL-2, IL-4 and IFN-γ).

Sontakke et al. (2007) compared the efficacy of the same *Boswellia serrata* extract (BSE) (WokVel™ containing 333 mg BSE thrice daily) with a selective COX-2 inhibitor (10 mg valdecoxib once daily) for six months in knee osteoarthritis patients (n=66) in a randomized, prospective, open label study. WOMAC pain and stiffness scores decreased significantly in the BSE treated group from the second month of intervention and onwards and these scores were significantly lower than with valdecoxib at the end of 7 months. In the valdecoxib group, WOMAC scores decreased significantly at the end of one month and persisted as long as the treatment continued. The onset of action with BSE was slower, but persisted even after treatment cessation, whereas valdecoxib acted faster but waned rapidly after intervention. BSE persistent effect was due presumably to the prevention of articular cartilage degradation, underlying osteoarthritis pathology. In an earlier study, *Boswellia serrata* extract (333 mg Wokvel™ capsule thrice daily) for 8 weeks reduced knee pain and swelling frequency and severity, and clinically improved knee function compared to placebo in knee osteoarthritis patients (18F/12M, 59 years mean age) (Kimmatkar et al. 2003). The treatment increased knee flexion, walking distance and improved knee movement range in this randomized double blind placebo controlled crossover (21 days washout period) study. Each Wokvel™ capsule contained 40% minimum boswellic acids [BA] (6.44% KBA, 2% AKBA, 18.51% β-BA, 8.58% A-BBA, 6.93% α-BA and 1.85% A-ABA). These BAs dose dependently inhibit leukotriene biosynthesis and reduce pro-inflammatory marker (5-LOX, 5-HETE [5-hydroxyeicosatetraenoic acid] and leukotriene B4 [LTB-4] levels. *Boswellia serrata* extract (4 tablets daily for 8 weeks) significantly improved WOMAC pain, stiffness and overall scores, VAS pain intensity score (30% improvement) and sleep efficiency (latency scores, wakefulness after sleep onset and number of awakenings) in healthy older adults (> 40 years) with persistent knee pain. The extract reduced pain synergistically with omega-3 fatty acid supplement (AvailOM®) in this randomized double blind and controlled clinical study (Pérez-Piñero et al. 2023).

Goni et al. (2021) compared the effects of Nucart VG tablet [(Gufic Bioscience Ltd) containing *Boswellia serrata* extract (600 mg) and glucosamine (750 mg)] with glucosamine sulphate (750 mg) on knee osteoarthritis in an open label, parallel group, randomized trial. Nucart VG (1 tablet twice daily post-meal for 3 months) significantly improved the health status (EuroQol-5D indicator, EQ-5D) and relieved pain of patients (21F/12M, 53 years median age, 23 kg/m^2 average BMI) with clinical unilateral or bilateral OA symptoms. Pain reduction (VAS and WOMAC [pain, stiffness and physical function] scores) was significant in the Nucart VG group compared to only glucosamine group even after the first month treatment (15 and 26% vs 8 and 17% for glucosamine sulfate). Moreover, the rate of pain reduction over treatment time was also steeper with Nucart VG compared to glucosamine sulfate (Goni et al. 2021). A boswellic acid formulation, MEBAGA (5 g methylsulfonylmethane and 7.2 mg boswellic acids) has also been compared for its efficacy to glucosamine sulphate (1.5 g) (Notarnicola et al. 2016). MEBAGA treatment (38F/17M, 59 years median age, 30 kg/m^2 average BMI)) improved visual analog pain scale (VAS, 40 vs 23%; 39 vs 21% for GS Lequesne pain index) and joint function (Lequesne index [LI], 32 vs 16% for GS) compared to glucosamine group (42F/15M) in arthritis (grade 3 Kellgren/Lawrence) patients.

Other formulations such as AflaB2® ([100 mg Aflapin® and 40 mg negative type II collagen], 1 capsule daily for 3 months) relieved pain and improved physical function in middle aged (23F/17M, 59 years mean age) knee osteoarthritis patients (Jain et al. 2020). VAS pain score and WOMAC pain, stiffness and physical function scores decreased (18, 11.5, 10.8 and 12.6%, respectively) after 5 days of treatment. These OA symptoms indicators decreased linearly with treatment duration in this pilot, observational, single arm, multicentre, open label, real world study. AflaB2® pain, inflammation and joint stiffness reduction was attributed to the dual action of its component Aflapin® and native type II collagen. Aflapin® acts on the mechanical pathway, whereas collagen mediates the immune pathway to control cartilage degradation and control OA symptoms (Jain et al. 2020). However, in primary human chondrocytes, Aflapin® increased glycosaminoglycans production indicating that *B. serrata* extract suppressed the cartilage matrix degradation in human cartilage explant model by inhibiting MMP-9 and -13 expressions (Ashruf and Ansari 2023 and references therein). Matrix metalloproteinase-13(MMP-13), a major collagenase is highly expressed in OA cartilage and promotes cartilage extracellular matrix (ECM) degradation. Osteoarthritis (OA) characterized by ECM causes joint pain and disability. Opera® (Gamfarma srl, Milan, Italy) a dietary supplement contains *Boswellia serrata* resin, α-lipoic acid, methylsulfonylmethane and bromolain (40, 240, 200 and 20 mg, respectively) in a single capsule, Opera ® capsule daily for 12 weeks reduced pain (~10% VAS scale) and neuropathic impairment in chemotherapy-induced peripheral neuropathy (CIPN) affected patients (19F/6M, 64 years median age, Karnofsky performance score [KSN] ≥70) (Desideri et al. 2017). Moreover, the dietary supplement improved CIPN symptoms in patients treated with neurotoxic chemotherapy without interaction and toxicity.

Curcuma longa and *Boswellia serrata* (combination) have been used to treat inflammatory, pain-related and degenerative disorders for ~ 4 millennia (Sethi et al. 2022 and references therein). Many clinical trials use *Boswellia serrata* resin extract or boswellic acids in combination with turmeric or curcuminoids, its active components. A systematic review reported that curcuminoid and boswellia formulations relieved pain and improved physical function effectively treating knee osteoarthritis compared to placebo (Bannuru et al. 2018). However, curcuminoids efficacy was not significantly different compared to NSAIDs, although gastrointestinal adverse events were less likely with curcuminoids. Four RCTs included in the review reported significantly beneficial functional outcomes of boswellia formulations compared to placebo with low overall quality of the evidence due to high bias risk. Boswellia formulations significantly reduced knee OA pain compared to placebo, although none of the studies compared these formulations with NSAIDs (Bannuru et al. 2018).

Haroyan et al. (2018) compared the efficacy of combined boswellic acid extract (Curamin®, 350 mg curcuminoids and 150 mg boswellic acid, 62F/5M) with curcuminoid complex extract (CuraMed®, 333 mg curcuminoids in 500 mg capsules, 60F/6M) to treat osteoarthritis in a three-arm, parallel group, randomized, double blind, placebo (65F/3M) controlled trial. Curamin® significantly improved the physical functions of patients (56 years median age, 29 kg/m^2 average BMI) with degenerative hypertrophic OA (Kellgren/Lawrence grade I-III) and relieved pain (WOMAC index and physical tests) compared to placebo after 3 months. CuraMed® treatment (daily dose of 3 capsules [1.5 g] orally for 12 weeks) also improved WOMAC joint pain index compared to placebo. However, the beneficial effect of Curamin®

was superior to CuraMed® presumably due to increased curcumin bioavailability and synergy of curcumin and boswellic acid. Curamin® capsule contains 350 mg BCM-95® and 150 mg *Boswellia serrata* gum resin (75% boswellic acids and 10% 3-O-acetyl-11-keto-boswellic acid [AKBA]). Each CuraMed® capsule contains 552–578 mg BCM-95® as a dry extract corresponding to 500 mg curcuminoids ([376 mg curcumin, 124 mg demethoxycurcumin and bisdemethoxycurcumin] and ~50 mg *Curcuma longa* volatile oil containing 22–23.4 mg aromatic turmerone) (Haroyan et al. 2018).

Curcuma longa and *Boswellia serrata* (CB) combined benefit to improve quality of life has been investigated in osteoarthritis patients (Pinzon et al. 2019). This controlled trial randomized middle age knee OA patients (Kellgren/Lawrence grade II or III) (63 years median age) into three groups. The treatment group (25F/9M) received *C. longa* and *B. serrata* extract (350 and 150 mg). The second group (26F/7M) was treated with NSAIDs (400 mg ibuprofen or 50 mg diclofenac). The third group (33F/5M) received the CB extract together with NSAIDs. Each medication was taken twice daily for 1 month. The CB with NSAIDs treatment (group 3) significantly improved quality of life, particularly pain/discomfort (P<0.001), mobility (P =0.002) and anxiety depression (P =0.008). The CB alone treatment significantly (P =0.013) improved usual activity, whereas pain/discomfort, anxiety/depression and usual activity improved significantly in NSAID group. CB extracts effectively reduced anxiety/depression, improved mobility, usual activity and self-care by 50%. It also had the least adverse events and reduced need of rescue medication compared to the other groups (Pinzon et al. 2019).

The strong analgesic activity of *Boswellia serrata* extract was recently demonstrated in effectively relieving exercise induced acute musculoskeletal pain in about 3 h (Rudrappa et al. 2022). The formulation of boswellia and turmeric extracts in sesame oil (Rhuleave-K, TBF) was evaluated for its analgesic effects in lowering exercise-induced acute musculoskeletal pain in healthy individuals. TBF (containing 266 mg curcuminoids and 10 mg acetyl keto-boswellic acid [AKBA] solubilized in sesame oil, Arjuna Natural Pvt. Ltd., Aluva, India) significantly reduced pain intensity and onset of pain relief and continuously improved pain relief. In this randomized, placebo controlled, double blind, single dose, single day, multicenter study, TBF (1 g) significantly (statistical and clinical) reduced acute pain intensity in adult participants (57F/57M, 37 years median age) with exercise induced acute musculoskeletal pain (≥ 5 numerical rating scale [NRS]). The significantly fast (~ 5 times perceptible pain relief [PPR] than placebo) onset of analgesia of the TBF group (99 vs 10% for placebo) indicated lower pain duration. Curcumin's analgesic (nociception) effect presumably occurred via antagonism of transient receptor potential vanilloid 1(TRPV1)-mediated pain. The analgesic activity of boswellia was attributed to boswellic acids, especially AKBA. Boswellic acids inhibit the transformation of arachidonic acid to leukotrienes via 5-lipoxygenase but can also enhance arachidonic acid liberation in human leukocytes and platelets. Hence, TBF's multiple natural anti-inflammatory ingredients can account for its analgesic effect to improve acute and neuroinflammation and modulate multiple pain pathways. The low number needed to treat [NNT (1.1)] for categorical pain relief scale [PRS] at rest, movement, and pressure indicates an excellent analgesic effect. TBF's significant fast pain relief onset (68.5 min PPR and 192 min meaningful pain relief [MPR]) indicates an effective and safe natural alternative to manage acute pain (Rudrappa et al. 2022). The same group previously showed that TBF (1 g daily for 7 days) reduced pain intensity to a level similar to acetaminophen with superior pain relief (over 8 times better McGill Pain Questionnaire) in healthy adults (42 years median age) with musculoskeletal pain (Rudrappa et al. 2020).

Chilelli et al. (2016) studied the effects of supplementing curcumin and *Boswellia serrata* gum resin for 3 months on plasma levels of oxidative stress markers, inflammation and glycation in healthy male cyclists. The supplemented nutraceutical tablet (Fitomuscle®, ForFarma, Rome, Italy) contained turmeric Pytosome® (50 mg turmeric corresponding to 10 mg curcumin) and *Boswellia* extract (140 mg corresponding to 105 mg boswellic acids). The supplement (one Fitomuscle® tablet daily) was administered to healthy cyclists (n = 25, 46 years median age, 24 kg/m^2 average BMI) along with a Mediterranean diet, and 22 cyclists were on the Mediterranean diet alone. The curcumin/*Boswellia serrata* resin gum extract supplementation significantly reduced oxidative stress (advanced glycation end-products [AGEs], -42 vs 4%) and lipid peroxidation (malondialdehyde [MDA], −0.10 vs −0.07 μmol/L) in chronically exercising master athletes. The benefit of curcumin/*Boswellia serrata* was attributed to their direct and significant effect on plasma lipid oxidation since no changes occurred in inflammatory status between the two groups (with and without supplementation). Advanced glycation products are known to accumulate with aging, particularly at the periphery of skeletal muscle fibers. However, the precise nature of AGE-induced protein modifications and their clinical significance in the aging process of skeletal muscle are currently unclear.

Clagen™ [Aflapin® (*Boswellia serrata* extract) + native type 2 collagen + Mobilee (hyaluronic acid [60–70%], polysaccharide (>10%), and collagen (>5%) + CureQlife (curcumin)] is considered a disease-modifying osteoarthritis drug (DMOAD). DMOADs have anti-inflammatory and analgesic properties to manage OA instead of NSAIDs inherent risk and adverse events (Kamat et al. 2023). Clagen™ (one capsule daily for 2 months) treatment significantly reduced VAS pain score (49%), improved range of motion (54% Goniometer knee extension; 16% knee injury and osteoarthritis outcome scores [KOOS] composite score) and KOOS for symptoms, function and quality of life (9.6, 9.8 and 7.8%, respectively) in knee osteoarthritis patients (35F/64M, 50 years median age, 24 kg/m^2 average BMI, Kellgren/Lawrence I-III radiological stages). Both curcumin and boswellia extract showed positive results in VAS and goniometer scale in this retrospective, observational study. Boswellic acid significantly reduces leukocytes infiltration in the knee joint, decreasing inflammation and preventing type II collagen degradation thereby relieving pain.

3.4.3 Mechanism of Action

Studies (animal and pilot clinical trials) support *B. serrata* gum resin extract (BSE) potential to treat various inflammatory diseases (inflammatory bowel disease [IBD], rheumatoid arthritis [RA], osteoarthritis [OA] and asthma) (Abdel-Tawab et al. 2011). In 2002, the European Medicines Agency classified (designated) BSE as an 'orphan drug' to treat peritumoral brain oedema. The pharmacological effects of BSE were mainly attributed to two boswellic acids (11-keto-β-boswellic acid [KBA] and 3-O-acetyl-11-keto-β-boswellic acid [AKBA]) that suppress leukotriene formation via 5-lipoxygenase [5-LOX] inhibition (Table 3.9). Two *Boswellia serrata* constituents, BBA and AKBBA synergistically exert anti-inflammatory/anti-arthritic activity to relieve pain and improve physical ability of knee osteoarthritis patients. These boswellic acids (AKBA and KBA) are also quality indicators of Indian frankincense (air dried *B. serrata* gum resin exudate) in European Pharmacopoiea 6.0 monograph. However, low AKBA and KBA

TABLE 3.9 Mechanism of action of β-boswellic acids against knee osteoarthritis

BOSWELLIC ACIDS	*ROLE/FUNCTION*
	INHIBITION
BBA	Cytokines (inflammatory interleukins: ILS[-1β,-6, -15, -17) and TNF-α
	Lipopolysaccharides (LPS)
	microsomal Prostaglandin E Synthase-1 (mPGES-1)
	Prostaglandin E2 (PGE2)
BBA, AKBBA	Cathepsin G (CatG)
	Reactive Oxygen Species (ROS)
	Nuclear factor-κB (NF-κB)
	Human leukocyte elastase (HLE)
	5-Lipoxygenase (5-LOX)
	High-sensitivity C-reactive protein (hs-CRP)
	Activation
BBA	Endothelial NOS (eNOS)/nitric oxide (NO)/
	cyclic guanosine 3′, 5′-monophosphate (cGMP) Pathways

Source: Adapted from Majeed et al. (2019); BBA—β-boswellic acid; AKBBA—3-O-acetyl-11-keto-β-boswellic acid

plasma concentrations (far below the effective concentration for *in vitro* bioactivity) failed to inhibit leukotriene formation in human whole blood. In contrast, plasma concentration of β-boswellic acid (β-BA) is high and abundant to inhibit microsomal prostaglandin E synthase-1 and the serine protease cathepsin G, enzymes that are potential targets for BSE anti-inflammatory properties (Abdel-Tawab et al. 2011).

In a phase IIa trial, BSE treatment (800 mg three times daily for 4 weeks) produced steady-state plasma concentrations of 6.35, 4.9, 0.33 and 0.04 μmol/L of β-BA, AβBA, KBA and AKBA, respectively in three patients. High fat meals increased KBA, AKBA and β-BA plasma concentrations 3, 6 and 6-fold, respectively. Earlier studies (1985–1990) indicated that H15 Gufic™ (400 mg *B. serrata* dry extract) effectively improved chronic arthritic symptoms (joint swelling and stiffness) without relieving pain. In clinical trials, BSE significantly relieved pain and improved physical function. Abdel-Tawab et al. (2011) speculated that the beneficial BSE effects observed in anti-inflammatory and arthritis studies of OA and RA are related to interference with PGE2 synthesis due to direct microsomal prostaglandin E synthase (mPGES-1) inhibition. These investigators suggested that Cathepsin G and mPGES-1 may serve as the major pharmacological targets for boswellic acid, especially for β-BA. Boswellic acids (BAs), the active principles of *B. serrata* modulate several key inflammation enzymes (5-lipoxygenase [5-LOX), cyclooxygenase-1 [COX-1], human leukocyte elastase [HLE], cathepsin G [catG], microsomal prostaglandin E synthase-1 [mPGES-1]), nuclear factor-kappaB (NF-κB) and inhibit cytokines (TNFα, IL-1β and IL-6) (Riva et al. 2016; Majeed et al. 2019).

Sanchez et al. (2022) investigated the mechanisms of action of *Curcuma longa* (CL) and *Boswellia serrata* (BS) extracts on primary human osteoarthritic chondrocytes at therapeutic plasma concentrations. Articular cartilage samples were obtained from 18 knee osteoarthritis patients (11F/7M, 65.6 years median age) at the time of total knee joint replacement surgery. Chondrocytes from ten patients (3F/7M, 61 and 75 years mean age for males and females, respectively) were used for RNA-sequence analysis. *Boswellia serrata* extract's anti-oxidant/detoxifying activities were associated with Nrf1 and PPARα pathways activation (Table 3.10). In contrast to *C. longa*, *B. serrata* extract modified fewer genes (156 differentially expressed genes [DEGs]) with genes involved in cartilage development and endochondrial ossification pathway and inflammatory response accounting for 37 and 26% of the total DEGs. Nrf1 genes (HMOX1 and metallothioneins) were most upregulated by *B. serrata* extract. *Curcuma longa* extract over-expressed anti-oxidative, detoxifying and cytoprotective genes involved in the Nrf2 pathway and suppressed pro-inflammatory cytokines and chemokines genes (Table 3.11). *Curcuma longa* modified 325 differentially expressed genes with those involved in inflammatory response and ossification pathway

TABLE 3.10 Genes modulated by *Boswellia serrata* extract

ROLE/FUNCTION	*GENE SYMBOL*	*GENE NAME*	*VARIATION*
Nrf1 pathway genes	HMOX1	Heme oxygenase 1	x 2.8
20 DEGs	MT1G	Metallothionein 1G	x 2.2
	MT1H	Metallothionein 1H	x 2.1
	MT1X	Metallothionein 1X	x 2.3
	MT1F	Metallothionein 1F	91%
	MT1M	Metallothionein 1M	86%
	FTL	Ferritin light chain	45%
Inflammatory response	CCRL2	C-C motif chemokine receptor like 2	165%
41 DEGs	CYP26B1	CytochromeP450 family 26 subfamily B member 1	110%
	NRROS	Negative regulator of reactive oxygen species	109%
	IL36RN	Interleukin 36 receptor antagonist	102%
	IL36B	Interleukin 36 beta	68%
	TNFRSF1B	TNF receptor superfamily member 18	66%
	IL6	interleukin 6	**–54%**

(Continued)

TABLE 3.10 (Continued)

ROLE/FUNCTION	*GENE SYMBOL*	*GENE NAME*	*VARIATION*
	TNFSF10	TNF superfamily member 10	**–54%**
	CCL7	C-C motif chemokine ligand 7	**–53%**
	CCL2	C-C motif chemokine ligand 2	**–45%**
	IL17RE	Interleukin 17 receptor E	**–45%**
Extracellular	SERPINA3	Serpin family A member 3	95%
13 DEGs	CTSL	Cathepsin L	60%
	MMP15	Matrix metallopeptidase 15	41%
	ADAMTS1	ADAM metallopeptidase thrombosponidin motif 1	**–41%**
	ADAMTS5	ADAM metallopeptidase thrombosponidin motif 5	**–40%**
Ossification Pathways	ANGPT2	Angiopoietin 2	133%
58 DEGs	LAMA3	Laminin subunit alpha 3	123%
	GREM1	Gremlin 1 DAN family BMP antagonist	123%
	COL21A1	Collagen type XXI alpha 1 chain	**–53%**
	CCN1	Cellular communication network factor 1	**–52%**
	COL3A1	Collagen type III alpha 1 chain	**–50%**
Autophagy Pathway	TRIM17	Tripartite motif containing 17	140%
24 DEGs	PLEKHM1	Pleckstrin homology and RUN domain containing 1	61%
	SQSTM1	Sequestosome 1	54%
	DEPP1	DEPP1 autophagy regulator	**–48%**
	EVA1A	Eva-1 homolog A, programmed cell death regulator	**–44%**

Source: Adapted from Sanchez et al. (2022); DEGs—Differentially expressed genes from specific pathways; Numbers denote DEGs modulated for each function/role

TABLE 3.11 Genes modulated by *Curcuma longa* extract

ROLE/FUNCTION	*GENE SYMBOL*	*GENE NAME*	*VARIATION*
Nrf2 Pathways	AKR1B10	Aldo-keto reductase family member B10	x 10
Detoxification Phase I	NQO1	NAD(P)H quinone dehydrogenase 1	x 3.6
8 DEGs	EPHX1	Epoxide hydrolase 1	x 2.7
Detoxification Phase III	SLCO2B1	Solute carrier organic anion transporter 2B1	x 4.2
2 DEGs	ABCB6	ATP binding cassette subfamily B member 6	x 3.7
Antioxidant GSH-based	GCLM	Glutamate-cysteine ligase modifier subunit	x 3
12 DEGs	GCLC	Glutamate-cysteine ligase catalytic subunit	x 2.1
	SLC7A11	Cysteine/glutamate transporter	x 2.3
Antioxidant TXN-based	TXNRD1	Thioredoxin reductase 1	x 2.7
NADPH regeneration	ME1	Malic enzyme 1	X 2.4
7 DEGs	PGD	Phosphogluconate dehydrogenase	X 2
Heme & Iron metabolism	HMOX1	Heme oxygenase 1	x 14.1
6 DEGs	FTL	Ferritin light chain	x 2.7
Chaperone/stress proteins	HSPA6	Heat shock protein family A (HSP70) member 6	X 8.6
22 DEGs	HSPA1A	Heat shock protein family A (HSP70) member 1A	X 3.3
	HSPA1B	Heat shock protein family A (HSP70) member 1B	X 3
	DNAJC6	Dnal heat shock protein (HSP40) member C6	X 2.7
	DNAJB4	Dnal heat shock protein (HSP40) member B4	X 2.2

(Continued)

TABLE 3.11 (Continued)

ROLE/FUNCTION	*GENE SYMBOL*	*GENE NAME*	*VARIATION*
	HSPB8	Heat shock protein family A (HSP70) member 8	X 3
Autophagy proteostatis	SQSTM1	Sequestosome 1	x 2.6
17 DEGs	FBXO30	F-box protein 30	X 2
Inflammatory response	CXCL6	C-X-C motif chemokine ligand 6	85.2%
111 DEGs	CXCL1	C-X-C motif chemokine ligand 1	82.3%
	IL6	Interleukin 6	81.5%
	GGT5	Gamma glutamyltransferase 5	80.6%
Extracellular degradation	HTRA3	Htr A serine peptidase 3	x 5
27 DEGs	MMP12	Matrix metallopeptidase 12	**–58.4%**
	ADAMTSL4	ADAMTS like 4	**–56.1%**
	MMP13	Matrix metallopeptidase 13	**–55.1%**
Ossification Pathways	ANGPT2	Angiopoietin 2	x 2.8
109 DEGs	GPC1	Glypican 1	63%
	LRRC15	Leucine rich repeat containing 15	**–74.6%**
	TNXB	Tenascin XB	**–65.5%**
	VCAN	Versican	**–65.2%**
	FGF1	Fibroblast growth factor 1	**–64.6%**

Source: Adapted from Sanchez et al. (2022); DEGs—Differentially expressed genes from specific pathways; Numbers denote DEGs modulated for each function/role

TABLE 3.12 Benefits of Boswellia and Curcuma combination

ROLE/FUNCTION	*GENE SYMBOL*	*GENE NAME*	*VARIATION (%)*
Pro-autophagy			
	PLEKHM1	Pleckstrin homology & RUN domain M1	64
	RUBCNL	Rubicon like autophagy enhancer	31
	SQSTM1	Sequestosome 1	28
	ATG2A	Autophagy related 2A	20
	FOXO1	Forkhead box O1	18
	LAMP2	Lysosomal associated membrane protein 2	16
Anti-oxidative			
	MT1X	Metallothionein 1X	100
	HMBG2	High mobility group box 2	31
	UCP2	Uncoupling protein 2	20
	LDLR	Low-density lipoprotein receptor	**–39**
	HMGCR	3-Hydroxy-3-methylglutaryl-CoA reductase	**–14**
Inflammation			
	GDF15	Growth differentiation factor 15	140
	IL-6	Interleukin 6	**–70**
	CCL2	C-C motif chemokine ligand 2	**–52**
	TLRs	Toll like receptors	**–20**
Anti-catabolism			
	ADAMTS-1	ADAM metallopeptidase with thrombosponidin motif 1	**–36**
	ADAMTS-5	ADAM metallopeptidase with thrombosponidin motif 5	**–28**

Source: Adapted from Sanchez et al. (2022). The variation is in comparison with *C. longa* alone

each contributing ~⅓ of the total. However, the highest upregulated genes were HMOX1, heme and iron metabolism, AKR1B10, Nrf2 pathway, and HSPA6, chaperone/stress protein genes.

Its anti-inflammatory effects were associated with GDF15 increase, reduction in cholesterol cell intake and fatty acid metabolism and suppressed Toll-like receptors (TLRs) activation. CL and BS down regulated ADAMTS [1, 5] and MMP [3, 13] genes expressions and protect cartilage due to their anti-oxidative, anti-inflammatory and anti-catabolic activities. *C. longa* and *B. serrata* have different time course effects on human OA chondrocytes to regulate antioxidative, detoxifying, anti-inflammatory, and anti-catabolic activities (Sanchez et al. 2022). For example, curcumin activates Nrf2 pathways (inducing HMOX1, antioxidant and cytoprotective gene expression), induces transcription of genes associated with iron/heme homeostasis, ROS detoxification, autophagy and metallothioneins. CL reduced cartilage catabolism in OA by inhibiting HTRA1 (serine protease involved in matrix degradation products) synthesis. *B. serrata* and *C. longa* combination prevents cartilage degradation by suppressing cartilage degradation (ADAMTS) genes and activating inflammatory response particularly inducing GDF15 gene expression (Table 3.12). The two extracts act on distinct pathways and with different time courses justifying their association to treat osteoarthritis (Sanchez et al. 2022).

3.5 TULSI—*OCIMUM SANCTUM*

Ocimum sanctum also known as *Ocimum tenuiflorum* (Tulsi) is an erect, multi-branch tall (30–60 cm) shrub with simple opposite strongly scented green or purple leaves. The ovate leaves have petiole (up to 5 cm long) and flowers are purple in elongate racemes in close whorls (Valsan et al. 2022 and references therein). Three generally popular *O. sanctum* forms are: Rama tulsi (green stems and leaves), Krishna tulsi (purple stems and leaves), and Vana tulsi (basic wild form) (Srivastava et al. 2018). Green *O. sanctum* variety (*Sri Tulasi*) had higher ursolic acid content than black variety (*Krishna Tulasi*) (0.35 and 0.48 vs ~0.26%, w/w) while oleanolic acid (0.17 and 0.22% w/w) was only present in the latter (Anandjiwala et al. 2006).

O. sanctum or tulsi is used in traditional or ethnobotany to reduce swelling and pain when applied to affected surface. The fresh juice from *O. sanctum* leaves is also applied to relieve headache and ear infection [myringitis] (Mandal et al. 2022). *Ocimum sanctum* dry leaf powder contained (%. w/w) moisture (2.26), water soluble extractives (18), alcohol soluble extractives (20.38), total ash (9.95), water soluble ash (3.24), acid insoluble ash (4.50), acid soluble ash (2.21), foreign organic matter (0.86) and eugenol (2.97) (Lalla et al. 2007). Several components of *O. sanctum* have great nutritional value; it is a good source of vitamin A (intake of 100 g fresh basil provides 175% of daily vitamin A requirement). It is also rich in iron (3.17 mg iron/100 g fresh leaves) (Mandal et al. 2022). Phytochemicals present in *O. sanctum* include 13 phenolic compounds (e.g., caffeic, chlorogenic, ocimumnaphthanoic, proyocatechuic and rosmarinic acids); 12 flavanoids (e.g., vicenin, orientin, cirsimaritin); 7 phenyl propanoids (e.g., eugenol, citrusin C, bieugenol); 7 neolignans (e.g., Tulsinols A-G); 3 coumarins (ocimarin, aesculetin and aesculin); 3 steroids (stigmasterol, campesterol and β-sitosterol); 12 terpenes (e.g., β-elemene, pinene, neral) (Mandal et al. 2022).

3.5.1 Composition

Saran et al. (2017) characterized ten *O. sanctum* plant accessions (Gujarat local selection) grown in western India and compared with check variety "Angana" (Uttar Pradesh) (Table 3.13). DOS-1, the basil accession had maximum dry leaf recovery (230 g/kg), total chlorophyll content (1.25 mg/g), carotenoid content (8.5 mg/g), oil content in green herbage (50 g/kg), oil and eugenol yield (73 and 67 kg/ha) and maximum eugenol elements (990 g/kg). Eugenol yield varied over four folds, whereas oil and essential oil

TABLE 3.13 Characteristic of ten *O. sanctum* accessions

TRAITS	*MIN*	*MAX*	*ANGANA*
Biological yield (kg/ha)	14,400	19,900	19,700
Green herbage yield (kg/ha)	10,100	16,300	16,900
Fresh leaf yield (kg/ha)	5500	7900	8000
Dry leaf yield (kg/ha)	1200	2300	1600
Eugenol yield (kg/ha)	16	67	44
Oil yield (kg/ha)	21	73	57
Leaf recovery (g/kg)	180	230	210
Essential oil yield (g/kg)	17	50	34
Eugenol element (g/kg)	730	990	770
Leaf area (cm^2)	7.0	12.6	9.5
Total chlorophyll (mg/g)	0.87	1.25	1.21
Chlorophyll A (mg/g)	0.71	1.00	0.95
Chlorophyll B (mg/g)	0.16	0.24	0.26
Carotenoids (mg/mL)	4.8	8.5	7.3

Source: Adapted from Saran et al. 2017

yields varied 3.5 and~3 folds among the plant accessions. The check variety "Angana" was superior only in green herbage and fresh leaf yields and chlorophyll B content compared to the Gujarati accessions. Eugenol was the predominant essential oil constituent followed by β-caryophyllene.

The Ocimum plant generates ~110 kg oil/hectare with basil oil price Rs. 650/kg and cultivation cost ~Rs. 23,546/ha (Srivastava et al. 2018). The oil contains phenols, aldehydes, tannins saponin and fats. The sweet basil is native to India, but also cultivated elsewhere (France, Italy, Bulgaria, Egypt, Hungary, South Africa and USA). Sweet basil (*Ocimum basilicum*) oil contains methyl chavicol, linalool and other minor constituents. India cultivates *O. basilicum* and *O. sanctum* on large scale, mostly for important essential oil in several states (Uttar Pradesh, Punjab, Haryana, and Madhya Pradesh). *O. sanctum* cultivars CIM-Ayu and CIM-Angana can produce 15–17 and 12–14 quintals of dry leaves per hectare (Srivastava et al. 2018).

Pandey et al. (2015) determined the polyphenolic contents of methanol leaf extracts from *O. sanctum* grown in four different geographical regions (Himachal Pradesh, Uttar Pradesh, Madhya Pradesh and West Bengal) in India (Table 3.14). A total of 8 samples were analyzed simultaneously by UPLC-ESI/MS/MS method, three each harvested in 2009 and 2011. The sample (OS-2) from Uttar Pradesh (2011) had the maximum amount of total phenolic acids because it had the highest concentration of protocatechuic, caffeic and ferulic acids. Ursolic acid was the most abundant constituent, whereas epicatechin and catechin were below detection level. Rosmarinic acid (35 folds), ferulic acid, caffeic acid and apigenin (~12 folds) content varied the most among the samples. Hakkim et al. (2007) reported higher sinapic acid and eugenol contents (54 and 70 mg/100 g, dm) and minimum rosmarinic acid (25 mg/100 g) for *O. sanctum* leaves from plants field grown in Coimbatore, Tamilnadu. The leaves also contained carnosic acid and isothymusin (19 and 14 mg/100 g) with high antioxidant activity (hydroxyl radical and superoxide activities ~70% at 1 mg/mL) (Hakkim et al. 2007).

Phenolics and flavonoids were major contributing markers among nine chemical classes of metabolites for species-specific variation in leaves from five *Ocimum* species including *O. sanctum* (Rai et al. 2023). *O. sanctum* samples had high amounts of flavonoids, phenolics, anthraquenones, sterols, sugars, and aromatic compounds compared to other *Ocimum* species. For example, apigenin, apigenin-7-*O*-glucuronide, kaemferol and cirsillineol were present in high concentrations in *O. sanctum* leaves. Moreover, *O. sanctum* had relatively high amounts of the phenolic metabolite coniferaldehyde (eugenol precursor) and protocatechuic acid (Rai et al. 2023). *Ocimum sanctum* fresh leaves and stems extract contain

TABLE 3.14 Polyphenolic contents (mg/100 g) of *Ocimum sanctum* leaf extracts

POLYPHENOLS	*MIN*	*MAX*
Phenolic acids		
Caffeic acid	14.0	167.3
Chlorogenic acid	30.8	35.0
Ellagic acid	10.9	83.0
Ferulic acid	35.7	436.7
Gallic acid	25.3	94.3
Protocatechuic acid	52.3	158.3
Rosmarinic acid	22.7	800
Sinapic acid	21.0	39.7
Vanillic acid	5.4	11.6
Total	233	1578
Flavonoids		
Apigenin	12.4	146
Chrysin	7.6	15.5
Kaempferol	20.4	22.6
Kaempferol-3-O-rutinoside	19.9	22.2
Luteolin	62.0	271.3
Quercetin	3.9	4.9
Quercetin dihydrate	16.5	21.6
Quercetin-3,4′-diglucoside	29.7	32.6
Rutin	9.5	29.8
Scutallerein	2.6	9.4
Total	214	533
Eugenol	4.6	18.6
Ursolic acid	210	1613

Source: Adapted from Pandey et al. 2015

antioxidant phenolic compounds (cisilineol, circimaritin, isothymusin, apigenin and rosmarinic acid) and appreciable amounts of eugenol. Two flavonoids (orientin and vicenin) and other compounds (ursolic acid, apigenin, luteolin, apigenin-7-*O*-glucuronide, luteolin-7-)-glucuronie, orientin and molludistin) have been isolated from leaf extract (Pandey and Mahduri 2010 and references therein).

O. sanctum phenolics (i.e., rosmarinic acid, apigenin, cirsimaritin, isothymusin, and isothymonin) exhibit antioxidant and anti-inflammatory activities and are effective ACE inhibitors (Uma Devi et al. 2000). The flavonoid, cirsimaritin possesses anti-inflammatory activity; the flavone C-glycoside, orientin offers neuroprotection, while Vicenin-2 is a potential anti-inflammatory constituent (Rastogi and Shasany 2018 and references therein).

3.5.2 Essential Oil Composition

Ocimum sanctum essential oil contains eugenol, methyl chavicol, camphor, linalool, and methyl cinnamate (Srivastava et al. 2018). Eugenol ranges between 40 and 70% of the essential oil composition in *O. sanctum* leaves. That eugenol concentration decreases significantly during the monsoon season, at times decreasing up to 0% from 80%. Essential oil composition of *O. sanctum* leaves is highly variable even within the same cultivars depending on the location and year of plant growth (Table 3.15).

TABLE 3.15 Composition (%) of *O. sanctum* essential oil

CULTIVARS	*EO (%V/W)*	*EUGENOL*	*METHYL EUGENOL*	*β-ELEMENE*	*E-CARYO-PHELLENE*	*GERMA-CRENE D*	*β-SELINENE*	*REFERENCE*
CIM-Ayu	0.36	68.0	4.7	4.1	8.4	0.3	0.8	Verma et al. (2013)
CIM-Angana	0.15	54.9	0.3	10.2	4.0	5.3	0.6	
CIM-Kanchan	0.53	1.5	79.7	4.5	5.0	nr	0.1	
Shyama	0.41	0.6	67.8	4.3	17.1	1.4		Awasthi and Dixit (2007)
Rama	0.43	46.2	nd	16.3	27.6	0.1		
CIM-Ayu	0.50	43.0		1.4	2.6	nr	29.1	Saroj and Krishna (2017)
CIM-Angana	0.40	78.3		1.0	1.5	nr	7.4	
Local Rama	0.30	23.5		0.7	0.9	nr	27.0	
Local Shyama	0.20	62.8		1.6	0.8	nr	11.3	
Krishna Tulsi	0.20	0.9	82.9	0.5	4.1	2.3		Joshi and Hoti (2014)
CIM-Ayu		67.4	t	11.0	7.3	2.4	0.1	Padalia and Verma (2011)
"Purple"		72.8	0.2	10.9	8.4	2.2	0.2	

EO—Essential oil values are on fresh weight basis, nr—not reported, nd—not detected, t—trace amount

Samples were collected from foot and mid-hill regions of north India (Purara, Bageshwar, Uttarakhand) and the essential oil was estimated on fresh weight basis and composition determined by GC/MS (Verma et al. 2013). Eighteen constituents were identified representing 93, 88 and 94% of total oil composition for CIM-Ayu, CIM-Angana and CIM-Kanchan cultivars, respectively.

Saroj and Krishna (2017) obtained their samples from local farmers of the semi-arid Hamirpur region, Uttar Pradesh where farmers are organic *O. sanctum* growers. These investigators identified only 11 compounds by GC/MS, including significant amount of Muurola-4,(10)-dien-8β-ol (17.8, 9.1, 28.7 and 18.6%), R-Cembrenene (2.5, 9.3, 3.3 and 0.14%) and Sabinene (6.3, 6.7, 0.4 and 0.2%) for cultivars CIM-Ayu, CIM-Angana, Local Rama and Shyama, respectively (Padalia and Verma 2011). Camphene, α-copaene and β-costal were other minor essential oils reported for these cultivars (Saroj and Krishna 2017). Eugenol and E-caryophyllene contents were similar, although β-elemene values were different for CIM-Ayu reported by Verma et al. (2013) and Padalia and Verma (2011). Hydro distilled essential oil composition of *O. sanctum* 'Green' (CIM-Ayu) and 'Purple' cultivars (grown in Uttarakhand) was similar; predominantly phenylpropanoids (68.1 and 73.5%) represented primarily by eugenol (67.4 and 72.8%). The other major classes were sesquiterpene hydrocarbons (~22%) comprised of β-elemene (~11%), β-caryophyllene (7.3 and 8.4%) and germacrene D (~2.3%). Both oils are considered 'eugenol' types (Padalia and Verma 2011). Major essential constituents of Krishna Tulsi from northwest Karnataka were methyl eugenol (82.9%), β-carophyllene (4.1%), borneol (2.4%), germacrene D (2.3%) and α-copane (1.9%). Phenyl derivative (83.8%) was the predominant constituent, followed by sesquiterpene hydrocarbons (11.1%), oxygenated monoterpenes (3.1%), monoterpene hydrocarbons (0.6%) and oxygenated sesquiterpene (0.3%) (Joshi and Hoti 2014). Cultivar Shyama consisted of 34 essential oil constituents comprising 98.3% of the oil containing methyl eugenol (67.8%) and E-caryophyllene (17.1%). Thirty nine constituents were identified in cultivar Rama comprising 98% of the oil consisting of eugenol, E-caryophellene and β-elemene (46.2, 27.6 and 16.3%, respectively) (Awasthi and Dixit 2007).

Flowering aerial part of *O. sanctum* (Krishna Tulsi) grown in northwest Karnataka, India, consisted primarily of phenyl derivative compounds (83.8%) rich essential oil. Twenty six essential oils constituents comprised 98.9% of the total oil with methyl eugenol as the major compound (82.9%) and sesquiterpene hydrocarbons (11.1%) comprised of 4.1% β-caryophyllene, 2.4% borneol, 2.3% germacrene D and 0.5% β-elemene (Dubey and Pandey 2018). Gupta et al. (2018) described characteristics and essential oil composition of some *O. sanctum* varieties. For example, CIM-Ayu is a high-yielding eugenol-rich oil producing variety

[83% eugenol, 7.5% β-elemene]) developed through mass selection. CIM-Angana (dark purple pigmented variety developed through half-sib selection) contains 40.4% eugenol, 14% β-elemene, 9% β-caryophyllene and 16.7% germacrene D. The essential oil of CIM-Kanchan (developed through selection) consists of 70% methyl eugenol, 7.6% β-elemene and 15.7% β-caryophyllene). A hybrid strain of *Ocimum gratissimum* known as "Clocimum" developed by recurrent selection and released by RRL (Regional Research Laboratory, Jammu) contains 60–65% eugenol. Another heterotic F1 Ocimum strain "Clocimum-3c" trade name "RRL-og-1" contains 90–95% eugenol (Prakash and Gupta 2005). Eugenol is known to inhibit cyclooxygenase (97% cyclooxygenase inhibitory activity at 1000 μM concentration) comparable to NSAIDs, ibuprofen, naproxen and aspirin (10, 10 and 1000 μM concentrations) (Pattanayak et al. 2010).

3.5.3 Analgesic Activity

Ocimum sanctum is recommended in Ayurvedic medicine to treat arthritis and other diseases. Its leaves are used for their analgesic, anti-inflammatory, antipyretic and other activities (antifungal, antibacterial, antioxidant, anticancer). Aqueous extract of *O. sanctum* also exerts immunomodulatory activity by inducing antibody production that enhances immune cells and haemoglobin production. The immunostimulant activity of the leaves is attributed to its main constituents (phenolic compounds, eugenol, methyl eugenol and caryophyllene) (Valsan et al. 2022 and references therein). Aqueous extract of fresh *O. sanctum* leaves (0.3, 0.6 or 0.9 mL equivalent to 0.21, 0.43 and 0.64 g leaf) injected intraperitoneally exerted analgesic activity in rats (Rathore and Jain 2013). The moderate dose (0.6 mL) increased reaction time (44% 'rat tail method') from initial value indicating long lasting (effective at 3 h) pain relief. Maximum pain relief occurred at the maximum dose (0.9 mL) with the highest increased reaction time (52%) with peak analgesic effect at 90 min and retained up to 3 h. The increase in mean reaction time was dose dependent (Rathore and Jain 2013).

Methanol extract and aqueous suspension of *Ocimum sanctum* leaves possess anti-inflammatory, analgesic and antipyretic activities attributed partly to inhibition of prostaglandins biosynthesis. *Ocimum* methanol extract (500 mg/kg) inhibited acute and chronic inflammation (carrageenan-induced pedal edema and granuloma pouch) equivalent to the response observed with sodium salicylate (300 mg/kg), a known prostaglandin biosynthesis inhibitor (Godhwani et al. 1987). Ethanol (80%) extract of *O. sanctum* leaves (500 mg/kg) had maximum analgesic effect (hot plate test of white albino mice) at 3 h post administration comparable to the standard drug Ketorolac (16 vs 15.3 sec post drug reaction time for Ketrorolac [2.5 mg/kg i.p.]). The extract inhibited carrageenan induced inflammation (30 vs 67% paw oedema for diclofenac sodium) with duration of action comparable to that of the standard drug diclofenac sodium (Hannan et al. 2011).

Petroleum ether extracted *O. sanctum* seed oil (3 mL/kg) had antipyretic activity (febrile response reduction in rats) comparable to aspirin attributed to its prostaglandin inhibitory activity (Singh et al. 2007). The analgesic activity of that oil (3 mL/kg) was demonstrated by its acetic acid induced writhing inhibition (~75%) comparable to aspirin (100 mg/kg) (Singh and Majumdar 1995). The oil also possessed significant anti-inflammatory property against phlogistic agents (i.e., carrageenan, prostaglandin [PGE2] and histamine). The seed oil analgesic activity is peripherally mediated resulting from the combined inhibitory effects of prostaglandin, histamine and acetylcholine. *O. sanctum* seed oil (3 mL/kg) has beneficial antiarthritic activity in all inflammation models (Freund's complete adjuvant-, formaldehyde-, and turpentine-induced arthritis) comparable to aspirin (100 mg/kg, i.p.) due to transaminase enzymes (glutamic-oxaloacetic transaminase [SGOT] and glutamic-pyruvic transaminase [SGPT]) inhibition (Singh et al. 2007). The beneficial effects of *O. sanctum* seed oil are attributed to its linolenic acid that inhibits arachidonic acid metabolism accounting for its analgesic and anti-inflammatory activities. The fixed oil blocks cyclooxygenase and lipoxygenase pathways of arachidonic acid metabolism. *Ocimum sanctum* oil possesses significant anti-inflammatory activity against carrageenan and other mediator-induced paw edema in rats. This effect is attributed to the dual inhibition of arachidonic acid metabolism due to selective inhibition of the cyclooxygenase and lipoxygenase pathways (Pandey et al. 2014).

Neuropathic pain is generally characterized by sensory abnormalities (i.e., dysesthesia [unpleasant abnormal sensation], hyperalgesia [increased response to painful stimuli], and allodynia [painful response to {non-noxious} stimuli that does not normally prove pain]). Among chemotherapeutic agents, vincristine uniquely produces predictable and uniform neurotoxicity in all patients even at therapeutic doses. Traditionally, *Ocimum sanctum* has been used as nerve tonic and to alleviate joint pain, headache and muscular pain particularly in South India. *O. sanctum* also has demonstrated ameliorative potential in attenuating sciatic nerve neuropathic pain (Kaur et al. 2010).

Treatment with *O. sanctum* (100 and 200 mg/kg p.o.) and its saponin rich fraction administered for 14 days significantly attenuated oxidative stress and calcium levels in vincristine-induced neuropathic pain in rats. Neuropathic pain relief effects of *O. sanctum* are attributed to attenuation of nerve injury inciting agent due to increased calcium and free radicals contents (Forouzanfar and Hosseinzadeh 2018; Kaur et al. 2015). This beneficial effect in neuropathic pain is attributed to *O. sanctum's* saponin rich fraction that includes pentacyclic triterpenoids (i.e., ursolic and oleanolic acids). The saponin rich fraction also attenuated cold allodynia, mechanical and thermal hyperalgesia significantly higher than the hydro-alcoholic extract at the same concentration. The same group later demonstrated that *O. sanctum* and its saponin rich fraction (100 and 200 mg/kg) for 14 days also attenuate peripheral neuropathic pain due to chronic constriction injury (Kaur et al. 2010). The saponin rich fraction showed greater potency in attenuating chronic constriction injury induced neuropathic pain similar to the previous study.

Improved transdermal ursolic acid (UA) delivery of *Ocimum sanctum* leaf extract (OLE) with loaded nano-structured lipid carriers (OLE-NLCs) treated arthritis as effectively as diclofenac gel (Ahmad et al. 2018). The prolonged UA release (up to 12 h) probably due to high entrapment efficiency (90%) was superior in drug permeation (~2.7 enhancement ratio) compared to commercial formulation. Supercritical carbon dioxide extracted (SFE; 200 Psi, 45 °C, 1.5 h) ocimum leaf was used to load the nanostructured lipid carriers by solvent evaporation followed by ultrasonication. The *in vitro* anti-arthritic OLEC-NLC gel formulation activity was comparable (87 vs 85% inhibition of protein denaturation) to those of diclofenac gel (1.16%). Moreover, the OLE-NLC gel formulation exhibited significantly higher analgesic activity (35 vs 20 sec thermal stimulus time) compared to control drug formulation due to higher skin penetration. The OLE-NLC gel formulation significantly inhibited the immune response in complete Freund's adjuvant (CFA) induced arthritic rats due to reduced vascular permeability caused by inflammatory mediator (i.e., PGE2) inhibition. Furthermore, the OLE-NLC gel formulation inhibited COX-1 and COX-2 enzymes and reduced IL-1 and TNF-α level. Radiological studies showed good arthritis recovery from the formulation attributable to its therapeutic efficacy and improved skin penetration (Ahmad et al. 2018).

Khanna and Bhatia (2003) evaluated the analgesic activity and potential mechanism of action of 70% ethyl alcohol extract (10–100 mg/kg) from *O. sanctum* leaves. The leaf extract (50 and 100 mg/kg) administered orally and intraperitoneally reduced glacial acetic acid-induced writhing and radiant heat induced tail flick in mice. The analgesic effects acted both centrally and peripherally. Moreover, the receptor antagonist opioid naloxone and the central noradrenaline depletory DSP-4 (*N*-2-chloroethyl-*N*-ethyl-2-bromobenzylamine) significantly reduced *O. sanctum*'s analgesic activity indicating its involvement in the neurotransmitter system. The serotonin depletory PCPA (D-parachloro-phenylalanine methyl ester hydrochloride) potentiated the analgesic activity of the extract in the tail flick test suggesting serotogenic involvement and interactions with other neurotransmitters in pain management (Khanna and Bhatia 2003).

Ethyl acetate fraction from methanol extract of *Ocimum sanctum* roots (STE) displayed the most potent anti-inflammatory, analgesic and anti-pyretic activity in a dose dependent manner presumably due to flavonoids presence (Kumar et al. 2015a). Analgesic profile of STE was evaluated using chemical (i.e., acetic acid and formalin) and thermal (i.e., hot plate and tail immersion) induced pain model. STE (100 and 300 mg/kg) inhibited carrageenan induced paw edema (46 and 61%), protected against thermally induced pain (31 and 51% vs 71% for diclofenac sodium, a standard NSAID), significantly reduced acetic acid induced writhing (34 and 53% vs 71% for 10 mg/kg diclofenac sodium), exerted analgesic activity (time spent paw licking) (22 and 52 vs 77% for diclofenac sodium) and anti-pyretic activity (temperature reduction) (37 and 47% vs 54% for paracetamol [150 mg/kg] after 5 h) (Kumar et al. 2015c). STE acted centrally and peripherally.

3.5.4 Human Clinical Trials

Jamshidi and Cohen (2017) reviewed the literature of human trials on *O. sanctum*'s clinical efficacy and safety. They identified 24 studies with therapeutic effects on metabolic disorders, cardiovascular disease, immunity and neurocognition. However, only 7 studies were considered high quality (Jadad scores 4–5). Nevertheless, published reports indicate that *O. sanctum* intervention can help normalize glucose, blood pressure and lipid profiles and improve psychological immunological stress and potentially prevent or reduce various health conditions. *O. sanctum* leaves powder (1 g daily for 30 days) supplementation significantly lowered blood glucose (21%), glycated serum proteins (11%), serum lipids [total cholesterol, low-density and very low density lipoprotein-cholesterol, and triglycerides], uronic acids (14%) and total amino acids (14%) in non-insulin-dependent diabetes mellitus (NIDDM) patients (17M/10F, 54 years median age, 25 kg/m^2 average BMI) compared to control group (5M/5F) (Rai et al. 1997). The leaf powder (1 g) provided 0.49 g carbohydrate, 0.23 g crude fiber, 0.20 g protein and 0.07 g fat. The NIDDM patients were diabetic for 3 to 12 years and continued their regular treatment (glibenclamide, glipizide, penformin and chloropropamide) during the study. The beneficial hypoglycaemic and hypolidaemic activity of the leaf powder was attributed to its bioactive components (i.e., fiber, saponins, eugenol and others) (Rai et al. 1997). Treatment with holy basil leaves significantly lowered fasting and postprandial blood glucose (17.6 and 7.3%) and urine glucose levels in NIDDM patients in a randomized placebo-controlled, single blind trial (Agrawal et al. 1996). Holixer™, a proprietary (patent pending [202241023540]) hydroalcoholic extract from *O. tenuiflorum* leaves (250 mg daily for 8 weeks) buffered physiological and subjective responses evidenced by significantly lower concentrations in salivary cortisol, blood pressure and subjective stress in stressed adults (34F/16M, 46 years median age, 26.5 kg/cm^2 average BMI) (Lopresti et al. 2022). The treatment reduced oxidative stress induced by the MAST—Maastricht acute stress (cold water 7 °C) test in this randomized, double blind placebo controlled trial.

Mondal et al. (2011) evaluated the immunomodulatory effects of *O. sanctum* leaves ethanol (70%) extract in a double blind, randomized controlled cross-over trial. The extract (300 mg capsule) was administered on empty stomach for four weeks followed by three weeks wash out period prior to crossover to the next intervention in twenty two young healthy volunteers (27 years median age, 23 kg/m^2 average BMI). The *O. sanctum* intervention significantly increased blood cytokine release (IFN-γ and IL-4) and lymphocytes (NK-cells) indicating an effective immune response. The beneficial immunomodulatory effects were attributed to flavonoids present in the extract (Mondal et al. 2011). *Ocimum sanctum* (Tulsi) compounds (i.e., ursolic acid, carnosol, rosmarinic acid, cirsilineol, apigenin, eugenol, and crisimaritin) upregulate IL-2 and IFN-γ, and downregulate IL-1β, that represent a major defense mechanism to assess T-cell dependent antibody response (Singh et al. 2012). However, analgesic effects of *Ocimum* have not been investigated in human clinical trials.

3.6 BLACK CUMIN—*NIGELLA SATIVA*

India is known to be the largest producer of Nigella in the world. The other producing countries are Sri Lanka, Bangladesh, Nepal, Egypt, Iraq and Pakistan. Indian cultivation of *Nigella sativa* spreads over several states (Punjab, Himachal Pradesh, Madhya Pradesh, Gangetic plains of Bihar and Bengal, Assam and Maharashtra) (Dubey et al. 2016). Area of production is estimated at ~9000 ha producing 7000–8000 tons of black cumin in India (Huchchannanavar et al. 2019). The global black seed market is expected to reach USD 31.06 million by 2027 (from 21.39 million in 2021), with 6.41% CAGR according to Marketwatch (2023).

In the Indian system of medicine, *Nigella sativa* seeds are used for various therapeutic activities like astringent, stimulant, diuretic, anthelmintic, jaundice, fever, dyspepsia, paralysis, piles and also in skin disease (Tiwari et al. 2019). *Nigella sativa* accession from India and Pakistan are very close from genetic diversity point of view. Moreover, Indian *Nigella sativa* accession is genetically diverse from those of Egypt and Oman based on UPGMA (unweighted pair group method with arithmetic mean) method using RAPD (random amplified polymorphic DNA) among eight germplasm from different geographies (Sudhir et al. 2016). Shariq et al. (2015) characterized seed traits of *Nigella sativa* germplasms collected from ten different states in India. Seeds varied in size (2.83–3 mm length, 1.66–2 mm width) and average weight (2.16–2.4 g).

3.6.1 Composition

Black seed oil is a natural pain killer and used as an antiseptic and analgesic remedy to treat joint pain and stiffness. *Nigella sativa* seed oil market size is estimated to surpass USD25 million in 2025 with demand for cosmetic applications and personal care projected to surpass 700 tons (Ramadan 2021). The seed oil consists of monoterpenes, carbonyl compounds, phenols, alcohols and esters (~46, 25, 1.7, 0.9 and 16%, respectively) (Dubey et al. 2016). Saxena et al. (2017) reported the fatty acid composition of twenty three *N. sativa* genotypes collected from different regions (i.e., Rajasthan, Uttar Pradesh, Madhya Pradesh, Punjab, Kota and Gujarat) in India and grown at the experimental farm in Ajmer, Rajasthan, India during the *rabi* season (Table 3.16). The main fatty acids were linoleic, palmitic and oleic acids. Polyunsaturated fatty acid was the dominant *N. sativa* oil component due to high linoleic acid. Myristic, heptadecanoic and arachidic acids were present at low levels (<1%) in all genotypes. Monounsaturated fatty acids varied significantly among the genotypes, although all except oleic acid content were below 1% level. The tert-butyl methyl ether (MTBE) extract of the Indian *N. sativa* cultivar 'Ajmer Nigella-1' grown in Andhra Pradesh (2013–2014) contained 2.39% cis-13,16-docosadienoic acid along with linoleic (68.14%), palmitic (12.14%), oleic (6.61%), stearic (1.9%) and myristic acids (0.21%) (Kalidasu et al. 2017). One *N. sativa* sample sourced from India contained the lowest linoleic acid (50.24%), along with palmitic, oleic, stearic and gondoic acids (10.83, 19.09, 2.47 and 2.25%, respectively) (Thilakarathne et al. 2018). Seed oil from the twenty three *N. sativa* genotypes exhibited free radical scavenging capacity (11–48%) due to the presence of antioxidants (2.4–14.8 μg BHT equivalent/mL), total phenolic (129–212 μg GAE/mL) and flavonoid (739–818 μg QE/mL) (Saxena et al. 2017).

TABLE 3.16 Fatty acid composition (%) of *Nigella sativa*

FATTY ACIDS (%)	*MIN*	*MAX*
Myristic acid	0.03	0.24
Palmitic acid	9.3	15.1
Palmitoleic acid	0.22	1.66
Margaric acid	0.21	0.79
10-Heptadecenoic acid	0.21	0.9
Stearic acid	2.02	3.91
Oleic acid	6.8	19.0
Linoleic acid	60.9	71.4
α-Linolenic acid	0.28	1.01
Arachidic acid	0.09	0.24
Eicosadienoic acid	2.98	8.09
ΣSaturated fatty acids	12.1	18.1
ΣMonounsaturated fatty acids	8.0	19.0
ΣPolyunsaturated fatty acids	65.1	77.1

Source: Adapted from Saxena et al. 2017

3.6.2 Essential Oil

Nigella sativa essential oil produces significant analgesic activity primarily due to the presence of active compounds (i.e., thymoquinone). Thymoquinone (18%) the major compound in *N. sativa* essential oil exhibited the most effective antifungal activity (7 and 8.5 MIC against *Penicillum digitatum* and *Colletotrichium gloeosporoides*) due to strong binding with fungal protein (i.e., α-tubulin) (Akansha et al. 2023). Black cumin seeds from Punjab had an average 0.66% (fresh weight basis) essential oil yield consisting of 48 compounds. The essential oil contained terpenoids (39.4%), cycloalkane derivatives (12.6%), aldehydes (10.6%), fatty acid esters (9.1%), phenols (7.5%), esters (7.5%) and other minor components. Predominant essential oils consisted of thymoquinone (18.1%), 2, 10-dimethyl-9-undecenal (10.6%), 1,1-dichloro-2, 2,3,3-tetramethyl-cyclopropane (9.6%) and *p*-cymene (5.6%) (Table 3.17). Higher *p*-cymene (36.4%) and thymoquinone (29.8%) have previously been reported for *N. sativa* essential oil. These differences in essential oil composition are primarily due to genetic and environmental factors (Akansha et al. 2023).

Major volatile compounds in Indian *N. sativa* seeds were *p*-cymene (41.8%), α-thujene (13.93%), thymoquinone (10.27%), methyl linoleate (4.02%), carvacrol (3.65%), β-pinene (2.96%), d-limonene (2.11%), 4,5-epoxy-1-isopropyl-4-methyl-1-cyclohexane (1.8%), sabinene (1.5%) and 4-terpineol (1.22%). Terpine hydrocarbons and terpene ketones were the major volatile oil classes (63 and 11%) (Kabir et al. 2020). Essential oil of *N. sativa* seeds purchased from Gorakhpur local market, Uttar Pradesh, consisted of 33 components (representing 90% of the total amount), primarily thymoquinone (37.6%), *p*-cymene (31.2%), α-thujene (5.6%), thymohydroquinone (3.4%) and longifolene (2%). The total phenolic content (Folin-Ciocalteau method) of *N. sativa* essential oil was 11.5 mg GAE/g with strong antibacterial activity (90% zone inhibition against *Fusarium moniliforme*) (Singh et al. 2014). Major *N. sativa* essential oil components of Indian origin comprised α-pinene, β-pinene, β-cymene, and 4-caranol and characterized by the highest β-chamigrene (9.54%) content. Principal component analysis classified the essential oil chemotype as 4-caranol/β-cymene/α-phelandrene/β-chamigrene (Dalli et al. 2021).

Forty seven volatile compounds have been identified in *N. sativa* seeds (from Mysore) extracted by supercritical carbon dioxide (SF1–28 MPa/50 °C), (SF2–12 MPa/40 °C) and hydrodistillation of SF1 [SFE-1] (HD SFE). The major compounds of SF1, SF2 and SFE-1 were thymoquinone (35, 33 and 38%, respectively), γ-terpinene (27.5, 13.2 and 12.9%), thymol (7.4, 5.3 and 17%), β-caryophyllene (2.9, 5.1 and 4.8%) and thymohydroquinone (1.2, 1.1 and 2.3%) (Tiruppur Venkatachallam et al. 2010). Thymoquinone content in *Nigella sativa* seeds varied from 42.88 µg/100 mg seed [4.28%](Ajmer Nigella-13) to 247.60 µg/100 mg seed [24.76%](Ajmer Nigella-6) among 40 accessions collected from different agro-climatic areas of India (14 Indian states) and grown in one experimental field (Ajmer).

Thymoquinone was undetected/below detection level in two genotypes, Ajmer Nigella-21 and Ajmer Nigella–27. The genotypes and elite varieties segregated into four distinct clusters based on their thymoquinone content (Ravi et al. 2023).

Aqueous, methanol and ethanol *N. sativa* extracts as well as *N. sativa* oil and thymoquinone can relieve pain mediated through mu-and kappa-opioid receptor. The pain relief effect could be due to the antioxidant and anti-inflammatory properties and microglia inhibitory activity (based on anti-nociceptive properties of thymoquinone demonstrated in animal studies) (Beheshti et al. 2016). Thymoquinone has been suggested as a potential treatment agent for osteoarthritis since it mediates signaling pathways in human osteoarthritis chondrocytes as well as several other pathways (Huseini et al. 2016). It is the primary bioactive component of *N, sativa* responsible for its anti-rheumatoid arthritis properties demonstrated in animal models. It reduces levels of inflammatory cytokines (IL-1β, IL-6, tumor necrosis factor-α [TNF-α], interferon-gamma [IFN-γ] and prostaglandin E2 [PGE2]), and oxidative stress markers (i.e., articular elastase, myeloperoxidase, lipoxygenase, nitric oxide [NO], cyclooxygenase-2[COX-2] and metalloproteinase 13[MMP-13]) and significantly increases anti-inflammatory cytokines (IL-10) and enzymes activities (i.e., superoxide dismutase [SOD], glutathione peroxidase [GPx] and catalase [CAT) (Hadi et al. 2021).

Black cumin seed extracts (BCSEs) can protect against collagen degradation because they are considered as antiglycation agents and collagenase inhibitors. Glycation is an internal chronic stress for collagen. Thymocid®, a commercially produced cold pressed BCSE is chemically standardized to thymoquinone

TABLE 3.17 Essential oil composition (%) of *N. sativa* seeds of Indian origin

VOLATILES	*PEAK AREA %*			
COMPOUNDS	*A*	*B*	*C*	*D*
Thymoquinone	18.07	10.27	37.6	
p-Cymene		41.8	31.4	14.22
β-Pinene		2.96	1.7	8.59
Carvacrol	1.13	3.65	1.4	
4-Terpineol (Terpinen-4-ol)	0.29	1.22	1.0	3.35
α-Thujene		13.93	5.6	
d-Limonene		2.11	1.0	
4-Caranol				18.81
α-Phellandrene				10.58
β-Chamigrene				9.54
Thymol				4.64
β-Phellandrene				4.16
Thymohydroquinone			3.4	
α-Pinene			1.4	
α-Terpinen				1.67
γ-Terpinene			0.2	6.38
Methyl linoleate	2.36	4.02		
Neodihydrocarveol				3.34
2, 10-Dimethyl-9-undecenal	10.55			
1,1-Dichloro-2,2,3,3-tetramethylcyclopropane	9.58			
Butylhydroquinone	5.58			
Trans-4-Methoxy thujane	4.29			
D-Longifolene	4.25		2.0	
Succinic acid, dec-4-enyl ethyl ester	3.72			
(5E)-2,3,5,8-Tetramethyl-1,5,9-decatriene	2.9			
1-Monopalmitin, 2 tms derivative	2.82			
3-Cyclopentylpropionic acid, 2-dimethylaminoethyl ester	2.57			
2,2,4-Trimethyl-3–(3,8,12,16-tetramethyl-heptadeca	2.4			
α-Longipinene				2.23
β-Cyclocitral				2.1
2,3,5-Trimethylhydroquinone (trimethylhydroquinone)	1.9			
Pulegone	1.83			
Sabinene		1.5		
1-Bromotetracosane	1.48			
Propyleneglycol monoleate	1.46			

Source: Data for samples A, B, C and D adapted from Akansha et al. 2023; Kabir et al. 2020; Singh et al. 2014; Dalli et al. 2021, respectively

(TQ) content (5.12%). Thymocid® (50, 100 and 300 μg/mL) reduced fructose induced advanced glycation end products (AGEs) in a BSA model dose dependently (17, 33 and 71%, respectively) and inhibited methylglyoxal (MGO) induced collagen cross-linking (45, 93 and 93%, respectively vs 12.5% for the positive control aminoguanidine at 100 μg/mL). Thymocid® (62.5–1000 μg/mL) also protected bovine type I collagen degradation from glycation induced cross linking (10.4–92.4% reduced collagenase activity)

(Li et al. 2020). This antiglycation and collagen protection of Thymocid® can potentially be helpful in osteoarthritis management.

Clinical studies (4 oral and 1 topical) confirm that thymoquinone in *N. sativa* fixed oil has valuable analgesic and inflammatory effects in rheumatoid arthritis patients via different pathways (Mehboubi et al. 2018). Poly (ε-caprolactone) based nanoparticles loaded with *N. sativa* seeds essential oil is used to enhance indomethacin's analgesic and anti-inflammatory effects (Soltane et al. 2021) due to effective encapsulation efficiency (70 and 84%) and drug loading (14 and 5.6% for indomethacin and *Nigella sativa*, respectively). Phospholipid vesicles (liposomes) have also been used as natural *N. sativa* crude oil nanocapsules to develop analgesic creams formula applicable in cosmetics (Soltane et al. 2021 and references therein).

3.6.3 Clinical Studies

Several animal studies demonstrate *N. sativa* analgesic activity mediated by its active components (i.e., essential oil primarily thymoquinone and fatty acids). Here, we only present human studies relevant to *N. sativa* pain relief. Nineteen *N. sativa* oil studies related to rheumatoid arthritis were included five human clinical trials in a systematic review (Khabbazi et al. 2020). One study included both human and animal models. *N. sativa* oil supplementation significantly reduced rheumatoid arthritis by reducing disease activity score 28 (DAS28) compared to placebo in 4 clinical trials (Hadi et al. 2016; Kheirouri et al. 2016; Gheita and Kenawy 2012; Mahdy and Gheita 2009).

Nigella sativa oil intervention (5 mL *N. sativa* oil syrup [2.5 mL cold pressed oil, 1.25 mL honey and 1.25 mL water] every 8 h for 1 month) significantly reduced WOMAC scores for pain (23 vs 4% for placebo), stiffness (33 vs 7%), function (30 vs 0.3%) and total (28 vs 1.3%) and visual analog pain scores (34 vs 9%) in primary knee osteoarthritis patients (Kellgren/Lawrence grade 1 or 2) (42F/16M, 60 years median age, 30 kg/m^2 average BMI) (Huseini et al. 2016). The placebo contained mineral oil instead of *N. sativa* oil and the group consisted of slightly older patients (45F/13M, 63 years median age). The *N. sativa* oil contained 0.97 mg/mL thymoquinone, along with fatty acids (59, 28 and 6.7% polyunsaturated, saturated and monounsaturated fatty acids, respectively). The intervention (~7.275 mg thymoquinone daily for one month) did not alter the secondary outcomes, liver function (ALT—alanine aminotransferase, AST—aspartate aminotransferase, ALP—alkaline phosphatase) and kidney function (creatinine and BUN—blood urea nitrogen). WOMAC and VAS reduction (20 and 15%) by the *N. sativa* intervention were clinically relevant since they were higher than minimal clinically important differences from baseline. The pain relief of the cold pressed *N. sativa* oil in knee osteoarthritis is presumed to be mediated by its bioactive components (i.e., thymoquinone and fatty acids). Thymoquinone's analgesic effect is due to activation of the supra-spinal opioid receptors and its anti-inflammatory action exerted by enzymes (i.e., COX and LOX) inhibition; these enzymes are also inhibited by unsaturated fatty acids. Moreover, the intervention significantly reduced the use of rescue medication (500 mg acetaminophen) (1.2 vs 3.1 acetaminophen tablet taken daily during trial) (Huseini et al. 2016). The fatty acid of cold pressed *N. sativa* oil consisted of linoleic (49.4%), palmitic (20.3%), oleic (5.8%) and stearic (5.5%) acids and small amounts (<1%) of palmitoleic, arachidic and myristic acids.

N. sativa seed powder (2 g daily for 12 weeks) showed insignificant difference in KOOS score results and rescue medication (500 mg acetaminophen tablets) use between treatment (24F/13, 55 years median age, 31 kg/m^2 average BMI) and placebo groups (34F/16M) in knee osteoarthritis (42 mean KOOS pain score) patients (Salimzadeh et al. 2017). The powdered *N. sativa* seeds were soaked in vinegar (1:2 w/v, 24 h), dried and re-powdered, sieved (14 mesh), then filled into two piece red opaque gelatin capsule. Processing eliminated thymoquinone from the crude *N. sativa* seed (1.7%) and increased longifolene (10.4 vs 3.3% for crude seed) and sesquiterpenoid (12.8 vs 4%) content without affecting fatty acid content. The insignificant *N. sativa* (processed seed) efficacy between the treatment and placebo group (62.9 vs 59.2 VAS, P = 0.31) presumably reflects the importance of thymoquinone to relieve pain in knee osteoarthritis patients.

Tuna et al. (2018) investigated black cumin oil's effect on elderly osteoarthritis patients (23F/7M, 68 years median age) suffering from knee pain (7.5 VAS pain score). Topical application (gentle rub without massaging) of black cumin oil (a dessert spoon [~11 mL]from 30 mL bottle) for 20–30 min 3 times weekly for 30 days effectively relieved knee osteoarthritis pain (7.5 vs 6.3 VAS pain score, -4%, $P < 0.001$) compared to insignificant change in the control group with routine prescription. In a crossover clinical trial, topical *N. sativa* oil application every 8 h for 1 weeks followed by 1 month washout period relieved pain in elderly patients (22F/18M, 70–76 years old) with knee osteoarthritis (Kooshki et al. 2016). The control group received three acetaminophen tablets (325 mg) for 3 weeks. *N. sativa* oil exhibited significantly higher pain relief intensity (4.23 vs 4.76 for acetaminophen, $P = 0.01$) than oral acetaminophen in elderly knee osteoarthritis patients with moderate pain. This confirms an earlier study where *N. sativa* oil significantly reduced pain in 40 women compared to placebo (Gheita and Kenawy 2012). In that study, cold pressed *Nigella sativa* oil (1 g daily for one month) after 1 month placebo (starch) treatment improved disease activity in 40 female RA patients (43 years median age, 28 kg/cm^2 average BMI). Patients were on combined therapy (16 mg/week methotrexate, 200 mg/day hydrochloroquine, 5 mg folic acid [after methotrexate injection] and 50 mg once or twice daily diclophenac sodium). *N. sativa* oil supplementation significantly reduced morning stiffness (44%), swollen joints (41%), VAS pain (12%), disease activity score (8.8%) and white blood cell count (12%) (Gheita and Kenawy 2012).

T-lymphocyte cells (CD4+, CD8+ and CD4+CD25+) are involved in the induction and/or protection of rheumatoid arthritis. Cold pressed *N. sativa* seed oil (1 g for 2 months) treatment significantly reduced serum high-sensitivity C-reactive protein (hs-CRP) level and disease activity scores of 28 joints (DAS28, 18.7%) and improved numerous swollen joints compared with baseline and placebo groups (n =20, 41 years median age, 24.5 kg/m^2 average BMI) in 23 female rheumatic arthritis patients (43 years median age, 25 kg/m^2 average BMI) (Kheirouri et al. 2016). *N. sativa* oil supplementation significantly reduced CD8+ T cell percentage that was positively associated with disease activity. All patients had the same medical treatment (10 mg methotrexate weekly, 200 mg hydroxychloroquine daily and prednisolone therapy) during this randomized, double blind, placebo-controlled, parallel-group clinical trial. This study demonstrates that *N. sativa* oil treatment improves disease activity via the immunomodulatory effect on T cells in rheumatic arthritis patients (Kheirouri et al. 2016).

Cold pressed *Nigella sativa* oil (3 g daily for 8 weeks containing 1.25 mg thymoquinone/100 mg oil) with calorie restricted diet reduced inflammatory biomarkers (TNF-α [41 vs 16% for placebo) and high sensitivity C-reactive protein [hs-CRP; 55 vs 21%]) in forty three obese women (41 years median age, 32 kg/m^2 average BMI) in a randomized double blind, placebo-controlled clinical trial (Mahdavi et al. 2016). However, interlukin-6 reduction (8.6 vs 2.4%, P=0.6) was insignificant compared to the placebo group receiving sunflower oil (with similar/comparable fatty acid content [57% linoleic acid and 28% oleic acid]) instead of *N. sativa* oil.

Nigella sativa oil [palmitic, oleic, linoleic and linolenic acids; 9.7, 20.9, 46 and 18.2%, respectively] (2 g with 100 mg vitamin E daily for 2 months) significantly reduced erythrocyte sedimentation rate [anti-inflammatory indicator; therapeutic effect towards rheumatoid arthritis], plasma prostaglandin E2 level [inflammatory mediator], erythrocyte superoxide dismutase activity [oxidation stress reduction], and plasma creatinine in 28 rheumatoid arthritis male adult patients. The oil improved rheumatoid arthritis presumably due to increased antioxidant state and simultaneous reduction of reactive oxygen species (Al-Okbi et al. 2000). Elevated serum creatinine level is associated with methotrexate therapy in patients with chronic inflammatory diseases. Oil administration reduced morning joints stiffness (~25%) caused by synovial congestion, effusions and thickening of joint capsules. *N. sativa* oil treatment (1 g daily for 8 weeks) increased serum IL-10 level and reduced oxidative stress [serum MDA and NO] and disease activity score [DAS28] compared with baseline in23 women (44 years median age. 25 kg/m^2 average BMI) with mild to moderate RA under (DMARDs) treatment (methotrexate, hydroxychloroquine and < 10 mg prednisone daily). However, differences between *N. sativa* oil and placebo groups were insignificant in cytokine [TNF-α] level, and antioxidant defense system [TAS, SOD, and catalase] in this randomized, double blind, placebo-controlled clinical trial (Hadi et al. 2016). NS oil intervention reduces oxidative stress in RA patients.

N. sativa oil (1 g daily for two weeks) after two weeks placebo (starch) significantly reduced mean DAS-28 (4.5%), Ritchie articular index [RAI] (10%), morning stiffness (28%) and WBC count (12%) in 36 female rheumatoid arthritis patients (42 years median age, 27 kg/m^2 average BMI); VAS pain also decreased by ~10%. The study indicated that *N. sativa* seed oil concomitantly with methotrexate therapy relieves pain, improves daily function and reduces joint inflammation including swelling and tenderness thereby improving quality of life for rheumatoid arthritis patients. These anti-rheumatoid arthritis benefits are attributed to bioactive constituents of *N. sativa* oil, thymoquinone in particular (Mahdy and Gheita 2009).

Three Iranian studies (Hadi et al. 2016; Kheirouri et al. 2016; Kooshki et al. 2016) and one in Egypt (Gheita and Kenawy 2012) demonstrate the therapeutic effects of *N. sativa* oil on rheumatoid arthritis. *N. sativa* has similar effects in rheumatic arthritis patients to those observed in animal studies. *N. sativa* oil (500 mg capsules daily for 8 weeks) significantly reduced MDA, NO, high-sensitivity C-reactive protein (hs-CRP) level, cytotoxic T-cells (CD8+), disease activity score-28 (DAS-28), and number of swollen joints compared with baseline and placebo (Hadi et al. 2016; Kheirouri et al. 2016). *N. sativa* oil also increased serum IL-10, CD4+ T-cell percentage, T regulatory cell percentage (CD4+CD25+ T-cell) levels and CD4+/CD8+ ratio. These studies indicate that *N. sativa* shifts T-helper lymphocyte cell type 1 (Th1) to type 2 (Th2) thereby improving RA. *N. sativa* affects the immunomodulatory system as evidenced in a study where *N. sativa* oil (1 g daily for 4 weeks) intervention significantly reduced DAS-28, joint inflammation and morning stiffness (Gheita and Kenawy 2012).

3.7 CURRENT AND FUTURE PERSPECTIVES

Eighty three families of medicinal plants (out of 242 plants from 184 genera) are used to relieve 25 types of pain in 19 indigenous tribes in four states of Western Himalayan region (Jammu and Kashmir, Himachal Pradesh, Ladakh, and Uttarakhand) (Balkrishna et al. 2022). *Artemisia* genus and *Saussurea costus* were the primary analgesic plant species used in powder, decoction, and paste forms by tribal people, in spite of the high analgesic potential of leaves and roots of these medicinal species. Ayurvedic preparations that rely heavily on native plants/crops can reduce pain and increase function in osteoarthritic people and help manage symptoms in type-2 diabetics according to the National Center for Complementary and Integartive Health (NCCIH)'s overview (Platkin et al. 2022).

Terminalia chebula, commonly called black myrobalan, ink tree or chebulic myrobalan is a medium to large deciduous tree from the combretaceae family. It grows in deciduous forests in Himachal Pradesh, Tamil Nadu, Kerala, Karnatka, Uttar Pradesh, Andhra Pradesh and West Bengal. The dried fruit is extensively used in Ayurvedic medicine because of its therapeutic potential to treat or prevent many diseases (Ashwini et al. 2011). Chebulagic acid isolated from *Terminalia chebula* is a strong antioxidant and inhibits COX-LOX. Kumar et al. (2015b) evaluated the analgesic activity of single oral dose of *Terminalia chebula* using hot air pain model in 12 healthy male adults (33 years median age, 23 kg/m^2 average BMI) in a randomized, double blind, placebo controlled cross over (after two weeks washout) trial. The standardized aqueous fruit extract of *Terminalia chebula* (1 g) [containing not less than ($\nless$) 15% chebulinic acid, $\nless$10% chebulagic acid and $\nless$15% other low molecular weight hydrolysable tannins] significantly increased mean pain threshold (20%) and tolerance time (17.5%) compared to baseline and placebo. The analgesic activity of *T. chebula* aqueous extract is attributed to its bioactive components known to inhibit many key constitutive and inducible enzymes (i.e., COX-2, 5-lipooxygenase, NO) involved in pain pathways.

Commiphoa mukul concentrated extract (3.5% guggulsterones) (1.5 g daily for 1 month) significantly improved WOMAC scores (total 29%, pain 32%, stiffness 27% and function 27%), VAS function (51%) and total symptom count (21%) in knee osteoarthritis (Kellgren/Lawrence ≥ 2 for one knee; 20F/10M, 65 years median age) patients (Singh et al. 2003). These improvements continued at follow-up (one month post treatment). Other secondary measures (i.e., 6 min walk) showed progressive benefits, but not statistically different from baseline. Guggulesterone is known to reduce the elevated inflammatory protein in arthritis. Abha gugglu is a polyherbal Ayurvedic medicine containing *Terminalia chebula, Commiphoa mukul* along with other herbs. Abha guggulu (AG) (3 g daily for 30 days) significantly improved WOMAC total score in twelve adults (40–70 years old) osteoarthritis patients. The improvement in this quasi-experimental single arm trial is attributed to synergistic action of the herbal bioactive compounds (Hedaoo et al. 2023).

Peganum harmala oil treatment (4 drops 3 times daily for 4 weeks) significantly relieved/lowered pain and improved function in knee osteoarthritis patients (24F/3M, 54 years median age, Kellgren/Lawrence grade II or III) (Abolhassanzadeh et al. 2015). Pain reduction of the *Peganum* treatment was over twice compared to the olive oil control (27F) group (53 vs 17% VAS scale and 38 vs 16% WOMAC). The treatment also significantly increased total and function WOMAC scores (38 vs 16% for the control group) without significant changes in WOMAC stiffness scores in this double blind controlled randomized clinical trial. The *P. harmala* oil contained 0.0025% harmaline and 0.057% harmine (based on HPLC analysis) responsible for the analgesic activity. These alkaloids have both central and peripheral antinociceptive activities presumably through opioid receptors) (Abolhassanzadeh et al. 2015).

Coriander (*Coriandrum sativum* L.) an ancient herb of the Apiaceae (Umbelliferae) family is the main ingredient (in the form of fruit or seeds) of curry powder. India is a major player in coriander cultivation, although coriander presumably originated in the Near East. It has also been used traditionally in India to treat digestive, urinary, and respiratory systems disorders (Arituluk 2022). Rajeshwari et al. (2012) investigated the antioxidant and antiarthritic activities of coriander leaf powder (5 g daily for 60 days) in osteoarthritis patients. *Coriandrum sativum* treatment significantly reduced elevated oxidative stress (i.e., lipid peroxidation, increased erythrocyte catalase activity, decreased serum β-carotene and vitamin C), glutathione content (GSH) and alkaline phosphatase activity and increased erythrocyte antioxidant enzymes (i.e., glutathione-S-transferase [GST]) and serum calcium levels. The treatment significantly reduced lipid peroxidation in erythrocyte (42 %) and plasma (20%), and improved kidney function [reduced serum creatinine (6 vs 27 % for control) and urea (11 vs 19%)], reduced glutathione (43 vs 6%) and glutathione-s-transferase (62 vs 4%). *C. sativum* controlled oxidative stress (increased β-carotene and vitamin C, erythrocyte antioxidant enzymes (GST), decreased lipid peroxidation and increased antiarthritic efficacy (decreased erythrocyte sedimentation rate [ESR]). These effects are attributed to the bioactivity of the various leaf compounds (i.e., phenolic acids, flavonoids, other secondary metabolites) (Rajeshwari et al. 2012).

Tetracyclic triterpenoids (cucurbitanes) and their glycosides (momordicine and momordicoside) are the main phytochemical constituents of *Momordica charantia.* Soo May et al. (2018) determined the effectiveness of *Momordica charantia* (bitter melon) supplementation (4.5 g daily for 3 months) in pain reduction, improvement of symptoms and the quality of life in knee osteoarthritis patients (23F/15M, 62 years median age, 27 kg/m^2 average BMI, Kellgren/Lawrence grades 1–3) in a single blind, randomized placebo controlled study. Patients were prescribed rescue medication (1 g acetaminophen tablet for mild to moderate pain, 200 mg celecoxib capsule for severe pain or 50 mg tramadol tablet if unable to tolerate celecoxib). *M. charantia* (entire bitter melon fruit) supplementation significantly reduced (~24%) the analgesic KOOS (knee injury and osteoarthritis outcome score), KOOS pain and improved symptoms and function and VAS pain score. These beneficial pain management effects and improved quality of life were achieved with reduced rescue medication. *M. charantia* supplementation significantly reduced pain and rescue medication intake, improved symptoms, daily activities and quality of life (Soo May et al. 2018).

Agarwal (2013) disclosed a synergistic phytochemical composition comprising *Curcuma longa* oil and *Curcuma longa* polysaccharide rich water extract (1 and 99% by weight, respectively) possessing analgesic and anti-inflammatory activity to manage chronic pain and treat osteoarthritis. The composition of the invention was evaluated for its effectiveness to treat painful osteoarthritis in a randomized, single blind, placebo controlled trial. Osteoarthritis patients (83F/37M) received either the composition (1 g daily), or glucosamine sulphate (1.5 g daily) or placebo (1 g daily) or combination of the composition and glucosamine sulphate for 42 days. The treatment significantly reduced WOMAC and VAS scores compared to placebo and use of rescue medication. Moreover, osteoarthritis signs and symptoms improved with glucosamine sulphate and its combination with the composition (Agarwal 2016).

Antony (2022) developed a dispersion of *Boswellia* and/or *Curcuma longa* extracts in sesame seed oil to enhance the analgesic and anti-inflammatory properties of the composition. One of the composition BSC-4 contained 54 g sesame lignans (from sesame oil), 300 g AKBA and 12 g KBA (from *Boswellia*), 884 g curcuminoids and 70 g turmeric oil (from *C. longa*) and 2000 ppm antioxidant. This composition was evaluated for its efficacy in adult patients with acute musculoskeletal pain in a randomized active controlled study. Patients (n=88) with musculoskeletal pain (≥ 5 score) were randomized (1:1), provided with BSC-4 or paracetamol and asked to rate their pain severity (numerical rating scale for pain [NRS]) before and half-hourly post-dose up to 6 h. BSC-4 significantly relieved pain at par with paracetamol at all tested times demonstrating that it can act as an alternative to paracetamol for pain (Antony 2022).

A similar turmeric-boswellia-sesame formulation (Arjuna Natural Pvt. Ltd., Aluva, India) relieved menstrual pain in a double-blind, randomized, placebo-controlled study in healthy women with primary dysmenorrhea (Agarwal and Chaudhary 2023). The active ingredient of the formulation were 95% turmeric extract (28% curcuminoids), 10% *Boswellia serrata* extract (10% AKBA) and 62% sesame oil. A single day single dose (1 g) of the formulation at the beginning of menstruation (n=30, 26 years median age) significantly relieved pain (over 12-fold) (18.9 vs 1.5 [placebo] mean total pain relief on the numerical rating scale) and pain intensity half-hourly post-dose until 6 h. Moreover, the treatment lowered total pain intensity (sum of the pain intensity at 6 h) over 20-fold compared to placebo (34 vs 1.7 rating for placebo). Furthermore, the treatment had significantly higher maximum (≥ 70%) mean total pain relief than placebo (80 vs 17% respondents). The treatment group (73%) rated the formulation as an excellent pain reliever based on the global evaluation categorical scale. The unique composition of the formulation relieves menstrual pain by selectively modulating hormones that sense pain as well as reducing the elevated prostaglandins caused by dysmenorrhea (Agarwal and Chaudhary 2023).

The Belgian company Fermedics (Brown 2023) has two traditional lactic acid fermented products (ashwagandha and curcumin) discussed in this chapter. Fermanolide™ (standardized to contain >1.5% total withanolides, >800 mg withanolide A) is fermented ashwagandha with increased withanolide A content (by ~70% post fermentation; 3 times more withanolide A than non-fermented ashwagandha) by metabolizing withaferine A, thereby eliminating genotoxic potential. It has high bioavailability, rapid absorption in the small intestine requiring small dose (150 mg) under normal circumstances (https://fermedics.com/ingredients/fermanolide/). Fermeric™ (standardized to contain >2.3% curcuminoids, >1.4% curcumin, >20 billion CFU/g paraprobiotics) is highly bioactive turmeric obtained by traditional lactic acid fermentation. This fermentation increases curcumin content (12.7%), absorption (17.2 and 6.3 times within 24 and 48 h) and apparent permeability, consequently its antioxidant activity. Lower daily doses (300–500 mg) are generally sufficient to provide high efficacy of bioactive substances than unfermented turmeric (https://fermedics.com/ingredients/fermeric/).

Many analgesic/natural pain relief herbs are available in the market in analgesic topical forms (i.e., essential oil, cream [for direct application at pain site], capsules, tinctures and tea) (Gautam et al. 2013). Many plant-derived compounds have been or are currently in preclinical and clinical trials; these plants include amla, ashok, aswagandha, brahmi, long pepper, sandalwood, gritkumari, and neem, where the secondary metabolites are widely used for health-related problems. Unfortunately, osteoarthritis (OA) in India increased from 23.46 to 62.35 million individuals in 19 years (1990–2019) (Singh et al. 2022). The age standardised prevalence of OA increased from 4895 to 5313 per 100,000 during that period. In 2019, OA was the 20th most common cause of years lived with disability (YLDs) in India accounting for ~1.5% of all YLDs with knee osteoarthritis as the most common OA form followed by hand OA. The prevalence, incidence, and YLDs for OA and knee OA were consistently higher in females than males with highest level of OA prevalence in Uttar Pradesh, Maharashtra and West Bengal. The burden and impact of osteoarthritis in India are substantial and is increasing. Preventive measures are required to reduce modifiable risk factors (i.e., obesity, injury) as well as therapeutic treatments without side effects. These therapeutic treatments must include the native analgesic plants, their extracts and/or their bioactive components in pharmaceutical/drug formats. Seventeen to eighteen thousand plant species have been documented with over 7000 used in folk medicine and documented AYUSH system of medicine (Ayurveda, Unani, Siddha and homeopathy). The consumption of nearly 242 species of medicinal and aromatic plants exceeds 100 metric tons/year.

Many indigenous medicinal plants have been evaluated for their analgesic activity, mostly in animal model primarily by the hot plate method and/or acetic acid induced writhing in mice. This chapter explored well known and some other native plants and/or their extracts that have been investigated in human clinical studies. We focused mostly on the analgesic activity of individual/single native plants and or their extract, even with large variation in secondary metabolites or bioactives. Analgesic effects of common polyherbal Ayurvedic pain relief formulations are not considered in this chapter since it is difficult to delineate the effects of individual plant or herb from the mixture. The therapeutic potential of many native crops is still unexplored to date for relieving pain and improving quality of life in humans.

REFERENCES

Abdel-Tawab, M., Werz, O., & Schubert-Zsilavecz, M. (2011). *Boswellia serrata* an overall assessment of *in vitro*, preclinical, pharmacokinetic and clinical data. *Clinical Pharmacokinetics*, *50*(6), 349–369.

Abolhassanzadeh, Z., Aflaki, E., Yousefi, G., & Mohagheghzadeh, A. (2015). Randomized clinical trial of peganum oil for knee osteoarthritis. *Journal of Evidence-based Complementary & Alternative Medicine*, *20*(2), 126–131. https://doi.org/10.1177/2156587214566867

Agarwal, A. (2016). Synergistic phytochemical composition and a process for preparation thereof. *US Patent 9,345,258 B2* issued 24 May, 2016. https://patents.google.com/patent/US9345258B2/en.

Agarwal, D., & Chaudhary, P. (2023). Effect of turmeric-boswellia-sesame formulation in menstrual cramp pain associated with primary dysmenorrhea—A double-blind, randomized, placebo-controlled study. *Journal of Clinical Medicine*, *12*, 3968. https://doi.org/10.3390/jcm12123968

Agarwal, K. A., Tripathi, C. D., Agarwal, B. B., & Saluja, S. (2011). Efficacy of turmeric (curcumin) in pain and postoperative fatigue after laparoscopic cholecystectomy: A double-blind, randomized placebo-controlled study. *Surgical Endoscopy*, *25*, 3805–3810.

Agrawal, P., Rai, V., & Singh, R. B. (1996). Randomized placebo-controlled, single blind trial of holy basil leaves in patients with noninsulin-dependent diabetes mellitus. *International Journal of Clinical Pharmacology and Therapeutics*, *34*(9), 406–409.

Agriwatch (2023, January 11). *Turmeric Market Sentiments*. https://www.agriwatch.com

Ahmad, A., Abuzinadah, M. F., Alkreathy, H. M., Banaganapalli, B., & Mujeeb, M. (2018). Ursolic acid rich *Ocimum sanctum* L leaf extract loaded nanostructured lipid carriers ameliorate adjuvant induced arthritis in rats by inhibition of COX-1, COX-2, TNF-α and IL-1: Pharmacological and docking studies. *PLoS ONE*, *13*(3), e0193451. https://doi.org/10.1371/journal.pone.093451

Ahmed Nasef, N., Thota, R. N., Mutukumira, A. N., Rutherfurd-Markwick, K., Dickens, M., Gopal, P., . . . & Garg, M. L. (2022). Bioactive yoghurt containing curcumin and chlorogenic acid reduces inflammation in postmenopausal women. *Nutrients*, *14*(21), 4619. https://doi.org/10.3390/nu14214619

Akansha, Kaushal, S., Arora, A., Heena, Sharma, P., & Jangra, R. (2023). Chemical composition and synergistic antifungal potential of *Nigella sativa* L. seeds and *Syzygium aromaticum* (L.) Merr. & LM Perry buds essential oils and their major compounds, and associated molecular docking studies. *Journal of Essential Oil Bearing Plants*, 1–24. https://doi.org/10.1080/0972060X.2023.2220348

Al-Askar, M., AlMubarak, A. M., Alqutub, M. N., Mokeem, S., Javed, F., Vohra, F., & Abduljabbar, T. (2022). Analgesic efficacy of *Curcuma longa* (Curcumin) after surgical periodontal therapy. *Oral Health & Preventive Dentistry*, *20*(1), 19–26.

Aleebrahim-Dehkordy, E., Nasri, H., Baradaran, A., Nasri, P., Tamadon, M. R., Hedaiaty, M., . . . & Rafieian-Kopaei, M. (2017). Medicinal plants, effective plant compounds (compositions) and their effects on stomach cancer. *International Journal of Preventive Medicine*, *8*, 96. https://doi.org/10.4103/ijpvm.IJPVM_4_17

Al-Okbi, S. Y., Ammar, N. M., Soroor, K. A., & Mohammed, D. A. (2000). Impact of natural oils supplements on disease activity and antioxidant state of Egyptian patients with rheumatoid arthritis. *Medical Journal of Islamic Academy of Sciences*, *13*(4), 161–171.

Amirchaghmaghi, M., Pakfetrat, A., Delavarian, Z., Ghalavani, H., & Ghazi, A. (2016). Evaluation of the efficacy of curcumin in the treatment of oral lichen planus: A randomized controlled trial. *Journal of Clinical and Diagnostic Research: JCDR*, *10*(5), ZC134–ZC137.

Anandaraj, M., Prasath, D., Kandiannan, K., Zachariah, T. J., Srinivasan, V., Jha, A. K., . . . & Maheswari, K. U. (2014). Genotype by environment interaction effects on yield and curcumin in turmeric (*Curcuma longa* L.). *Industrial Crops and Products*, *53*, 358–364.

Anandjiwala, S., Kalola, J., & Rajani, M. (2006). Quantification of eugenol, luteolin, ursolic acid, and oleanolic acid in black (Krishna Tulasi) and green (Sri Tulasi) varieties of *Ocimum sanctum* Linn. using high-performance thin-layer chromatography. *Journal of AOAC International*, *89*(6), 1467–1474.

Anil, A., Gujjari, S. K., & Venkatesh, M. P. (2019). Evaluation of a curcumin-containing mucoadhesive film for periodontal postsurgical pain control. *Journal of Indian Society of Periodontology*, *23*(5), 461–468.

Annie, S. L., Singh, D., & Umesha, C. (2022). Performance of turmeric (*Curcuma longa* L.) genotypes on growth and yield under Prayagraj agro-climatic zone. *International Journal of Environment and Climate Change*, *12*(11), 976–982.

Antony, B. (2022). Pharmaceutical composition made from hydrophobic phytochemicals dispersed in sesame oil to enhance bioactivity. *US Patent 2022/0133836 A1* issued 5 May. https://patents.google.com/patent/US20220133836A1/en

Arituluk, Z. C. (2022). *Coriandrum sativum* L. In F. T. G. Dereli et al. (eds.), *Novel Drug Targets with Traditional Herbal Medicines* (p. 151). Springer Nature. https://doi.org/10.1007/978-3-031-07753-1_12

Ashruf, O. S., & Ansari, M. Y. (2023). Natural compounds: Potential therapeutics for the inhibition of cartilage matrix degradation in osteoarthritis. *Life*, *13*(1), 102. https://doi.org/10.3390/life13010102

Ashwini, R., Gajalakshmi, S., Mythili, S., & Sathiavelu, A. (2011). *Terminalia chebula*-a pharmacological review. *Journal of Pharmacy Research*, *4*(9), 2884–2887.

Asteggiano, A., Curatolo, L., Schiavo, V., Occhipinti, A., & Medana, C. (2023). Development, validation, and application of a simple and rugged HPLC method for boswellic acids for a comparative study of their abundance in different species of boswellia gum resins. *Applied Sciences*, *13*(3), 1254. https://doi.org/10.3390/app13031254

Aswathi, A. P., Raghav, S. B., & Prasath, D. (2023). Assessment of genetic variation in turmeric (*Curcuma longa* L.) varieties based on morphological and molecular characterization. *Genetic Resources and Crop Evolution*, *70*(1), 147–158.

Awasthi, P. K., & Dixit, S. C. (2007). Chemical compositions of *Ocimum sanctum* Shyama and *Ocimum sanctum* Rama oils from the plains of Northern India. *Journal of Essential Oil Bearing Plants*, *10*(4), 292–296.

Balkrishna, A., Nag, S., Dabas, A., & Arya, V. (2022). Analgesic potential of medicinal plants used by the tribes of Western Himalayan region of India: A systematic review. *Medicinal Plants-International Journal of Phytomedicines and Related Industries*, *14*(3), 382–404. https://doi.org/10.5958/0975-6892.2022.00043.0

Bannuru, R. R., Osani, M. C., Al-Eid, F., & Wang, C. (2018). Efficacy of curcumin and Boswellia for knee osteoarthritis: Systematic review and meta-analysis. *Seminars in Arthritis and Rheumatism*, *48*(3), 416–429.

Barrett, M. L. (2019). Proprietary botanical ingredient scientific and clinical monograph for BCM-95®/Curcugreen™ turmeric rhizome curcumin extract preparation. *American Botanical Council*. www.herbalgram.org

Basu, P., Maier, C., & Basu, A. (2021). Effects of curcumin and its different formulations in preclinical and clinical studies of peripheral neuropathic and postoperative pain: A comprehensive review. *International Journal of Molecular Sciences*, *22*(9), 4666. https://doi.org/10.3390/ijms22094666

Beheshti, F., Khazaei, M., & Hosseini, M. (2016). Neuropharmacological effects of *Nigella sativa*. *Avicenna Journal of Phytomedicine*, *6*(1), 124–141.

Belcaro, G., Dugall, M., Luzzi, R., Ledda, A., Pellegrini, L., Cesarone, M. R., . . . & Errichi, M. (2014). Meriva® + glucosamine versus chondroitin+ glucosamine in patients with knee osteoarthritis: An observational study. *European Review for Medical and Pharmacological Science*, *18*(24), 3959–3963.

Bhakshu, L. M., & Ratnam, K. V. (2023). Phytochemistry of boswellia species. In T. Pullaiah, K. V. Ratnam, M. P. Swamy, & L. M. Bhakshu (eds.), *Frankincense–Gum Olibanum: Botany, Oleoresin, Chemistry, Extraction, Utilization, Propagation, Biotechnology, and Conservation* (pp. 69–82). CRC Press.

Bhat, J. A., Akther, T., Najar, R. A., Rasool, F., & Hamid, A. (2022). *Withania somnifera* (L.) Dunal (Ashwagandha); current understanding and future prospect as a potential drug candidate. *Frontiers in Pharmacology*, *13*, 1029123. https://doi.org/10.3389/fphar.2022.1029123

Bikshapathi, T., & Kumari, K. (1999). Clinical evaluation of Ashvagandha in the management of Amavata. *Journal of Research Ayurveda Science*, *20*(1–2), 46–53.

Brown, O. (2023). Fermedics: Enhancing botanical bioavailability with fermentation. *NutraIngredients Europe*, 25 September. https://nutraingredients.com/Article/2023/09/25

Caputi, F. F., Acquas, E., Kasture, S., Ruiu, S., Candeletti, S., & Romualdi, P. (2018). The standardized *Withania somnifera* Dunal root extract alters basal and morphine-induced opioid receptor gene expression changes in neuroblastoma cells. *BMC Complementary and Alternative Medicine*, *18*(1), 1–10.

Chandran, B., & Goel, A. (2012). A randomized, pilot study to assess the efficacy and safety of curcumin in patients with active rheumatoid arthritis. *Phytotherapy Research: PTR*, *26*(11), 1719–1725. https://doi.org/10.1002/ptr.4639

Chatterjee, S., Srivastava, S., Khalid, A., Singh, N., Sangwan, R. S., Sidhu, O. P., . . . & Tuli, R. (2010). Comprehensive metabolic fingerprinting of *Withania somnifera* leaf and root extracts. *Phytochemistry*, *71*(10), 1085–1094.

Chauhan, S., Mandliya, T., Jain, D., Joshi, A., Khatik, C. L., Singh, A., . . . & Jain, R. (2022). Early selective strategies for higher yielding bio-economic Indian ginseng based on genotypic study through metabolic and molecular markers. *Saudi Journal of Biological Sciences*, *29*(4), 3051–3061.

Chilelli, N. C., Ragazzi, E., Valentini, R., Cosma, C., Ferraresso, S., Lapolla, A., & Sartore, G. (2016). Curcumin and *boswellia serrata* modulate the glyco-oxidative status and lipo-oxidation in master athletes. *Nutrients*, *8*(11), 745. https://doi.org/10.3390/nu8110745

Chopra, A., Lavin, P., Patwardhan, B., & Chitre, D. (2000). Randomized double blind trial of an ayurvedic plant derived formulation for treatment of rheumatoid arthritis. *The Journal of Rheumatology*, *27*(6), 1365–1372.

Chopra, A., Lavin, P., Patwardhan, B., & Chitre, D. (2004). A 32-week randomized, placebo-controlled clinical evaluation of RA-11, an Ayurvedic drug, on osteoarthritis of the knees. *JCR: Journal of Clinical Rheumatology*, *10*(5), 236–245.

Chopra, A., Saluja, M., Kianifard, T., Chitre, D., & Venugopalan, A. (2018). Long term effectiveness of RA-1 as a monotherapy and in combination with disease modifying anti-rheumatic drugs in the treatment of rheumatoid arthritis. *Journal of Ayurveda and Integrative Medicine*, *9*(3), 201–208.

Chopra, A., Saluja, M., & Tillu, G. (2010). Ayurveda–modern medicine interface: A critical appraisal of studies of Ayurvedic medicines to treat osteoarthritis and rheumatoid arthritis. *Journal of Ayurveda and Integrative Medicine*, *1*(3), 190–198.

Chopra, A., Saluja, M., Tillu, G., Venugopalan, A., Sarmukaddam, S., Raut, A. K., . . . & Patwardhan, B. (2011). A randomized controlled exploratory evaluation of standardized Ayurvedic formulations in symptomatic osteoarthritis knees: A Government of India NMITLI Project. *Evidence-based Complementary and Alternative Medicine*, 724291. https://doi.org/10.1155/2011/724291

Choudhary, A. K., & Rahi, S. (2018). Organic cultivation of high yielding turmeric (*Curcuma longa* L.) cultivars: A viable alternative to enhance rhizome productivity, profitability, quality and resource-use efficiency in monkey–menace areas of north-western Himalayas. *Industrial Crops and Products*, *124*, 495–504.

Chumroenphat, T., Somboonwatthanakul, I., Saensouk, S., & Siriamornpun, S. (2021). Changes in curcuminoids and chemical components of turmeric (*Curcuma longa* L.) under freeze-drying and low-temperature drying methods. *Food Chemistry*, *339*, 128121. https://doi.org/10.1016/j.foodchem.2020.128121

Daily, J. W., Yang, M., & Park, S. (2016). Efficacy of turmeric extracts and curcumin for alleviating the symptoms of joint arthritis: A systematic review and meta-analysis of randomized clinical trials. *Journal of Medicinal Food*, *19*(8), 717–729.

Dalli, M., Azizi, S. E., Benouda, H., Azghar, A., Tahri, M., Bouammali, B., . . . & Gseyra, N. (2021). Molecular composition and antibacterial effect of five essential oils extracted from *Nigella sativa* L. seeds against multidrug-resistant bacteria: A comparative study. *Evidence-Based Complementary and Alternative Medicine*, *2021*. https://doi.org/10.1155/2021/6643765

Deshmukh, R. A., & Bagewadi, A. S. (2014). Comparison of effectiveness of curcumin with triamcinolone acetonide in the gel form in treatment of minor recurrent aphthous stomatitis: A randomized clinical trial. *International Journal of Pharmaceutical Investigation*, *4*(3), 138–141.

Desideri, I., Francolini, G., Becherini, C., Terziani, F., Delli Paoli, C., Olmetto, E., . . . & Livi, L. (2017). Use of an alpha lipoic, methylsulfonylmethane and bromelain dietary supplement (Opera®) for chemotherapy-induced peripheral neuropathy management, a prospective study. *Medical Oncology*, *34*, 46. https://doi.org/10.1007/s12032-014-0907-4

Dhanani, T., Shah, S., Gajbhiye, N. A., & Kumar, S. (2017). Effect of extraction methods on yield, phytochemical constituents and antioxidant activity of *Withania somnifera*. *Arabian Journal of Chemistry*, *10*, S1193–S1199.

Dhar, R. S., Verma, V., Suri, K. A., Sangwan, R. S., Satti, N. K., Kumar, A., . . . & Qazi, G. N. (2006). Phytochemical and genetic analysis in selected chemotypes of *Withania somnifera*. *Phytochemistry*, *67*(20), 2269–2276.

Di Pierro, F., Rapacioli, G., Di Maio, E., Appendino, G., Franceschi, F., & Togni, S. (2013). Comparative evaluation of the pain-relieving properties of a lecithinized formulation of curcumin (Meriva®), nimesulide, and acetaminophen. *Journal of Pain Research*, *6*, 201–205.

Di Pierro, F., Simonetti, G., Petruzzi, A., Bertuccioli, A., Botta, L., Bruzzone, M. G., Cuccarini, V., Fariselli, L., & Lamperti, E. (2019). A novel lecithin-based delivery form of Boswellic acids as complementary treatment of radiochemotherapy-induced cerebral edema in patients with glioblastoma multiforme: A longitudinal pilot experience. *Journal of Neurosurgical Sciences*, *63*(3), 286–291. https://doi.org/10.23736/S0390-5616.19.04662-9

Drobnic, F., Riera, J., Appendino, G., Togni, S., Franceschi, F., Valle, X., Pons, A., & Tur, J. (2014). Reduction of delayed onset muscle soreness by a novel curcumin delivery system (Meriva®): A randomised, placebo-controlled trial. *Journal of the International Society of Sports Nutrition*, *11*, 31. https://doi.org/10.1186/1550-2783-11-31

Dubey, P. N., Singh, B., Mishra, B. K., Kant, K., & Solanki, R. K. (2016). Nigella (*Nigella sativa*): A high value seed spice with immense medicinal potential. *Indian Journal of Agricultural Sciences*, *86*(8), 967–979.

Dubey, R., & Pandey, S. K. (2018). Medicinally important constituents of tulsi (*Ocimum* spp.). In A. Tewari & S. Tewari (eds.), *Synthesis of Medicinal Agents from Plants* (pp. 151–176). Elsevier Ltd. http://doi.org/10.1016/B978-0-08-102071-5.00007-6

Enting, R. H., Oldenmenger, W. H., Van Gool, A. R., van der Rijt, C. C., & Smitt, P. A. S. (2007). The effects of analgesic prescription and patient adherence on pain in a Dutch outpatient cancer population. *Journal of Pain and Symptom Management*, *34*(5), 523–531.

Fatima, N., Usharani, P., & Muralidhar, N. (2012). Randomised double blind placebo controlled study to evaluate the effect of CardiPro on endothelial dysfunction and biomarkers in patients with type 2 diabetes mellitus. *Innovative Journal of Medical and Health Science*, *2*(6), 122–128.

Fink, E. A., Xu, J., Hübner, H., Braz, J. M., Seemann, P., Avet, C., . . . & Gmeiner, P. (2022). Structure-based discovery of nonopioid analgesics acting through the α2A-adrenergic receptor. *Science*, *377*(6614), eabn7065. https://doi.org/10.1126/science.abn7065

Forouzanfar, F., & Hosseinzadeh, H. (2018). Medicinal herbs in the treatment of neuropathic pain: A review. *Iranian Journal of Basic Medical Sciences*, *21*(4), 347–358.

Francesco, D. P., Giuliana, R., Eleonora, A. D. M., Giovanni, A., Federico, F., & Stefano, T. (2013). Comparative evaluation of the pain-relieving properties of a lecithinized formulation of curcumin (Meriva®), nimesulide, and acetaminophen. *Journal of Pain Research*, 201–205.

Gautam, G. K., Vidhyasagar, G., Das, S., & Dwivedi, B. (2013). Comparative analgesic activity of selected medicinal plants from Indian origin. *International Journal of Pharmaceutical Sciences and Research*, *4*(7), 2726–2729.

Ghawte, S. A., Nikhat, S., Ahmad, J., & Mulla, G. (2014). *Withania somnifera* L. Dunal a potential herb for the treatment of rheumatoid arthritis. *Annals of Phytomedicine*, *3*(1), 98–102.

Gheita, T. A., & Kenawy, S. A. (2012). Effectiveness of *Nigella sativa* oil in the management of rheumatoid arthritis patients: A placebo controlled study. *Phytotherapy Research*, *26*(8), 1246–1248.

Godhwani, S., Godhwani, J. L., & Vyas, D. S. (1987). *Ocimum sanctum*: An experimental study evaluating its anti-inflammatory, analgesic and antipyretic activity in animals. *Journal of Ethnopharmacology*, *21*(2), 153–163.

Goni, V. G., Mishra, M., Kadam, S., & Gandhi, S. S. (2021). Effect of Nucart VG (*Boswellia serrata* in combination with veg glucosamine sulphate) in comparison with glucosamine sulphate to improve quality of life of knee osteoarthritis patients: A randomized controlled trial. *International Journal of Research in Orthopaedics*, *7*(5), 976–985.

Gota, P., Damle, N., Patil, S., Sing, S., Nandave, M., & Gota, V. (2015). Efficacy of solid lipid *Boswellia serata* particles (SLBSP) in osteoarthritis of knee. *Journal of Pharmaceutical Sciences & Technology Management*, *1*(2), 70–76.

Gupta, S., Srivastava, A., Shasany, A. K., & Gupta, A. K. (2018). Genetics, cytogenetics, and genetic diversity in the genus *Ocimum*. In A. K. Shasany & C. Kole (eds.), *The Ocimum Genome, Compendium of Plant Genomes* (pp. 73–87). Springer Nature Switzerland.

Hadi, V., Kheirouri, S., Alizadeh, M., Khabbazi, A., & Hosseini, H. (2016). Effects of *Nigella sativa* oil extract on inflammatory cytokine response and oxidative stress status in patients with rheumatoid arthritis: A randomized, double-blind, placebo-controlled clinical trial. *Avicenna Journal of Phytomedicine*, *6*(1), 34–43.

Hadi, V., Pahlavani, N., Malekahmadi, M., Nattagh-Eshtivani, E., Navashenaq, J. G., Hadi, S., . . . & Norouzy, A. (2021). *Nigella sativa* in controlling Type 2 diabetes, cardiovascular, and rheumatoid arthritis diseases: Molecular aspects. *Journal of Research in Medical Sciences: The Official Journal of Isfahan University of Medical Sciences*, *26*. https://doi.org/10.4103.jrms.JRMS_236_20

Hakkim, F. L., Shankar, C. G., & Girija, S. (2007). Chemical composition and antioxidant property of holy basil (*Ocimum sanctum* L.) leaves, stems, and inflorescence and their in vitro callus cultures. *Journal of Agricultural and Food Chemistry*, *55*(22), 9109–9117.

Hannan, J. M. A., Das, B. K., Uddin, A., Bhattacharjee, R., Das, B., Chowdury, H. S., & Mosaddek, A. S. M. (2011). Analgesic and anti-inflammatory effects of Ocimum sanctum (Linn) in laboratory animals. *International Journal of Pharmaceutical Sciences and Research*, *2*(8), 2121–2125.

Haroyan, A., Mukuchyan, V., Mkrtchyan, N., Minasyan, N., Gasparyan, S., Sargsyan, A., . . . & Hovhannisyan, A. (2018). Efficacy and safety of curcumin and its combination with boswellic acid in osteoarthritis: A comparative, randomized, double-blind, placebo-controlled study. *BMC Complementary and Alternative Medicine*, *18*(1), 7. https://doi.org/10.1186/s12906-017-2062-z

Hedaoo, M., Patil-Bhole, T., Sharma, R., & Mahajan, M. (2023). Exploratory quasi-experimental study of anti-arthritic activity of ayurvedic polyherbal formulation, Abha Guggulu in osteoarthritis patients. *Drug Metabolism and Personalized Therapy*. https://doi.org/10.1515/dmpt-2022-0187

Henrotin, Y., Malaise, M., Wittoek, R., De Vlam, K., Brasseur, J. P., Luyten, F. P., . . . & Dierckxsens, Y. (2019). Bio-optimized Curcuma longa extract is efficient on knee osteoarthritis pain: A double-blind multicenter randomized placebo controlled three-arm study. *Arthritis Research & Therapy*, *21*(1), 1–10. https://doi.org/10.1186/s13075-019-1960-5

Heyninck, K., Lahtela-Kakkonen, M., Van der Veken, P., Haegeman, G., & Berghe, W. V. (2014). Withaferin A inhibits NF-kappaB activation by targeting cysteine 179 in IKKβ. *Biochemical Pharmacology*, *91*(4), 501–509.

Huchchannanavar, S., Yogesh, L. N., & Prashant, S. M. (2019). The black seed *Nigella sativa*: A wonder seed. *International Journal of Chemical Studies*, *7*(3), 1320–1324.

Huseini, H. F., Kianbakht, S., Mirshamsi, M. H., & Zarch, A. B. (2016). Effectiveness of topical *Nigella sativa* seed oil in the treatment of cyclic mastalgia: A randomized, triple-blind, active, and placebo-controlled clinical trial. *Planta Medica*, *82*(4), 285–288.

Jain, A. V., Jain, K. A., & Vijayaraghavan, N. (2020). AflaB2® and osteoarthritis: A multicentric, observational, post-marketing surveillance study in Indian patients suffering from knee osteoarthritis. *International Journal of Research in Orthopaedics*, *7*(1), 110–115.

Jamshidi, N., & Cohen, M. M. (2017). The clinical efficacy and safety of Tulsi in humans: A systematic review of the literature. *Evidence-Based Complementary and Alternative Medicine*, *2017*. https://doi.org/10.1155/2017/9217567

Joshi, C., Gajbhiye, N., Phurailatpam, A., Geetha, K. A., & Maiti, S. (2010). Comparative morphometric, physiological and chemical studies of wild and cultivated plant types of *Withania somnifera* (Solanaceae). *Current Science*, 644–650.

Joshi, R. K., & Hoti, S. L. (2014). Chemical composition of the essential oil of *Ocimum tenuiflorum* L. (Krishna Tulsi) from North West Karnataka, India. *Plant Science Today*, *1*(3), 99–102.

Kabir, Y., Akasaka-Hashimoto, Y., Kubota, K., & Komai, M. (2020). Volatile compounds of black cumin (*Nigella sativa* L.) seeds cultivated in Bangladesh and India. *Heliyon*, *6*(10). https://doi.org/10.1016/j.heliyon.2020.e05343

Kalidasu, G., Reddy, G. S., Kumari, S. S., Kumari, A. L., & Sivasankar, A. (2017). Secondary volatiles and metabolites from *Nigella sativa* L. seed. *Indian Journal of Natural Products and Resources*, *8*(2), 151–158.

Kamat, Y. D., Das, B., Thakkar, K., Mahajan, M., & Das Sr, B. (2023). A retrospective observational study evaluating the synergistic effect of a novel combination of Alfapin+ native type 2 collagen+ Mobilee (hyaluronic acid)+ CurQlife (curcumin) nutraceuticals in the symptomatic improvement of knee osteoarthritis. *Cureus*, *15*(3). https://doi.org/10.7759/cureus.36123

Kanjilal, S., Gupta, A. K., Patnaik, R. S., & Dey, A. (2021). Analysis of clinical trial registry of India for evidence of anti-arthritic properties of *Withania somnifera* (Ashwagandha). *Alternative Therapies in Health and Medicine*, *27*(6), 58–66.

Karlapudi, V., Sunkara, K. B., Konda, P. R., Sarma, K. V., & Rokkam, M. P. (2023). Efficacy and safety of Aflapin®, a novel *boswellia serrata* extract, in the treatment of osteoarthritis of the knee: A short-term 30-day randomized, double-blind, placebo-controlled clinical study. *Journal of the American Nutrition Association*, *42*(2), 159–168.

Kaur, G., Bali, A., Singh, N., & Jaggi, A. S. (2015). Ameliorative potential of *Ocimum sanctum* in chronic constriction injury-induced neuropathic pain in rats. *Anais da Academia Brasileira de Ciências*, *87*, 417–429.

Kaur, G., Jaggi, A. S., & Singh, N. (2010). Exploring the potential effect of *Ocimum sanctum* in vincristine-induced neuropathic pain in rats. *Journal of Brachial Plexus and Peripheral Nerve Injury*, *5*(1), e3–e11. http://www.jbppni.com/content/5/1/3

Khabbazi, A., Javadivala, Z., Seyedsadjadi, N., & Mahdavi, A. M. (2020). A systematic review of the potential effects of *Nigella sativa* on rheumatoid arthritis. *Planta Medica*, *86*(7), 457–469.

Khanna, N., & Bhatia, J. (2003). Antinociceptive action of *Ocimum sanctum* (Tulsi) in mice: Possible mechanisms involved. *Journal of Ethnopharmacology*, *88*(2–3), 293–296.

Kheirouri, S., Hadi, V., & Alizadeh, M. (2016). Immunomodulatory effect of *Nigella sativa* oil on T lymphocytes in patients with rheumatoid arthritis. *Immunological Investigations*, *45*(4), 271–283.

Kia, S. J., Basirat, M., Mortezaie, T., & Moosavi, M. S. (2020a). Comparison of oral Nano-Curcumin with oral prednisolone on oral lichen planus: A randomized double-blinded clinical trial. *BMC Complementary Medicine and Therapies*, *20*(1), 1–7. https://doi.org/10.1186/s12906-020-03128-7

Kia, S. J., Basirat, M., Saedi, H. S., & Arab, S. A. (2021). Effects of nanomicelle curcumin capsules on prevention and treatment of oral mucosits in patients under chemotherapy with or without head and neck radiotherapy: A randomized clinical trial. *BMC Complementary Medicine and Therapies*, *21*(1), 1–11. https://doi.org/10.1186/s12906-021-03400-4

Kia, S. J., Mansourian, A., Basirat, M., Akhavan, M., Mohtasham-Amiri, Z., & Moosavi, M. S. (2020b). New concentration of curcumin orabase in recurrent aphthous stomatitis: A randomized, controlled clinical trial. *Journal of Herbal Medicine*, *22*, 100336. https://doi.org/10.1016/j.herbmed.2020.100336

Kia, S. J., Shirazian, S., Mansourian, A., Fard, L. K., & Ashnagar, S. (2015). Comparative efficacy of topical curcumin and triamcinolone for oral lichen planus: A randomized, controlled clinical trial. *Journal of Dentistry (Tehran, Iran)*, *12*(11), 789–796.

Kimmatkar, N., Thawani, V., Hingorani, L., & Khiyani, R. (2003). Efficacy and tolerability of *Boswellia serrata* extract in treatment of osteoarthritis of knee—a randomized double blind placebo controlled trial. *Phytomedicine*, *10*(1), 3–7. https://doi.org/10.1078/094471103321648593

Kooshki, A., Forouzan, R., Rakhshani, M. H., & Mohammadi, M. (2016). Effect of topical application of *Nigella sativa* oil and oral acetaminophen on pain in elderly with knee osteoarthritis: A crossover clinical trial. *Electronic Physician*, *8*(11), 3193–3197.

Kulkarni, P. D., Damle, N. D., Singh, S., Yadav, K. S., Ghante, M. R., Bhaskar, V. H., . . . & Gota, V. S. (2020). Double-blind trial of solid lipid *Boswellia serrata* particles (SLBSP) vs. standardized Boswellia serrata gum extract (BSE) for osteoarthritis of knee. *Drug Metabolism and Personalized Therapy*, *35*(2), 20200104. https://doi.org/10.1515/dmpt-2020-0104

Kulkarni, R. R., Patki, P. S., Jog, V. P., Gandage, S. G., & Patwardhan, B. (1991). Treatment of osteoarthritis with a herbomineral formulation: A double-blind, placebo-controlled, cross-over study. *Journal of Ethnopharmacology*, *33*(1–2), 91–95.

Kumar, A., Agarwal, K., Maurya, A. K., Shanker, K., Bushra, U., Tandon, S., & Bawankule, D. U. (2015a). Pharmacological and phytochemical evaluation of Ocimum sanctum root extracts for its antiinflammatory, analgesic and antipyretic activities. *Pharmacognosy Magazine*, *11*(Suppl 1), S217.

Kumar, C. U., Pokuri, V. K., & Pingali, U. (2015b). Evaluation of the analgesic activity of standardized aqueous extract of *Terminalia chebula* in healthy human participants using hot air pain model. *Journal of Clinical and Diagnostic Research: JCDR*, *9*(5), FC01–FC04. https://doi.org/10.7860/JCDR/2015/11369.5916

Kumar, G., Srivastava, A., Sharma, S. K., Rao, T. D., & Gupta, Y. K. (2015c). Efficacy & safety evaluation of Ayurvedic treatment (Ashwagandha powder & Sidh Makardhwaj) in rheumatoid arthritis patients: A pilot prospective study. *Indian Journal of Medical Research*, *141*(1), 100–106.

Kumar, G., Srivastava, A., Sharma, S. K., Rao, T. D., & Gupta, Y. K. (2015). Efficacy & safety evaluation of Ayurvedic treatment (Ashwagandha powder & Sidh Makardhwaj) in rheumatoid arthritis patients: a pilot prospective study. *Indian Journal of Medical Research, 141*(1), 100–106.

Kumar, S., Singh, R., Gajbhiye, N., & Dhanani, T. (2018). Extraction optimization for phenolic-and withanolide-rich fractions from Withania somnifera roots: Identification and quantification of withaferin A, 12-deoxywithastromonolide, and withanolide A in plant materials and marketed formulations using a reversed-phase HPLC–photodiode array detection method. *Journal of AOAC International, 101*(6), 1773–1780.

Kunnumakkara, A. B., Hegde, M., Parama, D., Girisa, S., Kumar, A., Daimary, U. D., . . . & Aggarwal, B. B. (2023). Role of turmeric and curcumin in prevention and treatment of chronic diseases: Lessons learned from clinical trials. *ACS Pharmacology & Translational Science*, *6*(4), 447–518.

Kuptniratsaikul, V., Dajpratham, P., Taechaarpornkul, W., Buntragulpoontawee, M., Lukkanapichonchut, P., Chootip, C., . . . & Laongpech, S. (2014). Efficacy and safety of *Curcuma domestica* extracts compared with ibuprofen in patients with knee osteoarthritis: A multicenter study. *Clinical Interventions in Aging*, 451–458.

Kuptniratsaikul, V., Thanakhumtorn, S., Chinswangwatanakul, P., Wattanamongkonsil, L., & Thamlikitkul, V. (2009). Efficacy and safety of Curcuma domestica extracts in patients with knee osteoarthritis. *The Journal of Alternative and Complementary Medicine*, *15*(8), 891–897

Lalla, J., Hamrapurkar, P., & Singh, A. (2007). Quantitative HPTLC analysis of the eugenol content of leaf powder and a capsule formulation of *Ocimum sanctum*. *JPC-Journal of Planar Chromatography-Modern TLC*, *20*(2), 135–138.

Lepcha, T. O., Das, S., Dutta, B., & Medda, P. S. (2022). Effect of curing methods on quality and drying characteristics of turmeric. *Journal of Spices & Aromatic Crops*, *31*(1), 65–74. https://doi.org/10.25081/josac.2022.v31.i1.7698

Li, H., DaSilva, N. A., Liu, W., Xu, J., Dombi, G. W., Dain, J. A., . . . & Ma, H. (2020). Thymocid®, a standardized black cumin (*Nigella sativa*) seed extract, modulates collagen cross-linking, collagenase and elastase activities, and melanogenesis in murine B16F10 melanoma cells. *Nutrients*, *12*(7), 2146. https://doi.org/10.3390/nu12072146

Lopresti, A. L., Smith, S. J., Malvi, H., & Kodgule, R. (2019). An investigation into the stress-relieving and pharmacological actions of an ashwagandha (*Withania somnifera*) extract: A randomized, double-blind, placebo-controlled study. *Medicine*, *98*(37), e17186. http://doi.org/10.1097/MD.0000000000017186

Lopresti, A. L., Smith, S. J., Metse, A. P., & Drummond, P. D. (2022). A randomized, double-blind, placebo-controlled trial investigating the effects of an *Ocimum tenuiflorum* (Holy Basil) extra(Holixer™) on stress, mood, and sleep in adults experiencing stress. *Frontiers in Nutrition*, *9*, 965130. https://doi.org/10.3389/fnut.2022.965130

Madhu, K., Chanda, K., & Saji, M. J. (2013). Safety and efficacy of *Curcuma longa* extract in the treatment of painful knee osteoarthritis: A randomized placebo-controlled trial. *Inflammopharmacology*, *21*, 129–136.

Mahboubi, M., Kashani, L. M. T., & Mahboubi, M. (2018). *Nigella sativa* fixed oil as alternative treatment in management of pain in arthritis rheumatoid. *Phytomedicine*, *46*, 69–77.

Mahdavi, R., Namazi, N., Alizadeh, M., & Farajnia, S. (2016). *Nigella sativa* oil with a calorie-restricted diet can improve biomarkers of systemic inflammation in obese women: A randomized double-blind, placebo-controlled clinical trial. *Journal of Clinical Lipidology*, *10*(5), 1203–1211.

Mahdy, A., & Gheita, T. (2009). Beneficial effects of *Nigella sativa* seed oil as adjunct therapy in rheumatoid arthritis. *Journal of the Egyptian Society of Toxicology*, *41*, 31–37.

Mahto, H., Mahato, D., & Sharma, H. P. (2022). Phytoconstituent estimation and LC-MS studies of field grown *Withania somnifera* (L.) Dunal root extract in Jharkhand & Bihar. *Research Journal of Chemistry and Environment*, *26*(11), 135–141.

Majeed, M., Majeed, S., Narayanan, N. K., & Nagabhushanam, K. (2019). A pilot, randomized, double-blind, placebo-controlled trial to assess the safety and efficacy of a novel *Boswellia serrata* extract in the management of osteoarthritis of the knee. *Phytotherapy Research*, *33*(5), 1457–1468. https://doi.org/10.1002/ptr.6338

Mandal, A. K., Poudel, M., Neupane, N. P., & Verma, A. (2022). Phytochemistry, pharmacology, and applications of *Ocimum sanctum* (Tulsi). In M. H. Masoodi & M. U. Rehman (eds.), *Edible Plants in Health and Diseases*. Springer. https://doi.org/10.1007/978-981-16-4959-2_4

Manifar, S., Obwaller, A., Gharehgozloo, A., Boorboor Shirazi Kordi, H. R., & Akhondzadeh, S. (2012). Curcumin gel in the treatment of minor aphthous ulcer: A randomized, placebo-controlled trial. *Journal of Medicinal Plants*, *11*(41), 40–45.

Mankowski, R. T., Sibille, K. T., Leeuwenburgh, C., Lin, Y., Hsu, F. C., Qiu, P., . . . & Anton, S. D. (2023). Effects of curcumin C3 complex® on physical function in moderately functioning older adults with low-grade inflammation—A pilot trial. *The Journal of Frailty & Aging*, *12*(2), 143–149.

Mansourian, A., Bahar, B., Moosavi, M. S., Amanlou, M., & Babaeifard, S. (2017). Comparison of the efficacy of topical triamcinolone in orabase and curcumin in orabase in oral graft-versus-host disease. *Journal of Dentistry (Tehran, Iran)*, *14*(6), 313–320.

Marketwatch (2023). Global black seed oil market insight [2023–2030], published 5 June. https://www.giiresearch.com/report/vmr1392343-global-black-seed-oil-market-research-report.html

Maulina, T., Purnomo, Y. Y., Tasman, S. G. R., Sjamsudin, E., & Amaliya, A. (2022). The efficacy of curcumin patch as an adjuvant therapeutic agent in managing acute orofacial pain on the post-cleft lip and cleft palate surgery patients: A pragmatic trial. *European Journal of Dentistry*. https://doi.org/10.1055/s-0042-1750802

McCutcheon, A. (2018). *Adulteration of Boswellia Serrata*. Austin, TX: Botanical adulterants prevention program. *Botanical Adulterants Bulletin*, 1–7. https://umb.herbalgram.org/media/5ymjxzdi/bapp-babs-boswellia-cc-v2.pdf

Meghana, M. V. S., Deshmukh, J., Devarathanamma, M. V., Asif, K., Jyothi, L., & Sindhura, H. (2020). Comparison of effect of curcumin gel and noneugenol periodontal dressing in tissue response, early wound healing, and pain assessment following periodontal flap surgery in chronic periodontitis patients. *Journal of Indian Society of Periodontology*, *24*(1), 54–59.

Menon, M. K., & Kar, A. (1971). Analgesic and psychopharmacological effects of the gum resin of *Boswellia serrata*. *Planta medica*, *19*(2), 333–341.

Merolla, G., Dellabiancia, F., Ingardia, A., Paladini, P., & Porcellini, G. (2015). Co-analgesic therapy for arthroscopic supraspinatus tendon repair pain using a dietary supplement containing *Boswellia serrata* and *Curcuma longa*: A prospective randomized placebo-controlled study. *Musculoskeletal Surgery*, *99*, 43–52.

Mishra, S., Bishnoi, R. S., Maurya, R., & Jain, D. (2020). *Boswellia serrata* ROXB.– A bioactive herb with various pharmacological activities. *Asian Journal of Pharmaceutical and Clinical Research*, *13*(11), 33–39.

Mohsenzadeh, A., Karimifar, M., Soltani, R., & Hajhashemi, V. (2023). Evaluation of the effectiveness of topical oily solution containing frankincense extract in the treatment of knee osteoarthritis: A randomized, double-blind, placebo-controlled clinical trial. *BMC Research Notes*, *16*(1), 28. https://doi.org/10.1186/s13104-023-06291-5

Mondal, S., Varma, S., Bamola, V. D., Naik, S. N., Mirdha, B. R., Padhi, M. M., . . . & Mahapatra, S. C. (2011). Double-blinded randomized controlled trial for immunomodulatory effects of Tulsi (*Ocimum sanctum* Linn.) leaf extract on healthy volunteers. *Journal of Ethnopharmacology*, *136*(3), 452–456. https://doi.org/10.1016/j.jep.2011.05.012

Mugilan, R., Jayaswal, R., Sowmya, R., Vincent, S. S., Vaishali, K., & Prasad, K. (2020). Effect of curcumin on healing of extraction sockets in type II diabetic patients – A pilot study. *Journalof Evoution of Medical and Dental Sciences*, *9*, 1045–1049. https://doi.org/10.14260/jemds/2020/225

Mukadam, S., Ghule, C., Girme, A., Shinde, V. M., Hingorani, L., & Mahadik, K. R. (2023). A simple HPTLC approach of quantification of serratol and tirucallic acid with boswellic acids in *Boswellia serrata* by validated densitometric method with MS/MS characterization. *Journal of Chromatographic Science*. https://doi.org/10.1093/chromsci/bmad012

Murthy, M. N. K., Gundagani, S., Nutalapati, C., & Pingali, U. (2019). Evaluation of analgesic activity of standardised aqueous extract of *Withania somnifera* in healthy human volunteers using mechanical pain model. *Journal of Clinical and Diagnostic Research*, *13*(1), FC01–FC04. https://doi.org/10.7860/JCDR/2019/37590.12441

Nakagawa, Y., Mukai, S., Yamada, S., Matsuoka, M., Tarumi, E., Hashimoto, T., . . . & Nakamura, T. (2014). Short-term effects of highly-bioavailable curcumin for treating knee osteoarthritis: A randomized, double-blind, placebo-controlled prospective study. *Journal of Orthopaedic Science*, *19*(6), 933–939.

Nalina, L., Rajamani, K., Shanmugasundaram, K. A., & Boomiga, M. (2022). Breeding and conservation of medicinal plants in India. In *Medicinal and Aromatic Plants of India* (Vol. 1, pp. 201–236). Springer International Publishing.

Nalini, P., Manjunath, K., SunilKumarReddy K., & Usharani, P. (2013). Evaluation of the analgesic activity of standardized aqueous extract of *Withania somnifera* in healthy human volunteers using Hot Air Pain Model. *Research Journal of Life Sciences*, *1*(2), 1–6.

Neelam, A., Om, P., Tripathi, S., & Pant, A. K. (2015). *Curcuma longa* L: Elemental and nutritional profiling of fifty accessions from Uttarakhand region in India. *Asian Journal of Chemistry*, *27*(11), 4160–4166.

Nile, S. H., Liang, Y., Wang, Z., Zheng, J., Sun, C., Nile, A., . . . & Kai, G. (2022). Chemical composition, cytotoxic and pro-inflammatory enzyme inhibitory properties of *Withania somnifera* (L.) Dunal root extracts. *South African Journal of Botany*, *151*, 46–53.

Notarnicola, A., Maccagnano, G., Moretti, L., Pesce, V., Tafuri, S., Fiore, A., & Moretti, B. (2016). Methylsulfonylmethane and boswellic acids versus glucosamine sulfate in the treatment of knee arthritis: Randomized trial. *International Journal of Immunopathology and Pharmacology*, *29*(1), 140–146.

Padalia, R. C., & Verma, R. S. (2011). Comparative volatile oil composition of four Ocimum species from northern India. *Natural Product Research*, *25*(6), 569–575.

Pal, C. P., Singh, P., Chaturvedi, S., Pruthi, K. K., & Vij, A. (2016). Epidemiology of knee osteoarthritis in India and related factors. *Indian Journal of Orthopaedics*, *50*(5), 518–522. https://doi.org/10.4103/0019-5413.189608

Pal, K., Chowdhury, S., Dutta, S. K., Chakraborty, S., Chakraborty, M., Pandit, G. K., . . . & Mandal, S. (2020). Analysis of rhizome colour content, bioactive compound profiling and ex-situ conservation of turmeric genotypes (*Curcuma longa* L.) from sub-Himalayan terai region of India. *Industrial Crops and Products*, *150*, 112401. https://doi.org/10.1016/j.indcrop.2020.112401

Panahi, Y., Hosseini, M. S., Khalili, N., Naimi, E., Simental-Mendía, L. E., Majeed, M., & Sahebkar, A. (2016). Effects of curcumin on serum cytokine concentrations in subjects with metabolic syndrome: A post-hoc analysis of a randomized controlled trial. *Biomedicine & Pharmacotherapy*, *82*, 578–582.

Panahi, Y., Rahimnia, A. R., Sharafi, M., Alishiri, G., Saburi, A., & Sahebkar, A. (2014). Curcuminoid treatment for knee osteoarthritis: A randomized double-blind placebo-controlled trial. *Phytotherapy Research*, *28*(11), 1625–1631

Panda, S. K., Nirvanashetty, S., Parachur, V. A., Mohanty, N., & Swain, T. (2018). A randomized, double blind, placebo controlled, parallel-group study to evaluate the safety and efficacy of Curene® versus placebo in reducing symptoms ofk OA. *BioMed Research International*, *2018*, 5291945. https://doi.org/10.1155/2018/5291945

Pandey, A. K., Singh, P., & Tripathi, N. N. (2014). Chemistry and bioactivities of essential oils of some *Ocimum* species: An overview. *Asian Pacific Journal of Tropical Biomedicine*, *4*(9), 682–694.

Pandey, G., & Madhuri, S. (2010). Pharmacological activities of *Ocimum sanctum* (tulsi): A review. *International Journal of Pharmaceutical Sciences Review and Research*, *5*(1), 61–66.

Pandey, R., Chandra, P., Srivastava, M., Mishra, D. K., & Kumar, B. (2015). Simultaneous quantitative determination of multiple bioactive markers in *Ocimum sanctum* obtained from different locations and its marketed herbal formulations using UPLC-ESI-MS/MS combined with principal component analysis. *Phytochemical Analysis, PCA*, *26*(6), 383–394. https://doi.org/10.1002/pca.2551

Panknin, T. M., Howe, C. L., Hauer, M., Bucchireddigari, B., Rossi, A. M., & Funk, J. L. (2023). Curcumin supplementation and human disease: A scoping review of clinical trials. *International Journal of Molecular Sciences*, *24*(5), 4476. https://doi.org/10.3390/ijms24054476

Park, J., & Ernst, E. (2005, April). Ayurvedic medicine for rheumatoid arthritis: A systematic review. In *Seminars in Arthritis and Rheumatism* (Vol. 34, No. 5, pp. 705–713). WB Saunders.

Pattanayak, P., Behera, P., Das, D., & Panda, S. K. (2010). *Ocimum sanctum* Linn. A reservoir plant for therapeutic applications: An overview. *Pharmacognosy Reviews*, *4*(7), 95.

Patwardhan, B. (1996). Method of treating musculoskeletal disease and a novel composition therefor. *U.S. Patent 5,494,668* issued 27 February. https://patents.google.com/patent/US5494668A/en

Paultre, K., Cade, W., Hernandez, D., Reynolds, J., Greif, D., & Best, T. M. (2021). Therapeutic effects of turmeric or curcumin extract on pain and function for individuals with knee osteoarthritis: A systematic review. *BMJ Open Sport & Exercise Medicine*, *7*(1), e000935. http://doi.org/10.1136/bmjsem-2020-000935

Pellegrini, L., Milano, E., Franceschi, F., Belcaro, G., Gizzi, G., Feragalli, B., . . . & Giacomelli, L. (2016). Managing ulcerative colitis in remission phase: Usefulness of Casperome®, an innovative lecithin-based delivery system of *Boswellia serrata* extract. *European Review for Medical and Pharmacological Sciences*, *20*(12), 2695–2700.

Pérez-Piñero, S., Muñoz-Carrillo, J. C., Victoria-Montesinos, D., García-Muñoz, A. M., Andreu-Caravaca, L., Gómez, M., . . . & López-Román, F. J. (2023). Efficacy of *Boswellia serrata* extract and/or an omega-3-based product for improving pain and function in people older than 40 years with persistent knee pain: A randomized double-blind controlled clinical trial. *Nutrients*, *15*(17), 3848. https://doi.org/10.3390/nu15173848

Pingali, U., Pilli, R., & Fatima, N. (2014). Effect of standardized aqueous extract of *Withania somnifera* on tests of cognitive and psychomotor performance in healthy human participants. *Pharmacognosy Research*, *6*(1), 12–18.

Pinzon, R. T., Sanyasi, R.L. R., Pramudita, E. A., & Periska, S. D. (2019). The benefit of *Curcuma longa* and *Boswellia serrata* to improve quality of life in osteoarthritis patients: A randomized controlled trial. *International Journal of Research in Orthopaedics*, *5*(6), 1005–1014.

Platkin, C., Cather, A., Butz, L., Garcia, I., Gallanter, M., & Leung, M. M. (2022). *Food as Medicine: Overview and Report: How Food and Diet Impact the Treatment of Disease and Disease Management*. Center for Food as Medicine and Hunter College NYC Food Policy Center, 30 March. foodmedcenter.org and nycfoodpolicy.org

Polumackanycz, M., Petropoulos, S. A., Śledziński, T., Goyke, E., Konopacka, A., Plenis, A., & Viapiana, A. (2023). *Withania somnifera* L.: Phenolic compounds composition and biological activity of commercial samples and its aqueous and hydromethanolic extracts. *Antioxidants*, *12*(3), 550. https://doi.org/10.3390/antiox12030550

Prakash, P., & Gupta, N. (2005). Therapeutic uses of *Ocimum sanctum* Linn (Tulsi) with a note on eugenol and its pharmacological actions: A short survey. *Indian Journal of Physiological Pharmacology*, *49*(2), 125–131.

Pullaiah, T. (2023). Botany of boswellia. In T. Pullaiah, K. V. Ratnam, M. P. Swamy, & L. M. Bhakshu (eds.), *Frankincense–Gum Olibanum: Botany, Oleoresin, Chemistry, Extraction, Utilization, Propagation, Biotechnology, and Conservation* (pp. 11–40). CRC Press.

Rai, A. K., Khan, S., Kumar, A., Dubey, B. K., Lal, R. K., Tiwari, A., . . . & Ch, R. (2023). Comprehensive metabolomic fingerprinting combined with chemometrics identifies species-and variety-specific variation of medicinal herbs: An *Ocimum* study. *Metabolites*, *13*(1), 122. https://doi.org/10.3390/metabo13010122

Rai, V., Mani, U. V., & Iyer, U. M. (1997). Effect of *Ocimum sanctum* leaf powder on blood lipoproteins, glycated proteins and total amino acids in patients with non-insulin-dependent diabetes mellitus. *Journal of Nutritional & Environmental Medicine*, *7*(2), 113–118.

Rajeshwari, C. U., Siri, S., & Andallu, B. (2012). Antioxidant and antiarthritic potential of coriander (*Coriandrum sativum* L.) leaves. *e-SPEN Journal*, *7*(6), e223–e228.

Ramadan, M. F. (2021). *Black Cumin (Nigella sativa) Seeds: Chemistry, Technology, Functionality, and Applications*. Springer Nature. https://doi.org/10.1007/978-3-030-48798-0

Ramakanth, G. S., Uday Kumar, C., Kishan, P. V., & Usharani, P. (2016). A randomized, double blind placebo controlled study of efficacy and tolerability of *Withania somnifera* extracts in knee joint pain. *Journal of Ayurveda Integrated Medicine*, *7*(3), 151–157.

Ramarao, P., Rao, K. T., Srivastava, R. S., & Ghosal, S. (1995). Effects of glycowithanolides from *Withania somnifera* on morphine-induced inhibition of intestinal motility and tolerance to analgesia in mice. *Phytotherapy Research*, *9*(1), 66–68.

Rastogi, S., & Shasany, A. K. (2018). Ocimum genome sequencing—A futuristic therapeutic mine. In A.K. Shasany and C. Kole (eds.), *The Ocimum Genome, Compendium of Plant Genomes*, 127–148. https://doi.org.10.1007/978-3-319-97430-9_10

Rastogi, S., Kalra, A., Gupta, V., Khan, F., Lal, R. K., Tripathi, A. K., . . . & Shasany, A. K. (2015). Unravelling the genome of Holy basil: An "incomparable" "elixir of life" of traditional Indian medicine. *BMC Genomics*, *16*(1), 1–15.

Rathore, R., & Jain, S. (2013). An experimental study of analgesic effect of medicinal plant Tulsi (*Ocimum sanctum*). *Studies on Ethno-Medicine*, *7*(1), 27–30.

Ravi, Y., Vethamoni, I. P., Saxena, S. N., Velmurugan, S., Santanakrishnan, V. P., Raveendran, M., . . . & Harsh, M. (2023). Guesstimate of thymoquinone diversity in *Nigella sativa* L. genotypes and elite varieties collected from Indian states using HPTLC technique. *Open Life Sciences*, *18*(1), 20220536. https://doi.org/10.1515/biol-2022-0536

Rawat, P., Kumar, S., & Malik, J. K. (2023). *Boswellia serrata*: Herbal remedy to reduce inflammation. *Research Journal of Medicine and Pharmacy*, *2*(1), 8–14.

Revathy, S., Elumalai, S., & Antony, M. B. (2011). Isolation, purification and identification of curcuminoids from turmeric (*Curcuma longa* L.) by column chromatography. *Journal of Experimental Sciences*, *2*(7), 21–25.

Rezaei, F., Eikani, M. H., Nosratinia, F., & Bidaroni, H. H. (2023). Optimization of ethanol-modified subcritical water extraction of curcuminoids from turmeric (*Curcuma longa* L.) rhizomes: Comparison with conventional techniques. *Food Chemistry*, *410*, 135331. https://doi.org/10.1016/j.foodchem.2022.135331

Riva, A., Morazzoni, P., Artaria, C., Allegrini, P., Meins, J., Savio, D., . . . & Abdel-Tawab, M. (2016). A single-dose, randomized, cross-over, two-way, open-label study for comparing the absorption of boswellic acids and its lecithin formulation. *Phytomedicine*, *23*(12), 1375–1382.

Rolfe, V., Mackonochie, M., Mills, S., & MacLennan, E. (2020). Turmeric/curcumin and health outcomes: A meta-review of systematic reviews. *European Journal of Integrative Medicine*, *40*, 101252. https://doi.org/10.1016/j.eujim.2020.101252

Rudrappa, G. H., Chakravarthi, P. T., & Benny, I. R. (2020). Efficacy of high-dissolution turmeric-sesame formulation for pain relief in adult subjects with acute musculoskeletal pain compared to acetaminophen: A randomized controlled study. *Medicine*, *99*(28). http://doi.org/10.1097/MD.0000000000020373

Rudrappa, G. H., Murthy, M., Saklecha, S., Kare, S. K., Gupta, A., & Basu, I. (2022). Fast pain relief in exercise-induced acute musculoskeletal pain by turmeric-boswellia formulation: A randomized placebo-controlled double-blinded multicentre study. *Medicine*, *101*(35), e30144. http://doi.org/10.1097/MD.0000000000030144

Ryan, J. L., Heckler, C. E., Ling, M., Katz, A., Williams, J. P., Pentland, A. P., & Morrow, G. R. (2013). Curcumin for radiation dermatitis: A randomized, double-blind, placebo-controlled clinical trial of thirty breast cancer patients. *Radiation Research*, *180*(1), 34–43. https://doi.org/10.1667/RR3255.1

Sabinsa. (1991). Boswellin®. https://sabinsa.com/products-and-services/standardized-phyto-extracts

Sahebkar, A., & Henrotin, Y. (2016). Analgesic efficacy and safety of curcuminoids in clinical practice: A systematic review and meta-analysis of randomized controlled trials. *Pain Medicine (Malden, Mass.)*, *17*(6), 1192–1202. https://doi.org/10.1093/pm/pnv024

Sahoo, A., Kar, B., Jena, S., Dash, B., Ray, A., Sahoo, S., & Nayak, S. (2019). Qualitative and quantitative evaluation of rhizome essential oil of eight different cultivars of *Curcuma longa* L. (Turmeric). *Journal of Essential Oil Bearing Plants*, *22*(1), 239–247.

Salimzadeh, A., Ghourchian, A., Choopani, R., Hajimehdipoor, H., Kamalinejad, M., & Abolhasani, M. (2017). Effect of an orally formulated processed black cumin, from Iranian traditional medicine pharmacopoeia, in relieving symptoms of knee osteoarthritis: A prospective, randomized, double-blind and placebo-controlled clinical trial. *International Journal of Rheumatic Diseases*, *20*(6), 691–701.

Sanchez, C., Zappia, J., Lambert, C., Foguenne, J., Dierckxsens, Y., Dubuc, J. E., . . . & Henrotin, Y. (2022). *Curcuma longa* and *Boswellia serrata* extracts modulate different and complementary pathways on human chondrocytes *in vitro*: Deciphering of a transcriptomic study. *Frontiers in Pharmacology*, *13*. https://doi.org/10.3389/fphar.2022.931914

Saran, P. L., Tripathy, V., Saha, A., Kalariya, K. A., Suthar, M. K., & Kumar, J. (2017). Selection of superior *Ocimum sanctum* L. accessions for industrial application. *Industrial Crops and Products*, *108*, 700–707.

Saroj, T., & Krishna, A. (2017). A comparison of chemical composition and yiels of essential oils from shoot system parts of *Ocimum sanctum* found in semi-arid region of Uttar Pradesh. *Agrotechnology*, *6*(3), 172. https://doi.org/10.4172/2168-9881.1000172

Saxena, A., Mendoza, T., & Cleeland, C. S. (1999). The assessment of cancer pain in north India: The validation of the Hindi Brief Pain Inventory—BPI-H. *Journal of Pain and Symptom Management*, *17*(1), 27–41.

Saxena, S. N., Rathore, S. S., Diwakar, Y., Kakani, R. K., Kant, K., Dubey, P. N., . . . & John, S. (2017). Genetic diversity in fatty acid composition and antioxidant capacity of *Nigella sativa* L. genotypes. *LWT*, *78*, 198–207.

Schmiech, M., Lang, S. J., Werner, K., Rashan, L. J., Syrovets, T., & Simmet, T. (2019). Comparative analysis of pentacyclic triterpenic acid compositions in oleogum resins of different Boswellia species and their in vitro cytotoxicity against treatment-resistant human breast cancer cells. *Molecules*, *24*(11), 2153. https://doi.org/10.3390/molecules24112153

Sengupta, K., Alluri, K. V., Satish, A. R., Mishra, S., Golakoti, T., Sarma, K. V., . . . & Raychaudhuri, S. P. (2008). A double blind, randomized, placebo controlled study of the efficacy and safety of 5-Loxin® for treatment of osteoarthritis of the knee. *Arthritis Research & Therapy*, *10*(4), 1–11. http://arthritis-research.com/content/10/4/R85

Sengupta, K., Krishnaraju, A. V., Vishal, A. A., Mishra, A., Trimurtulu, G., Sarma, K. V., . . . & Raychaudhuri, S. P. (2010). Comparative efficacy and tolerability of 5-Loxin and Aflapin® against osteoarthritis of the knee: A double blind, randomized, placebo controlled clinical study. *International Journal of Medical Sciences*, *7*(6), 366–377. https://doi.org/10.7150/ijms.7.366

Sethi, V., Garg, M., Herve, M., & Mobasheri, A. (2022). Potential complementary and/or synergistic effects of curcumin and boswellic acids for management of osteoarthritis. *Therapeutic Advances in Musculoskeletal Disease*, *14*. https://doi.org/10.177/1759720X221124545

Shankar, B. A., Yadav, M. K., Kumar, M., & Burman, V. (2022). Differential gene expression analysis under salinity stress in the selected turmeric (*Curcuma longa* L.) cultivars for curcuminoid biosynthesis. *Research Square*. https://doi.org/10.21203/rs.3.rs-1951563/v1

Shariq, I. M., Israil, A. M., Iqbal, A., & Brijesh, P. (2015). Morpho-physiological characterization of seeds and seedlings of *Nigella sativa* Linn.: Study on Indian germplasm. *International Research Journal of Biological Sciences*, *4*(4), 38–42.

Siddiqui, M. Z. (2011). *Boswellia serrata*, a potential antiinflammatory agent: An overview. *Indian Journal of Pharmaceutical Sciences*, *73*(3), 255–261.

Singh, A., Das, S. K., Chopra, A., Danda, D., Paul, B. J., March, L., . . . & Antony, B. (2022). POS1425. The burden of osteoarthritis across the states of India, 1999–2019: Findings from the global burden of disease study 2019. *Annals of Rheumatic Diseases*, 1055–1056. https://doi.org/10.1136/annrheumdis-2022-eular.2361

Singh, B. B., Mishra, L. C., Vinjamury, S. P., Aquilina, N., & Shepard, N. (2003). The effectiveness of *Commiphora mukul* for osteoarthritis of the knee: An outcomes study. *Alternative Therapies in Health & Medicine*, *9*(3), 74–79.

Singh, D. P., Tripathi, P. K., Tripathi, S., Verma, N. K., Chandra, V., & Roshan, A. (2012). Phytochemical constituents and pharmacological activities of *Ocimum sanctum* (Tulsi): A review. *Journal of Pharmaceutical Research and Clinical Practice*, *2*(1), 118–126.

Singh, S., Das, S. S., Singh, G., Schuff, C., de Lampasona, M. P., & Catalan, C. A. (2014). Composition, in vitro antioxidant and antimicrobial activities of essential oil and oleoresins obtained from black cumin seeds (*Nigella sativa* L.). *BioMed Research International*, *2014*. http://doi.org/10.1155/2014/918209

Singh, S., & Majumdar, D. K. (1995). Analgesic activity of *Ocimum sanctum* and its possible mechanism of action. *International Journal of Pharmacognosy*, *33*(3), 188–192.

Singh, S., Taneja, M., & Majumdar, D. K. (2007). Biological activities of *Ocimum sanctum* L. fixed oil—An overview. *Indian Journal of Experimental Biology*, *45*, 403–412.

Singhal, S., Hasan, N., Nirmal, K., Chawla, R., Chawla, S., Kalra, B. S., & Dhal, A. (2021). Bioavailable turmeric extract for knee osteoarthritis: A randomized, non-inferiority trial versus paracetamol. *Trials*, *22*(1), 1–1. https://doi.org/10.1186/s13063-021-05053-7

Smith, D. I., & Trần, H. (eds.). (2022). *Pathogenesis of Neuropathic Pain: Diagnosis and Treatment*. Springer Nature.

Smith, T., Resetar, H., & Morton, C. (2022). US sales of herbal supplements increase by 9.7% in 2021. *HerbalGram*, *136*, 42–69.

Soltane, R., Mtat, D., Chrouda, A., Alzahrani, N., Al-Ghamdi, Y. O., El-Desouky, H., & Elbanna, K. (2021). *Nigella sativa* seed extract in green synthesis and nanocomposite. In M. F. Ramadan (ed.), *Black Cumin (Nigella Sativa) Seeds: Chemistry, Technology, Functionality, and Applications* (pp. 179–190). Food Bioactive Ingredients. https://doi.org/10.1007/978-3-030-48798-0_12

Sontakke, S., Thawani, V., Pimpalkhute, S., Kabra, P., Babhulkar, S., & Hingorani, L. (2007). Open, randomized, controlled clinical trial of *Boswellia serrata* extract as compared to valdecoxib in osteoarthritis of knee. *Indian Journal of Pharmacology*, *39*(1), 27–29.

Soo May, L., Sanip, Z., Ahmed Shokri, A., Abdul Kadir, A., & Md Lazin, M. R. (2018). The effects of *Momordica charantia* (bitter melon) supplementation in patients with primary knee osteoarthritis: A single-blinded, randomized controlled trial. *Complementary Therapies in Clinical Practice*, *32*, 181–186. https://doi.org/10.1016/j.ctcp.2018.06.012

Spice Board of India (2023). http://www.indianspices.com/sites/default/files/majorspicewise

Srivastava, R. K., Kumar, S., & Sharma, R. S. (2018). *Ocimum* as a promising commercial crop. In A. K. Shasany & C. Kole (eds.), *The Ocimum Genome, Compendium of Plant Genomes* (pp. 1–8). Springer Nature. https://doi.org/10.1007/978-3-319-97430-9_1

Srivastava, R. K., Kundu, A., Pradhan, D., Jyoti, B., Chokotiya, H., & Parashar, P. (2021). A comparative study to evaluate the efficacy of curcumin lozenges (TurmNova®) and intralesional corticosteroids with hyaluronidase in management of oral submucous fibrosis. *The Journal of Contemporary Dental Practice*, *22*(7), 751–755.

Srivastava, S., Saksena, A. K., Khattri, S., Kumar, S., & Dagur, R. S. (2016). Curcuma longa extract reduces inflammatory and oxidative stress biomarkers in osteoarthritis of knee: A four-month, double-blind, randomized, placebo-controlled trial. *Inflammopharmacology*, *24*, 377–388.

Sterniczuk, B., Rossouw, P. E., Michelogiannakis, D., & Javed, F. (2022). Effectiveness of curcumin in reducing self-rated pain-levels in the orofacial region: A systematic review of randomized-controlled trials. *International Journal of Environmental Research and Public Health*, *19*(11), 6443. https://doi.org/10.3390/ijerph19116443

Sudhir, S. P., Kumarappan, A., Malakar, J., & Verma, H. N. (2016). Genetic diversity of *Nigella sativa* from different geographies using RAPD markers. *American Journal of Life Sciences*, *4*(6), 175–180.

Thilakarathne, R. C. N., Madushanka, G. D. M. P., & Navaratne, S. B. (2018). Determination of composition of fatty acid profile of Ethiopian and Indian black cumin oil (*Nigella sativa*). *International Journal of Food Science and Nutrition*, *3*(3), 1–3.

Tiruppur Venkatachallam, S. K., Pattekhan, H., Divakar, S., & Kadimi, U. S. (2010). Chemical composition of *Nigella sativa* L. seed extracts obtained by supercritical carbon dioxide. *Journal of Food Science and Technology*, *47*, 598–605.

Tiwari, P., Jena, S., Satpathy, S., & Sahu, P. K. (2019). *Nigella sativa*: Phytochemistry, pharmacology and its therapeutic potential. *Research Journal of Pharmacy and Technology*, *12*(7), 3111–3116.

Tuna, H. I., Babadag, B., Ozkaraman, A., & Alparslan, G. B. (2018). Investigation of the effect of black cumin oil on pain in osteoarthritis geriatric individuals. *Complementary Therapies in Clinical Practice*, *31*, 290–294.

Twycross, R.G. (1984). Analgesics. *Postgraduate Medical Journal*, 60, 876–880.

Uma Devi, P., Ganasoundari, A., Vrinda, B., Srinivasan, K. K., & Unnikrishnan, M. K. (2000). Radiation protection by the ocimum flavonoids orientin and vicenin: Mechanisms of action. *Radiation Research*, *154*(4), 455–460.

Uritu, C. M., Mihai, C. T., Stanciu, G. D., Dodi, G., Alexa-Stratulat, T., Luca, A., . . . & Tamba, B. I. (2018). Medicinal plants of the family Lamiaceae in pain therapy: A review. *Pain Research and Management*, 1–44. https://doi.org/10.1155/2018/7801543

Usharani, P., Fatima, N., Kumar, C. U., & Kishan, P. V. (2014). Evaluation of a highly standardized *Withania somnifera* extract on endothelial dysfunction and biomarkers of oxidative stress in patients with type 2 diabetes mellitus: A randomized, double blind, placebo controlled study. *International Journal of Ayurveda and Pharma Research*, *2*(3), 22–32.

Valsan, A., Athulya Bose, G. K., Amrutha, C. S., Anil Kumar, A. K., Jayamol, K. V., & Kumar, S. (2022). Preliminary phytochemical screening of indigenous medicinal plants *Ocimum tenuiflorum*, *Ocimum basilicum*, and *Ocimum gratissimum*. *Research Journal of Agricultural Sciences*, *13*(4), 925–930.

Verma, R. S., Padalia, R. C., Chauhan, A., & Thul, S. T. (2013). Exploring compositional diversity in the essential oils of 34 Ocimum taxa from Indian flora. *Industrial Crops and Products*, *45*, 7–19.

Vishal, A. A., Mishra, A., & Raychaudhuri, S. P. (2011). A double blind, randomized, placebo controlled clinical study evaluates the early efficacy of Aflapin in subjects with osteoarthritis of knee. *International Journal of Medical Sciences*, *8*(7), 615–622. https://doi.org/10.7150/ijms.8.615

Vitali, M., Naim Rodriguez, N., Pironti, P., Drossinos, A., Di Carlo, G., Chawla, A., & Gianfranco, F. (2019). ESWT and nutraceutical supplementation (Tendisulfur Forte) vs ESWT-only in the treatment of lateral epicondylitis, Achilles tendinopathy, and rotator cuff tendinopathy: A comparative study. *Journal of Drug Assessment*, *8*(1), 77–86. https://doi.org/10.1080/21556660.2019.1605370

Yu, G., Xiang, W., Zhang, T., Zeng, L., Yang, K., & Li, J. (2020). Effectiveness of Boswellia and Boswellia extract for osteoarthritis patients: A systematic review and meta-analysis. *BMC Complementary Medicine and Therapies*, *20*(1), 1–16. https://doi.org/10.1186/s12906-020-02985-6

Orchid Wealth of India for Immunity Development

4

Lakshman Chandra De

4.1 INTRODUCTION

Orchidaceae family is the most diverse, advanced, evolved and perhaps the oldest flowering plants (originated ~100 million years ago) growing in different ecosystems and habitats. The rich Indian diversity is used in Ayurveda, Tibetan and Siddha medicine. Over 800 species are found in three hot spots in India: Eastern Himalayas, Western Ghats and Indo-Burma region. Moreover, five states in North-East India have orchids as their state flower because of their sacred value and cultural importance in society (Badaya et al. 2016). Orchids produce flowers and fruits (capsules) and each fruit contains thousands of very small (~470–560 μm long, 80–129 μm wide, ~6 μg weight) seeds. However, only 1% of the seed germinate naturally, but embryo culture or embryo rescue can elevate that germination rate to 99%. From a historical perspective, about 50 orchid species became endangered and threatened because of large scale imports and unavailability of modern hybrids in India. Interestingly, the imported hybrids (primarily from Thailand, Taiwan, Japan, Australia and USA) became segregated populations unqualified as varietal standards as per Indian Seed Act, 1966 and PPV & FR Act, 2001. Native gene pool (indigenous species) varieties obviously adapt to local conditions (G x E) and performs well under pests and disease stress. These new varieties developed using native orchid resources help to provide genetic control of traits and construct genetic maps thereby alleviating pressure for habitual natural collections from forest areas.

The breeding programmes at the National Research Centre for Orchids in Sikkim have developed inter-generic, primary, secondary and tertiary hybrids. Three varieties (B.S. Basnet [*Cymbidium*], KungaGyatso [*Aranda*] and V. Nagaraju [*Dendrobium*]) have been released for cultivation after the first flowering of primary hybrid from first indigenous cross. The first orchid hybrid (*Calanthe dominyi*) flowered in 1856 and since then many artificial hybrids have been produced at intergeneric and interspecific level. Over 125,000 hybrids (~10, 000 annual average) have been registered to date (Badaya et al. 2016).

Orchidaceae is the second largest family of flowering plants (~7% of the total flowering plants), although some species of the family risk extinction (Pal et al. 2022). ICAR-National Research Center for Orchids, Sikkim, established in 1996 provides research support to orchid growers in India, preserves orchid germplasm and develops sustainable use of orchid biodiversity. The center has collected and preserved 3150 accessions of 352 native orchid species (the germplasm of ~400 species across the country [83

DOI: 10.1201/9781003352341-4

of which are rare, endangered, and threatened (RET) and 52 species are of medicinal interest]. Globally, the largest orchid genera are *Bulbophyllum*, *Epidendrum*, *Dendrobium* and *Pleurothallis* (2000, 1500, 1400 and 1000 species, respectively). Orchids have a wide distribution range in India with ~1350 species belonging to 186 genera occurring in eight orchid habitats. The Northeastern states hold ~876 species in 151 genera contributing 70% of the country's orchid wealth. However, many of these species (~400) are endemic and rare with high ornamental value (Pal et al. 2022).

The medicinal value of orchids is recorded in ancient Sanskrit scriptures (~250–300 BC) as remedy for many ailments. Many orchid genera are used in conventional health care systems; however, *Eulophia campestris*, *Orchis latifolia*, and *Vanda roxburghii* have medicinal properties of interest to the scientific community. Other orchid species (*Habenaria acuminata*, *H. susannae*, *Orchis latifolia*, *Pholidata articulata* and *Satyrium* species) are consumed as nutritious food in Nagaland. Leaves and new shoots of *Cymbidium* are also used as food by many tribes in Nagaland (Pal et al. 2022).

The traditional medicinal system in India uses about 150 orchid species. In tribal or folk medicine, *Geodorum densiflorum* is used to treat ephemeral fever, *Soccolabium papillosum* (*Acampe praemorsa*) for bone fracture and body ache and *Bulbophyllum leopardinum* against sunstroke and diabetes. Fifteen of the 396 orchid species in Nagaland (Northeastern Himalaya state) are used by local practitioners to treat various diseases (rheumatism, cholera, nervous disorder and tuberculosis) as well as antimicrobial agent and antidotes to snake and insect bites (Pal et al. 2022). Terrestrial orchids of the Himalayas (1200–4000 m altitude) are part of the Astavarga group called 'Jeevaka' (strengthens vitality and immunity). These orchids known for their immunomodulating properties are: *Microstylis muscifera* Ridl. (*Malaxis muscifera*), *Habenaria edgeworthii* Hook f. ex Colt (*Platanthera edgeworthii*), and *Habenaria intermedia* (*H. arietina*) (Pal et al. 2022).

Till date, 29,199 species have been accepted (Govaerts et al. 2017). Orchids, one of the best-known plant groups in the global horticultural and cut flower trades, are also grown and traded for various purposes (e.g., ornamental plants, medicinal products and food). The medicinal orchids belong mainly to the genera: *Calanthe, Coelogyne, Cymbidium, Cypripedium, Dendrobium, Ephemerantha, Eria, Galeola, Gastrodia, Gymnadenia, Habenaria, Ludisia, Luisia, Nevilia* and *Thunia* (Szlachetko 2001). Orchids are commercially used in Chinese and South Asian traditional medicine systems (Leon and Lin 2017). The most prominently used orchids in traditional Chinese medicines are various *Dendrobium* spp. used to make the drug shi-hu [particularly *D. catenatum* Lindl. (including *D. officinale* Kimura & Migo), *D. loddigesii* Rolfe, *D. moniliforme* (L.) Sw. and *D. nobile* Lindl.)] (Leon and Lin 2017; Teoh 2016). Other orchids are also used including *Gastrodia elata* Blume tubers, *Bletilla striata* (Thunb.) Rchb.f.) rhizomes, *Anoectochilus* spp. rhizomes and stems, *Cremastra appendiculata* (D. Don) Makino, *Pleione bulbocodioides* (Franch.) Rolfe and *P. yunnanensis* (Rolfe) Rolfe corms (Leon and Lin 2017; Teoh 2016). They are also popularly used in some African traditional medicine (e.g. *Vanilla madagascariensis* Rolfe in Madagascar (Randriamiharisoa et al. 2015), North American folk medicine (e.g. *Cypripedium acaule* Aiton and *C. parviflorum* Salisb. [6], and the Unani medicine system [e.g. *Dactylorhiza hatagirea* (D. Don) *Soó Vanda tessellata* (Roxb.) Hook.ex G. Don, *Cymbidium bicolor* Lindl., and *Ipsea speciosa* Lindl. (Jayaweera and Fosberg 1980; Thakur and Dixit 2007; Khajuria et al. 2017).

4.2 ORCHIDS USES IN AYURVEDIC MEDICINE

In Ayurvedic medicine, 'Astavarga' a rejuvenating herbal formulation is derived from 8 herbs including some orchids (i.e., Jivak (*Microstylis wallichii*), Rishbhaka (*Habenaria acuminata*), riddhi (*H. intermedia*) and vriddhi (*H. edgeworthii*) (Handa 1986; Singh and Duggal 2009). *Flickingeria macraei* is used in 'Ayurveda' by the name 'Jeevanti' as an astringent agent, aphrodisiac and in the treatment of asthma and bronchitis (Kirtiker and Basu 1975). Other commonly used orchid drugs in the Ayurvedic system are salem (*Orchis latifolia* and *Eulophia latifolia*), jewanti (*Dendrobium alpestre*), shwethuli and rasna (*Acampe papillosa* and *Vanda tessellata*). In 'Sushrutasamhita' *Orchis latifolia* underground tuber is used in the drug 'munjatak' to relieve cough. *Vanda roxburghii* leaveshave been prescribed in the ancient Sanskrit literature for external application for rheumatism, ear infections, fractures and diseases of the nervous system. Nepal's Ayurvedic trade has 94 orchid species (Acharya and Rokaya 2010; Subedi et al. 2013), including *Crepidium acuminatum* (D.Don) Szlach., *Habenaria intermedia* D.Don, *Herminium edgeworthii* (Hook.f. ex Collett) X.H. Jin, Schuit., Raskoti & Lu Q. Huang and *Malaxis muscifera* (Lindl.) Kuntze (Hossain 2009; Dhyani et al. 2010; Khajuria et al. 2017). *Eulophia* spp. are also widely used medicinally in different parts of India [*E. dabia* (D.Don) Hochr., *E. spectabilis* Suresh in D.H. Nicolson, C.R. Suresh & K.S. Manilal (*E. nuda* Lindl.) (Jalal et al. 2014) and *D. hatagirea* is used to treat various ailments (Pant and Rinchen 2012).

In some Malaya regions women boil *Nervilia aragoana* leaves and drink the liquid immediately after childbirth as a precaution against possible post-natal sickness. *Corymborchis longiflora*, *Tropidia curculigoides* and *Acriopsis javanica* are reported as febrifuges in treating malaria.

4.3 ORCHIDS CHEMISTRY

Orchid phytochemicals can be classified as stilbenoids (stilbene, bibenzyls, phenanthrenes, 9,10-dihydrophenanthrenes, phenanthraquinones, 9,10-dihydrophenanthraquinones, phenanthropyrans and pyrones, 9,10-dihydrophenanthropyrans and pyrones, fluorenones), anthraquinones, pyrenes, coumarins, flavonoids, anthocyanins and anthocyanidins, chroman derivatives, lignans, simple benzenoid compounds, terpenoids (monoterpenes, sesquiterpenes, diterpenes, triterpenes), steroids, alkamines, amino acids, mono- and dipeptides, alkaloids and higher fatty acids. Dendrobium species produce various secondary metabolites such as phenanthrenes, bibenzyls, fluorenones and sesquiterpenes, and alkaloids are responsible for their wide ranging medicinal properties. Besides, numerous phenanthrenes compounds isolated from Dendrobium species are dihydrophenanthrene, ephemeranthoquionone, shihunidine, shihunine, dendrophenol, moscatilin, moscatin, denfigenin, defuscin, amoenumin, crepaditin, rotundatin, cumulatin, and gigantol. Some other orchid genera like *Eulophia, Cypripedium, Gastrodia, Bletilla, Bulbophyllum, Anoectochilus, Arundina, Eria, Malaxis, Habenaria, Vanda*, and *Vanilla* are enriched with different important phytochemicals (Hossain 2011) (Table 4.1).

4.4 PHARMACOLOGICAL PROPERTIES OF ORCHIDS

Antimicrobial: *Vanilla planifolia, Galeola foliata, Cypripedium macranthos var. rebunense, Spiranthes mauritianum, Bletilla striata*

Anti-inflammatory: *Anoectochilus formosanus, Gastrodia elata, Dendrobium moniliforme, Pholidota chinensis, Vanda roxburghii*

Anti-oxidant: *Anoectochilus formosanus, Anoectochilus roxburghii, Dendrobium moniliforme, D. nobile, Gastrodia elata.*

Antidiabetic: *Anoectochilus formosanus, Dendrobium candidum*

Antihepatotoxic: *Anoectochilus formosanus, Goodyera* species

Neuroprotective: *Coeloglossum viride, Gastrodia elata*

Anti-viral: Cymbidium hybrid, *Epipactis helleborine, Listera ovata, Gastrodia elata*

Antipyretic: *Dendrobium moniliforme*

Anticancer/Anti-tumor: *Anoectochilus formosanus, Bletilla striata, Bulbophyllum kwangtungense, Dendrobium chrysanthum, Dendrobium fimbriatum, Dendrobium nobile, Ephemerantha ionchophylla, Gastrodia elata, Spiranthes australis, Bulbophyllum odoratissimum*

TABLE 4.1 Some chemical constituents from orchids

SL.NO	*CHEMICAL NAME*	*PLANT SOURCE*	*SL.NO*	*CHEMICAL NAME*	*PLANT SOURCE*
1.	Aeridin	*Aerides crispum*	36.	Cypripedin, Cryptostylin	*Cypripedium calceolus, Cypripedium macranthum*
2.	Agrostophyllin	*Agrostophyllum brevipes*	37.	Defuscin, Dendroflorin, Dengidsin, Kaempferol, Naringenin, Taraxerol	*Dendrobium auranticum var. denneanum*
3.	Annoquinone	*Cypripedium macranthum*	38.	Dendrobine, Denbinobin, Dendrobinobine, Dendroside A, D, E, F, G, Dendronobiloside A, Nobilin Dand ENobilone	*Dendrobum nobile*
4.	Arundinin, Isoarundinin-I, II, Arundin	*Arundina graminifolia*	39.	Dendrocandin A, B, C, D, E, F, G, H, I	*Dendrobium candidum*
5.	Batatasin III	*Epidendrum rigidum*	40.	Dendrocrepine	*Dendrobium crepidatum*
6.	Blestrianol A, B, C, Bletilol-A, B Blestrin A, B, C, D	*Bletilla striata*	41.	Dendrochrysanene, Erianin	*Dendrobium chrysotoxum*
7.	Bulbophythrin A, B, 3,7-Dihydroxy-2-4-6-trimethoxyphenanthrene,	*Bulbophyllum odoratissimum*	42.	Dendromoniliside A, B, C, Moniliformin	*Dendrobium monoliforme*
8.	Callosmin, Imbricatin, Orchinol	*Agrostophyllum callosum*	43.	Dendroprimine, Hygrine	*Dendrobium primulinum*
9.	Calanthoside, Isatin, Indican, Glucoindican	*Calanthe discolor and C. liukiuensis*	44.	Denthyrsin, Denthyrsinone, Denthyrsinine, Denthyrsinol, Hircinol	*Dendrobium thyrsiflorum*
10.	Chysin A, B	*Chysis bractescens*	45.	9,10-Dihydro-2,5-Dihydroxy-3, 4 dimethoxy phenanthrene, Erianthridin, Fimbrinol A	*Maxillaria densa*
11.	Chrysotobibenzyl, Chrysotoxin	*Dendrobium aurantiacum*	46.	2,3-Dimethoxy-9,10-dihydrophenanthrene-4,7 diol	*Epidendrum rigidum*
12.	Cirrhopetalanthin	*Cremastra appendiculata*	47.	Ephemeranthrone, Lonchophylloid A, B, 3-methylgigantol	*Ephemerantha ionchophylla*
13.	Coelonin, 3,7-Dihydroxy-2,4,8-trimethoxyphenanthrene	*Coelogyne elata, Pholidota yunnanensis*	48.	Flaccidin, Flaccidinin, Oxoflaccidin, Isooxoflaccidin	*Coelogyne flaccida*
14.	Coeloginanthrin Coeloginanthridin CombretastatinC-1, Coelogin	*Coelogyne cristata*	49.	Gigantol	*Cymbidium giganteum Epidendrum rigidum., Scaphyglottis livida, Dendrobium aurantiacum var. denneanum*
15.	Confusarin, Coumarin	*Dendrobium aurantiacum*	50.	Gymconopin A, B, D	*Gymnadenia conopsea*

(Continued)

TABLE 4.1 (Continued)

SL.NO	*CHEMICAL NAME*	*PLANT SOURCE*	*SL.NO*	*CHEMICAL NAME*	*PLANT SOURCE*
16.	Crepidine, Crepidamine	*Dendrobium crepidatum*	51.	Gymnopusin	*Bulbophyllum gymopus*
17.	Cumulatin, Densiflorol A	*Bulbophyllum kwangtungense*	52.	Homoeridictyol, Scoparone, Dendroflorin	*Dendrobium densifiorum*
18.	Isoamoenylin, Amoenylin	*Dendrobium amoenum*	53.	Quercetin	*Dendrobium tosaense*
19.	Kuramerine	*Liparis kurameri*	54.	Shihunine, Shihunidine	*Dendrobium loddigesii*
20.	Lusianthridin	*Nidema boothii Lindl.*	55.	Sinensol A, B, C, D, E, F, Spirasineol B, Spiranthol-C, Spiranthoquinone	*Spiranthes sinensis var. amoena*
21.	Malaxin	*Malaxis congesta*	56.	Tristin	*Bulbophyllum triste, Dendrobium aurantiacum var. denneanum*
22.	Moscatin, Moscatilin	*Dendrobium moschatum, Dendrobium aurantiacum var. denneanum, Dendrobium loddiesii*	57.	Thunalbene	*Thunia alba*
23.	N-methylpiperidine	*Vandopsis longicaulis*	58.	Aloifol-I, Cymbinodin-A, B	*Cymbidium aloifolium*
24.	Nudol, Eulophiol	*Eulophia nuda*	59.	Erianin, Erianthridin	*Eria carinata*
25.	Ochrone-A, B Ochrolic acid, Ochrolon	*Coelogyne ochracea*	60.	Pendulin	*Cymbidium pendulum*
26.	Phalaenopsine	*Phalaenopsis mannii, Phalaenopsis equestris, Phalaenopsis ambilis*	61.	Agrostonin, Agrostonidin, Callosin, Callosumin	*Agrostophyllum callosum*
27.	Pholidotol A, B	*Pholidota chinensis*	62.	Flavanthridin	*Eria flava*
28.	Pieradine	*Dendrobium pierardii, Dendrobium aphyllum*	63.	Shancilin, Shanciol C, D, E, Sanjidin A, B	*Pleione bulbocodioides*
29.	Plicatol B	*Dendrobium plicatile*	64.	Flavidin *Flavidinin*	*Coelogyne flavida*
30.	Parviflorin	*Vanda parviflora*	65.	Benzaldehyde	*Zygopetalum mackayi*
31.	Tessalatin	*Vanda tessellata*	66.	Vanillyl methyl ether, Piperidinic acid	*Vanilla planifolia*
32.	Dengibsin, Dengibsinin	*Dendrobium gibsonii*	67.	Cycloartenol	*Catteya sp*
33.	Loroglossin	*Orchis maculata 0. incarnata, 0. latifolia*	68.	Parishin, Parishin B, C, Gastrol	*Gastrodia elata*
34.	Kinsenoside	*Anoectochilus formosanus*	69.	Heptacosane, Octacosano	*Vanda roxburghii*
35.	Habenariol	*Habenaria repens*	70.	Kaempferol-7β-D-glucopyranoside, Isorhamnetin-3-O β-D-glucopyranoside, Quercetin	*Anoectochilus roxburghii*

4.5 ORCHIDS WITH IMMUNOMODULATORY ACTION

Bletilla striata: Tuber is used to treat tuberculosis and haemorrhage. In China and Japan, it is used in wound healing, ulcers, inflammation, and haemostatic and as immunomodulator (He et al. 2017).

Corallorhiza maculata: Dried stems are used to restore blood in pneumonia patients in America and Europe (Hossain 2011).

Corymborchis longiflora: In Malayasia, it is used as febrifuge in treating malaria (Hossain 2011).

Dactylorhiza hatagirea: Tuber is used in burning sensation during urination, general debility, cough and cold, while decoction of tuber mixed with sugar is used as a drink in tuberculosis and effective against impotency (Panda and Mandal 2013).

Dendrobium aurantiacum: In China, herb is used as antipyretic, immunomodulatory, anti-ageing and in eye disorders.

Dendrobium candidum: Herb is used to strengthen stomach capacity, promote body fluid; used in the treatment of cataract, throat inflammation and immune boosters (Wang et al. 2014).

Dendrobium chrysanthum: Powdered dry leaves are used to treat eye related problems, skin diseases, as immunomodulator and antipyretic (Gutiérrez 2010).

Dendrobium denudans: In Tibet, Amchi people use the stem for cough, cold, nasal block and tonsillitis. The Nepali folk healers use it as tonic to increase the strength of old people and children (Panda and Mandal 2013).

Dendrobium nobile: Sesquiterepenes glycosides with alloaromadendrane, emmetin and picrotoxane types aglycones are isolated from stems of *Den. nobile*. These compounds exhibit immunomodulatory activity (Ye et al. 2002).

Eulophia ochreata: Tubers are used to combat general fatigue, boost immunity, and treat constipation, fever, skin diseases, wounds, tumours, boils, sunburns, cuts, injury and abdominal pain.

Habenaria edgeworthii: Leaves and tubers are used in blood and skin diseases, coughs, cold, asthma, leprosy, gout, general debility and as brain tonic and rejuvenator (Jalal et al. 2008).

Malaxis muscifera: Powdered bulbs are used to treat male fertility, while decoction is used in fever (Panda and Mandal 2013).

Satyrium nepalense: In Sikkim, tubers are used for reducing cold, cough and fever and mixed with yak ghee, used as aphrodisiac. Plant is used to proper child development and growth (Panda and Mandal 2013).

4.6 HEALTH BENEFITS

Some compounds isolated from orchids demonstrate their potential physiological benefits. For example, alkaloids, primarily dendrobine from *Dendrobium nobile* stem ethanol (95%) extract improve cognitive deficits, attenuate neuroinflammation and beta amyloid (Aβ) accumulation in Alzeimer's disease (AD) mice model (Li et al. 2022). The alkaloids are potential therapeutic agents to prevent and treat AD since they inhibit LPS-induced NOD-like receptor family 3 (NLRP3) inflammasome activation, release pro-inflammatory cytokines (IL-1β and IL-8) in the hippocampus (sensitive hippocampal CA1 region that causes cell death) and protects neuronal injury and working memory impairment. The alkaloids induce neuroprotection by ameliorating NLRP3-mediated pyroptosis (programmed cell death associated with cellular inflammatory response) (Li et al. 2022).

Paudel et al. (2022) list 36 anticancer compounds isolated from *Dendrobium* species. These compounds are from bibenzyl, phenanthtrene and fluorenone groups (26, 8 and 2 compounds, respectively). The anticancer mechanism is mediated via inhibition of cancer cell proliferation, apoptosis, induction, metastasis suppression and angiogenesis. Five compounds (**5**, **17**, **23**, **25** and **30**) discussed below have been investigated for their *in vitro* and *in vivo* anticancer effects on different cancer cell lines.

Compound **5**—4,5,4'-trihydroxy-3,3'-dimethoxybibenzyl from *D. ellipsophyllum* is highly cytotoxic on lung cancer cells (H23, H460 and H292) and upregulates tumor repressor protein p53 significantly increasing early and late apoptosis. Erianin (compound **17**) isolated from *D. chrysotoxum* induces apoptosis in T47D cells by attenuating Bcl-2 expression and activating caspase signaling as well as suppressing CDKs causing cell cycle arrest. It inhibits HeLa cell growth, induces apoptosis and cell cycle arrest at the G2/M phase and increases Bax and caspase-3 expressions. Gigantol (compound **23**) inhibits lung cancer (H292 and H460) cell migration, downregulates Cav-1 and activates Akt and Cdc-42, thereby suppressing filopodia formation. It increases EMT markers including N-cadherin, vimentin and slug causing significant suppression of protein kinase B, extracellular signal-regulated kinase, and Cav-1 survival pathways. Moscatilin (compound **25**) (100 mg/kg) significantly suppresses breast cancer metastasis to the lungs and reduces the number of metastatic lung nodules and lung weight without toxic effects. It also impedes angiogenesis by suppressing the activation of VEGY receptor 2 (Flk-1/KDR) and c-Raf-MEK1/2-ERK1/2 signals. Denbinobin (compound **30**) isolated from *D. nobile* and *D. moniliforme* induces human glioblastoma (GBM) cell apoptosis through IκB kinase inactivation, followed by Akt and fork head in rhabdomyosarcoma dephosphorylation and caspase-3-activation signaling cascade. It also induces apoptosis in lung and colorectal cancer via Akt inactivation, Bad activation, mitochondrial dysfunction, apoptosis-inducing factor releasing and DNA damage. It increases tubulin polymerization levels and deregulates Bcr-Abl signaling to inhibit human leukemia (K562) cell proliferation (Paudel et al. 2022).

4,5,4'-Trihydroxy-3–3'-dimethoxybibenzyl (TDB) extracted from *Dendrobium ellipsophyllum* inhibits human lung cancer cells by suppressing the AKT/GSK-3β signaling pathway. It also modulates adipocyte differentiation that regulates obesity by limiting G0/G1 phase progression, deactivating the AKT/GSK-3β signaling pathway and attenuating adipogenic regulators. TDB is proposed as a potential therapeutic agent against obesity (Khine et al. 2022).

4.7 VALUE ADDED AND HEALTH BENEFITS

Value addition in floriculture increases the economic value and consumer appeal of any floral commodity. In floriculture, value is added through genetic changes, processing or diversification. Orchid is a highly diversified flower crop. Indigenous species of *Aerides, Bulbophyllum, Calanthe, Coelogyne, Cymbidium, Paphiopedilum, Rhyncostylis, Renanthera* and *Vanda* are used as breeding materials, dry flowers, potted plants and herbal medicines for value addition (Table 4.2). They are adapted to diversified climate grown both epiphytically and terrestrially. Orchids are grown organically with locally available resources. Many orchids can be grown on rocks and logs for placing in the landscape. A beautiful colour scheme can be developed with *Cymbidium* and *Dendrobium* orchids. Orchid hybrids of *Cymbidium, Dendrobium, Vanda, Phalaenopsis, Oncidium, Cattleya, Paphiopedilum, Mokara, Aranda, Renantanda* etc. with different colour and forms are used as cut flowers, floral display and exhibits.

Tribal people of North-eastern hill region use wild orchids for various folk medicines as orchids are rich in alkaloids, flavonoids, glycosides, carbohydrates and other phytochemicals. Fragrant orchids like *Aerides multiflorum, Aerides odoratum, Cattleya maxima, Coelogyne cristata, Coelogyne ochracea, Dendrobium chrysotoxum, Lycaste, Oncidium spaceolatum, Rhyncostylis retusa,* and *Zygopetalum intermedium* are delightful in outdoor living areas. Leaves, tubers and pseudobulbs of different species are used for edible purposes. Vanilla a major spice crop and source of vanillin comes from *Vanilla planifolia. Anoectochilus* leaves are used as vegetables in Indonesia and Malayasia. *Cymbidium maladimum* and *Dendrobium speciosum* pseudobulbs and *Microtis uniflora* and *Caladenia carnea* tubers are edible. Miniature cymbidiums can be used as value added packed items. Bright flowers of orchid genera like *Dendrobium, Cymbidium, Paphiopedilum Cattleya, Pholidota* etc. can be used for drying. Among orchids, *Cymbidium, Dendrobium* and *Phalaenopsis* are excellent for wedding counter-pieces.

TABLE 4.2 Value addition to 75 Indian orchid species

SL. NO.	*NAME*	*HABITAT*	*VALUE ADDITION*	*PHOTOGRAPHS*
1.	*Acampe praemorsa*	Epiphytic	Crushed root used as tonic. Paste applied externally to treat rheumatism and pains. Paste taken orally to cure arthritis	A close-up captures a cluster of intricately detailed flowers, each featuring light-colored petals with darker markings in the form of stripes.
2.	*Aerides multiflorum*	Epiphytic	Ideal for pots and slat baskets. Leaf paste applied to treat cuts, wounds and earaches. Also used in tonic preparation. As parents for developing intergeneric hybrids	A potted plant with long, slender leaves and cascading clusters of small flowers
3.	*Aerides odorata*	Epiphytic	Leaf is antibacterial. Leaf paste is used to treat cuts and wounds, heal boils in ears and noses. The whole plant is used in tuberculosis, joint pain and swellings. Seeds used for wound healing. As parents for developing intergeneric hybrids (*Aeridopsis*, *Aeridovanda*).	A potted plant labeled as "Aerides odoratum," with the identification number "NOAC#2277". The plant has numerous small flowers arranged in cascading clusters and is placed on a metal grid surface.
4.	*Agrostophyllum brevipes*	Epiphytic	Rich in agrostophyllin. Tuber powder is used to prepare jaundice medicine.	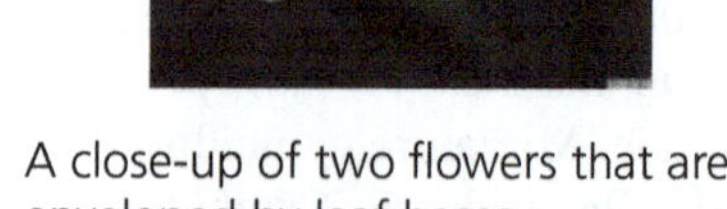 A close-up of two flowers that are enveloped by leaf bases.

(*Continued*)

TABLE 4.2 (Continued)

SL. NO.	*NAME*	*HABITAT*	*VALUE ADDITION*	*PHOTOGRAPHS*
5.	*Agrostophyllum callosum*	Epiphytic	Rich in Agrostonin, Agrostonidin, Callosin, Callosumin. Tuber powder is used in inflammation, diabetes, wound and skin disorders.	 A close-up image features a plant with long, narrow leaves and small, clustered flowers.
6.	*Anoectochilus roxburghii*	Terrestrial	Grown as attractive foliages. Rich in Kaempferol-7β-D-glucopyranoside, Isorhamnetin-3-O β-D-glucopyranoside, Quercetin. Infusion of whole plant consumed to control fever, lung disease, hypertension and to treat tuberculosis	 A close-up image captures four prominently veined, ovate-shaped leaves.
7.	*Anthogonium gracile*	Terrestrial	Ground tuber applied to cracking heels for quick pain relief. Paste used to treat bone fracture, dislocation and boils	 A close-up image features several small flowers with a bell-like shape, where the petals curve outward at the edges. Surrounding the flowers are long, narrow leaves.
8.	*Arachnis cathcartii*	Epiphytic	Used for wedding decorations, flower arrangements. Used as parent for developing intergeneric hybrids (i.e., Aranda, Aranthera, Arachnopsis)	 A close-up view of orchid flowers, highlighting their large, textured petals and intricate central structures. Surrounding the flowers are elongated, smooth leaves.

(Continued)

TABLE 4.2 (Continued)

SL. NO.	*NAME*	*HABITAT*	*VALUE ADDITION*	*PHOTOGRAPHS*
9.	*Arundina graminifolia*	Terrestrial	Used as a garden plant for landscaping. Loose flowers are ideal for drying and decorations and ornaments. Rich in Arundinin, Isoarundinin-I, II, and Arundin. Roots are used to relieve bodyache. Scrapped bulbous stem applied on foot-heels to treat cracks	 The plants feature long, slender leaves and tall stems with clusters of small flowers at the top.
10.	*Ascocentrum ampullaceum*	Epiphytic	Species are ideal for bamboo basket culture. Used as parent for developing intergeneric hybrids (Ascocenda, Kagawara, Mokara, Vascostylis, Robinara, Knudsonara)	 A close-up image features a potted plant with long, narrow leaves and clusters of small flowers near their base.
11.	*Bulbophyllum careyanum*	Epiphytic	Decoction of leaf powder consumed to get fast recovery after childbirth and to cause abortion. Fresh pulp of pseudobulb used in burns. Ideal for basket culture.	 A close-up image of a plant with a single elongated leaf and a bulbous base from which roots emerge. A curved stem extends outward, ending in a cluster of small buds or flowers.
12.	*Bulbophyllum odoratissimum*	Epiphytic	Rich in Bulbophythrin A, and B, 3,7-Dihydroxy-2-4-6-tri-methoxyphenanthrene. Infusion or decoction of whole plant used to treat tuberculosis, chronic inflammation and fracture	 A close-up of a flower with numerous elongated, thin petals radiating outward from the center.

(*Continued*)

TABLE 4.2 (Continued)

SL. NO.	NAME	HABITAT	VALUE ADDITION	PHOTOGRAPHS
13.	*Calanthe sylvatica*	Terrestrial	Flower juice is used to stop nose or gum bleeding. Leaves used to treat cold and cough. Ideal for bed culture and cut flowers.	 A close-up captures a plant with large leaves and a flowering stem. Each flowering stem holds multiple small flowers, each with elongated petals.
14.	*Calanthe triplicata*	Terrestrial	Ideal for bed culture in landscaping. Root and flower are used as analgesic during diarrhea and teeth cavities. Pseudobulb is used to treat gastrointestinal disorders. Used as parent for developing hybrids such as *Ghita Norby*.	 A close-up image showcases a cluster of white flowers, each with delicate, elongated petals that radiate symmetrically from the center.
15.	*Calanthe masuca*	Terrestrial	Used as cut flowers. Used to treat acne and inflammatory cysts. As parent for developing hybrids like Tydares Sieboca.	 A potted plant with large leaves and a tall stem, topped with a cluster of small flowers. The leaves have visible tears and holes.

(Continued)

TABLE 4.2 (Continued)

SL. NO.	*NAME*	*HABITAT*	*VALUE ADDITION*	*PHOTOGRAPHS*
16.	*Cephalanthera longifolia*	Terrestrial	Rhizomes are used as appetizer, tonic and wound healer. Powder used to increase vigour, vitality and alleviate impotency.	 A close-up image showcases a flowering plant with a tall central stem, bearing multiple small blossoms in a vertical arrangement. The flowers appear delicate, with slightly curved petals.
17.	*Cleisostoma williamsonii*	Epiphytic	Leaves and stems are used to heal fractured bones and also consumed during diarrhea.	 A close-up image features small flowers with delicate, radiating petals, arranged on thin stems
18.	*Coelogyne corymbosa*	Lithophytic or epiphytic	Ideal for hanging baskets. Pseudobulb juice applied in wounds and burns as analgesic. Paste applied on forehead to relieve headache.	 A close-up image presents two flowers with elongated petals and sepals, arranged symmetrically around a central column containing the reproductive structures.

(*Continued*)

TABLE 4.2 (Continued)

SL. NO.	NAME	HABITAT	VALUE ADDITION	PHOTOGRAPHS
19.	*Coelogyne cristata*	Lithophytic or epiphytic	Ideal for hanging baskets and pot culture. Rich in Coeloginanthrin, Coeloginanthridin, Combretastatin C-1, Coelogin. Pseudobulbs are used as an aphrodisiac to overcome constipation. Juice applied in wounds and boils. Gums exuded from pseudobulbs used to treat sores.	 A photograph captures several flowers growing through a wire fence. The flowers have elongated petals and are accompanied by long, narrow leaves.
20.	*Coelogyne fimbriata*	Lithophytic or epiphytic	Ideal for basket and pot culture. Pseudobulb paste or juice is consumed to relieve headache, fever and indigestion. Pulp applied over burnt skin. Powder form used as ingredients in tonic preparation.	 A close-up image features an orchid with elongated petals and a distinctive fringed lip, showcasing its intricate floral structure.
21.	*Coelogyne flaccida*	Lithophytic or epiphytic	Ideal for basket and pot culture. Rich in Flaccidin, Flaccidinin, Oxoflaccidin, Isooxoflaccidin. Pseudobulb paste is applied externally to relieve frontal headache and fever. Juice taken during indigestion.	 A close-up image features a potted plant with several small, star-shaped flowers, positioned in a container with multiple drainage holes.

(Continued)

TABLE 4.2 (Continued)

SL. NO.	NAME	HABITAT	VALUE ADDITION	PHOTOGRAPHS
22.	*Coelogyne ovalis*	Lithophytic or epiphytic	Ideal for basket and pot culture. Pseudobulb is an aphrodisiac. The whole plant is effective against cough, urinary infections and eye disorders.	 A close-up image features several flowers and buds at the tip with elongated leaves.
23.	*Corymborkis veratrifolia*	Terrestrial	Leaf juice is used as an emetic. Also used during constipation and fever in children	 A close-up image features a plant with large, broad leaves and clusters of small flowers, arranged at the nodes.
24.	*Cremastra appendiculata*	Terrestrial	Rich in Cirrhopetalanthin. Root powder is used in snakebite. Stem used in dental caries.	 A close-up image features two flowering plants with elongated stems and multiple narrow, drooping petals arranged around them.

(Continued)

TABLE 4.2 (Continued)

SL. NO.	NAME	HABITAT	VALUE ADDITION	PHOTOGRAPHS
25.	*Crepidium acuminatum*	Terrestrial	Root powder used in burns. Pseudobulb used to treat bronchitis, cold, cough, fever, tuberculosis and weakness. Pseudobulb also used as tonic and male aphrodisiac.	 A close-up image showcases a plant with small flowers attached to a central stem, surrounded by dense foliage. The flowers appear to be in various stages of blooming
26.	*Cymbidium aloifolium*	Epiphytic	Used as pot plants. Rich in Aloifol-I, Cymbinodin-A, B. Rhizome is purgative, also used in bone fracture. Powdered rhizome is consumed as tonic. Root paste used to cure rheumatism and nervous disorders. Pseudobulb is used as demulcent agent. Seed powder used for healing wounds. Leaves are used in boils and fever. Whole plant is used in weakness of eye, burns and sores.	 A close-up image features a potted plant with long, pointed leaves and small flowers emerging from the base. The flowers are visible in various stages, from buds to fully opened flowers,
27.	*Cymbidium ensifolium*	Epiphytic	Used as cut flower and pot plants. As parent for developing miniature hybrids. Root decoction used to treat gonorrhoea. Flower decoction is used in eye sore disorders.	 A close-up image captures a potted plant with elongated leaves and flowers emerging from a central stem.

(*Continued*)

TABLE 4.2 (Continued)

SL. NO.	NAME	HABITAT	VALUE ADDITION	PHOTOGRAPHS
28.	*Cymbidium iridioides*	Epiphytic	Used as pot plants and parent for developing standard hybrids. Fresh juice of leaf used as blood clotting in fresh wounds. Pseudobulb and root powder are used as tonic and eaten during diarrhea.	 A group of potted orchids with flowers, arranged on a metal grid structure. The orchids have delicate petals and slender stems, with leaves.
29.	*Dactylorhiza hatagirea*	Terrestrial	Tuber used in burning sensation during urination, general debility, cough and cold, while decoction of tuber mixed with sugar is used as a drink in tuberculosis and effective against impotency.	 A close-up of a plant with a cluster of small flowers densely packed along the stem and broad, serrated leaves frame the flowers.
30.	*Dendrobium chrysanthum*	Epiphytic	Ideal for pot culture. Loose flowers are ideal for freezing drying. Stem used as tonic to nourish stomach, enhance immune system, promote the production of body fluid and reduce fever. Leaf is used as antipyretic and mild skin diseases.	 A branch with small, tubular flowers arranged in clusters, accompanied by elongated leaves.

(*Continued*)

TABLE 4.2 (Continued)

SL. NO.	*NAME*	*HABITAT*	*VALUE ADDITION*	*PHOTOGRAPHS*
31.	*Dendrobium chrysotoxum*	Epiphytic	Ideal for pot culture and drying. Rich in Dendrochrysanene, Erianin. The extract of stem and flower used as tonic. Leaf extract used as antipyretic.	 A cluster of yellow flowers along with elongated leaves is resting on a wooden surface.
32.	*Dendrobium densiflorum*	Epiphytic	Ideal for pot culture and drying. Rich in Homoeridictyol, Scoparone, Dendroflorin. Pseudobulb cures vomiting and quenches thirst. Lowers fever of phlegm. Rejuvenates degenerated power of the digestive system caused due to lack of digestive fluids and appetite loss. Pulps of the pseudobulbs are used in boils, pimples and other skin eruptions. Leaf paste used in fractured bones, sprains and inflammations.	 Several potted plants placed on a metal mesh table. The plants have broad leaves with white spots, and one of them displays a cluster of yellowish flowers hanging down.
33.	*Dendrobium devonianum*	Epiphytic	Ideal for pot culture and drying. Dried stems used as an immune system enhancer.	 A hanging orchid with white flowers and long stems are set against a black fabric backdrop.

(Continued)

TABLE 4.2 (Continued)

SL. NO.	*NAME*	*HABITAT*	*VALUE ADDITION*	*PHOTOGRAPHS*
34.	*Dendrobium fimbriatum*	Epiphytic	Ideal for pot culture. The whole plant used in liver upset and nervous debility. Pseudobulb is considered as aphrodisiac, stimulant and demulcent. Pseudobulb used in liver upsets, nervous debility, fever and as tonic. Leaf used in fractured bones.	 A cluster of orchids growing among various stems and leaves in a greenhouse. The flowers have five petals each with a dark center.
35.	*Dendrobium longicornu*	Epiphytic	Ideal for pot culture, drying and tree mounting. Plant juice was used to relieve fever. Boiled roots used to feed livestock suffering from cough.	 A close-up of a flowering plant with several delicate blossoms and elongated leaves.
36.	*Dendrobium moniliforme*	Epiphytic	Ideal for wooden log culture. Rich in Dendromoniliside A, B, C and Moniliformin. Used as anti-inflammatory, antioxidant and anti-pyretic.	 A tree trunk with a cluster of small, star-shaped flowers growing directly on it. The flowers are light-coloured and densely packed together.

(*Continued*)

TABLE 4.2 (Continued)

SL. NO.	NAME	HABITAT	VALUE ADDITION	PHOTOGRAPHS
37.	*Dendrobium moschatum*	Epiphytic	Loose flowers are ideal for freeze drying. Rich in Moscatin, Moscatilin. Pseudobulb decoction is consumed during general body weakness. Pseudobulb paste is used to treat fractured and dislocated bones. Leaf juice used during earache.	 A close-up view of a potted plant with numerous flowers. The flowers have elongated petals and a central dark area.
38.	*Dendrobium nobile*	Epiphytic	Used as a pot plant and parent for developing new hybrids. Rich in Dendrobine, Denbinobin, Dendrobinobine, Dendroside A, D, E, F and G, Dendronobiloside A, Nobilin D and E, Nobilone. Stem is considered as tonic, vigour, aphrodisiac, stomachic, analgesic and anti-ageing. Stem decoction used to reduce salivation, thirst, tongue dryness, night sweating, menstrual pain and nightfall. Powdered stem used in dyspepsia, parched and thirsty mouth, fever, anorexia, pulmonary tuberculosis and lumbago. The whole plant is used in nervous disorders. Seeds applied to freshly cut wounds for quick healing.	 A potted plant with numerous flowers at the end of all branches. The flowers have light-colored petals with dark centers.
39.	*Dendrobium transparens*	Epiphytic	Ideal for pot culture. Pseudobulb paste is used to treat fractured and dislocated bones.	 A cluster of white flowers with dark centers attached to a thin stem, positioned in front of a chain-link fence.

(*Continued*)

TABLE 4.2 (Continued)

SL. NO.	NAME	HABITAT	VALUE ADDITION	PHOTOGRAPHS
40.	*Epipactis helleborine*	Terrestrial	Tuber used to treat insanity, gout, headache and stomach ache. The whole plant is used in nervous disorders.	 A close-up of a delicate flower with multiple textured petals and a central structure that appears to have a small round object on top.
41.	*Eulophia dabia*	Terrestrial	Rhizomes are used as appetizers, tonic and aphrodisiac. Used in treatment of purulent cough and heart trouble. Infusion given to infants to reduce cough, cold and as blood purifier.	 A close-up of a flower with multiple small blossoms arranged along a central stem.
42.	*Eulophia ochreata*	Terrestrial	Tubers are used to combat general fatigue, boost immunity, and treat constipation, fever, skin diseases, wounds, tumours, boils, sunburns, cuts, injury and abdominal pain.	 A close-up view of several orchid flowers with delicate petals and intricate details.
43.	*Eulophia spectabilis*	Terrestrial	Tuber is an appetizer, aphrodisiac and used as a tonic and blood-purifier. Also useful in tuberculosis glands on neck, tumors and bronchitis. Decoction of tuber is used to treat pinworm and roundworm infections. Leaf paste used in skin diseases.	 A close-up of an orchid plant with several buds and flowers arranged along a central stem.

(Continued)

TABLE 4.2 (Continued)

SL. NO.	NAME	HABITAT	VALUE ADDITION	PHOTOGRAPHS
44.	*Geodorum densiflorum*	Terrestrial	Tubers are used as poultice in wounds, skin diseases, carbuncles and insect bites. Also used to cure dysentery, diabetes, improving fertility in men and to regularize menstrual cycle in women.	 A close-up view of a potted plant with broad, textured leaves and a delicate, white flower.
45.	*Goodyera repens*	Terrestrial	Tuber paste externally applied to treat syphilis. Extract taken as blood purifier during irregular menstruation. The whole plant used against eye disorders and snake bites.	 A close-up view of a plant stem with multiple small, fuzzy flowers densely arranged along it. The flowers appear white and have a delicate, hairy texture.
46.	*Gymnadenia orchidis*	Terrestrial	Powered tubers are used to treat cuts, wounds, gastritis, diarrhea, liver and urinary disorders. Milk decoction is taken as aphrodisiac and tonic. Rejuvenates bodily strength, increases sperm count and restores diminished kidney heat.	 A close-up of a flowering plant with numerous small flowers clustered along a central stem. The petals curve outward in a tubular shape.

(*Continued*)

TABLE 4.2 (Continued)

SL. NO.	*NAME*	*HABITAT*	*VALUE ADDITION*	*PHOTOGRAPHS*
47.	*Habenaria dentata*	Terrestrial	The whole plant considered as analgesic, aphrodisiac, disinfectant and anti-rheumatic. Used to treat urinary and orthopedic problems.	 A close-up of a cluster of flowers with fringed petals, radiating outward in a circular pattern.
48.	*Habenaria furcifera*	Terrestrial	Tuber paste is applied on cuts, wounds and insect bites. Also used as tonic to improve body fluid.	 A close-up view of a plant stem with small, delicate flowers and buds attached along its length.
49.	*Herminium lanceum*	Terrestrial	Extracts of plant given in suppressed urination.	 A close-up view of a plant stem with small flowers arranged along its length. The petals extend outward in a curved shape.
50.	*Liparis nervosa*	Epiphytic	Tubers are used to treat stomachache, malignant ulcers	 A close-up view of a plant stem with small flowers featuring elongated petals.

(Continued)

TABLE 4.2 (Continued)

SL. NO.	*NAME*	*HABITAT*	*VALUE ADDITION*	*PHOTOGRAPHS*
51.	*Liparis odorata*	Epiphytic	Leaf juice is used to treat cancerous ulcer, gangrene and burns. Stem used to cure throat cancer.	 A close-up view of a plant with multiple flowers and buds growing along a vertical stem.
52.	*Luisia trichorrhiza*	Epiphytic	Paste of whole plant applied externally to cure muscular pain and orally consumed to treat jaundice.	 A close-up view of a plant's flowers and buds along its stem. The small, delicate blossoms have partially opened petals.
53.	*Luisia tristis*	Epiphytic	Leaf juice is used to treat chronic wounds, boils and burns.	 A close-up view of a small flower with translucent petals and a soft highlighted central structure.
54.	*Malaxis muscifera*	Terrestrial	Pseudobulbs are useful in sterility, seminal weakness, dysentery, fever and general debility as a tonic. Powdered form used to treat male infertility and decoction in fever and burning limbs.	 A close-up view of various plants with a tall, slender plant featuring a spike-like flower structure at the center. Surrounding foliage consists of leaves in different shapes and sizes.

(*Continued*)

TABLE 4.2 (Continued)

SL. NO.	NAME	HABITAT	VALUE ADDITION	PHOTOGRAPHS
55.	*Monomeria barbata*	Lithophytic	Tubers are used in treating coughs, pulmonary tuberculosis and trauma.	 A close-up of a plant stem with multiple small flowers and buds in various stages of bloom.
56.	*Oberonia caulescens*	Epiphytic	Tubers are used in liver ailments	 A close-up view of a flower spike with numerous tiny flowers densely packed along a central stem. Each small blossom has multiple petals and distinct reproductive structures.
57.	*Oberonia falconeri*	Epiphytic	The whole plant is used in bone fractures	 A close-up view of a tree trunk with elongated leaves and long, thin, spiky structures hanging down.

(*Continued*)

TABLE 4.2 (Continued)

SL. NO.	*NAME*	*HABITAT*	*VALUE ADDITION*	*PHOTOGRAPHS*
58.	*Otochilus porrectus*	Terrestrial	The whole plant is used as tonic and also in the treatment of sinusitis and rheumatism	 A close-up of a plant with small, delicate flowers arranged along a stem. The elongated petals extend outward.
59.	*Papilionanthe teres*	Epiphytic	Paste of whole plant applied externally to treat high fevers and dislocated bones. Juice used in burning sensation and to reduce fever, cough and cold.	 A greenhouse filled with various potted plants. In the foreground, there is a plant with vibrant flowers.
60.	*Paphiopedilum sp*	Semi-terrestrial	Indigenous species viz. *Paph. fairreanum, Paph. hirsutissimum, Paph. insigne, Paph. spicerianum, Paph. venustum, Paph. villosum* and *Paph. druyri* are used as pot plants, cut flowers and parents for developing multi-stemmed hybrids and inter-specific hybrids.	 Several potted plants are arranged on a wooden table, with some displaying long, narrow leaves and hanging flowers.

(*Continued*)

TABLE 4.2 (Continued)

SL. NO.	*NAME*	*HABITAT*	*VALUE ADDITION*	*PHOTOGRAPHS*
61.	*Phaius tankervilleae*	Terrestrial	Ideal for bed culture in landscaping. Tuber is tonic. Tuber paste is used to reduce swelling of gout, pain of fractured bones, redness and swelling. Also eaten during dysentery. Tubers and leaves poultice are used to heal infected wounds and boils.	 A plant with large, textured leaves and tall stems bearing buds and flowers.
62.	*Pholidota articulata*	Epiphytic	The whole plant used as tonic. Root powder is used to treat cancer. Fruit juice used to treat skin ulcers and skin eruptions. Pseudobulb applied to treat dislocated bones.	 A close-up of a plant with elongated leaves and a cluster of small, bell-shaped flowers hanging from the stem.
63.	*Pholidota chinensis*	Epiphytic	Pseudobulb extract is used in toothache, stomachache and inflammation.	 A potted plant with broad, elongated leaves and clusters of small, densely grouped flowers that cascade outward from the base of the foliage.

(*Continued*)

TABLE 4.2 (Continued)

SL. NO.	NAME	HABITAT	VALUE ADDITION	PHOTOGRAPHS
64.	*Pholidota imbricata*	Epiphytic	Pseudobulb juice is applied to relieve nasal, abdominal and rheumatic pain. Paste used to reduce fever, pain and swelling during arthritis. Powder used as tonic.	 A plant with elongated clusters of small flowers hanging downward in a dense, spiral pattern along the stems and broad, smooth leaves.
65.	*Pholidota pallida*	Epiphytic	Juice applied to relieve nasal, abdominal and rheumatic pains. Powder used to treat insomnia.	 An epiphytic plant growing on a tree trunk. The plant features elongated leaves and clusters of small, tubular flowers arranged densely along vertical stems.
66.	*Platanthera latilabris*	Terrestrial	Pseudobulb is used as blood purifier	 A close-up view of a plant stem adorned with small, delicate flowers.

(Continued)

TABLE 4.2 (Continued)

SL. NO.	*NAME*	*HABITAT*	*VALUE ADDITION*	*PHOTOGRAPHS*
67.	*Pleione humilis*	Terrestrial/ lithophyte	Dried powder of pseudobulb is used as tonic. Paste used in cut and wounds.	 A close-up of delicate flowers with elongated white petals and fringed edges, growing on a mossy surface.
68.	*Pleione maculata*	Terrestrial/ lithophyte	Rhizomes are used in liver complaints and stomach ailments.	 A close-up view of several flowers with elongated, pointed petals and a distinct central lip adorned with dark stripes and spots.
69.	*Pleione praecox*	Terrestrial/ lithophyte	Dried powder of pseudobulb is tonic. Paste used in cut and wounds.	 A close-up of several flowers with elongated petals and a central structure with serrated edges.

(*Continued*)

TABLE 4.2 (Continued)

SL. NO.	NAME	HABITAT	VALUE ADDITION	PHOTOGRAPHS
70.	*Polystachya concreta*	Epiphytic	Pseudobulb is used to treat arthritis	 A plant with elongated leaves and a tall central stem bearing small flowers at the top.
71.	*Renanthera imschootiana*	Epiphytic	Ideal as cut flowers and parents for developing intergeneric hybrids such as Aranthera, Renantanda, Renanthopsis, Renanstylis and Sappanara. Leaf paste is used in skin diseases.	 A potted orchid with a tall, slender stem and several broad leaves at its base. The flowers are arranged at the top of the stem.
72.	*Rhynchostylis retusa*	Epiphytic	Ideal for pot plants, festive decorations and flower arrangements. Used as parents for developing hybrids such as Rhynchorides, Neostylis, Renanstylis, Rhyncovanda, Rhyncovola and Rhyncolaeliocattleya. The paste of leaves and roots used in rheumatism. Leaf juice is used in constipation, gastritis, acidity and as emollient. Root juice is used in cuts and wounds. Root decoction of used in menstrual pain, arthritis, cuts and wounds. Dry flower is used as emetic.	 A potted plant with broad leaves and long, cascading flower spikes densely packed with small individual flowers that hang downward.

(Continued)

TABLE 4.2 (Continued)

SL. NO.	NAME	HABITAT	VALUE ADDITION	PHOTOGRAPHS
73.	*Satyrium nepalense*	Terrestrial	Tubers are used in diarrhea, dysentery and malaria. Tubers eaten as aphrodisiac and growth supplement for children. Juice is used externally in cut and wounds. Powder is tonic and used to reduce cold, cough and fever.	 An orchid plant with long elongated tubers having bud-like structures and long leaves.
74.	*Vanda coerulea*	Epiphytic	Ideal for cut flowers and pot plants. Used as parents for developing inter-generic hybrids like Aranda, Ascocenda, Renantanda, Kagawara, Mokara, Aeridovanda, Vascostylis and Robinara. Leaf juice is expectorant and used in eye diseases, diarrhoea, loose motion, dysentery and external skin diseases.	 A potted orchid plant with long, slender leaves and multiple flowering stems.
75.	*Vanda cristata*	Epiphytic	Ideal for pot plants. Leaf juice is expectorant and used in tonsillitis, bronchitis, dry cough and general weakness as tonic. Leaf paste applied to cuts and wounds. Root paste is used in cuts, wounds, boils and dislocated bones.	 An epiphytic orchid growing on a tree trunk. The orchid has elongated leaves and small and delicate flowers.

4.8 COSMECEUTICAL APPLICATIONS

Orchids are also used in the cosmetic industry because of their relevant properties to prevent and/or treat skin dryness, skin wrinkles and skin ageing (Kanlayavattanakul and Lourith 2022). These adverse skin effects are due to oxidants, radicals, inflammatory mediators, enzymes responsible for dryness and skin hyper-pigmentation. Orchids used for astringency or tonic effects are associated with their anti-inflammatory activities that help prevent and/or treat skin dryness and oxidative stress induced cellular inflammatory lesions. Moreover, orchid's antioxidant activities attenuate oxidative stress in dermal cells surplus and overproduction of skin melanin pigments. Some orchid species are already commercialized in cosmetic industry. Some Dendrobium (second largest genus in the Orchidaceae family) species have stronger inhibitory activities against mushroom tyrosinase than kojic acid (IC_{50} 57–112 vs 152 μg/mL) because of their phenolics and flavonoids contents. The anthocyanin rich ethanol (70%) extract of *Dendrobium Sonia* strongly inhibits collagenase, elastase, and tyrosinase that prevent collagen and elastin degradation occurring in skin dullness. The extract also suppresses cellular melanogenesis; these effects were attributed to the presence of ten phenolics (primarily sinapic and ferulic acids) and three anthocyanin (pelargonidin, cyanidin and keracyanin) constituents. Compounds isolated from *Dendrobium denneannum* stem exert potent anti-inflammatory effects. These compounds ([phenantheroide] 2,5-dihydro-4-methoxy-phenanthrene, 2-O-β-D-glucopyranoside [cucurbetacin] and 5-methoxy-2,4,7,9S-tetra-hydroxy-9–10-dihydrophenanthrene) suppress iNOS by p38, JNK, MAPK and IκBα inhibition through the MPKs and NF-κB pathways (Kanlayavattanakul and Lourith 2022).

Malaxis acuminata (both wild and *in vitro* derived plants) methanol leaf and stem extracts strongly inhibit skin ageing related enzymes (collagenase, elastase, tyrosinase and xanthine oxidase) and protect against UV-B and UV-A radiations *in vitro* with high sun protection factor. The extracts also possess anti-inflammatory activities (5-lipoxygenase and hyaluronidase) with considerable radical scavenging antioxidant activity (Bose et al. 2017).

A Spanish company in Barcelona has developed cultured *Calanthe discolor* (terrestrial orchid native to Japan) stem cell line from a greenhouse plant (Provital 2018). Adipocytes derived stem cells (ASCs) have beneficial effects on aging skin cells, primarily on dermal fibroblasts by increasing their antioxidant defense and extracellular matrix protein production and inhibition of metalloproteinase production. Moreover, cell proliferation and migration increase thereby aiding wound healing. Thus, ASCs influence wound healing, skin regeneration and photoaging. *Calanthe discolor* stem cell extract was incubated *in vitro* with human ASCs; orchistem induces the production of growth factors in the ASCs by increasing TGFβ1 production (50%). Orchistem also attenuates the inflammation processes, supports skin regeneration, and invigorates fibroblasts proliferation and migration. Orchistem as an active ingredient reduces sagging and redefine face contours suitable for anti-aging products intended to improve skin firmness, according to Provital (2018).

4.9 CURRENT AND FUTURE PERSPECTIVES

India is one of the orchid-rich (~1350 species in about 185 genera) countries in the world giving rise to rich biodiversity in several growing/geographical environments. Most species are ornamental (~200), while 55 species are considered medicinally important. However, orchids have the least production area and the minimum contribution in the turnover of floriculture products estimated at Rs. 717 crore businesses for 285,000 ha of floriculture crops with 2,284,000 tons of loose flowers and 947,000 tons cut flowers (APEDA 2023-24). The Government of India has initiated developmental programs to ensure native germplasm conservation and promote sustainable development of the orchid industry through research and development programmes. India produces and registers only about 200 orchid hybrids; however, tissue-culture true-to-type hybrid clones biotechnology approach has been the foremost contributor that can be commercially exploited. Commercial potential of orchids in India has its strength in rich orchid diversity/germplasm, diverse Agroclimate for orchid growing, technical capacity in growing, propagation, biotechnology and greenhouse technology, cheap labor and high-end increasing consumer market. Weakness of the industry is due to inadequate quantity and quality of planting materials, lack of market-driven approach in plant and flower production, low local hybrid production, inconsistent R&D backup with technical innovations, low production of planting material, inadequate quantity and quality of cut flowers to meet market demands, inadequate training and extension programmes and lack of community involvements (Hegde 2020).

Pharmacological studies on orchids indicate the immense potential of these plants in treatment of conditions such as neurodegenerative disorders, anticonvulsive, anticancer, antidiabetic, viral diseases and others. However, gaps in studies carried out are apparent which need to be bridged to exploit the full medicinal potential of orchids. Orchids have recently been proved to be a rich storehouse of chemical constituents with promising anti-tumor, anti-cancer and anti-inflammatory activities as revealed in modern biology-based studies. Investigations in progress can identify new biomolecules that confirm the usefulness of traditional remedies to develop new therapeutics. Orchid's species have recently been targeted for many investigations related to their chemical, biological, pharmacological, and medical properties. Traditional use of orchids preparation of Yin tonic in the Chinese, Tibetian and Ayurvedic medicine needs to be revised considering modern science of health and diseases. It is true that many people of developing countries from rural areas now prefer traditional medicines over synthetic ones because of many side effects, low production cost, easy availability and wide effectiveness. Meanwhile, consumers in developed countries are becoming disillusioned with modern health systems and are seeking alternatives. Since herbal medicines serve the health needs of about 80% of the world's population, and orchids containing numerous bioactive phytochemicals can be used as a promising source of medicine.

REFERENCES

Acharya, K. P., and Rokaya, M. B. (2010). Medicinal orchids of Nepal: Are they well protected? *Our Nature*, 8, 82–91.

APEDA (2023–24). Agricultural and Processed Food Products Export Development Authority, Ministry of Commerce and Industry, Government of India.

Badaya, V. K., Dadheech, A., Dhoot, M., and Khan, I. (2016). Orchid: Its commercialization and varietal development. *Indian Farmer*, 3(8), 563–568.

Bose, B., Choudhury, H., Tandon, P., and Kumaria, S. (2017). Studies on secondary metabolite profiling, anti-inflammatory potential, *in vitro* photoprotective and skin-aging related enzyme inhibitory activities of *Malaxisacuminata*, a threatened orchid of nutraceutical importance. *Journal of Photochemistry and Photobiology B: Biology*, 173, 686–695.

Dhyani, A., Nautiyal, B. P., and Nautiyal, M. C. (2010). Importance of Astavarga plants in traditional systems of medicine in Garhwal, Indian Himalaya. *International Journal of Biodiversity Science, Ecosystem Services & Management*, 6, 13–19.

Govaerts, R., Bernet, P., Kratochvil, K., Gerlach, G., Carr, G., Alrich, P., Pridgeon, A. M., Pfahl, J., Campacci, M. A., Holland Baptista, D., Tigges, H., Shaw, J., Cribb, P., George, A., Kreuz, K., and Wood, J. J. (2017). *World Checklist of Orchidaceae. Kew: Facilitated by the Royal Botanic Gardens.* Available at: http://apps.kew.org/wcsp/ (accessed 23 March 2017).

Gutiérrez, R. M. P. (2010). Orchids: A review of uses in traditional medicine, its phytochemistry and pharmacology. *Journal of Medicinal Plants Research*, 4(8), 592–638.

Handa, S. S. (1986). Orchids for drugs and chemicals. In: S. P. Vij (ed.), *Biology, Conservation and Culture of Orchids*. New Delhi: East West Press, pp. 89–100.

He, X., Wang, X., Fang, J., Zhao, Z., Huang, L., Guo, H., and Zheng, X. (2017). *Bletilla striata*: Medicinal uses, phytochemistry and pharmacological activities. *Journal of Ethnopharmacology*, 195, 20–38.

Hegde, S. N. (2020). Status of Orchid Industry in India. In S. M. Khasim, S. N. Hegde, M. T. González-Arnao and K. Thammasiri (eds.), *Orchid Biology: Recent Trends & Challenges*. Springer Nature, Singapore Pte Ltd., pp. 11–20. https://doi.org/10.1007/978-981-32-9456-1_2.

Henkel, A. (1906). *Wild Medicinal Plants of the United States* (No. 89). Washington, DC: US Department of Agriculture, Bureau of Plant Industry, Government Printing Office.

Hossain, M. M. (2009). Traditional therapeutic uses of some indigenous orchids of Bangladesh. *Medicinal and Aromatic Plant Science and Biotechnology*, 42, 101–106.

Hossain, M. M. (2011). Therapeutic orchids: Traditional uses and recent advances—An overview. *Fitoterapia*, 82, 102–140.

Jalal, J. S., Jayanthi, J., and Kumar, P. (2014). *Eulophia spectabilis*: A high value medicinal orchid under immense threat due to overexploitation for medicinal use in Western Ghats, Maharastra. *The MIOS Journal*, 15, 9–15.

Jalal, J. S., Kumar, P., and Pangtey, Y. P. S. (2008). Ethnomedical orchids of Uttarakhand, Western Himalaya. *Ethnobotanical Leaflets*, 12, 1227–1230.

Jayaweera, D. M. A., and Fosberg, F. R. (1980). *A Revised Handbook to the Flora of Ceylon—Complete Set*. A.A Balkema.

Kanlayavattanakul, M., and Lourith, N. (2022). Orchid extracts and cosmetic benefits. In J.-M. Mérillon and H. Kodja (eds.), *Orchids Phytochemistry, Biology and Horticulture: Fundamentals and Applications*. Springer International Publishing, pp. 609–626. https://doi.org/10.1007/978-3-030-38392-3_22

Khajuria, A. K., Kumar, G., and Bisht, N. S. (2017). Diversity with ethnomedicinal notes on orchids: A case study of Nagdev forest range, Pauri Garhwal, Uttarakhand, India. *Journal of Medicinal Plants*, 5, 171–174.

Khine, H. E. E., Sungthong, R., Sritularak, B., Prompetchara, E., and Chaotham, C. (2022). Untapped pharmaceutical potential of 4, 5, 4′-trihydroxy-3, 3′-dimethoxybibenzyl for regulating obesity: A cell-based study with a focus on terminal differentiation in adipogenesis. *Journal of Natural Products*, 85(6), 1591–1602.

Kirtiker, K. R., and Basu, B. D. (1975). *Indian Medicinal Plants*, 2nd ed., Vol. 4. Bishen Singh Mohendra Pal Singh.

Leon, C., and Lin, Y. L. (2017). *Chinese Medicinal Plants, Herbal Drugs and Substitutes: An Identification Guide*. Kew Publishing, p. 816. https://doi.org/978-1-84246-387-1

Li, D. D., Fan, H. X., Yang, R., Li, Y. Y., Zhang, F., and Shi, J. S. (2022). *Dendrobium Nobile* Lindl. alkaloid suppresses NLRP3-mediated pyroptosis to alleviate LPS-induced neurotoxicity. *Frontiers in Pharmacology*, 1403. https://doi.org/10.3389/fphsr.20220846541

Pal, R., Babu, P. K., and Dayamma, M. (2022). Indian Orchid germplasm: Conservation and utilization. In S. K. Datta and Y. C. Gupta (eds.), *Floriculture and Ornamental Plants*, pp. 359–387. https://doi.org/10.1007/978-981-15-3518-5_13

Pal, R., Meena, N. K., Dayamma, M., and Singh, D. R. (2022). Ethnobotany and recent advances in Indian medicinal orchids. In J.-M. Mérillon and H. Kodja (eds.), *Orchids Phytochemistry, Biology and Horticulture: Fundamentals and Applications*. Springer, pp. 361–387. https://doi.org/10.1007/978-3-030-38392-3_26-1

Panda, A. K., and Mandal, D. (2013). The folklore medicinal orchids of Sikkim. *Ancient Science of Life*, 33(2), 92–96. https://doi.org/10.4103/0257-7941.139043

Pant, S., and Rinchen, T. (2012). *Dactylorhiza hatagirea*: A high value medicinal orchid. *Journal of Medicinal Plants Research*, 6, 3522–3524.

Paudel, M. R., Bhattarai, H. D., and Pant, B. (2022). Traditionally used medicinal *Dendrobium*: A promising source of active anticancer constituents. In J.-M. Mérillon and H. Kodja (eds.), *Orchids Phytochemistry, Biology and Horticulture: Fundamentals and Applications*. Springer, pp. 389–414. https://doi.org/10.1007/978-3-030-38392-3_16

Provital (2018). *Calanthe Discolor*, 17 September. Available at: www.weareprovital.com

Randriamiharisoa, M. N., Kuhlman, A. R., Jeannoda, V., Rabarison, H., Rakotoarivelo, N., Randrianarivony, T., Raktoarivony, F., Randrianasolo, A., and Bussmann, R. W. (2015). Medicinal plants sold in the markets of Antananarivo, Madagascar. *Journal of Ethnobiology and Ethnomedicine*, 11, 60. https://doi.org/10.1186/s13002-015-0046-y

Singh, A., and Duggal, S. (2009). Medicinal orchids: An overview. *Ethnobotanical Leaflets*, 13, 351–363.

Subedi, A., Kunwar, B., Choi, Y., Dai, Y., van Andel, T., Chaudhary, R. P., de Boer, H. J., and Gravendeel, B. (2013). Collection and trade of wild-harvested orchids in Nepal. *Journal of Ethnobiology and Ethnomedicine*, 9, 64. http://www.ethnobiomed.com/content/9/1/64

Szlachetko, D. (2001). Genera et species Orchidalium. 1. *Polish Botanical Journal*, 46, 11–26.

Teoh, E. S. (2016). *Medicinal Orchids of Asia*, vol. 16, no. 4. Springer. https://doi.org/10.1007/978-3-319-24274-3

Thakur, M., and Dixit, V. K. (2007). Aphrodisiac activity of *Dactylorhiza hatagirea* (D.Don) Soo in male albino rats. *Evidence-Based Complementary and Alternative Medicine*, 4, 29–31.

Wang, Y., Liu, D., Chen, S., Wang, Y., Jiang, H., and Yin, H. (2014). A new glucomannan from Bletilla striata: Structural and antifibrosis effects. *Fitoterapia*, 92, 72–78.

Ye, Q., Qin, G., and Zhao, W. (2002). Immunomodulatory sesquiterpene glycosides from *Dendrobium nobile*. *Phytochemistry*, 61, 885–890.

Anti-COVID-19 Agents from Native Indian Crops

5

Mani Divya and Sekar Vijayakumar

5.1 INTRODUCTION

Severe acute respiratory syndrome coronavirus 2 (*SARS-CoV-2*) spreads horrifyingly often from person to person. The virus has already spread around the globe, causing devastating economic and social effects for millions of people (Barbosa and de Carvalho Junior 2021). Since the World Health Organization (WHO) recently affirmed an unprecedented coronavirus epidemic, there has been a worldwide surge in scientific interest in coronaviruses (Mani et al. 2020). The human population has been greatly diminished by this virus. So far, the disease has been contained using various anti-COVID-19 agents, including medicinal and traditional plants (Raza et al. 2021). Nonetheless, vaccines are being researched and developed globally as the best way to prevent SARS-CoV-2 infection (Zhou et al. 2020). In an effort to counteract SARS-CoV-2 and its symptoms, scientists around the world are actively searching for new treatments and vaccines. Although developing vaccines would be ideal, it still requires significant time and resources. Recent research (Long et al. 2020) demonstrates the importance of the immune system in combating viral infection. As a result, the immune response of hosts should be boosted by any promising new therapeutic interventions.

SARS-CoV-2 has very high nucleotide sequence identity (79.7%) with SARS-CoV-1. The Spike protein, one of the four human coronavirus envelopes, promotes host attachment and virion-cell membrane fusion during infection, thereby determining the host range and tissue tropism. Therapeutic interventions against human coronaviruses (CoVs) aim to either activate the host defense and immune system or block viral life cycle events (transmission, cell binding, replication, assembly, and enzymes involved in the synthesis of viral components). Three strategies are generally used to search for therapeutics against CoVs: evaluate existing broad-spectrum antiviral drugs, identify lead molecules or phytochemicals against viral or host proteins via *in silico* studies, and design or develop rational drugs based on genomic information and COVID-19 pathological characteristics (Meng et al. 2022). Ayurvedic medicines practiced in the Indian subcontinent for centuries are used to treat viral diseases by preparing herbal formulations containing active pharmacological agents and phytochemicals with physiological benefits. Several Indian Ayurvedic herbs, spices, and their bioactive phytochemicals have been investigated for their potential prophylactic and therapeutic use against COVID-19 (Maurya and Sharma 2022).

DOI: 10.1201/9781003352341-5

Plants and other natural sources have been the backbone of modern medicine (Atanasov et al. 2015). These herbal remedies are extremely significant in the modern era of medicine because of their potential to extend life by boosting the immune system. It is common practice to examine the effectiveness of traditional medicine extracts against viruses. Treatment strategies based on conventional medicinal products have only recently been used for the anticipation and action of COVID-19 in some select Asian countries (China, India, Japan, and South Korea) (Ang et al. 2020). Nontoxicity, biocompatibility, biodegradability, and antiviral properties are just some of the many benefits of these compounds. Biopolymers (e.g., sodium alginate, chitosan, and gums) are compounds that have garnered interest in the medical community for use in vaccination and delivery system development (Mallakpour et al. 2021). Plants and their metabolites have great potential as vaccine component against both newly developing and pandemic pathogens (Mosafer et al. 2019; Yang et al. 2018).

Generally, effective medications can be developed by mimicking the structure of natural chemicals that demonstrate the required action. Lin et al. (2014) provided a brief overview of naturally occurring antiviral agents acting against general coronaviruses, and Pang et al. (2020) and Lu (2020) both provided overviews of COVID-19 therapies, but natural therapeutics were only briefly mentioned without delving into the energetic compounds or their mechanisms of action.

There has been prior discussion of the value of agricultural diversification and the incorporation of neglected crops with potential as functional foods into the mainstream. Plants may provide nutrients for human health and help fight severe illnesses like COVID, but there has been little progress in finding the best prospects of underused crops and using agricultural development technologies to create superior varieties (Tiozon et al. 2021). The main widely grown crops are those used for human consumption, such as wheat (Xing Shi 2020), rice (de Lima Lessa et al. 2020), maize (Cheah et al. 2020), cassava (Okwuonu et al. 2021), and sweet potato (Siwela et al. 2020), or horticultural crops like strawberries (Budke et al. 2020). The effects of abiotic and biotic pressures on crop growth and production have been well documented, and they are only expected to worsen as a result of global climate change. It is therefore important to focus on improving crop tolerance to stress by developing new methods and creating better cultivars. However, most crop development efforts are concentrated on cereal grains like rice, wheat, maize, and others.

Mayes et al. (2012) proposed five criteria to identify the best possible underused crops' suitability to the region: economic price of the item, value-added flavor and texture, and decreased needs for agricultural products. Additionally, extracts and chemicals with antiviral properties are being developed that can potentially stop virus transmission or infection (Huang et al. 2020a). Plant-based vaccines and low-cost antiviral proteins are now under development (Mahmood et al. 2020)., These standards, according to the present pandemic, highlight minor millets, also known as tiny millets, to provide the poorest and most vulnerable people of society with food security and possibly even save their lives (Muthamilarasan and Prasad 2021). Naja and Hamadeh (2020) presented a multi-level action plan during the epidemic to encourage optimal nutrition at the individual, local, and governmental levels. Evidence supporting food-borne SARS-CoV-2 transmission has been fully examined, as well as plant bioactive chemicals and their mechanism by which they aid the human immune system to fight infection (Galanakis 2020; Panyod et al. 2020).

Besides the currently recommended COVID-19 prophylaxis, randomized regulated population-based testing might also assess the efficacy of plant extracts of various therapeutic crops, natural antiviral substances, and chemicals. Plant-based medications frequently originate from a pharmacological repertoire of organic molecules with antiviral properties and various mechanisms of action. Natural plants and their extracts have been utilized for centuries in alternative medicine to treat viral infections (Ganjhu et al. 2015). The scientific databases provide many studies on the anthelmintic, antiviral, antifungal, and antibacterial qualities of medicinal plants and crops with a wide range of ethnobotanical origins (Ben-Shabat et al. 2019; Zambounis et al. 2020; Rios and Recio 2005). In addition, the links between medicinal plants' health benefits and several specific secondary metabolites that are present may be the cause of some effects. Natural drug discovery against SARS-CoV-2 may be aided by the data presented here, concurrent with the COVID-19 pandemic vaccination against infectious diseases. In this chapter, we discuss the research on the medicinal properties of plant-derived metabolites, focusing on their effectiveness against viruses. Natural antiviral medicines are favored over synthetic ones because of their superior efficacy, lower risk of side effects, and lower cost. However, due to their low bioavailability, many of these natural drugs fail to pass clinical trials. Plants are a rich resource for chemodiversity, and it is possible that some of the compounds may be used to protect people from the deadly effects of COVID-19. As such, the purpose of this chapter is to compile and organize information about the medicinal history and plant parts used for antiviral potential of indigenous Indian crops.

5.2 COVID-19

The coronavirus outbreak has now reached pandemic proportions and spread to other nations. These coronaviruses are enclosed RNA viruses that belong to the *Coronaviridae* family, *Nidovirales* order, which includes four genera: *Alphacoronavirus*, *Betacoronavirus*, *Gammacoronavirus*, and *Deltacoronavirus* (Chen et al. 2020). *Alphacoronaviruses* and *betacoronaviruses* are connected with infections in mammalian species, but *gammacoronaviruses* and *deltacoronaviruses* have a broader variety of hosts. Recently, in December 2019, a novel strain of human coronavirus dubbed severe acute respiratory syndrome coronavirus 2 (SARS-CoV-2) was identified; this strain is accountable for the fragmentary epidemic coronavirus disease 2019 (COVID-19) (World Health Organization: Naming the Coronavirus Disease [COVID-19] and the Virus That Causes It). The coronavirus, COVID-19, is a pathogenic virus that primarily affects the human respiratory system as well as erstwhile organs, rapidly becoming a major public health problem (Shereen et al. 2020). SARS-CoV-2 is a novel strain that has never been discovered in people before. However, there have been earlier outbreaks of other beta coronavirus varieties, including the most recent Middle East respiratory illness (MERS-CoV), which was identified for the first time in 2012, and the severe acute respiratory syndrome (SARS-CoV-1), which occurred in 2002 and 2003. Currently, no particular therapy has been established for SARS-CoV-2 infection; nevertheless, various medications have shown apparent usefulness in inhibiting the disease's virus. Formally designated as 2019-nCoV by the World Health Organization (WHO), 2019-nCoV has been identified as the disease's etiological agent, and phylogenetic analysis indicates that it is distinct from previously identified coronaviruses (Drosten et al. 2003). The new coronavirus illness (nCoV) is a global pandemic danger since December 2019 (Qiao 2020). Coronaviruses are enclosed, single-stranded, positive-sense RNA viruses that prey on the respiratory system in humans. It has been demonstrated that viruses can sicken both people and animals (Wang et al. 2020). In order to create effective therapies, research has concentrated on elucidating the nature of the illness in response to the rising number of fatalities. Currently, there is no known therapy for COVID–19. The pandemic of COVID-19 necessitates quick manufacture of reagents, vaccines, monoclonal antibodies (mAbs), and medicines. In terms of SARS-CoV-2 immunization, ensuring fair access for all individuals in their respective nations is a significant obstacle. The catastrophic effect of this pandemic, particularly in poor countries, necessitates the development of low-cost protein technologies to manufacture SARS-CoV-2 vaccine candidates with the aim of controlling and preventing COVID–19.

Most research has recently investigated the feasibility of generating SARS-CoV-2 vaccine in plants. The fast and exponential spread of SARS-CoV-2 has placed a serious economic and health cost worldwide, particularly in middle- and low-income countries, necessitating the development of effective COVID-19 vaccines or treatment interventions. Children, people in good health, and people of all ages are being affected by the global coronavirus disease 2019 (COVID-19) pandemic brought on by the SARS-CoV-2 coronavirus (Grubaugh et al. 2018).

The emergence of COVID-19, the world's third major respiratory disease pandemic, has emphasized the world's economic equilibrium. The rising frequency of transferable illnesses has had a profound impact on the lives of millions of individuals (Liao et al. 2017). COVID-19 causes acute respiratory distress syndrome, anemia, heart damage, and secondary infections in patients. Antibiotics like azithromycin, cephalosporins, carbapenems, quinolones, tigecycline, and vancomycin; antivirals like lopinavir, oseltamivir, remdesivir, and ritonavir; and corticosteroids like dexamethasone and methylprednisolone have been frequently utilized to treat COVID-19 patients (Dubey et al. 2022).

5.3 ANTI-COVID-19 AGENTS

This continuous mutation of SARS-CoV-2 has been generally cited as the cause of increased coronavirus infections. Herbal remedies that target the host receptor ACE-2 may be useful in the prevention and treatment of SARS-CoV-2 infection because it infects cells through this receptor, unlike SARS-CoV. Natural compounds, such as herbs and mushrooms, have had significant antiviral and anti-inflammatory effects in the past. Therefore, natural compounds can potentially serve as efficient therapeutics against COVID-19 (Shahzad et al. 2020). Inhibitory characteristics have been reported in many edible medicinal plants, including *Cassia occidentalis* (Khan and Kumar 2019), *Cynara scolymus* (Elsebai et al. 2016), and *Punica granatum* (Tito et al. 2021). Rudravanti, also known as *Cressa cretica* (Linn), *Convolvulaceae* family, contains antitussive qualities, as well as bronchodilatory and mast cell-stabilizing characteristics, that has been used historically for cough and other respiratory ailments (Priyashree et al. 2010, 2012). Because it has antiviral activity against respiratory syncytial virus, the rhizome of *Zingiber officinale*, often known as ginger (Ardraka), has been used as a treatment for bronchitis and colds (Chang et al. 2013). Clove, also known as *Syzygium (S.) aromaticum*, is a member of the conventional flavor category and is rich in several phytochemicals, such as sesquiterpenes, hydrocarbons, monoterpenes, phenolic compounds, and others. These substances have various pharmacological effects and are used to preserve food (El-Saber Batiha et al. 2020).

The World Health Organization (WHO) estimates that roughly 230 million confirmed cases were documented in only a single day. As of the 31st of January, 2022, over 8 billion doses of vaccination had been distributed around the globe. In order to battle the pandemic, many medications have been repurposed, and vaccines are given the green light based on an emergency use approval procedure (Chavda and Apostolopoulos 2021).

5.4 INDIAN CROPS AGAINST COVID-19

Spread of infection while an epidemic is ongoing can be stopped by convincing antiviral herbs. Every civilization ever has made use of the healing properties of plants utilized for their medicinal purposes. Most of the worldwide population in developing and low-income nations rely on traditional medicine for basic health care (Kumar and Navaratnam 2013). Several pharmacological interest have been exhibited in plants extensively. Approximately 45,000 plant species have been employed in traditional medical systems in India, which has a rich and diverse plant collection (Pal and Jain 1998). Vaccines derived from plants are undergoing evaluation in clinical trials for influenza, hepatitis B, and other communicable diseases (Salazar-Gonzalez et al. 2015). More than 85% of people in underdeveloped nations take these drugs for health reasons (Sivasankari et al. 2014). More research is required to implement contemporary plant-based treatments and explore the antiviral ability of coronavirus in humans with native Indian crops (Figure 5.1 & Table 5.1).

Antiviral agents of human corona virus with native Indian crops.

Scientists employed binding mode algorithms to search drug catalogues for active ingredients during the SARS-CoV-2 pandemic (Zhang et al. 2020). PubMed and Google Scholar search was combined with numerous significant traditional Persian medical manuscripts, including *The Book of Al-Havi*, *The Canon of Medicine*, *Zakhireh-i-Kharazmshahi*, *Qarabadine-Kabir*, *Tohfat ol Moemenin*, and

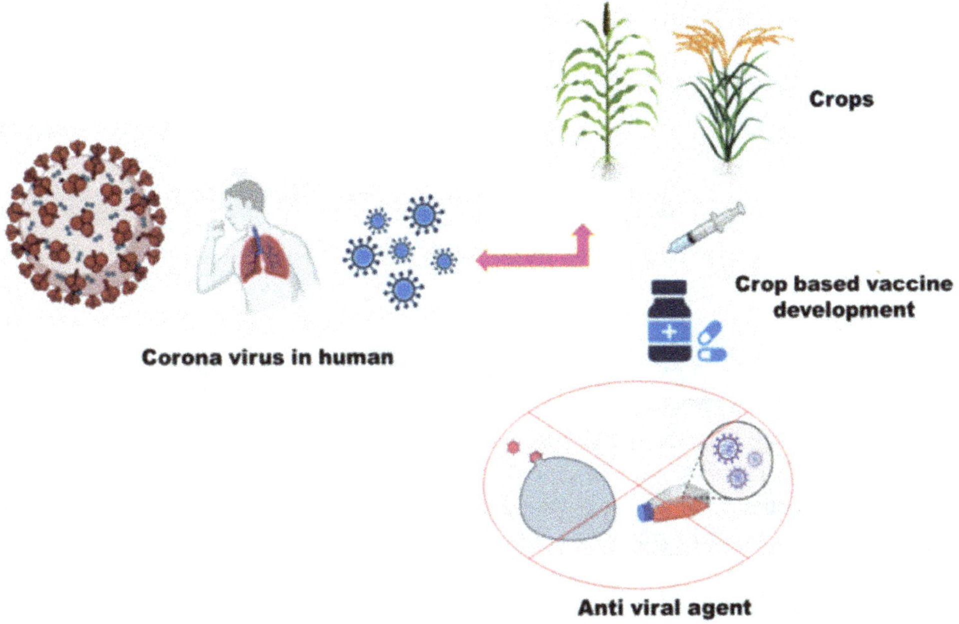

FIGURE 5.1

TABLE 5.1 Indian plants and its compounds effect against viral infections

INDIAN PLANTS	*COMPOUNDS*	*FUNCTIONS*	*REFERENCES*
Ocimum sanctum, *Curcuma longa* L, *Tinospora cordifolia*, *Piper nigrum* L, *Zingiber officinale*, *Syzygium aromaticum* (L.), *Elettaria cardamomum* (L.), *Citrus limon* (L.), *Withania somnifera*,	Kaempferol Stigmasterol Chavicine	Phytochemicals against SARS-CoV-2 main protease, spike protein, human ACE2 and furin proteins	Maurya and Sharma (2022)
Tinospora cordifolia	Aqueous extract	Effective against SARS-CoV-2	Kumar et al. (2020b)
Piper longum	Ayurvedic supplementary ingredient	Antiviral activity	Pandey et al. (2013)
Cocculus hirsutus	Coclaurine, Lirioresinol	Effective against SARS-CoV-2	Rajan et al. (2022)
Garcinia cambogia	Quercetin Vitexin, etc., Compound 5 Remdesivir	Effective against COVID-19	Aati et al. (2022)
Glycyrrhiza glabra	–	Inhibits viral replications of SARS-CoV	Nourazarian et al. (2016)
Vitex trifolia *Indigofera tinctoria* *Cassia alata* *Leucas aspera* *Gymnema sylvestre*	–	Effect against corona virus	Vimalanathan et al. (2009)
Sphaeranthus indicus	–	Decreases inflammatory cytokines in SARS-CoV	Alam et al. (2002)

Makhzan-ol-ol-Vieh, to generate a natural science dataset of crops with antiviral potential. The search was conducted using phrases pertaining to medicinal plants used to treat respiratory ailments (Ameri et al. 2015; Buso et al. 2020).

Antiviral plant metabolites may eliminate viruses and protect the respiratory system from illness (Huang et al. 2020b; Yang et al. 2018). Together with pharmacological network analysis, these novel methodologies were used to define physiologically active chemicals from *Juniperus communis* fruits and plants including *Thymus vulgaris*, *Curcuma longa*, *Rosmarinus officinalis*, *Ocimum basilicum*, *Melissa officinalis*, and *Mentha piperita* (Sampangi-Ramaiah et al. 2020). Quercetin from onions (*Allium cepa*), apples (*Malus domestica*), green tea (*Camellia sinensis*), and buckwheat (*Fagopyrum esculentum*) binds and inhibits the 6LU7 and 6Y2E proteases of SARS-CoV-2 (Zhang et al. 2020; Sampangi-Ramaiah et al. 2020; Lee et al. 2015).

India's pulses production index for fiscal year 2020 was about 164, a 64% increase compared to fiscal year 2008. In that year, the total agricultural production index was 141 (Chauhan et al. 2016). In terms of production and consumption, Bengal gram, red gram, pea, mung bean, urad bean, and lentils were among the most significant pulse crops with high protein content. The most common forms of these pulses are whole, split dhal, and flour. Pulses are scarce in South Indian cuisine owing to a lack of knowledge about their high nutritional value and practical applications, as well as the presence of anti-nutritional components, their frequent availability, and the absence of sufficient storage facilities for pulses (Bessada et al. 2019).

A little shrub with yellow blossoms, the ashwagandha plant is indigenous to North Africa and India. The traditional Indian herb, ashwagandha, also referred to as "Indian winter cherry" boosts energy while lowering anxieties and stress, and boosting the immune system. Extracts or powder from the plant's root or leaves are used to treat numerous illnesses (Dar et al. 2015). A fruit resembling a gourd that is endemic to tropical and subtropical areas, pumpkin is well known for its high nutritional composition. There are three economically important species of pumpkin, including *Cucurbita maxima*, *Cucurbita pepo*, and *Cucurbita moschata*, and the United States, China, Russia, and India are the world's leading producers. Pumpkin has indeed been strongly featured on the vegetable list due to its immune-boosting function and is preferred throughout and after the COVID-19 epidemic (Komarayanti et al. 2020).

Several Ayurvedic herbs have been used to manage SARS-CoV-2 infection. The Ministry of AYUSH (Ayurveda, Yoga and Naturopathy, Unani, Siddha & Homeopathy), Government of India, has recommended an Ayurvedic formulation composed of the following extracts to protect against COVID-19: *Tinospora cordifolia*, *Adhatoda vasica*, *Zingiber officinale*, *Piper longum*, *Hygrophila auriculata*, *Tragia involucrata*, *Plectranthus amboinicus*, *Anacyclus pyrethrum*, *Terminalia chebula*, *Saussurea costus*, *Syzygium aromaticum*, *Andrographis paniculata*, *Clerodendrum serratum*, *Sida acuta*, and *Cyperus rotundus*. SARS-CoV-2 protease and/or ACE2 inhibitor are the primary targets of these Ayurvedic antiviral compounds. Some Ayurvedic polyherbal formulations can be used as vaccine adjuvants to increase vaccine potency and immune responses since many COVID-19 symptoms mostly affect the respiratory, immune, and other physiological systems (Sarkar and Mukhopadhyay 2022). Ayurvedic adjuvants stimulate and potentiate cellular or humoral immune responses by activating cell signaling pathways and can benefit COVID-19 patients to avoid or prevent cytokine storm activation (Sarkar and Mukhopadhyay 2022). Most single or polyherbal formulations in Ayurvedic medicine modulate the immune response by increasing cytokine (IL-2 and IFNγ) production and enhancing natural killer and T cell activities. *Tinospora cordifolia* is the most frequently used intervention in COVID-19 preventive (n = 29) and recovery (n = 16) Clinical Trials Registry—India (CTRI) registered Ayurveda AYUSH studies for COVID-19 (Bhapkar et al. 2022).

Borse et al. (2021) explored the immunomodulatory and anti–SARS-CoV-2 potential of Ayurvedic *Rasayana* phytoconstituents from *Withania somnifera* (Ashwagandha) root, *Tinospora cordifolia* (Guduchi) stem, and *Asparagus racemosus* (Shatavari) root. Ayurvedic preparations (hydroalcoholic and aqueous extracts) generated 31 UHPLC-PDA-MS identified compounds (11 *W. somnifera*, 10 *T. cordifolia*, and 10 *A. racemosus* phytochemicals) with antiviral and immunomodulatory activities based on molecular docking and simulations. Muzanzagenin, the main drug-like molecule from *A. racemosus*, had the best docking score (~−6.0 kcal/mol) against all three SARS-CoV-2 protein targets. All *T. cordifolia* and *W. somnifera* phytochemicals except Ashwagandhanolide, Withanoside IV, and V are drug-like substances against SARS-CoV-2 replication. Moreover, the *Rasayana* phytoconstituents positively modulate several immune pathways, boosting the immune system, treating the inflammasome, regulating vascular permeability, and maintaining physiological homeostasis (Borse et al. 2021).

Nesari et al. (2022) evaluated the efficacy of "AYURAKSHA" kit as a post-interventional strategy to boost immunity and quality of life against COVID-19 in a nonrandomized, controlled, prospective prophylactic interventional trial (Identifier: CTRI/2020/05/025171). The AYURAKSHA kit is an immune-boosting prophylaxis to prevent COVID-19 infection. The kit contains three products: Sanshamani Vati (*Tinospora cordifolia* preparation, 500 mg BD after lunch and dinner), AYUSH Kadha (3 g/day decoction [4:2:2:1 ratio] of four medicinal herbs: Tulsi [*Ocimum sanctum*], Cinnamon [*Cinnamomum zeylanicum*], Ginger [*Zingiber officinale*], and black pepper [*Piper nigrum*]), and Anu Talia oil for nasal applications (two drops in each nostril, twice daily). Delhi police participants (n = 80) receiving the AYURAKSHA kits were evaluated for IgG COVID-19 positivity percentage, immune status questionnaire (ISQ), the WHO Quality of Life Brief Version (QOL BREF) scores, and hematological parameters as secondary outcomes compared to the control group (n = 66). The AYURAKSHA treatment for 2 months significantly reduced the percentage of COVID-19 IgG positivity (17.5% vs 39.4%, $P = 0.003$) and COVID-19 infection risk (55.6%) compared to the control group. Moreover, the control group showed a significant decline (~21% change of the cellular immune responses) in lymphocyte subsets CD3+, CD4+, and CD8+ levels, confirming their higher COVID-19 infection severity (lymphopenia) than the trial group. The AYURAKSHA kit is therefore associated with reduced COVID-19 positivity with a better quality of life (Nesari et al. 2022). The kit reduces the risk of liver abnormalities risk and maintains cytokine and normal blood sugar levels.

Ayush-64 is an Ayurvedic polyherbal drug developed and patented by the Central Council of Research in Ayurvedic Sciences (CCRAS). Ayush-64 tablet is prepared from a combination of four Ayurvedic herbs: *Alstonia scholaris*, *Swertia chirayita*, *Picrorhiza kurroa*, and *Caesalpinia crista* for their anti-inflammatory and immune-modulatory activities, *Agastha Hareetaki* for its rejuvenating and beneficial lung capacity, and *Anu Talia* (nasal insufflation therapy). Ayush-64 (2 tablets 500 mg each, thrice daily) was administered to 36 mild to moderate COVID-19 patients (19M/17F, median age 43 years) after food for either 8 or 14 days in a single-arm pilot study (Singh et al. 2021). Participants (86.1%) recovered clinically after 14 days, with 75% clinical recovery within 7 days. Moreover, 69.4% of participants turned negative for SARS-CoV-2 (nasal or throat swab evaluated by RT-PCR test) by the 15th day (50% negative on the 8th day). Ayush-64 is considered a safe treatment option in mild to moderate COVID-19 cases. It significantly facilitates clinical improvement based on the duration of clinical recovery and negative SARS-CoV-2 conversion without any adverse events (Singh et al. 2021).

Srivastava et al. (2021) performed a retrospective three-month (May–July 2020) study of 70 confirmed mild COVID-19 cases treated through Ayurvedic management. The Ayurvedic management consisted of AYUSH-64 tablet (1 g twice daily) and *Agasthya Hareetaki Rasayana* (5 g twice daily) after food and 2 drops of *Anu Talia* nasal administration in the morning before food for 10 days. The Ayurvedic interventions showed a higher clinical recovery rate in managing mild COVID-19 symptomatic cases than other clinical treatment (94.3% vs 69.5% in Chandigarh, India). Body temperature, pulse rate, and COVID-19-related symptoms (fever, headache, nasal discharge [circulatory illness], cough, sore throat, and dyspnea [cardiorespiratory disorders]) declined after the 3rd treatment day.

Wanjarkhedkar et al. (2022) evaluated the efficacy of an Ayurveda regime containing *Dasamoolkaduthrayan Kashaya* and *Guluchyai Kwatham* as an add-on to Standard of Care in mild and moderate COVID-19-infected patients to determine their symptom resolution rate. These formulations (*Dasamoolkaduthrayan Kashaya* and *Guluchyai Kwatham*), obtained from aqueous extracts of several medicinal plants, are manufactured as GMP-certified tablets (900 and 600 mg, respectively). They have been in the market for over two decades, primarily used to treat respiratory tract diseases. The treatment (n = 60, 44 years mean age) (two tablets of each) administered after meals for 7 days significantly alleviated breathlessness (53–1.6% vs 46–28% of patients for the control group) and ageusia or gustatory dysfunction (75–3.3% vs 46–1.3% of patients for the control group). The treatment also reduced the median hospital stay for COVID pharyngitis patients (5 vs 7 days for the control group). The anti-COVID-19 activity of the treatment is due to the known analgesic, antipyretic, antiviral, and antiplatelet aggregation activities of *Dasamoolkaduthrayan Kashaya* and *Guluchyai Kwatham* (Wanjarkhedkar et al. 2022).

Amla (*Phyllantus emblica*) fruit powder (2 g daily every 12 hours) administered for 10 days in COVID-19-positive (RT-PCR test) patients (12M/18F, 48 years median age) significantly reduced fever, cough severity, breath shortness, myalgia, and length of hospital stay (4.44 vs 7.18 days for control group) and improved oxygen saturation (SpO_2) and C-reactive protein (CRP) levels in a randomized, double-blind controlled trial. These ameliorative effects of clinical signs are due to *P. emblica*'s antiviral activity in the viral replication process by binding to 3-chymotrypsin-like protease receptor binding site (Varnasseri et al. 2022).

Maurya and Sharma (2022) explored phytochemicals from Kadha (used to control various respiratory disorders) Ayurvedic medicine potential to inhibit different stages of SARS-CoV-2 infection and other coronavirus target proteins using molecular docking and molecular dynamic simulation approach. Kadha phytochemicals have significant binding affinity with different CoVs proteins, suggesting their potential to control viral infection and multiplication in host cells. Most of these phytochemicals present in Ashwagandha (*Withania somnifera*), Giloy (*Tinospora cordifolia*), Tulsi (*Ocimum sanctum*), Clove (*Syzygium aromaticum*), and black pepper (*Piper nigrum*) have significant anti-inflammatory properties and interact with most of the druggable proteins (SARS-CoV-2 Mpro; human ACE2 receptor; SARS-CoV-2 NP; and Nsp3 protein). Thus, regular consumption of Ayurvedic Kadha can reduce the inflammatory response, boost individual immunity, and reduce the risk of CoVs infection (Maurya and Sharma 2022). Kadha has been evaluated for its efficacy as a primary therapy to protect healthcare workers during the COVID-19 pandemic (Alam et al. 2021). The observational study (ClinicalTrials.gov NCT04387643) evaluated 52 participants for their physical and psychological capacity to cope with distress after 30 days (March 1–April 2, 2020). Results of the study have not been reported. *Guduchi GhanVati* aqueous *T. cordifolia* extract was evaluated as a primary therapy to manage confirmed asymptomatic or mildly asymptomatic COVID-19 patients (ClinicalTrials.gov NCT04480398). The treatment group, 40 individuals, were given two Guduchi GhanVati tablets (500 mg each) orally, twice each day after meals for 28 days. The control group (51 individuals), provided with conventional and standard therapy, showed reduced symptoms (11.7%) after an average of 1.8 days. Viral clearance of the treatment group at day 14 was significantly superior to that of the control (100% vs 82.3%). Moreover, the Guduchi GhanVati treatment reduced average hospital stay (by ½ from 12.8 days for control to 6.4 days) (Alam et al. 2021 and references therein). Guduchi GhanVati (1 g; 2 tablets, 2 times/day, 14 days) was an effective primary therapy to manage COVID-19 infection in 40 patients in an open-label single-arm feasibility trial. All patients tested negative for COVID-19 after 14 days with 32.5 and 95% viral clearance on days 3 and 7, respectively (Kumar et al. 2020a, 2020b).

Giloy or Guduchi (*Tinospora cordifolia*) in combination with "Pippali or Indian Long Pepper" (*Piper longum*) is also effective against COVID–19. The efficacy of this Ayurvedic combination (*T. cordifolia* [300 mg] and *P. longum* [75 mg] twice daily) was evaluated as a primary treatment for 28 confirmed asymptomatic to mildly symptomatic COVID-19 patients in a placebo-controlled randomized trial (ClinicalTrials.gov Identifier: NCT04621903). Patients recovered completely after one-week treatment compared to control (100% vs 60% for control group), and the probability of delayed healing process decreased significantly (by 40%) from infection in the treatment group (Devpura et al. 2021). *Piper longum* is known to enhance the bioavailability and absorption of other Ayurvedic active components.

A community-based participatory study evaluated the effectiveness of Ayurvedic intervention as supportive care for mild to moderate COVID-19 patients (ClinicalTrials.gov Identifier: NCT04716647). SARS-CoV-2-confirmed patients (28 people) were given oral tablets of *Withania somnifera* (250 mg–5 g), *Tinospora cordifolia* (0.5–1 g), and *Ocimum sanctum* (0.5–1 g) based on age, weight, and symptom severity. The treatment relieved symptoms and positively impacted patients' recovery within seven days based on negative nasopharyngeal swab test (Kulkarni et al. 2021). Two other Ayurvedic therapies were evaluated for efficacy as supportive care for flu-like symptoms of COVID-19 people in self-isolation (ClinicalTrials.gov Identifiers: NCT04345549 and NCT04351542). However, the precise interventions and results of the studies were undisclosed (Alam et al. 2021 and references therein).

Black cumin (*Nigella sativa*) is another herbal plant used in Ayurvedic medicine, and its seed forms part of the Indian spice cuisine. Its anti-COVID-19 efficacy is due to reduced SARS-CoV replication in cell cultures because of its primary bioactive components: thymoquinone, α-heredin, and nigelledine. These phytochemicals have high binding affinity with SARS-CoV-2 molecular enzyme and protein targets. *N. sativa* ground seed powder capsule (80 mg/kg/day) together with natural honey (1 g/kg/day) were evaluated as co-therapy in treating COVID-19 in a phase 3 randomized, placebo-controlled, open-label, add-on, cohort study (ClinicalTrial.gov Identifier: NCT04347382). The 14-day treatment alleviated disease symptoms, reduced mortality among severe patients, and improved overall clinical score with substantially earlier viral clearance compared to control (Ashraf et al. 2020). By day 6, patients in the treatment group resumed their normal activity—63.6% of those with moderate and 28% with severe COVID-19 infection. *N. sativa* seed oil capsule (500 mg, twice daily for 10 days) was evaluated as primary therapy in treating mild COVID-19 in a phase 2 open-label non-randomized controlled pilot trial (ClinicalTrial.gov Identifier: NCT04401202). Patients (n = 86, 53% male, average age 36 years) receiving *N. sativa* oil recovered faster (10.7 vs 12.3 days), with a higher recovery rate (62% vs 36%) than the control group (Koshak et al. 2021).

Chopra et al. (2022) designed a prospective, randomized, double-blind, parallel-group, placebo-controlled, two-arm exploratory study on healthy volunteers administered Ashwagandha seven days after first or second COVISHIELD™ vaccination (Identifier: CTRI/2021/06/034496). The investigators propose to evaluate Ashwagandha (1 tablet standardized aqueous extract equivalent to 4 g Ashwagandha root powder per day for 24 weeks) effects on safety, immunogenicity, and protection in healthy COVISHIELD™ vaccinated population since the single-dose vaccine has 76% efficacy after 22–90 days.

Ashwagandha is also being investigated in a randomized controlled multi-centric clinical study for its prophylactic activity against COVID-19 in high-risk health care workers compared to hydroxychloroquine (Identifiers: CTRI/2020/8/027163, CTRI/2020/05/025332).

Muralikrishna (2020) evaluated the intake of Ashwagandha tablet (500 mg twice daily for one month) to prevent COVID-19 in healthy individuals in an observational study. The clinical trial is an open-label randomized controlled prospective interventional community-based study on healthy subjects. Rangnekar et al. (2020) assessed the efficacy of herbal extracts in boosting innate immunity of patients with COVID-19 infection in a randomized two-arm parallel double-blind controlled exploratory trial for 30 days. The study involved two capsules (investigational products 1, 400 and 450 mg of edible starch [placebo]) each to be taken twice daily for 15 days, and one capsule of product 2 from day 16 to be consumed orally twice daily for another 15 days. The investigational products 1 and 2 consisted of herbal extracts (a blend of water and CO_2 extracts) of *Zingiber officinale*, *Embella ribes*, *Glycyrrhiza glabra*, *Terminalia chebula*, *Tinospora cordifolia*, *Asparagus racemosus*, *Embelica officinalis*, *Piper longum*, and calcined zinc. The herbal extract effectively reduced viral load, acting as an immunomodulatory with safe and early recovery.

5.5 CURRENT AND FUTURE PERSPECTIVE

Several strategies have been and continue to be deployed to increase the potency, efficacy, and/or safety of Ayurvedic herbal plant products to safeguard against human coronaviruses. For example, glycyrrhizin, one of liquorice root's saponins, inhibits SARS-associated coronavirus replication (300–600 mg/L EC_{50} value during and after virus adsorption, respectively). It completely blocked virus replication of infected cells at 4000 mg/L and is safer than the antiviral drug ribavirin, which is responsible for hemolysis and drastic hemoglobin reduction in SARS patients (Cinatl et al. 2003). Glycyrrhizin potency in blocking virus entry into the cells is further enhanced (10-fold) by 2-acetamido-β-$_{D}$-glucopyranosylamine addition to its glycoside chain because coronaviruses spike proteins (S-proteins) are heavily glycosylated (Hoever et al. 2005). Moreover, glycyrrhizin disappears from blood plasma of treated subjects after oral administration (100 mg), since it metabolizes to glycyrrhetinic acid by β-$_D$-glucuronidase present in the intestinal microbiome (Chen et al. 2004). Similarly, computational biology studies of curcumin, the most notable polyphenolic compound from turmeric (*Curcuma longa*), demonstrate that its keto form (keto-curcumin) preferentially attenuates SARS-CoV-2 viral infection by blocking the bioactive drug target residues based on its docking score, binding energy and molecular dynamics simulation (Shanmugarajan et al. 2020).

A comprehensive literature search review (until the end of May 2020) from in *vitro cell-* and non-cell-based studies highlights the mechanisms of action and bioactive compounds responsible for the anti-COVID-19 potential (Prasansuklab et al. 2021). For example, *Curcuma longa* polyphenol curcumin inhibits the SARS-CoV papain-like protease activity and ACE activity. *Psoralea corylifolia* (used in Ayurvedic medicine) chalcone and flavonoids also inhibit the SARS-CoV papain-like protease activity. *Clitoria ternatea* (flavonoid) inhibits (32% inhibition) the ion channel activity of SARS-CoV3a protein in a cell-based assay. Flavonoids (cinnatannin B1, procyanidin A2, and procyanidin B1) from cinnamon possess antiviral activity against SARS-CoV in a cell-based assay.

Some Ayurvedic polyherbal formulations can be used as vaccine adjuvants to increase vaccine potency and immune responses, since many COVID-19 symptoms mostly affect the respiratory, immune, and other physiological systems (Sarkar and Mukhopadhyay 2022). Ayurvedic adjuvants stimulate and potentiate cellular or humoral immune responses by activating cell signaling pathways and can benefit COVID-19 patients to avoid/prevent cytokine storm activation. Withanolides from ashwagandha (*Withania somnifera*) hot water root extract have been patented as a vaccine adjuvant to considerably enhance adjuvant activity (Jadhav et al. 2012). Other *Withania somnifera* active compounds, such as withanone, block SARS-CoV-2 entry by interrupting electrostatic interactions between the viral receptor binding domain (RBD) and host cell angiotensin-converting enzyme 2 (ACE2), and inhibiting the virus' main protease (Mpro). *Tinospora cordifolia* bioactive compounds (berberine, magnoflorine, tinocordiside, and isocolumbin) have higher binding affinity to SARS-CoV-2 surface glycoprotein than antiviral medicines (Favipiravir, Lopinavir/Ritonavir, and Remdesivir). Terpenoids and polyphenols from crude hydroalcoholic *Ocimum sanctum* extract and natural compounds (gedunin and epoxyazadiradione) from neem (*Azadirachta indica*) also exert their antiviral activities due to high binding efficacy with viral proteins (Sarkar and Mukhopadhyay 2022).

Withania somnifera phytochemicals (withanone and withaferin A) target viral main protease (M^{Pro}) and host transmembrane (TMRSS2), and glucose-related protein 78 (GPR78) indicating their potential as viral entry inhibitors. These *W. somnifera* phytochemicals interact with TMRSS2 and blocks viral entry into the host cell, since withanone downregulates TMRSS2 transmission while withaferin A binds to GRP78 (involved in the MRSA infection). Balkrishna et al. (2021a) prepared a withanone-enriched methanol extract without withaferin A to resolve the complex immunomodulation associated with SARS-CoV-2 infection. The extract efficiently ameliorates human-like pathological responses induced in humanized (human alveolar epithelial A549 cells xenotransplant) zebrafish by SARS-CoV-2 recombinant spike (S) protein. It very strongly inhibits and disrupts the binding of SARS-CoV-2 spike protein with angiotensin-converting enzyme (IC_{50} 0.3 ng/mL), attenuates SARS-CoV-2 protein-induced kidney necrosis, and prevents skin hemorrhaging in humanized zebrafish. Moreover, the extract is safe due to its low cytotoxicity and can easily be standardized for clinical trials.

Balkrishna et al. (2021b) developed a tri-herbal (*Withania somnifera*, *Tinospora cordifolia*, and *Ocimum sanctum*) aqueous extract, Coronil, that rescued humanized zebrafish from SARS-CoV-2-induced pathologies. Coronil effectively ameliorated the cytokine response mounted during SARS-CoV-2 infection by acting at various levels of physiological systems. It prevented SARS-CoV-2 S-protein-mediated viral entry into A549 cells by inhibiting spike protein–ACE2 interactions.

Novel sustainable green technologies can increase the yield, purity, and safety and enhance the bioavailability and efficacy of anti-coronavirus phytochemicals. For example, zinc oxide nanoparticles (ZnO NPs), similar to those developed to stabilize *Azadirachta indica* gum (Vijayakumar et al. 2021a) can be used to increase potency, anti-inflammatory and other physiological bioactivities, and bioavailability of

A. indica antiviral compounds. Similarly, potency, bioactivity, and bioavailability of black cumin (*Nigella sativa*) antiviral compounds can be enhanced using silver nanoparticles (Vijayakumar et al. 2021b). In addition to conventional medical treatments and preventative vaccinations, adopting a healthy diet that is mostly composed of pharmaceutical products could prove to be the most effective method for fending against sickness, particularly the immune-related pandemic known as COVID-19. It is essential for people to include plant foods into their diets on a regular basis to keep their bodies in a state of equilibrium and to strengthen their immune systems. Combinatorial chemistry techniques have demonstrated significant advantages to the drug discovery process. This is based on the growing acceptance that the chemical diversity of natural products is fundamental to provide starting scaffolds for future drugs, and the fact that combinatorial chemistry techniques are being used more frequently. It is necessary to have a multidisciplinary strategy combining molecular diversity from natural product sources in addition to combinatorial native crops as the most effective solution to drug discovery and development. It is intended that researchers would use the information offered here as a guide in the process of producing safe and effective anti-coronavirus treatment medicines that are derived from naturally occurring substances as well as traditional crops unique to the area. SARS-CoV-2 is still spreading rapidly around the world, despite the efforts of research organizations located all over the world that are dedicated to the creation of vaccines and medications for therapy. As a result, few recently published studies discuss the cultivation of plant crops that have the potential to ameliorate respiratory illnesses and pneumonia, in addition to reducing the likelihood of a COVID-19 epidemic. Novel approaches to disease prevention and the stimulation of effective immune responses need to be investigated. Native Indian crops have the potential to play a significant role in this scenario.

5.6 CONFLICT OF INTEREST

The authors declare no conflict of interest.

REFERENCES

Aati HY, Ismail A, Rateb ME, AboulMagd AM, Hassan HM, Hetta MH. (2022). *Garcinia cambogia* phenolics as potent anti-COVID-19 agents: Phytochemical profiling, biological activities, and molecular docking. *Plants*, 11(19):2521. https://doi.org/10.3390/plants11192521

Alam G, Wahyuono S, Ganjar IG, Hakim L, Timmerman H, Verpoorte R. (2002). Tracheospasmolytic activity of viteosin-A and vitexicarpin isolated from *Vitex trifolia*. *Planta Medica*, 68(11):1047–1049.

Alam S, Sarker MM, Afrin S, Richi FT, Zhao C, Zhou JR, Mohamed IN. (2021). Traditional herbal medicines, bioactive metabolites, and plant products against COVID-19: Update on clinical trials and mechanism of actions. *Frontiers in Pharmacology*, 12:671498. https://doi.org/10.3389/fphar.2021.671498

Ameri A, Heydarirad G, Mahdavi Jafari J, Ghobadi A, Rezaeizadeh H, Choopani R. (2015). Medicinal plants contain mucilage used in traditional Persian medicine (TPM). *Pharmaceutical Biology*, 53(4):615–623.

Ang L, Lee HW, Choi JY, Zhang J, Lee MS. (2020). Herbal medicine and pattern identification for treating COVID-19: A rapid review of guidelines. *Integrative Medicine Research*, 9(2):100407. https://doi.org/10.1016/j.imr.2020.100407

Ashraf S, Ashraf S, Ashraf M, Imran MA, Kalsoom L, Siddiqui UN, Ghufran M, Majeed N, Farooq I, Habib Z, Hilal A. (2020). Therapeutic efficacy of Honey and *Nigella sativa* against COVID-19: A multi-center randomized controlled clinical trial (HNS-COVID-PK). *medRxiv*. https://doi.org/10.1101/2020.10.30.20217364

Atanasov AG, Waltenberger B, Pferschy-Wenzig EM, Linder T, Wawrosch C, Uhrin P, Temml V, Wang L, Schwaiger S, Heiss EH, Rollinger JM. (2015). Discovery and resupply of pharmacologically active plant-derived natural products: A review. *Biotechnology Advances*, 33(8):1582–614.

Balkrishna A, Haldar S, Singh H, Roy P, Varshney A. (2021b). Coronil, a tri-herbal formulation, attenuates spike-protein-mediated SARS-CoV-2 viral entry into human alveolar epithelial cells and pro-inflammatory cytokines production by inhibiting spike protein-ACE-2 interaction. *Journal of Inflammation Research*, 14:869–884. https://doi.org/10.2147/JIR.S298242

Balkrishna A, Pokhrel S, Singh H, Joshi M, Mulay VP, Haldar S, Varshney A. (2021a). Withanone from *Withania somnifera* attenuates SARS-CoV-2 RBD and host ACE2 interactions to rescue spike protein induced pathologies in humanized zebrafish model. *Drug Design, Development and Therapy*, 11:1111–1133.

Barbosa JR, de Carvalho Junior RN. (2021). Polysaccharides obtained from natural edible sources and their role in modulating the immune system: Biologically active potential that can be exploited against COVID–19. *Trends in Food Science & Technology*, 108:223–235.

Ben-Shabat S, Yarmolinsky L, Porat D, Dahan A. (2020). Antiviral effect of phytochemicals from medicinal plants: Applications and drug delivery strategies. *Drug Delivery and Translational Research*, 10(2):354–367. https://doi.org/10.1007/s13346-019-00691-6

Bessada SM, Barreira JC, Oliveira MB. (2019). Pulses and food security: Dietary protein, digestibility, bioactive and functional properties. *Trends in Food Science & Technology*, 93:53–68.

Bhapkar V, Sawant T, Bhalerao S. (2022). A critical analysis of CTRI registered AYUSH studies for COVID–19. *Journal of Ayurveda and Integrative Medicine*, 13(1):100370. https://doi.org/10.1016/j.jaim.2020.10.012

Borse S, Joshi M, Saggam A, Bhat V, Walia S, Marathe A, Sagar S, Chavan-Gautam P, Girme A, Hingorani L, Tillu G. (2021). Ayurveda botanicals in COVID-19 management: An in silico multi-target approach. *PLoS ONE*, 16(6):e0248479. https://doi.org/10.1371/journal.pone.0248479

Budke C, Thor Straten S, Mühling KH, Broll G, Daum D. (2020). Iodine biofortification of field-grown strawberries–Approaches and their limitations. *Scientia Horticulturae*, 269:109317. https://doi.org/10.1016/j.scienta.2020.109317

Buso P, Manfredini S, Reza Ahmadi-Ashtiani H, Sciabica S, Buzzi R, Vertuani S, Baldisserotto A. (2020). Iranian medicinal plants: From ethnomedicine to actual studies. *Medicina*, 56(3):97. https://doi.org/10.3390/medicina56030097

Chauhan JS, Singh BB, Gupta S. (2016). Enhancing pulses production in India through improving seed and variety replacement rates. *Indian Journal of Genetics and Plant Breeding*, 76(4):410–419.

Chavda VP, Apostolopoulos V. (2021). Mucormycosis–An opportunistic infection in the aged immunocompromised individual: A reason for concern in COVID–19. *Maturitas*, 154:58–61.

Cheah ZX, O'Hare TJ, Harper SM, Kochanek J, Bell MJ. (2020). Zinc biofortification of immature maize and sweetcorn (*Zea mays* L.) kernels for human health. *Scientia Horticulturae*, 272:109559. https://doi.org/10.1016/j.scienta.2020.109559

Chen F, Chan KH, Jiang Y, Kao RY, Lu HT, Fan KW, Cheng VC, Tsui WH, Hung IF, Lee TS, Guan Y. (2004). *In vitro* susceptibility of 10 clinical isolates of SARS coronavirus to selected antiviral compounds. *Journal of Clinical Virology*, 31(1):69–75.

Chen Y, Liu Q, Guo D. (2020). Emerging coronaviruses: Genome structure, replication, parthenogenesis. *Journal of Virology*, 92, 418–423.

Chopra A, Chavan-Gautam P, Tillu G, Saluja M, Borse S, Sarmukaddam S, Chaudhuri S, Rao BC, Yadav B, Srikanth N, Patwardhan B. (2022). Randomized, double blind, placebo controlled, clinical trial to study ashwagandha administration in participants vaccinated against COVID-19 on safety, immunogenicity, and protection with COVID-19 vaccine–a study protocol. *Frontiers in Medicine*, 9:761655. https://doi.org/10.3389/fmed.2022.761655

Cinatl J, Morgenstern B, Bauer G, Chandra P, Rabenau H, Doerr H. (2003). Glycyrrhizin, an active component of liquorice roots, and replication of SARS-associated coronavirus. *The Lancet*, 361(9374):2045–2046.

Dar NJ, Hamid A, Ahmad M. (2015). Pharmacologic overview of *Withania somnifera*, the Indian Ginseng. *Cellular and Molecular Life Sciences*, 72(23):4445–4460.

de LimaLessa JH, Raymundo JF, Corguinha AP, Martins FA, Araujo AM, Santiago FE, de Carvalho HW, Guilherme LR, Lopes G. (2020). Strategies for applying selenium for biofortification of rice in tropical soils and their effect on element accumulation and distribution in grains. *Journal of Cereal Science*, 96:103125. https://doi.org/10.1016/j.jcs.2020.103125

Devpura G, Tomar BS, Nathiya D, Sharma A, Bhandari D, Haldar S, Balkrishna A, Varshney A. (2021). Randomized placebo-controlled pilot clinical trial on the efficacy of ayurvedic treatment regime on COVID-19 positive patients. *Phytomedicine*, 84:153494. https://doi.org/10.1016/phytomed.2021.153494

Drosten C, Günther S, Preiser W, Van Der Werf S, Brodt HR, Becker S, Rabenau H, Panning M, Kolesnikova L, Fouchier RA, Berger A. (2003). Identification of a novel coronavirus in patients with severe acute respiratory syndrome. *New England Journal of Medicine*, 348(20):1967–1976.

Dubey S, Biswas P, Ghosh R, Chatterjee S, Dubey MJ, Chatterjee S, Lahiri D, Lavie CJ. (2022). Psychosocial impact of COVID–19. *Diabetes & Metabolic Syndrome: Clinical Research & Reviews*, 14(5):779–788.

El-Saber Batiha G, Alkazmi LM, Wasef LG, Beshbishy AM, Nadwa EH, Rashwan EK. (2020). *Syzygium aromaticum* L. (Myrtaceae): Traditional uses, bioactive chemical constituents, pharmacological and toxicological activities. *Biomolecules*, 10(2):202. https://doi.org/10.3390/biom10020202

Elsebai MF, Mocan A, Atanasov AG. (2016). Cynaropicrin: A comprehensive research review and therapeutic potential as an anti-hepatitis C virus agent. *Frontiers in Pharmacology*, 7:472. https://doi.org/10.3389/fphar.2016.00472

Galanakis CM. (2020). The food systems in the era of the coronavirus (COVID-19) pandemic crisis. *Foods*, 9(4):523. https://doi.org/10.3390/foods9040523

Ganjhu RK, Mudgal PP, Maity H, Dowarha D, Devadiga S, Nag S, Arunkumar G. (2015). Herbal plants and plant preparations as remedial approach for viral diseases. *Virus Disease*, 26(4):225–236.

Grubaugh ND, Ladner JT, Lemey P, Pybus OG, Rambaut A, Holmes EC, Andersen KG. (2019). Tracking virus outbreaks in the twenty-first century. *Nature Microbiology*, 4(1):10–19.

Hoever G, Baltina L, Michaelis M, Kondratenko R, Baltina L, Tolstikov GA, Doerr HW, Cinatl J. (2005). Antiviral activity of glycyrrhizic acid derivatives against SARS– coronavirus. *Journal of Medicinal Chemistry*, 48(4):1256–1259.

Huang J, Tao G, Liu J, Cai J, Huang Z, Chen JX. (2020a). Current prevention of COVID-19: Natural products and herbal medicine. *Frontiers in Pharmacology*, 11:588508. https://doi.org/10.3389/fphar.2020.588508

Huang J, Wu L, Ren X, Wu X, Chen Y, Ran G, Huang A, Huang L, Zhong D. (2020b). Traditional Chinese medicine for corona virus disease 2019: A protocol for systematic review. *Medicine*, 99(35):e21774. https://doi.org/10.1097/MD.0000000000021774

Jadhav SKR, Patel KA, Dholakia BB, Khan BM.(2012). Structural characterization of a flavonoid glycosyltransferase from *Withania somnifera*. *Bioinformation*, 8(19):943–949.

Khan MY, Kumar V. (2019). Mechanism & inhibition kinetics of bioassay-guided fractions of Indian medicinal plants and foods as ACE inhibitors. *Journal of Traditional and Complementary Medicine*, 9(1):73–84.

Khan T, Khan MA, Ullah N, Nadhman A. (2021). Therapeutic potential of medicinal plants against COVID-19: The role of antiviral medicinal metabolites. *Biocatalysis and Agricultural Biotechnology*, 31:101890. https://doi.org/10.1016/j.bcab.2020.101890

Komarayanti S, Suharso W, Herrianto E. (2020). Local fruits and vegetables of jember district that can increase immunity during the Covid-19 pandemic. *Budapest International Research in Exact Sciences (BirEx) Journal*, 2(4):492–508.

Koshak AE, Koshak EA, Mobeireek AF, Badawi MA, Wali SO, Malibary HM, . . . & Madani TA. (2021). *Nigella sativa* for the treatment of COVID-19: An open-label randomized controlled clinical trial. *Complementary Therapies in Medicine*, 61:102769. https://doi.org/10.1016/j.ctim.2021.102769

Kulkarni V, Sharma N, Modi D, Kumar A, Joshi J, Krishnamurthy N. (2021). A community-based participatory research to assess the feasibility of Ayurveda intervention in patients with mild-to-moderate COVID–19. *MedRxiv*. https://doi.org/10.1101/2021.01.20.21250198

Kumar A, Prasad G, Srivastav S, Gautam VK, Sharma N. (2020a). A retrospective study on efficacy and safety of *Guduchi Ghan Vati* for Covid-19 asymptomatic patients. *MedRxiv* 2020.07.23.20160424. https://doi.org/10.1101/2020.07.23.20160424

Kumar A, Prasad G, Srivastav S, Gautam VK, Sharma N. (2020b). Efficacy and safety of *Guduchi Ghan vati* in the management of asymptomatic COVID-19 infection: An open label feasibility study. *MedRxiv* 2020.09.20.20198515. https://doi.org/10.1101/2020.09.20.20198515

Kumar VS, Navaratnam V. (2013). Neem (*Azadirachta indica*): Prehistory to contemporary medicinal uses to humankind. *Asian Pacific Journal of Tropical Biomedicine*, 3(7):505–514.

Lee H, Lei H, Santarsiero BD, Gatuz JL, Cao S, Rice AJ, Patel K, Szypulinski MZ, Ojeda I, Ghosh AK, Johnson ME. (2015). Inhibitor recognition specificity of MERS-CoV papain-like protease may differ from that of SARS-CoV. *ACS Chemical Biology*, 10(6):1456–1465.

Liao Y, Xu B, Wang J, Liu X. (2017). A new method for assessing the risk of infectious disease outbreak. *Scientific Reports*, 7(1):1–2.

Lin LT, Hsu WC, Lin CC. (2014). Antiviral natural products and herbal medicines. *Journal of Traditional and Complementary Medicine*, 4(1):24–35.

Long QX, Liu BZ, Deng HJ, Wu GC, Deng K, Chen YK, Liao P, Qiu JF, Lin Y, Cai XF, Wang DQ. (2020). Antibody responses to SARS-CoV-2 in patients with COVID–19. *Nature Medicine*, 26(6):845–848.

Lu H. (2020). Drug treatment options for the 2019-new coronavirus (2019-nCoV). *Bioscience Trends*, 14(1):69–71.

Mahmood N, Nasir SB, Hefferon K. (2020). Plant-based drugs and vaccines for COVID–19. *Vaccines*, 9(1):15. https://doi.org/10.3390/vaccines9010015

Mallakpour S, Azadi E, Hussain CM. (2021). Chitosan, alginate, hyaluronic acid, gums, and β-glucan as potent adjuvants and vaccine delivery systems for viral threats including SARS-CoV-2: A review. *International Journal of Biological Macromolecules*, 182:1931–1940.

Mani JS, Johnson JB, Steel JC, Broszczak DA, Neilsen PM, Walsh KB, Naiker M. (2020). Natural product-derived phytochemicals as potential agents against coronaviruses: A review. *Virus Research*, 284:197989. https://doi.org/10.1016/j.virusres.2020.097989

Maurya DK, Sharma D. (2022). Evaluation of traditional ayurvedic Kadha for prevention and management of the novel Coronavirus (SARS-CoV-2) using *in silico* approach. *Journal of Biomolecular Structure and Dynamics*, 40(9):3949–3964.

Mayes S, Massawe FJ, Alderson PG, Roberts JA, Azam-Ali SN, Hermann M. (2012). The potential for underutilized crops to improve security of food production. *Journal of Experimental Botany*, 63(3):1075–1079.

Meng J, Li R, Zhang Z, Wang J, Huang Q, Nie D, Fan K, Guo W, Zhao Z, Han Z. (2022). A review of potential therapeutic strategies for COVID–19. *Viruses*, 25;14(11):2346. https://doi.org/10.3390/v14112346

Mosafer J, Sabbaghi AH, Badiee A, Dehghan S, Tafaghodi M. (2019). Preparation, characterization and in vivo evaluation of alginate-coated chitosan and trimethylchitosan nanoparticles loaded with PR8 influenza virus for nasal immunization. *Asian Journal of Pharmaceutical Sciences*, 14(2):216–221.

Muralikrishna DC (2020). *Observational Study of Ashwagandha Tablet Intake as a Preventive Measure in Pandemic of—COVID-19—an Open Label, Randomized, Controlled, Prospective, Interventional, Community-Based Clinical Study on Healthy Subjects* (Clinical Trials Registry-India Identifier: CTRI/2020/05/025166). https://ctri.nic.in/Clinicaltrials/showallp.php?mid1=43553&EncHid=&userName=025166

Muthamilarasan M, Prasad M. (2021). Small millets for enduring food security amidst pandemics. *Trends in Plant Science*, 26(1):33–40.

Naja F, Hamadeh R. (2020). Nutrition amid the COVID-19 pandemic: A multi-level framework for action. *European Journal of Clinical Nutrition*, 74(8):1117–1121.

Nesari T, Kadam S, Vyas M, Huddar VG, Prajapati PK, Rajgopala M, More A, Rajgopala S, Bhatted S, Yadav RK, Mahanta V. (2022). AYURAKSHA, a prophylactic Ayurvedic immunity boosting kit reducing positivity percentage of IgG COVID-19 among frontline Indian Delhi police warriors: A non-randomized controlled intervention trial. *Frontiers in Public Health*, 10:920126. https://doi.org/10.3389/fpubh.2022.920126

Nourazarian SM, Nourazarian A, Majidinia M, Roshaniasl E. (2016). Effect of root extracts of medicinal herb *Glycyrrhiza glabra* on HSP90 gene expression and apoptosis in the HT-29 colon cancer cell line. *Asian Pacific Journal of Cancer Prevention*, 16(18):8563–8566.

Okwuonu IC, Narayanan NN, Egesi CN, Taylor NJ. (2021). Opportunities and challenges for biofortification of cassava to address iron and zinc deficiency in Nigeria. *Global Food Security*, 28:100478. https://doi.org/10.1016/j.gfs.2020.100478

Pal DC, Jain SK. (1998). *Tribal Medicine*. Naya Prokash, 206, Bidhan Sarani, Calcutta, India, p. 316.

Pandey MM, Rastogi S, Rawat AK. (2013). Indian traditional ayurvedic system of medicine and nutritional supplementation. *Evidence-Based Complementary and Alternative Medicine*, 2013:376327. https://doi.org/10.1155/2013/376327

Pang J, Wang MX, Ang IY, Tan SH, Lewis RF, Chen JI, Gutierrez RA, Gwee SX, Chua PE, Yang Q, Ng XY. (2020). Potential rapid diagnostics, vaccine and therapeutics for 2019 novel coronavirus (2019-nCoV): A systematic review. *Journal of Clinical Medicine*, 9(3):623. https://doi.org/10.3390/jcm9030623

Panyod S, Ho CT, Sheen LY. (2020). Dietary therapy and herbal medicine for COVID-19 prevention: A review and perspective. *Journal of Traditional and Complementary Medicine*, 10(4):420–427.

Prasansuklab A, Theerasri A, Rangsinth P, Sillapachaiyaporn C, Chuchawankul S, Tencomnao T. (2021). Anti-COVID-19 drug candidates: A review on potential biological activities of natural products in the management of new coronavirus infection. *Journal of Traditional and Complementary Medicine*, 11(2):144–157. https://doi.org/10.1016/j.jtcme.2020.12.001

Priyashree S, Jha S, Pattanayak SP. (2010). A review on *Cressa Cretica Linn.*: A halophytic plant. *Pharmacognesy Review*, 4:161–166.

Priyashree S, Jha S, Pattanayak SP. (2012). Bronchodilatory and mast cell stabilising activity of *Cressa cretica* L.: Evaluation through *in vivo* and *in vitro* experimental models. *Asian Pacific Journal of Tropical Medicine*, 5(3):180–186.

Qiao J. (2020). What are the risks of COVID-19 infection in pregnant women? *The Lancet*, 395(10226): 760–762.

Rajan M, Prabhakaran S, Prusty JS, Chauhan N, Gupta P, Kumar A. (2022). Phytochemicals of *Cocculus hirsutus* deciphered SARS-CoV-2 inhibition by targeting main proteases in molecular docking, simulation, and pharmacological analyses. *Journal of Biomolecular Structure and Dynamics*, 41(15):7406–7420. https://doi.org/10.1080/07391102.2022.2121758

Rangnekar H, Patankar S, Suryawanshi K, Soni P. (2020). Safety and efficacy of herbal extracts to restore respiratory health and improve innate immunity in COVID-19 positive patients with mild to moderate severity: A structured summary of a study protocol for a randomised controlled trial. *Trials*, 21(1):943. https://doi.org/10.1186/s13063-020-04906-x

Raza ZA, Taqi M, Tariq MR. (2022). Antibacterial agents applied as antivirals in textile-based PPE: A narrative review. *The Journal of the Textile Institute*, 113(3):515–526.

Rios JL, Recio MC. (2005). Medicinal plants and antimicrobial activity. *Journal of Ethnopharmacology*, 100(1–2):80–84.

Salazar-González JA, Angulo C, Rosales-Mendoza S. (2015). Chikungunya virus vaccines: Current strategies and prospects for developing plant-made vaccines. *Vaccine*, 33(31):3650–3658.

Sampangi-Ramaiah MH, Vishwakarma R, Shaanker RU. (2020). Molecular docking analysis of selected natural products from plants for inhibition of SARS-CoV-2 main protease. *Current Science*, 118(7):1087–1092.

San Chang J, Wang KC, Yeh CF, Shieh DE, Chiang LC. (2013). Fresh ginger (*Zingiber officinale*) has anti-viral activity against human respiratory syncytial virus in human respiratory tract cell lines. *Journal of Ethnopharmacology*, 145(1):146–151.

Sarkar PK, Mukhopadhyay CD. (2022). Mechanistic insights from the review and evaluation of ayurvedic herbal medicines for the prevention and management of COVID-19 patients. *Journal of Herbal Medicine*, 32:100554. https://doi.org/10.1016/j.hermed.2022.100554

Shahzad F, Shahzad U, Fareed Z, Iqbal N, Hashmi SH, Ahmad F. (2020). Asymmetric nexus between temperature and COVID-19 in the top ten affected provinces of China: A current application of quantile-on-quantile approach. *Science of the Total Environment*, 736:139115. https://doi.org/10.1016/j.scitotenv.2020.139115

Shanmugarajan D, Prabitha P, Kumar BP, Suresh B. (2020). Curcumin to inhibit binding of spike glycoprotein to ACE2 receptors: Computational modelling, simulations, and ADMET studies to explore curcuminoids against novel SARS-CoV-2 targets. *RSC Advances*, 10(52):31385–31399.

Shereen MA, Khan S, Kazmi A, Bashir N, Siddique R. (2020). COVID-19 infection: Emergence, transmission, and characteristics of human coronaviruses. *Journal of Advanced Research*, 24:91–98.

Singh NR, Madan A, Yadav B, Gupta A, Rana RK, Wetal VR, Dubey N, Jameela S, Singhal R, Khanduri S, Sharma BS. (2021). Effect of an Ayurvedic intervention (Ayush-64) in mild to moderate COVID-19: An exploratory prospective single arm clinical trial. https://doi.org/10.31219/osf.io/azb9t

Sivasankari B, Anandharaj M, Gunasekaran P. (2014). An ethnobotanical study of indigenous knowledge on medicinal plants used by the village peoples of Thoppampatti, Dindigul district, Tamilnadu, India. *Journal of Ethnopharmacology*, 153(2):408–423.

Siwela M, Pillay K, Govender L, Lottering S, Mudau FN, Modi AT, Mabhaudhi T. (2020). Biofortified crops for combating hidden hunger in South Africa: Availability, acceptability, micronutrient retention and bioavailability. *Foods*, 9(6):815. https://doi.org/10.3390/foods9060815

Srivastava S, Singh H, Muralidharan S, Mohan R, Chaudhary S, Rani P, Payyappalli U, Srikanth N. (2021). A retrospective analysis of Ayurvedic clinical management of mild COVID-19 patients. *Journal of Research in Ayurvedic Sciences*, 5(2):80–86. https://doi.org/10.4103/jras.jras_15_21

Tiozon RN, Fernie AR, Sreenivasulu N. (2021). Meeting human dietary vitamin requirements in the staple rice via strategies of biofortification and post-harvest fortification. *Trends in Food Science and Technology*, 109:65–82.

Tito A, Colantuono A, Pirone L, Pedone E, Intartaglia D, Giamundo G, Conte I, Vitaglione P, Apone F. (2021). Pomegranate peel extract as an inhibitor of SARS-CoV-2 spike binding to human ACE2 receptor (*in vitro*): A promising source of novel antiviral drugs. *Frontiers in Chemistry*, 9:638187. https://doi.org/10.3389/fchem.2021.638187

Varnasseri M, Siahpoosh A, Hoseinynejad K, Amini F, Karamian M, Yad MJ, Cheraghian B, Khosravi AD. (2022). The effects of add-on therapy of *Phyllanthus emblica* (Amla) on laboratory confirmed COVID-19 Cases: A randomized, double-blind, controlled trial. *Complementary Therapies in Medicine*, 65:102808. https://doi.org/10.1016/j.ctim.2022.102808

Vijayakumar S, Divya M, Vaseeharan B, Chen J, Biruntha M, Silva LP, Duran-Lara EF, Shreema K, Ranjan S, Dasgupta N. (2021b). Biological compound capping of silver nanoparticle with the seed extracts of blackcumin (*Nigella sativa*): A potential antibacterial, antidiabetic, anti-inflammatory, and antioxidant. *Journal of Inorganic and Organometallic Polymers and Materials*, 31:624–635.

Vijayakumar S, Divya M, Vaseeharan B, Ranjan S, Kalaiselvi V, Dasgupta N, Chen J, Durán-Lara EF. (2021a). Biogenic preparation and characterization of ZnO nanoparticles from natural polysaccharide *Azadirachta indica*. L. (neem gum) and its clinical implications. *Journal of Cluster Science*, 32:983–993.

Vimalanathan S, Ignacimuthu S, Hudson JB. (2009). Medicinal plants of Tamil Nadu (Southern India) are a rich source of antiviral activities. *Pharmaceutical Biology*, 47(5):422–429.

Wang J, Zhou M, Liu F. (2020). Exploring the reasons for healthcare workers infected with novel coronavirus disease 2019 (COVID-19) in China. *Journal of Hospital Infection*, 105(1):100–101.

Wanjarkhedkar P, Sarade G, Purandare B, Kelkar D. (2022). A prospective clinical study of an Ayurveda regimen in COVID 19 patients. *Journal of Ayurveda and Integrative Medicine*, 13(1):100365. https://doi.org/10.1016/j.jaim.2020.10.008

Xing Shi Gan. (2020). Decoction against Coronavirus Disease 2019 (COVID-19): In silico and experimental study. *Pharmacological Research*, 157:104820. https://doi.org/10.1016/j.phrs.2020.104820

Yang M, Yang T, Jia J, Lu T, Wang H, Yan X, Wang L, Yu L, Zhao Y. (2018). Fabrication and characterization of DDAB/PLA-alginate composite microcapsules as single-shot vaccine. *RSC Advances*, 8(24):13612–13624.

Zambounis A, Sytar O, Valasiadis D, Hilioti Z. (2020). Effect of photosensitisers on growth and morphology of *Phytophthora citrophthora* coupled with leaf bioassays in pear seedlings. *Plant Protection Science*, 56(2):74–82.

Zhang DH, Wu KL, Zhang X, Deng SQ, Peng B. (2020). In silico screening of Chinese herbal medicines with the potential to directly inhibit 2019 novel coronavirus. *Journal of Integrative Medicine*, 18(2):152–158.

Zhou X, Jiang X, Qu M, Aninwene GE, Jucaud V, Moon JJ, Gu Z, Sun W, Khademhosseini A. (2020). Engineering antiviral vaccines. *ACS Nano*, 14(10):12370–12389.

Therapeutic Benefits of Holy Basil (Tulsi)

6

Vandita Anand, Divya Singh, and Anjana Pandey

6.1 INTRODUCTION

6.1.1 Historical Perspective

India is the home of Ayurveda, a 5000-year-old system of ancient Hindu medicine. The origin of Ayurvedic medicine is credited to Dhanvantari, the divine doctor in Hindu mythology. The Vedas, especially the *Atharvaveda*, are the source of Ayurveda, which emphasizes the use of locally grown plant-based remedies to heal illnesses.

Tulsi, "Queen of herbs," is known for its religious and spiritual sanctity and is described as a sacred and medicinal plant in ancient literature. The traditional Ayurvedic and Unani systems of herbal medicine and holistic health played a significant role by Charaka in the Charaka Samhita, an Ayurvedic classic. Ancient Rishis identified tulsi as one of India's most potent medicinal plants thousands of years ago. Sanskrit, which translates to "matchless one," is where the name Tulsi originates. Tulsi, also known as Vishnupriya, denotes the one who pleases Lord Vishnu. It is a commonly growing, revered plant that originated in its natural state (Vana tulsi). Tulsi was the first natural herbal concoction to be used as a medication in China. Tulsi has existed since 4000–5000 B.C. in ancient literature (Vishwabhan et al. 2011).

It is a member of the Lamiaceae family of the *Ocimum* genus, which is known for its therapeutic properties. Tulsi comes in two primary varieties: black (Krishna tulsi) and green (Rama tulsi), both of which have comparable chemical components (Das et al. 2006).

It is a fragrant, upright, heavily branched plant that grows to a mature height (30–60 cm), with simple, opposite, elliptic, oblong, obtuse or acute fragrant leaves. The leaves are whole or sub-serrate or dentate-edged and grow up to 5 cm long. The tiny, purple to scarlet blooms of tulsi grow in tight clusters on cylinder-shaped spikes. At the foot of each flower cluster are heart-shaped bracts without stalks and a hairless sepal cup. Flowers seldom exceed 5 mm in length, with a bearded base of the hairy calyx tube. The fruits are tiny, and the seeds range in color from yellow to reddish (Buddhadev et al. 2014).

6.1.2 Latest Production

Tulsi finds its application in the manufacture of aromatic substances and essential oils. Two tulsi varieties are cultivated on a large scale in India; they are Sri Tulsi (green type) and Krishna Tulsi (purple type).

DOI: 10.1201/9781003352341-6

Other varieties (Drudriha, Babi, Tukashmiya, Vana and Kapoor) tulsi are also available for farming in India. Tulsi farms produce approximately 10,000 kg of fresh leaves, amounting to 10–20 kg of essential oil per hectare. Yields are higher under irrigation (20 tonnes of herbage and 40 kg oil/ha). Flower production is about 3–4 tonnes/ha, producing around 25–32 kg of oil. Steam distillation is the preferred process to obtain oil from young inflorescence or the whole herb because it is less time-consuming. The dried leaves (bunched and put in propylene bags) are generally used by food processing industries, and the market for dried leaves determines the quality (taste, flavor, moisture content, and appearance) of the product (Jagdish 2019).

6.1.3 Breeding

In Asia's tropical regions, the *Ocimum* genus is home to 50 to 150 species of plants and shrubs.

1. *Ocimum basilicum*—Sweet Basil.
2. *Ocimum sanctum*—Tulsi and Holy Basil.
3. *Ocimum minimum*—This is a dwarf variety of tulsi.
4. *Ocimum gratissimum*—Ram tulsi and Shrubby Basil.
5. *Ocimum americanum*—Kali tulsi, Mummry, and Hairy Basil.

Improved Varieties:

- R.R.L.O.C.-11
- R.R.L.O.C.-12
- R.R.L.O.C.-14

Ocimum has undergone a thorough crop development program at the Regional Research Laboratory in Jammu, and newer, more promising varieties have emerged that are now advised for commercial production. The varieties are:

O. canum RRL-01
O. americanum RRL-02
O. viride RRL-08
O. gratissimum RRL-08
O. basilicum RRL-07
O. basilicum RRL-011

Synthesized Amphidiploid of Ocimum RRL-015

Varieties like RRL-011, Vikarsudha, Kusumohak, and CIM-Saumya have been released from different institutes.

India exports about 5000 tons of *Ocimum* dry leaves (herbs), its products, essential oil, and its derivatives/chemical constituents (eugenol, methyl eugenol, linalool, methyl chavicol, germacrene A and D, elemicin, β-elemene, and (Z)-ocimene) annually to Europe (Lal et al. 2018). The Council of Scientific and Industrial Research–Central Institute of Medicinal and Aromatic Plants (CSIR-CIMAP), Lucknow, is actively involved in genetic enhancement of the *Ocimum* species by breeding to develop better plant types with high yield and quality. CIM-CIMAP developed CIM Ayu, a high-yielding (200 ql/ha herb yield), eugenol-rich (84% eugenol, 111 kg/ha oil yield) oil-producing *O. sanctum* variety, and CIM Angana, a high-yielding (181 ql/ha herb yield), dark purple pigmented *Shyam tulsi* (40% eugenol, 92 kg/ha oil yield).

Pandey et al. (2015) developed and validated a rapid, sensitive UPLC-ESI/MS/MS method to simultaneously determine bioactive markers in *Ocimum sanctum* leaves (Table 6.1). Bioactive contents varied significantly in methanol extracts of leaves from different regions. Predominant phenolic acids were

TABLE 6.1 Content of *Ocimum sanctum* methanol extracts of dried leaves from four geographic regions (2009–2013)

ANALYTE (MG/G)	*HP2013*	*UP-2009–11*	*MP-9–11*	*WB-9–11*	*RANGE*	*OVERALL*
PHENOLIC ACIDS	*OS1*	*OS2-3-4*	*OS5–6*	*OS7–8*	*RANGE*	*AVERAGE*
Gallic	0.32	0.81	0.28	0.28	0.25–0.94	0.48
Protocatechuic	0.83	1.19	0.83	0.59	0.52–1.58	0.90
Chlorogenic	0.32	0.31	0.33	0.31	0.31–0.35	0.32
Caffeic	1.01	1.10	0.56	0.22	0.14–1.67	0.73
Ellagic	0.11	0.53	0.14	0.16	0.11–0.83	0.29
Ferulic	0.36	2.53	0.39	1.48	0.36–4.37	1.46
Sinapic	0.23	0.24	0.32	0.26	0.21–0.40	0.26
Rosmarinic	8.00	3.87	0.79	0.23	0.23–8.00	2.71
Vanillic	0.11	0.11	0.11	0.08	0.05–0.12	0.10
Total	11.28	10.69	3.75	3.62	2.33–15.78	7.26
Flavonoids						
Epicatechin	BDL	0.09	0.08	0.09	BDL–0.09	0.09
Catechin	BDL	0.05	0.05	0.06	BDL–0.06	0.05
Quercetin-3,4′-diglucoside	0.30	0.31	0.30	0.30	0.30–0.33	0.30
Rutin	0.30	0.14	0.16	0.18	0.10–0.30	0.18
Kaemferol-3-O-rutinoside	0.22	0.20	0.20	0.21	0.20–0.22	0.21
Scutellarein	0.05	0.04	0.06	0.07	0.03–0.09	0.05
Quercetin	0.04	0.05	0.04	0.04	0.04–0.05	0.04
Luteolin	1.53	1.12	1.71	0.66	0.62–2.71	1.20
Quercetin dihydrate	0.17	0.17	0.20	0.17	0.17–0.22	0.18
Apigenin	0.44	0.79	1.42	0.18	0.12–1.46	0.75
Kaemferol	0.23	0.21	0.21	0.21	0.20–0.23	0.21
Chrysin	0.09	0.12	0.10	0.13	0.08–0.16	0.11
Total	3.37	3.19	4.36	2.24	2.14–5.33	3.27
Eugenol	0.05	0.13	0.14	0.13	0.05–0.19	0.12
Ursolic acid	5.05	13.95	7.38	2.68	2.10–16.13	8.38

Adapted from Pandey et al. (2015); *Ocimum santum* samples were from Himachal Pradesh collected in 2013 (HP2013), Uttar Pradesh [mean value for 3 years—2009 to 2011] (UP-2009–11), Madhya Pradesh [mean value for 2 years—2009 and 2011] (MP-9–11) and West Bengal [mean value for 2 years—2009 and 2011] (WB-9–11); BDL—below detection limit

rosmarinic and ferulic acids, whereas luteolin was the major flavonoid. Leaves from Uttar Pradesh and Himachal Pradesh had the most, and those from West Bengal had the least, phenolic acids. Total flavonoid content was highest in Madhya Pradesh leaves and lowest in West Bengal leaves. Ursolic acid content was highly variable among the samples.

Shukla et al. (2021) developed an elite melatonin-rich *Ocimum sanctum* germplasm line. Eighty germplasm lines established under *in vitro* conditions were screened for their antioxidant potential (DPPH bioassay; 27.3–1452 μmol Trolox equivalents/g), with 5 lines showing high-antioxidant values (1227–1452 μmol Trolox equivalents/g). Phytohormones, neurotransmitters, melatonin, and serotonin levels were quantified in field-grown and greenhouse-grown tissues of a selected clonal micropropagated line, Vrinda. Melatonin content was consistent in both conditions, with higher serotonin levels under field conditions. Micropropagated tulsi lines had high phenolic and antioxidant activity (67–93 mg GAE/g and 6742–7718 μmol Trolox equivalents/g for mature plants). The melatonin and serotonin content of *Vrinda* leaves were 327 and 686 ng/g, respectively.

6.2 USES

6.2.1 Indian Food Products

Basil seeds are widely used in several Asian nations, including Iran and India, for cultural, nutritional, medicinal, and cosmetic purposes. The seeds are often consumed in beverages (Sharbat) and frozen sweets (Faloodeh) in these nations for aesthetic reasons and as a source of nutritional fiber (Munir et al. 2017; Cherian 2019). A drink containing up to 0.3% basil seeds had good sensory qualities like taste, texture, and acceptability according to Munir et al. (2017). Moreover, the drink's fiber and protein contents were higher than those of the control drink, and it contained significant amounts of minerals and phenolic compounds.

Basil seed mucilage has been investigated for its functional and nutritional benefits and is used in various products (low-sodium meat products, sponge cakes, pudding, ice cream, and low-fat yogurt) due to its interactions with milk protein and as a fat replacement (75% reduction) in sponge cakes. The mucilage enhances the rheological characteristics, alleviates syneresis, produces high gel strengths (Song and Kim 2019) of the food products, and improves the physicochemical and sensory qualities of bread and other bakery goods (Israr et al. 2017).

Standard procedures were used to determine the phytochemical composition of methanol, ethyl acetate, petroleum ether, and chloroform extracts of *Ocimum sanctum* seeds from Solan, Himachal Pradesh (Sharma et al. 2021). The methanol extract had the highest total phenolic content (757 mg GAE/g) compared to those reported in methanol extract of leaves (112 mg GAE/g dw vs 365 mg GAE/g for ethanol extract). Flavonoid content (283 mg RE/g) was also high in the methanol extract. Maximum flavonoid, phenolic, saponin, glycoside, terpenoid, steroid, tannin, and alkaloid contents were observed in the methanol and chloroform seed extracts (Table 6.2).

The methanol extract exhibited the highest DPPH radical scavenging activity due to the presence of higher total phenolic and flavonoid contents and the highest antimicrobial activity against *E. coli*, *K. pneumoniae*, *S. aureus*, and *B. subtilis*. Moreover, the methanol *Ocimum sanctum* seed extract contained borneol, caryophyllene, and caryophyllene oxide (12.63%, 3.87%, and 2.17% peak area), along with eight

TABLE 6.2 *Ocimum sanctum* seed composition

PHYTOCHEMICALS	*METHANOL*	*CHLOROFORM*	*ETHYL ACETATE*	*PETROLEUM ETHER*
Phenols (mg GAE/g)	757	535	183	139
Tannins (mg GAE/g)	362	237	258	156
Flavonoids (mg RE/g)	283	266	190	230
Saponins (mg DIO/g)	102	101	11.9	67.8
Glycosides (mg DIG/g)	296	242	67	193
Terpenoids (mg LE/g)	280	210	152	76
Sterol (mg CHO/g)	494	155	297	58
Alkaloids (%) seeds	24.2	27.8	21.2	20.8

Adapted from Sharma et al. (2021); GAE—gallic acid equivalent, RE—rutin equivalent, DIO—diosgenin, DIG—digitoxin, LE—linalool equivalent, CHO—cholesterol

other identified compounds. The antimicrobial (antibacterial and antifungal) activity in *O. sanctum* seed extracts is presumably due to the presence of secondary metabolites and characteristic chemical profile (Sharma et al. 2021). The seed is considered an abundant source of fixed oil (18%–22%), sitosterol, polysaccharides, and mucilage (Hiltunen et al. 1999). Linoleic acid is the main fatty acid (66%) of the seed oil (Naji-Tabasi and Razavi 2017) and together with α-linolenic acid (15.7%) may presumably be responsible for its anti-inflammatory, anticoagulant, hypotensive, chemopreventive, antihypercholesterolemic and immunomodulatory activities (Singh and Chaudhuri 2018 and references therein).

6.2.2 Basil and Basil Oil Are Used in Many Other Foodstuffs

Like other herbs, basil may be used in a variety of cuisines. A book on using herbs in cooking by Law (1975) includes recipes for sandwiches, soups, biscuits, and salads, and the use of herbs to garnish other food products.

Basil is a crucial component in the soups paysanne, turtle soup, and oxtail soup. Additionally, cheese and cheese dishes go nicely with basil. Basil is frequently used to flavor a variety of vegetables. The indigestible carbohydrates in peas and beans create flatulence, which is lessened by adding basil to recipes with those ingredients. Basil may also be used to flavor baked goods and desserts. For instance, basil pairs nicely with apple products like apple pie, apple sauce, and apple jellies.

According to Bassiouny et al. (1990), pure ether extracts of basil (0.02% w/w in dough) may be used instead of traditional antioxidants in soda cracker cookies in Egypt; the extracts had no impact on the organoleptic qualities. Basil applied to food concentrates or chocolate was proven by Russian scientists (Fomicheva et al. 1982) to prevent peroxides production as well stall already developed peroxides.

6.2.2.1 *Meat and Fish*

Basil oil is used to extend the shelf life of different products due to its antibacterial and antioxidant properties. There are Russian patents covering the use of basil oil or comminuted basil in the production of sausages and other meat products (Guseinov et al. 1992). Basil oil improves taste and color intensity and reduces microbial contamination (Dinarieva et al. 1984). Sausage taste improved when basil was added as microencapsulated oleoresin, which permeated even the intramuscular lipids in the sausages (Flint and Seal 1985).

According to Polic and Nedeljkovic (1978) basil causes off-taste in poultry meat. Basil, however, is a component of the traditional Lebanese cuisine known as stuffed turkey (Salah 1977), and many textbooks feature recipes for basil-infused fowl dishes. Beef stews and bouillons, pig, meatballs, shish kebabs, lamb meals, inner organ dishes, chicken, and turkey are all suggested to be flavored with basil. Basil makes canned beef stew taste handmade. All fish recipes, including stuffings, go well with basil, which can also be used in other seafood meals.

6.2.2.2 *Butter and Cheese*

A new "fresh table dairy speciality" created by the Italian company Yomo is being marketed under the name "Belgioioso." It is a spreadable, creamy product with traditional Mediterranean flavors like basil and has 18.5 g of fat, 9 g of protein, and 3 g of carbohydrate (905 kJ; 216 kcal/100 g) (Maiocchi 1995). Basil extract gives spiced butter its pleasant flavor. Basil or basil oil has been used as an ingredient in cheeses, such as "Brodinskii", a processed cheese produced in Russia (Samodurov et al. 1991), and "Pre Monde," a cheddar-type cheese produced in the USA, which is a condiment composed of dried sweet basil and cheese.

Basil is a common ingredient in sauces that are used to flavor entrées and side dishes; pesto is one of the most well-known varieties, and its main components are garlic, basil, and olive oil. It goes well with pasta dishes, soups, dips, and other foods. Pesto originated in Genoa, Italy. It may be made at home but is also sold commercially, and it is especially well-liked in Mediterranean nations (Dziezak 1991). The use of basil for sauces and dressings includes basil honey, basil yogurt, basil tomato, basil anchovies, and basil Parmesan sauce.

6.2.2.3 Beverages

The global holy basil iced tea market is estimated at 108.33 million USD (2022) and is projected to reach 177.94 million USD by 2030 (6.4% CAGR) during 2023–2030 (Virtue Market Research 2023). The growth is fueled by the natural stress-reducing properties of holy basil. It is rich in minerals (iron, zinc, calcium, manganese) and vitamins (A, C, and K). Vitamin K supports heart health and bone strength, and a half cup of fresh holy basil provides almost the recommended daily vitamin K intake. Moreover, the growth is also spurred by the potential of holy basil in diabetes prevention. Regular holy basil iced tea consumption, known for its anti-inflammatory properties, improves pancreatic β-cell function and insulin secretion, lowering the risk of type 2 diabetes. North America accounted for the major market share. Global holy basil iced tea market key players include several Indian companies (Hector Beverages Private Ltd., Chill Incorporated, Organic India Private Ltd., Vahdam Teas Private Ltd., and Teas of All Nations).

Basil is a common ingredient in alcoholic drinks, including bitters, liquors, and spirits. Basil is partly responsible for the exquisite scent of chartreuse liquor. Russians have patented a technique for enhancing the organoleptic qualities and storage durability of a carbonated fermented milk beverage by adding a blend of coriander, basil, and fennel essential oils to a salt solution of whey (Askerova et al. 1993). Reikhan is a non-alcoholic beverage that has also been granted a patent in Russia. Basil leaves and stems are infused in water at 95%–100 °C, filtered, combined with sugar and citric acid, and chilled (Kerimov 1993). Even grape and high-eugenol basil remnants may be used to make a non-alcoholic beverage concentrate, according to the Russian patent. Basil (1–40 g/L), fresh, dried, or frozen, is also employed in alcoholic drinks that are either sweet or dry and made with spirits, garlic, or lemon according to a German patent. Alcoholic beverages like Dushanbinskaya bitters use basil seeds instead of leaves in their recipes (Bagdasarov et al. 1978). The screening of 20 naturally occurring aromatic herbs in Ukraine resulted in the creation of four novel soft drinks using wormwood, mint, fennel, borage, and basil. These drinks include a considerable amount of vitamin C, thiamine, and riboflavin and have outstanding sensory qualities. The herbal antibacterial components in the drinks give them a 10–12 days shelf life.

6.2.2.4 Foods

Sweet basil or its extracts and oils are frequently combined with other spices and herbs in various foods, including pickled vegetables, sauces, vinegar, ice creams, mustard, and confectionery goods. Spice mixtures frequently include basil. A salt alternative for dietetic usage that contains basil but doesn't include sodium salts has been patented by Russians. Basil leaf flavoring is patented in Great Britain and Germany. A product called Spice Grains was first offered in 1974 by Norda Inc. in New York. It contained mono- and diglyceride antioxidants, food coloring, and herb oil made from natural herbs like basil and was used in the confectionery and snack industries (Anon 1974).

6.2.2.5 Microcapsule for Oil Storage

Basil oil is safe to use in the food industry, has a longer shelf life, and doesn't experience hygiene issues as a result of microbial contamination when kept in microcapsules. Sheen and Tsai (1991) found that when it came to oil retention, basil microcapsules were more stable than those of ginger or garlic. The highest oil stability and the slowest rate of disintegration were achieved by double-walled microcapsules (Sheen and Tsai 1991). This capsule had good sensory characteristics, perfect water solubility, and was microbiologically safe.

There is now a basil flavor that can be sprayed onto foodstuff. Basil extract is obtained using a centrifuge from the pulverized material and is then pressure-stored in bottles according to an Italian patent (Antonucci-Tarolla 1988). Product manufacturers may obtain a "fresh taste" using herb concentrates to give dishes the genuine flavor of fresh herbs. Since microwave cooking times are so brief, they are especially well suited to meals where quick flavor creation is required.

Sweet basil (*Ocimum basilicum*) extracts, flavorings, oils, spices, and oleoresins are on the American Food and Drug Administration's GRAS list, which denotes that they are widely acknowledged to be safe. Holy basil is considered safe for most people (GRAS) by the FDA, although it may interact with certain medications and should be used with caution in certain circumstances.

6.3 MEDICINAL PROPERTIES

Humans have used plants as medicines for thousands of years. *Ocimum* species like *Ocimum sanctum* and *Ocimum basilicum* are members of the *Lamiaceae* family (Shrinet et al. 2021). In Ayurvedic medicine, leaves, seeds, and roots are used to cure various diseases. Numerous nutrients and bioactive substances are found in *Tulsi*. It is grown widely in damp soil from the Himalayas to the Andaman and Nicobar Islands, and in various regions of Asia and Africa, and is used in food, beverages, and medicines. *Tulsi*'s therapeutic qualities depend on the type of soil and rainfall frequency. In Asia's tropical regions, around 150 species of the *Ocimum* genus are grown (Shah et al. 2018). Different plant species of the *Ocimum* genus may have similar morphologies, yet they can have significantly different genotypes and chemical profiles.

6.3.1 Phytonutrients

Tulsi is a rich source of phytonutrients. *Ocimum sanctum* contains vitamin A, vitamin C, β-carotene, chlorophyll, and minerals. The leaves contain vitamin C and carotene (83 and 2.5 μg/100 g), total carotenoid (19.8 g/100 g), and thiamine and riboflavin (300 and 60 μg/100 g) (Singh and Chaudhuri 2018 and references therein).

Ocimum sanctum [Ram Tulsi, green] is a highly mineral-rich species compared to other *Ocimum* species (*O. basilicum* [Shyma, black Tulsi], *O. gratissimum* [Krishna Tulsi, Deona], *O. canum* [Dulal Tulsi], and *O. americanum* [Ban Tulsi, Mamari]) (Tripathi et al. 2015). The high mineral abundance in *O. sanctum* leaves may presumably be one reason for its highest cultivation and utmost popularity over other *Ocimum* species. *O. sanctum* leaves had higher Ca, Mg, and Na content compared to those of *O. basilicum*, while K was comparatively higher in *O. basilicum* (Table 6.3). These elements and K maintain the electrolytic balance of the human body, whereas Mg, K, and Ca enable the use of *O. sanctum* leaves for diabetic and heart patients.

6.3.2 Phenolics

Chlorogenic acid, caffeic acid, vanillic acid, naphthanoic acid, and methyl salicylic glucoside are among the phenolic compounds in the *Ocimum sanctum* plant and are recovered from the plant's aerial portions (Skaltsa et al. 2019). HPLC analysis demonstrated the presence of gallic acid ethyl ester, 4-hydroxybenzoic acid, protocatechuic acid, gallic acid methyl ester, vanillin, and 4-hydroxybenzaldehyde (Jahanger et al. 2022). Dried leaves from hydroalcoholic extract of *Ocimum sanctum* leaves had high luteolin-7-O-glucoside, catechin, and resveratrol (430, 356, and 270 mg/100 g, respectively) (Table 6.4).

TABLE 6.3 Mineral composition (mg/kg) in dried leaves of *Ocimum* species

MINERALS	*O. SANCTUM*	*O. BASILICUM*	*O. GRATISSIMUM*	*O. AMERICANUM*
	RAM (GREEN)	*SHYMA*	*KRISHNA*	*BAN*
Calcium	15,634	13,556	19,428	14,502
Magnesium	5118	4553	7714	3022
Sodium	7157	6582	256	459
Potassium	10,521	11,255	2816	47,990
Boron	29	19	20	35
Copper	18	9.4	7.6	9.6
Iron	325	228	597	254
Manganese	40	27	35	39
Phosphorous	3690	3806	2892	3080
Zinc	37	31	21	24

Adapted from Tripathi et al. (2015)

TABLE 6.4 Composition (mg/100 g) of polyphenolic compounds

PHENOLIC ACIDS	*AMOUNT*	*FLAVONOIDS*	*AMOUNT*
Caffeic acid	55.68	Apigenin	132
Ferulic acid	7.17	Catechin	356
Gallic acid	4.02	Hesperetin	164
4-Hydroxy benzoic acid	12.61	Kaemferol	57.03
2-Hydroxy cinnamic acid	10.19	Luteolin	64.1
Protocatechuic acid	23.53	Luteolin-7*O*-glucoside	430
Sinapic acid	18.58	Myricetin	43.7
Polyphenols		Naringenin	6.43
Ellagic acid	18.14		
Resveratrol	270		

Adapted from Kondapalli et al. (2022)

O. sanctum extract (850 mg/kg) increased blood hemoglobin, high-density cholesterol (HDL), and insulin levels; decreased total cholesterol, triglycerides, and serum IL-6 levels in female Sprague-Dawley rats. The extract also exhibited a prebiotic effect (change in caecal *Lactobacillus*, reduced *Firmicutes*) comparable to FOS (Kondapalli et al. 2022).

6.3.3 Flavonoids

Methoxy flavonoids and their glycosides (cirsimartin, isothymusin, vicenin, luteolin, isovitexin, and isoorientin) and orientin are among the key components of the *Ocimum sanctum* plant (Kelm et al. 2000). Other flavones found by atmospheric pressure chemical ionization mass spectrometry (APCIMS) are salvigenin, cirsumaritin, crisilineol, isothymusin, apigenin, and eupatorium (Grayer et al. 2002).

6.3.4 Neolignans and Coumarins

Neolignan components such as Tulsinol A to Tulsinol G, produced through the polymerization of eugenol content, were found and identified in the methanolic extracts of *Ocimum sanctum* (Suzuki et al. 2009).

On the other hand, the tulsi plant produces three coumarin substances, aeculetin, aesculin, and ocimarin (Singh and Chaudhuri 2018 and references therein).

6.3.5 Steroids

Stigmasterol, campesterol, sitosterol, and sitosterol-3-O-D-glucopyranoside are the steroid substances that are isolated from the stem and leaves of *Ocimum sanctum* (Gupta et al. 2007).

6.3.6 Essential Oil

The leaves of *Ocimum sanctum* are an abundant source of volatile oil, generally with 71% eugenol and 20% methyl eugenol. Other reported volatile oil constituents include sesquiterpene hydrocarbons (caryophyllene and carvacrol), rosmalinic acid, oleanolic acid, caryophyllene, linalool, borneol, nerol, terpinene, 4-decylaldehydes, selinene, pinenes, and camphene. The essential oil constituents vary based on agro-climatic factors (growing region, cultivation, species, cultivars, growing season, and light exposure) and harvesting conditions (age or position of the chosen leaves and harvesting techniques).

Raina et al. (2013) evaluated the essential oils of 32 *Ocimum sanctum* germplasm accessions collected from different North Indian regions (all from Uttar Pradesh except four from Uttarakhand). Table 6.5 shows the actual essential oil composition of only two promising *Ocimum sanctum* chemotypes (eugenol and methyl eugenol). The essential oils (0.13–0.45% v/w content on a fresh weight basis) of fresh herbage at full bloom stage varied in eugenol (1.94–60.2%), methyl eugenol (0.87–82.98%), β-caryophyllene (4.13–44.6%), and β-elemene (0.76–32.41%). The two eugenol- and methyl eugenol-rich chemotypes contained high (>55%) eugenol or methyl eugenol (>70%). Eugenol is responsible for the therapeutic effects of tulsi, whereas methyl eugenol is a high-value aroma compound used as a flavoring agent.

Ocimum sanctum collected from three different sites (variable altitude—214 m, 232 m, and 411 m) in Uttarakhand, North India at the vegetative and full bloom stage (Table 6.5) contained β-elemene

TABLE 6.5 Essential oil composition

	EUGENOL	METHYL EUGENOL	VEGETATIVE	BLOOM	RED	WHITE
COMPOUNDS	(% ESSENTIAL OILS)				(μG/ML)	
Linalool	0.42		0.02–0.13	0.06–0.1		1.09
Borneol	0.59	0.11	0.12	0.14–0.17	7.79	2.8
Eugenol	60.21	0.86	2.3–18.3	1.2–45.1		1.5
β-Elemene	5.06	0.25	29.3–57.9	26.5–37.8	65.7	2.79
Methyl eugenol	0.26	82.98	0.25	0.16–0.25	684	98.4
β-Caryophyllene	25.95	13.69	8.7–39.3	15.7–24.6	146	
α-Humulene	2.03	1.03	0.63–2.47	0.97–1.45	13.4	1.42
γ-Muurolene		0.07			2.41	
Germacrene D	3.36	0.09	0.04–17.6	0.05–15.1	0.55	1.23
β-Selinene	0.37		0.18–1.07	0.26–0.39		
α-Selinene	0.54		0.41–0.99	0.31–0.51		0.4
β-Bisabolene	0.57		0.05–0.09	0.03–0.08		
Caryophyllene oxide	0.12	0.28	0.75–4.0	0.99–2.88		
References	Raina et al. (2013)		Rana et al. (2021)		Tangpao et al. (2018)	

(26.5–57.9%), eugenol (1.2–45.1%), (E)-β-ocimene (0.01–17.7%), germacrene D (0.04–17.6%), (E)-caryophyllene (8.7%–39.3%), and α-copaene (1.24%–7.1%) as the main components of essential oil. The oil had high sesquiterpene hydrocarbons (52.3%–79.8%). β-Elemene was the major compound (57.9%) during the first harvest, whereas eugenol (45.1%) predominated during the second harvest (Rana et al. 2021). Fresh leaves of *Ocimum sanctum* (cvs. Green—CIM-Ayu and Purple) grown in Uttarakhand, India at blooming stage had similar essential oil composition (Padalia and Verma 2011). The predominant phenylpropanoids (68.1% and 73.5% for green and purple cultivars, respectively) were represented by eugenol (67.4% and 72.8%), followed by sesquiterpene hydrocarbons (22.4% and 22.6%; primarily β-elemene [11% and 10.9%], β-caryophyllene [7.3% and 8.4%], and germacrene D [2.4% and 2.2%]). Both oils were characterized as eugenol types.

Fresh *Ocimum sanctum* red (var. Shyama) and white (var. Rama) leaves from plants grown at Chiang Mai University, Thailand were evaluated for their essential oil composition (Tangpao et al. 2018). Maximum hydro-distilled oil yield of red and white varieties was ~0.4% and ~0.33%, respectively. Major components of red *O. sanctum* leaves (Table 6.5) were methyl eugenol, β-caryophyllene, α-cubebene, β-elemene, α-copaene, and α-humulene (684, 146, 105, 65.7, 20.7, and 13.4 μg/mL, respectively). Methyl eugenol, α-cubebene, and α-copaene (98.4, 9.9, and 4.7 μg/mL, respectively) were the principal components of white *O. sanctum* leaves (Tangpao et al. 2018).

A study analyzed seasonal variations in essential oils of leaves (purple variety) from *Ocimum sanctum* grown (~800 m altitude) in Northwest Karnataka, India, collected throughout one year (May 2014—April 2015). Yield of essential oil varied (0.14–0.21% w/w), with 23 to 39 identified constituents comprising 95% of the total oil. Methyl eugenol was the most abundant compound (average 70%; 55.5%–76.8% range), followed by β-caryophyllene (10.4%), α-ylangene (4.2%), γ-muurolene (3.6%), and borneol (2%). Methyl eugenol content was optimal in November and lowest at maturity in April when accumulation of sesquiterpenes increased and phenylpropanoids decreased (Joshi and Sharma 2021).

Saran et al. (2017) characterized eleven *Ocimum sanctum* accessions grown in Gujarat, western India at three different harvesting stages for 21 traits and six essential oil components. Oil content, oil, and eugenol yields varied (17–50 g/kg, 21–73, and 16–67 kg/ha, respectively). At maturity, fresh leaf essential oil (g/kg) consisted of eugenol (730–940), azulene (0–110), β-caryophyllene (0–182), α-cubebene (0–125), δ-3-carene (0–27), and α-humulene (0–8). On average, dry leaf yield was 1843 kg/ha, containing 73 kg/ha oil from harvests (2015–2016) priced at US$2.32 and 38.88/kg for dry leaf and oil in the local market.

Ocimum sanctum essential oil (specific gravity 0.92) obtained from the Centre of Aromatic Plants (Dehradun, India) consisted of 41 identified compounds (relative abundance ≥2.5% of the total) (Salvi et al. 2022). The most abundant essential oil compounds were: α-citral (14.75%), cyclohexane, 1-ethenyl-1-methyl-2,4-bis(1-methylethenyl)-,[1S–(1α,2β,4β)] (11.81%), caryophyllene (8.91%), phenol, 2-methoxy-3-(2-propenyl)- (8.52%), α-amorphene (7.78%), γ-muurolene (7.43%), and 2H-1-benzopyran, 3,4,4a,5,6,8a-hexahydro-2,5,5,8a-tetramethyl-(2α, 4aα, 8aα) (4.52%). The oil showed potent antibacterial activity in the 12.5% to 25% range and effectively scavenged DPPH (9.48 ng/mL EC_{50}). The abundance of phenol, 2-methoxy-3-(2-propenyl)- (8.5%) and γ-muurolene (7.4%) are presumed to be responsible for its high antibacterial activity. Furthermore, the high antimicrobial and antioxidant activities are attributed to the presence of α- and β-pinene and β-caryophyllene oxide and β-caryophyllene (Salvi et al. 2022).

The essential oil of store (Heritage Bio-natural Systems Pvt. Ltd., Hyderabad) bought *Ocimum sanctum* dried leaves consisted primarily of methyl eugenol (13.96%), 1,3,5-cycloheptatriene (11.75%), 2,4-diamino-5-[3,4]-p-chlorobenz (9.25%), E-4-[-2-benzothiazolyl)ethenyl (7.11%), eugenol, N,N-bis(4-chlorobenzylidene)benzene (5.61%), and caryophyllene oxide (3.64%) (Kondapalli et al. 2022). Devendran and Balasubramanian (2011) identified ten compounds in *O. sanctum* hydroalcoholic leaf extract with eugenol, caryophyllene, and cyclohexane 1,2,4-triethenyl as major compounds (43.9%, 26.5%, and 15.3%, respectively). Moreover, the amount of eugenol was associated with antioxidant activity. Crude methanol *O. sanctum* leaf extract contained eugenol (3.31%), caryophyllene (3.61%), germacrene (1.52%), 9i, 215-octadecatrienoic acid methyl ester (1.88%), hexadecanoic acid methyl ester (2.30%), and phenol-2-methoxy-4-(1-propenyl) (3.31%) (Arulraj et al. 2014).

Ocimum sanctum grown in central European (Czech Republic) conditions in dense spacing produced average annual dry biomass (5963 kg/ha) containing 1% essential oil with a major proportion of linalool (34–39%), estragole (22–26%), t-methyl cinnamate (22%–28%), and rosmarinic acid (1.05–6.34 mg/g) (Pavela et al. 2023). Other minor essential oils were identified in this linalool–estragole chemotype, including eucalyptol, ocimene, camphor, terpinen-4-ol, bornyl acetate, eugenol, and c-methyl cinnamate.

6.4 MEDICINAL PROPERTIES

6.4.1 Antistress

More free radicals are produced due to stress, which harms human tissues and various vital organs. Adaptogens or anti-stress agents are substances that promote generalized resistance and physical endurance. Recent research has demonstrated that *O. sanctum* possesses strong adaptogenic qualities. Tulsi leaves can help avoid and lessen both physical and mental stress. Tulsi is a rejuvenator, lowering stress levels, calming the body, and enhancing memory (Prakash and Gupta 2005). Administration of an oral dose of ethanolic tulsi extract for 7 days to mice boosted their adrenaline and noradrenaline levels and reduced dopamine and serotonin levels.

Tulsi can be used twice a day to take advantage of its sedative effects. A study discovered that *O. sanctum*'s ethanolic extract aids in reducing the shift in plasma corticosterone levels that is brought on by exposure to noise stress. This finding supports tulsi's anti-stress properties.

In another study, rabbits were given fresh leaves of *O. sanctum*, where oxidative stress was reduced, resulting in alleviation of plasma superoxide dismutase (SOD) loss (23%) and reduced glutathione (29%) levels. *O. sanctum*'s anti-stress effects are aided by its antioxidant capabilities (Jyoti et al. 2007).

6.4.2 Adaptogen

Tulsi is an excellent source of adaptogenic qualities, which help reduce erratic mood swings and promote mental calm and clarity. The two most important adaptogen components in tulsi's chemical makeup, eugenol and caryophyllene, help reduce corticosterone levels, which are the main source of stress. Additionally, it improves memory and reduces the likelihood of mental health issues brought on by aging. On the other hand, both ursolic and oleanolic acids serve as adaptogens and are highly effective at reducing stress levels (Devi 2001).

6.4.3 Antidiabetic

Ocimum sanctum leaves exhibit antidiabetic properties (Khan et al. 2012). *Ocimum sanctum* aqueous extract lowers blood glucose levels in induced hyperglycemic tilapia (*Oreochromis niloticus*) (Arenal et al. 2012). Due to their ability to suppress carbohydrate-hydrolyzing enzymes, both floral and leafy sections of plants can be employed in alternative nutritional therapy, particularly for managing diabetes. Tetracyclic triterpenoids isolated from aerial parts of *Ocimum sanctum* are claimed to have a similar antidiabetic effect. When compared to untreated diabetic rats, the aerial portion of *Ocimum sanctum* test compounds dramatically lowers elevated serum glucose levels. It also reverses the cholesterol, triglyceride, low-density lipoprotein (HDL), and high-density lipoprotein (LDL) values (Mahomoodally et al. 2012). Therefore, diabetes and metabolic stress can be treated with *Ocimum sanctum*. It has been demonstrated that tulsi and neem lower blood sugar levels in people with diabetes (Patil et al. 2012).

6.4.4 Anti-Inflammatory

In an experimental study, the inflammatory activity in rats was demonstrated using 500 mg methanolic extract of *Ocimum sanctum*. The cyclooxygenase and lipoxygenase routes of arachidonic acid metabolism can be blocked by essential oil and linolenic acid present in tulsi (Kelm et al. 2000). As a result, they exhibit anti-inflammatory properties against prostaglandin II and leukotrienes, which cause rat edema. In rats with carrageenan-induced paw edema, *O. sanctum*'s aqueous extract (200 mg/kg or 400 mg/kg) showed considerable efficacy. This demonstrated the superior effect of *O. sanctum* over standard medication indomethacin. Moreover, the oil's ability to inhibit lipoxygenase and block histamine contributes to tulsi's antiulcer properties (Singh et al. 2007). Consuming tulsi leaf (OS Linn.) on an empty stomach increases immunity (Mondal et al. 2011). The anti-inflammatory properties of *Ocimum sanctum* seed oil are due to combined suppression of arachidonate metabolism and the addition of antihistaminic activity (Singh et al. 2007). The seed oil also exhibits antipyretic properties due to prostaglandin inhibition and analgesic efficacy with peripheral action, anticoagulant, hypotensive, and immune-modulating effects.

Tulsi is primarily used in immune-based therapy to treat poor mothering, manage ecto- and endoparasites, and improve fertility and bone setting. Tulsi also demonstrates immuno-modulating characteristics, such as modifications in cytokine production, histamine release, immunoglobulin secretion, class switching, cellular coreceptor expression, lymphocyte expression, and phagocytosis (Upadhayay et al. 2012).

6.4.5 Antifungal

The essential oil produced by steam distilling (1.1% w/v) *O. gratissimum*'s aerial parts exhibit antifungal properties. It inhibits the growth of all the fungi examined, including the phytopathogens *Botryospaeria rhodina*, *Rhizoctonia* sp., and two strains of *Alternaria* sp. (Prabuseenivasan et al. 2006). Tulsi shows antifungal activity against *Trichophyton rubrum*, *Microsporum canis*, *M. gypseum*, and *T. mentagrophytes*. Moreover, hexane extract of *O. gratissimum* with eugenol is particularly efficient against the dermatophyte *Trichophyton rubrum*, the most prevalent dermatophyte in Brazil (Bandeira et al. 2017).

A *Colletotrichum* species isolated from rotten tomatoes was evaluated using ethanol, hot water, and cold-water extracts of *O. gratissimum*. Hot water extract showed the highest inhibition zone, followed by ethanolic extract, and cold-water extract had the lowest inhibition zone (Orji et al. 2015).

6.4.6 Antiviral

Joshi et al. (2014) evaluated the antiviral activity of *O. sanctum* 95% ethanol extract against the 2009 swine flu pandemic (H1N1pdm) virus in Madin-Darby canine kidney (MDCK) cells through different virus inhibition assays. The extract had very low toxicity (726 µg/mL) and dose-dependently inhibited the virus by strongly impairing and reducing viral protein expression and infectivity, reducing virus yield through inhibition of viral replication. The maximum inhibitory effect was obtained with 150 µg/mL extract at the early stages (2–4 hpi infection) of virus replication. The extract markedly reduced genomic viral mRNA and viral protein and additionally exhibited immunomodulatory effects. Molecular docking of apigenin among the top five *O. sanctum* phytocompounds (oleanic acid, vicenin-2, apigenin, stigmasterol, and urcolic acid) with H1N1 proteins showed the strongest binding energy with good ADMET property compared to oseltamivir and zanamivir (Alhazmi 2015).

Umashankar et al. (2021) reported several *O. sanctum* leaf extract studies demonstrating its therapeutic, prophylactic, and virucidal activities. Leaf extract showed therapeutic activity against the H9N2 virus in an *in ovo* model by reducing infection level. Tulsi crude extracts or its individual isolated compounds exert wide-spectrum antiviral activity against HSV, adenovirus, coxsackievirus B, and enteroviruses. Ursolic acid exerts the strongest activity against HSV, ADV-8, CVB1, and EV71. Recent molecular docking studies demonstrate that tulsinol A–G and dihydro-dieuginol B are potent inhibitors of SARS coronavirus main protease (Mpro) and papain-like protease (PLpro), indicating that *O. sanctum* can be used as a preventive against CoV, due to its immunomodulatory, ACE2 blocking, and viral replication inhibition properties (Umashankar et al. 2021 and references therein). Rajan et al. (2021) screened 11 selected Ayurvedic herbs, including *Ocimum sanctum*, for *in silico* docking with the active site of the crystal structure of SARS-CoV-2 virus main protease (PDB ID: 6LU7). Apigenin from *O. sanctum* had high binding constant (–5.27 binding energy) and inhibition activity (140 μmol/L inhibition constant) against PDB ID: 6LU7 protease, indicating good antiviral property. This is supported by quantitative data analysis (citation frequency and use value) demonstrating that *Ocimum tenuiflorum* was among the most frequently used plant species (0.924 citation frequency and 0.022 use value) by Bangladeshi respondents in the prevention and management of COVID-19 symptoms (Islam et al. 2021).

6.4.7 Antibacterial

Since *Ocimum* sp. essential oils contain carvacrol, methyl eugenol, and caryophyllene, it is believed that they possess antibacterial characteristics (Zeedan and Abdalhamed 2021). *Ocimum sanctum* extract was found to be equally efficient against pathogenic Gram-positive and Gram-negative bacteria (Khosla 1995). *Ocimum sanctum* leaf oil, aqueous, alcoholic, and chloroform extracts were tested for their antibacterial properties against *E. coli, P. aeruginosa, S. typhimurium,* and *S. aureus*. *Streptococcus mutans* was sensitive to tulsi (OS) extract, although *Klebsiella pneumoniae* and *Escherichia coli* isolates were resistant to it at higher concentrations (6.25–25 mg/mL). Tulsi extracts in solvents and water have demonstrated antibacterial action against *S. aureus,* which is multi-drug resistant (Dahiya and Purkayastha 2012). Eugenol, methyl eugenol, linalool, and 1,8-cineole all demonstrated exceptional cytotoxicity to *Candida species* and Tulsi (OS Linn.) oils (Agarwal and Nagesh 2010). According to Mahomoodally et al. (2010), unripe tulsi fruit extract from the *Lamiaceae* family was incredibly effective against a resistant strain of *Staphylococcus aureus*. When coupled with chloramphenicol (C) and trimethoprim (Tm), its leaf extract shows strong antibacterial activities against drug-resistant *S. enterica* serovar *typhi (S. typhi)* (Joshi et al. 2011a).

6.4.8 Anticancer

Tulsi has demonstrated apoptosis and reduced proliferation in several studies of prostate, breast, pancreatic, oral, and lung cancer cells. These studies indicate that the anticancer activities of *O. sanctum* extracts or its essential oil generally rely on upregulating apoptotic genes and simultaneously promoting survival genes. Magesh et al. (2009) investigated the antitumor mechanism of *O. sanctum* ethanol extract against human nonsmall cell lung carcinoma (NSCLC) A549 cells *in vitro* and the Lewis lung carcinoma (LLC) animal model. The extract was cytotoxic against A549 cells, increased the sub-G1 population, and induced apoptosis by PARP (poly [ADP-ribose] polymerase) cleavage, cytochrome C release into cytosol, and simultaneous caspase-9 and caspase-3 protein activation, increased Bax/Bcl-2 ratio, Akt phosphorylation, and ERK (extracellular signal-regulated kinase) inhibition. The extract dose-dependently suppressed the growth of LLC inoculated onto C57BL/6 mice.

Ethanol *O. sanctum* leaf extract inhibited the proliferation, migration, invasion, and induced apoptosis of pancreatic cancer (PC) cells (*AsPC-1, MiaPaCa*, and *Capan-1*) *in vitro* by downregulating activated ERK-1/2, FAK, and p65 (subunit of NF-κB) (Shimizu et al. 2013). Intraperitoneal injections of the aqueous extract significantly inhibited the growth of orthotopically transplanted PC cells *in vivo* by upregulating metastasis-inhibiting genes (*E-cadherin*) and inducing apoptosis (*BAD*) genes and simultaneously promoting survival genes (*Bcl-2* and *Bcl-xL*). Apparently, both ethanol extracts and essential oil of *O. sanctum* leaves significantly inhibit the aggressiveness of PC cells and inhibit the growth of orthotopically implanted PC cells. Ethanol (70%) *O. sanctum* leaf extract significantly inhibits the growth of LNCaP prostate cancer cells during *in vitro* treatment without affecting normal HPrEC cell viability (Dhandayuthapani et al. 2015). The extract induces apoptosis in LNCaP cells by downregulating Bcl-2 expression, significantly increasing PARP cleavage level, and activating caspase-9 and caspase-3, leading to DNA fragmentation and cell death. Aqueous *O. sanctum* (Rama and Krishna Tulsi) leaves exhibited significant cytotoxic effect against oral cancer (mouth epidermal carcinoma cells [KB cells]) cell line (Shivpuje et al. 2015).

Monga et al. (2011) investigated the anti-melanoma activity of 50% alcoholic aqueous leaf extracts from various *Ocimum* species. Orally ingested leaf extract (200 mg/kg) significantly decreased tumor volume, increased average body weight, and increased mouse survival rates. In order to mediate anticancer effects, *Ocimum* sp. contains phytochemicals like eugenol, rosmarinic acid, apigenin, retinal, luteolin, sitosterol, and carnosic acid (Baliga et al. 2013). These phytochemicals also increase antioxidant activity, alter gene expression, induce apoptosis, and inhibit angiogenesis and metastasis in chemically induced skin cancer. Tulsi's aqueous extract and the bio-organic flavonoids, orintin and vicenin, protect mice against radiation sickness and reduce mortality. Only the tumor-enhancing effects of the radiation are protected from healthy tissues. Additionally, some important phytochemicals, such as eugenol, rosmarinic acid, apigenin, and carnosic acid, shield DNA against damage caused by radiation (Baliga et al. 2013).

Essential oil (250 μg/mL) from *Ocimum sanctum* leaves significantly reduced the number of migrated cancer cells and suppressed matrix metalloproteinase (MMP-9) activity in lipopolysaccharide (LPS)-induced inflammatory cells (Manaharan et al. 2014). Inhibition of MMP-9 activity generally reduces inflammation and prevents cancer progression and metastasis. Hence, *O. sanctum* essential oil is potentially useful in developing novel therapeutic strategies for inflammation associated cancer (Manaharan et al. 2014). The same investigators extended their study to elucidate the anticancer and apoptosis mechanisms of essential oil from *Ocimum sanctum* leaves (Manaharan et al. 2016). *O. sanctum* essential oil dose-dependently inhibited MCF-7 (Michigan Cancer Foundation-7) cell proliferation (170 μg/mL IC_{50}) by inducing apoptosis, upregulating the apoptotic genes *p53* and *Bid* and elevating the *Bax/Bcl-2* ratio. Moreover, *O. sanctum* essential oil had superior cytotoxic and apoptotic effects than resveratrol, a polyphenol that induces apoptosis and inhibits growth of several cell lines (Manaharan et al. 2016).

The lignin rabdosin isolated from aqueous methanol *Ocimum sanctum* leaf extract is highly cytotoxic against human cancer cell lines MCF-7, SKBR3, and HCT-116 with very low cytotoxicity against peripheral blood mononuclear cells (PBMCs) (Flegkas et al. 2019).

6.4.9 Hepatoprotection

Some *Ocimum sanctum* phytochemicals, including eugenol, carvacrol, ursolic acid (UA), caryophyllene, and rosmarinic acid, have shown anti-inflammatory, gastrointestinal, and hepatoprotective properties (Prakash 2005). Lahon and Das (2011) investigated the hepatoprotective activity of *Ocimum sanctum* alcoholic leaf extract against paracetamol-induced liver injury in albino rats and concluded that it synergized with silymarin and had a strong hepatoprotective effect. *Ocimum sanctum* extracts may lessen oxidative stress and hepatic steatosis by increasing glutathione peroxidase and catalase levels in the liver (Lahon and Das 2011). The oil possesses anti-inflammatory characteristics due to dual reduction of arachidonate metabolism and the addition of antihistaminic action. Its anti-pyretic qualities are due to prostaglandin inhibition and peripherally acting analgesic action. Studies on animals have demonstrated the oil's value in reducing arthritis and joint swelling brought on by turpentine oil and formaldehyde adjuvants (Singh et al. 2007).

6.4.10 Immunomodulatory Activity

Tulsi leaf consumption on an empty stomach enhances immunity (Mondal et al. 2011). Its ethanolic extract exhibits immunomodulatory properties. Tulsi is mostly utilized in immune-based therapy to manage poor mothering, ecto- and endoparasite control, improvement of fertility, and bone setting. Furthermore, it exhibits immune-modulating properties, including changes in cytokine secretion, histamine release, immunoglobulin secretion, class switching, cellular co-receptor expression, lymphocyte expression, and phagocytosis (Upadhayay et al. 2012). According to Dutta et al. (2007), tulsi leaf extract (DTLE) is protective against genotoxicants. Aqueous extract of *Ocimum sanctum* at oral doses of 100 and 200 mg/kg/day increased the production of RBC, WBC, hemoglobin, and antibodies in rats without changing the metabolic parameters (Jeba et al. 2011).

6.4.11 Antilipidemic Efficacy

People are now more frequently diagnosed with atherosclerosis, hyperlipidemia, and other associated disorders (Amrani et al. 2006). *O. basilicum* aqueous extract can reduce acute hyperlipidemia brought on by triton WR-1339 in rats by lowering LDL cholesterol, triglycerides, and total cholesterol levels. Fresh tulsi leaves lower overall cholesterol levels when given to rabbits for 28 days (Rachmawati and Muhammad 2021).

6.4.12 Antifertility

The antifertility effect of tulsi leaves is due to their high ursolic acid content, which exerts anti-estrogenic activity associated with the cessation of male fertility and inhibition of female ovulation. This substance may be a safe and efficient choice when paired with other potent antifertility medications. Tulsi leaves, which lessen the Sertoli cells' activity, prevent the creation of sperm in males (Kumar et al. 2022). In *in vitro* studies, *O. canum* leaves were found to have anti-implantation properties in albino rats. Ursolic acid has an anti-sterility property, while tulsi leaves have antiandrogenic properties. Benzene extract of *O. sanctum* administered to albino rats reduces the number and movement of sperm (Deshmukh et al. 2015).

6.4.13 Antipyretic

Rats receiving the typhoid-paratyphoid A/B vaccine-induced pyrexia were used to assess the antipyretic efficacy of *Ocimum sanctum* fixed oil. The antipyretic efficacy of the oil on administration was seen in the significantly reduced febrile response. The oil's antipyretic efficacy was on par with aspirin when administered at 3 mL/kg. The fixed oil also has prostaglandin inhibitory action, which apparently accounts for its antipyretic effects (Singh and Majumdar 1995).

6.5 HUMAN STUDIES AND CLINICAL TRIAL ON TULSI

6.5.1 Animal Studies

In animal models, *Ocimum sanctum* (OS) has been shown to have neuroprotective, stress-relieving, and cognitive-improving properties. OS leaf extracts are acknowledged for their calming and relaxing properties. Apigenin, ursolic acid, luteolin, orientin, isoorientin, stigmasterol, vicenin-2, isovitexin, vitexin, chlorogenic acid, aesculetin, aesculin, galuteolin, circineol, gallic acid, protocatechuic acid, vanillin acid, chlorogenic acid, and apigenin-7-O-glucuronide are all present in the alcohol extract of OS leaves. EtOS's psychopharmacological effects were first assessed, and they revealed that it exhibited anticonvulsant, anti-anxiety, and antidepressant effects (Sakina et al. 1990). Ethanolic extract of OS (EtOS) administration exhibited a positive impact on cognitive activity in animal models. It improved learning and memory in an Alzheimer's rat model and reduced the amnesic consequences of cognitive impairment in rats (Giridharan et al. 2011). EtOS also demonstrated positive results in cases of cerebral reperfusion injury and cerebrovascular insufficiency (Ahmad et al. 2012). Aqueous *O. sanctum* extract (300 mg/kg i.p) showed anticataleptic action in experimentally induced Parkinsonism rats and improved mice performance in the rota-rod chimney test, indicating muscle rigidity reduction. This antiparkinsonian activity exhibited by aqueous *O. sanctum* extract is ascribed to linalool, which modulates central neurotransmission (Joshi et al. 2011).

The effects of acute and chronic noise stress on numerous parameters, including plasma corticosterone (Sembulingam et al. 1997), WBC count, and neutrophil function (Archana and Namasivayam 2000), were normalized in albino rats treated with EtOS. According to a mouse swimming performance test, high EtOS doses also produced stimulant and/or anti-stress effects on the neurological system that were comparable to those of desipramine (Maity et al. 2000). In albino rats, OS-containing formulations (EuMil) reduced changes in glucose tolerance, the suppression of male sexual activity, and cognitive impairment in a dose-dependent manner similar to ginseng (Muruganandam et al. 2002). Rats exposed to restraint stress had their behavior altered, but aqueous extracts of OS and *C. sinensis* prevented this (Tabassum et al. 2010). Several behavioral investigations assessed the anti-anxiety and antidepressant effects of EtOS, (Chatterjee et al. 2011). OS extracts were used in formulations that helped rats with chronic stress-related neurochemical disturbances. When exposed to electroshock stress, noradrenaline (NA), dopamine (DA), and 5-hydroxytryptamine (5HT) concentrations in the frontal cortex, pons-medulla, hypothalamus, and hippocampus all returned to normal (Samson et al. 2006). Similar outcomes were obtained by administering 70% EtOS, which normalized the neurotransmitter levels in specific brain regions affected by noise stress (Ravindran et al. 2005).

The aqueous extract of *Ocimum sanctum* (*O. sanctum*) leaf was studied for its immunotherapeutic potential in treating bovine sub-clinical mastitis (SCM). After intramammal infusion of aqueous *O. sanctum* leaf extract, somatic cell count (SCC), total bacterial count (TBC), milk differential leukocyte count (DLC), phagocytic activity and phagocytic index, as well as leukocyte lysosomal enzymes like myeloperoxidase and acid phosphatase content, were assessed. The outcomes showed that the *O. sanctum* aqueous extract treatment decreased TBC, raised neutrophil and lymphocyte numbers, and improved phagocytic activity and phagocytic index. Similar to this, animals given the extract had considerably more lysosomal enzymes in their milk polymorphonuclear cells (PMNs). According to the findings, *O. sanctum* (leaf) crude aqueous extract contains several biologically active components that are antibacterial and immunomodulatory in nature. As a result, the current work supports the therapeutic use of medicinal herbs and also highlights the potential of widely accessible nontoxic medicines to boost breast immunity (Mukherjee et al. 2005).

In underdeveloped nations, malaria is a serious public health issue that results in at least 247 million infections and 1 million fatalities per year (WHO 2008). The two predominant malaria species in India are *Plasmodium falciparum* and *Plasmodium vivax*, which account for almost one fifth (22.6%) of clinical episodes of *Plasmodium falciparum* and 42% of episodes of *Plasmodium vivax* worldwide (Hay et al. 2010). One of India's malaria-endemic states, Odisha accounts for 23% of all malaria cases, 50% of malaria deaths nationwide, and 40% of cases in the state are caused by *Plasmodium falciparum* (Mahapatra et al. 2012). Odisha's malaria status is worse than that of sub-Saharan Africa (Narain 2008). According to a different study, malaria can be prevented by using 16 traditional plant species from 12 families. The three districts of Odisha's healers most frequently reported using *Andrographis paniculata*, *Azadirachta indica*, *Nyctanthes arbor-tristis*, *Ocimum sanctum*, *Piper nigrum*, and *Zingiber officinale* as malaria preventatives. The majority of the treatments were utilized as decoctions.

6.5.2 Human Studies

Jamshidi and Cohen (2017) conducted a comprehensive review of human studies that reported clinical outcomes after tulsi ingestion of 24 independent clinical studies. The studies were classified according to three main clinical domains: metabolic disorders, neurocognitive or mood conditions, and immunity and infections (15, 4, and 5 studies, respectively). Only three out of 24 studies were high quality (4–5 Jadad scores), two of which examined neurocognitive effects and one reported on immunity. Results of the 24 human studies indicate that tulsi is a safe herbal intervention that can help normalize glucose, blood pressure, and lipid profiles and deal with psychological and immunological stress. Seventeen clinical trials reported on metabolic conditions (10 reporting on type 2 diabetes or metabolic syndrome with six randomized placebo-controlled clinical trials). Only three out of 24 studies were high quality (4–5 Jadad scores), two of which examined neurocognitive effects and one reported on immunity. The three studies were: lipid profile reduction (Mondal et al. 2012), immunomodulation (Mondal et al. 2011) (both 5 Jadad score), viral infection (Sampath et al. 2015) (5 Jadad score) and Saxena et al. (2012) (4 Jadad score).

The cardioprotective effect of tulsi was evaluated in a randomized, double-blind, controlled trial in healthy volunteers (n = 22, 27 years mean age, 23 kg/m^2 average BMI) for 11 weeks (Mondal et al. 2012). Dried *O. sanctum* ethanolic (70%) leaf extract (300 mg capsule) was administered on an empty stomach once daily for 4 weeks, followed by a washout (3 weeks) period prior to crossover to placebo (sucrose). Tulsi intervention significantly (P = 0.003) reduced elevated cholesterol in individuals (n = 6) with apparently normal medical history but high cholesterol and triglyceride levels. However, this decrease did not continue when subjects crossed over to the placebo group.

The immunomodulatory role of *Ocimum sanctum* ethanol (70%) leaf extract was demonstrated in a double-blind randomized controlled crossover trial in healthy young adults (Clinical Trial Registry of India—CTRI/2009/091/000350) (Mondal et al. 2011). The study involved 22 healthy young adults (mean age 27 years, placebo–*O. sanctum* sequence [n = 12], *O. sanctum*–placebo sequence [n = 10]) receiving *O. sanctum* ethanol extract (300 mg capsules on an empty stomach daily for 4 weeks) with a 3-week washout period. *Ocimum sanctum* intervention significantly increased IFN-γ and IL-4 (interferon-γ and interleukin-4) levels, T-helper and NK-cell percentages after 4 weeks without any toxic effects. Moreover, a triglyceride reduction trend was observed in 6 hypercholesterolemic subjects. The immunomodulatory effects, expressed as increased cytokine levels, are presumably due to flavonoids present in the extract (Mondal et al. 2011).

A double-blind randomized controlled study was designed to investigate the cognition-enhancing and anti-stress potential of ethanolic (70%) *O. sanctum* leaf extract (ursolic acid >2.7% w/w) in healthy young men (n = 40, 27 years mean age, 23 kg/m^2 average BMI) (Sampath et al. 2015). *O. sanctum* treatment (300 mg capsule once daily for 30 days) significantly improved Stenberg reaction time and error rate (38% for placebo vs 66% error rate change for intervention). The intervention reduced salivary cortisol (14%) after 15 days, although the inter-group comparisons were not significant. The extract improved short-term memory (Stenberg reaction time and error rate) and cognitive flexibility (Stroop task).

Saxena et al. (2012) evaluated the efficacy of OciBest, an *Ocimum tenuiflorum* whole plant extract, in symptomatic control of general stress in a randomized, double-blind, placebo-controlled study. Adult stressed patients (48 years average age) received either placebo (45M/37F) or OciBest (400 mg capsules thrice daily, 44M/32F) for six weeks. The severity of stress-related symptoms was self-evaluated biweekly by patients. OciBest significantly (P ≤ 0.05) reduced mean scores of all stress symptoms (headache, palpitation at rest, frequent GI symptoms, frequent sleep problems, and forgetfulness). OciBest overall effect size was 1.6 times (39%) higher than placebo and provided effective relief from frequent exhaustion, forgetfulness, sleep, and sexual problems. These adaptogenic effects of OciBest could be ascribed to neuroprotective, immunostimulant, free radical scavenging and plasma cortisol lowering effects of *O. sanctum* (Saxena et al. 2012 and references therein).

The metabolic effects of *O. sanctum* extract were evaluated in 30 overweight/obese subjects in a randomized, parallel group, open label pilot study (Satapathy et al. 2017). Young obese adults (n = 16, 21 years median age, 25 kg/m^2 average BMI) received one 250 mg *O. sanctum* extract twice daily for 8 weeks, while the control group (n = 14) received no intervention. *O. sanctum* supplementation significantly improved body weight, BMI, plasma insulin, and insulin resistance in the intervention group and lipid parameters (22% HDL-C increase compared to the control group). *O. sanctum* extract exhibited antidiabetic and hypolipidemic effects in type 2 diabetic patients. Aqueous *O. sanctum* leaves extract (5 mL daily for 2 months) significantly lowered total cholesterol (142 to 132 mg/dL) and LDL (91 to 85 mg/dL), but increased HDL (25 to 27 mg/dL) level (cardioprotective lipid) in type 2 diabetic patients (≥140 mg/dL fasting plasma glucose, 51 years median age, 27.5 k/m^2 average BMI) (Dineshkumar et al. 2010).

Venu Prasad (2014) investigated the anti-fatigue effects of *O. sanctum*-enriched bar in young (n = 15, 18–30 years) healthy volunteers in a randomized controlled study. The volunteers were instructed to take two *O. sanctum*-enriched bars (25 g) morning and evening before undergoing the treadmill test for 10 days. The *O. sanctum* treatment enhanced performance by significantly increasing the final VO_2 max (29.5 to 43.4 mL/kg/min) and lowering serum creatinine kinase and lactic acid accumulation. The lactic acid levels before and after treadmill exercise were 9.7 and 29.4 mg/g for control vs 9.4 and 15.1 mg/g for *O. sanctum*-supplemented bar. Moreover, 13 out of 15 volunteers in the control bar group were positive for HHV-6 virus, whereas viral DNA was absent in almost 50% of the volunteers in the *O. sanctum*-supplemented group. The *O. sanctum*-enriched bar consisted of lyophilized powders of tulsi leaves ethanol (70%) extract (20 g/kg of oat/resin/peanut/skim milk mixture).

Lyophilized ethanol (70%) extract of *Ocimum sanctum* leaves (500 mg/capsule, twice daily after meals for 60 days) significantly attenuated generalized anxiety disorders and its associated stress and disorders in adult subjects (21M/14F, 38.4 years average age) (Bhattacharyya et al. 2008). *O. sanctum* intervention reduced the baseline anxiety score index (19.2% and 34.2% after 30 and 60 days), stress index (11.5% and 27.5%), and depression index (13.2 and 30.8%). Moreover, the treatment significantly improved [adaptogenic activity] adjustment willingness (10.6 and 25.1%) and attention (16.3 and 33.9% on day 30 and 60, respectively) compared to subjects' baseline scores. These benefits are due to *O. sanctum*'s potential action in regulating the hypothalamo-hypophyseal-adrenocortical (HHA) axis, particularly during stress-related disorders in humans. *Ocimum sanctum* ethanolic leaf extract (300 mg capsules) administered for 30 days significantly improved reaction time (Sternberg task), percent of correct responses in facilitation task (Stroop task), in a double-blind randomized controlled trial involving 30 volunteers (Suneetha et al. 2012). Moreover, the treatment improved attention, indicating that *Ocimum sanctum* possesses memory-enhancing effects.

An open prospective multicenter study was carried out in patients with multiple ophthalmic disorders, notably acute dacryocystitis, conjunctivitis, conjunctival xerosis (dry eye), degenerative conditions (pterygium or pinguecula), and postoperative cataract patients with a herbal eye drop preparation (Ophthacare) comprising basic principles of different herbs which have been traditionally used in the Ayurvedic system of medicine since time immemorial. They included *Meldespumapum, Carum copticum, Cinnamomum camphora, Emblica officinalis, Terminalia belirica, Ocimum sanctum*, and *Curcuma longa*. Thus, according to the reports, these plants have antibacterial and anti-inflammatory properties (Biswas et al. 2001). A prescription containing Swasari Ras, Giloy Ghanvati, Ashwagandha, Tulsi Ghanvati, and Anu Taila was created based on the principles of Ayurveda that can symptomatically treat COVID–19. Swasari Ras, one of the medications in the consortium under study, is a traditional Ayurvedic therapy advised for respiratory diseases. It is often used in the traditional Indian medical system to treat rhinitis, excessive mucus production, bronchitis, and other extremely uncomfortable respiratory disorders including asthma (Balkrishna et al. 2020). The two other medications included in the treatment plan, Giloy Ghanvati and Tulsi Ghanvati, are made from aqueous extracts of *Tinospora cordifolia* and *Ocimum sanctum*, respectively. These two herbs are well-known for their ability to reduce inflammation and boost the immune system.

6.6 HEALTH BENEFITS

6.6.1 Oral Health

Ocimum sanctum is described as a promising agent for oral health care management since it reduces gum inflammation caused by periodontal diseases such as periodontitis and gingivitis (Fernandes et al. 2020). *In vitro* tests demonstrate the antibacterial activity of *Ocimum sanctum* extract (as toothpastes, mouthwash, and gels) against oral pathogenic bacteria, *Streptococcus mutans* and *Enterococcus faecalis*. There are two main *Ocimum sanctum* subspecies: *Ocimum sanctum* L. green (Tulsi) and *Ocimum sanctum* purple (Krishna Tulsi) in India. These subspecies differ in essential oil composition. The green subspecies contains eugenol, limonene, and (*E*)-caryophellene, whereas eugenol and (*E*)-caryophellene predominate in the purple subspecies. These essential oil components exhibit antibacterial activity by varied mechanistic pathways. For example, eugenol disrupts membrane primarily by inhibiting ATPase activity and several virulence factors at sub-inhibitory concentrations. Carvacrol also disrupts membrane, inhibits ATPase activity, destabilizes membrane, and reduces proton motive force (Fernandes et al. 2020).

Ocimum sanctum extract often has comparable/superior antibacterial activity against periodontal pathogens to established gold standards (Chlorhexidine, Triclosan, and CetylPyridinum Chloride) due to the presence of eugenol. Essential oils from *Ocimum* have been extensively investigated for their capacity to control oral cariogenic bacteria and biofilm-forming microorganisms (Fernandes et al. 2020). Various forms of *Ocimum* have been used for its anticariogenic activity, including chewing the whole leaves that significantly reduces *Streptococcus mutans* colony counts in saliva of young (9–12 years) clinical volunteers. In another study, chewing *Ocimum sanctum* (Tulsi) leaves increased saliva pH after 30 min chewing due to reduction in acidic conditions necessary for cariogenesis induction and halitosis. Tulsi has also been used in gingivitis and root canal disinfection and to treat oral mucosal fibrosis, confirming the potential of its extract as adjuvant therapy to manage oral submucosal fibrosis.

Novel drug delivery formulations with *Ocimum sanctum* have been applied in dentistry. For example, resorbable *Ocimum sanctum* (10% w/w)-loaded polyvinyl vinyl acetate nanofibers have been evaluated for periodontitis. Supramolecular complexation with β-cyclodextrin has been used to protect *O. sanctum* essential oil against degradation. The β-cyclodextrin complex with *O. sanctum* exerts anti-inflammatory activity that can be applicable in periodontitis. Silver nanoparticles with *O. sanctum* have very high polydispersity and stability with enhanced antimicrobial activity against *Escherichia coli* and *Staphylococcus aureus* (Fernandes et al. 2020).

Ocimum sanctum is a suitable herbal candidate for oral squamous cell carcinoma (OSCC) treatment because components (quercetin, rutin, and tannic acid) from defatted seed extracts show the best binding with EGFR, β2-AR, and keapl/nrf2 receptors in docking studies (Sharma et al. 2020). Gupta et al. (2014) evaluated the effectiveness of *Ocimum sanctum* on dental plaque and gingival inflammation compared with gold standard chlorhexidine and normal saline. *Ocimum sanctum* mouthwash prepared from ethanol (100%) leaves extract (10 mL 4% *O. sanctum* mouthwash, twice daily for 30 days) was as effective as chlorhexidine in reducing plaque and gingivitis in this placebo-controlled, triple-blind randomized control trial of volunteer medical students (n = 36, 23 years mean age). Mean plaque and gingival index decreased (3.0 vs 2.5 and 2.2 vs 1.4) on the 30th day compared to baseline. These anti-gingivitis and anti-plaque effects of *O. sanctum* extract confirm previous studies and were attributed to the anti-inflammatory and antibacterial activities of isolated compounds, particularly phenolic compounds, eugenol, and carvacrol (Gupta et al. 2014). Ahirwar et al. (2018) evaluated the antimicrobial efficacy of *Ocimum sanctum* essential oil compared to that of triple antibiotic paste (TAP—ciprofloxacin, minocycline, and metronidazole) in microbiological samples from 40 children's (4–9 years) root canals of primary molars. The antibacterial efficacy of *Ocimum sanctum* treatment was not significantly different from that of TAP.

6.6.2 Eye Health

Commercial aqueous *Ocimum sanctum* leaf extract (Dabur Pharmaceuticals Ltd., Ghaziabad, India) (150 µg/mL, treated for 30 min at 37 °C with 5% CO_2 and 95% air) protected human lens epithelial cells (HLEC) from H_2O_2 insult and maintained their normal architecture (Halder et al. 2009). The epithelial cells were obtained from normal transparent lenses of donors' (20–40 years) eyes within 5–8 h of death. The extract significantly reduced cytoplasmic vacuole size, preventing chromatin condensation and clumping in the cell nuclei. Moreover, it protected against apoptotic cell death of the lens epithelial cells, thereby maintaining cellular integrity even in the presence of oxidative stress. The protective effect of the aqueous *Ocimum sanctum* leaf extract is attributed to its antioxidant properties due to free radical scavenging activity, particularly by the radioprotective water-soluble flavonoids, orientin and vicenin ((Halder et al. 2009).

A formulation containing *Ocimum sanctum* aqueous extract (0.1%) with hydroxypropyl methylcellulose (0.25% w/v) (one eye drop twice daily) significantly delayed the onset and progression of galactose cataract compared to control (0.75 vs 1 and 2.81 vs 4 opacity index for control on 7th and 30th day, respectively). However, the opacity index was considerably reduced with 0.15% *O. sanctum* aqueous extract + 0.01% *C. longa* extract + 0.25% hydroxypropyl methylcellulose (1.79 vs 4 for control). The anti-cataract activity of the herbal formulation is attributed to its antioxidant activity (reduced lipid peroxidation [TBARS], increased glutathione [GSH] levels, superoxide dismutase [SOD], glutathione peroxidase [GPx], catalase [CAT], and glutathione-S-transferase [GST] enzyme activities) in isolated rat lenses exposed to osmotic/oxidative stress (Gupta et al. 2003).

6.6.3 Antituberculosis

Essential oil from *Ocimum sanctum* leaves has anti-*Mycobacterium tuberculosis* effect due to growth inhibition of *Mycobacterium tuberculosis* strain H37Rv and nine other clinical isolates (4 isoniazid- and rifampicin-resistant, 2 rifampicin-resistant, and 2 isoniazid-resistant mycobacterial strains) (Jayapal et al. 2021). The minimal inhibitory concentration was 3 µL (2.93 µg) for H37Rv and 1.5 to 6 µL (1.47–5.87 µg) for the clinical isolates. The anti-*Mycobacterium tuberculosis* effect is attributed to the diverse composition of the essential oil, each of which may have a different mode of anti-mycobacterial action.

6.6.4 Antimalrial

The essential oil of *O. sanctum* also has insecticidal and larvicidal activities against mosquitoes. Two natural antimalarial products (trans- and cis-diol moiety) were fractionated from ethanol extract of *O. sanctum* bark. Their corresponding derivatives (Formula I and II, respectively) possess antimalarial activity superior to, and less toxic than, commonly used antimalarial drugs chloroquine and mefloquine and are potent against both chloroquine-sensitive (D-6) and -resistant (W-2) malarial strains. Malaria can be treated by administering a therapeutically effective amount of Formula I compound according to Zhu (2010). Ethanolic *Ocimum sanctum* root extract (800 mg/kg) showed *in vivo* maximum antiplasmodial activity (3.2% parasitemia) against *Plasmodium berghei*, the causative malarial agent in mice, on the 4th day (Singh et al. 2010). This antimalarial activity is attributed to the presence of alkaloid and/or flavonoid constituents in *O. sanctum* roots and leaves.

6.6.5 Wound Healing

Walia et al. (2014) disclose an invention that provides a wound healing matrix exhibiting antibiotic, anti-inflammatory, antioxidant, and biocompatible properties on the wound surface. The matrix (bandage) is coated with green synthesized silver nanoparticles with *tulsi* (*Ocimum sanctum*) plant extract, curcumin particles, and zwitterionic chitosan. The nanoparticles (~56 nm) are obtained using silver nitrate and aqueous *O. sanctum* plant extract. The matrix has antibiotic, anti-inflammatory, antioxidant, analgesic, hemostatic, anticancer, and immune-modulating properties and induces fibroblast and epithelial cell proliferation.

6.6.6 Hair/Scalp Treatment

Ocimum sanctum oil obtained from leaves acts synergistically with zinc pyrithione to exhibit antimicrobial activity against *Malassezia* yeasts, particularly *Malassezia furfur*, the main cause of dandruff (Chandra 2006). Eugenol is the *O. sanctum* oil compound responsible for this synergistic effect. *Lactobacillus plantarum*-fermented aerial part of *Ocimum sanctum* (2% w/w) shampoo inhibited *Malassezia furfur*, the dandruff-causative and scalp-irritating agent. Moreover, the antifungal activity against *M. furfur* (656 strain) was comparable to that of ketoconazole [an antifungal medicine] (Punyoyai et al. 2018). Fermented *O. sanctum* was rich in total soluble protein (65 mg/100 mL) presumably derived from the release of the bacterial starter culture. The antibacterial and antifungal activities of the fermented *O. sanctum* product are attributed to these proteins (Punyoyai et al. 2018).

6.6.7 Mental Health

Paste made from ground *Ocimum sanctum* leaves exerts antipsychotic effect in male Wistar rats and albino mice. The paste (2, 4, and 8% w/w) orally administered for 15 days with a specially prepared diet induced catalepsy, significantly decreased locomotor activity, brain dopamine levels, and ketamine-induced stereotyped behavior, comparable to the antipsychotic drug Olanzapine (Renu and Milind 2015). The antipsychotic activity of *O. sanctum* leaves paste is due to antidopaminergic activity attributable to its antioxidant activity and perhaps the presence of alkaloids, tannins, steroids, and glycosides presumably responsible for the psychopharmacological action. The paste constituents inhibit dopaminergic neurotransmission and potentially block dopamine D2 receptor (Renu and Milind 2015).

6.6.8 Antiulcer

Aqueous *Ocimum sanctum* leaves extract (100 mg/kg fed orally) and melatonin (20 mg/kg, i.p. injection) completely inhibited piroxicam-induced gastric mucosal ulceration and significantly decreased oxidative stress (reduced LPO and increased gastric GSH) in rats. Moreover, the *O. sanctum* extract and melatonin combination mitigated the changes in gastric antioxidant enzyme (Cu-Zn SOD, Mn SOD, gastric peroxidase, glutathione peroxidase, and catalase) activities. The gastroprotective effect of the combined extract–melatonin is through antioxidant mechanism (Basu et al. 2013). In this regard, *Vrinda* obtained from a germplasm line by micropropagation is a high-antioxidant phytohormone (*melatonin* and *serotonin*) *Ocimum sanctum* plant with consistent melatonin levels (327 and 342 ng/g for leaf; 365 and 379 ng/g for root), respectively, in both field and greenhouse conditions (Shukla et al. 2021).

6.6.9 Cardioprotection-Heart Health

Aqueous methanol (50%) *Ocimum sanctum* leaves extract (maximal effect at 150 mg/kg body weight) significantly reduced TBARS, NFκB expression, COX-2 (cyclooxygenase-2) and 5-LOX (5-lipoxygenase) activities, LTB_4 (leukotrienes) and TXB_2 (thromboxane) concentrations, and cardiac markers (phospholipases [A, C, and D] and phospholipid content), increased SOD activity, and normalized hsCRP levels in myocardial-induced Dawley albino male rats (Kavitha et al. 2015). These cardioprotective effects of *O.*

sanctum leaves extract are due to the presence of its high phenolic content (16%) that reduces oxidative stress, thereby suppressing NFκB and arachidonic acid catabolism, resulting in reduced inflammation.

6.6.10 Radioprotection

Aqueous *Ocimum sanctum* leaves extract protects mouse bone marrow against radiation clastogenesis and stem cell lethality. The extract (10 mg/kg/daily) for 5 consecutive days prior to whole-body radiation produced higher stem cell survival than the well-known synthetic radioprotector WR-2721. Two flavonoids, orientin and vicenein, isolated from the extract, increased 30-day survival of lethally irradiated mice. Orientin is as effective as, and vicenein better than, WR-2721 in reducing the number of aberrant cells, presumably due to their strong free radical scavenging activity *in vitro* and anti-lipid peroxidation *in vivo* at very low concentrations. In fact, the flavonoids protect chromosomes at a lower dose (50 μg/kg) than the synthetic compounds without any systemic toxicity in mice (Devi et al. 1998).

6.7 CURRENT AND FUTURE PROSPECTS

Since humans have used "tulsi" for a very long time, it is a strong herb. According to research in contemporary medicine, tulsi may be a helpful treatment for issues including ulcers, high cholesterol, type 2 diabetes, obesity, and weakened/suppressed immune systems (from conditions like cancers and AIDS). It is without a doubt thought of as a natural energizer. There have been certain instances when research has purified specific plant parts, which have been analyzed to discover their chemical make-up and biopharmacological effects.

The traditional uses of tulsi in Ayurveda result from some inherent qualities in wide tulsi varieties, such as essential oils including the anti-inflammatory component eugenol and different acids with anti-inflammatory and antioxidant effects. These characteristics can lend credence to suggestions that tulsi is an effective Ayurvedic remedy for numerous ailments. Tulsi is helpful because it contains phytochemicals that have antioxidant, anti-infective, and immunological properties, like alpha-linolenic acid. More research is needed, though, to be more particular about this. In the future, this plant will be beneficial in treating various health issues. This will change people's decisions regarding herbal medicines and open up new lines of investigation for the pharmaceutical and nutraceutical industries.

REFERENCES

Agarwal, P., & Nagesh, L. (2010). Evaluation of the antimicrobial activity of various concentrations of Tulsi (*Ocimum sanctum*) extract against *Streptococcus mutans*: An in vitro study. *Indian Journal of Dental Research*, *21*(3), 357–359.

Ahirwar, P., Shashikiran, N. D., Sundarraj, R. K., Singhla, S., Thakur, R. A., & Maran, S. (2018). A clinical trial comparing antimicrobial efficacy of "essential oil of *Ocimum sanctum*" with triple antibiotic paste as an intracanal medicament in primary molars. *Journal of Indian Society of Pedodontics and Preventive Dentistry*, *36*(2), 191–197.

Ahmad, A., Khan, M. M., Raza, S. S., Javed, H., Ashafaq, M., Islam, F., . . . & Islam, F. (2012). *Ocimum sanctum* attenuates oxidative damage and neurological deficits following focal cerebral ischemia/reperfusion injury in rats. *Neurological Sciences*, *33*, 1239–1247.

Alhazmi, M. I. (2015). Molecular docking of selected phytocompounds with H1N1 proteins. *Bioinformation*, *11*(4), 196–202.

Amrani, S., Harnafi, H., Bouanani, N. E. H., Aziz, M., Caid, H. S., Manfredini, S., . . . & Bravo, E. (2006). Hypolipidaemic activity of aqueous *Ocimum basilicum* extract in acute hyperlipidaemia induced by triton WR-1339 in rats and its antioxidant property. *Phytotherapy Research: An International Journal Devoted to Pharmacological and Toxicological Evaluation of Natural Product Derivatives*, *20*(12), 1040–1045.

Anon (1974). Oils of spices used to create new flavour material. *Candy Ind*, *139*, 45.

Antonucci-Tarolla, M. (1988). System for aromatizing foods. *European Patent Application*, EP0259274A1.

Archana, R., & Namasivayam, A. (2000). Effect of *Ocimum sanctum* on noise induced changes in neutrophil functions. *Journal of Ethnopharmacology*, *73*(1–2), 81–85.

Arenal, A., Martín, L., Castillo, N. M., de la Torre, D., Torres, U., & González, R. (2012). Aqueous extract of *Ocimum tenuiflorum* decreases levels of blood glucose in induced hyperglycemic tilapia (*Oreochromis niloticus*). *Asian Pacific Journal of Tropical Medicine*, *5*(8), 634–637.

Arulraj, J., Shanmugaiah, V., & Lakshmanan, N. (2014). Studies on phytochemical analysis and antimicrobial activity of Tulsi (*Ocimum sanctum Linn*) leaf extract against human and fish pathogens. *International Journal of Advanced Life Sciences (IJALS)*, *7*(1), 27–34.

Askerova, A., Guseinov, I., Azimov, A., Dmitrieva, N., & Shamsizade, R. (1993). Manufacture of the carbonated fermented milk beverage, Airan. *USSR Patent, SU*, 1796122.

Bagdasarov, S., Nazyrova, R., Airopet'yants, B., & Voitenko, T. (1978). Composition of ingredients for "Dushanbinskaya" bitter lemon. *USSR Patent*, 587151.

Baliga, M. S., Jimmy, R., Thilakchand, K. R., Sunitha, V., Bhat, N. R., Saldanha, E., . . . & Palatty, P. L. (2013). *Ocimum sanctum* L (Holy Basil or Tulsi) and its phytochemicals in the prevention and treatment of cancer. *Nutrition and Cancer*, *65*(sup1), 26–35.

Balkrishna, A., Solleti, S. K., Singh, H., Tomer, M., Sharma, N., & Varshney, A. (2020). Calcio-herbal formulation, Divya-Swasari-Ras, alleviates chronic inflammation and suppresses airway remodelling in mouse model of allergic asthma by modulating pro-inflammatory cytokine response. *Biomedecine & Pharmacotherapie*, *126*, 110063. https://doi.org/10.1016/j.biopha.2020.110063

Bandeira Jr, G., Pês, T. S., Saccol, E. M., Sutili, F. J., Rossi Jr, W., Murari, A. L., . . . & Baldisserotto, B. (2017). Potential uses of *Ocimum gratissimum* and *Hesperozygis ringens* essential oils in aquaculture. *Industrial Crops and Products*, *97*, 484–491.

Bassiouny, S., Hassanien, R., Abd El Razik, A., & El Kayati, M. (1990). Efficiency of antioxidants from natural sources in bakery products. *Food Chemistry*, *37*, 297–305.

Basu, A., Mukherjee, D., Ghosh, A. K., Mitra, E., Firdaus, S. B., Ghosh, D., . . . & Bandyopadhyay, D. (2013). Melatonin augments the protective effects of aqueous leaf homogenate of Tulsi (*Ocimum sanctum* L.) against piroxicam-induced gastric ulceration in rats. *Asian Journal of Pharmaceutical and Clinical Research*, *6*(2), 123–132.

Bhattacharyya, D., Sur, T. K., Jana, U., & Debnath, P. K. (2008). Controlled programmed trial of Ocimum sanctum leaf on generalized anxiety disorders. *Nepal Medical College Journal*, *10*(3), 176–179.

Biswas, N. R., Gupta, S. K., Das, G. K., Kumar, N., Mongre, P. K., Haldar, D., & Beri, S. (2001). Evaluation of Ophthacare® eye drops—a herbal formulation in the management of various ophthalmic disorders. *Phytotherapy Research*, *15*(7), 618–620.

Buddhadev, S. G., Buddhadev, S. S., & Mehta, N. D. (2014). A review article on *Ocimum sanctum* Linn. *International Peer-Reviewed Ayurveda Journal*, *2*(2), 1–6.

Chandra, L. (2006). Hair and/or scalp treatment composition. *U.S. Patent 0013796 A1*, issued 19 January.

Chatterjee, M., Verma, P., Maurya, R., & Palit, G. (2011). Evaluation of ethanol leaf extract of Ocimum sanctum in experimental models of anxiety and depression. *Pharmaceutical Biology*, *49*(5), 477–483.

Cherian, R. (2019). Health benefits of basil seeds. *International Journal of Scientific Research in Science Engineering and Technology*, *6*(2), 511–515. https://doi.org/10.32628/IJRSET1962145

Dahiya, P., & Purkayastha, S. (2012). Phytochemical screening and antimicrobial activity of some medicinal plants against multi-drug resistant bacteria from clinical isolates. *Indian Journal of Pharmaceutical Sciences*, *74*(5), 443–450. https://doi.org/10.4103/0250-474X.108420

Das, S. K., & Vasudevan, D. M. (2006). Tulsi: The Indian holy power plant. *Natural Products Repository*, *5*(4), 279–283.

Deshmukh, A. S., Deshmukh, G. B., & Shirole, P. D. (2015). *Ocimum sanctum*: A medicinal gift from nature. *International Journal of Pharmacognosy*, *2*(12), 550–559.

Devendran, G., & Balasubramanian, U. (2011). Qualitative phytochemical screening and GC-MS analysis of *Ocimum sanctum* L. leaves. *Asian Journal of Plant Science & Research*, *1*, 44–48.

Devi, P. U. (2001). Radioprotective, anticarcinogenic and antioxidant properties of the Indian holy basil, *Ocimum sanctum* (Tulasi). *Indian Journal of Experimental Biology*, *39*, 185–190.

Devi, P. U., Bisht, K. S., & Vinitha, M. (1998). A comparative study of radioprotection by *Ocimum* flavonoids and synthetic aminothiol protectors in the mouse. *The British Journal of Radiology*, *71*(847), 782–784.

Dhandayuthapani, S., Azad, H., & Rathinavelu, A. (2015). Apoptosis induction by *Ocimum sanctum* extract in LNCaP prostate cancer cells. *Journal of Medicinal Food*, *18*(7), 776–785.

Dinarieva, G., Solntseva, G., Belusova, E., Kandilov, N., Shukyurov, N., Kocharin, A., & Guseinov, V. (1984). Spice essential oil composition for use in sausage production. *USSR Patent*, SU1069755A.

Dineshkumar, B., Analava, M., & Manjunatha, M. (2010). Antidiabetic and hypolipidaemic effects of few common plants extract in type 2 diabetic patients at Bengal. *International Journal of Diabetes and Metabolism*, *18*(2), 59–65.

Dutta, D., Devi, S. S., Krishnamuthi, K., Kumar, K., Vyas, P., Muthal, P. L., . . . & Chakrabarti, T. (2007). Modulatory effect of distillate of *Ocimum sanctum* leaf extract (Tulsi) on human lymphocytes against genotoxicants. *Biomedical and Environmental Sciences*, *20*(3), 226–234.

Dziezak, J. D. (1991). Getting savvy on sauces. *Food Technology (Chicago)*, *45*(6), 84–87.

Fernandes, T., D'souza, A., & Sawarkar, S. P. (2020). *Ocimum sanctum* L: Promising agent for oral health care management. In: D. N. Chauhan, P. R. Singh, K. Shah, & N. S. Chauhan (eds.), *Natural Oral Care in Dental Therapy* (pp. 259–269). Beverly, MA: Scrivener Publishing LLC. https://doi.org/10.1002/9781119618973.ch16.

Flegkas, A., Milosević Ifantis, T., Barda, C., Samara, P., Tsitsilonis, O., & Skaltsa, H. (2019). Antiproliferative activity of (-)-rabdosiin isolated from *Ocimum sanctum* L. *Medicines*, *6*(1), 37. https://doi.org/10.3390/medicines6010037

Flint, F., & Seal, R. (1985). The sausage seasoning scene. *Food Manufacturing*, *60*, 43, 45.

Fomicheva, L., Keller, E., Gulyaev, V., Roenko, T., & Koptyaeva, I. (1982). Activity of vegetable additives used for food concentrates. *Konservnaya i Ovoshchesushil naya promyshlennost'*, *1*, 39–41.

Giridharan, V. V., Thandavarayan, R. A., Mani, V., Dundapa, T. A., Watanabe, K., & Konishi, T. (2011). *Ocimum sanctum* Linn. leaf extracts inhibit acetylcholinesterase and improve cognition in rats with experimentally induced dementia. *Journal of Medicinal Food*, *14*(9), 912–919.

Grayer, R. J., Kite, G. C., Veitch, N. C., Eckert, M. R., Marin, P. D., Senanayake, P., & Paton, A. J. (2002). Leaf flavonoid glycosides as chemosystematic characters in Ocimum. *Biochemical Systematics and Ecology*, *30*(4), 327–342.

Gupta, D., Bhaskar, D. J., Gupta, R. K., Karim, B., Jain, A., Singh, R., & Karim, W. (2014). A randomized controlled clinical trial of *Ocimum sanctum* and chlorhexidine mouthwash on dental plaque and gingival inflammation. *Journal of Ayurveda and Integrative Medicine*, *5*(2), 109–116.

Gupta, P., Yadav, D. K., Siripurapu, K. B., Palit, G., & Maurya, R. (2007). Constituents of Ocimum sanctum with antistress activity. *Journal of Natural Products*, *70*(9), 1410–1416.

Gupta, S. K., Joshi, S., Srivastava, S., & Trivedi, D. (2003). Herbal ophthalmic formulation for preventing cataract. *World Patent WO 03/080091 A1*, issued 2 October, World Intellectual Property Organization.

Gupta, S. K., Srivastava, S., Trivedi, D., Joshi, S., & Halder, N. (2005). *Ocimum sanctum* modulates selenite-induced cataractogenic changes and prevents rat lens opacification. *Current Eye Research*, *30*(7), 583–591.

Guseinov, V., Shukyurov, N., Movsum-zade, A., Ibragimov, F., Sadykhova, R., Askerova, K., & Alieva, D. (1992). Manufacture of sausages. *USSR Patent*, SU 1722371.

Halder, N., Joshi, S., Nag, T. C., Tandon, R., & Gupta, S. K. (2009). *Ocimum sanctum* extracts attenuate hydrogen peroxide induced cytotoxic ultrastructural changes in human lens epithelial cells. *Phytotherapy Research*, *23*(12), 1734–1737.

Hay, S. I., Sinka, M. E., Okara, R. M., Kabaria, C. W., Mbithi, P. M., Tago, C. C., . . . & Godfray, H. C. J. (2010). Developing global maps of the dominant *Anopheles* vectors of human malaria. *PLoS Medicine*, *7*(2), e1000209. https://doi.org/10.1371/journal.pmed.1000209

Hiltunen, R. (1999). Chemical composition of Ocimum species. In *Basil* (pp. 74–82). CRC Press

Islam, A. R., Ferdousi, J., & Shahinozzaman, M. (2021). Previously published ethno-pharmacological reports reveal the potentiality of plants and plant-derived products used as traditional home remedies by Bangladeshi COVID-19 patients to combat SARS-CoV–2. *Saudi Journal of Biological Sciences*, *28*(11), 6653–6673.

Israr, T., Rakha, A., Rashid, S., Shehzad, A., Ahmed, W., & Sohail, M. (2017). Effect of basil seed gum on physico-chemical and rheological properties of bread. *Journal of Food Processing and Preservation*, *41*(5), e13128. https://doi.org/10.1111/jfpp.13128

Jagdish (2019). Tulsi farming project report (Basil). Cultivation Economics. https://www.agrifarming.in/tulsi-farming-project-report-basil-cultivation-economics

Jahanger, M. A., Patra, K. K., Kumari, S., Singh, A., Manika, N., Srivastava, R. P., . . . & Singh, L. (2022). A glance at the phytochemical and ethno-pharmacological understanding of four Ocimum species. *Current Pharmaceutical Biotechnology*, *24*(9), 1094–1107.

Jamshidi, N., & Cohen, M. M. (2017). The clinical efficacy and safety of Tulsi in humans: A systematic review of literature. *Evidenced-Based Complementary and Alternative Medicine*, 9217567. https://doi.org/10.1155/2017/9217567

Jayapal, V., Raj, C. V., Muthaiah, M., Chadha, V. K., Brammacharry, U., Selvaraj, S., & Easow, J. M. (2021). In-vitro anti-*Mycobacterium tuberculosis* effect of essential oil of *Ocimum sanctum* L. (Tulsi/Basil) leaves. *Indian Journal of Tuberculosis*, *68*(4), 470–473.

Jeba, C. R., Vaidyanathan, R., & Rameshkumar, G. (2011). Immunomodulatory activity of aqueous extract of *Ocimum sanctum* in rat. *International Journal of Pharmacy & Biomedical Research*, *2*(1), 33–38.

Joshi, B., Sah, G. P., Basnet, B. B., Bhatt, M. R., Sharma, D., Subedi, K., . . . & Malla, R. (2011a). Phytochemical extraction and antimicrobial properties of different medicinal plants: *Ocimum sanctum* (Tulsi), *Eugenia caryophyllata* (Clove), *Achyranthes bidentata* (Datiwan) and *Azadirachta indica* (Neem). *Journal of Microbiology and Antimicrobials*, *3*(1), 1–7.

Joshi, G., Sharma, S., Acharya, J., & Parida, M. (2014). Assessment of *In vitro* antiviral activity of *Ocimum sanctum* (Tulsi) against pandemic swine flu H1N1 virus infection. *World Research Journal of Antimicrobial Agents*, *3*(1), 62–67.

Joshi, R. K., & Sharma, A. K. (2021). Determination of seasonal variation of volatile organic constituents of the leaves of traditional herb *Ocimum sanctum* Linn. *Indian Journal of Pharmaceutical Sciences*, *83*(4), 750–757.

Joshi, S. V., Bothara, S. B., & Surana, S. J. (2011b). Evaluation of aqueous extract of Ocimum sanctum in experimentally induced Parkinsonism. *Journal of Chemical and Pharmaceutical Research*, *3*(1), 478–487.

Jyoti, S., Satendra, S., Sushma, S., Anjana, T., & Shashi, S. (2007). Antistressor activity of *Ocimum sanctum* (Tulsi) against experimentally induced oxidative stress in rabbits. *Methods and Findings in Experimental and Clinical Pharmacology*, *29*(6), 411–416.

Kavitha, S., John, F., & Indira, M. (2015). Amelioration of inflammation by phenolic rich methanolic extract of *Ocimum sanctum* Linn. leaves in isoproterenol induced myocardial infarction. *Indian Journal of Experimental Biology*, *53*, 632–640.

Kelm, M. A., Nair, M. G., Strasburg, G. M., & DeWitt, D. L. (2000). Antioxidant and cyclooxygenase inhibitory phenolic compounds from *Ocimum sanctum* Linn. *Phytomedicine*, *7*(1), 7–13.

Kerimov, T. (1993). Manufacture of the non-alcoholic beverage Reikhan. *USSR Patent, SU1796122*.

Khan, V., Najmi, A. K., Akhtar, M., Aqil, M., Mujeeb, M., & Pillai, K. K. (2012). A pharmacological appraisal of medicinal plants with antidiabetic potential. *Journal of Pharmacy & Bioallied Sciences*, *4*(1), 27–42

Khosla, M. K. (1995). Sacred tulsi (Ocimum sanctum l.) In traditional medicine and pharmacology. *Ancient Science of Life*, *15*(1), 53–61.

Kondapalli, N. B., Hemalatha, R., Uppala, S., Yathapu, S. R., Mohammed, S., Venkata Surekha, M., . . . & Bharadwaj, D. K. (2022). *Ocimum sanctum, Zingiber officinale,* and *Piper nigrum* extracts and their effects on gut microbiota modulations (prebiotic potential), basal inflammatory markers and lipid levels: Oral supplementation study in healthy rats. *Pharmaceutical Biology*, *60*(1), 437–450.

Kumar, R., Saha, P., Lokare, P., Datta, K., Selvakumar, P., & Chourasia, A. (2022). A systemic review of *Ocimum sanctum* (Tulsi): Morphological characteristics, phytoconstituents and therapeutic applications. *International Journal for Research in Applied Sciences and Biotechnology*, *9*(2), 221–226.

Lahon, K., & Das, S. (2011). Hepatoprotective activity of *Ocimum sanctum* alcoholic leaf extract against paracetamol-induced liver damage in Albino rats. *Pharmacognosy Research*, *3*(1), 13–18. https://doi.org/10.4103/0974-8490.79110

Lal, R. K., Gupta, P., Chanotiya, C. S., & Sarkar, S. (2018). Traditional plant breeding in Ocimum. In A. K. Shasany & C. Kole (eds.), *The Ocimum Genome, Compendium of Plant Genomes* (pp. 89–98). Springer Nature Switzerland. https://doi.org/10.1007/978-3-319-7450-9_7

Law, D. (1975). *Herbs for Health and Flavour*. J. Bartholomew.

Magesh, V., Lee, J. C., Ahn, K. S., Lee, H. J., Lee, H. J., Lee, E. O., . . . & Kim, S. H. (2009). *Ocimum sanctum* induces apoptosis in A549 lung cancer cells and suppresses the *in vivo* growth of Lewis lung carcinoma cells. *Phytotherapy Research*, *23*(10), 1385–1391.

Mahapatra, N., Marai, N., Dhal, K., Nayak, R. N., Panigrahi, B. K., Mallick, G., . . . & Kerketta, A. S. (2012). Malaria outbreak in a non endemic tribal block of Balasore district, Orissa, India during summer season. *Tropical Biomedicine*, *29*(2), 277–285.

Mahomoodally, M. F., Gurib-Fakim, A., & Subratty, A. H. (2010). Screening for alternative antibiotics: An investigation into the antimicrobial activities of medicinal food plants of Mauritius. *Journal of Food Science*, *75*(3), M173–M177.

Mahomoodally, M. F., Subratty, A. H., Gurib-Fakim, A., Choudhary, M. I., & Nahar Khan, S. (2012). Traditional medicinal herbs and food plants have the potential to inhibit key carbohydrate hydrolyzing enzymes *in vitro* and reduce postprandial blood glucose peaks *in vivo*. *The Scientific World Journal*, *1*, 285284. https://doi.org/10.1100/2012/285284

Maiocchi, G. (1995). Ultraspreadable. *Latte*, *20*, 490–493.

Maity, T. K., Mandal, S. C., Saha, B. P., & Pal, M. (2000). Effect of *Ocimum sanctum* roots extract on swimming performance in mice. *Phytotherapy Research, 14*(2), 120–121.

Manaharan, T., Thirugnanasampandan, R., Jayakumar, R., Kanthimathi, M. S., Ramya, G., & Ramnath, M. G. (2016). Purified essential oil from *Ocimum sanctum* Linn. Triggers the apoptotic mechanism in human breast cancer cells. *Pharmacognosy Magazine, 12*(Suppl 3), S327–S331. https://doi.org/10.4103/0973-1296.185738

Manaharan, T., Thirugnanasampandan, R., Jayakumar, R., Ramya, G., Ramnath, G., & Kanthimathi, M. S. (2014). Antimetastatic and anti-inflammatory potentials of essential oil from edible *Ocimum sanctum* leaves. *The Scientific World Journal, 2014*. http://doi.org/10.1155/2014/239508

Mondal, S., Mirdha, B., Padhi, M., & Mahapatra, S. (2012). Dried leaf extract of Tulsi (*Ocimum sanctum* Linn) reduces cardiovascular disease risk factors: Results of a double blinded randomized controlled trial in healthy volunteers. *Journal of Preventive Cardiology, 1*(4), 177–181.

Mondal, S., Varma, S., Bamola, V. D., Naik, S. N., Mirdha, B. R., Padhi, M. M., . . . & Mahapatra, S. C. (2011). Double-blinded randomized controlled trial for immunomodulatory effects of Tulsi (*Ocimum sanctum* Linn.) leaf extract on healthy volunteers. *Journal of Ethnopharmacology, 136*(3), 452–456.

Monga, J., Sharma, M., Tailor, N., & Ganesh, N. (2011). Antimelanoma and radioprotective activity of alcoholic aqueous extract of different species of *Ocimum* in C57BL mice. *Pharmaceutical Biology, 49*(4), 428–436.

Mukherjee, R., Dash, P. K., & Ram, G. C. (2005). Immunotherapeutic potential of *Ocimum sanctum* (L) in bovine subclinical mastitis. *Research in Veterinary Science, 79*(1), 37–43.

Munir, M., Qayyum, A., Raza, S., Siddiqui, N. R., Mumtaz, A., Safdar, N., . . . & Bashir, S. (2017). Nutritional assessment of basil seed and its utilization in development of value added beverage. *Pakistan Journal of Agricultural Research, 30*(3), 266–271.

Muralikrishnan, G., Pillai, S. K., & Shakeel, F. (2012). Protective effects of *Ocimum sanctum* on lipid peroxidation and antioxidant status in streptozocin-induced diabetic rats. *Natural Product Research, 26*(5), 474–478.

Muruganandam, A. V., Kumar, V., & Bhattacharya, S. K. (2002). Effect of poly herbal formulation, EuMil, on chronic stress-induced homeostatic perturbations in rats. *Indian Journal of Experimental Biology, 40*(10), 115–160.

Naji-Tabasi, S., & Razavi, S. M. A. (2017). Functional properties and applications of basil seed gum: An overview. *Food Hydrocolloids, 73*, 313–325.

Narain, J. P. (2008). Malaria in the South-East Asia region: Myth & the reality. *The Indian Journal of Medical Research, 128*(1), 1–3.

Orji, J. O., Nwuzo, A. C., Ejikeugwu, P. C., Ugbo, E. N., Moses, I. B., Nwakaeze, E. A., & Nwankwo, C. P. (2015). Antifungal activities of *Ocimum gratissimum* and *Gongronema latifolium* leaves on *Colletotrichum* species isolated from spoilt tomatoes. *International Journal of Pharmaceutical Science Invention, 4*(5), 42–45.

Padalia, R. C., & Verma, R. S. (2011). Comparative volatile oil composition of four Ocimum species from northern India. *Natural Product Research, 25*(6), 569–575.

Pandey, R., Chandra, P., Srivastava, M., Mishra, D. K., & Kumar, B. (2015). Simultaneous quantitative determination of multiple bioactive markers in Ocimum sanctum obtained from different locations and its marketed herbal formulations using UPLC-ESI-MS/MS combined with principal component analysis. *Phytochemical Analysis, 26*(6), 383–394.

Patil, R. S., Kokate, M. R., & Kolekar, S. S. (2012). Bioinspired synthesis of highly stabilized silver nanoparticles using Ocimum tenuiflorum leaf extract and their antibacterial activity. *Spectrochimica Acta Part A: Molecular and Biomolecular Spectroscopy, 91*, 234–238.

Pavela, R., Kaffková, K., Smékalová, K., Vrchotová, N., Bednář, J., & Tříska, J. (2023). Biomass yield potential of Tulsi (*Ocimum sanctum* L.) in European conditions. *Industrial Crops and Products, 194*, 116365. https://doi.org/10.1016/j.indcrop.2023.116365

Polic, M., & Nedeljkovic, L. (1978). Effect of addition of spices on the flavour of chicken. *Tehnoligija Mesa, 19*, 359–362.

Prabuseenivasan, S., Jayakumar, M., & Ignacimuthu, S. (2006). *In vitro* antibacterial activity of some plant essential oils. *BMC Complementary and Alternative Medicine, 6*(1), 39. https://doi.org/10.1186/1472-6882-6-39

Prakash, P., & Gupta, N. (2005). Therapeutic uses of *Ocimum sanctum* Linn (Tulsi) with a note on eugenol and its pharmacological actions: A short review. *Indian Journal of Physiology and Pharmacology, 49*(2), 125–131.

Punyoyai, C., Sirilun, S., Chantawannakul, P., & Chaiyana, W. (2018). Development of antidandruff shampoo from the fermented product of *Ocimum sanctum* Linn. *Cosmetics, 5*(3), 43. https://doi.org.10.3390/cosmetics5030043

Rachmawati, E., & Muhammad, R. F. (2021, May). The ethanolic extract of holy basil leaves (*Ocimum sanctum* L.) attenuates atherosclerosis in high fat diet fed rabbit. In *AIP Conference Proceedings* (vol. 2353, no. 1, p. 030113). AIP Publishing LLC.

Raina, A. P., Kumar, A., & Dutta, M. (2013). Chemical characterization of aroma compounds in essential oil isolated from "Holy Basil" (*Ocimum tenuiflorum* L.) grown in India. *Genetic Resources and Crop Evolution, 60*, 1727–1735.

Rajan, M., Gupta, P., & Kumar, A. (2021). Promising antiviral molecules from ayurvedic herbs and spices against COVID–19. *Chinese Journal of Integrative Medicine*, *27*(4), 243–244.

Rana, L., Tewari, G., & Pande, C. (2021). Aroma profile of the aerial parts of *Ocimum sanctum* L. harvested at vegetative and full blooming stages from three altitudes of North India. *Journal of Essential Oil Bearing Plants*, *24*(3), 408–420.

Ravindran, R., Rathinasamy, S. D., Samson, J., & Senthilvelan, M. (2005). Noise-stress-induced brain neurotransmitter changes and the effect of Ocimum sanctum (Linn) treatment in albino rats. *Journal of Pharmacological Sciences*, *98*(4), 354–360.

Renu, K., & Milind, P. (2015). Antipsychotic potentials of *Ocimum sanctum* leaves. *International Journal of Pharmaceutical Sciences and Drug Research*, *7*(1), 46–51.

Sakina, M. R., Dandiya, P. C., Hamdard, M. E., & Hameed, A. (1990). Preliminary psychopharmacological evaluation of *Ocimum sanctum* leaf extract. *Journal of Ethnopharmacology*, *28*(2), 143–150.

Salah, N. (1977). *Arab World Cook Book*. Nahda Salah Publisher.

Salvi, P., Kumar, G., Gandass, N., Kajal., Verma, A., Rajarammohan, S., Rai, N., & Gautam, V. (2022). Antimicrobial potential of essential oils from aromatic plant *Ocimum* sp.; A comparative biochemical profiling and in-silico analysis. *Agronomy*, *12*(3), 627. https://doi.org/10.3390/agronomy12030627

Samodurov, V., Dolgoshchinova, V., Tarasyuk, V., Pruidze, G., Kraevaya, N., & Yakhnev, N. (1991). Composition for manufacture of "Brodinskii" processed cheese. *USSR Patent*, SU1690657.

Sampath, S., Mahapatra, S. C., Padhi, M. M., Sharma, R., & Talwar, A. (2015). Holy basil (*Ocimum sanctum* Linn.) leaf extract enhances specific cognitive parameters in healthy adult volunteers: A placebo controlled study. *Indian Journal of Physiology and Pharmacology*, *59*(1), 69–77.

Samson, J., Rathinasamy, S. D., Ravindran, R., Senthilvelan, M. (2006). Biogenic amine changes in brain regions and attenuating action of Ocimum sanctum in noise exposure. *Pharmacology, Biochemistry, and Behavior*, *83*(1), 67–75.

Saran, P. L., Tripathy, V., Saha, A., Kalariya, K. A., Suthar, M. K., & Kumar, J. (2017). Selection of superior *Ocimum sanctum* L. accessions for industrial application. *Industrial Crops and Products*, *108*, 700–707.

Satapathy, S., Das, N., Bandyopadhyay, D., Mahapatra, S. C., Sahu, D. S., & Meda, M. (2017). Effect of Tulsi (*Ocimum sanctum* Linn.) supplementation on metabolic parameters and liver enzymes in young overweight and obese subjects. *Indian Journal of Clinical Biochemistry*, *32*, 357–363.

Saxena, R. C., Singh, R., Kumar, P., Negi, M. P. S., Saxena, V. S., Geetharani, P., . . . & Venkateshwarlu, K. (2012). Efficacy of an extract of *ocimum tenuiflorum* (OciBest) in the management of general stress: A double-blind, placebo-controlled study. *Evidence-based Complementary and Alternative Medicine*, *2012*. https://doi.org/10.1155/2012/894509

Sembulingam, K., Sembulingam, P., & Namasivayam, A. (1997). Effect of Ocimum sanctum Linn on noise induced changes in plasma corticosterone level. *Indian Journal of Physiology and Pharmacology*, *41*(2), 139–143.

Shah, S., Rastogi, S., & Shasany, A. K. (2018). Genomic resources of ocimum. In *The Ocimum Genome* (pp. 99–110). Springer.

Sharma, S., Kumari, A., Dhatwalia, J., Guleria, I., Lal, S., Upadhyay, N., . . . & Kumar, A. (2021). Effect of solvents extraction on phytochemical profile and biological activities of two *Ocimum* species: A comparative study. *Journal of Applied Research on Medicinal and Aromatic Plants*, *25*, 100348. https://doi.org/10.1016/j.jarmap.2021.100348

Sharma, Y., Bharadwaj, M., Srivastava, N., Kaur, A., Kumar, M., Agarwal, M., . . . & Bala, K. (2020). In vitro antioxidant activity of defatted seed extracts of *Ocimum sanctum* on rat PC-12 cells and its inhibitory efficacy with receptors of oral squamous cell carcinoma. *Industrial Crops and Products*, *154*, 112668. https://doi.org/10.1016/j.indcrop.2020.112668

Sheen, L., & Tsai, S. (1991). Studies on spray-dried microcapsules of ginger, basil, and garlic essential oils. *Journal of Chinese Agricultural Chemical Society*, *29*, 226–237.

Shimizu, T., Torres, M. P., Chakraborty, S., Souchek, J. J., Rachagani, S., Kaur, S., . . . & Batra, S. K. (2013). Holy Basil leaf extract decreases tumorigenicity and metastasis of aggressive human pancreatic cancer cells in vitro and in vivo: Potential role in therapy. *Cancer Letters*, *336*(2), 270–280. https://doi.org/10.1016/j.canlet.2013.03.017

Shivpuje, P., Ammanangi, R., Bhat, K., & Katti, S. (2015). Effect of *Ocimum sanctum* on oral cancer cell line: An *in vitro* study. *The Journal of Contemporary Dental Practice*, *16*(9), 709–714.

Shrinet, K., Singh, R. K., Chaurasia, A. K., Tripathi, A., & Kumar, A. (2021). Bioactive compounds and their future therapeutic applications. In *Natural Bioactive Compounds* (pp. 337–362). Academic Press.

Shukla, M. R., Kibler, A., Turi, C. E., Erland, L. A., Sullivan, J. A., Murch, S. J., & Saxena, P. K. (2021). Selection and micropropagation of an elite melatonin rich Tulsi (*Ocimum sanctum* L.) germplasm line. *Agronomy*, *11*(2), 207. https://doi.org//agronomy110202207

Singh, D., & Chaudhuri, P. K. (2018). A review on phytochemical and pharmacological properties of Holy basil (Ocimum sanctum L.). *Industrial Crops and Products*, *118*, 367–382.

Singh, N., Hoette, Y., & Miller, R. (2010). *Tulsi: The Mother Medicine of Nature* (2nd ed., pp. 28–47). Lucknow: International Institute of Herbal Medicine.

Singh, S., & Majumdar, D. K. (1995). Anti-inflammatory and antipyretic activities of *Ocimum sanctum* fixed oil. *International Journal of Pharmacognosy*, *33*(4), 288–292.

Singh, S., Taneja, M., & Majumdar, D. K. (2007). Biological activities of *Ocimum sanctum* L. fixed oil—An overview. *Indian Journal of Experimental Biology*, *45*, 403–412.

Skaltsa, H., Tzakou, O., & Singh, M. (1999). Note Polyphenols of *Ocimum sanctum* from Suriname. *Pharmaceutical Biology*, *37*(1), 92–94. https://doi.org/10.1076/phbi.37.1.92.6318

Song, K. Y., & Kim, Y. S. (2019). Effect of mucilage extracted from Basil (*Ocimum basilicum* L.) seeds on physico-chemical and rheological properties in low-fat milk protein gel. *Journal of Food Processing and Preservation*, *43*(11), e14191. https://doi.org/10.1111/jfpp.14191

Suneetha, S., Talwar, A., Mahapatra, S. C., & Sharma, R. (2012). Effects of *Ocimum sanctum* Linn (OS) leaf extract on stress, memory and attention in healthy humans. *Planta Medica*, *78*(11), PF58. https://doi.org/10.1055/s-0032-1320605

Suzuki, A., Shirota, O., Mori, K., Sekita, S., Fuchino, H., Takano, A., & Kuroyanagi, M. (2009). Leishmanicidal active constituents from Nepalese medicinal plant Tulsi (*Ocimum sanctum* L.). *Chemical and Pharmaceutical Bulletin*, *57*(3), 245–251.

Tabassum, I., Siddiqui, Z. N., & Rizvi, S. J. (2010). Effects of *Ocimum sanctum* and *Camellia sinensis* on stress-induced anxiety and depression in male albino *rattus norvegicus*. *Indian Journal of Pharmacology*, *42*(5), 283–288.

Tangpao, T., Chung, H., & Sommano, S. (2018). Aromatic profiles of essential oils from five commonly used Thai basils. *Foods*, *7*(11), 175. https://doi.org/10.3390/foods7110175.

Tripathi, D. K., Pathak, A. K., Chauhan, D. K., Dubey, N. K., Rai, A. K., & Prasad, R. (2015). An efficient approach of laser induced breakdown spectroscopy (LIBS) and ICAP-AES to detect the elemental profile of Ocimum L. species. *Biocatalysis and Agricultural Biotechnology*, *4*(4), 471–479.

Umashankar, V., Deshpande, S. H., Hegde, H. V., Singh, I., & Chattopadhyay, D. (2021). Phytochemical moieties from Indian traditional medicine for targeting dual hotspots on SARS-CoV-2 spike protein: An integrative in-silico approach. *Frontiers in Medicine*, *8*, 672629. https://doi.org/10.3389/fmed.2021.672629

Upadhayay, U. P. P. D. D., Ewam, P. C. V. V., Ewam, U. P. C. V. V., & Sansthan, G. A. (2012). Immunomodulatory and therapeutic potentials of herbal, traditional/indigenous and ethnoveterinary medicines" Mahima,"Anu Rahal," Rajib Deb,"Shyma K. Latheef," Hari Abdul Samad. *Pakistan Journal of Biological Sciences*, *15*(16), 754–774.

Venu Prasad, M. P. (2014). *Antifatigue and Neuroprotective Properties of Selective Species of Ocimum L.* PhD Thesis, University of Mysore. http://hdl.handle.net/10603/70183

VirtueMarket Research (2023). *Holy Basil Iced Tea Market Size (2022–2030).* https://virtuemarketresearch.com/report/holy-basil-iced-tea-market, accessed October 8, 2023.

Vishwabhan, S., Birendra, V. K., & Vishal, S. (2011). A review on ethnomedical uses of *Ocimum sanctum* (Tulsi). *International Research Journal of Pharmacy*, *2*(10), 1–3.

Walia, P., Walia, A., & Talwar, T. (2014). A multifunctional natural would healing matrix. *World Patent 147638 A1*, issued 25 September. World Intellectual Property Organization.

WHO (2008). World malaria report 2008, 215 p. World Health Organization.

Zeedan, G. S. G., & Abdalhamed, A. M. (2021). Antiviral effects of plant extracts used in the treatment of important animal viral diseases. *World*, *11*(4), 521–533.

Zhu, S. (2010). Natural product derivatives with antimalarial activity. *U.S. Patent 7851508 B2*, issued 14 December.

Custard Apple and Its By-products

Nutritional, Bioactive, Medicinal Potential, and Functional Food Development

7

Seyashree Hazra, Najmun Nahar, Tanmay Sarkar, and Runu Chakraborty

7.1 INTRODUCTION

The custard apple plant, native to tropical America, *Annona squamosa* L., sometimes called "sugar apple" or "sweetsop", grows in several tropical nations worldwide. Custard apple seeds were apparently introduced by the Portuguese in the 16th century, based on archaeological investigations of botanical remains in Uttar Pradesh (UP) (Pokharia 2008). However, earlier records have been found for remains of custard apple (fruit coat and seeds) at other archaeological sites in Punjab (Kushana Period, 100–300 AD) and UP (Early Iron Age, 1300–700 BC). A delicious fruit, custard apple was once utilized in South China as a traditional remedy to heal malignant wounds and has insecticidal activity (Chen et al. 2012). It is therefore evident that custard apple, along with other economically important plants, was grown much earlier than the Portuguese arrival in India.

The more desirable and nutritionally relevant part is the ripe fruit, abundant in minerals and vitamins (Olesen and Muldoon 2012). *A. squamosa* belongs to the Annonaceae family and *Annona* genus, all under the Plantae kingdom (Mondal et al. 2018). The custard apple is a bush-type tree that mainly grows in low-altitude regions. There are 1200 species in the Annonaceae family, which has 126 genera and are all found in tropical and subtropical regions. The most typical place to find custard apples (CA) is in the backyard gardens of the coastal towns in America. It is grown in the Malay

DOI: 10.1201/9781003352341-7

Archipelago, India, Sri Lanka, the Philippines, Polynesia, Australia, and major regions in Africa (Bapat et al. 2020; Pinto et al. 2005). Traditional plant *Annona squamosa* is mainly found in Tamil Nadu, India (Varadharajan et al. 2012). The flowers of CA are grouped with green color and white flesh within. Flowers contain sweet and delightful fragrances, unlike leaves, which have an unpleasant aroma. Good pollination requires low temperature and high humidity with slightly acidic soil (pH 6.0–6.5) (Adil 2019). *A. squamosa* is a well-liked fruit native to the equatorial regions of India, where a season lasts only three months out of the year. This heart-shaped fruit, popularly known as "sweet sop," weighs around 150 g and has exceptionally rough skin. The pulp is creamy, delightful, and flavorful when fully ripe. It is typically consumed as a fruit and is widely used to make ice creams and beverages (Chikhalikar et al. 2000). Plants are propagated by grafting or from seed stock (Bapat et al. 2020). Between 2004–05 and 2017–18, the total fruit production in India climbed from 50.9 to 97.35 million tonnes. The total cultivated area and production of CA in the year 2017–2018 were 46,000 ha and 401,000 MT, respectively. The total export of apples in India was valued at 527,712 lakhs in 2015–2016 (Bala et al. 2018). Custard apples are annually farmed in India on 29.87 thousand hectares, producing 228.37 million tonnes (Chandel et al. 2018). The major custard apple-producing states in India are Assam, Telangana, Madhya Pradesh, Bihar, Maharashtra, Uttar Pradesh, Odisha, Rajasthan, Andhra Pradesh, and Tamil Nadu (Gautam et al. 2021).

The biosphere is seriously threatened when conventional mineral oil-based lubricants are released into the environment. Mineral oils have long been present in the environment due to their low degradability and are expected to be swiftly replaced with eco-friendly, biodegradable lubricants. The globe does not produce enough edible oils, although numerous plant species whose seeds are currently underused and untapped can be used to manufacture biodiesel in a nation like India. A potential food source for the development of bio-lubricants in India is non-edible oil seeds. The fatty acid composition of custard apple seed oil is appropriate for biological lubricants and has the capacity to replace mineral-based lubricants available worldwide (Ashokkumar et al. 2020).

Compared to grapefruit, CA has higher vitamin C levels, 35–42 mg/100 g (Marahatta et al. 2019). Thiamine, potassium, and dietary fiber are essential nutrients present in CA. Several polyketides are derived from the bark (acetogenin and squamone), and the seed (squamocin, acetogenin, annonacin, and annotemoyin 1 and 2). Many phytochemicals such as anonaine, coryeline, aporphine, norcorydine, isocorydine, and glaucine have also been obtained from the plant's stems, leaves, and roots (Pandey and Barve 2011). These and several other phytochemicals suggest that custard apples have antimalarial, antibacterial, anticancer, anthelmintic, antidiabetic, and hepatoprotective properties. Various CA parts have different uses; for example, CA leaves include vermicidal protein characteristics that can alleviate and/or treat abscesses, bug bites, malignant tumors, and other skin conditions. In tribal areas, crushed leaves are used as a remedy for ulcers and sores and are inhaled to cure hysteria and fainting spells. Bark scrapings from roots are used to relieve toothaches, and powdered seeds are used to eradicate fleas and head lice. However, extreme care must be exercised to prevent seed flour from entering the eyes, which can be painful (Singh et al. 2019). In India, a mixture of salt and crushed ripe fruits is used to treat tumors. The roots and bark are extremely astringent. The bark extract is administered as a sedative to provide relief from diarrhea. The rhizome is a radical treatment for dysentery and other illnesses due to its potent purgative activity (Bhattacharya and Chakraverty 2016). All tests were run using methanolic fruit extracts (fresh and dried pulp and seeds). The methanolic extract of custard apple pulp has good ABTS and FRAP antioxidant potential. GC/MS analysis of pulp extracts indicated the presence of fatty acids, alcohols, alkanes, aldehydes, alkenes, and ketones, some of which have medicinal applications and are also used in food preparations, such as 2-octanol, which is useful as a flavoring compound as well as in perfumes. In diabetic patients, D-mannitol works as a sweetener. Heptanoic acid is included in the manufacture of fragrances (Bhardwaj et al. 2014). Custard apple peel has astringent qualities and tannins, which are used to formulate herbal medicines (Lydia et al. 2017). Research suggests that custard apples offer hepatoprotective, anti-genotoxic, anti-malaria, and antitumor properties (Singh et al. 2019). In India, custard apple is commonly used as a tonic to improve blood health, reduce vomiting, and treat cancer. It is also used as a vermicide (a medication used to kill worms) and for skin conditions and wound healing. The leaves, bark, root, and fruit of the tree are used in various traditional medicine preparations. However, more scientific research is needed to confirm the safety and efficacy of these traditional uses (Shami 2017). This chapter reviews CA's nutritional, botanical, and health benefits and its conversion of agricultural wastes into value-added products.

7.2 BOTANICAL DETAILS

Annona squamosa (Figure 7.1; Table 7.1) is probably the most drought-resistant plant in the Annonaceae family because it grows and produces poorly in areas with abundant rain. The tree thrives in sand, thick loam with well-drained soil, oolitic limestone, and tropical or almost tropical climates (Bapat et al. 2020).

Altitude 700–2500 m
Texture Soft
Fruit 150–500 mg
Length 6–10 m
Seed Toxic (not in use)
Leaves Ovate and pointed at the apex

7.2.1 Leaves

A semi-deciduous custard apple tree (3–7 meters high) has branches that are unevenly spaced and a broad, open crown. Its exterior bark is light brown, and its inner layer is light yellow and slightly bitter, with visible leaf scars. The leaves are glabrate or nearly so, lanceolate or oblong-lanceolate, 6–17 × 3–6 cm, and pale green on both surfaces (Saha 2011). Young leaf edges are subtly hairy and can have slightly uneven sides (Bhattacharya and Chakraverty 2016).

7.2.2 Flowers

The custard apple's flowering season lasts from the beginning of spring until the beginning of summer, and in places with high moisture levels, it lasts year-round. Proteaginous, actinomorphic, and penicillate blooms are present. The flower (up to 2–4 cm long) consists of three deteriorated sepals and six degenerated petals. There are two whorls made up of six petals, and the inner whorl's petals either degenerate into tiny scales or vanish entirely. Pistils are numerous on the conical receptacle at the flower's center, while the perimeter is surrounded by numerous stamens (Olesen and Muldoon 2012). Numerous crowded stamens are white and less than 16 mm long (Saha 2011).

7.2.3 Fruits

The tree starts producing fruit when 3–4 years old. In India, fruiting occurs from July to August. Custard apples have a sweet, sugary flavor. The dense rind is covered in knobby segments that soften as it ripens and crack open, emitting a delicious perfume. Normal ripening occurs between 15 and 30 °C (Olesen and Muldoon 2012). The average diameter and length of rounded, heart-shaped, ovate, or conical fruits are 5 to 7.5 cm and 6 to 10 cm, respectively. Several factors, like cultivar, pollination, and nutrition, affect the development of fruit size, which varies between 120 and 330 g. The white pulp resembles custard and has a delicious sweet-sour flavor (Pinto et al. 2005).

TABLE 7.1 Different varieties of *Annona* species

BOTANICAL NAME	*COMMON NAME*	*REGION OF CULTIVATION*	*PHYSICAL CHARACTERISTICS*	*REFERENCES*
Annona glabra L.	Pond apple, Alligator apple	The Bahamas, Florida, South, and Central America, West Africa, Caribbean	Fruits rectangular to oval in shape, typically 2.5–3.5 inches across and 3–5 inches long. Mature fruit is deliciously scented with fleshy, floury, and solid pulp. It has a yellowish-orange hue.	Dilrukshi et al. (2020); Lonarkar et al. (2021)
Annona reticulate L.	Bullock's heart, Ramphal, Custard apple	West Indies, Southern India, South America, Burma, and Bengal	Heart-shaped and edible, yellow-colored fruits are delicious. They change color to yellowish red when ripe.	Jamkhande and Wattamwar (2015)
Annona diversifolia L.	Ilama	Southern Florida	Whitish green outer layer with white or pink flesh	Noonan (1953)
Annona squamosa L.	Custard Apple, sugar apple	Deccan Plateau, Central India, Maharashtra and Islands of west India, Northern region of South America, Nepal	The fruit weighs between 120 and 330 g and is between 5 and 7.5 cm in diameter and 6 to 10 cm long. It is spherical, elliptical, or angular.	Bala et al. (2018); Marahatta et al. (2019); Pinto et al. (2005)
Annona muricata L.	Soursop, Guanabana	Central America, Yunnan in China.	Oblong, spherical fruit with an irregular shape; color changes from light green to bright green during ripening. Weighs up to 4.5 kg, 10–30 cm long.	Bonavia et al. (2004); Gajalakshmi et al. (2012); Noonan (1953)
Annona senegalensis Pers.	Wild soursop	Senegal, Africa	The unripe fruit is whitish green and changes color to yellow or orange when ripe.	Pinto et al. (2005); Saha (2011)
Annona cherimolia Mill.	Cherimoya	Peru, Andes, Central America, and Mexico	Usually angular, oval-shaped, and with an uneven surface.	Bonavia et al. (2004)

7.2.4 Seed

The fruit consists of 35 to 45 ovoid-shaped, dark-colored seeds varying from 0.6 to 0.8 cm in width and 1.5 to 2.0 cm in height (Bhattacharya and Chakraverty 2016; Pinto et al. 2005).

7.2.5 Stems

CA stems are cylindrical with a distinctive odor and bitter flavor. Thick outer cork cells develop upon maturation (Bhattacharya and Chakraverty 2016).

FIGURE 7.1 Different part of custard apple tree

7.3 NUTRITIONAL POTENTIAL

CA is an enriched source of nutrients that are briefly mentioned here (Table 7.2). *A. squamosa* pulp tastes delicious partly due to its sugar content (28%), predominantly sucrose (2.53%), followed by dextrose (5.05%) and laevulose (0.04%). The pulp has a low glycemic index and a mild glycemic load in spite of the high sugar content. It also has fragrant flavors, according to Patel and Kumar (2008). The pulp has significant levels of Fe, Ca, thiamine, amino acids, potassium, ascorbic acid, carotene, riboflavin, magnesium, niacin, and dietary fibers, as well as aliphatic ketones (i.e., palmitone). *Annona* fruit contains considerable amounts of carbohydrates, protein, and energy (Albuquerque et al. 2016; Saha 2011) and is an excellent source of minerals (i.e., Ca, K, Na, P, Mg, and Zn) (Akomolafe and Ajayi 2015; Sarkar et al. 2022). *A. squamosa* leaves contain micronutrients in varying amounts, such as Mg, P, Zn, and Cu (65.65, 43.1, 0.46, and 0.31 mg/100 g, respectively), and selenium is present in trace amounts (Varadharaj et al. 2014). Potassium is present in the highest amount in fruit pulp and seed (371 and 356 mg/100 g). The fruit pulp of custard apple is a rich source of potassium, calcium, sodium, and magnesium, and contains a minimum amount of Fe, Zn, and Cu compared to the seed. Storage (9 days) reduces the fruit pulp's ascorbic acid concentration and increases its total sugar, sucrose, and reducing sugar contents (Abdualrahman et al. 2019). The primary nutritional and bioactive components in *Annona squamosa* seeds include carotenoids, proteins, cyclopeptides, acetogenins, polysaccharides, alkaloids, fatty acids, ascorbic acid, and tocopherols (Kumari et al. 2022). The average granular size of *A. squamosa* fruit is generally 4.84 µm, tested by SEM (scanning electron microscope). The starch of sweetsop has remarkable swelling ability and solubility compared to that of soursop. It also has a very low gelatinization temperature, suggesting a weakened grainy structure. Sweetsop starch has a low pasting temperature, high paste clarity, viscosity peaks and breakdown, lower setback, and improved freeze-thaw stability. It also has low gelatinization temperature and high freeze-thaw resilience similar to waxy corn. Hence, this fruit can be used as a thickener in frozen dishes due to its physiochemical properties (Nwokocha and Williams 2009).

TABLE 7.2 Nutritional composition of different *A. squamosa* parts

FRUITS PART	*ENERGY (KCAL/100 G)*	*SUGAR (%)*	*CARBO-HYDRATE (%)*	*PROTEIN (%)*	*FAT (%)*	*MOISTURE (%)*	*CRUDE FIBERS (%)*	*ASH (%)*	*REFERENCES*
Peel	–	–	86.75	3.3	7.78	3.5	–	2.17	Shehata et al. (2021)
	–	–	–	–	–	10.9	–	10.4	Shivamathi et al. (2019)
	–	3.17	–	0.21	0.24	9.11	–	1.12	Nwokocha and Williams (2009)
	–	–	19.9	1.7	0.4	–	–	–	Adil (2019)
	–	–	21.95	3.78	0.22	–	–	4.58	Marahatta et al. (2019)
	92.9	–	20.41	–	0.14	70.8	–	0.57	Bhardwaj et al. (2014)
	–	28	–	–	–	–	–	–	Singh et al. (2019)
Pulp	93.8		21.1	1.96	0.18		5.3		Albuquerque et al. (2016)
	51.8 kJ/100 g	–	10.8	1.13	0.79	78.5	5.9	2.84	Abdualrahman et al. (2019)
	–	–	23.5	1.6	–	–	–	–	Lydia et al. (2017)
	242.4	–	30.3	–	–	70	–	7.5	Hassan et al. (2008)
	–	–	23.9	1.6	0.3	73.3	–	–	Patil et al. (2011)
	–	20.9	–	2.8	0.39	74.6	3.3	1.05	Bala et al. (2018)
	–	–	–	17.8	–	–	–	–	Panadare et al. (2020)
Seed	463.6	–	12.45	–	–	–	36.3	2.78	Hassan et al. (2008)
	434.5 KJ/100 g	–	21.8	18.3	30.4	6.65	17.6	5.24	Abdualrahman et al. (2019)
	375	–	10.32	17.4	29.4	9.67	32.5	0.81	Amoo et al. (2008)

7.3.1 Carbohydrates

Annona squamosa fruits are highly nutrient-dense, and their high levels of crude fiber and low lipids place them in the healthy fruits category. Such fruits are quite diverse and have many health advantages. They are, therefore, perfect for those who favor low-sugar diets. When ingested in appropriate quantities, CA fruit can benefit the keto diet, which means consuming less than 30 grams of carbs daily (Marahatta et al. 2019). The study identified the phytochemicals and antioxidant components in *Annona squamosa* fruits, seeds, and leaves cultivated in Egypt (Giza, Mansoura, Alexandria, and Menofia). At Menofia, Alexandria, and Giza, *Annona squamosa* had the highest sugar concentrations in the following order: leaves > seeds > fruits. In contrast, Alexandria and Menofia contain the highest concentrations of glucose in their fruits, seeds, and leaves, respectively. Enormous amounts of fructose were found in the seeds, fruits, and leaves of *A. squamosa* in Menofia and Alexandria. The highest fruit fructose level was found in the Giza and Mansoura regions, followed by seeds and leaves. Fruit was composed of the highest sucrose levels in Giza and Mansoura, whereas seeds had the highest sucrose in Menofia and Alexandria (Shukry et al. 2019).

Sweetness of 29 custard apple genotypes from Maharashtra varied significantly when evaluated as relative levels of fruit total soluble solids (19.3–28.5 °Brix) and pulp acidity (0.18%–0.34%). Moreover, total and reducing sugars also varied significantly among the genotypes (15.7%–26.2% and 14.1%–24.3%) (Ghawade et al. 2018). Total soluble sugar ranged from 19% to 26% for 21 custard apple germplasm accessions collected during an exploration program from six districts of Maharashtra (Dikshit et al. 2008). Another study of 25 custard apple genotypes collected from different locations in Gujarat showed significant variation in fruit total soluble solids (26.6–32.6 °Brix), acidity (0.20%–0.30%), total sugars (12.5%–17.6%), and ascorbic acid (18.25–38.24 mg/100 g of fruit juice) (Yadav et al. 2017).

7.3.2 Protein

Research on *Annona squamosa* protein has focused primarily on the seed because of its relatively higher content than other parts of the fruit. For example, *A. squamosa* seeds and pulp contained 18.34% and 1.13% crude protein (N × 6.25) (Abdualrahman et al. 2019). Similar protein content (15%) was reported for *A. squamosa* seeds from Southeastern Brazil that also contained 56.4% carbohydrates, 20.36% lipids, and 1.74% ash (Luzia and Jorge 2012). Short-term storage (120 days) reduces seed protein content (~45%), presumably due to proteolytic enzyme activities (Martinez-Maldonado et al. 2015). Two proteins (21 and 28 kDa) have been isolated and purified from *A. squamosa* seeds. These ribosomal inactivating proteins (RIPs) have lectin-binding properties and possess remarkable antimicrobial, antimutagenic, and cytotoxic activity, enabling their use to treat various diseases (Dhanraj et al. 2020).

Glutamic acid, aspartic acid, and arginine are the most abundant amino acids in *A. squamosa* seeds (13.5, 7.9, and 6.08 g/100 g protein, respectively) (Abdualrahman et al. 2019). Two of these amino acids (arginine and glutamine), along with histidine, are known to have strong effects on the immune functions in the body. These amino acids account for ~30% of the total amino acids. The lysine/arginine ratio (0.64), a determinant of cholesterolemic and atherogenic effects of a protein, is twice that of hempseed proteins, but less lipidemic and atherogenic than soybean protein with a lysine/arginine ratio of 0.88 (Oomah 2022). The aromatic amino acids (AAAs) (6.88 g/100 g protein) (only phenylalanine and tyrosine in this study) are associated with protein and other essential secondary metabolite synthesis. Some of these metabolites mediate nervous signal transmission and quench reactive oxygen species in the brain (Oomah 2022). The sum of some amino acids (histidine, lysine, methionine, proline, tryptophan, and tyrosine) (15.7 g/100 g protein) reflects the antioxidant amino acids that increase the potency of most food-derived peptides. Sulfur amino acids (4.12 g/100 g protein) mediate several crucial cellular mechanisms including redox balance and methylation. The branched-chain amino acids (valine, leucine, and isoleucine) (12.65 g/100 g protein, ~16% of total amino acids) are major players in preventing future risks of type 2 diabetes and other vascular or inflammatory stress-inducing pathologies

(Abolbaghaei et al. 2018). The low Fisher ratio (BCAA/AAA = 1.85) of *A. squamosa* seed protein is a desirable trait for functional foods to meet specific physiological needs. In defatted *A. squamosa* seeds, the amino acids (g/100 g protein) in descending order were glutamic acid (0.995), leucine (0.845), arginine (0.704), aspartic acid (0.684), phenylalanine-tyrosine (0.671), valine (0.642), alanine (0.594), isoleucine (0.464), lysine (0.407), glycine (0.392), threonine (0.324), serine (0.299), histidine (0.139), and methionine-cysteine (0.106) (Mariod et al. 2010). The Fisher and lysine/arginine ratios (2.91 and 0.58) of the defatted seeds were higher and lower, respectively, compared to the seed.

Eight cyclic peptides (cyclosquamosins A, B, D, E, H, and I; squamins A and B; and cherimolyacylopeptide B) were isolated from *A. squamosa* seeds. Cyclosquamosin D exerts strong anti-inflammatory activity by inhibiting inflammatory cytokines (TNF-α and IL-6) production from LPS-stimulated murine macrophage (J774A.1) cells presumably by binding to toll-like receptors TLR2/4 (Yang et al. 2008). Cyclosquamosin B has vasorelaxant effect primarily attributed to calcium influx inhibition through voltage-dependent calcium channels and can be useful to treat cerebral vasospasm and hypertension and improve peripheral circulation for vascular diseases (Morita et al. 2006). Fanzlizhicyclopeptides A and B exert potent anthelmintic activity against earthworms and anti-dermatophytic activities against fungi (*Trichophyton mentagrophytes* and *Microsporum audouinii*) (Dahiya and Dahiya 2021). According to the study (Panadare et al. 2020), custard apple seeds were once considered trash but later were found to be a useful source of protein and oil. Two extraction techniques were applied to extract the protein and seed oil: Ultrasound-Assisted Three Phase Partitioning (UTPP) and Ultrasound Pre-treatment followed by Three Phase Partitioning (UPTPP). TPP extracted 17.8% of protein and 30.16% of oil from custard apple seeds. UPTPP was a superior extraction technique to the other one.

7.3.3 Fat

Fat content of *A. squamosa* seeds (22.2%–30.4%, w/w) has high unsaturated fatty acids (71%–76% of total fatty acids (Mariod et al. 2010; Mariod et al. 2017; Abdualrahman et al. 2019; Luzia and Jorge 2012; Chen et al. 2016). The seed oil consists of palmitic (9.9%–20.3%), stearic (4.3%−13.6%), oleic (38.8%–56.5%), linoleic (20.5%–32.9%), linolenic (1.1%–2.2%), arachidonic (1.1%−1.5%), and behenic (0.1%–2%) acids. The seed kernel (67.7% of the dried seed) contains 22.2% oil (dry weight basis) with 12.1% palmitic, 22.9% linoleic, 47.4% oleic, 13.6% stearic, and 2.3% heneicosanoic acids (Rana 2015). The seed oil has a slight almond-like odor and a very mild and pleasant taste. It apparently promotes hair growth by follicle stimulation, inhibiting scalp inflammation due to improved blood circulation, preventing hair fall, and facilitating nutrient absorption (Singh et al. 2019). The oil inhibits hepatoma cells (H_{22} transplantation tumor model) growth with a 54% maximum inhibitory rate in mice. This antitumor effect is presumably due to the primary oil constituents (unsaturated fatty acids and acetogenins, 41 mg/g motrin equivalents) that suppress interleukin-6 (IL-6), Janus kinase (Jak), and phosphorylated signal transducers and activators of transcription (p-Stat3) (Chen et al. 2016).

7.3.4 Vitamins and Minerals

Custard apple from 25 genotypes collected from different locations in Gujarat varied significantly in minerals (Mg, 21.2–38.7; K, 257–296; Na, 4.3–15.3; and Ca, 15.4–21.4 mg/100 g fruit) and ascorbic acid (18.25–38.24 mg/100 g fruit juice) (Yadav et al. 2017). Indonesian *A. squamosa* (also known as Srikaya) fruits contained 2.3% ash and minerals (mg/kg dry matter) [Ca, 326; K, 8562; Mg, 1230; P, 1531; S, 555; B, 7.3; Cd, 0.066; Cr, 0.446; Fe, 26.4; Mn, 3.5; Mo, 0.178; Ni, 0.555; and Zn, 4.2] (Gökbel et al. 2015). Generally, the pulp has higher minerals, except magnesium, and trace element content than the seed. The pulp contains higher primary minerals than the seed, including calcium, iron, copper, zinc, and manganese. Only magnesium content was higher in seed (730 vs 518 ppm for the pulp). Secondary elements, including chromium, nickel, and molybdenum, were higher in pulp than the seed, except for tin (0.827 ppm) and selenium (0.873 ppm) (Bhardwaj et al. 2014).

The highest mineral contents were also reported in *A. squamosa* (var. Balady), except potassium, which was highest (116 mg/kg) in the peel (Shehata et al. 2021). However, the same report showed the highest mineral contents—except iron, potassium, and sodium (highest in the peel)—of a hybrid variety (*A. cherimoya* × *A. squamosa*).

Seed and pulp are good sources of calcium, followed by iron and zinc. Negligible amounts of trace elements like selenium, molybdenum, and cadmium are present in pulp and seed (Bhardwaj et al. 2014). Leaf mineral composition (μmol/g dry weight) varied among four varieties of *A. squamosa*: K (252–387), Na (61–95), Ca (79–172), and Fe (37–50) (Shukry et al. 2019). The mineral contents (sodium, calcium, iron, and potassium) of the various sections of *Annona squamosa* varied depending on locations in Egypt. Compared to other fruits like bananas and apples, K+ was the mineral that predominated in *A. squamosa* fruits and was present in higher concentrations (Shukry et al. 2019). Vitamin C is well proven to enhance immunity and maintain healthy blood vessels and connective tissue. Vitamin C is present in the highest amount in *A. squamosa* (39 mg/100 g), followed by *Citrus paradise, Mangifera indica, Annona reticulata, Ananus comosus, Annona cherimola, Persea Americana* (Marahatta et al. 2019). Luzia and Jorge (2012) investigated tocopherol isomers in the oil derived from soursop and sugar apple seed. Soursop contained 32.66 mg/kg total tocopherol, predominantly in the γ-isomer (76.5% of total tocopherol). However, δ-tocopherol was reported as the main isomer (70% of total tocopherol) in *A. squamosa* seed oils, with high total tocopherol content (155 and 166 mg/kg) (Mariod et al. 2010).

7.4 PHYTOCHEMICAL ANALYSIS

Custard apple seeds and peel extracts revealed the presence of many bioactive compounds, including lignans, terpenoids, phenylpropanoids, organic acids, sugars, and flavonoids (Tables 7.3 and 7.4).

7.4.1 Alkaloids

Alkaloids from the *Annona* genus possess many pharmacological activities. Some (i.e., liriodenine, anonaine, and asimilobine) have antibacterial activity against *Staphylococcus epidermidis*, while others like (+)-xylopine and isocoreximine have demonstrated significant anti-cancer activity against specific cell lines (Nugraha et al. 2019). Total phenolics, flavonoid concentration, and phytochemical components of *A. squamosa* seed were investigated in chloroform, petroleum ether, methanol, and ethyl acetate extracts using established laboratory techniques. The presence of substantial to moderate levels of phytochemicals, including flavonoids, coumarins, alkaloids, and terpenoids, can be linked to the plant's potential for considerable medical benefits (Biba et al. 2013). *Annona squmosa* leaves were ground and progressively extracted using various solvents. Initial phytochemical analysis of the extracts showed the presence of alkaloids, flavonoids, saponins, steroids, tannins, and carbohydrates. The number of substances in the various extracts was qualitatively determined using TLC and HPTLC techniques, and alkaloids were identified in the methanol extract (Agrawal et al. 2012). Alkaloid constituents separated from Indian-grown *Annona squamosa* leaves consisted of lanuginosine, (+)-corydine, (+)-*O*-methylarmepavine, (+)-norcorydine, (+)-norisocorydine, (+)-glaucine, (+)-anonaine, xylopine, and (+)-norcorydine. The seed additionally contained asimilobine, (+)-reticuline, (–)-nornuciferine or (–)-N-methylasimilobine, and (–)-anonaine (Nugraha et al. 2019). The LC chromatograms and MS data for the major alkaloid extract from *A. squamosa* leaves identified several alkaloids, including corydine, sanjoinine, norlaureline, norcodeine, oxanalobine, and aporphine, based on their retention times (1.562, 1.741, 1.896, 41.668, 43.7, and 45.569 min, respectively). The alkaloid fraction had high DPPH radical scavenging activity with low IC_{50} value (Shami 2017).

Other phytochemical tests on various *A. squamosa* plant sections indicated the presence of numerous bioactive components, including annonaceous acetogenins (ACGs), diterpenes (DITs), cyclopeptides, and alkaloids (ALKs). Nineteen alkaloids (primarily aporphine type) have been isolated and identified mostly from *A. squamosa* leaves, stems, and roots. These alkaloids have antihypertensive, antispasmodic, antihistaminic, and vasodilatory properties (Ma et al. 2017). The methanol extract of *A. squamosa* raw and dried pulp and seeds contains 1.25 g and 0.97 g/100 g DM alkaloids, respectively, followed by saponins (Bhardwaj et al. 2014). Three aporphine alkaloids isolated from *A. squamosa* bark were identified as N-nitrosoxylopine, roemerolidine, and duguevalline (Johns et al. 2011). These alkaloids possess antiplasmodial activities representing a potential source of antimalarial compounds. The annonaceous hybrid atemoya is a cross between *Annona squamosa* L. and *Annona cherimola* Mill. Seven alkaloids, including three oxoaporphine (liriodenine, lanuginosine, and lysicamine), two proaporphine (stepharine, pronuciferine), and two aporphine (anonaine and asimilobine), were discovered as a result of the phytochemical research using various spectrometric techniques, namely MS and NMR (Rabêlo et al. 2015). *A. squamosa* seeds are abundant in alkaloids and higher than those from shade-dried leaves (3.77% vs 0.48%, dry weight) (Larrota and Baquero 2018; Kavitha 2015). Four alkaloids (coclaurine, higenamine, magnoflorine and stepharine) are common to both seeds and leaves (Avula et al. 2018). Rabêlo et al. (2014) reviewing alkaloids in *Annona* species from 1930 to 2013 reported 15 alkaloids identified in Indian *A. squamosa.* Almost all alkaloids were present in leaves, except liriodenine in stembark, while others were extracted from stems, roots, or bark. The aporphines (anonaine, corydine, glaucine, and isocorydine) and higenamine were identified in bark, root, leaves, and stem; aporphine, dienone, norcorydine, norisocorydine, norlaureline, and roemerine in both leaves and stem; and lanuginosine, (+)-*O*-Methylarmepavine, and (–)-xylopine only in leaves. Nineteen alkaloids from *A. squamosa* stem and leaves have antispasmodic, bronchodilatory, antihypertensive, and antihistaminic activities beneficial for respiratory diseases (Dash et al. 2020). Anonaine, a benzylisoquinoline alkaloid, has wide-ranging biological activities, including anticancer, vasorelaxation, antioxidant, antiparasitic, and antimicrobial activities. These activities are mediated by different mechanisms: nitric oxide and reactive species production, glutathione reduction, caspases and apoptosis-related proteins activation, DNA damage and human lung, and cervical cancer cell growth inhibition (Dash et al. 2020).

Alkaloids present in ethanol extract (5% suspension) from *A. squamosa* defatted seeds apparently depressed the central nervous system, enhanced pentobarbitone-induced sleep, elevated pain threshold, and prevented electroshock-induced seizures in albino rats without eliciting anti-inflammatory effects (Saluja and Santani 1994). In a case-controlled study, chronic exposure to neurotoxic alkaloid benzyltetrahydroisoquinoline present in *A. squamosa* was associated with atypical Parkinsonism and progressive supranuclear palsy in Parkinson patients (12M/10F, 65 years median age) in Guadeloupe, the French West Indies (Caparros-Lefebvre and Elbaz 1999). *A. squamosa* herbal tea, considered a purgative is often (daily or weekly) used by many people in the French West Indies because of its sedative or hypnotic effects, presumably due to the opioid precursor benzyltetrahydroisoquinoline. This alkaloid has many pharmacological activities, including dopaminergic antagonism and serotoninergic agonism, associated with antidepressive properties (Caparros-Lefebvre and Elbaz 1999).

Three aporphine alkaloids (liriodenine, lysicamine, and lanuginosine) isolated from methanol extracts of *A. squamosa* leaves exert cytotoxicity against human adult T-cell leukemia/lymphoma (MT-1 and MT-2) tumor cells (among 245 screened extracts) (Nakano et al. 2013). Liriodenine induced apoptosis presumably by arresting G1/S cell cycle via nitric oxide- and p53-mediated pathways. Isoquinoline alkaloid extract from aerial parts of *A. squamosa* has high anticancer activity (0.01–100 µg/mL concentration range) against colon cancer cells (HCT116) and human breast cancer cells (MCF-7) (Regassa et al. 2022). Moreover, two benzylisoquinoline alkaloids [{(6, 7-dimethoxy-1-(α-hydroxy-4-methoxybenzyl)-2-methyl-1} and 2, 3, 4-tetrahydroisoquinolin (1) or 1, 2, 3, 4-tetrahydroisoquinoline] and coclaurine exert antitumor activity against colon cancer (HCT116) cells and human breast cancer cell lines (MCF-7) (Dev and Joseph 2021). The antitumor activity of these *A. squamosa* alkaloids against human breast cancer cells (MCF-7) indicates some similarity with the well-known vinca alkaloids. Alkaloids (aporphine, corydine, norcodeine, norlaureline, and sanjoinine) extracted from *A. squamosa* leaves exhibit antibacterial activity against *S. aureus, E. coli, B. cereus, P. aeruginosa, H. pylori*, and MRSA, and antioxidant activities (DPPH radical scavenging and superoxide dismutase activities). The antibacterial activity was due to changes in *B. cereus* cell wall morphology (swelling and rupture), cells lysis, and apoptosis (Shami 2017).

Higenamine is on the World Anti-Doping Agency (WADA) prohibited list as a banned substance for athletes (under S3 group beta-2 agonists) starting January 1, 2017. It is generally used in dietary supplements (often unlabeled) for weight loss. One such supplement contained 18.93 ng/mg higenamine (Stajić et al. 2017). In China, higenamine is often used to treat respiratory complications and coughs, and to enhance cardiovascular health because it activates the cardiac β-adrenoceptor, causes vasodilation, and prevents platelet aggregation by α-adrenoceptor interaction (Zhao et al. 2022). It is an adrenergic receptor agonist with potential benefits for athletes since it promotes fat breakdown and thermogenesis, muscle mass growth, and increases energy and concentration. However, it also increases heart rate and blood pressure and may be unsafe. In humans, higenamine is methylated, sulfated, and glucuronidated to generate 32 metabolites detectable in urine 6 hours after oral administration (Zhao et al. 2022).

A. squamosa bark was used to isolate three recognized aporphine alkaloids, namely N-nitrosoxylopine, duguevalline, and roemerolidine. With no discernible cytotoxicity, duguevalline and roemerolidine exhibited moderate antiplasmodial activity, except for N-nitrosoxylopine, which showed antiplasmodial activity with cytotoxicity. The antimalarial action of substances isolated from *A. squamosa* has sufficient potential to be considered for further in-depth mechanistic studies to gain insights into the mechanism of action and resistance in order to develop new therapeutic agents (Johns et al. 2011). Liriodenine (AAR-01), norushinsunine (AAR-02), and reticuline (AAR-03), three aporphine alkaloids, were extracted from the roots of *Annona reticulata*. ^{1}H NMR and ^{13}C NMR methods and mass spectrum analysis were used to determine the compounds' structures. Cytotoxicity of all extracted compounds was assessed by the MTT method against the cancer A-549, K-562, HeLa, and MDA-MB cell lines as well as healthy cell lines (Vero cells). All isolated compounds (at 5, 10, and 20 g/mL dilutions) had substantial dose-dependent cytotoxic activity against all cancer cell lines, whereas AAR-02 demonstrated potent cytotoxicity against carcinoma cell lines with (7.4–8.8 g/mL) IC_{50} values. Additionally, none of the separated chemicals had as strong of an impact on Vero cell lines as they did on cancer cell lines. The isoquinoline moiety in the structures of the three aporphine alkaloids liriodenine (AAR-01), norushinsunine (AAR-02), and reticuline (AAR-03) may be the reason for their notable cytotoxicity (Suresh et al. 2012).

7.4.2 Essential Oils

The primary essential oil constituent in the Annonaceae (Table 7.3) is monoterpene hydrocarbons in the fruit and seed, sesquiterpene hydrocarbons in the leaf, and oxygenated sesquiterpenes in the bark and root (Fournier et al. 1999). The most commonly identified compounds are the monoterpenes α-pinene, limonene, β-pinene, and *p*-cymene, and the sesquiterpenes β-caryophyllene and caryophyllene oxide. One *A. squamosa* fruit study identified 46 compounds (83.3% of essential oil), primarily monoterpenes and sesquiterpenes (14.1% and 51.4%).

The major compounds (in decreasing proportional order >2%) were spathulenol, germacrene D, elemene, β-pinene, methyl isoeugenol, α-terpinene, bornyl acetate, *p*-cymene, and γ-cadinene. A second *A. squamosa* fruit study identified 35 compounds containing the same amount of monoterpene but a higher sesquiterpene level (14.1% and 74.5%). The major compounds were τ-cardinol, τ-muurolol, spathulenol, α-copaene, α-terpineol, globulol, farnesol, and viridiflorol (Fournier et al. 1999). These investigators also reported two essential oil studies on *A. squamosa* peel, the first one containing 0.1% oil with 33 identified compounds, mainly α-pinene, β-pinene, limonene, β-farnesene, (E)-β-ocimene, and bornyl acetate. Forty-six compounds were identified in the second *A. squamosa* peel, accounting for 82.2% of identified compounds with monoterpenes and sesquiterpenes (33.5% and 40.5%). The major compounds (in decreasing proportional order >2%) were spathulenol, bornyl acetate, myrtenol, borneol, pinocarveol, and α-campholenic aldehyde.

Wong and Khoo (1993) identified 47 volatile compounds (92.3% of GC area) in *A. squamosa* (custard apple) fruit from Malaysia (9 mg/kg fruit pulp), consisting primarily of terpenoids (98.3% of fruit volatiles). Monoterpenoids accounted for 91.2% of the total volatiles, abundant in terpinen-4-ol (70.5%) and α-terpineol (10%). The most abundant sesquiterpenes (65.6%) were *epi*-α-cardinol (37%), *t*-muurolol (12.4%), spathulenol (6.8%), α-copaene (5.2%), globulol (4%), farnesol (3.5%), and viridiflorol (2.1%). The 13 identified monoterpenoids were most abundant in α-terpineol (5.1%), bornyl acetate (1.8%), 1,8-cineole (1.4%), and β-pinene (1.3%). Pulp from *A. squamosa* fruits grown in the Brazilian Amazon comprised 27 mono- and sesquiterpenes, comprising 99.1% of the oil. The most abundant monoterpenes (75.5% of the oil) were α-pinene (25.3%), sabinene (22.7%), limonene (10.1%), (*E*)-β-ocimene (7.2%), and the sesquiterpenes spathulenol (6.3%), germacrene D (6%), myrecene (4.2%), and bicyclogermacrene (3.5%), and α-cardinol (2.3%) (Andrade et al. 2001).

The essential oil content (109 mg/kg fruit pulp) of a Cuban *A. squamosa* (custard apple) cultivar consisted of 49 volatile compounds, primarily mono- and sesquiterpenoids (51.7% and 19.9%, respectively)

TABLE 7.3 Phytochemicals from various part of *A. squamosa* plant

PARTS	*PHYTOCHEMICALS*	*REFERENCES*
Leaves	Quercetin 3-*O* glucoside	Panda and Kar (2007)
	Sesquiterpenic hydrocarbons (trans-caryophyllene, germacrene D, bicyclogermacrene, humulene), phytol and squalene	Al-Nemari et al. (2020)
	Phytosterols, glycosides, phenols, saponins, alkaloids, and flavonoids	Kumar et al. (2021)
	Cytohexapeptides, acetogenins, annonaceous acetogenins	Gajalakshmi et al. (2012)
	Borneol, linalool, eugenol, farnesol, and geraniol	Patel and Kumar (2008)
	Alkaloid, isomeric hydroxyl ketones	Singh et al. (2019)
Stems	Annomosin A, annosquamosin C, D, E, F, and G	Yang et al. (2002)
	Annosqualine, dihydrosinapoyltyramine	Yang et al. (2004)
	Camphene, eugenol, camphor, car-3-ene, carvone, caryphyllene, farnesol, geraniol, 16-hetriacontanone, borneol	Bhattacharya and Chakraverty (2016)
Roots	Alkaloids-liriodenine, oxoanalobine	González-Esquinca et al. (2004)
	Hexacontanol, limonine, higemamine, and isocorydine	Bhattacharya and Chakraverty (2016)
Bark	1H-Cycloprop(e)azulene, kaur-16-ene, caryophyl-lene oxide, bisabolene epoxide, germacrene D, and bisabolene	Chavan et al. (2006)
	Squamone and acetogenins	Singh et al. (2019)
Seeds	Squamocin, cholesteryl glucopyranoside, annotemoyin-1, and 2	Rahman et al. (2005)
	Liriodenine (AS-1), moupinamide (AS-2), squamosamide (AS-8), daucosterol (AS-12)	Yang et al. (1992)
	Squamostanal-A,	Araya et al. (1994)
	Annosquamosin A	Tan and Zhou (2006)
	Samaquasine, acetogenin, annonacin and annonastatin	Singh et al. (2019)
	3′- Methoxynobiletin, p-anisaldehyde, 40-hydroxy-3,4,5-trimethoxystilbene, schisanhenol, schisandrin	Du et al. (2021)
Pulp	Dasycarpidan-1-methanol	Karthikeyan et al. (2016)
	3′-*O*-Methylviolanone, mMatairesinol, isopimpinellin, 2-methoxy-5-prop-1-enylphenol, 4-hydroxybenzaldehyde, 30-hydroxy-3,4,5,40-tetramethoxystilbene	Du et al. (2021)

(Pino 2000). Other compound classes were diterpenes (5.9%), oxygenated monoterpenoids (4.9%), acids (1.7%), oxygenated sesquiterpenoids (1.4%), and aliphatic esters (0.2%). Pino et al. (2003) identified 180 flavor compounds from fruit pulps of four commercial Cuban custard apple cultivars. The major constituents (mg/kg) were: α-pinene (50–100), β-pinene (71–97), myrcene (12–20), limonene (14–20), terpinen-4-ol (1–22), and germacrene D (10–27). Fruits with the highest terpenoid content had the highest custard-like and overall fruity aroma intensity. The presence of many terpenoids contributes to the unique custard apple flavor.

The essential oil of *A. squamosa* pericarp is light yellow (0.26% w/w, yield) and consists of 59 compounds, with (-)-spathulenol as the most abundant sesquiterpene, accounting for almost a third (32.5%) of the oil (Chen et al. 2017). The oil exhibits hepatoprotective effects (anti-hepatoma activity) and antitumor activity against human SMCC-7721 hepatoma cells due to increased cell apoptosis and pronounced cell cycle arrest. Twenty-two compounds (96.3% of identified volatiles) have been identified in *A. squamosa* root, consisting of 20.6% monoterpenes and 75.7% sesquiterpenes. The major compounds were (in decreasing proportional order >2%) β-caryophyllene, α-pinene, α-humulene, α-gurjunene, limonene, γ-elemene, caryophyllene oxide, camphene, β-eudesmol, and α-eudesmol (Fournier et al. 1999).

Research in India on *A. squamosa* essential oil has focused mainly on leaves, highlighting the regional differences. The essential oil (0.08% yield) of *A. squamosa* green leaves (collected locally in Bangalore, India) consists of β-caryophyllene (50%), α-pinene (7.3%), an unidentified monocyclic terpene (4.4%), and two bicyclic sesquiterpene hydrocarbons (5% each of the total oil) (Rai and Muthana 1954). *A. squamosa* leaves from Trivandrum, South India, yielded 0.14% essential oil on hydrodistillation containing 28 identified compounds. The major components (%) were: β-cedrene (23.3), β-caryophyllene (14.1), (E,E)-farnesol (7), cadina-1,4-diene (6.9), allo-aromadendrene (5.5), calamenene (5.1), β-elemene (4.5), α-bisabolene (3.2), α-humulene (3.1), β-bisabolene (2.4), and methyl eugenol (2.3) (Joy and Rao 1997). The essential oil composition of *A. squamosa* leaves from the North Indian plains (Lucknow) consisted of 18 compounds accounting for 86% of the oil. Sesquiterpene hydrocarbon was the major constituent (76%), along with oxygenated sesquiterpenes (7.1%) and monoterpene hydrocarbons (2.5%). The major oil constituents (%) were: β-caryophyllene (22.9), germacrene D (21.3), bicyclogermacrene (8.5), β-elemene (7.8), γ-cadinene (6.7), and α-muurolol (5.7) (Garg and Gupta 2005). The leaf essential oil of *A. squamosa* from the lower Himalaya region consists of 43 constituents, accounting for 88.6% of the total identified oil composition. The oil is classified into sesquiterpene hydrocarbons (63.4% of total identified), oxygenated sesquiterpenes (21.8%), monoterpene hydrocarbons (2%), and oxygenated monoterpenes (1.4%). Major constituents (%) of the oil were: (E)-caryophyllene (15.9), γ-cadinene (11.2), epi-α-cadinol (9.4), (Z)-caryophyllene (7.3), γ-muurolene (5.4), α-humulene (5.2), viridiflorene (5), α-cadinol (3.9), aromadendrene (2.9), δ-cadinene (2.9), α-cadinene (2.9), (2Z, 6Z)-farnesal (2.2), and caryophyllene oxide (2.1) (Verma et al. 2016). A total of 41 constituents have been identified from leaf essential oils of three *Annona* species (*A. cherimola, A. muricata, and A. squamosa*) grown in Kerala, India (Joseph and Dev 2021). The species differed considerably in composition; for example, mono- and sesquiterpenoids were almost equally distributed in *A. muricata*, whereas in *A. cherimola and A. squamosa,* sesquiterpene hydrocarbons dominated over sesquiterpenoids. *A. squamosa* leaf essential oil consists of 43.7% sesquiterpene hydrocarbons, 17.3% oxygenated sesquiterpenes, 15.9% monoterpene hydrocarbons, and 3.6% diterpenoids. Major identified (80.5% of total essential oil) compounds in *A. squamosa* were β-caryophyllene (11.9%), α-pinene (8.2%), β-pinene (6.5%), epi-α-cadinol (7.3%), δ-amorphene (4.9%), α-cadinol (3.6%), (E,E)-geranyl linalool (3.6%), and Z-caryophyllene (3.1%) (Joseph and Dev 2021).

Meira et al. (2015) identified 23 compounds in *A. squamosa* leaves from southeast Brazil with high abundance (95.5%) of sesquiterpenes: 27.4% (E)-caryophyllene, 17.1% germacrene D, 10.8% bicyclogermacrene, 7.3% (Z)-caryophyllene, 6.2% β-elemene, 5.7% α-humulene, 4.3% epi-α-cadinol, 4.2% γ-cadinene, and 4.1% δ-elemene. The essential oils exert potent trypanocidal and antimalarial activities, inducing ultrastructural changes in cell membranes and mitochondria, resulting in necrotic parasite (*Trypanosoma cruzi*) death.

According to phytochemical investigations, borneol, linalool, eugenol, geraniol, and farnesol are all present in the extract. These substances may act singly or collectively to exert antibacterial action. For the experiment, two Gram-positive and two Gram-negative microorganisms were chosen. The essential oil produced by steam distillation was used to identify the volatile components from *Annona squamosa* bark. The six primary components (%) were: 1H-cycloprop(e)azulene (3.46), caryophyllene oxide (29.38), germacrene D (11.44), bisabolene epoxide (3.64), bisabolene (4.48), and kaur-16-ene (19.13). The oil's antibacterial potential was also examined, showing notable antimicrobial action against *B. subtilis* and *S. aureus* (Chavan et al. 2006). The essential oil derived from the hydrodistillation of *A. squamosa* leaves collected from Lucknow, India, included δ-cadinene, limonene, terpinolene, α-muurolol, (Z)-nerolidol bicyclogermacrene, β-elemene, β-bourbonene, γ-cadinene, and α-copaene (Kumar et al. 2021). Supercritical CO_2 was used to extract both volatile and non-volatile oils from custard apple seed flour. The following parameters were maintained during the process: 25 MPa pressure, 318 K temperature, and 2.5 mL/min flow rate for non-volatile compound extraction for 60 min, producing a maximum 0.288 g/g yield. The ideal extraction parameters for volatile oils were 15 MPa pressure, 308 K temperature, and 1.5 mL/min flow rate for an extraction time of 60 minutes (Panadare et al. 2021).

Volatile organic compounds rich in terpenes such as pinene, linalool, limonene, and caryophyllene (present in *A. squamosa*) can benefit well-being and mental health, including neurological and neuropsychiatric disorders (Weston-Green et al. 2021). These terpenes have therapeutic efficacy similar to existing commercial medications for several indications, including analgesics, anti-inflammatories, anti-anxiety, and anti-depressant drugs. β-caryophyllene has potential therapeutic use in insulin resistance, type 2 diabetes, obesity, hyperlipidemia, and diabetic complications because of its demonstrated antioxidant, anti-inflammatory, organoprotective, and antihyperglycemic properties (Hashiesh et al. 2020).

7.4.3 Polyphenolic Compounds

Phenolic compounds have been investigated in *A. squamosa* fruit and other plant parts such as leaves, root, and bark using various extraction solvents and methods. The most recent study provides the full panel of phenolics from *A. squamosa* aqueous methanol peel extract related to antibacterial activity (Tlais et al. 2023). Phenolic content of water-extracted *A. squamosa* fruit parts (peel and pulp [outer, middle, and inner]) did not vary significantly (353–370 mg GAE/L). However, aqueous methanol (70%) extracted a higher amount of phenolics from the peel (1043 mg GAE/L), with no significant difference among the pulp (363–374 mg GAE/L). Flavan-3-ols was the most abundant of the 41 phenolic compounds identified in the aqueous methanol extract from the peel. Procyanidins were present in high concentration (~400 mg/L), consisting primarily of type B isomers (1, 3, and 2), dimer, trimer, and tetramer (57, 52, and 63 mg/L, respectively). Epicatechin (76.4 mg/L) was the major flavan-3-ol derivative, followed by catechin (923 mg/L) and lower amounts of catechin derivatives. The major peel flavonols were rutin and calabricoside (12.6 and 7.14 mg/L), whereas lariciresinol-glucopyranoside (9.4 mg/L) was the most abundant phenolic acid, along with lower amounts of quinic and vanillic acids. Antibacterial activity against *Staphylococcus aureus* correlated positively only with procyanidin type B (isomer 1) and rutin (Tlais et al. 2023).

The skin of the common tropical fruit known as the sugar apple (*Annona squamosa* L.) is discarded in municipal trash. Response surface methodology (RSM) has been used successfully to improve phenolic extraction from food. An ultrasound-assisted extraction technique was developed to extract antioxidant compounds from sugar apple peel. A central composite design was used to adjust solvent percentage (13.2%–46.8%), temperature (43.2–76.8 °C), and ultrasonic time (33.2–66.8 min). These conditions (20:1 mL/g solvent-to-solid ratio, 32.68% acetone, 67.23 °C for 42.54 min under ultrasonic irradiation) are optimal for extracting the maximal yield of total phenolic content (26.81 mg GAE/g fresh weight) (Deng et al. 2015). El-Chaghaby et al. (2014) investigated the effects of various extraction protocols on the antibacterial and antioxidant capacities of *Annona squamosa* L. leaves. Boiling water, 80% methanol, 50% acetone, and 50% ethanol were the four solvents selected for the investigation. Acetone and boiling water produced the highest extraction yields, with acetone being the most and water the least effective solvent for extracting total phenolic. The solvent extracts showed effective antibacterial activity compared to the raw fruit peel extract. Raw leaves contain limited amounts of total phenol content (48.85 and 17.96 mg GAE/g) in methanol and ethanol extracts (Ibrahim et al. 2020). However, dried leaves' ethanolic and methanolic fractions (117.2 and 112.9 mg GAE/g) have the highest phenol content as compared with other parts. The methanol extract of fresh and dried *A. squamosa* pulp contained comparable total phenolic content (546 and 537 µg GAE) (Bhardwaj et al. 2014).

The seed has the highest total phenolic (70–284 µg GAE/g dry sample) and flavonoid (36–112.7 µg QE/g dry sample) contents compared to the peel and pulp in aqueous extract (followed by 30 min heating at 70 °C). Antibacterial activity is highly expressed in the peel extract compared to those from the seed or pulp against *Bacillus subtilis*, *Candida albicans*, *Escherichia coli*, *Klebsiella pneumoniae*, *Salmonella senftenberg*, and *Staphylococcus aureus*. However, the seed extract has the highest anticancer activity compared to those from the peel or pulp. These activities presumably depend on the phenolic components present in *A. squamosa* aqueous seed extract such as ferulic, *o*-coumaric, and *p*-coumaric acids (5.08, 49.02, and 1.96 mg/100 g, respectively), and gallic, *p*-hydroxybenzoic, salicylic, and syringic acids (Shehata et al. 2021). Total phenolic content of *A. squamosa* fruits varies (82–747 mg gallic acid/100 g fresh weight), with the highest concentration in the dehydrated peel compared to that of the pulp or seed (12.1 vs 1.5 vs 0.3 gallic acid/100 g dry weight) (Leite et al. 2020). However, methanol extract of *A. squamosa* seed has higher total phenolics (32.5 vs 2.2 µg GAE/mg extract) and flavonoid contents (893 vs 247 µg QE/g extract) than the pulp (Leite et al. 2020).

Solvent and extraction conditions (parameters) affect total phenolic and flavonoid contents of *A. squamosa* seeds. For example, ethanol extracts higher amounts of flavonoids and phenolics than methanol or water (189 vs 184 vs 85 mg quercetin equivalent/100 g dry weight and 234 vs 232 vs 114 mg gallic acid equivalent/100 g dry weight for flavonoid and phenolics, respectively) (Nguyen et al. 2020). However, Leite et al. (2020) reported considerably higher total phenolic and flavonoid contents (32.53 µg GAE/mg and 893 µg QE/g extract) in *A. squamosa* methanol seed extract. Phenolics and flavonoids are present in the ethanol (ESE) and methanol (MSE) seed extracts compared to other extracts (petroleum ether, acetone) (Alaqeel et al. 2023). The free radical DPPH scavenging activity was 9 and 9.8 µg/mL EC_{50} compared to 11.7 µg/mL for ascorbic acid. The superoxide radical scavenging activities were 84%, 79%, and 90% for ethanol and methanol extracts and ascorbic acid (400 µg/mL), respectively. ESE and MSE have higher reducing power than ascorbic acid. The free radical scavenging activity (hydrogen peroxide method) (400 µg/mL) was 11, 15, and 24 µg/mL (EC_{50}) for ESE, MSE, and ascorbic acid, respectively. Only the methanol extract has higher nitric oxide activity than ascorbic acid (EC_{50} 39.5 vs 42.5 µg/mL) (Alaqeel et al. 2023). ESE (200 µg) was cytotoxic against MCF-7 breast cancer cells (IC_{50} 15.5 vs 24.8 µg for coumarin; selectivity index [SI] 12.8 vs 7.9 for coumarin) and more effective (7.5 SI) against HepG2 cancer cells. Selectivity index (>3) suppresses cancer cells; however, SI value of 3 is regarded as hazardous to normal cells (Alaqeel et al. 2023).

Five solvents (hexane, acetone, chloroform-methanol, ethanol, and water) were used to extract total phenolic and flavonoid from *A. squamosa* seeds (Kothari and Seshadri 2010). Total phenolic contents of microwave-assisted extracts were (in descending order): 243, 209, 172, 30, and 24.5 g GAE/g for chloroform-methanol, water, ethanol, acetone, and hexane extracts, respectively. Total flavonoid contents for the extracts were: 42.4, 32.7, 23.2, 9.9, and 5.7 mg QE/g dry extract for ethanol, acetone, chloroform-methanol, hexane, and water extracts, respectively. Interestingly, the water and ethanol extracts exert the highest antioxidant activities (778 and 427 g GAE/g dry extract for total antioxidant activity, and 3202 and 1926 g ascorbic acid equivalent antioxidant activity (AEAC)/100 g of dry extract for DPPH radical scavenging capacity) (Kothari and Seshadri 2010). Petroleum ether seed extract contains total phenolics, flavonoid, and flavonols (251 µg GAE, 339 µg CE, 606 µg/100 mg extract, respectively) (Bhardwaj et al. 2014).

Solvents such as acetone, methanol, water, and ethanol extract different amounts of phenolics from *A. squamosa* pulp (1.43, 1.02, 0.96, and 0.86 mg/g fresh weight, respectively). The methanol pulp extract contains gallic, caffeic, *p*-coumaric, and ferulic acids, vanillin, and catechol (Jagtap and Bapat 2012). The methanol pulp extract exerts good antioxidant activity because of its phenolic compounds: 537 µg GAE/100 mg total phenolic, 205 µg CE/100 mg total flavonoid, and 350 µg/100 mg flavonol (dry weight)

(Bhardwaj et al. 2014). Baskaran et al. (2016) identified 16, 15, and 13 free, bound, and esterified phenolic compounds, respectively, in *A. squamosa* ethanol (70%) fruit pulp extract. The major polyphenols were catechin, epicatechin, epigallocatechin gallate, caffeic, *p*-coumaric, ferulic, gallic, and sinapic acids. The free phenolic compounds consist of procyanidin B2, procyanidin trimer, citric, hydroxyadipic, malic, quinic, trans-aconitic, and 4-(β-D-glucopyranosyloxy) benzoic acids. The bound phenolics comprised 7-hydroxycoumarin 7-glucoside (skimmin), dihydroxyquercetin, ptocyanidin B2, xanthotoxol acetate, citric, malic, and *p*-decycloxybenzoic acids. Phenolic compounds identified from soluble esters (esterified phenolics) were gallic, protocatechuic, caffeic, p-coumaric, and sinapic acids as well as the flavonols catechin and epicatechin.

The phenolic content of *A. squamosa* peel is significantly influenced by the extracting solvent (Table 7.4). Acetone is the best solvent to extract total phenolic compounds, followed by methanol, water, and ethanol (1.43, 1.02, 0.96, and 0.86 mg GAE/g fresh weight, respectively). The acetone extract exhibits high antioxidant activity (96% DPPH), but the lowest *N*, *N*-dimethyl-*p*-phenylendiamide (DMPD) scavenging activity (12% vs 35% for other solvents) (Jagtap and Bapat 2012). Ethanol (95%) extract of freeze-dried peel (~⅓ of the fruit) for ten *A. squamosa* cultivars grown in Thailand varied in total phenolic (34–140 mg GAE/g dry sample), catechin (678–1805 µg catechin/g dry sample), and gallic acid (210–1980 µg GA/g dry sample) contents (Manochai et al. 2018). The peel also varied in content of catechins (µg/g): catechin (193–491), epicatechin [EC] (182–713), epigallocatechin [EGC] (33–840), epigallocatechin gallate [EGCG] (0.4–32), and epicatechin gallate [ECG] (from undetected to 327). Generally, the individual catechins followed the same trend, except for two cultivars: epigallocatechin > epicatechin > epicatechin gallate > epigallocatechin gallate (Manochai et al. 2018).

The total phenolics (243 mg GAE/g) and flavonoid (83 mg QE/g) contents of *A. squamosa* leaves (96% ethanol extract) correlate with its antioxidant activity (133 and 65 µg/mL IC_{50} for DPPH and ABTS scavenging activities) (Nguyen et al. 2020). The methanol (80%) extracts from *A. squamosa* root and bark contain higher total phenolic contents than dried leaves (~170 vs 97 g GAE/kg). However, the dried leaves have higher flavonoid content than the root and bark (223 vs ~105 mg rutin equivalent (RE)/g extract) (Mariod et al. 2012). However, HPLC analysis showed higher total phenolic content in the leaves compared to the bark and root (0.47 vs 0.30 vs 0.04 mg/g dry weight); these values are lower than those obtained by the Folin-Ciocalteu method (94 vs 168 vs 172 mg/g dry weight) (Mariod et al. 2012). The leaves contained chlorogenic, hydroxybenzoic, gallic, ferulic, and p-coumaric acids, and vanillin (3.51, 0.58, 0.25, 0.14, 0.011, and 0.26 µg/g dry weight, respectively). Predominant phenolic acid constituents in the bark were gallic, chlorogenic, and p-coumaric, and vanillin (0.35, 2.14, 0.28, 0.14, and 2.8 µg/g dry weight, respectively). Moreover, the bark and the leaves have the same amount of syringic acid (0.12 µg/g dry weight). The leaves exert lower antioxidant activity (ORAC values) than the root or bark (30 vs 65 vs 85 µM of Trolox) (Mariod et al. 2012).

7.4.4 Flavonoids

Another significant class of phenolic chemicals is flavonoids, which are crucial to plant biological processes. Anthocyanins, dihydroflavonols, dihydrochalcones, isoflavonoids, flavanones, flavones, flavonols, and flavanols were among the subdivisions of flavonoids that made up about 50% of the total chemicals identified and characterized using MS spectrum and literature study. Custard apple peel had TPC and TFC (61.7 mg GAE/g and 0.42 mg QE/g, respectively) based on LC-ESI-QTOF-MS/MS assessment with 42 identified flavonoids in this experiment (Du et al. 2021). Most of the compounds, including glycosylated flavonoids like calabricoside A (flavan-3-ol), poncirin (flavanone), and miconioside A have already been identified. Methanol extract of raw and dried *A. squamosa* pulp contained 88 and 205 µg CE/100 mg of total flavonoid and flavonol (Bhardwaj et al. 2014).

The total phenolics (243 mg GAE/g) and flavonoid (83 mg QE/g) contents of *A. squamosa* leaves (96% ethanol extract) correlate with its antioxidant activity (133 and 65 µg/mL IC_{50} for DPPH and ABTS scavenging activities) (Nguyen et al. 2020). The methanol (80%) extracts from *A. squamosa* root and bark

TABLE 7.4 Phytochemicals in different *A. squamosa* parts

	COMPOUNDS	*EXTRACTANT*	*DETECTION*	*REFERENCES*
Peel				
Phenolic compounds	Saponins, phenols, terpenoids	Water	UV spectrophotometer	Lydia et al. (2017)
	Catechin, epicatechin, and chlorogenic acid	Methanol	HPLC	Can-Cauich et al. (2017)
Sugar	Lactose, sucrose, galactose, glucose	Water	HPLC-MS	Chandraju et al. (2012)
Bioactive compound	Dasycarpidan-1-methanol, acetate(ester), 6-Octadecenoic acid, (Z). Eicosanoic acid	Distilled water	GC-MS	Karthikeyan et al. (2016)
Phytochemical compounds	Quinic and citric acids, calabricoside A, quercetin derivatives, rutin, catechin	Ethanol/water	HPLC-ESI-q-TOF-MS	Rojas-García et al. (2022)
Pulp				
Phenolic compounds	Dasycarpidan-1-methanol, isomenthone, methyl 2-cholorohexadecanoate	Distilled water	GC-MS	Karthikeyan et al. (2016)
	Rosmanol	Ethanol	LC-ESI-QTOF-MS/MS, HPLC-PDA	Du et al. (2021)
Flavonoids	Myricetin 3-*O*-glucoside, procyanidin trimer C1			
Volatile composition	Linalool, terpenoid compounds, α- and β-pinene, butylbutonate, spathulenol	1:1Dichloromethane: n-pentane	GC-MS	Shashirekha et al. (2008)
	(E)-β-ocimene, α-cubebene, D-germacrene, α-phellandrene, bornyl acetate, δ-cadinene	HS-SPME	GC-MS	de Sousa Galvão et al. (2020)
Seed				
Phenolic compounds	Carnosic acid	Ethanol	LC-ESI-QTOF-MS/MS, HPLC-PDA	Du et al. (2021)
	Acids (gallic, syringic, β hydroxyl benzoic, ellagic, ferulic, salicylic, benzoic)	Water	RP-HPLC	Shehata et al. (2021)
Flavonoids	Cyanidins, Quercetin	Ethanol	LC-ESI-QTOF-MS/MS, HPLC-PDA	Du et al. (2021)
Phytochemical compounds	Quinic acid, methyl-kaempferol-hexoside, quercetin rutinoside, poncirin	Ethanol/water	HPLC-ESI-q-TOF-MS	Rojas-García et al. (2022)

contain higher total phenolic contents than dried leaves (~170% vs 97% g GAE/kg). However, the dried leaves have higher flavonoid content than the root and bark (223% vs ~105% mg rutin equivalent [RE]/g extract) (Mariod et al. 2012).

7.4.5 Acetogenins, Chemical Characteristics, Bioactivity, and Therapeutic Uses

Annona cherimola Mill., or the custard apple, a species of the Annonaceae family, is widely used in traditional medicine as a valuable source of bioactive compounds. A unique class of secondary metabolites derived from this family are Annonaceous acetogenins, lipophilic polyketides considered to be among the most potent antitumor compounds. This review provides an overview of the chemical diversity, isolation procedures, bioactivity, modes of application and synthetic derivatives of acetogenins from *A. cherimola* Mill (Durán et al. 2021). Acetogenins (ACGs) are a group of naturally occurring compounds that have been isolated only from plants of the Annonaceae family. Alkaloids are the second most important chemical compounds found in the Annonaceae family. Squamostanin-A and Squamostanin-B, two novel Annonaceous acetogenins, were discovered in a 95% ethanol extract of *Annona squamosa* seeds. These two ACGs exert different anticancer effects by lowering the ATP level via suppressing the NADH oxidase of tumor cells' plasma membrane and complex I of mitochondria (Yang et al. 2009). Numerous human diseases, including cancer, diabetes, and neurological disorders, are known to be triggered by oxidative stress brought on by reactive oxygen intermediates, such as superoxide and hydrogen peroxide, which result in apoptotic cell death. A new class of substances known as Annonaceous acetogenins isolated from the defatted seeds of *Annona squamosa* exhibit strong antimicrobial, antiparasitic, pesticidal, and cell growth inhibiting properties (Pardhasaradhi et al. 2005). *A. squamosa* ACGs have C32 and C34 branched-chain fatty acids with a γ-lactone end and exert various health benefits including anticancer, antimicrobial, antitumor, antioxidant, antiparasitic, and pesticidal effects. Ten Annonaceous acetogenins (AAs), including squamocins-G, H, J, K, L, and M, squamostatin-A, squamocin, annotemoyin-1, and solamin, were isolated from the seed. Chemical characterization of each compound was done by nuclear magnetic resonance (NMR). Squamocin-H and G are also known as asimicin and bullatacin, respectively. Squamocin, squamocin-G, and squamostatin-A were prevalent out of the ten AAs extracted from the seed in this experiment, whereas the remaining compounds were present in small amounts (Dang et al. 2011). Six novel Annonaceous acetogenins, including annosquacins A–D, annosquatin A, and annosquatin B, were discovered in the ethanol fraction of custard apple seed. The Mosher analysis was used to define the precise configuration of the six compounds. Compound 1 was isolated as a white powder and named Annosquacin A($C_{35}H_{62}O_6$). Compounds 1–4, composed of an adjacent bistetrahydrofuran (bis-THF), were identified by NMR (Chen et al. 2012). Other varieties of *Annona* species also showed numerous amounts of acetogenins. The leaves of *Annona montana* were used to extract montanacin F, a unique Annonaceous acetogenin with a terminal lactone unit that showed cytotoxic activity (Wang et al. 2002).

Annomolin and annocherimolin are two novel acetogenins extracted from *Annona cherimola* seeds. Annomolin is composed of a mono-THF with two flanking hydroxyls and has 1,2-diol at C-7/8 of the aliphatic chain. Annocherimolin differs from annomolin by a double bond at C-21/22. These two compounds demonstrated different cytotoxic potencies in different cancer cell lines (Kim et al. 2001). García-Aguirre et al. (2008) investigated the cytotoxic potential of three isomeric acetogenins (Ace) isolated from *Annona cherimola* seeds in mice. The ^{1}H-NMR spectrum identified an Ace ending with γ-lactone, THF ring, hydroxyl group, and connected along with an aliphatic chain. HPLC analysis revealed the existence of three Ace in an 8:1:1 ratio. Acetogenins obtained from *Annona cornifolia* ethanol seed extract demonstrated potent DPPH radical scavenging activity. Ten acetogenins were recovered in their pure form. Circular dichroism was used to determine the absolute configuration of two acetogenins at the C-36 position. Acetogenins 4-desoxylongimicin B (3) and folianin A (2), both are novel compounds, although squamocins M (4) and L (5), bulatacin (9), and asimicin (10) have already been identified in *Annona squamosa* (Lima et al. 2010).

7.5 BENEFICIAL HEALTH POTENTIAL

Compared to expensive, pharmaceutical medications, green medicine is reliable and safe. Screening medicinal plants for possible biological activity is thus necessary. Numerous researches have indicated that various parts of *A. squamosa* have therapeutic potential (Tables 7.5 and 7.6).

7.5.1 Antidiabetic Properties

A. squamosa is the most studied *Annona* species for antidiabetic activity. The leaves have traditionally been used for diabetes, ulcer, hypertension, wounds, hemorrhage, and epilepsy (Ansari et al. 2022a). Agarwal et al. (2014) evaluated the antidiabetic effect of *A. squamosa* leaves extract on 24 diabetic patients (18–65 years) in a randomized, open-label controlled study. Patients enrolled for 3 months after a 1-month washout were randomized into study and control groups and evaluated every 4 weeks for clinical and biochemical parameters (body mass index, waist–hip ratio, blood pressure, fasting blood glucose, HbA1C, lipid profile, and insulin level). *A. squamosa* leaves extract (10 g) administered orally twice daily before meals significantly ($P < 0.001$) improved body mass index, fasting blood glucose level, HbA1C, triglyceride, and HDL. Changes were less significant ($P < 0.01$) in total cholesterol, LDL, and insulin levels. These hypoglycemic and antidiabetic effects of *A. squamosa* leaves extract are presumably due to enhanced insulin level from pancreatic islets, increased glucose utilization in muscles, and blood sugar maintenance (Agarwal et al. 2014). This is the only study to date on *A. squamosa* antidiabetic effect in humans, although the leaf constituent(s) responsible for this effect were not identified. However, Panda and Kar (2007) demonstrated that the antidiabetic effect of *A. squamosa* leaves was mediated by the flavonoid quercetin-3-*O*-glucoside. Isoquercetin or quercetin-3-*O*-glucoside obtained from *A. squamosa* methanol (80%) leaves extract ameliorated diabetes mellitus by reducing serum glucose concentration and increasing insulin level in alloxan-treated Wistar male rats and was marginally better than the reference drug glibenclamide. Moreover, the extracted quercetin-3-*O*-glucoside (15 mg/kg/day for 10 days) inhibited oxidative damage in renal and hepatic tissues by attenuating lipid peroxidation (LPO) and concomitantly increasing superoxide dismutase (SOD) and glutathione (GSH) contents in rats (Panda and Kar 2007).

Aqueous and ethanol *A. squamosa* leaves extracts also induce similar antidiabetic effects in streptozotocin-induced diabetic Wistar rats and alloxan-induced diabetic albino rabbits (Gupta et al. 2005a) without identifying the constituent(s) responsible for such effects. The aqueous leaves extract enhanced insulin levels from pancreatic islets, increased muscle glucose utilization, inhibited hepatic glucose output, and reversed abnormal lipid profile in diabetic animals. Administration of *A. squamosa* fruit pulp extract (2.5 and 5 g/kg body weight [bw]) improved glucose tolerance, urine sugar and protein, and eliminated glycohemoglobin in diabetic rabbits. The ethanol extract reduced fasting blood glucose (53%), urine sugar (75%), total cholesterol (49%), low-density lipoprotein (72%), and triglycerides levels (29%) and increased high-density lipoprotein (30%). These effects are also useful in controlling metabolic syndrome (Gupta et al. 2005b).

A. squamosa aqueous leaves extract (350 mg/kg) containing rutin and quercetin controlled elevated blood glucose levels (hyperglycemia) in high fat diet and streptozocin-induced type-2 diabetes in male Sprague Dawley rats (Kaur et al. 2013). This hypoglycemic action of the extract was probably mediated by pancreatic insulin secretion from β-cell islet or enhanced blood glucose transport to the peripheral tissues. Furthermore, the extract reduces Glipizide (an insulin secretion enhancing and utilization drug) dosage (25–50%) and improves diabetic complications from lipid peroxidation and antioxidant systems in experimental type-2 diabetic rats (Kaur et al. 2013). Hexane leaves extract and its active constituent quercetin-3-*O*-glucoside exert antidiabetic activity by modulating insulin signaling through protein-tyrosine-phosphatase 1B (PTP1B) inhibition. The antidiabetic effects were attributed to activation of insulin secretion and/or free radical scavenging activity of its active compounds. *A. squamosa* hexane extract (200 and 400 mg/kg body weight) increased insulin level comparable to glimepiride (1 mg/kg) in streptozotocin-induced diabetic rats. Moreover, the extract inhibited α-glucosidase activity comparable to acarbose (10 mg/kg). This hypoglycemic effect is presumably due to insulin secretion and inhibition of glucose-metabolizing enzymes (Quílez et al. 2018).

A. squamosa ethanol (70%) leaves extract (200 μL at 100 mg/kg bw) promotes wound healing in streptozocin-induced male diabetes Wistar rats (Ponrasu and Suguna 2012). The extract enhances new collagen production and helps its crosslinking by increasing/restoring hexosamine and uronic acid in the collagen matrix, stimulates cell proliferation, and protein synthesis. Topical administration of *A. squamosa* ethanol extract promotes various stages of wound healing, such as fibroplasia, collagen synthesis, wound contraction, and epithelialization. The high amount of phenolics (264 mg GAE/g) is attributed to the increased wound healing effects of *A. squamosa* leaves extract because of its free scavenging activity (Ponrasu and Suguna 2012). The antidiabetic effect of ethanol (80%, v/v) leaves extract (EEAS) is associated with its phenolic compounds (rutin and procyanidins) and acetogenin squafosacin G (Ansari et al. 2022b). The extract dose-dependently increases insulin secretion *in vitro*, presumably mediated via K_{ATP} and Ca^{2+} ion channel-independent pathways, inhibits glycation, glucose absorption, and dipeptidyl peptidase IV (DPP-IV) enzyme activity, and enhances glucose uptake and pancreatic insulin secretion in clonal BRIN BD11 β-cells and mouse islet cells. Moreover, the extract attenuated sucrose and glucose absorption and serum glucose levels in non-diabetic rats, whereas it reduced plasma DPP-IV levels in high fat-fed (HFF) obese Sprague Dawley male rats. EEAS insulin glycation reduction *in vitro* indicates its potential against insulin resistance and diabetes-associated complications (Ansari et al. 2022b). The antidiabetic effects of *A. squamosa* can also be attributed to the presence of β-caryophellene (a peroxisome proliferator-activated receptor [PPAR-γ] agonist) in leaves extracts because of its antihyperglycemic properties, insulin resistance, and hypolipidemic activity in streptozotocin-induced type-2 diabetic rats and its role in diabetic nephropathy and wound healing (Hashiesh et al. 2020).

7.5.2 Anticancer Activity

A. squamosa leaf extracts (100 μg/mL; methanol, acetone or water) strongly induced apoptosis in MCF-7 and MDA-MB-231 breast cancer cells (Al-Nemari et al. 2022). Organic extracts (methanol and acetone) induced 100% MCF-7 cell death compared to 60% cell death for aqueous extract. However, the aqueous extract was most potent in inducing apoptosis in MCF-7 cells (60 and 32% necrotic and apoptotic cells) due to Bcl-2 suppression and Bax activation. Aqueous extract (300 mg/kg directly injected in the tumor) reduced tumor size (6-fold) in dimethylbenzanthracene (DMBA)-induced mammary tumor in female Wistar albino rats due to p53 suppression, proliferation, and increased apoptosis in tumor tissues. These effects were attributed to acetogenins and/or other bioactive constituents (Al-Nemari et al. 2022).

Alkaloids from aerial parts of *A. squamosa* have anticancer activity (0.01–100 μg/mL) on liver, breast, and colon cancer cell lines. For example, coclaurin is cytotoxic against human breast cancer (MCF-7), colon cancer (HT116), and liver cancer (HEPG-2) cells (IC_{50} values of 15.3, 8.2, and 1.7 μg/mL, respectively) (Regassa et al. 2022). Bullatacin isolated from *A. cherimola* inhibits tumor growth (by ~65% at 15 μg/kg dose) in mice bearing HepS (hepatocellular carcinoma) and S180 sarcoma xenografts. Oral annonacin (10 mg/kg) administration significantly reduces lung cancer (by 58%) in hybrid mice (BDF-1) models (Al Kazman et al. 2022). Anonaine, a benzylisoquinoline alkaloid from *A. squamosa* leaves, has anticancer activity presumably mediated by apoptosis-related protein activation, DNA damage, and human lung and cervical cancer cell growth inhibition (Dash et al. 2020). *A. squamosa* alkaloids (ethanol extract from defatted aerial part) benzylisoquinoline alkaloids [{(6,7-dimethoxy-1-(α-hydroxy-4-methoxybenzyl)-2-methyl-1} and 2,3,4-tetrahydroisoquinolin (1) or 1,2,3,4-tetrahydroisoquinoline] and coclaurine show antitumor activity in colon cancer (HCT116) cells and human breast cancer cell lines (MCF-7) (Dev and Joseph 2021).

The acetogenin annoreticin from *A. squamosa* leaves induces apoptosis, whereas annosquatin A and B exert anticancer activity by reactive oxygen induction in human breast cancer (MCF-7) cells. Annosquacins A–D and annosquatin A, B from *A. squamosa* seeds also exhibit cytotoxic activity in human breast cancer (MCF-7) cells (Al-Ghazzawi 2019). Other major anticancer acetogenins reported from *Annona* species are annotacin, bulatacin, folianin, isolongimicin, reticulactin, and squamocin. For example, squamosin prevents human leukemia cell line proliferation resulting in apoptosis by caspase 3 activation. Another acetogenin, ascimicin, is cytotoxic to and can inhibit tumor (9KB nasopharynx, A549 lung, and HT-29 colon) cells (Regassa et al. 2022).

Table 7.5 demonstrates the anticancer effects of *A. squamosa* seed alkaloids and acetogenins evaluated on several cell lines and carcinoma. These components generally exhibit cytotoxic effects, inhibit cell proliferation, and regulate gene expression that modulates important pathways. Quilez et al. (2018) summarized the antitumoral neoplasms of *Annona* species including *A. squamosa*. Ethanol *A. squamosa* seed extract (containing 12, 15-cis-squamosin A and bullatacin) exerts high antitumor activity against breast (MCF-7) and hepatoma (HepG2) cancer cells and inhibits (70% maximum inhibitory rate) H_{22} hepatoma cell growth *in vivo* in transplanted tumor mice. This extract has been proposed as a potential candidate for a novel anti-liver cancer drug.

Crude ethanol and ethyl acetate *A. squamosa* leaves extract exerts significant anticancer activity on human epidermoid carcinoma (KB-3–1) and colon cancer (HCT-116) cell lines. *A. squamosa* bark extracts, aqueous and particularly ethanol (95%), have antigenotoxic effects in 7,12-dimethylbenz(a)anthracene (DMBA)-induced genotoxicity/carcinoma in golden Syrian hamsters. The antitumor activity of aqueous and organic *A. squamosa* seed extracts is attributed to their active compound squamocin that induces apoptosis in tumor cells through oxidative stress. The antitumor effects of annonaceous acetogenins against human hepatocarcinoma are due to G phase cell cycle arrest and apoptosis induction through mitochondrial and receptor pathways. The bark acetogenin, bullacin B, is highly potent (10 and up to a million times more potent than Adriamycin [Doxorubicin chemotherapy medication used to treat cancer]) against different human cell lines (Quilez et al. 2018).

TABLE 7.5 Anticancer effects of *A. squamosa* seed alkaloids and acetogenins

COMPOUND	*CANCER CELL LINES*	*FUNCTION/EFFECTS*
Annosquatin A, B	Human breast (MCF-7), leukemia (K-562) Colon carcinoma (COLO-205) (MCF-7), lung (A-549)	Induce ROS generation, Decrease glutathione levels Regulate Bcl-2 and PS externalization Exhibit high cytotoxic selectivity
(-)-Anonaine	H-22 solid tumor cell	Inhibits IL-6/Jak/Stat3 pathway Increases caspase-3 expression
Asimilobine	Colon cancer (WiDr) cell	Induces G2/M phase and apoptosis
Annosquatin III Dieporeticenin B, Squamocin	SMMC 7721/T, MCF-7/ADR	Exert inhibitory activity
Squamoxinone D	Nasopharyngeal (KB) cancer, Brest (MCF-7), lung (A-549), Leukemia (K-562)	Inhibits proliferation
Squamocin-I, II, III, Squamoxinone D	Hepatocellular (BEL 7402), Gastric (BGC 803), lung cancer Hep G2, SMMC 7721	Exhibit cytotoxic activity

Source: Adapted from Joseph and Dev (2023)

Annona squamosa seeds have been used to treat "malignant sore" (cancer). Petroleum ether extract (Soxhlet, 80 °C, 90 min; 23% oil yield) contained palmitic, linoleic, oleic, and stearic acids (9.9, 20.5, 56.5, and 9.1% of total fatty acids, respectively) and 41 mg/g total acetogenins. Oral administration of the seed oil inhibited (54% maximum inhibitory rate) H_{22} tumor cell growth in mice. This antitumor effect of the seed oil was due to reduced interleukin-6 (IL-6), janus kinase (Jak), and phosphorylated signal transducers and activators of transcription (p-Stat3) expression via IL-6/Jak/Stat3 pathway. The tumor suppression was attributed to the main seed oil constituents, the unsaturated fatty acids, and the acetogenins (Chen et al. 2016). *A. squamosa* seed contains ribosome inactivating (RIP) protein (25 μg) that cuts double-stranded supercoiled pUC19 DNA into nicked circular DNA (Sismindari and Mubarika 1998). RIP from the seeds has cytotoxic activity against gastric cancer cells (IC_{50} 43 μg/mL).

Extracts have been encapsulated into nanoparticles to investigate cytotoxicity efficiency. Chitosan nanoparticles (535 nm) loaded with *A. squamosa* leaves (96% ethanol) extract have potent cytotoxicity (IC_{50} 292 μg/mL) against human colon cancer (WiDr) cell lines due to significant caspase-3 upregulated expression and G2/M phase cell cycle arrest (Fadholly et al. 2019). Methanolic (70%) *A. squamosa* leaf extracts have potent hepato-renal protective effects against Ehrlich ascites carcinoma (Abd-Elghany et al. 2022). The extract and its loaded noisome (distinctive drug stable nanocarriers that effectively encapsulate drugs) attenuated inflammatory markers (TNF-α, NFκB, and COX-2 levels) and improved oxidative stress (SOD, GSH) activity in tumor, liver, and kidney tissues of Ehrlich ascites carcinoma-bearing mice. The extract-loaded niosomes were more effective at shrinking tumors than the free extract. The hepato-renal protective effect of *A. squamosa* leaf extract is attributed to its antioxidant and scavenging activities of its phenolic compounds and flavonoids (Abd-Elghany et al. 2022). *A. squamosa* seed oil cytotoxicity is improved (~7 fold, IC_{50} 1.3 vs 8.7 ng/mL for untreated oil) by TPGS (D-α-tocopheryl polyethylene glycol succinate) stabilized nanoparticles (~194 nm particle size) against mouse breast cancer 4T1 cells (Ao et al. 2022).

The nanoparticles reduced tumor volume (~500 mm^3 compared to 1500 mm^3 for negative control [saline]), increased antitumor activity (70 vs ~52% inhibition rate and 0.75 vs ~1.2 g tumor weight for untreated oil). Moreover, the seed oil nanoparticles had great potential as tumor-targeted delivery vehicles with higher (~1.5-fold) relative tumor-targeting index (RTTI) of acetogenins than that of acetogenins delivered alone (Ao et al. 2022). Ethanol (70%) *A. squamosa* fruit extract and its synthesized silver nanoparticles (6.63 nm average particle size) expressed high cytotoxicity against prostate adenocarcinoma (PC3), ovary adenocarcinoma (SKOV 3), and human cervical cancer (HeLa) cell lines. The most potent cytotoxicity was against HeLa and SKOV 3 cells (IC_{50} 0.001–1.6 μg/mL) and for PC3 cells (IC_{50} 1.7 vs 3.5 μg/mL for the ethanol extract). The extract and nanoparticles induced apoptosis (~80% compared to control) in all cancer cells due to S phase cell cycle arrest (Mokhtar et al. 2022). The anticancer activity is attributed to the 114 bioactive compounds (77 acetogenins, 20 diterpenes, and 17 other compounds) identified in the ethanol extract. The acetogenin squamocin P isolated from *A. squamosa* has significantly higher anticancer activity against human hepatocarcinoma (SMMC 7721/T), multidrug-resistant breast cancer (MCF-7/ADR), and human non-small lung cancer (A549/T) cell lines than cisplatin (IC_{50} 0.44 vs 199 μM, 3.34 vs 179 μM, and 6.32 vs 219 μM for cisplatin, respectively) (Nugraha et al. 2019).

A. squamosa leaf extracts are cytotoxic and reduce proliferation in colon cancer cell (HCT-116 and Lovo) lines (Al-Nemari et al. 2020). The leaf extracts (100 μg/mL) have significant cytotoxic and antiproliferative effects on breast cancer cell lines MCF-7 and MDA-MB–231. The extracts significantly triggered apoptosis and hindered wound closure (Al-Nemari et al. 2022).

Acetogenins, unique to the custard apple fruit family, are responsible for the anticancer effects. Breast and colon cancer cell types were examined using acetogenins (Kelloff et al. 2008). Flavonoids present in the methanolic *Annona squamosa* leaves extract were the main metabolites believed to have a significant impact on the cytotoxic activity of T47D cancer cell line (Amin et al. 2019). Montanacin F, a unique Annonaceous acetogenin derived from *Annona montana* leaves, showed cytotoxic activity against Lewis lung carcinoma (Wang et al. 2002). Annomolin and annocherimolin are two unique cytotoxic acetogenins obtained from *Annona cherimolia* seeds. Annomolin demonstrated cytotoxic selectivity for the human prostate tumor cell line, while annocherimolin was effective against breast and colon cancer cell lines (Kim et al. 2001). Squamostanin-A and Squamostanin-B, two novel annonaceous acetogenins, were discovered in a 95% EtOH extract of *Annona squamosa* seeds. The cytotoxicity of both the compounds was investigated against the human tumor cell lines of human colon adenocarcinoma (HCT), human prostate adenocarcinoma (PC-3), human breast carcinoma (MCF-7), and human lung carcinoma (A-549) by using the traditional MTT method. These acetogenins showed a wide spectrum of cytotoxic potential and can be considered as alternatives for future generation of antitumor medication (Yang et al. 2009).

7.5.3 Antibacterial Properties

Custard apple exerts several physiological activities (Table 7.6). The aqueous and methanolic extract of the *Annona muricata* leaves was tested for resistance to different microorganisms (i.e., *Proteus vulgaris*, *Escherichia coli*, *Bacillus subtilis*, *Klebsiella pneumoniae*, *Salmonella typhimurium*, *Streptococcus pyogenes*, *Staphylococcus aureus*, and *Enterobacter aerogenes*).

TABLE 7.6 Health beneficial potential of custard apple

HEALTH EFFECTS/ REFERENCES	*EXPERIMENTAL CONDITIONS*	*KEY FINDINGS*
Anticarcinogenic properties/ Al-Nemari et al. (2022)	Twelve female (5–8 weeks old, 110 ± 20 g) Wistar albino rats were divided into three equal groups. Dimethylbenzanthracene, a potent carcinogen, was subcutaneously injected into the pectoral region of two sets of rats (groups I and II), causing mammary tumors. Corn oil was administered to group I animals after eight weeks of injection. Group II rats were injected with *A. squamosa* aqueous extract suspended in corn oil straight into the tumor. A solitary dose (300 mg/kg) was administered every 48 hours for four weeks.	*A. squamosa* treatment significantly reduced tumor size compared to the control group. The treatment greatly inhibited tumor growth and progression.
Antioxidant properties/Panda and Kar (2007)	Male albino Wistar rats were intraperitoneally injected with freshly prepared streptozotocin (55 mg/kg body weight) after fasting overnight. Rats were classified into four groups: Control group received only buffer; the second was the diabetic control group. Group 3 was treated with protamine-zinc insulin, and group 4 was diabetic rats fed aqueous *A. squamosa* solution (300 mg/kg body weight daily) for 30 days.	*A. squamosa* normalized total hemoglobin and HbA1c levels in diabetic rats. TC, HDL-C, and LDL-C levels also returned to the normal range.
Hepatoprotective property/Zahid et al. (2020)	Sprague Dawley rats were exposed to 50% alcohol (12 mL/kg) for eight days resulting in liver damage. Ethanol extract of *A. squamosa* seed (EEAS) (200 and 400 mg/kg po) was administered for eight consecutive days.	EEAS reduced hepatocellular necrosis, promoted cell regeneration, and repair toward normal function.
Antibacterial property/Biba et al. (2013)	A chloroform *A. squamosa* seed extract (ASCH) was added to cell culture containing Gram-positive (*Bacillus subtilis* and *Staphylococcus aureus*) and Gram-negative bacteria *(Salmonella typhi, Proteus mirabilis, Escherichia coli,* and *Klebsiella pneumoniae).*	ASCH suppressed bacterial growth after incubation.
Anti-Inflammatory Activity/Dellai et al. (2010); Yang et al. (2008)	Cyclopeptides: cyclosquamosin D (A1) and met-cherimolacyclopeptide B (B) and their analogues were derived from *Annona squamosa* seeds. They were examined for their ability to block pro-inflammatory cytokine production in lipopolysaccharide-stimulated macrophage J774A.1 cell line.	Cyclopeptides significantly reduced inflammation by inhibiting IL-6 and TNF-α release.

(Continued)

TABLE 7.6 (Continued)

HEALTH EFFECTS/ REFERENCES	*EXPERIMENTAL CONDITIONS*	*KEY FINDINGS*
Anti-HIV propertyWu et al. (1996)	H9 lymphocytes (3.5 × 10^6 cells/mL) were incubated (1 h, 37 °C) with or without HIV–1. In the presence or absence of *A. squamosa*'s methanol extract, cells were thoroughly washed and redissolved (2 × 10^5 cells/mL final concentration). Cells were counted to estimate uninfected cultures' cell density after incubation (3 days, 37 °C) for drug's toxicity. The quantity of virus released into the medium of cultures infected with HIV was measured using a p24 antigen capture assay.	16β,17-dihydroxy-*ent*-kauran-19-oic acid, obtained from *A. squamosa*'s methanol extract, was efficacious against HIV in H9 lymphocyte cells.

The most vulnerable Gram-positive bacteria were *S. aureus* and *B. subtilis*, and the most susceptible were *P. vulgaris* and *K. pneumoniae* among Gram-negative bacteria. Treatment for various bacterial infectious ailments, including pneumonia, diarrhea, urinary tract infections, and even some skin conditions, uses *A. muricata* leaf extract (Gajalakshmi et al. 2012). *Pseudomonas aeruginosa* was most effectively inhibited by methanol extract (MIC: 130 µg/mL), followed by petroleum ether extract (MIC: 165 µg/mL), and methanol extract against *E. coli* (MIC: 180 µg/mL). The investigation showed the antibacterial activity of specific bioactive compounds (borneol, linalool, farnesol, eugenol, and geraniol) (Patel and Kumar 2008). GC/MS identified primary volatile components of *A. squamosa* L. bark as 1H-Cycloprop(e)azulene (3.5%), bisabolene (4.5%), caryophyllene oxide (29.4%), germacrene D (11.4%), kaur-16-ene (19%), and bisabolene epoxide (3.6%). The essential oil showed significant antibacterial action against *B. subtilis* and *S. aureus* (Chavan et al. 2006). Dasycarpidan-1-methanol, acetate (ester), a single chemical isolated from the peel and the pulp of custard apple, is principally responsible for the antibacterial activity and is most effective against the following particular microbes: *S. aureus*, *E. coli*, *Bacillus*, *Proteus*, and *Klebsiella* species (Karthikeyan et al. 2016).

The methanol leaf extract (ALS–6.6% yield) containing alkaloids (anonaine, asimilobine, liriodenine, nornuciferine, and reticuline) has broad-spectrum antibacterial activity effective against both Gram-negative and Gram-positive bacteria. It exerts bactericidal activity against *S. aureus* (an important human pathogen) and bacteriostatic effects against *Enterococcus faecalis* (medically opportunistic pathogen) and *K. pneumoniae* (common cause of hospital and community infections). The antimicrobial activity is associated with bacterial membrane destabilization (biofilm disruption, elevated membrane permeability, and significant cell viability reduction) induced by constituents of the leaf extract (alkaloids) (Pinto et al. 2017).

Anonaine has high antimicrobial activity against *Kocuria rhizophila* and *Staphylococcus epidermidis* similar to chloramphenicol (MIC 50 µg/mL). *A. squamosa* methanol seed extract has higher antimicrobial activity than neomycin against *E. coli* and *S. aureus* (MIC 50 vs 313 µg/mL). The fruit extract exerts antimicrobial activity against multidrug-resistant MRSA (MIC 5000 µg/mL) (Nugraha et al. 2019).

Vijayalakshmi and Nithiya (2015) evaluated the antibacterial activity of ethanol, acetone, and aqueous *A. squamosa* fruit extracts against *S. aureus*, *P. aeruginosa*, *K. pneumoniae*, *E. coli*, *S. typhi*, *S. pyogenes*, and *Aspergillus niger*. The fruit extracts have maximum activity against Gram-negative than Gram-positive bacteria, attributed to their bioactive phenolic compounds (flavonoids, tannins). A similar study investigated the antibacterial activity of *A. squamosa* leaves extracted with acetone, methanol, and water (1:10 w/v) against *B. subtilis*, *S. aureus*, *E. faecalis*, *E. coli*, *P. aeruginosa*, *K. pneumoniae*, and *S. typhimurium* (*LT2*) (Al-Nemari et al. 2020). The methanolic extract inhibited all bacterial growth, whereas the acetone extracts exerted antibacterial activity against both Gram-positive and Gram-negative bacterial strains, except *S. aureus*. Moreover, the methanol extract inhibited *S. aureus* more effectively than the standard antibiotic tetracycline (16.5 vs 14.8 mm inhibition zone). The aqueous extract was more sensitive to Gram-negative than Gram-positive bacteria, with moderate antibacterial effect against *E. faecalis*. The antibacterial activity of leaves extracts was attributed to their multitude of bioactive compounds, primarily benzoquinone alkaloid and sesquiterpene hydrocarbons. Alkaloids (aporphine, corydine, norcodeine, norlaureline and sanjoinine) extracted from *A. squamosa* leaves exhibit antibacterial activity against *S. aureus*, *E. coli*, *B. cereus*, *P. aeruginosa*, *H. pylori*, and MRSA, and antioxidant activities (DPPH radical scavenging and superoxide dismutase activities). The antibacterial activity was due to changes in *B. cereus* cell wall morphology (swelling and rupture), cell lysis, and apoptosis (Shami 2017). Ali and Jafar (2022a) investigated the antibacterial activity of *A. squamosa* pulp (ethanol extract) mouthwash against *S. aureus* and *E. coli* compared with chlorhexidine. The extract has antibacterial activity against *S. aureus* and *E. coli* obtained from saliva samples from 21 healthy children (7–9 years old). The antibacterial effect of the extract at 10 mg/mL was comparable to chlorhexidine, indicating its potential use as an active herbal mouthwash. Moreover, the extract has minimal to no cytotoxicity on human dermal fibroblast neonate cells and can be a safe substitute for chlorhexidine gluconate (0.12%) to prevent or treat oral bacteria-induced diseases (Ali and Jafar 2022b).

Aqueous *A. squamosa* seed extract synthesized copper oxide nanoparticles (~40 nm) were more effective against the Gram-negative *E. coli* than the Gram-positive *Microbacterium testaceum* bacteria (21 vs 17 mm zone of inhibition), presumably due to changes induced in the bacterial cell wall. The nanoparticles effectively inhibited the Gram-negative plant pathogenic bacteria *Xanthomonas oryzae*, the major cause of bacterial blight in rice crop (Singh et al. 2021). Silver nanoparticles (6.63 nm average particle size) of *A. squamosa* ethanol (70%) fruit extract had higher antibacterial activity than the ethanol extract (0.09–0.75 mg/mL vs 4–16.5 mg/mL) against *S. aureus*, *E. coli*, and *P. aeruginosa*, with the highest activity and the lowest MIC (0.093 mg/mL) against *Candida albicans*. Moreover, the nanoparticles acted synergistically with gentamicin to increase antibacterial activity only against *P. aeruginosa* and *E. coli* (Mokhtar et al. 2022). Silver nanoparticles synthesized from *A. squamosa* aqueous leaf (35–90 nm) and fruit (15–50 nm) extracts elicited antibacterial activity against Gram-negative *E. coli* and *P. aeruginosa*, particularly at high (100 μg/mL) concentration. The antibacterial activity is attributed to changes in bacterial cell wall induced by the bioactive components of the extracts in addition to the antibacterial effects of silver ions (Malik et al. 2022). Ribosomal inactivating protein (RIP) extracted from *Annona squamosa* seeds was remarkably effective against *E. coli*, *K. pneumoniae*, *P. aeruginosa*, *S. aureus*, and *B. subtilis* bacteria, especially at 30 μg/mL. The antibacterial activity was attributed to the presence of lectin in the protein that induces changes in the bacterial cell wall (Dhanraj et al. 2020).

Through LC-MS analysis, specific alkaloids including corydine, sanjoinine, norlaureline, norcodeine, oxanalobine, and aporphine were identified, and scanning electron microscopy (SEM) observed the mode of action of these alkaloids present in the leaves of *A. squamosa*. The SEM studies provided visual evidence of the morphological changes in the bacterial cells (*B. cereus*) caused by the alkaloids, such as swelling, rupture, lysis, and apoptosis, which supported the antibacterial activity. The study also found strong antioxidant activity, suggesting that these alkaloids may have potential as a new source of antimicrobial agents and antioxidants (Shami 2017). Phenolic compounds have a strong correlation with antimicrobial activity, particularly rutin and procyanidin dimer type B, which exert inhibitory effects on bacteria (Tlais et al. 2023). The study claimed that peel extract had better antimicrobial activity than pulp extract. Gram-positive bacteria like *S. saprophyticus*, *S. aureus*, *L. monocytogenes*, and *B. megaterium* were sensitive to peel extract, while among LAB only *Weissella cibaria* and *Lactococcus lactis* were inhibited.

7.5.4 Antioxidant Activities

Annona squamosa seeds from Thiruvananthapuram district (Kerala, India) were evaluated for their antioxidant content and activities. Different seed extracts exhibited good antioxidant activity and scavenging potential against free radicals measured by ABTS, DPPH, FRAP, super oxide, and nitric oxide radical scavenging assays (Vikas et al. 2017). The potential of *Annona squamosa* leaves to scavenge free radicals was investigated using various antioxidant screening models. The radical cation (ABTS) was scavenged to a maximum of 99% in the ethanolic extract at 1000 µg/mL, followed by the stable radicals DPPH and nitric oxide radical. However, the extract had limited anti-lipid peroxidation capability and superoxide radical scavenging ability in rat-brain homogenate. The results support *A. squamosa*'s antioxidant properties (Pandey and Barve 2011). *A. squamosa* aqueous seed extract demonstrated the highest level of radical scavenging activity (Kothari and Seshadri 2010). The maximum quantity of TPC (total phenolic content) was identified in the *A. squamosa* roots and bark. *Annona squamosa* leaf extracts demonstrated strong antioxidant properties by DPPH (1,1-diphenyl-2-picrylhydrazyl) free radical scavenging abilities (7.81–25 µg/mL) and by oxygen radical absorbance capacity and MTT assay (Mariod et al. 2012).

A. squamosa methanol fruit extract exerts considerable antioxidant activity due to high polyphenolic and flavonoid contents. Oxygen radical absorbance capacity (ORAC) and MTT assays indicate good antioxidant activity for different plant parts (leaves, bark, roots, and seeds). The petroleum ether seed extract (0.2 µg/mL) displays higher lipoxygenase inhibitory activity (52.7%) than the crude ethanol extract of fruit pulp and seeds. *A. squamosa* ethanol leaf extract (1 mg/mL) has strong radical scavenging activity (99% ABTS, 99% ABTS, 90% DPPH, and 74% NO). Most of the 11 ent-kauranes from fresh *A. squamosa* stems significantly attenuate superoxide anion generation by human neutrophils and inhibit nitric oxide generation in response to the macrophage activator, formyl-L-methionyl-L-leucyl-L-phenylalanine (fMLP/CB) (known to inhibit tumor necrosis factor-alpha [TNF-α] production) (Quílez et al. 2018).

7.5.5 Hepatoprotective Property

The ethanolic extract of bark and leaf of *Annona squamosa* was tested for its capacity to cure carbon tetrachloride-induced liver injury in Wistar rats. The extract (450 mg/kg oral dose) significantly increased protection by reducing the serum levels of alanine and aspartate aminotransferase, malondialdehyde, total bilirubin, and aminotransferase equivalent, a measure of liver lipid peroxidation. These histopathological findings were combined with a biochemical examination of a hepatic segment. The extract's functionality was also equivalent to the well-known hepatoprotective silymarin (Sonkar et al. 2016). Custard apple's hepatoprotective activity was studied in diethylnitrosamine (DEN) induced carcinogenic liver injury in Swiss albino mice. *A. squamosa* extract reduced alkaline phosphatase (ALP), Alpha-fetoprotein (AFP) (only in serum), Glutamyl Oxaloacetate Transaminase (GOT), Total and Direct Bilirubin (both in serum and tissue), Acid Phosphatase (ACP), Glutamyl Pyruvate Transaminase (GPT), and Alpha-fetoprotein (AFP) (only in serum) values in the DEN-induced rats. However, total protein content decreased in DEN-induced animals and elevated in DEN-induced groups that also received *Annona squamosa* extract, respectively. Histopathology supports *Annona squamosa*'s hepatoprotective properties. As a result, the study suggested using this plant in herbal remedies for hepatoprotective function (Sobiya Raj et al. 2009). *A. squamosa* combined with anti-tubercular medication and *Pongamia pinnata* significantly reduced serum glutamate pyruvic transaminase (SGPT), serum glutamate oxaloacetate transaminase (SGOT), and tissue malondialdehyde (MDA) levels in rat liver damage induced by anti-tubercular medications (Derangula et al. 2022).

A. squamosa ethanol seed extract (200 and 400 mg/kg/day) was administered to alcohol (50%)-induced (12 mL/kg po each) liver injury in Sprague Dawley rats for 8 days. The extract dose dependently increased serum hepatic enzymes (alanine transaminase, aspartate transaminase, alkaline phosphatase, lactate dehydrogenase), and antioxidant (SOD, CAT, GSH) levels, reduced malondialdehyde (MDA) levels, attenuated hepatocellular necrosis, and repaired cells to their normal state. These effects are attributed to the hepatoprotective and antioxidant activities of the extract due to the presence of phytochemicals (i.e., phenolic compounds, flavonoids, steroids, terpenoids, and alkaloids) (Zahid et al. 2020). Essential oil obtained by hydro distilling *A. squamosa* pericarps containing 59 compounds exerted anti-hepatoma activity (IC_{50} < 55 µg/mL) against human hepatoma SMMC-7721 cell line. The oil induced pro-apoptosis (shrunken or broken cell nuclei) and altered cell cycle (75% G1 phase cell cycle arrest, reduction in S and G2 phase populations). The hepatoprotective effect was attributed to the essential oil, particularly the high (32.5%) amount of (-)-spathulenol (Chen et al. 2017). *A. squamosa* methanol leaf extract (250 and 500 mg/kg body weight) protected against isoniazid and rifampicin (drugs used in tuberculosis treatment) induced oxidative liver injury in rats. The extract prevented increase in serum liver enzyme markers (alanine aminotransferase, aspartate aminotransferase, alkaline phosphatase, and gamma glutamyl transpeptidase). The hepatoprotective effect of the extract was attributed to the presence of leaf flavonoids that exert strong antioxidant activity. Apparently, aqueous and ethanol leaf extracts also have similar hepatoprotective effects (Uduman et al. 2011).

7.5.6 Insecticidal Activity

Acetogenins from *A. squamosa* extracts are apparently responsible for their biological insecticidal properties. Over 400 annonaceous acetogenins, including their isomers, have been identified so far. The target organism is killed by ATP deprivation using a succession of natural C-35/C-37 products (Mondal et al. 2018).

7.5.7 Antimalarial Activity

The antimalarial activity of *A. squamosa* is mostly focused on its larvicidal effect. The fruit peel has been suggested as a potential approach to control medically important malaria vectors (*Anopheles subpictus* and *Culex quinquefasciatus*). Similarly, ethyl acetate and methanol bark extracts can control the instar larvae of malaria vectors (*A. subpictus*, *A. stephensi*, *C. quinquefasciatus*, and *C. tritaeniorhynchus*). *A. squamosa* methanolic leaf extract has potent antiplasmodial activity in 3D7 and Dd2 chloroquine-sensitive strains (IC_{50} 2 and 30 μg/mL). The ethanol leaf extract has the highest antiplasmodial activity against both chloroquine-sensitive Pf3D7 and chloroquine-resistant PfINDO strains of *Plasmodium falciparum* among medicinal plants, thereby corroborating its use against malaria in India (Quilez et al. 2018). The stem and root bark extracts are generally more toxic toward malarial larvae (*Anopheles gambiae*; 50% mortality at 24 and 21 μg/mL, respectively). These antimalarial effects may be partly due to essential oil since *A. squamosa* essential oil inhibited *Plasmodium falciparum* erythrocyte stages, *Trypanosoma cruzi* epimastigotes and trypomastigotes (IC_{50} values 14.7, 16.2, and 12.7 μg/mL, respectively) (Nugraha et al. 2019). However, alkaloids and acetogenins isolated from *A. squamosa* leaves also possess potent anti-leishmanial effects against *Leishmania chagasi* promastigote (IC_{50} 23–38 μg/mL) and amastigote (IC_{50} 14–29 μg/mL) (Quilez et al. 2018).

7.5.8 Antiulcer Activity

Krishnaiah et al. (2013) investigated the protective effects of *A. squamosa* ethanol leaf extract against acetic acid–induced ulcerative colitis in Wistar rats (human intestinal inflammation simulation). The extract (250 and 500 mg/kg), administered for 3 consecutive days before and 7 days post induction, reduced mucosal epithelium damage and reverted biochemical parameters (endogenous antioxidants [SOD, GSH, and catalase], mucosal malondialdehyde, and myeloperoxidase). *A. squamosa* extract attenuates acetic acid–induced ulcerative colitis comparable to sulfasalazine by reducing oxidative stress in rats. The extract significantly reduced crypt abscess formation, gross colon mucosal inflammation, wet weight of inflamed colonic tissue, and diarrhea.

In an extensive study, Yadav et al. (2011) fractionated ethanol extract of *A. squamosa* twigs to elucidate compounds responsible for its antiulcer activity. The extract (50 mg/kg), along with its chloroform and hexane soluble fractions (20 mg/kg), was evaluated against cold restraint (CRU), pyloric ligation (PL), aspirin (ASP), alcohol (AL)–induced gastric ulcer in Sprague Dawley rats and histamine (HE)–induced duodenal ulcer in guinea pigs, and compared with standard drugs omeprazole (10 mg/kg) and sucralfate (500 mg/kg). *A. squamosa* twig ethanol extract and its fractions (chloroform and hexane) reduced ulcer formation better than or comparable to the standard drugs in CRU, PL, and HE models and elicited antisecretory activity *in vivo* through reduced free acidity, total acidity, and pepsin in PL. The gastroprotective effect of *A. squamosa* extracts is attributed to gastric hydrogen potassium ATPase inhibition and simultaneous strengthening of the mucosal defense system. Three alkaloids, (+)-*O*-methylarmepavine, N-methylcordaldine, and isocorydine purified from the chloroform fraction, inhibited gastric hydrogen potassium ATPase, reduced gastrin secretion in the ethanol-induced ulcer model, confirming their antisecretory potential and potent antiulcer activity (Yadav et al. 2011).

7.5.9 Other Activities

7.5.9.1 Anti-inflammatory Activity

Two compounds (caryophyllene oxide and 18-acetoxy-ent-kaur-16-ene) isolated from *A. squamosa* petroleum ether bark extract exhibited significant analgesic and anti-inflammatory effects. These effects occurred at 50 mg/kg for the petroleum extract and 12.5 and 25 mg/kg for 18-acetoxy-ent-kaur-16-ene. The diterpenoid 16β,17-dihydroxy-ent-kaurane-19-oic acid (DDKA) from *A. squamosa* stems inhibits respiratory bursts and degranulation of human neutrophils through the inhibition of cytosolic calcium mobilization. DDKA is a known anti-HIV agent and possesses antiplatelet and antithrombotic activities that involve anticoagulatory effects and extracellular adenosine 3',5'-cyclic monophosphate (cAMP) induction (Quilez et al. 2018).

7.5.9.2 Neuroprotection

Aqueous or ethanolic *A. squamosa* leaf extract (300 mg/kg daily for two months) attenuated oxidative stress (MDA and NO; increased SOD and GSH), NFκB and acetylcholine esterase levels, and caspase 3 activity in aluminum chloride–induced neuro-inflammation in albino male rat brains. Moreover, the extracts increased brain-derived neurotrophic factor (BDNF), thereby maintaining neurons in the central nervous system; BDNF expression is inversely correlated with the occurrence of Alzheimer's disease (Hendawy et al. 2019).

7.5.9.3 Metabolic Disorders

A. squamosa methanol leaf extract inhibited angiotensin-converting enzyme (IC_{50} 9.3 vs 6.4 μg/mL for Captopril) and pancreatic lipase (only moderately) (IC_{50} 6.9 vs 0.94 μg/mL for Orlistat) due to the presence of GC-MS identified bioactive compounds (luteolin, gallic acid, hexadecanoic acid methyl ester, and longifolene). These compounds have antihypertensive and lipase inhibitory properties, enabling *A. squamosa* use in hypertension and obesity management (Habib et al. 2022).

7.5.9.4 Anti-anxiolytic Effect

The anxiolytic potential of *A. squamosa* fruit juice has been evaluated in stressed and unstressed mice using anxiety models, viz. light–dark, elevated plus maze, and hole-board models (Sharma and Parle 2016). *Annona squamosa* fruit juice (3, 6, and 9% v/v, p.o.) administered for 14 days significantly enhanced the time spent in the lit compartment in the light–dark model, number of entries and time spent in the open arms in the elevated plus maze model, and head dip counts in the hole-board model in unstressed and stressed mice. *Annona squamosa* effectively attenuated blood plasma nitrite levels, thereby suggesting enhancement in scavenging of free radicals. Interestingly, brain GABA and serotonin levels were markedly increased indicating *A. squamosa* anxiolytic effect (Sharma and Parle 2016).

7.5.9.5 Miscellaneous Activities

Custard apples, which contain vitamin A, promote good skin. Vitamin-rich custard apple fruit peel and pulp improve vision and eyesight. Magnesium in custard apples assists in maintaining the body's water balance, eliminating different acids, and lowering the risk of arthritis (Adil 2019). In several animal models, the blended extract of *Nigella sativa* and *Annona squamosa* was investigated and found to have anti-arthritic, anti-inflammatory, and analgesic properties. The combined extract significantly decreased bone erosion, pannus development, AST, ALT, and TP levels, the swelling of the hind paws, and body weight (Manvi et al. 2011).

Two phytoparasitic nematodes, *Bursaphelenchus xylophilus* and *Meloidogyne incognita*, were successfully eradicated by the methanol *Annona squamosa* seeds extract. The chemical structures of ten annonaceous acetogenins (AAs) were determined by nuclear magnetic resonance. Squamocin-G exhibited strong nematicidal action against *M. incognita*. Squamocin, squamocin-G, and squamostatin-A exerted antifungal effects against *Phytophthora infestans*, the cause of tomato late blight. Squamostatin-A also successfully stopped the spread of *Puccinia recondita*–related wheat leaf rust. The seeds of *A. squamosa* and their bioactive AAs might serve as a viable supplementary source of a botanical nematicide and fungicide to manage various plant diseases (Dang et al. 2011).

Allopurinol, with its adverse side effects, is used to treat gout by acting as a xanthine oxidase inhibitor (XOI). While sugar apple ethanol extract can function as XOI with lesser side effects. Additionally, an *in silico* investigation of three sugar apple compounds revealed that rutin, kaempferol-3-rutinoside, and quercetin-3-glucoside can bind to the RBD of the S protein of COVID 19, indicating that the three compounds have anti-COVID-19 action (Muna et al. 2021).

A persistent inflammatory skin condition called psoriasis is characterized by an incomplete keratinization process and quick keratinocyte proliferation. It was previously discovered that ethanolic fraction from three medicinal herbs, namely *Curcuma longa*, *Alpinia galangal*, and *Annona squamosa*, have anti-psoriatic action using a HaCaT keratinocyte cell line as an *in vitro* model. The expression of psoriasis signaling indicators CD40 and NF-B1 were both considerably reduced by an *Annona squamosa* extract (Saelee et al. 2011).

Epidermal growth factor receptor (EGFR) is essential in normal skin development and function and also a molecular target for psoriasis therapy. EGFR is upregulated in active psoriatic plaques and has a functional role in psoriatic hyperplasia. *A. squamosa* ethanolic leaf extract (3.15 and 1.575 μg/mL) significantly reduced EGFR mRNA expression in HaCaT keratinocyte cell line and inhibited EGFR protein expression (Ronpirin et al. 2014). This effect is similar to the anti-psoriatic drug dithranol and erlotinib, the EGFR tyrosine kinase commonly used in lung cancer treatment and psoriasis in patients. The anti-psoriatic effect of *A. squamosa* leaf extract is attributed to its abundance in quercetin along with phenylated flavones that exhibit anti-psoriatic activity (Ronpirin et al. 2014).

Annona squamosa leaf extract (ASLE) or its niosomal-entrapped formulation was evaluated for its UVA radiation protective activity on the skin of the rat's dorsal area using spectrophotometry and transmission electron microscopy imaging. Niosomal-entrapped ASLE, a preparation created using nanotechnology, can be utilized as an effective photoprotector (sunscreen) against the harmful UVA radiation effects (Mohamad et al. 2022).

7.6 INCORPORATION/FOOD APPLICATIONS

The great nutritional and therapeutic value of CA has led to its long-standing application in the development of various food products (Table 7.7).

TABLE 7.7 Custard apple as a functional food development agent

PRODUCT/REFERENCE	*METHOD OF PREPARATION*	*KEY FINDINGS*
Chicken breast fillets Kadam et al. (2020)	Uniformly sized and shaped chicken breast fillets were submerged for 10 min in 0.1, 0.3, and 0.5% aqueous hydro-ethanolic (50%) solutions of custard apple peel extract (CPeE). Pure water was the negative and BHT (100 ppm solution) the positive control. Prepared fillets stored under refrigeration (4 ± 1 °C) in LDPE bags.	250 to 1250 mg/mL CPeE showed antibacterial activity. Total phenol content and ascorbic acid content improved. Color, smell, and taste were pleasantly accepted till the ninth storage day.
Basundi Bawale et al. (2018)	Custard apple pulp was added at 5, 10, and 15% to prepare the basundi.	Moisture and sucrose content increased by 5.5 % and 1.41 % compared to the control group. Protein, ash, fat, total solids and carbohydrate decreased. Basundi with 10% CA pulp addition had the best acceptance.
Whey based beverage M El-Aidie et al. (2020)	Whey and pulp juice from custard apples were added (25:75, 75:25, and 50:50, weight/weight). They were then inoculated with 1% *L. casei* and stored for 21 days at 5 ± 1 °C.	Vitamin A and C increased at 135 and 173% than the control. Growth of probiotic bacteria increased, and coliform, including other spore-forming bacterial growth decreased. 75:25 was the most acceptable ratio.
Pudding Vidanarachchi et al. (2021)	CA peel powder was added (0%, 10%, 15%, 20%, and 25%) to prepare pudding. Milk powder was added to the rest of the amount.	Addition of 25% peel powder produced the highest amount of energy (19%), fat (0.30%), and moisture (5.8%). The sensory score was excellent for treatment (25% peel)
Pound Cake de Souza and Schmiele (2021)	Fructooligosaccharides (FOS), soy protein hydrolysate (SPH), and custard apple puree (CAP) were used to prepare cake at varying percentages.	The best result was obtained by adding 94.94% FOS, 100% SPH, and 40.98% CAP. with 72% sensory acceptability rating. The cake had low glycemic load (4.7%). and moderate glycemic index (60%)

Custard apples are consumed fresh; however, they soften quickly at maturity and are extremely prone to rotting due to high perishability. Jain et al. (2019) reviewed post-harvest technologies to minimize loss and preserve the custard apple fruits. Fruit shelf life can be significantly extended by storage at 15–20 °C under low oxygen and ethylene and 10% carbon dioxide. Custard apples have good acceptability in various value-added products (i.e., juice, ice cream, toffee, milkshake, vinegar). The fruits are an excellent substrate to manufacture high-quality vinegar because customers praised the vinegar for its fragrance. In addition to raising the custard apple's economic and nutritional worth, this also gives India a method to use the fruit. Future research in the field of pulp value addition has a significant amount of potential (Jain et al. 2019).

Custard apple pulp was packed in 200 gauge polyethylene and processed with 100 ppm ascorbic acid to deter browning reaction. Then, it was frozen in an alcohol bath at –25 °C before storage in a deep freeze at –18 °C. Three cryoprotectants were investigated with a control sample devoid of any cryoprotectant used for comparison. Both the control and pulp containing the cryoprotectants were monitored for the ascorbic acid degradation half-life. Preparation (III) [1% propylene glycol + 2% glycerol + 10% maltodextrin + 10% glucose syrup] produced a substance that could be poured even while being frozen and significantly prolonged half-life (75.33% vs 34.82% days for the control) (Chikhalikar et al. 2000).

Probiotic fruit yogurts were prepared from three varieties of *Annona* species: *Annona muricata*, *Annona squamosa*, and *Annona reticulata*. Fruit pulp added to yogurts significantly increased their antioxidant activity. Titratable acidity increased, and pH decreased considerably in the fruit pulp incorporated yogurt with 412 mg GAE/100 g total phenolic content. *A. squamosa* fruit pulp yogurt scored the highest ratings for flavor, aroma, color, and overall acceptance by the panelists. According to this study, *A. squamosa* can be used to develop yogurt with good sensory qualities, antioxidant characteristics, and a microbial count that will last during storage (Senadeera et al. 2018). Custard apple carbonated beverages were prepared with a rich source of protein (0.7%), carbohydrates (12.9%), and fat (0.18%), compared to other artificial drinks (Patil et al. 2011).

Annona squamosa leaves were washed thoroughly, dried, ground, and packed in a tea bag. Thirty-two volunteers used sensory assessment to assess the prepared tea. The product underwent a quality assessment utilizing a questionnaire based on the Hedonic scale to detect and rank its degree of acceptance. The panelists preferred the 6-gram per serving Atis (*Annona squamosa*) tea for its color, taste, and smell, giving it a very high rating for approval. Additionally, the three treatments varied significantly in color, flavor, and aroma, with the amount of Atis tea used affecting these characteristics.

In another study, the nutritional composition of custard apple bagasse (CAB) flour was determined after extracting the juice. The flour contained 5.7% protein, 76.4% carbohydrate, 5.35% fat, and 10.4% moisture with 376 kcal/100 g energy. CAB flour is also a rich mineral (P, K, Cu, Fe, Mg, Mn, Zn, and Ca) source containing a high concentration of phenolic compounds. The customers responded favorably to the CAB flour's inclusion in cookie formulation, expressing an increased preference for some formulations of the product compared to cookies made entirely of wheat flour. The results showed that flour produced from custard apple pulp bagasse can be used in food preparation to improve nutritional characteristics (Souza et al. 2018).

A sensory study was conducted on custard apple peel extract incorporated in chicken breast fillets. For this purpose, the particular sized and shaped chicken breast fillets were derived from fresh chicken birds. Then, they were submerged in 0.1, 0.3, and 0.5% aqueous solutions of hydro-ethanolic (50%) custard apple peel extract (CPeE) for 10 minutes. 100 ppm of BHT and distilled water were used for the positive and negative controls, respectively. Low-density polyethylene (LDPE) bags were used for storage under refrigeration (4 ± 1 °C). CPeE-incorporated meat samples proved to be satisfactory in terms of taste, smell, and appearance until the ninth day of storage with enhanced total phenol content and ascorbic acid. CPeE-incorporated chicken breast fillets also showed good antimicrobial activity against *P. putida*, *E. fecalis*, *E. coli*, and *S. putrefaciens* at a concentration ranging from 250 to 1250 mg/mL (Kadam et al. 2020).

Fresh and mature custard apple powder was developed and added in various ratios (10 to 25%) to pudding mixture formulas. Physicochemical and sensory properties were evaluated for the pudding mixtures. Sensory analysis revealed that the 20% custard apple powder-fortified pudding received great marks for its appearance, flavor, and general appeal compared to other treatments. Custard apple flour (20%) and 80% milk were combined to create pudding with increased nutritive and sensory qualities. The findings of the research can be utilized to create a custard apple powder-based pudding mixture with high acceptance and a wealth of nutrients (Vidanarachchi et al. 2021).

Response surface methodology was used to explore the application of fructooligosaccharides (FOS), custard apple puree (CAP), and soy protein hydrolysate (SPH) as partial or full substitutes for sugar, wheat flour, and powdered whole milk, respectively. The best preparation for low-carbohydrate (carb) pound cake has a desirability score of 93.7% and replaces wheat flour, sucrose, and powdered whole milk with 41% CAP, 94.9% FOS, and 100% SPH. The number of digestible carbs was reduced by 61%. The starch level and nonreducing sugar content were significantly reduced due to wheat flour's substitution and sucrose's partial substitution. The low-carb pound cake, with a low glycemic load (4.7%) and moderate glycemic index (60%), was considered expertly acceptable with a 72% sensory acceptability grade (de Souza et al. 2018).

7.7 CURRENT AND FUTURE PERSPECTIVES

This chapter focuses on the nutritional potentials, antioxidants, and other chemical compounds obtained from different parts of *A. squamosa*. Their health benefits and incorporation in food products are evidenced by various studies highlighted above. For example, studies on anti-psoriatic activity can be a gateway to the development of cosmeceuticals based on *A. squamosa* by targeting the expression of epidermal growth factor receptor EGFR biomarker. *Annona squamosa* leaf extract can be used to develop anti-aging cream according to Mahawar et al. (2019) due to antioxidant activity.

A similar study (Omar et al. 2020) suggests that *Annona muricata* leaf extract with strong antioxidant activity can be used in cosmetic cream. Moreover, *Annona squamosa* leaf (methanol) extract effectively protects skin against UVA irradiation (Mohamad et al. 2022). The anti-aging skin rejuvenating effects of *Annona squamosa* are due to the presence of acetogenins according to Brownell et al. (2022). Moreover, N-acyltryptamine present in ethanol extract of *A. squamosa* fruit is very effective against atopic dermatitis (Somei 2021).

Many plant species are currently underutilized and unexplored in India. However, research on custard apple in India dates back to 1954 and maybe even earlier. Studies in India were the first to report the single randomized clinical study on diabetic subjects (Agarwal et al. 2014) as well as other pioneering research on the mechanism of action and health benefits of *A. squamosa* on chronic diseases. Besides, custard apple is used in traditional folk medicine for lice, diabetes, ulcer, hypertension, wounds, hemorrhage, and epilepsy. Young *A. squamosa* leaves often are part of polyherbal Indian traditional medicine, for example, in combination with black pepper (*Piper nigrum*) as an antidiabetic still in use today. *A. squamosa* also appears in many Ayurvedic medicinal formulations and often is sourced from traditional "*local Sitaphal*" growing wild in the forests in India. Custard apple is also cultivated and grown in many states with Maharashtra and Madhya Pradesh growing and producing the most fruit (~65% and 25% of total area and production, respectively). Custard apple is a profitable crop in Western Maharashtra with a 1.30 cost-benefit ratio indicating its growth potential, particularly in rainfed areas, according to the economic analysis of Gawali et al. (2016). Several reviews highlight value addition of *A. squamosa* plant by-products from the fruit, seed, for example, as well as leaves to reduce agricultural waste for improved circular economy. A good example is applications in catalysis and cell bioimaging using *A. squamosa* fruits to prepare biogenic fluorescent nitrogen-doped carbon dots (Dhanush et al. 2022). Apparently, *A. squamosa* seed oil is available all over India and pyrolysis has been used to produce an alternative fuel for internal combustion engines with increased brake thermal efficiency and lower nitric oxide emission (Kumar and Periyasamy 2019). The fatty acid composition of custard apple seed oil is appropriate for biological lubricant and has the capacity to replace mineral-based lubricants available worldwide (Ashokkumar et al. 2020).

A. squamosa plant and its parts (fruits, leaves, bark, twigs, and roots) have been investigated because of their abundance of secondary metabolites including alkaloids, acetogenins, and essential oils. Some of these secondary metabolites have been used to synthesize nanoparticles to increase their bioavailability and bioactivities as natural insecticides or pesticides to enhance rice crop production. It is foreseeable that *A. squamosa* will be a common commodity with extensive use in functional food and nutraceuticals as well as in other non-food industries.

7.8 ACKNOWLEDGEMENTS

This study was funded by The University Grant Commission, New Delhi, India.

7.9 CONFLICT OF INTEREST

The authors declare no conflict of interest.

REFERENCES

Abd-Elghany, A. A., Ahmed, S. M., Masoud, M. A., Atia, T., Waggiallah, H. A., El-Sakhawy, M. A., & Mohamad, E. A. (2022). *Annona squamosa* L. extract-loaded niosome and its anti-ehrlich ascites' carcinoma activity. *ACS Omega*, *7*(43), 38436–38447.

Abdualrahman, M. A. Y., Ma, H., Zhou, C., Yagoub, A. E. A., Ali, A. O., Tahir, H. E., & Wali, A. (2019). Postharvest physicochemical properties of the pulp and seed oil from Annona squamosa L. (Gishta) fruit grown in Darfur region, Sudan. *Arabian Journal of Chemistry*, *12*(8), 4514–4521. https://doi.org/10.1016/j.arabjc.2016.07.008

Abolbaghaei, A., Oomah, B. D., Tavakoli, H., & Hosseinian, F. (2018). Influence of branched chain amino acids on insulin sensitivity and the mediator roles of short chain fatty acids and gut hormones: A review. *Journal of Food Bioactives*, *2*, 1–15.

Adil, M. (2019). Qualitative features and therapeutic value of custard apple fruit at refregiration storage. *International Journal of Food and Allied Sciences*, *4*(1), 12–17. https://doi.org/10.21620/ijfaas.2018112-17

Agarwal, V., Bhadu, I., Goyal, S., Agrawal, R. P., & Goyal, D. (2014). PO409 Efficacy & safety of *Annona squamosa* (sugar apple) on diabetic subjects: A randomized, open label controlled study. *Diabetes Research and Clinical Practice*, *106*, S254. https://doi.org/10.1016/S0168-8227(14)70703-X

Agrawal, M., Agrawal, Y., Itankar, P., Patil, A., Vyas, J., & Kelkar, A. (2012). Phytochemical and HPTLC studies of various extracts of *Annona squamosa* (Annonaceae). *International Journal of PharmTech Research*, *4*(1).

Akomolafe, S. F., & Ajayi, O. B. (2015). A comparative study on antioxidant properties, proximate and mineral compositions of the peel and pulp of ripe *Annona muricata* (L.) fruit. *International Food Research Journal*, *22*(6), 2381–2388.

Alaqeel, N. K., Almalki, W. H., Binothman, N., Aljadani, M., Al-Dhuayan, I. S., Alnamshan, M. M., . . . & Tarique, M. (2023). The inhibitory and anticancer properties of *Annona squamosa* L. seed extracts. *Brazilian Journal of Biology*, *82*, e268250. https://doi.org/10.1590/1519-6984.268250

Albuquerque, T. G., Santos, F., Sanches-Silva, A., Oliveira, M. B., Bento, A. C., & Costa, H. S. (2016). Nutritional and phytochemical composition of *Annona cherimola* Mill. fruits and by-products: Potential health benefits. *Food Chemistry*, *193*, 187–195. https://doi.org/10.1016/j.foodchem.2014.06.044

Al-Ghazzawi, A. M. (2019). Anti-cancer activity of new benzyl isoquinoline alkaloid from Saudi plant *Annona squamosa*. *BMC Chemistry*, *13*(1), 1–6. https://doi.org/10.1186/s13065-019-0536-4.

Ali, S. H., & Jafar, Z. J. (2022a). Comparison of antibacterial efficacy of *Annona squamosa* mouthwash with chlorhexidine for children. *Journal of Research in Medical and Dental Science*, *10*(1), 87–94.

Ali, S. H., & Jafar, Z. J. (2022b). *In vitro* cytotoxic effect of *Annona squamosa* pulp extract as a mouthwash for children on human normal cell line. *Journal of Baghdad College of Dentistry*, *34*(1), 60–66.

Al Kazman, B. S., Harnett, J. E., & Hanrahan, J. R. (2022). Traditional uses, phytochemistry and pharmacological activities of Annonacae. *Molecules*, *27*(11), 3462.

Al-Nemari, R., Al-Senaidy, A., Semlali, A., Ismael, M., Badjah-Hadj-Ahmed, A. Y., & Ben Bacha, A. (2020). GC-MS profiling and assessment of antioxidant, antibacterial, and anticancer properties of extracts of *Annona squamosa* L. leaves. *BMC Complementary Medicine and Therapies*, *20*(1), 1–14. https://doi.org/10.1186/s12906-020-03029-9

Al-Nemari, R., Bacha, A. B., Al-Senaidy, A., Almutairi, M. H., Arafah, M., Al-Saran, H., . . . & Semlali, A. (2022). Cytotoxic effects of *Annona squamosa* leaves against breast cancer cells via apoptotic signaling proteins. *Journal of King Saud University—Science*, *34*(4), 102013. https://doi.org/10.1016/j.jksus.2022.102013

Alvionita, M., & Oktavia, I. (2019). Bioactivity of flavonoid in ethanol extract of *Annona squamosa* L. fruit as xanthine oxidase inhibitor. In *IOP Conference Series: Materials Science and Engineering, 546*(6), 062003. IOP Publishing.

Amin, M., Saleh, I., & Hidayat, R. (2019). The anticancer activity of Srikaya Leaves fraction (*Annona squamosa* L.): An in vitro study. *Bioscientia Medicina: Journal of Biomedicine and Translational Research*, *3*(4), 39–44. https://doi.org/10.32539/bsm.v3i4.102

Amoo, I. A., Emenike, A. E., & Akpambang, V. O. E. (2008). Compositional evaluation of Annona cherimoya (Custard Apple) fruit. *Trends in Applied Sciences Research*, *3*(2), 216–220. https://doi.org/10.3923/tasr.2008.216.220

Andrade, E. H. A., Maria das Graces, B. Z., Maia, J. G. S., Fabricius, H., & Marx, F. (2001). Chemical characterization of the fruit of *Annona squamosa* L. occurring in the Amazon. *Journal of Food Composition and Analysis*, *14*(2), 227–232. https://doi.org/10.1006/jfca.2000.0968

Ansari, P., Akther, S., Khan, J. T., Islam, S. S., Masud, M., Samim, R., . . . & Abdel-Wahab, Y. H. (2022a). Hyperglycaemia-linked diabetic foot complications and their management using conventional and alternative therapies. *Applied Sciences*, *12*(22), 11777. https://doi.org.10.3390/app122211777

Ansari, P., Hannan, J. M. A., Seidel, V., & Abdel-Wahab, Y. H. (2022b). Polyphenol-rich leaf of *Annona squamosa* stimulates insulin release from BRIN-BD11 cells and isolated mouse islets, reduces (CH_2O) digestion and absorption, and improves glucose tolerance and GLP-1 (7–36) levels in high-fat-fed rats. *Metabolites*, *12*(10), 995. https://doi.org/10.3390/metabo1200995

Ao, H., Lu, L., Li, M., Han, M., Guo, Y., & Wang, X. (2022). Enhanced solubility and antitumor activity of *Annona squamosa* seed oil via nanoparticles stabilized with TPGS: Preparation and *in vitro* and *in vivo* evaluation. *Pharmaceutics*, *14*(6), 1232. https://doi.org/10.3390/pharmaceutics14061232

Araya, H., Hara, N., Fujimoto, Y., & Sahai, M. (1994). Squamostanal-A, apparently derived from tetrahydrofuranic acetogenin, from *Annona squamosa*. *Bioscience, Biotechnology, and Biochemistry*, *58*(6), 1146–1147. https://doi.org/10.1271/BBB.58.1146

Ashokkumar, S., Elanthiraiyan, A., Shagu, S., Srinivasan, M., & Deepan, K. B. (2020, December). Comparison and analysis of custard apple seed oil with engine lubricant (Bio Lubricant). *IOP Conference Series: Materials Science and Engineering*, *993*(1), 012008. https://doi.org/10.1088/1757-899X/993/1/012008

Avula, B., Bae, J. Y., Majrashi, T., Wu, T. Y., Wang, Y. H., Wang, M., . . . & Khan, I. A. (2018). Targeted and non-targeted analysis of annonaceous alkaloids and acetogenins from *Asimina* and *Annona* species using UHPLC-QToF-MS. *Journal of Pharmaceutical and Biomedical Analysis*, *159*, 548–566.

Bala, S., Nigam, V. K., Singh, S. S., Kumar, A., & Kumar, S. (2018). Evaluation of nutraceutical applications of *Annona squamosa* L. based food products. *Journal of Pharmacognosy and Phytochemistry*, *SP1*, 827–831.

Bapat, V. A., Jagtap, U. B., Ghag, S. B., & Ganapathi, T. R. (2020). Molecular approaches for the improvement of under-researched tropical fruit trees: Jackfruit, guava, and custard apple. *International Journal of Fruit Science*, *20*(3), 233–281. https://doi.org/10.1080/15538362.2019.1621236

Baskaran, R., Pullencheri, D., & Somasundaram, R. (2016). Characterization of free, esterified and bound phenolics in custard apple (*Annona squamosa* L) fruit pulp by UPLC-ESI-MS/MS. *Food Research International*, *82*, 121–127. https://doi.org/10.1016/j.foodres.2016.02.001

Bawale, S. M., Narwade, S. G., & Kamble, N. S. (2018). Studies on preparation of basundi blended with custard apple pulp. *The Pharma Innovation Journal*, *8*(9), 307–309.

Bhardwaj, A., Satpathy, G., & Gupta, R. K. (2014). Preliminary screening of nutraceutical potential of *Annona squamosa*, an underutilized exotic fruit of India and its use as a valuable source in functional foods. *Journal of Pharmacognosy and Phytochemistry*, *3*(2), 172–180.

Bhattacharya, A., & Chakraverty, R. (2016). The pharmacological properties of *Annona squamosa* Linn: A review. *International Journal of Pharmacy and Engineering*, *4*(2), 692–699.

Biba, V. S., Lakshmi, S., Dhanya, G. S., & Remani, P. (2013). Phytochemical analysis of *Annona squamosa* seed extracts. *International Research Journal of Pharmaceutical and Applied Sciences*, *3*(4), 29–31.

Bonavia, D., Ochoa, C. M., Oscar Tovar, S., & Palomino, R. C. (2004). Archaeological evidence of cherimoya (*Annona cherimolia* Mill.) and guanabana (Annona muricata L.) in ancient Peru. *Economic Botany*, *58*(4), 509–522.

Brownell, L. A., Chu, M., Corneliusen, B., Hong, M. F., Hwang, J. H., Hyun, E. J., . . . & Ylmam, M. (2022). Skin care compositions and methods of use thereof. *U.S. Patent 11,446,225 B2*, issued 20 September.

Can-Cauich, C. A., Sauri-Duch, E., Betancur-Ancona, D., Chel-Guerrero, L., González-Aguilar, G. A., Cuevas-Glory, L. F., . . . & Moo-Huchin, V. M. (2017). Tropical fruit peel powders as functional ingredients: Evaluation of their bioactive compounds and antioxidant activity. *Journal of Functional Foods*, *37*, 501–506. https://doi.org/10.1016/j.jff.2017.08.028

Caparros-Lefebvre, D., & Elbaz, A. (1999). Possible relation of atypical parkinsonism in the French West Indies with consumption of tropical plants: a case-control study. *The Lancet*, *354*(9175), 281–286.

Chandel, S. S., Dikshit, S. N., & Sharma, H. G. (2018). Collection and evaluation of custard apple (*Annona squamosa* L.) genotypes in Chhattisgarh plains. *Journal of Pharmacognosy and Phytochemistry*, Published Online, 149–152.

Chandraju, S., Mythily, R., & Kumar, C. C. (2012). Qualitative chromatographic analysis of sugars present in non-edible rind portion of custard apple (*Annona squamosa* L.). *Journal of Chemical and Pharmaceutical Research*, *4*(2), 1312–1318.

Chavan, M. J., Shinde, D. B., & Nirmal, S. A. (2006). Major volatile constituents of *Annona squamosa* L. bark. *Natural Product Research*, *20*(8), 754–757. https://doi.org/10.1080/14786410500138823

Chen, Y., Chen, Y., Shi, Y., Ma, C., Wang, X., Li, Y., . . . & Li, X. (2016). Antitumor activity of *Annona squamosa* seed oil. *Journal of Ethnopharmacology*, *193*, 362–367.

Chen, Y. Y., Peng, C. X., Hu, Y., Bu, C., Guo, S. C., Li, X., . . . & Chen, J. W. (2017). Studies on chemical constituents and anti-hepatoma effects of essential oil from *Annona squamosa* L. pericarps. *Natural Product Research*, *31*(11), 1305–1308.

Chen, Y., Chen, J., Wang, Y., Xu, S., & Li, X. (2012). Six cytotoxic annonaceous acetogenins from Annona squamosa seeds. *Food Chemistry*, *135*(3), 960–966. https://doi.org/10.1016/j.foodchem.2012.05.041

Chikhalikar, N. V., Sahoo, A. K., Singhal, R. S., & Kulkarni, P. R. (2000). Studies on frozen pourable custard apple (*Annona squamosa* L.) pulp using cryoprotectants. *Journal of the Science of Food and Agriculture*, *80*(9), 1339–1342.

Dahiya, R., & Dahiya, S. (2021). Natural bioeffective cyclooligopeptides from plant seeds of Annona genus. *European Journal of Medicinal Chemistry*, *214*, 113221.

Dang, Q. L., Kim, W. K., Nguyen, C. M., Choi, Y. H., Choi, G. J., Jang, K. S., . . . & Kim, J. C. (2011). Nematicidal and antifungal activities of annonaceous acetogenins from *Annona squamosa* against various plant pathogens. *Journal of Agricultural and Food Chemistry*, *59*(20), 11160–11167. https://doi.org/10.1021/jf203017f

Dash, S., Kar, B., Sahoo, N., & Pattnaik, G. (2020). Annonaine an alkaloid from leaves of custard apple (*Annona squamosa*): A comprehensive review on its phytochemicals and pharmacological activities. *Asian Journal of Chemistry*, *32*(8), 1824–1836.

Dellai, A., Maricic, I., Kumar, V., Arutyunyan, S., Bouraoui, A., & Nefzi, A. (2010). Parallel synthesis and anti-inflammatory activity of cyclic peptides cyclosquamosin D and Met-cherimolacyclopeptide B and their analogs. *Bioorganic and Medicinal Chemistry Letters*, *20*(19), 5653–5657. https://doi.org/10.1016/j.bmcl.2010.08.033

Deng, G. F., Xu, D. P., Li, S., & Li, H. B. (2015). Optimization of ultrasound-assisted extraction of natural antioxidants from sugar apple (*Annona squamosa* L.) peel using response surface methodology. *Molecules*, *20*(11), 20448–20459. https://doi.org/10.3390/molecules201119708

Derangula, S. S. R., Muthiah, N. S., Surendra, B. V., Reddy, K. S., Somashekar, H. S., Sukumar, E., & Prabhu, K. (2022). Comparative study on hepatoprotective activity of *Pongamia pinnata* (PP) & *Annona squamosa* (AS) leaf extracts against anti-tubercular drugs (isoniazid & rifampin) induced hepatotoxicity in rats. *International Journal of Health Sciences*, 1686–1697. https://doi.org/10.53730/ijhs.v6nS2.5174

de Sousa Galvão, M., de Santana, K. L., Nogueira, J. P., Neta, M. T. S. L., & Narain, N. (2020). Method optimization study on isolation of volatile compounds by headspace solid-phase microextraction (HS-SPME) from custard apple (*Annona squamosa* L.) pulp. *Journal of Analytical Sciences, Methods and Instrumentation*, *10*(3), 59–77. https://doi.org/10.4236/jasmi.2020.103005

de Souza, A. R., & Schmiele, M. (2021). Custard apple puree, fructooligosaccharide and soy protein hydrolysate as alternative ingredients in low carb pound cake. *Journal of Food Science and Technology*, *58*(9), 3632–3644. https://doi.org/10.1007/s13197-021-05155-9

Dev, A. A., & Joseph, S. M. (2021). Anticancer potential of Annona genus: A detailed review. *Journal of the Indian Chemical Society*, *98*(12), 100231. https://doi.org/10.1016/j.jics.2021.100231

Dhanraj, S. R., Vennila, J. J., & Dhanraj, M. (2020). Pharmacological investigation of ribosome inactivating protein (RIP)–like protein extracted from *Annona squamosa* L. seeds. *Journal of King Saud University-Science*, *32*(7), 2982–2988.

Dhanush, C., Aravind, M. K., Ashokkumar, B., & Sethuraman, M. G. (2022). Synthesis of blue emissive fluorescent nitrogen doped carbon dots from *Annona squamosa* fruit extract and their diverse applications in the field of catalysis and bio-imaging. *Journal of Photochemistry and Photobiology A: Chemistry*, *432*, 114097.

Dikshit, N., Bharad, S. G., & Badge, M. P. (2008). Diversity in custard apple germplasm collections from Maharastra, India. *Indian Journal of Plant Genetic Resources*, *21*(1), 95–96.

Dilrukshi, M. K. D. T., Dharmadasa, R. M., Abeysinghe, D. C., & Abhayagunasekara, A. V. C. (2020). Selection of superior quality Annona species by means of bioactive compounds and antioxidant capacity. *World Journal of Agricultural Research*, *8*(2), 39–44. https://doi.org/10.12691/wjar-8-2-3

Du, J., Zhong, B., Subbiah, V., Barrow, C. J., Dunshea, F. R., & Suleria, H. A. R. (2021). Lc-esi-qtof-ms/ms profiling and antioxidant activity of phenolics from custard apple fruit and by-products. *Separations*, *8*(5), 62. https://doi.org/10.3390/separations8050062

Durán, A. G., Gutiérrez, M. T., Mejías, F. J., Molinillo, J. M., & Macías, F. A. (2021). An overview of the chemical characteristics, bioactivity and achievements regarding the therapeutic usage of acetogenins from *Annona cherimola* Mill. *Molecules*, *26*(10), 2926. https://doi.org/10.3390/molecules26102926

El-Aidie, S. A. M., Hatem, H. O., Ewis, A. M., & El-Nimer, A. M. (2020). Production of nutraceutical probiotic whey–based beverage fortified with *Annona squamosa* L (custard Apple) fruit. *Journal of Food and Dairy Sciences*, *11*(6), 171–177. https://doi.org/10.21608/jfds.2020.106390

El-Chaghaby, G. A., Ahmad, A. F., & Ramis, E. S. (2014). Evaluation of the antioxidant and antibacterial properties of various solvents extracts of Annona squamosa L. leaves. *Arabian Journal of Chemistry*, *7*(2), 227–233. https://doi.org/10.1016/j.arabjc.2011.06.019

Fadholly, A., Proboningrat, A., Iskandar, R. P. D., Rantam, F. A., & Sudjarwo, S. A. (2019). In vitro anticancer activity *Annona squamosa* extract nanoparticle on WiDr cells. *Journal of Advanced Pharmaceutical Technology & Research*, *10*(4), 149–154.

Fournier, G., Leboeuf, M., & Cavé, A. (1999). Annonaceae essential oils: A review. *Journal of Essential Oil Research*, *11*(2), 131–142.

Gajalakshmi, S., Vijayalakshmi, S., & Devi, R. V. (2012). Phytochemical and pharmacological properties of *Annona muricata*: A review. *International Journal of Pharmacy and Pharmaceutical Sciences*, *4*(2), 3–6.

García-Aguirre, K. K., Zepeda-Vallejo, L. G., Ramón-Gallegos, E., Alvárez-González, I., & Madrigal-Bujaidar, E. (2008). Genotoxic and cytotoxic effects produced by acetogenins obtained from *Annona cherimolia* Mill. *Biological and Pharmaceutical Bulletin*, *31*(12), 2346–2349. https://doi.org/10.1248/bpb.31.2346

Garg, S. N., & Gupta, D. (2005). Composition of the leaf oil of *Annona squamosa* L. from North Indian Plains. *Journal of Essential Oil Research*, *17*, 257–258.

Gautam, D., Jain, S. K., Bhatnagar, P., Meena, N., & Chhipa, H. (2021). Utilization of custard apple pulp for preparation of blended nectar. *Indian Journal of Horticulture*, *78*, 229–235. https://doi.org/10.5958/0974-0112.2021.00033.5

Gawali, A. S., Kumbhar, J. S., & Yadav, D. B. (2016). Economic analysis of custard apple (*Annona squmosa* L.) production in Western Maharashtra. *Journal of Tree Sciences*, *35*(1), 30–33.

Ghawade, P. M., Supe, V. S., Shete, M. B., & Idate, G. M. (2018). Biochemical characterization of custard apple genotypes. *Journal of Pharmacognosy and Phytochemistry*, *7*(4), 636–638.

Gökbel, H., Harmankaya, M., & Özcan, M. M. (2015). Determination of metal, non-metal and heavy metal contents of some tropical fruits growing in Indonesia. *Quality Assurance and Safety of Crops & Foods*, *7*(4), 545–549.

González-Esquinca, A., Morales, M., & Toriz, F. (2004). Oxoaporphine alkaloids in Guatteria diospyroides Baill. and *Annona squamosa* L. (Annonaceae):(with 1 figure). *International Journal of Experimental Botany*, *73*, 53–55.

Gupta, R. K., Kesari, A. N., Watal, G., Murthy, P. S., Chandra, R., Maithal, K., & Tandon, V. (2005a). Hypoglycaemic and antidiabetic effect of aqueous extract of leaves of *Annona squamosa* (L.) in experimental animal. *Current Science*, *88*(8), 1244–1254.

Gupta, R. K., Kesari, A. N., Watal, G., Murthy, P. S., Chandra, R., & Tandon, V. (2005b). Nutritional and hypoglycemic effect of fruit pulp of *Annona squamosa* in normal healthy and alloxan-induced diabetic rabbits. *Annals of Nutrition and Metabolism*, *49*(6), 407–413. https://doi.org/10.1159/000088987

Habib, M. R., Igarashi, Y., & Rabbi, M. A. (2022). *In vitro* inhibitory activity of *Annona squamosa* leaves against enzymes associated with metabolic disorders. *Journal of Research in Pharmacy*, *26*(5), 1272–1280.

Hashiesh, H. M., Meeran, M. F. N., Sharma, C., Sadek, B., Kaabi, J. A., & Ojha, S. K. (2020). Therapeutic potential of β-caryophyllene: A dietary cannabinoid in diabetes and associated complications. *Nutrients*, *12*, 2963. https://doi.org/10.3390/nu12102963

Hassan, L. G., Muhammad, M. U., Umar, K. J., & Sokoto, A. M. (2008). Comparative study on the proximate and mineral contents of the seed and pulp of sugar apple (*Annona squamosa*). *Nigerian Journal of Basic and Applied Sciences*, *16*(2), 179–182.

Hendawy, O. M., ELBana, M. A., Abdelmawlla, H. A., Maliyakkal, N., & Mostafa-Hedeab, G. (2019). Effect of *Annona squamosa* ethanolic and aqueous leave extracts on aluminum chloride-induced neuroinflammation in albino rats. *Biomedical and Pharmacology Journal*, *12*(4), 1723–1730.

Ibrahim, F., Jaber, A., Ibrahim, G., & Cheble, E. (2020). Antioxidant activity and total phenol content of different parts from Lebanese *Annona squamosa* L. *International Journal of Pharmacy and Pharmaceutical Sciences*, *12*(8), 100–105. https://doi.org/10.22159/ijpps.2020v12i8.36992

Jagtap, U. B., & Bapat, V. A. (2012). Antioxidant activities of various solvent extracts of custard apple (*Annona squamosa* L.) fruit pulp. *Nutrafoods*, *11*, 137–144.

Jain, C., Champawat, P. S., Mudgal, V. D., Madhu, B., & Jain, S. K. (2019). Post-harvest processing of custard apple (*Annona squamosa* L.): A review. *International Journal of Chemical studies*, *7*(3), 1632–1637.

Jamkhande, P. G., & Wattamwar, A. S. (2015). *Annona reticulata* Linn. (Bullock's heart): Plant profile, phytochemistry and pharmacological properties. *Journal of Traditional and Complementary Medicine*, *5*(3), 144–152. https://doi.org/10.1016/j.jtcme.2015.04.001

Johns, T., Windust, A., Jurgens, T., & Mansor, S. M. (2011). Antimalarial alkaloids isolated from *Annona squamosa*. *Phytopharmacology*, *1*(3), 49–53.

Joseph, S. M., & Dev, A. R. A. (2021). Analysis of essential oil constituents of three Annona species growing in Kerala, India. *International Journal of Research in Ayurveda Pharmacy*, *12*(5), 45–48.

Joseph, S. M., & Dev, A. R. A. (2023). Annonaceae: Tropical medicinal plants with potential anticancer acetogenins ans alkaloids: In: K. Arunachalam, X. Yang and S. P. Sasidharan (eds.), *Bioprospecting of Tropical Medicinal Plants*. Springer Nature Switzerland AG, pp. 565–588. https://doi.org/10.1007/978-3-031-28780-0_22

Joy, B., & Rao, J. M. (1997). Essential oil of the leaves of *Annona squamosa* L. *Journal of Essential Oil Research*, *9*(3), 349–350.

Kadam, B. R., Ambadkar, R. K., Somkuwar, A. P., Kurkure, N. V., Bonde, S. W., Chaudhari, S. P., & Waskar, V. S. (2020). Effect of custard apple (Annona squamosa) seed extract on quality of chicken breast fillets. *Asian Journal of Dair & Food Research, 329*(1).

Kalidindi, N., Thimmaiah, N. V., Jagadeesh, N. V., Nandeep, R., Swetha, S., & Kalidindi, B. (2015). Antifungal and antioxidant activities of organic and aqueous extracts of *Annona squamosa* Linn. leaves. *Journal of Food and Drug Analysis*, *23*(4), 795–802. https://doi.org/10.1016/j.jfda.2015.04.012

Karthikeyan, K., Abitha, S., & Kumar, V. S. (2016). Identification of bioactive constituents in peel, pulp of prickly custard apple (*Annona muricata*) and its antimicrobial activity. *International Journal of Pharmacognosy and Phytochemical Research*, *8*(11), 1833–1838.

Kaur, R., Afzal, M., Kazmi, I., Ahamd, I., Ahmed, Z., Ali, B., . . . & Anwar, F. (2013). Polypharmacy (herbal and synthetic drug combination): A novel approach in the treatment of type-2 diabetes and its complications in rats. *Journal of Natural Medicines*, *67*, 662–671.

Kavitha, V. V. (2015) Evaluation of antioxidant, carotene and alkaloid content in *Spinach oleracea* and *Annona squamosa*. *International Journal of Pharma and Bio Sciences*, *6*(2), B1071–B1076; ISSN:0975–6299.

Kelloff, G. J., Hawk, E. T., & Sigman, C. C. (Eds.). (2008). *Cancer Chemoprevention: Strategies for Cancer Chemoprevention, 2*. Springer Science & Business Media.

Kim, D. H., Ma, E. S., Suk, K. D., Son, J. K., Lee, J. S., & Woo, M. H. (2001). Annomolin and annocherimolin, new cytotoxic annonaceous acetogenins from *Annona cherimolia* seeds. *Journal of Natural Products*, *64*(4), 502–506.

Kothari, V., & Seshadri, S. (2010). Antioxidant activity of seed extracts of *Annona squamosa* and Carica papaya. *Nutrition and Food Science*, *40*(4), 403–408. https://doi.org/10.1108/00346651011062050

Krishnaiah, V. C., Saleem, T. M., Sujatha, D., Pusphakumari, B., & Anitha, G. (2013). Putative antioxidant property of *Annona squamosa* on acetic acid induced ulcerative colitis. *Journal of Advanced Pharmacy Education & Research*, *3*(2), 177–186.

Kumar, M., Changan, S., Tomar, M., Prajapati, U., Saurabh, V., Hasan, M., . . . & Mekhemar, M. (2021). Custard apple (*Annona squamosa* L.) leaves: Nutritional composition, phytochemical profile, and health-promoting biological activities. *Biomolecules*, *11*(5), 614. https://doi.org/10.3390/biom11050614

Kumar, P. S., & Periyasamy, S. (2019). Performance, combustion and emission characteristics of a diesel engine fuelled with the *Annona Squamosa* seed cake pyrolysis oil. *International Advanced Research Journal in Science, Engineering and Technology*, *6*(5), 39–46.

Kumari, N., Prakash, S., Kumar, M., Radha, Zhang, B., Sheri, V., . . . & Lorenzo, J. M. (2022). Seed waste from Custard Apple (*Annona squamosa* L.): A comprehensive insight on bioactive compounds, health promoting activity and safety profile. *Processes*, *10*(10), 1–20. https://doi.org/10.3390/pr10102119

Larrota, H. R., & Baquero, L. C. P. (2018). Antioxidant activity of ethanolic extracts and alkaloid fractions from seeds of three species of Annona. *Pharmacologyonline*, *2*, 206–218.

Leite, D. O., de FA Nonato, C., Camilo, C. J., de Carvalho, N. K., da Nobrega, M. G., Pereira, R. C., & da Costa, J. G. (2020). Annona genus: Traditional uses, phytochemistry and biological activities. *Current Pharmaceutical Design*, *26*(33), 4056–4091.

Lima, L. A. S., Pimenta, L. P., & Boaventura, M. A. D. (2010). Acetogenins from *Annona cornifolia* and their antioxidant capacity. *Food Chemistry*, *122*, 1129–1138. https://doi.org/10.1016/j.foodchem.2010.03.100

Lonarkar, A. G., Rathod, K. S., Ambadkar, R. K., Patil, P. S., & Dhagare, L. G. (2021). Shelf life of chicken samosa incorporated with custard apple (*Annona squamosa*) peel powder. *Journal of Meat Science*, *16*(1and2), 17–22. https://doi.org/10.5958/2581-6616.2021.00005.0

Luzia, D. M. M., & Jorge, N. (2012). Soursop (*Annona muricata* L.) and sugar apple (*Annona squamosa* L.): Antioxidant activity, fatty acids profile and determination of tocopherols. *Nutrition and Food Science*, *42*(6), 434–441. https://doi.org/10.1108/00346651211277690

Lydia, D. E., John, S., Swetha, V. K., & Sivapriya, T. (2017). Investigation on the antimicrobial and antioxidant activity of custard apple (*Annona reticulata*) peel extracts. *Research Journal of Pharmacognosy and Phytochemistry*, *9*(4), 241–247. https://doi.org/10.5958/0975-4385.2017.00045.0

Ma, C., Chen, Y., Chen, J., Li, X., & Chen, Y. (2017). A review on *Annona squamosa* L.: Phytochemicals and biological activities. *American Journal of Chinese Medicine*, *45*(5), 933–964. https://doi.org/10.1142/S0192415X17500501

Mahawar, V., Patidar, K., & Joshi, N. (2019). Development and evaluation of herbal antiaging cream formulation containing *Annona squamosa* leaf extract. *Asian Journal of Pharmceutical and Clincal Research*, *12*, 210–214.

Malik, M., Iqbal, M. A., Malik, M., Raza, M. A., Shahid, W., Choi, J. R., & Pham, P. V. (2022). Biosynthesis and characterizations of silver nanoparticles from *Annona squamosa* leaf and fruit extracts for size-dependent biomedical applications. *Nanomaterials*, *12*(4), 616. https://doi.org/10.3390/nano12040616

Manochai, B., Ingkasupart, P., Lee, S. H., & Hong, J. H. (2018). Evaluation of antioxidant activities, total phenolic content (TPC), and total catechin content (TCC) of 10 sugar apple (*Annona squamosa* L.) cultivar peels grown in Thailand. *Food Science and Technology*, *38*(Suppl.1), 294–300.

Manvi, F. V., Nanjawade, B. K., & Shing, S. (2011). Pharmacological screening of combined extract of *Annona squamosa* and *Nigella sativa*. *International Journal of Pharma and Bio Sciences*, *2*(2), 520–529.

Marahatta, A. B., Aryal, A., & Basnyat, R. C. (2019). The phytochemical and nutritional analysis and biological activity of *Annona squamosa* Linn. *International Journal of Herbal Medicine*, *7*(4), 19–28.

Mariod, A. A., Abdelwahab, S. I., Elkheir, S., Ahmed, Y. M., Fauzi, P. N. M., & Chuen, C. S. (2012). Antioxidant activity of different parts from *Annona squamosa*, and *Catunaregam nilotica* methanolic extract. *Acta Scientiarum Polonorum Technologia Alimentaria*, *11*(3), 249–258.

Mariod, A. A., Elkheir, S., Ahmed, Y. M., & Matthäus, B. (2010). *Annona squamosa* and *Catunaregam nilotica* seeds, the effect of the extraction method on the oil composition. *Journal of the American Oil Chemists' Society*, *87*(7), 763–769.

Mariod, A. A., Mirghani, M. E. S., & Hussein, I. H. (2017). *Unconventional Oilseeds and Oil Sources*, pp. 145–150. Academic Press.

Martinez-Maldonado, F. E., Miranda-Lasprilla, D., Magnitskiy, S., & Melgarejo, L. M. (2015). Germination, protein contents and soluble carbohydrates during storage of sugar apple seeds (*Annona squamosa* L.). *Journal of Applied Botany and Food Quality*, *88*(1), 308–313.

Meira, C. S., Guimarães, E. T., Macedo, T. S., da Silva, T. B., Menezes, L. R., Costa, E. V., & Soares, M. B. (2015). Chemical composition of essential oils from *Annona vepretorum* Mart. and *Annona squamosa* L. (Annonaceae) leaves and their antimalarial and trypanocidal activities. *Journal of Essential Oil Research*, *27*(2), 160–168.

Mohamad, E. A., Ahmed, K. A., & Mohammed, H. S. (2022). Evaluation of the skin protective effects of niosomal-entrapped *Annona squamosa* against UVA irradiation. *Photochemical & Photobiological Sciences*, 1–11.

Mokhtar, F. A., Selim, N. M., Elhawary, S. S., Abd El Hadi, S. R., Hetta, M. H., Albalawi, M. A., . . . & Ibrahim, R. M. (2022). Green biosynthesis of silver nanoparticles using *Annona glabra* and *Annona squamosa* extracts with antimicrobial, anticancer, apoptosis potentials, assisted by in silico modeling, and metabolic profiling. *Pharmaceuticals*, *15*(11), 1354.

Mondal, P., Biswas, S., Pal, K., & Ray, D. P. (2018). *Annona squamosa* as a potential botanical insecticide for agricultural domains: A review. *International Journal of Bioresource Science*, *5*(1), 81–89. https://doi.org/10.30954/2347-9655.01.2018.11

Morita, H., Iizuka, T., Choo, C. Y., Chan, K. L., Takeya, K., & Kobayashi, J. I. (2006). Vasorelaxant activity of cyclic peptide, cyclosquamosin B, from *Annona squamosa*. *Bioorganic & Medicinal Chemistry Letters*, *16*(17), 4609–4611.

Muna, I. N., Suharti, S., Muntholib, M., & Subandi, S. (2021). Powder preparation of sugar apple (*Annona squamosa* L.) and analyzing its potencies as anti-gout and anti-COVID–19. *AIP Conference Proceedings*, *2353*(1), 030096. https://doi.org/10.1063/5.0052676

Nakano, D., Ishitsuka, K., Kamikawa, M., Matsuda, M., Tsuchihashi, R., Okawa, M., . . . & Kinjo, J. (2013). Screening of promising chemotherapeutic candidates from plants against human adult T-cell leukemia/lymphoma (III). *Journal of Natural Medicines*, *67*(4), 894–903.

Nguyen, M. T., Nguyen, V. T., Le, V. M., Trieu, L. H., Lam, T. D., Bui, L. M., . . . & Danh, V. T. (2020). Assessment of preliminary phytochemical screening, polyphenol content, flavonoid content, and antioxidant activity of custard apple leaves (*Annona squamosa* Linn.). In *IOP Conference Series: Materials Science and Engineering, 736*(6), 062012. IOP Publishing.

Noonan, J. C. (1953). Review of investigations on the Annona species. *In Proceedings of the Florida State Horticulture Society*, *66*, 205–210.

Nugraha, A. S., Damayanti, Y. D., Wangchuk, P., & Keller, P. A. (2019). Anti-Infective and anti-cancer properties of the Annona species: Their ethnomedicinal uses, alkaloid diversity, and pharmacological activities. *Molecules*, *24*(23), 1–31.

Nwokocha, L. M., & Williams, P. A. (2009). New starches: Physicochemical properties of sweetsop (*Annona squamosa*) and soursop (*Anonna muricata*) starches. *Carbohydrate Polymers*, *78*(3), 462–468. https://doi.org/10.1016/j.carbpol.2009.05.003

Olesen, T., & Muldoon, S. J. (2012). Effects of defoliation on flower development in atemoya custard apple (*Annona cherimola* Mill. *A. squamosa* L.) and implications for flower-development modelling. *Australian Journal of Botany*, *60*(2), 160–164. https://doi.org/10.1071/BT11299

Omar, N. A., Abd Razak, F. S., Tang, C. H., Helal Uddin, A. B. M., Sarker, M. Z. I., Janakiraman, A. K., & Liew, K. B. (2020). Development and evaluation of a cosmetic cream formulation containing *Annona muricata* leaves extract. *Latin American Journal of Pharmacy*, *40*(2), 258–266.

Oomah, B. D. (2022). Hempseed: A functional food source. In: R. Campos-Vega and B. Dave Oomah (eds.), *Molecular Mechanisms of Functional Food*. John Wiley & Sons, Ltd., 269–356.

Panadare, D. C., Dialani, G., & Rathod, V. K. (2021). Extraction of volatile and non-volatile components from custard apple seed powder using supercritical CO2 extraction system and its inventory analysis. *Process Biochemistry*, *100*, 224–230. https://doi.org/10.1016/j.procbio.2020.09.030

Panadare, D. C., Gondaliya, A., & Rathod, V. K. (2020). Comparative study of ultrasonic pretreatment and ultrasound assisted three phase partitioning for extraction of custard apple seed oil. *Ultrasonics Sonochemistry*, *61*, 104821.

Panda, S., & Kar, A. (2007). Antidiabetic and antioxidative effects of *Annona squamosa* leaves are possibly mediated through quercetin-3-O-glucoside. *BioFactors*, *31*(3–4), 201–210. https://doi.org/10.1002/BIOF.5520310307

Pandey, N., & Barve, D. (2011). Phytochemical and pharmacological review on *Annona squamosa* Linn. *International Journal of Research in Pharmaceutical and Biomedical Sciences*, *2*(4), 1404–1412.

Pardhasaradhi, B. V. V., Reddy, M., Ali, A. M., Kumari, A. L., & Khar, A. (2005). Differential cytotoxic effects of *Annona squamosa* seed extracts on human tumour cell lines: Role of reactive oxygen species and glutathione. *Journal of Biosciences*, *30*, 237–244.

Patel, J. D., & Kumar, V. (2008). Annona squamosa L.: Phytochemical analysis and antimicrobial screening. *Phytochemical Analysis*, *1*(1), 34–38.

Patil, S. R., Kurhekar, S. P., & Patil, R. R. (2011). Study on development of custard apple carbonated beverage. *International Journal of Processing and Post Harvest Technology*, *2*(1), 56–58.

Pino, J. A. (2000). Volatile components of Cuban Annona fruits. *Journal of Essential Oil Research*, *12*(5), 613–616.

Pino, J. A., Marbot, R., & Fuentes, V. (2003). Characterization of volatiles in bullock's heart (*Annona reticulata* L.) fruit cultivars from Cuba. *Journal of Agricultural and Food Chemistry*, *51*, 3836–3839.

Pinto, A. D. Q., Cordeiro, M. C. R., De Andrade, S. R. M., Ferreira, F. R., Filgueiras, H. D. C., Alves, R. E., & Kinpara, D. I. (2005). *Annona species*. https://doi.org/10.1111/j.1744-7348.1947.tb06347.x

Pinto, N. C., Silva, J. B., Menegati, L. M., Guedes, M. C. M., Marques, L. B., Silva, T. P., . . . & Fabri, R. L. (2017). Cytotoxicity and bacterial membrane destabilization induced by *Annona squamosa* L. extracts. *Anais da Academia Brasileira de Ciências*, *89*, 2053–2073.

Pokharia, A. K. (2008). Palaeoethnobotanical record of cultivated crops and associated weeds and wild taxa from Neolithic site, Tokwa, Uttar Pradesh, India. *Current Science*, 248–255.

Ponrasu, T., & Suguna, L. (2012). Efficacy of *Annona squamosa* on wound healing in streptozotocin-induced diabetic rats. *International Wound Journal*, *9*(6), 613–623. https://doi.org/10.1111/j.1742-481X.2011.00924.x

Quílez, A. M., Fernández-Arche, M. A., García-Giménez, M. D., & De la Puerta, R. (2018). Potential therapeutic applications of the genus Annona: Local and traditional uses and pharmacology. *Journal of Ethnopharmacology*, *225*, 244–270.

Rabêlo, S. V., Araújo, C. S., Costa, V. C. O., Tavares, J. F., Silva, M. S., Barbosa-Filho, J. M., & Almeida, J. R. G. S. (2014). Occurrence of alkaloids in species of the genus Annona L. (Annonaceae): A review. In S. K. Brar, S. Kaur & G. S. Dhillon (eds.), *Nutraceuticals and Functional Foods: Natural Remedy*. Nova Science Publishers, Inc., 41–60.

Rabêlo, S. V., Costa, E. V., Barison, A., Dutra, L. M., Nunes, X. P., Tomaz, J. C., . . . & Almeida, J. R. G. da S. (2015). Alkaloids isolated from the leaves of atemoya (*Annona cherimola* × *Annona squamosa*). *Revista Brasileira de Farmacognosia*, *25*(4), 419–421. https://doi.org/10.1016/j.bjp.2015.07.006

Rahman, M. M., Parvin, S., Haque, M. E., Islam, M. E., & Mosaddik, M. A. (2005). Antimicrobial and cytotoxic constituents from the seeds of *Annona squamosa*. *Fitoterapia*, *76*(5), 484–489. https://doi.org/10.1016/j.fitote.2005.04.002

Rai, C., & Muthana, M. S. (1954). Essential oil from the leaves of "*Annona squamosa* Linn". *Journal of the Indian Institute of Science*, *36*(2), 117–121.

Rana, V. S. (2015). Fatty oil and fatty acid composition of *Annona squamosa* Linn. seed kernels. *International Journal of Fruit Science*, *15*(1), 79–84.

Regassa, H., Sourirajan, A., Kumar, V., Pandey, S., Kumar, D., & Dev, K. (2022). A review of medicinal plants of the Himalayas with anti-proliferative activity for the treatment of various cancers. *Cancers*, *14*(16), 3898. https://doi.org/10.3390/cancers14163898

Rojas-García, A., Rodríguez, L., Cádiz-Gurrea, M. de la L., García-Villegas, A., Fuentes, E., Villegas-Aguilar, M. D. C., . . . & Segura-Carretero, A. (2022). Determination of the bioactive effect of custard apple by-products by in vitro assays. *International Journal of Molecular Sciences*, *23*(16), 9238. https://doi.org/10.3390/ijms23169238

Ronpirin, C., Charueksereesakul, T., Thongrakard, V., & Tencomnao, T. (2014). Effects of ethanolic extract of *Annona squamosa* L. leaves on the expression of EGFR. *Journal of Chemical and Pharmaceutical Research*, *6*(4), 791–797.

Saelee, C., Thongrakard, V., & Tencomnao, T. (2011). Effects of Thai medicinal herb extracts with anti-psoriatic activity on the expression on NF-κB signaling biomarkers in HaCaT keratinocytes. *Molecules*, *16*(5), 3908–3932. https://doi.org/10.3390/molecules16053908

Saha, R. (2011). Pharmacognosy and pharmacology of *Annona squamosa*: A review. *International Journal of Pharmacy and Life Sciences*, *2*(10), 1183–1189.

Saluja, A. K., & Santani, D. D. (1994). Pharmacological screening of an ethanol extract of defatted seeds of *Annona squamosa*. *International Journal of Pharmacognosy*, *32*(2), 154–162.

Sarkar, T., Salauddin, M., Roy, A., Sharma, N., Sharma, A., Yadav, S., . . . & Simal-Gandara, J. (2022). Minor tropical fruits as a potential source of bioactive and functional foods. *Critical Reviews in Food Science and Nutrition*, 1–45. https://doi.org/10.1080/10408398.2022.2033953

Senadeera, S. S., Prasanna, P. H. P., Jayawardana, N. W. I. A., Gunasekara, D. C. S., Senadeera, P., & Chandrasekara, A. (2018). Antioxidant, physicochemical, microbiological, and sensory properties of probiotic yoghurt incorporated with various Annona species pulp. *Heliyon*, *4*(11), e00955. https://doi.org/10.1016/j.heliyon.2018.e00955

Shami, A. M. (2017). The effect of alkaloidal fraction from *Annona squamosa* L. against pathogenic bacteria with antioxidant activities. *Pharmaceutical Sciences*, *23*(4), 301–307.

Sharma, K., & Parle, M. (2016). *Annona squamosa* as an antianxiety agent: Effects on behavioural and brain chemical changes. *World Journal of Pharmacy and Pharmaceutical Sciences*, *5*(10), 730–743.

Shashirekha, M. N., Baskaran, R., Rao, L. J., Vijayalakshmi, M. R., & Rajarathnam, S. (2008). Influence of processing conditions on flavour compounds of custard apple (*Annona squamosa* L.). *LWT—Food Science and Technology*, *41*(2), 236–243. https://doi.org/10.1016/j.lwt.2007.03.005

Shehata, M. G., Abu-Serie, M. M., El-Aziz, A., Mohammad, N., & El-Sohaimy, S. A. (2021). Nutritional, phytochemical, and in vitro anticancer potential of sugar apple (*Annona squamosa*) fruits. *Scientific Reports*, *11*(1). https://doi.org/10.1038/s41598-021-85772-8

Shivamathi, C. S., Moorthy, I. G., Kumar, R. V., Soosai, M. R., Maran, J. P., Kumar, R. S., & Varalakshmi, P. (2019). Optimization of ultrasound assisted extraction of pectin from custard apple peel: Potential and new source. *Carbohydrate Polymers*, *225*, 115240. https://doi.org/10.1016/j.carbpol.2019.115240

Shukry, W. M., Galilah, D. A., Elrazek, A. A., & Shapana, H. A. (2019). Mineral composition, nutritional properties, vitamins, and bioactive compounds in *Annona squamosa* L. grown at different sites of Egypt. *Series of Botany and Environmental Science*, *1*, 7–22.

Singh, P., Singh, K. R., Singh, J., Das, S. N., & Singh, R. P. (2021). Tunable electrochemistry and efficient antibacterial activity of plant-mediated copper oxide nanoparticles synthesized by *Annona squamosa* seed extract for agricultural utility. *RSC Advances*, *11*(29), 18050–18060

Singh, Y., Bhatnagar, P., & Thakur, N. (2019). A review on insight of immense nutraceutical and medicinal potential of custard apple (*Annona squamosa* Linn.). *International Journal of Chemical Studies*, *7*(2), 1237–1245.

Sismindari, A. H., & Mubarika, S. (1998). *In vitro* cleavage of supercoiled double stranded DNA by crude extract of *Annona squamosa* L. *Majalah Farmasi Indonesia*, *9*, 146–152.

Sobiya Raj, D., Vennila, J. J., Aiyavu, C., & Panneerselvam, K. (2009). The hepatoprotective effect of alcoholic extract of *Annona squamosa* leaves on experimentally induced liver injury in Swiss albino mice. *International Journal of Integrative Biology*, *5*(3), 182–186.

Somei, M. (2021). Method for reducing itching in atopic dermatitis. *U.S. Patent 11,065,229*, issued 20 July.

Sonkar, N., Yadav, A. K., Mishra, P. K., Jain, P. K., & Rao, C. V. (2016). Evaluation of hepatoprotective activity of *Annona squamosa* leaves and bark extract against carbon tetrachloride liver damage in wistar rats. *World Journal of Pharmacy and Pharmaceutical Sciences*, *5*, 1353–1360. https://doi.org/10.20959/wjpps20168-7418

Souza, F. T. C., Santos, E. R., Silva, J. D. C., Valentim, I. B., Rabelo, T. C. B., Andrade, N. R. F. D., & Silva, L. K. D. S. (2018). Production of nutritious flour from residue custard apple (*Annona squamosa* L.) for the development of new products. *Journal of Food Quality*, *2018*. https://doi.org/10.1155/2018/5281035

Stajić, A., Anđelković, M., Dikić, N., Rašić, J., Vukašinović-Vesić, M., Ivanović, D., & Jančić-Stojanović, B. (2017). Determination of higenamine in dietary supplements by UHPLC/MS/MS method. *Journal of Pharmaceutical and Biomedical Analysis*, *146*, 48–52.

Suresh, H., Shivakumar, B., & Shivakumar, S. I. (2012). Cytotoxicity of aporphine alkaloids from the roots of *Annona reticulata* on human cancer cell lines. *International Journal of Plant Research*, *2*(3), 57–60. https://doi.org/10.5923/j.plant.20120203.02

Tan, N. H., & Zhou, J. (2006). Plant cyclopeptides. *Chemical Reviews*, *106*(3), 840–895. https://doi.org/10.1021/cr040699h

Tlais, A. Z. A., Rantsiou, K., Filannino, P., Cocolin, L. S., Cavoski, I., Gobbetti, M., & Di Cagno, R. (2023). Ecological linkages between biotechnologically relevant autochthonous microorganisms and phenolic compounds in sugar apple fruit (*Annona squamosa* L.). *International Journal of Food Microbiology*, *387*, 110057. https://doi.org/10.1016/j.ijfoodmicro.2022.110057

Uduman, T. S., Sundarapandian, R., Muthumanikkam, A., Kalimuthu, G., Parameswari, S. A., Vasanthi Srinivas, T. R., & Karunakaran, G. (2011). Protective effect of methanolic extract of *Annona squamosa* Linn in isoniazid-rifampicin induced hepatotoxicity in rats. *Pakistan Journal of Pharmaceutical Sci*ence, *24*(2), 129–134.

Van Zonneveld, M., Scheldeman, X., Escribano, P., Viruel, M. A., Van Damme, P., Garcia, W., . . . & Hormaza, J. I. (2012). Mapping genetic diversity of cherimoya (*Annona cherimola* Mill.): Application of spatial analysis for conservation and use of plant genetic resources. *PLoS ONE*, *7*(1), 29845. https://doi.org/10.1371/journal.pone.0029845

Varadharaj, V., Janarthanan, U. D., Krishnamurthy, V., & Synnah, J. (2014). Assessment of phytonutrients in the ethanolic leaf extract of *Annona squamosa* (L.). *World Journal of Pharmacy and Pharmaceutical Sciences*, *3*(12), 725–732.

Varadharajan, V., Janarthanan, U. K., & Krishnamurthy, V. (2012). Physicochemical, phytochemical screening and profiling of secondary metabolites of *Annona squamosa* leaf extract. *World Journal of Pharmaceutical Research*, *1*(4), 1143–1164.

Verma, R. S., Joshi, N., Padalia, R. C., Singh, V. R., Goswami, P., & Chauhan, A. (2016). Characterization of the leaf essential oil composition of *Annona squamosa* L. from foothills of north India. *Medicinal Aromatic Plants (Los Angel)*, *5*(270), 2167–0412.

Vidanarachchi, V. A. D. S., Jemziya, M. B. F., Wijewardane, R. M. N. A., & Rifath, M. R. A. (2021). Development and quality evaluation of instant pudding mixture fortified with powder of custard apple *(Annona muricata L.)*. http://ir.lib.seu.ac.lk/handle/123456789/5880

Vijayalakshmi, R., & Nithiya, T. (2015). Antimicrobial activity of fruit extract of *Annona squamosa* L. *World Journal of Research in Pharmacy and Pharmaceutical Sciences*, *4*, 1257–1267.

Vikas, B., Akhil, B. S., Remani, P., & Sujathan, K. (2017). Free radical scavenging properties of *Annona squamosa*. *Asian Pacific Journal of Cancer Prevention: APJCP*, *18*(10), 2725–2732. https://doi.org/10.22034/APJCP.2017.18.10.2725

Vikas, B., Nehru, J., & Botanic, T. (2013). Antibacterial activity of *Annona squamosa* seed extract. *International Journal of Pharmacy and Technology*, *5*(3), 5651–5659.

Wang, L. Q., Min, B. S., Li, Y., Nakamura, N., Qin, G. W., Li, C. J., & Hattori, M. (2002). Annonaceous acetogenins from the leaves of *Annona montana*. *Bioorganic & Medicinal Chemistry*, *10*(3), 561–565.

Weston-Green, K., Clunas, H., & Jimenez Naranjo, C. (2021). A review of the potential use of pinene and linalool as terpene-based medicines for brain health: Discovering novel therapeutics in the flavours and fragrances of cannabis. *Frontiers in Psychiatry*, *12*, 583211. https://doi.org/10.3389/fpsyt.2021.583211

Wong, K. C., & Khoo, K. H. (1993). Volatile components of Malaysian Annona fruits. *Flavour and Fragrance Journal*, *8*(1), 5–10.

Wu, Y. C., Hung, Y. C., Chang, F. R., Cosentino, M., Wang, H. K., & Lee, K. H. (1996). Identification of ent-16β,17-dihydroxykauran-19-oic acid as an anti-HIV principle and isolation of the new diterpenoids annosquamosins A and B from *Annona squamosa*. *Journal of Natural Products*, *59*(6), 635–637. https://doi.org/10.1021/np960416j

Yadav, D. K., Singh, N., Dev, K., Sharma, R., Sahai, M., Palit, G., & Maurya, R. (2011). Anti-ulcer constituents of *Annona squamosa* twigs. *Fitoterapia*, *82*(4), 666–675.

Yadav, V., Singh, A. K., Singh, S., & Rao, V. A. (2017). Variability in custard apple (*Annona squamosa*) genotypes for quality characters from Gujarat. *Indian Journal of Agricultural Sciences*, *87*(12), 1627–1632.

Yang, H. J., Li, X., Zhang, N., Chen, J. W., & Wang, M. Y. (2009). Two new cytotoxic acetogenins from *Annona squamosa*. *Journal of Asian Natural Products Research*, *11*(3), 250–256. https://doi.org/10.1080/10286020802682916

Yang, X. J., Xu, L. Z., Sun, N. J., Wang, S. C., & Zheng. (1992). Studies on the chemical constituents of *Annona squamosa*. *Yao Xue Xue Bao-Acta Pharmaceutica Sinica*, *27*(3), 185–190.

Yang, Y. L., Chang, F. R., & Wu, Y. C. (2004). Annosqualine: A novel alkaloid from the stems of *Annona squamosa*. *Helvetica Chimica Acta*, *87*(6), 1392–1399. https://doi.org/10.1002/HLCA.200490127

Yang, Y. L., Chang, F. R., Wu, C. C., Wang, W. Y., & Wu, Y. C. (2002). New ent-kaurane diterpenoids with anti-platelet aggregation activity from *Annona squamosa*. *Journal of Natural Products*, *65*(10), 1462–1467. https://doi.org/10.1021/np020191e

Yang, Y. L., Hua, K. F., Chuang, P. H., Wu, S. H., Wu, K. Y., Chang, F. R., & Wu, Y. C. (2008). New cyclic peptides from the seeds of *Annona squamosa* L. and their anti-inflammatory activities. *Journal of Agricultural and Food Chemistry*, *56*(2), 386–392. https://doi.org/10.1021/jf072594w

Zahid, M., Arif, M., Rahman, M. A., & Mujahid, M. (2020). Hepatoprotective and antioxidant activities of *Annona squamosa* seed extract against alcohol-induced liver injury in Sprague Dawley rats. *Drug and Chemical Toxicology*, *43*(6), 588–594. https://doi.org/10.1080/01480545.2018.1517772

Zhao, X., Yuan, Y., Wei, H., Fei, Q., Luan, Z., Wang, X., Xu, Y., & Lu, J. (2022). Identification and characterization of higenamine metabolites in human urine by quadrupole-orbitrap LC-MS/MS for doping control. *Journal of Pharmaceutical and Biomedical Analysis*, *214*, 114732. https://doi.org/10.1016/j.jpba.2022.114732

Nutraceutical Potential and Applications of Citrus Fruit Waste in Food Industries

8

An Overview

Rachna Gupta, Shweta Suri, Anupama Singh, and Prabhat K. Nema

8.1 INTRODUCTION

Citrus fruits that relate to the family *Rutaceae* are widely grown, consumed, and processed everywhere in the world [1]. Fruits like orange (*Citrus sinensis*), lime (*Citrus aurantiifolia*), lemon (*Citrus limon*), mandarin (*Citrus reticulata*), and grapefruit (*Citrus paradisi*) are known globally for their high market potential [2]. Fruits belonging to the citrus family are well-recognized for their unique sweet and sour taste, lavishing aroma, fresh scent, and high nutritional content, especially vitamin C content [3]. In addition, these citrus plants are highly adaptable to different climatic and soil conditions, which are responsible for their efficient farming in tropical, subtropical, and temperate areas [4]. Furthermore, citrus fruits possess excellent nutraceutical, therapeutic, and medicinal properties owing to their enormous bioactive content. Besides, these fruits are composed of water (75%–90%), sugar (6%–9%), and the rest contain fiber, pectin, minerals, and essential oils. Citrus fruits also contain carotenoids and flavonoids [5] (Figure 8.1).

Typically, around 75% of citrus fruits, including grapefruits, mandarins, lemons, and oranges, are consumed fresh, while approximately 25% are processed by the food industry into foodstuffs like citrus juices, jams, jellies, marmalades, and dehydrated citrus foods. This processing generates significant amounts of waste [6].

High levels of processing, treatment, and consumption produce enormous masses of solid organic waste [7]. The citrus waste that is produced by the food sector each year accounts for

DOI: 10.1201/9781003352341-8

FIGURE 8.1 Bioactive constituent of citrus fruits [1]

almost half of the volume of fresh fruit [8], and approximately 120 million tonnes of citrus waste is produced from citrus juice processing every year [9]. This citrus juice processing waste has low pH, high moisture, and organic contents, and is not suitable for landfill disposal under European regulations [5]. Conversely, greater amounts of citrus waste are typically dumped into nearby landfills or waterways or incinerated, causing environmental pollution, lower dissolved oxygen levels in contaminated water, and health risks [9]. Besides, citrus waste is employed as livestock or animal feed, but when included in excess in animals' diets, it can affect their digestive systems and cause an illness called rumen parakeratosis, which suggests its low end-use effectiveness as animal feed [10]. Nonetheless, the citrus wastes (~40%–50% of the overall fruit weight) consist of pomace (membrane), seeds, peels (albedo and flavedo), and pulp [7]. Citrus waste contributes to the potent family of nutraceuticals comprising biologically active constituents, for example, phenols, flavonoids, carotenoids, essential oils, and dietary fiber. Citrus processing by-products contain good antioxidant, hypoglycemic, anti-inflammatory, anti-analgesic, anti-cancerous, and hypotonic activities. Also, in a bioeconomic context, citrus waste can be transformed into biofuels as well as other non-food commodities like biofilms, packaging material, or encapsulating agents [11]. Citrus peels constitute a major portion of citrus waste, although they contain natural pigments, essential oils, flavonoids, carbohydrate-soluble sugars, flavoring compounds, and phytochemicals that have great potential value. Apart from the citrus peels, the seeds obtained from citrus fruits, the non-edible part, contain excellent oil content (20%–40% by weight). Citrus pomace is another by-product generated after citrus fruits are used to make fruit juice or several other citrus-based food products [4]. This citrus pomace can be used to prepare fiber-rich bakery food products. It can also be used as a nanoparticle to enrich the supply of nutrients. Apart from the food uses of citrus pomace, it can also be used as a biofertilizer and bioadsorbent [1, 6].

This chapter provides an in-depth account of several nutraceutical/therapeutic properties (antioxidant, hypoglycemic, anti-inflammatory, anticarcinogenic, and antibacterial), as well as different traditional and green extraction methods to recover biologically active components found in the peels, seeds, and pomace of citrus fruit. In addition, the chapter outlines investigations on utilizing citrus by-products as food additives, prebiotics, encapsulating agents, and edible packaging agents, as displayed in Figure 8.2.

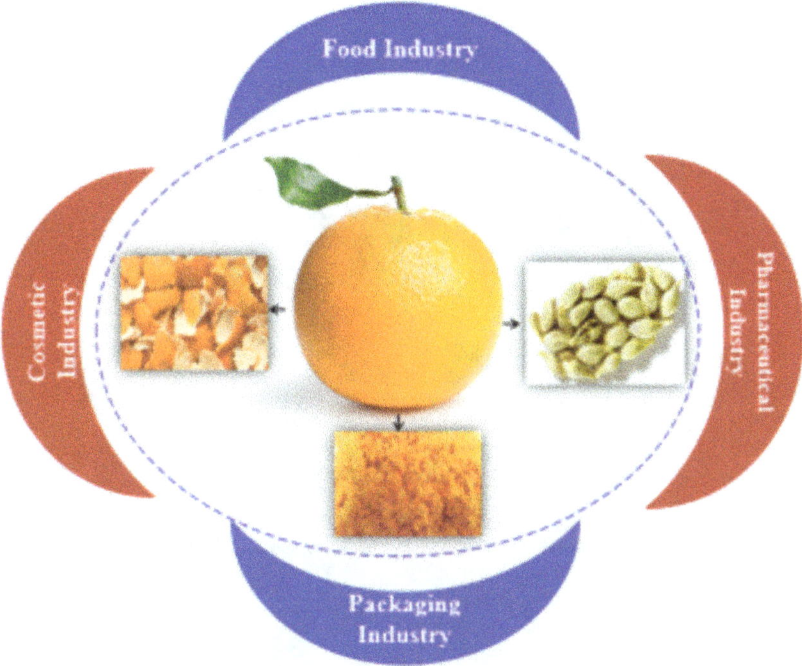

FIGURE 8.2 Effective valorization of citrus processing waste in various processing sectors [1]

8.2 CURRENT STATUS OF CITRUS PROCESSING INDUSTRIES

Overall, citrus fruits represent nearly 98% of total industrial fruit crops, whereas oranges alone account for about 82% of all citrus fruit production [12, 13]. In 2020, the citrus fruit worldwide production was about 158.49 million tonnes, with the Asian continent contributing the maximum output (47.7%), followed by Africa (43.7%), the United States (8.1%), Europe (0.4%), and Oceania (0.1%) (Figure 8.3) [14].

Among citrus-cultivating nations, China produces the highest (44.63 million tonnes) amount of citrus fruits, leading to 28.16% of the global cultivation in 2020, while Brazil, India, Mexico, and other nations each have considerable citrus fruit production [14] (Figure 8.4).

The production of citrus fruit in India was 14 million tonnes in 2020, with approximately 1.07 million hectares engaged in citrus fruit cultivation [4]. According to the report published by ICAR-Central Citrus Research Institute, Nagpur, India, the cultivation of citrus fruits plays a robust role in the Indian fruit industry, contributing to 13% of the country's annual fruit production. India generally sells citrus fruits overseas to nearby nations; however, approximately 4%–5% of citrus fruits are processed annually in the food sector [15], but large amounts of lemons and limes are used to produce pickles in the unorganized sectors, which is usually unaccounted for.

Furthermore, reports indicate the extensive use of citrus fruits since the creation of juice processing businesses in the US states of California and Florida in the 20th century [16]. About 18% of the worldwide production of citrus fruits is used for industrial purposes [17], particularly to produce juice. Besides, these fruits are likewise used in the canning sector to produce canned foods such as marmalade and mandarin fractions, in addition to the recovery of biologically active constituents like flavonoids and essential oils [18].

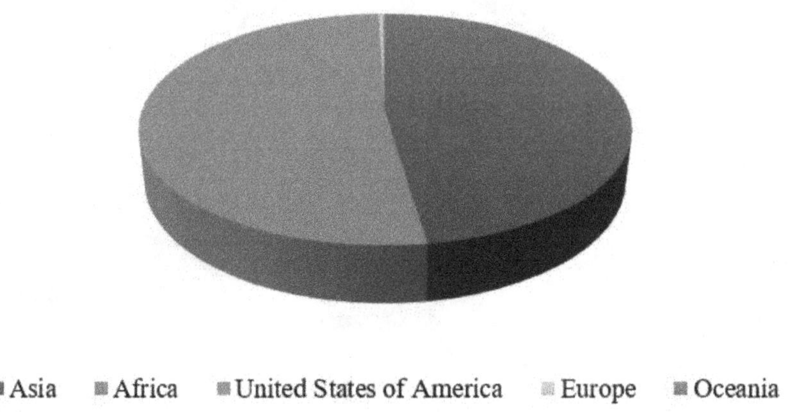

FIGURE 8.3 Citrus fruit production in different regions of the world [14]

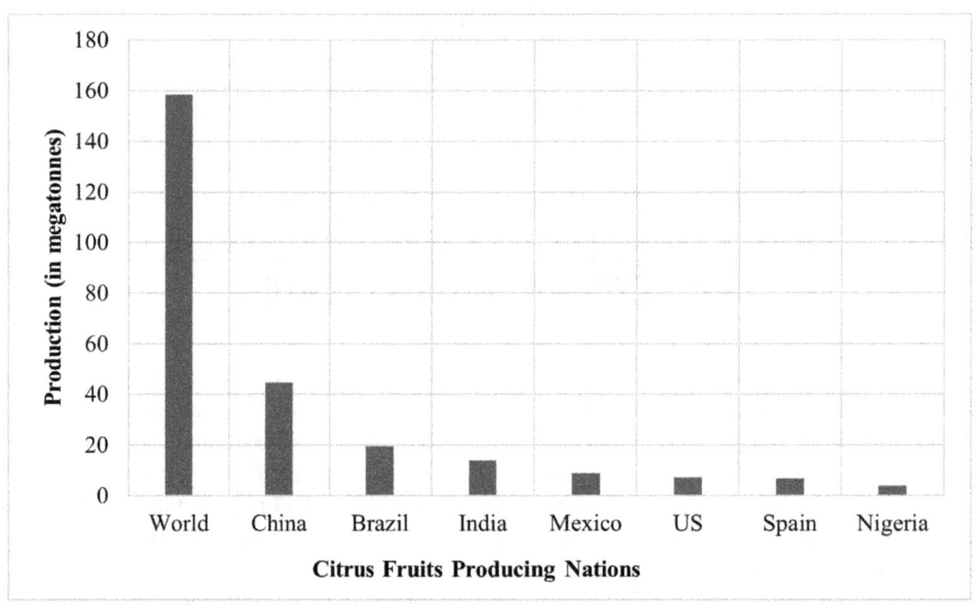

FIGURE 8.4 Global citrus fruits production [14]

8.3 CITRUS WASTES (PEEL, POMACE, AND SEED)

The citrus processing by-products (flavedo, peels, albedo, pith, and seed) are principal components, which include various unit processes and technologies. Table 8.1 enlists different uses of citrus waste in varying sectors. In a morphological context, flavedo (exocarp of fruit/outermost cover), as well as albedo (the spongy sheet of parenchyma cells found aligned with flavedo), are the exterior layers of the citrus fruit [19]. Flavedo consists of oil glands that include enzymes, phytochemicals (chlorophyll, flavonoids, carotenoids, as well as phenolic components), paraffin waxes, fatty acids, along with some steroids/triterpenoids. The chlorophyll-holding flavedo cells found within the chloroplast change during ripening by carotenoid compounds, primarily xanthophyll. After ripening, this significant hormonally regulated process causes the fruit to turn from green to yellow. The abundance of essential oils is rooted in the multicellular entities with spherical or pyriform profiles in the flavedo's inner layer. The inside citrus peel layer, called albedo, located near the flavedo, is abundant in gelatin, lignin, and dietary fibers [20]. Citrus fruits' endocarp (pulp/flesh) is typically made up of segments or carpels, which are divided by a membrane of tiny epidermal chains containing numerous juice sacs, termed vesicles or seeds. The axis of the fruit is the principal white, spongy tissue, which resembles the albedo layer. The "rag" of the isolated juice is the collective term for the core plus segments.

8.3.1 Citrus Peel

Around 50%–55% of the weight of fruit, or most of the citrus waste, consists of citrus peels comprised of two structural layers (flavedo and albedo). Citrus peel is abundant in essential oils, dietary fiber, polyphenolic chemicals, poly-methoxylated flavonoids (*hesperidin, naringin, nobiletin,* and *tangerine*), and other bioactive components [21]. Citrus peel essential oils, particularly α-terpinolene, α-pinene, and D-limonene, are in high demand due to their utilization in the food, beverage, perfumery, and cosmetic segments. The substantial quantity of soluble sugars, for example, fructose, glucose, and sucrose, present in citrus peel can be used to produce bioethanol. Besides, the cellular mass of the citrus peel consists of polysaccharides like cellulose, hemicellulose, pectin, arabinose, as well as galacturonic acid. The presence of soluble as well as insoluble carbohydrates in citrus peel implies its potential for creating value-added products through metabolic processes [22]. The main challenge in safely using citrus waste for food purposes is the application of pesticides, which are applied to safeguard citrus fruits from harmful molds and insects prior to, as well as after, harvest [23]. Citrus peels frequently have a larger exposure to fungicides (4468 mg/kg) compared to the entire fruit; nevertheless, a minute amount of these compounds is usually transferred from peel to pulp [24].

8.3.2 Citrus Seeds

Citrus seeds (20%–40% of the citrus fruit's weight) are the only part that cannot be consumed due to their high oil content [25]. Citrus seed oil may be used to generate biodiesel, according to literature. Citrus seeds correspond to 30%–35% of the citrus processing by-products. Whole and dehulled citrus seed flours contain crude fat (52%), carbohydrates (28.5%), crude fiber (5.5%), and crude protein (3.1%) on a dry weight basis [26].

8.3.3 Citrus Pomace

The pomace obtained from the citrus fruit is a by-product or residue from processing citrus fruit juice [27]. More than 50% of citrus fruits are discarded during the production of commercial citrus juice, resulting in considerable ecological problems caused by the large quantity of citrus pomace [28]. Efficient pomace management is necessary to avoid environmental issues. Furthermore, citrus pomace contains high levels of sugar, pectin, dietary fibers, and essential oils, which are valuable for various industries [29].

8.4 CHARACTERIZATIONS OF CITRUS WASTE COMPONENTS

Figures 8.5a, b, and c illustrate the anatomical and quantitative physical constitution of citrus fruit, its chemical constituents, and non-edible components of citrus fruits, often known as citrus by-products. Citrus waste contains ash, fat, protein, soluble sugar, starch, fiber containing hemicellulose, cellulose, and pectin, as well as various bioactive substances [30]. These residues can potentially be used to produce ethanol, nutrient-rich fiber components in confectionery products, macro- and micronutrient extraction from by-products, fortified nutrient-rich animal feed, and bio-oil and essential oil [31]. The industry's net production and profits increase because of these alternatives, which also help to prevent environmental damage. Citrus waste includes a notable level of soluble pectin (42.5 g/100 g), along with sugars (16.9 g/100 g), hemicelluloses (10.5 g/100 g), and cellulose (9.21 g/100 g). This significant amount of soluble, as well as insoluble, carbohydrates have great potential for creating high-value products [22]. Citrus peel

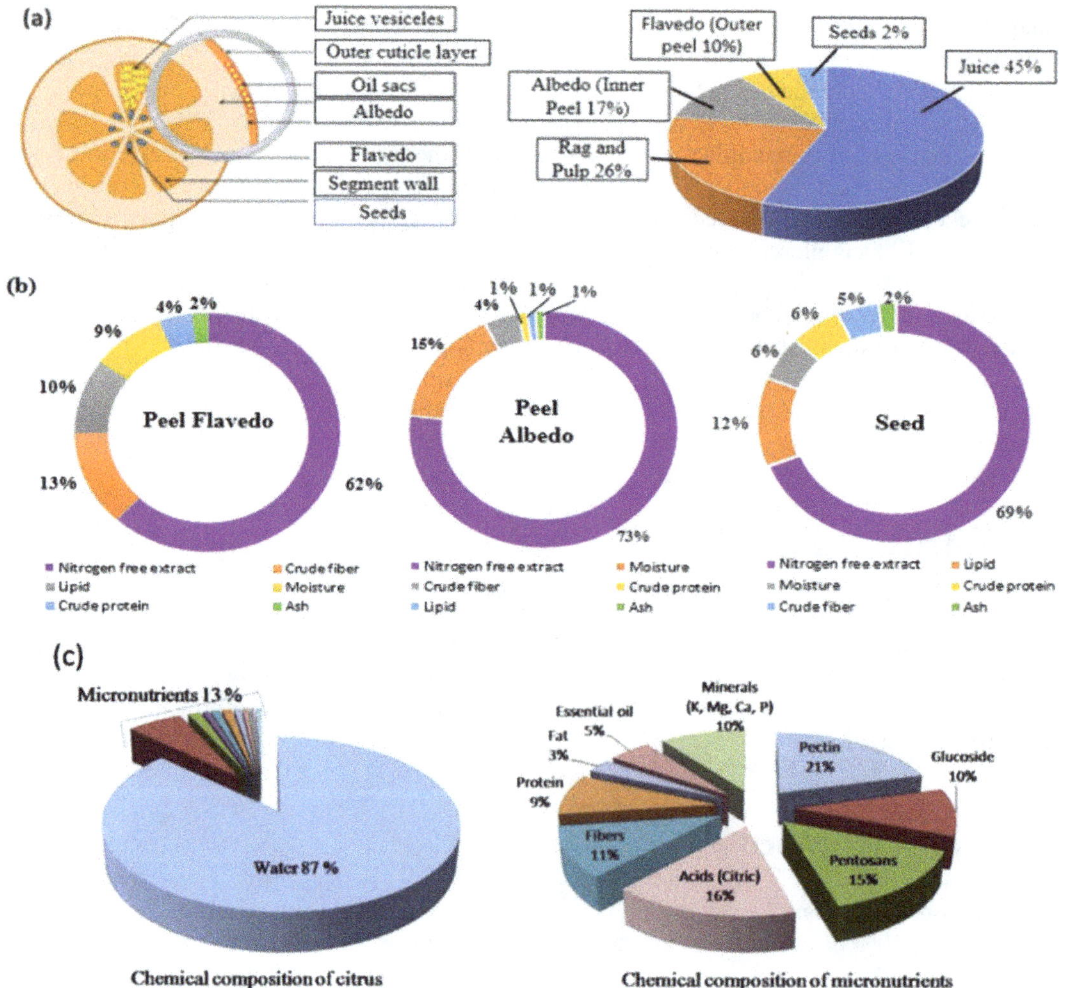

FIGURE 8.5 (a) Anatomical view of citrus fruits along with their physical components; (b) composition of peel flavedo, albedo, and seeds, (c) chemical/micronutrient components of citrus fruits [35, 36].

TABLE 8.1 Proximate analysis of citrus by-products

PARAMETERS (G/100 G)	*CITRUS PEELS*	*CITRUS PULP*	*CITRUS SEEDS*
Moisture content[a]	76.015	85.7	–
Crude protein[b]	8.12	8.6	3.1
Crude fiber[b]	57.0	4.9	5.50
Crude fat[b]	0.80	4.9	52.0
Total ash[b]	3.17	6.5	2.5
Source	[32]	[33]	[34]

[a] % wet basis
[b] g/100 g dry basis

has high levels of soluble sugars (46.2 g/100 g db), *viz.*, glucose, fructose, and sucrose. The rind consists of insoluble cellulose, pectin, and hemicelluloses, with high levels of galacturonic acid, galactose, and arabinose, along with trace amounts of xylose, rhamnose, and glucose. The valuable essential oils found in the peel, for instance, α-terpinolene, D-limonene, and α-pinene, have great commercial potential, as they can be utilized in various products like food, beverages, fragrances, perfumes, cosmetics, and more [32]. Table 8.1 shows the proximate composition of citrus waste products.

8.5 THERAPEUTIC BENEFITS OF CITRUS BY-PRODUCTS

The essential health benefits of safe phytochemicals include powerful antioxidants, antimutagenic properties, and anticancer effects [37]. In general, citrus fruits contain high levels of flavonoids and vitamin C, which are important for a healthy diet and can help prevent various diseases [38]. Various edible citrus fruits/veggies contain secondary plant compounds, for example, flavonoids and phenolics (gallic and ferulic acids). Flavonoids have been applied in medicine for preserving vascular health, as treatments for osteoporosis, and as agents to protect the liver. Numerous flavonoids exhibit anticancer properties demonstrated in both *in vitro* and *in vivo* studies [39]. These flavonoids have been exploited in the development of effective anti-inflammatory medications [40] because they can block the development of inflammatory mediators (e.g., prostaglandins, leukotrienes, and nitric oxide). Ascorbic acid and dehydroascorbic acid, two types of vitamin C, are both active and water-soluble antioxidants [41]. This vitamin's capacity to act as a hydrogen donor inhibits free radicals, providing its antioxidant properties [42]. The other vital element of *Citrus* species is carotenoids, which can prevent photo-oxidation damage [43]. The bioavailability and consumption of flavonoids and vitamin C correlate directly with their health benefits. Chronic toxicity can result from synthetic antioxidants; therefore, plant-derived natural antioxidants become more significant. Natural, efficient, and harmless antioxidants called polyphenols possess free-radical scavenging activity with no negative outcomes [44]. Hesperidin and its aglycone are essential in avoiding disorders linked to inflammation and oxidative stress. Some important therapeutic properties of citrus waste are discussed below.

8.5.1 Anti-inflammatory

Inflammatory cytokines, for example, TNF-α, interleukin-1, and interleukin-6, along with molecular mediators such as iNOS and COX-2, are intricately controlled by the body to trigger a complex inflammatory response. Additionally, these inflammatory cytokines are involved in the development of numerous chronic inflammatory conditions, for instance, multiple sclerosis, Alzheimer's, Parkinson's disease, as well as colon cancer [45]. PPAR-γ agonist activity was found in orange peel extract, which also blocked COX-2 expression and PGE2 production in UVB-induced HaCaT cells.

8.5.2 Anticancer Activity

Uncontrolled cell division and development cause numerous disorders known as cancer. It is believed that abnormal cells prefer glucose as their main metabolic substrate and energy source. The negative impact of oxidative stress, including DNA changes and cancer proliferation, can be prevented by eating foods rich in antioxidants like phenolic compounds or other types of phytochemical antioxidants. Citrus flavonoids are considered to be safe and nontoxic drugs that can control tyrosine kinases and, in turn, regulate apoptosis and possess antiproliferative qualities. Flavanones can reduce tumor growth and halt cell division, inducing cancer cell death through death receptors and caspase-mediated mitochondrial pathways in both laboratory and animal studies [47]. Flavonoids in citrus fruits can reduce the risk of oral cancer development, and their cancer-fighting effects have been observed in rodents [48]. Research has shown that hesperidin has anticancer properties that can combat various types of cancer, such as breast and colon cancers. Essential oils are currently acknowledged as agents with anticarcinogenic properties. Moreover, cancer prevention and treatment can be positively affected by dietary fiber and vitamin C [49]. By blocking protein kinase B, flavonoids in *Citrus aurantium* L. extract induced apoptosis in human leukemia cells. Furthermore, the same fruit extract caused apoptosis in non-small-cell lung cancer cells (A549), halting their proliferation [50]. The citrus *Platy mamma* extract's flavonoids showed a similar impact on these cell lines with added anti-inflammatory and antiangiogenic qualities. However, it did not regulate cancer cell apoptosis, although it still may be able to protect from the disease [51].

8.5.3 Antidiabetic Activity

The metabolism of insulin and glucose is significantly impacted by the flavonoid hesperidin. The main problem in diabetes mellitus is an altered blood glucose metabolism. By activating glucose-regulating enzymes, hesperidin can decrease glucose and fat levels, thus lowering the risks of diabetes-related complications due to its antioxidant properties and ability to suppress proinflammatory cytokines [52]. Citrus flavonoids (e.g., hesperidin, naringin, nobiletin, and neo-hesperidin) can lower hyperglycemia in HepG2 cells by reducing amylase enzyme ability to break down starch. Furthermore, they also enhanced hepatic glycolysis and glycogen synthesis, indicating the possibility of preventing hyperglycemia [53].

8.5.4 Antimicrobial Activity

Studies show that citrus flavonoids possess antibacterial effects. Hesperetin and naringenin showed the most potent antibacterial impact on *Helicobacter pylori* strains [49]. Another research indicated that bergamot extract is more useful in fighting Gram-negative bacteria [54]. Additional research has also confirmed the antibacterial effects of naringin and its derivatives on Gram-positive bacteria [55]. *C. aurantium* flower possesses extensive antibacterial effects and considerable antioxidant properties. It can, therefore, be used as an antibacterial agent for functional foods and health applications. Flavanones are thought to have antibacterial properties because they break down bacterial membranes and interact with microbial enzymes to produce DNA in bacteria.

8.6 BIOACTIVE COMPOUNDS IN CITRUS WASTE AND THEIR FOOD INDUSTRY APPLICATIONS

8.6.1 Polyphenols

Citrus waste comprises several bioactive complexes, for instance, polyphenols, mainly phenolic acids, as well as flavonoids [57]. Remarkably, citrus processing waste contains more polyphenols than the edible citrus fruits portion [58]; so, instead of treating them as waste, they can be used to recover valuable bioactive components, enabling a circular economy. Flavonoids, namely flavones, flavonols, flavanones, isoflavones, and anthocyanidins, are the major polyphenols in citrus peels [1]. Several researchers have studied the efficiency of green extraction methods compared with conventional phenol extraction from citrus peels. The ultrasonically assisted phenol extraction from citrus waste showed 26.5% maximum yield, which is 1.77 times more than microwave extraction. Polyphenolic compounds were extracted from three *Citrus* mandarin hybrid varieties (*Clemenvilla*, *Ortanique*, and *Nadorcott*) through ultrasound-based extraction techniques. The ultrasonic process (400 W power, 80% duty cycle, 40 °C) improved the physicochemical profile, amount of bioactive constituents, and antioxidant properties of tangerine peel extracts [59]. In another study, polyphenolic compounds from peels of kinnow (*Citrus reticulata*) and mosambi (*Citrus limetta*) extracted by ultrasonic-assisted extraction were superior to the maceration technique. The yield for kinnow and mosambi mandarins was 5.85% and 12.95% by ultrasound-aided extraction, and 5.2% and 12.2% by maceration [60]. Additionally, the peel waste of fresh kinnow (*Citrus reticulata*) contained 31.05 mg GAE/g, and 25.91 mg quercetin equivalent/g of total phenol and flavonoid contents, along with 56.6% DPPH radical scavenging activity [61], and even higher (66.5%) DPPH radical scavenging activity [62]. Variations in polyphenolic levels and antioxidant effectiveness in kinnow peels are attributed to different citrus fruit types, regions, harvest periods, extraction methods, and measurement tools [63].

8.6.2 Essential Oils

This is a volatile mixture of aromatic fluids that is often present in seed oil sacs and citrus peels [64]. Essential oils recovered from citrus by-products are employed in the flavoring industry as well as for fragrances [65]; however, the oil yield, as well as citrus peel essential oil constituents, varies based on extraction techniques. Supercritical fluid extracted higher essential oil (limonene) yield from lemon peel than solvent-free microwave extraction procedure (4.5% vs 1.26%) [66, 67]. Likewise, limonene extraction from orange peel (*Citrus unshiu* Kuno) utilizing the supercritical carbon dioxide method showed superior limonene yield (30.65% vs 13.16%) at higher pressure (300 vs 100 bars) [68].

Citrus seeds are known as a powerful reservoir of essential oils. The variety of citrus fruit determines the citrus seed oil content (20%–40% by weight). The citrus seed oils from varieties, for instance, *C. sinensis*, *C. reticulata*, and *C. paradisi*, are considered to be beneficial oils due to their composition of palmitic, oleic, and linoleic acids, along with traces of stearic as well as linolenic acids [1, 2, 69]. Citrus seed oil contains 98.6% fatty acids, including high levels of linoleic, palmitic, and oleic acids (43.7%, 21.9%, and 21.3%, respectively), lower amounts of linolenic, stearic, and cis-vaccenic acids (5%, 4%, and 2%, respectively), and small quantities of arachidic, eicosanoid, and behenic acids (0.4%, 0.1%, and 0.2%, respectively), together with tocopherol (13.7 mg/kg) [70]. Currently, only a few local producers are involved in producing citrus essential oils on a large scale using leftover peels and seeds. The need for citrus essential oil is growing yearly because of the healing advantages of the oil extracted from citrus seeds and peels [16]. Furthermore, citrus seed waste can be used for processing and effectively utilizing citrus seed oil extraction.

8.6.3 Carotenoids

Citrus by-products are rich in carotenoids that can be employed as natural pigments in the food and beverage manufacturing sector [71]. Mandarin variety of citrus fruits showed a carotenoid content of 200 mg/100 g dry weight [72]. Mandarin (*Citrus reticulata* L.) pericarp extract was used as a coloring agent to manufacture bakery products (cakes, bread, and others), where the baked goods containing citrus seed flour showed a significant increase in carotenoid content [73].

8.6.4 Pectin

Pectin obtained from citrus fruit peels is applied in the food manufacturing sector as a gelling compound for making jellies and jams [74]. Around 85% of the globally marketable pectin is recovered from peel waste attained through citrus processing operations. The fresh and dried citrus fruit peels consist of 1.5%–3% and 9%–18% pectin [65]. Pectin can be extracted from citrus by-products by several procedures.

Pectin from *Citrullus lanatus* peel has been extracted using the microwave-assisted method with varying liquid-solid ratios (1:10–1:30 g/ml), microwave power (160–480 W), as well as microwave irradiation time (60–180 s) [75]. A pectin extraction yield of 25.8% was achieved using microwave power (477 W), exposure time (128 s), and a solid-liquid ratio (1:20) [76]. A similar pectin amount was recovered (27.6% yield) from mandarin peels by sequential microwave-assisted solvent and used to prepare jams [77].

8.6.5 Dietary Fiber

The demand for dietary fiber has increased both as a dietary supplement and as a raw material in natural/processed foods. These dietary fibers aid in enhancing the volume of the diet. The fiber in citrus peels is rich in soluble as well as insoluble fiber, known for its health benefits and nutritional value [65]. Various methods, including enzymatic process, ultrasound, dry/wet processing, acid-alkali treatment, and a combination of techniques, are used to extract fibers from citrus peels [78]. Steam explosion and diluted acid soaking were employed to extract fibers from *Citrus sinensis* rind. Orange has a high soluble fiber content of 33.7% [79]. In addition, researchers investigated the extra benefits of dietary fiber from kinnow peels by adding it to make baked goods. Citrus peel fiber is packed with nutrients and has functional properties that offer great nutraceutical advantages. Research has indicated that enriching foods with fiber enhances emulsification and prolongs shelf life [80]. Fiber in kinnow peel is easily identifiable by its unique microstructure and high levels of soluble fiber [81].

8.7 CONVENTIONAL AND NOVEL TECHNIQUES FOR ISOLATION OF BIOACTIVE COMPONENTS

Innovative "green" methods have the potential to extract bioactive components from citrus waste for various industries, including food, pharmaceuticals, nutraceuticals, and cosmetics. Generally, bioactive components must be isolated from the fruit cells for extraction. This involves several processes (i.e., cleaning, drying, and grinding) without destroying or distorting the desired functional components. Dichloromethane is used as a solvent when the target chemical is lipophilic; higher-polarity solvents (methanol or ethanol) are used when the target molecule is hydrophilic. It is crucial to focus on the function of the extraction methods, since the required molecules may be polar, nonpolar, or heat-labile [82]. Different extraction methods, for example, Soxhlet, hydro-distillation, maceration, and so on, along with innovative green techniques like pulse electric field (PEF), microwave-assisted extraction (MAE), ultrasound-assisted extraction (UAE), subcritical fluid extraction (SFE), and enzyme-assisted extraction (EAE), have been employed to isolate phytochemicals from citrus waste. These include essential oils, pectin, dietary fibers, and polyphenols [83]. Each method has its own benefits and drawbacks. The traditional methods use hazardous solvents, high temperatures, and long extraction times. In contrast, nontraditional (novel) techniques resolve the solvent toxicity issue, use less solvent, and require shorter extraction time [84].

Additionally, green extractions use less energy and preserve the extracted components better than conventional procedures [85]. Many product development procedures, such as encapsulation and extrusion, are used to separate, purify, and introduce the extracted substances to food systems. The basic steps in the analytical method for detecting bioactive compounds in citrus waste involve sampling, preparing samples, extracting target compounds, purifying crude extracts, identifying qualitatively, and quantitatively determining using instrumental techniques (Figure 8.6).

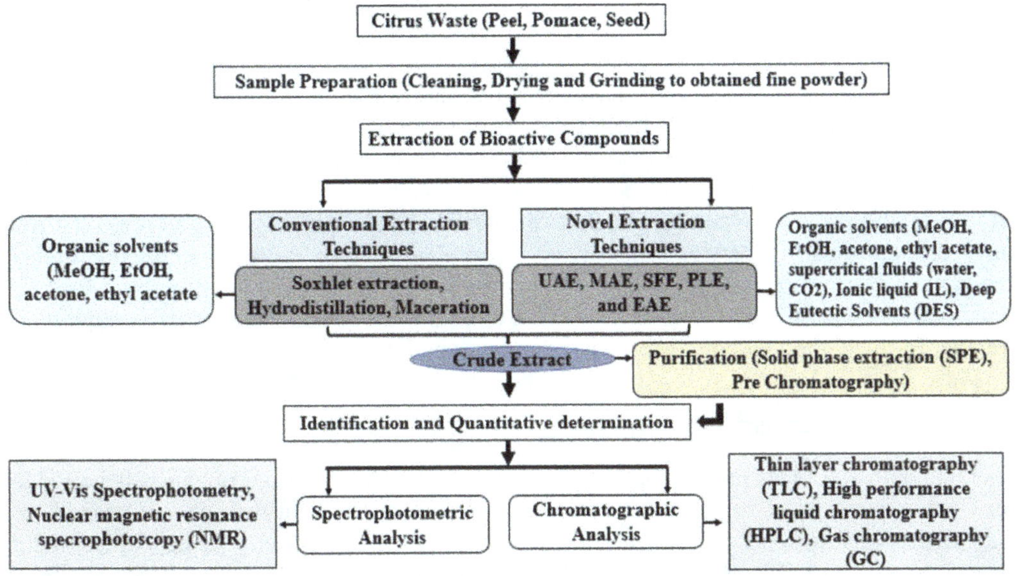

FIGURE 8.6 Steps and analytical procedures to extract and identify citrus waste bioactive substances. Abbreviations: pulse electric field (PEF), microwave-assisted extraction (MAE), ultrasound-assisted extraction (UAE), subcritical fluid extraction (SFE), and enzyme-assisted extraction (EAE); solid-phase extraction (SPE); thin-layer chromatography (TLC); gas chromatography (GC); high-performance liquid chromatography (HPLC); nuclear magnetic resonance spectroscopy (NMR); methanol (MeOH); ethanol (EtOH); ionic liquid (IL); deep eutectic solvent (DES) [85]

8.7.1 Conventional Extraction

Traditional extraction methods used to separate bioactive phytochemicals from various food matrices include maceration, hydro-distillation, and solvent extraction. The efficiency of a technique changes based on the polarity of the specific compound being targeted and the solvent used for extraction. Designing a definite method that can effectively extract and separate all target chemicals is difficult because of variations in polarity. An ideal solvent should possess minimal toxicity, low boiling point, ability to extract complex molecules from food, and enhance rapid mass transfer. Besides the solvent, it is important to also keep track of the impacts of factors, for instance, extraction duration, extract quality, and temperature on extract yield.

8.7.1.1 Maceration

The most popular method to obtain phytochemicals/essential oils from citrus waste is maceration. Raw material is crushed into smaller particles while being macerated to increase its surface area, as well as ensure proper blending with the solvent. Stirring during maceration helps in extracting substances by accelerating the spread and removing the concentrated solution from the sample's outer layer; this technique has been utilized for a considerable period to manufacture bioactive compounds and essential oils [86]. Maceration of different citrus peels resulted in high antioxidant activity (43% and 22.5%) and total phenolic content (TPC) (23 and 19 mg GAE/g peels) in kinnow and mousambi peel extract [6]. Extraction yield of bioactive chemicals depends on solvent type and concentration during solvent extraction. Extraction using different solvents resulted in varying yields from kinnow peels: the highest (18.5% and 15.6%) with 80% and 50% ethanol, and the lowest (5.12%) with 100% ethyl acetate [87].

8.7.1.2 Hydro-distillation

Hydro-distillation is a frequently used method to extract bioactive phytochemicals from citrus by-products. Hydro-distillation is utilized to extract the volatile part of the sample without the need for organic solvents, using distilled water and usually lasting 6 to 8 hours. This method relies on three fundamental physicochemical methods like hydro-diffusion, hydrolysis, as well as thermal decomposition. Chemicals can deteriorate the components during extraction at high temperatures, limiting the application of this method [88]. Hydro-distillation can physically extract and separate both volatile and nonvolatile organic molecules simultaneously. Azeotropic distillation is used to remove the volatile organic components from the matrix, after which they are condensed, gathered, and sorted in a Florentin flask. Boiling water embedded with the matrix in the alembic is utilized to obtain the soluble nonvolatile organic molecules. However, hydro-distillation requires a lot of time and energy [89].

8.7.1.3 Solvent Extraction (Soxhlet)

The apparatus is filled with a minute quantity of dry material, and the solvent is continuously heated to a boil and then condensed while passing over the sample. Solvent extraction has historically been used to separate thermally stable analytes from solid matrices that aren't very soluble in solvents. Long extraction durations and the requirement for huge volumes of solvent are two significant disadvantages of Soxhlet extraction, despite its dependability and ability to perform continuous extraction. Maximum amount of total phenolic content (TPC) was extracted using ethanol (72%, 121%, 74%, 60%, and 118% from grapefruit, mandarin, orange, Meyer lemon, and from Yen Ben lemon, respectively). Furthermore, a 2-hour solvent extraction procedure at 85 °C produced a 1.51% yield from flavonoids extraction from pomelo peels [90].

8.7.2 Nonconventional/Novel Techniques

These methods, referred to as "cold extraction," have no impact on the stability of the extracted molecules and use less energy [91, 92]. The new and innovative tactics mentioned in the following section can address many of the issues related to traditional methods. These eco-friendly extraction methods aim to obtain a high extraction rate, enhance energy efficiency, optimize mass and heat transfer, lower equipment size, as well as minimize processing stages. The intention of utilizing these techniques is to safeguard the surroundings and its resources [93, 94].

8.7.2.1 Ultrasound-assisted Extraction (UAE)

Sound waves that arise between 20 kHz and 100 MHz are described as ultrasound. Cavitation is a phenomenon that includes the development, expansion, and collapse of bubbles and is caused by UAE (Figure 8.7) [86]. UAE is a powerful extraction technique that works well with various samples and analytes. Ultrasound disintegrates plant cell walls, accelerating heat and mass transfer, and thereby enabling easy release of target compounds from natural sources [95]. In ultrasonic extraction, the two main physical processes are washing the cell content after cell wall disintegration and diffusion from the cell wall. Several variables, including sonicator temperature and time, pressure, and frequency, can alter ultrasound activity [96]. When compared to alternative extraction methods, this procedure is easier to apply, adaptable, flexible, and economical. Ultrasound is a useful technique for extracting several biomaterials and compounds, including proteins, essential oils, polysaccharides, peptides, chemicals, dyes, and colorants [97].

There are two possible outcomes: direct and indirect. The intensity of ultrasonography is often 100 times higher when handled straight to the medium, with no barrier, for example, a probe device [98]. When using an ultrasonic water bath for indirect sonication, the vibrations must travel through the water and strike the sample. Because ultrasonic energy improves yield, speeds up kinetics, and boosts mass transfer coefficient, it is a tried-and-true method for removing bioactive compounds.

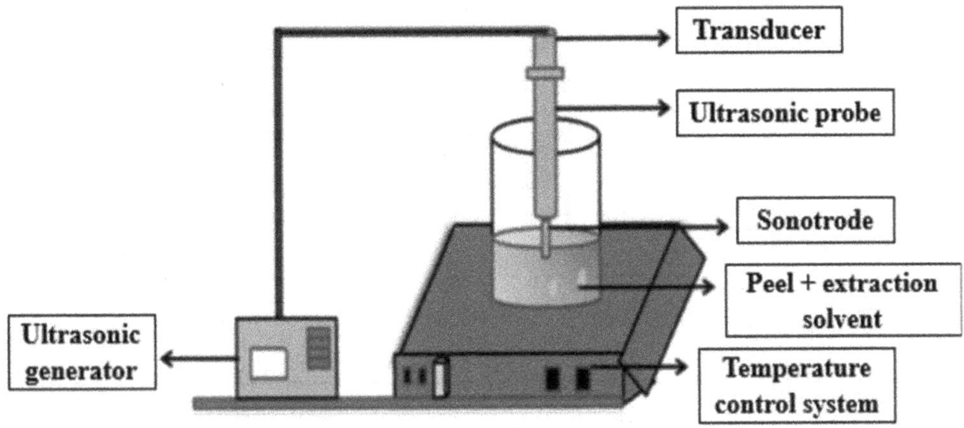

FIGURE 8.7 Ultrasound-assisted extraction system [99]

Orange peels' trans-β-carotene extraction efficiency is significantly improved by utilizing ultrasonic technology [100]. Extraction yield increased with decreasing particle size, peaking at 0.28 mm. On the other hand, since the cell walls were already cracked all through the fine grinding process, extracting trans-β-carotene from too fine powders did not lead to a higher yield. Results showed that extraction yields benefited from the combined effects of heat treatment and cavitation, especially at higher temperatures between 25 and 45 °C. These benefits were shown to be positively linked with the intensity of ultrasonication.

There was significant temperature, time, and amplitude (at 20 kHz) sensitivity in the orange peel lutein extraction process using ultrasonication. Maximum lutein content was obtained after 34 min extraction at 43 °C and 33% amplitude [26]. The kinnow peel extract had 48.23% antioxidant activity, 28.3 mg GAE/g polyphenols, and 4.4 mg CE/g flavonoids, according to an analysis conducted as part of comparative research on the ultrasonication-based isolation of phenolic components from numerous citrus peels. On the other hand, the extract from mousambi peel contained 2.07 mg CE/g flavonoids and 22 mg GAE/g polyphenols, along with 39.7% antioxidant activity. The use of ultrasonography was linked to noticeably higher phenolic component synthesis than maceration extract [60].

8.7.2.2 Microwave-assisted Extraction

Microwave-assisted extraction (Figure 8.8) is a favorable method for isolating bioactive compounds from citrus by-products. With this technique, which employs microwave radiation to create heat, the samples are heated either directly or indirectly. Amid 300 MHz and 300 GHz, the electromagnetic wave (wavelength 1 cm–1 m) and the coupled electric and magnetic forces make up the microwave spectrum. However, 915 to 2450 MHz and 12 to 20 cm are the most commonly employed frequencies and wavelengths in MAE, respectively [101]. The microwave radiation applied to the sample causes localized heating from the inside to the outside, which causes ions and molecules to rotate and move [75]. Ionic conduction and dipole rotation are the primary procedures for mass and heat transfer between cells in a matrix. The separation of molecules and tissues brought about by water evaporation allows for the isolation of both volatile, as well as nonvolatile, compounds. When exposed to microwaves, the cell wall degrades, letting the bioactive molecules inside leak out of the cell [102]. Rapid heating and component extraction from the product are caused by the moisture in the biomass and the microwave radiation [103].

By decreasing the temperature gradient and increasing energy flow, this non-contact heat source can produce more efficient heating. Different components, such as antioxidants, flavorings, pigments, and other organic compounds, can be productively separated by employing this technology [104].

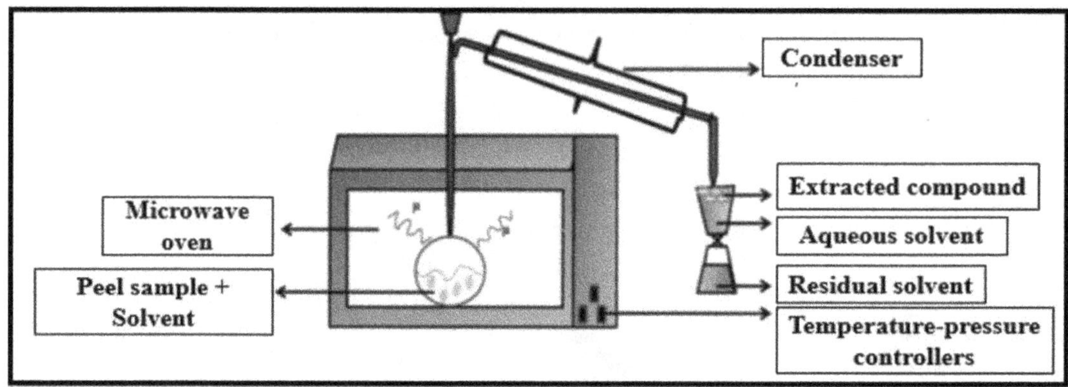

FIGURE 8.8 Microwave-assisted extraction system [99]

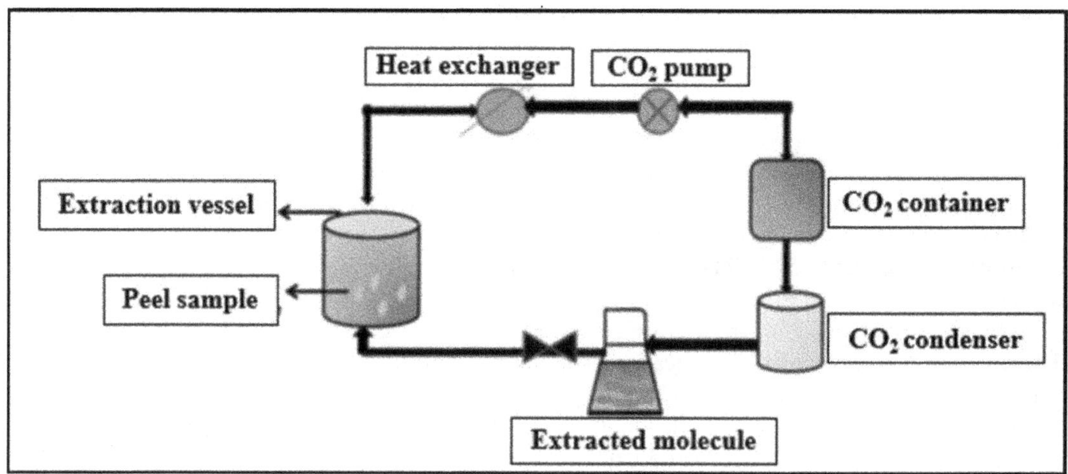

FIGURE 8.9 Supercritical fluid extraction system [99]

Mandarin peels (*C. inshiu*) were used to extract hesperidin and narirutin (5.86 and 1.31 g/100 g) using aqueous ethanol and microwave-aided extraction [105]. Total flavonoids (637.6 mg QE/100 g) and total polyphenols (131.6 mg/100 g) were present in the aqueous methanol extract [106]. TPC content increased at 400 to 500 W microwave power, but then decreased with each consecutive increase, reaching its lowest value at 800 W, most likely as a result of polyphenols' breakdown at higher microwave powers. Likewise, extracts exposed to microwave radiation for prolonged periods degrade due to their breakdown, which lowers extraction effectiveness for extraction times greater than 125 seconds [107]. Using solid by-products from orange juice processing, vacuum-assisted MAE also improved phenolic components' extraction on a large scale [108].

When compared to traditional heating methods, microwave equipment accelerates reactions, increases yield, and improves purity. Its use can occur with or without the addition of a solvent, and can simply, safely, and repeatably attain high temperatures [109].

8.7.2.3 Supercritical Fluid Extraction (SFE)

Changes in pressure and temperature during supercritical extraction cause the gas and liquid phases to mix, changing the gas in the supercritical fluid (Figure 8.9). Convection in the supercritical solvent phase is the main mode of transport in this mass transfer activity [110]. Small quantities of material may be used for the extraction, which is a quick and selective method that doesn't require any additional processing [110]. A further significant benefit is the potential for direct integration with analytical chromatographic techniques like gas chromatography (GC) and supercritical fluid chromatography (SFC) [93].

The chemical components in the solid matrix are first solubilized and subsequently split in the supercritical solvent in the two stages of this procedure. The solvent dissolves the constituent elements of the matrix by penetrating the tightly packed bed.

Accordingly, the pressure drops and the temperature rises, and the solvent subsequently emerges from the extractor as a solvent-free extract [93]. Low viscosity, as well as low surface tension, allow supercritical fluids to readily disperse within a solid matrix and swiftly penetrate the solid, respectively, increasing extraction efficiency. These beneficial features of supercritical fluids contribute to their enhanced flexibility. A fluid's force may be altered by varying the extraction pressure due to the relationship between density and solubility [111].

Supercritical extraction is commonly employed to separate nonpolar bioactive components due to the nonpolar character of the solvents used in this method. Modifiers can be used to extract polar chemicals like flavonoids; the extracts' ability to dissolve the modifier depends on its density. As a result, the extraction process makes use of various supercritical fluids [93].

While temperature, pressure, and the use of co-solvents all impact SFE performance, carbon dioxide (CO_2) is the most ecologically friendly supercritical fluid for efficiently extracting bioactive components since it experiences low process degradation or loss [112].

The solvent power associated with density changes affects SFE efficiency. SFE is an effective alternative to conventional methods because it uses CO_2 instead of vast volumes of organic solvents and doesn't require sample preparation. With 20% ethanol applied as a co-solvent, the maximum naringin yields (35.3, 44, and 19.9 mg/g in orange, tangerine, and lemon peel extracts) were achieved in liquid CO_2 extraction (20 MPa and 20 °C) and supercritical extraction (30 MPa and 60 °C). Additionally, the extract exhibited the most antiradical activity (31.8–59.5 mol TE/g) against DPPH and ABTS+ [113].

8.7.2.4 Subcritical Water Extraction

Subcritical water extraction (SWE) (Figure 8.10) is cost-effective compared to conventional extraction methods; this fast-acting technology yields high-quality extraction products while also being ecologically benign [114]. SWE, often referred to as pressurized hot water extraction or superheated water extraction, uses water at pressures up to 22.1 MPa (higher than vapor saturation) and temperatures between 100 °C to 374 °C (critical temperature) to maintain the liquid state of the water molecules throughout the process [115]. Water has a density of 1000 kg/m^3 and a dielectric constant of around 80, making it a polar solvent.

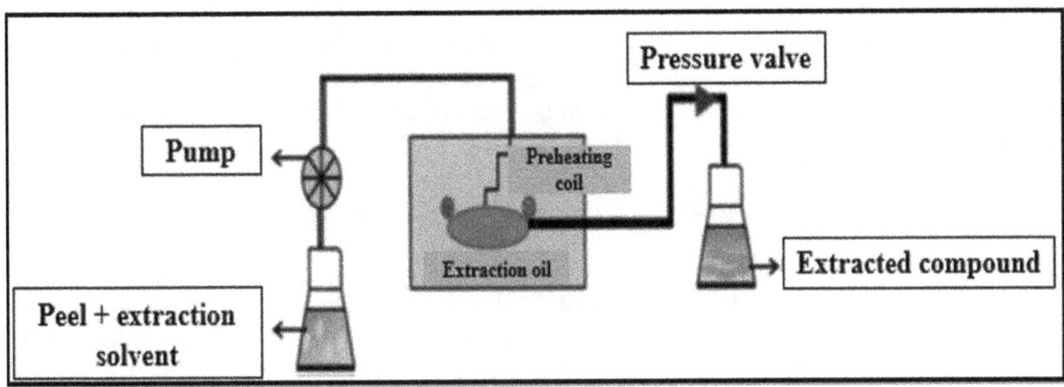

FIGURE 8.10 Subcritical water extraction system [99]

Because the dielectric constant of water decreases with increasing temperature due to the weakening of hydrogen bonds, water may be used as a material reaction medium.

Water has a density of 79.9% at ambient temperature and atmospheric pressure [116]. The amount of water in a liquid state can be lowered to 27%–32.5% by increasing the temperature to 250 °C at 5 MPa pressure. At room temperature, the densities of methanol (32.5%) and ethanol (27%), respectively, are equal to those of water (27%) [117]. The latter makes it possible for water to interact with polar compounds, lowering the binding force and enabling the dissolution of material in water at higher temperatures and pressure.

The ionic water constant (Kw), which is approximately three times larger than at ambient temperature, increases as the reaction temperature rises. The enhanced reactivity of water with increasing H+ and OH− concentrations in the aqueous medium makes it appropriate for use as an acid or basic catalyst in hydrolysis processes. Therefore, subcritical water treatment can be used to hydrolyze organic waste and remove its constituent elements [118]. In SWE, solute desorption at high temperatures and pressure is the initial stage, after which the extracted chemicals diffuse into the solvent. The extracted solutions are moved to a collecting container after elution from the extraction cell. The main factors influencing the SWE process are temperature and length of the reaction, as well as the ratio of solids to water [119]. Bioactive compounds become more soluble and have a better mass transfer rate at elevated treatment temperatures. However, because the extraction solvent's viscosity and surface tension decrease with temperature, the selected compounds may deteriorate at elevated temperatures. Consequently, each particular scenario should have its specific process temperature and time, which are largely determined by the characteristics of the finished product [120]. To extract bioactive flavonoids, a semi-continuous synthesis of *C. unshiu* peel was conducted utilizing subcritical water (SW). The different SW extracts were examined for their flavonoid yields, antioxidant capability, and enzyme-inhibitory properties [121].

8.7.2.5 Enzyme-Assisted Extraction

Enzyme-assisted extraction is typically utilized for bioactive substances that are difficult to extract using traditional techniques because they are tightly linked to the cell wall. The bioactive materials inside the cell plasma leak out of the cell when enzymes disintegrate the cell wall [122]. Bioactive compounds have been extracted from citrus by-products using several common enzymes [57]. Enzymes connect to the polysaccharide-lignin network that constitutes the active site of the cell wall during the extraction process, hydrolyzing the lipid body and polysaccharide structures by breaking down glycoside linkages in the cell wall and proteolytic bonds in the intermediate lamella [123]. Through the cell wall, the bioactive materials in the cell and middle lamella are therefore released.

The selection of extraction enzymes is contingent upon the process of isolating the target compounds, which may include breaking down the cell wall, pectin, or polysaccharides [124]. enzyme-assisted extraction of polyphenols and other bioactive compounds is influenced by both enzyme concentration and extraction time. When enzyme-assisted extraction is combined with supercritical extraction, polyphenol yields from pomegranate peels can be increased. Pectinase, protease, and cellulase were used in a 25:25:50 ratio at 3.8% concentration, 49 °C, 85 min treatment, and 6.7 pH to obtain the highest yield of extract containing phenolic compounds [125].

8.7.2.6 Pulsed Electric Field Assisted Extraction (PEF)

PEF extraction (Figure 8.11) is used to increase extraction yield by electroporating the cell membrane. An electric potential that permeates the cell membrane divides molecules according to their charge as a result of holes that emerge, increasing their permeability [96]. One of the PEF applications that has drawn the most scientific interest is the recovery of intracellular active compounds from plant food sources, food wastes, and by-products. This is mostly due to its capacity to weaken and tear cell membranes, allowing enclosed substances to be released [95].

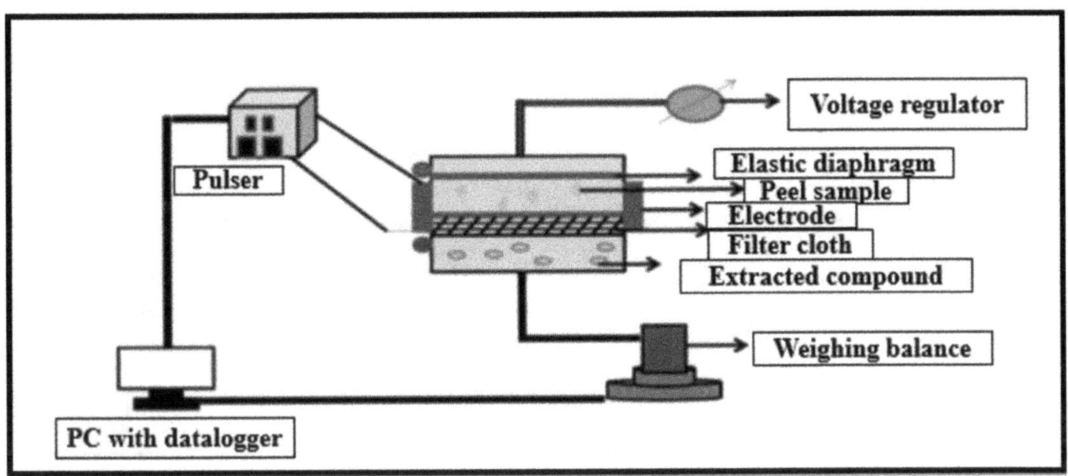

FIGURE 8.11 Pulsed electric field extraction system [99]

A suitable product handling system contains a PEF production unit with a high-voltage generator, a pulse generator, a treatment chamber, and several monitoring and regulating devices forming part of a typical pumpable fluid treatment system [126].

Process variables such as temperature, material characteristics, pulse number, field intensity, and input energy are determined by how well the therapy works [86]. PEF uses renewable plant resources, and agri-solvents instead of water or other solvents, reduces energy consumption, and optimizes unit operations to produce extracts of extraordinary quality and purity [127]. The water-based pulsed electric field approach improves the extraction of various substances, as well as the temperature, rate of extraction, and amount of solvents used [96]. This technology facilitates phase separation and purification by allowing for the particular acquisition of important compounds without harming the treated matrix [128].

Large voltages may be produced by electrically pulsed discharges, which makes them useful for recovering food waste and by-products [129]. By shortening the duration and temperature of the extraction processes, pulsed electrical energy can both boost production and enhance the quality of extracted compounds [127]. Phenolic compound synthesis and antioxidant activity were increased with pulsed electric fields, with maximum cell disintegration index attained at 60 μs [130], and lower extraction times without the use of organic solvents.

The effects of pulsed electric field treatment were investigated to extract bioactive chemicals continuously at 85 °C between 12.4 and 38.4 kV/cm [131]. The permeabilization of the membrane facilitated faster component extraction than with traditional treatment, increasing the yield of bioactive metabolites.

8.7.2.7 Ultrasound-assisted Enzymatic Extraction (UAEE)

Few studies have combined ultrasound and enzyme-aided extraction methods, known as UAEE (Figure 8.12), for leaching out phytochemicals [132, 133]. It is a combination of two related extraction techniques that enable additional benefits. Enzymes support the retrieval of specific bioactive molecules during enzyme-assisted extraction while disrupting and obliterating cell membranes. However, unlike cell walls, the cellular matrix cannot be hydrolyzed by enzymes [134]. In contrast to EAE, the bubbling effect that UAEE produces can greatly dislocate and rupture the medium to facilitate enzymatic response and the consequent discharge of targeted molecules. Enzyme-assisted extraction commonly uses different physical techniques, including shaking, to improve mass transmission. The ultrasonic-aided technique is the best option among these, as it can enhance mass transfer via the matrix medium's interior as well as exterior [135]. Moreover, the contact area between the matrix phases rises when ultrasonic power increases. This can expose more substrates to the enzymes in an enzyme-assisted system and result in

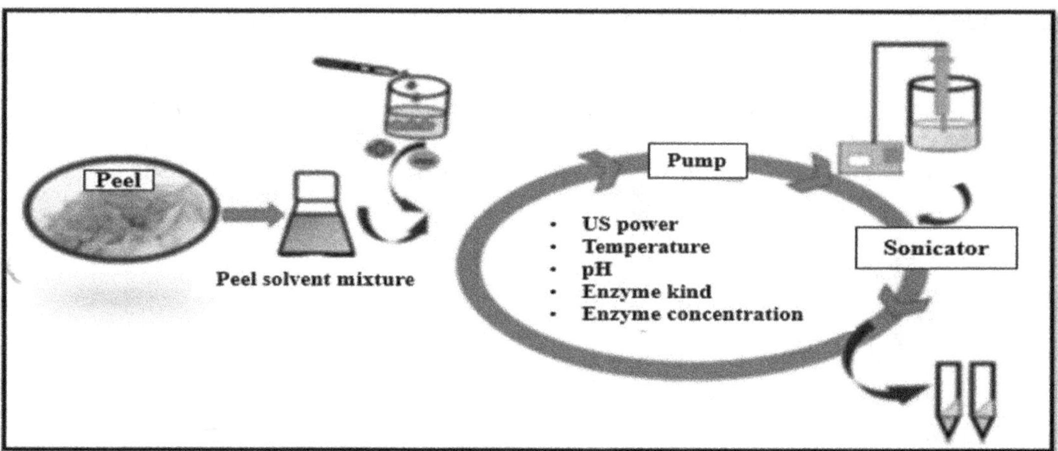

FIGURE 8.12 Ultrasound-assisted enzymatic extraction system [99]

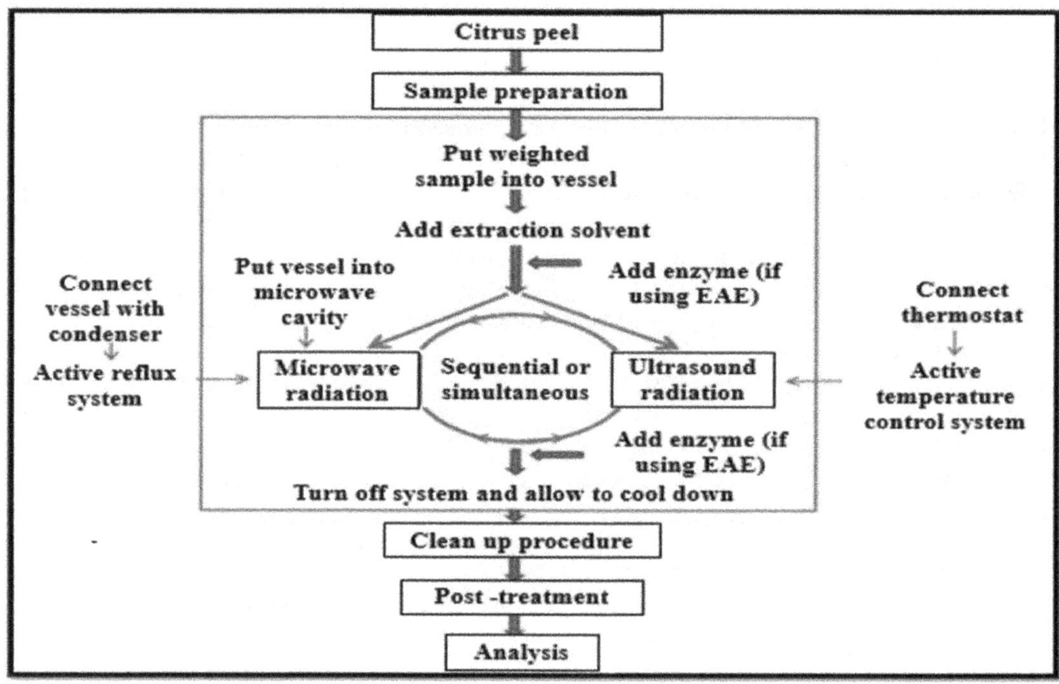

FIGURE 8.13 Synergistic effects of UAE (Ultrasonic-assisted extraction), MWAE (Microwave-assisted extraction), and EAE (Enzymatic assisted extraction) [99]

higher production of the desired bioactive metabolites. UAEE accelerates phytochemical extraction from different plant tissue components, increasing the extract yield of bioactive compounds from plant materials (leaves, fruits, and peels) [136].

8.7.2.8 Microwave-assisted Enzymatic Extraction (MWAEE)

Microwave-assisted extraction (Figure 8.13) has currently been identified as an acceptable technology due to its advantages of improved extraction efficacy, ease of handling, low energy consumption, and low solvent consumption. Enzymolysis and microwave treatment are used in MWAEE to break down and improve the permeability of cell walls, making it easier to transfer the right molecules into the solvent inside the targeted cell [137]. The enzymatic process can be carried out either before or after microwave treatment. Samples should be cooled to a specified temperature to prevent immobilization of previously used enzymes before microwave treatment. Grapefruit peel was used to extract soluble dietary fiber using MWAEE, enzyme (cellulase: 3000 U/g), and microwave technology combined [138]. MWAEE has significant potential for the use of dietary fiber isolated from grape peel in the functional foodservice industry because it not only improves the structural qualities of dietary fiber, but also improves functional traits, particularly by showing a greater ability to bind water, cholesterol, oil, and nitrite ion.

8.7.2.9 Microwave-Ultrasonic Assisted Extraction (MWUAE)

Microwave-Ultrasonic assisted extraction (MWUAE) is the best and most likely strategy for removing obstacles (Figure 8.13). The process can be used as a cost-effective extraction method due to quick sample preparation, enhanced efficacy, short extraction times, and high yields [139]. The US can break down cell membranes, improve mass transfer, increase contact area, solvent quantity that enters the sample medium, and boost the solubility of soluble substances [140].

In MWUAE, the connection between MW and US energy may be quite variable. Throughout the extraction process, MW or US waves periodically break, and the power on/off modules that are employed for the irradiations can harmonize or self-regulate. When a zero refluxing unit is attached to the extraction unit to enable enhanced low-boiling-point solvents (e.g., methanol, ethanol), the harmonized method is usually employed with an increase in energy flow into the treatment unit. The evaporation rate may be decreased, and the solvent temperature can be regulated by the intermittent process [141]. MW or US energy can be produced by two different pieces of equipment: one for ultrasound and the other for microwave radiation. One piece of equipment must be treated first, and then the other equipment is treated again. Concurrent radiation can produce higher energy and then transfer heat to the extraction medium within a predetermined time, but integrating simultaneous irradiation in one container requires a specially designed setup. Ultrasonic waves were sent into the microwave oven's chamber by a probe that was designed into a MWUAE device. A safety device configuration is necessary while creating the MWUAE apparatus. Notwithstanding the possibility of irreversible harm to the probe's basic architecture, the materials listed above have a maximum reliable working power of just 90 W. Furthermore, the use of MWUAE has never been hindered by the higher power value created by the probe material, as the simultaneous radiation produced by the MWUAE unit uses less power than ultrasonic or microwave irradiation. MAE and UAE were found to be the optimal parameters for producing the maximum yield when performing the sequential MWUAE method on a *Citrus maxima* fruit peel and citric acid [142].

8.8 UTILIZATION OF CITRUS PROCESSING WASTE IN FOOD INDUSTRIES

Citrus processing waste is used in different food and non-food sectors due to its tremendous health-promoting properties. The jam, jellies, and marmalade industries, as well as flavor/cosmetics industries that employ citrus processing waste, have been in operation for many years.

8.8.1 Food Additives

Food additives are compounds that give varied sensory (i.e., taste, aroma, color, flavor), along with the nutritional quality of the foodstuffs. Increased attention has been paid to the inclusion of naturally derived additives attained from food processing waste because of enhanced awareness about the consumption of healthy and organic food ingredients. Food processing waste consists of bioactive compounds that have good antioxidant action, which can be natural substitutes for artificial food additives [44].

The advent of technology has allowed various citrus processing wastes like peels, seed oil, and pomace to be treated and added to baked food products, thereby increasing the nutritional quotient of prepared food and enhancing their use. For example, muffins are made with 10% and 15% fiber obtained from orange bagasse. Muffins' incorporation of high-fiber orange bagasse (15%) contained a good amount of fiber (15.3%) and low fat (15%) with a low glycemic index.

Citrus by-products like peels in both fermented as well as unfermented forms have been used as food additives in bread dough, where a combination of 4% and 6% unfermented peel dried at 150 °C and 100 °C, respectively, showed higher organoleptic potential [46]. In a trial, sponge cakes were developed by combining different quantities of powder obtained from the citrus peel (0, 10%, 20%, 30%, and 40%). Substituting 10% powder obtained from citrus peel for sponge cake was suitable for boosting the nutritional content along with the sensory characteristics of the cake [47]. In addition, biscuits have been prepared by substituting wheat flour with 3%, 6%, and 9% mandarin peel flour. Addition of 6% citrus peel powder improved ash/mineral (1.32%), fiber (0.85%), carotenoids (69 µg/g), ascorbic acid (1.5 mg/100 g), polyphenols (2150 µg GAE or gallic acid equivalent/g), and total antioxidative activity (24.5%) of cookies compared with non-substituted control [48]. Another important application of citrus peels in the food industry is the use of citrus peel pectin as an emulsifier, thickener, stabilizer, texturizer, and gel-forming agent in the development of fruit jams, jellies, and low-fat/low-sugar products [19]. Furthermore, citrus peels are also used as raw material in the development of probiotic beverages, ice cream, and additional foodstuffs [1].

8.8.2 Prebiotics

Prebiotics are often referred to as "indigestible food ingredients that benefit the host by explicitly carrying out cellular division or action of varied microorganisms in the colon, and thereby improving host health" [49]. Prebiotics are becoming important in everyday life. Although the citrus peel pectin is used to make jams, jellies, as well as marmalades, the pectin in citrus peels displays effective prebiotic action. The pectin-based oligosaccharides in citrus fruit processing waste are dynamic prebiotic constituents with improved antimicrobial and microbiota-modulating ability [74]. Current research suggests the possibility of using citrus peel by-products in the preparation of prebiotic functional foods with enhanced biological functions. Pectic oligosaccharides attained by partial hydrolysis of pectin are emergent prebiotics resulting from agricultural waste (i.e., citrus peel waste) [50, 51].

8.8.3 Encapsulating Agent/Edible Packaging Materials

Another use of citrus by-products is in the development of packaging material. Biodegradable packaging can be synthesized using citrus peels. *Citrus limon* and *Citrus aurantifolia* peels have been used to develop edible coatings for fresh strawberries [54]. In addition, an edible coating was prepared through pectin extracted from lemon peel as well as sweet potato starch (3:1 ratio), together with titanium oxide (0.5%–2.0%) nanoparticles [55]. Besides, grapefruit peel powder (*Citrus grandis*) and tea polyphenols (10%) were used to develop bioactive edible packaging films by casting. The film had an excellent moisture barrier, with antibacterial and antioxidant properties and strong mechanical action. The developed bioactive edible packaging can likewise be a green substitute for 30-day soybean oil packaging [56].

8.9 CURRENT AND FUTURE PERSPECTIVES

In 2017, India was the second-largest global producer of lemon and limes (2,364,000 tonnes) after Mexico (2,528,174 tonnes) and the third-largest orange producer (7,647,000 tonnes) after Brazil and China. It had been anticipated that citrus natural goods would primarily come from Northeast India [143].

Kinnow mandarin (*Citrus reticulata*) produced (~414,090 tonnes) in Punjab generates 25% peel, 23% residue, and 2% seeds, along with 50% juice. Kinnow peels contain on average 22.5% total solids, 1.4% acidity, 6.2% total sugars, 6% reducing sugars, 1.9% pectin, 0.8% fat, 0.7% ash, 41.6 mg/100 g ascorbic acid, 13.7 mg/100 g carotenoids, 7.4 mg/100 g β-carotene, 0.420 mg/g naringin, and 4.7 mg/g limonin. Kinnow peels (flavedo) were shredded and frozen (-18 °C), and part of it was blanched-frozen (kept in boiling water for 4 min before freezing). Plain ice cream prepared with 1%, 3%, and 5% untreated and blanched-frozen kinnow peels had increased total solids (~37%), ascorbic acid (2.9–6 mg/100 g), and naringin (46.6–81.1 μg/g) with increasing peel addition. Ice cream with 3% unblanched frozen kinnow peel had the best mean sensory scores (8.1 overall acceptability score) [144]. The most recent study [145] showed that 4% phenolic extract (heat-assisted hydroalcoholic [20% ethanol], 90 °C, 15 min) from mandarin (*Citrus reticulata*) peels supplementation increased total flavonoid content (2.2 mg/g), DPPH antioxidant activity (37 mg/g IC_{50}), and phenolic (quercetin, caffeic acid, and hesperidin [33.3, 0.75, and 83 μg/g]) contents of wheat bread. Moreover, most phenolics, except hesperidin, degraded completely after *in vitro* digestion. The duodenal hesperidin concentration doubled (159 vs 77 μg/g for the control) in 4% enriched bread, indicating a beneficial health impact to consumers.

The microwave method (900 watts, 8 min) has been used to prepare carbon dots from *Citrus limon* peel to generate the highest yield (60% carbon, 9% hydrogen, and 8.2% nitrogen, 56% quantum yield similar to the standard quinine sulfate in 0.1 M H_2SO_4 dye [54% at 340 nm]). The carbon dots from *Citrus limon* peel were spherical, in relatively uniform size (1.5–3.5 nm), well dispersed, and had great photostability. The carbon dots can detect Fe^{3+} with complete (100%) recovery [146].

Citrus fruit waste is used commercially to produce rhamnose, the basic raw material for the synthesis of strawberry-flavored furaneol (2,5-dimethyl-4-hydroxy-3(2H)-furanone). Citrus wastes are used for enzyme production; for example, α-amylase and pectinase can be produced from orange waste and orange peel using *Aspergillus niger*, and filter-paperase (FPase) from kinnow waste by *Trichoderma reesei*. High pectinase enzyme yield and activity (6800 IU/g) are produced when orange peel extract is used along with peptone in submerged fermentation. Orange peel waste is one of the best-known and studied raw materials for the production of clouding agents [147].

Several patents have been issued using citrus waste to develop novel products. For example, citrus waste is pretreated to produce pasteurized treated citrus waste solids and water containing peel oil components. Pasteurization efficiently disrupts cell structure and removes volatile peel oil components from citrus processing waste. Ground citrus waste is preheated (55–99 °C, jacketed screw conveyor) and heated at elevated temperature (100–160 °C, 20 bar barometric pressure) to concentrate and separate peel oil. The deoiled citrus waste (<1500 ppm limonene) is cooled before enzyme (cellulolytic, hemicellulolytic, and pectinolytic enzyme mixture) hydrolysis. The citrus waste solids can be fermented with enzymes to produce ethanol or other products [148]. A process is described to convert citrus by-products (i.e., peel and juice from juice sacs attached to the peel) into dietary fiber, naringin, juice, and oil [149]. The process involves peels pressing to release the juice from vesicles, peels grinding in water to form a slurry, peel color and oil removal by flotation, peel decolorization and debittering by boiling and washing before obtaining dietary fiber. Oil from citrus by-products is obtained by mechanically pressing (5–15 psig) citrus peels, grinding (0.1–10 mm) to form a slurry, flushing the slurry to remove flavonoids and citrus oil, debittering the peel particles (150–230 °F), peel pressing to remove water solubles (flavonoids, citrus oil, and flavor), and collecting the citrus oil by dewatering the slurry and debittering the peel particles.

A microwave-assisted hydrothermal low-temperature (biorefinery) process is described to isolate pectin, D-limonene, flavonoid, soluble monosaccharide, and cellulose from citrus waste [150]. The biorefinery process is described as a novel cascade-type valorization of citrus waste for conversion and separation into bio-derived chemicals and materials using low-temperature hydrothermal microwave treatment. Orange peel is separated into limonene, pectin, monosaccharides, and cellulose in a single process step. Waste orange peels and pith were cut into small pieces before microwave hydrothermal treatment (200 °C, 10 min under continuous stirring), followed by four separation steps. The first step separated (filtered) the gel (devoid of pectin), dewatered (by solvent exchange with ethanol and acetone), followed by drying to obtain mesoporous cellulose (19 nm pore diameter, 0.38 cm^3/g pore volume; 11.6% dry weight). Organic compound (α-terpineol/bio-oil; 14% dw) is separated using ethyl acetate, and pectin (11.6% dw) is precipitated from the residue using aqueous ethanol (1:3 ratio). The water-ethanol solution was dried after pectin separation to obtain a mixture of sugars (25.6% dw).

A homogeneous, uniformly smooth product of micronized orange pomace (<250 microns) is used to increase orange juice viscosity (~10-fold), mouthfeel, and fiber content. The beverage (8 oz) comprised about 20% weight of the micronized orange pomace and contained 5 g fiber (1:2 soluble to insoluble fiber ratio). The orange pomace contained 72%–82% moisture, 4.2%–8.5% insoluble fiber, 2.8%–5.8% soluble fiber, 6.95%–14.22% total fiber, 40.9–87.5 ppm limonin, 2562–4380 ppm hesperidin, 5.3%–7.4% sugars, and 20–45 mg vitamin C/100 g [151]. Citrus waste is treated with limewater and acidified (50% HCl solution) to isolate/separate hesperidin, and polysaccharides are precipitated from the aqueous extract with ethyl alcohol (96%), followed by filtration, purification, and drying. The polysaccharides exert antiulcer activity [152].

Manak Citrus Products Private Limited (incepted in 1955) is a well-known Indian manufacturer offering dried lime peels, lime juice, and lime oil. The dried lime peel (12 kg package) can be used for cosmetics, tea flavor, and Ayurvedic preparation. The company also offers cold-pressed lime oil and distilled lime oil (5 L aluminum bottle) [153]. A new company, "Wonky Fruit," uses fruit waste (peels, seeds, and even flesh) from the lemon industry to create a wide range of cosmetic products, from lip balm to fragrant candles containing lemon waste. Cold-pressed lime and lemon seed oils and peel-extracted oil are used as emollients in their cosmetics [154].

Citrus by-products possess a good reservoir of bioactive compounds that could be recovered in different ways. Many studies have stated that citrus waste extraction yield is improved by the application of green/nonconventional extraction techniques such as supercritical fluid, high pressure, microwave, ultrasonication, as well as pulsed electric field extraction. Nevertheless, with progressive studies toward the efficient utilization of citrus waste, the usage of all these processes on an industrial scale is yet to be realized. Therefore, industry-focused research must be done for improved waste utilization, financial growth, and environmental sustainability. Extraction kinetics studies are also important to understand the exact mechanism of mass transfer, which could be beneficial for industrial-scale applications.

In addition, focus is required on the threats or challenges in implementing green extraction methods at the industry level for citrus waste recovery and in integrated biofiltration industries. Besides, meta-research that combines active harmonization, as well as collaboration among citrus-cultivating agriculturists, citrus juice processing industries, and industrial investigators, is important to comprehend the future of comprehensive biomass utilization for making a safe ecological system for upcoming generations.

New extraction techniques can result in larger extraction yields with short extraction times, better product quality, and fewer environmental issues. Research on the fusion of these novel extraction technologies is becoming more and more popular today. To summarize, citrus waste is very important in a variety of fields, including the pharmaceutical, cosmetic, and food industries. Citrus waste should receive more attention because of the enormous amount requiring management, value-added processing, and optimized profit for a circular economy. Citrus biomass can be used as a non-synthetic conditioner for the manufacturing of compost, specifically in agricultural areas, improving the nutritional content and organic matter of agricultural soil. It can also be used as animal feed. The presence of various bioactive compounds, including dietary fiber, limonoids, pectins, flavonoids, carotenoids, and essential oils, are all potential sources of promoting health benefits.

Based on compiled reviews, the food industry can benefit economically by adequately utilizing citrus waste with appropriate processing technologies. Since citrus waste is full of bioactive components, new innovative approaches can be explored to valorize this waste to obtain phenols, pigments, essential oil, fiber, pectin, flavonoids, and other phytonutrients for further use in food industries, especially the baking industry. More detailed studies are needed to study the extraction techniques, optimization of extraction parameters, and extraction kinetics concerning quality.

REFERENCES

1. Suri, S., Singh, A., & Nema, P. K. (2022). Current applications of citrus fruit processing waste: A scientific outlook. *Applied Food Research*, 2(1), 100050. https://doi.org/10.1016/j.afres.2022.100050
2. Rashid, U., Ibrahim, M., Yasin, S., Yunus, R., Taufiq-Yap, Y. H., & Knothe, G. (2013). Biodiesel from *Citrus reticulata* (mandarin orange) seed oil, a potential non-food feedstock. *Industrial Crops and Products*, 45, 355–359. https://doi.org/10.1016/j.indcrop.2012.12.039
3. Ledesma-Escobar, C. A., & de Castro, M. D. L. (2014). Towards a comprehensive exploitation of citrus. *Trends in Food Science & Technology*, 39(1), 63–75. https://doi.org/10.1016/j.tifs.2014.07.002
4. Suri, S., Singh, A., & Nema, P. K. (2021). Recent advances in valorization of citrus fruits processing waste: a way forward towards environmental sustainability. *Food Science and Biotechnology*, 30(13), 1601–1626. https://doi.org/10.1007/s10068-021-00984-y
5. Santiago, B., Moreira, M. T., Feijoo, G., & González-García, S. (2020). Identification of environmental aspects of citrus waste valorization into D-limonene from a biorefinery approach. *Biomass and Bioenergy*, 143, 105844. https://doi.org/10.1016/j.biombioe.2020.105844
6. Panwar, D., Saini, A., Panesar, P. S., & Chopra, H. K. (2021). Unraveling the scientific perspectives of citrus by-products utilization: Progress towards circular economy. *Trends in Food Science & Technology*, 111, 549–562. https://doi.org/10.1016/j.tifs.2021.03.018; Yadav, V., Sarker, A., Yadav, A., Miftah, A. O., Bilal, M., & Iqbal, H. M. (2022). Integrated biorefinery approach to valorize citrus waste: A sustainable solution for resource recovery and environmental management. *Chemosphere*, 293, 133459. https://doi.org/10.1016/j.chemosphere.2021.122459
7. Mahato, N., Sharma, K., Sinha, M., Baral, E. R., Koteswararao, R., Dhyani, A., . . . & Cho, S. (2020). Biosorbents, industrially important chemicals and novel materials from citrus processing waste as a sustainable and renewable bioresource: A review. *Journal of Advanced Research*, 23, 61–82. https://doi.org/10.1016/j.jare.2020.01.007
8. Wadhwa, M., & Bakshi, M. P. S. (2013). Utilization of fruit and vegetable wastes as livestock feed and as substrates for generation of other value-added products. *Rap Publication*, 4(2013), 67.
9. Lotito, A. M., De Sanctis, M., Pastore, C., & Di Iaconi, C. (2018). Biomethanization of citrus waste: Effect of waste characteristics and of storage on treatability and evaluation of limonene degradation. *Journal of Environmental Management*, 215, 366–376. https://doi.org/10.1016/j.jenvman.2018.03.057
10. Martin, M. A., Siles, J. A., Chica, A. F., & Martin, A. (2010). Bio-methanization of orange peel waste. *Bioresource Technology*, 101(23), 8993–8999. https://doi.org/10.1016/j.biortech.2010.06.133
11. Kuna, A., Sowmya, M., Sahoo, M. R., Mayengbam, P. D., Dasgupta, M., & Sreedhar, M. (2018). Value addition and sensory evaluation of products made from underutilized Kachai Lemon (*Citrus jambhiri*) Lush. fruits. *Journal of Pharmacognosy and Phytochemistry*, 7(5), 3032–3036.
12. Marín, F. R., Soler-Rivas, C., Benavente-García, O., Castillo, J., & Pérez-Alvarez, J. A. (2007). By-products from different citrus processes as a source of customized functional fibres. *Food Chemistry*, 100(2), 736–741. https://doi.org/10.1016/j.foodchem.2005.04.040
13. FAOSTAT. (2022). Food and agriculture data. *Food and Agriculture Organization*. Rome: FAO. Available from https://www.fao.org/faostat/en/#data/QCL (Accessed 5 August 2022).
14. ICAR-Central Citrus Research Institute. (2019). *Annual Report*. Indian Council of Agricultural Research, Nagpur, Maharashtra, India. Available from https://ccri.icar.gov.in/ccringp/
15. Berk, Z. (2016). Chapter 1-Introduction: History, production, trade, and utilization. In Z. Berk (Ed.), *Citrus Fruit Processing* (pp. 1–8). San Diego, CA: Academic Press.
16. FAO. (2017). *Citrus Fruit Fresh and Processed—Statistical Bulletin 2016*. Available from http://www.fao.org/publications/card/en/c/534798b4-2ee5-4626-84b1-0090df36dd69/ (Accessed 4 August 2022).
17. Izquierdo, L., & Sendra, J. M. (2003). Citrus fruits composition and characterization. In B. Caballero, L. Trugo, & P. Finglas (Eds.), *Encyclopedia of Food Sciences and Nutrition* (p. 6000). Oxford: Academic Press.

18. Chavan, P., Singh, A. K., & Kaur, G. (2018). Recent progress in the utilization of industrial waste and by-products of citrus fruits: A review. *Journal of Food Process Engineering*, 41, e12895. https://doi.org/10.1111/jfpe.12895
19. Mamma, D., & Christakopoulos, P. (2013). Biotransformation of citrus by-products into value added products. *Waste and Biomass Valorization*, 5, 529–549. https://doi.org/10.1007/s12649-013-9250-y
20. Wang, L., Xu, H., Yuan, F., Fan, R., & Gao, Y. (2015). Preparation and physicochemical properties of soluble dietary fiber from orange peel assisted by steam explosion and dilute acid soaking. *Food Chemistry*, 185, 90–98. https://doi.org/10.1016/j.foodchem.2015.03.112
21. Rivas, B., Torrado, A., Torre, P., Converti, A., & Dominguez, J. M. (2008). Submerged citric acid fermentation on orange peel auto hydrolysate. *Journal of Agricultural and Food Chemistry*, 56, 2380–2387. https://doi.org/10.1021/jf073388r
22. Ortelli, D., Edder, P., & Corvi, C. (2005). Pesticide residues survey in citrus fruits. *Food Additives and Contaminants*, 22, 423–428. https://doi.org/10.1080/02652030500089903
23. Calvaruso, E., Cammilleri, G., Pulvirenti, A., Lo Dico, G. M., Lo Cascio, G., Giaccone, V., Vitale Badaco, V., Cipri, V., Alessandra, M. M., Vella, A., Macaluso, A., Di Bella, C., & Ferrantelli, V. (2020). Residues of 165 pesticides in citrus fruits using LC-MS/MS: A study of the pesticides distribution from the peel to the pulp. *Natural Product Research*, 34, 34–38. https://doi.org/10.1080/14786419.2018.1561682
24. Rosa, A., Era, B., Masala, C., Nieddu, M., Scano, P., Fais, A., Porcedda, S., & Piras, A. (2019). Supercritical CO2 extraction of waste citrus seeds: Chemical composition, nutritional and biological properties of edible fixed oils. *European Journal of Lipid Science and Technology*, 121, 1800502. https://doi.org/10.1002/ejlt.201800502
25. Saini, A., Panesar, P. S., & Bera, M. B. (2020). Valuation of *Citrus reticulata* (kinnow) peel for the extraction of lutein using ultrasonication technique. *Biomass Conversion and Biorefinery*, 1–9. https://doi.org/10.1007/s13399-020-00605-4/
26. Wang, L., Lee, W. W., Yang, H. W., Ryu, B. M., Cui, Y. R., Lee, S. C., Lee, T. G., & Jeon, Y. J. (2018). Protective effect of water extract of citrus pomace against AAPH-induced oxidative stress in vitro in Vero cells and in vivo in zebrafish. *Preventive Nutrition and Food Science*, 23, 301–308. https://doi.org/10.3746/pnf.2018.23.4.301
27. Majerska, J., Michalska, A., & Figiel, A. (2019). A review of new directions in managing fruit and vegetable processing by-products. *Trends in Food Science & Technology*, 88, 207–219. https://doi.org/10.1016/j.tifs.2019.03.021
28. Papoutsis, K., Pristijono, P., Golding, J. B., Stathopoulos, C. E., Scarlett, C. J., Bowyer, M. C., & Van Vuong, Q. (2016). Impact of different solvents on the recovery of bioactive compounds and antioxidant properties from lemon (*Citrus limon* L.) pomace waste. *Food Science and Biotechnology*, 25, 971–977. https://doi.org/10.1007/s10068-016-0158-8
29. Wadhwa, M., & Bakshi, M. P. S. (2013). *Utilization of Fruit and Vegetable Wastes as Livestock Feed and as Substrates for Generation of Other Value-Added Products*. Rome, Italy: Food and Agriculture Organization.
30. Martin, M. A., Siles, J. A., Chica, A. F., & Martín, A. (2010). Biomethanization of orange peel waste. *Bioresource Technology*, 101, 8993–8999. https://doi.org/10.1016/j.biortech.2010.06.133
31. M'hiri, N., Ioannou, I., Ghoul, M., & Mihoubi, N. B. (2015). Proximate chemical composition of orange peel and variation of phenols and antioxidant activity during convective air drying. *Journal of New Sciences, Agriculture and Biotechnology*, 9, 881–890.
32. Ani, P. N., & Abel, H. C. (2018). Nutrient, phytochemical, and antinutrient composition of Citrus maxima fruit juice and peel extract. *Food Science and Nutrition*, 6, 653–658. https://doi.org/10.1002/fsn3.604
33. Akpata, M. I., & Akubor, P. I. (1999). Chemical composition and selected functional properties of sweet orange (*Citrus sinensis*) seed flour. *Plant Foods Human Nutrition*, 54, 353–362. https://doi.org/10.1023/A:1008153228280
34. Oikeh, E., Oriakhi, K., & Omoregie, E. (2013). Proximate analysis and phytochemical screening of *Citrus sinensis* fruit wastes. *The Bioscientists*, 1(2), 164–170.
35. Khan, U. M., Sameen, A., Aadil, R. M., Shahid, M., Sezen, S., Zarrabi, A., . . . & Butnariu, M. (2021). Citrus genus and its waste utilization: a review on health-promoting activities and industrial application. *Evidence-Based Complementary and Alternative Medicine*, 2021, 1–17. https://doi.org/10.1155/2021/2488804/
36. Espın, J. C., Garcia-Conesa, M. T., & Tomas-Barberan, F. A. (2007). Nutraceuticals: facts and fiction. *Phytochemistry*, 68, 2986–3008.
37. Wilmsen, P. K., Spada, D. S., & Salvador, M. (2005). Antioxidant activity of the flavonoid hesperidin in chemical and biological systems. *Journal of Agricultural and Food Chemistry*, 53(12), 4757–4761. https://doi.org/10.1021/jf0502000
38. Di Carlo, G., Mascolo, N., Izzo, A. A., & Capasso, F. (1999). Flavonoids: Old and new aspects of a class of natural therapeutic drugs. *Life Sciences*, 65(4), 337–353.

39. Reddy, L., Odhav, B., & Bhoola, K. D. (2003). Natural products for cancer prevention: A global perspective. *Pharmacology & Therapeutics*, 99(1), 1–13. https://doi.org/10.1016/S0163-7258(03)00042-1
40. McCullough, M. L., Peterson, J. J., Patel, R., Jacques, P. F., Shah, R., & Dwyer, J. T. (2012). Flavonoid intake and cardiovascular disease mortality in a prospective cohort of US adults. *American Journal of Clinical Nutrition*, 95(2), 454–464. https://doi.org/10.3945/ajcn.111.016634
41. Halliwell, B. (1996). Commentary oxidative stress, nutrition and health. Experimental strategies for optimization of nutritional antioxidant intake in humans. *Free Radical Research*, 25(1), 57–74. https://doi.org/10.3109/10715769609145656
42. Garg, A., Garg, S., Zaneveld, L. J. D., & Singla, A. K. (2001). Chemistry and pharmacology of the citrus bioflavonoid hesperidin. *Phytotherapy Research*, 15(8), 655–669. https://doi.org/10.1002/ptr.1074
43. Beecher, G. R. (1999). Phytonutrients' role in metabolism: Effects on resistance to degenerative processes. *Nutrition Reviews*, 57(9), 3–6.
44. Heiss, E., Herhaus, C., Klimo, K., Bartsch, H., & Gerhauser, C. (2001). Nuclear factor kappa B is a molecular target for sulforaphane-mediated anti-inflammatory mechanisms. *Journal of Biological Chemistry*, 276, 32008–32015.
45. Norihiro, Y., Takahiro, F., Hitoshi, M., Takeshi, O., Kunio, S., & Ron, H. (2014). Orange peel extract, containing high levels of polymethoxy flavonoid, suppressed UVB-induced COX-2 expression and PGE2 production in HaCaT cells through PPAR-γ activation. *Experimental Dermatology*, 23, 18–22. https://doi.org/10.1111/exd.12394
46. Hwang, S. L., Shih, P. H., & Yen, G. C. (2012). Neuroprotective effects of citrus flavonoids. *Journal of Agricultural and Food Chemistry*, 60(4), 877–885. https://doi.org/10.1021/jf204452y
47. Aranganathan, S., Selvam, J. P., & Nalini, N. (2008). Effect of hesperetin, a citrus flavonoid, on bacterial enzymes and carcinogen-induced aberrant crypt foci in colon cancer rats: A dose-dependent study. *Journal of Pharmacy and Pharmacology*, 60, 1385–1392. https://doi.org/10.1211/jpp.60.10.0015
48. Moon, S. H., Lee, J. H., Kim K. T., et al. (2013). Antimicrobial effect of 7-O-butylnaringenin, a novel flavonoid, and various natural flavonoids against *Helicobacter pylori* strains. *International Journal of Environmental Research and Public Health*, 10(11), 5459–5469. https://doi.org/10.3390/ijerph10115459
49. Park, K. (2012). Induction of the cell cycle arrest and apoptosis by flavonoids isolated from Korean *Citrus aurantium* L. in nonsmall-cell lung cancer cells. *Food Chemistry*, 135, 2728–2735. https://doi.org/10.1016/j.foodchem.2012.06.097
50. Nagappan, A., Lee, H. J., Saralamma V. V. G., et al. (2016). Flavonoids isolated from *Citrus platymamma* induced G2/M cell cycle arrest and apoptosis in A549 human lung cancer cells. *Oncology Letters*, 12(2), 1394–1402. https://doi.org/10.3892/ol.2016.4793/
51. Yamamoto, M., Suzuki, A., & Hase, T. (2008). Short-term effects of glucosyl hesperidin and hesperetin on blood pressure and vascular endothelial function in spontaneously hypertensive rats. *Journal of Nutritional Science & Vitaminology*, 54, 95–98. https://doi.org/10.3177/jnsv.54.95
52. Shen, W., Xu, Y., & Lu, Y. H. (2012). Inhibitory effects of citrus flavonoids on starch digestion and antihyperglycemic effects in HepG2 cells. *Journal of Agricultural and Food Chemistry*, 60, 9609–9619. https://doi.org/10.1021/jf3032556
53. Zhang, Y., Wang, J. F., Dong J., et al. (2013). Inhibition of α-toxin production by subinhibitory concentrations of naringenin controls *Staphylococcus aureus* pneumonia. *Fitoterapia*, 86, 92–99. https://doi.org/10.1016/j.fitote.2013.02.001
54. Degirmenci H., & Erkurt, H. (2019). Relationship between volatile components, antimicrobial and antioxidant properties of the essential oil, hydrosol and extracts of *Citrus aurantium* L. flowers. *Journal of Infection and Public Health*, 13(1), 58–67. https://doi.org/10.1016/j.jiph.2019.06.017
55. Huang, S. M., Tsai, S. Y., Lin, J. A., Wu, C. H., & Yen, G. C. (2012). Cytoprotective effects of hesperetin and hesperidin against amyloid beta-induced impairment of glucose transport through downregulation of neuronal autophagy. *Molecular Nutrition & Food Research*, 56, 601–609. https://doi.org/10.1002/mnfr.201100682
56. Sharma, K., Mahato, N., Cho, M. H., & Lee, Y. R. (2017). Converting citrus wastes into value-added products: Economic and environmently friendly approaches. *Nutrition*, 34, 29–46. https://doi.org/10.1016/j.nut.2016.09.006
57. Balasundram, N., Sundram, K., & Samman, S. (2006). Phenolic compounds in plants and agri-industrial by-products: Antioxidant activity, occurrence, and potential uses. *Food Chemistry*, 99(1), 191–203. https://doi.org/10.1016/j.foodchem.2005.07.042
58. Anticona, M., Blesa, J., Lopez-Malo, D., Frigola, A., & Esteve, M. J. (2021). Effects of ultrasound-assisted extraction on physicochemical properties, bioactive compounds, and antioxidant capacity for the valorization of hybrid Mandarin peels. *Food Bioscience*, 42, 101185. https://doi.org/10.1016/j.fbio.2021.101185
59. Saini, A., Panesar, P. S., & Bera, M. B. (2019). Comparative study on the extraction and quantification of polyphenols from citrus peels using maceration and ultrasonic technique. *Current Research in Nutrition and Food Science*, 7(3), 678.

60. Suri, S., Singh, A., & Nema, P. K. (2022). Infrared drying of Kinnow (*Citrus reticulata*) peel waste: Kinetics and quality characterization. *Biomass Conversion and Biorefinery*, 1–12. https://doi.org/10.1007/s13399-022-02844-z/
61. Rafiq, S., Singh, B., & Gat, Y. (2019). Effect of different drying techniques on chemical composition, color and antioxidant properties of kinnow (*Citrus reticulata*) peel. *Journal of Food Science and Technology*, 56(5), 2458–2466. https://doi.org/10.1007/s13197-019-03722-9/
62. Goulas, V., & Manganaris, G. A. (2012). Exploring the phytochemical content and the antioxidant potential of Citrus fruits grown in Cyprus. *Food Chemistry*, 131(1), 39–47. https://doi.org/10.1016/j.foodchem.2011.08.007
63. El Asbahani, A., Miladi, K., Badri, W., Sala, M., Addi, E. A., Casabianca, H., . . . & Elaissari, A. (2015). Essential oils: From extraction to encapsulation. *International Journal of Pharmaceutics*, 483(1–2), 220–243. https://doi.org/10.1016/j.ijpharm.2014.12.069
64. Berk, Z. (2016). By-products of the citrus processing industry. In Z. Berk (Ed.), *Citrus Fruit Processing* (pp. 219–233). San Diego: Academic press. https://doi.org/10.1016/b978-0-12-803133-9.00010-2
65. Golmakani, M. T., & Moayyedi, M. (2016). Comparison of microwave-assisted hydrodistillation and solvent-less microwave extraction of essential oil from dry and fresh *Citrus limon* (Eureka variety) peel. *Journal of Essential Oil Research*, 28(4), 272–282. https://doi.org/10.1080/10412905.2016.1145606
66. Lopresto, C. G., Meluso, A., Di Sanzo, G., Chakraborty, S., & Calabrò, V. (2019). Process-intensified waste valorization and environmentally friendly d-limonene extraction. *Euro-Mediterranean Journal for Environmental Integration*, 4(1), 1–12. https://doi.org/10.1007/s41207-019-0122-0/
67. Safranko, S., Ćorković, I., Jerković, I., Jakovljević, M., Aladić, K., Šubarić, D., & Jokić, S. (2021). Green extraction techniques for obtaining bioactive compounds from mandarin peel (*Citrus unshiu* var. Kuno): Phytochemical analysis and process optimization. *Foods*, 10(5), 1043. https://doi.org/10.3390/foods10051043
68. Waheed, A., Mahmud, S., Saleem, M., & Ahmad, T. (2009). Fatty acid composition of neutral lipid: Classes of citrus seed oil. *Journal of Saudi Chemical Society*, 13(3), 269–272. https://doi.org/10.1016/j.jscs.2009.10.007
69. Juhaimi, F. A., Matthäus, B., Özcan, M. M., & Ghafoor, K. (2016). The physico-chemical properties of some citrus seeds and seed oils. *Zeitschrift für Naturforschung C*, 71(3–4), 79–85. https://doi.org/10.1515/znc-2016-0004
70. Mahato, N., Sharma, K., Sinha, M., & Cho, M. H. (2018). Citrus waste derived nutra-/pharmaceuticals for health benefits: Current trends and future perspectives. *Journal of Functional Foods*, 40, 307–316. https://doi.org/10.1016/j.jff.2017.11.015
71. M'hiri, N., Ioannou, I., Ghoul, M., & MihoubiBoudhrioua, N. (2017). Phytochemical characteristics of citrus peel and effect of conventional and nonconventional processing on phenolic compounds: A review. *Food Reviews International*, 33(6), 587–619. https://doi.org/10.1080/87559129.2016.1196489
72. Ordóñez-Santos, L. E., Esparza-Estrada, J., & Vanegas-Mahecha, P. (2021). Ultrasound-assisted extraction of total carotenoids from mandarin epicarp and application as natural colorant in bakery products. *LWT*, 139, 110598. https://doi.org/10.1016/j.lwt.2020.110598
73. Foti, P., Ballistreri, G., Timpanaro, N., Rapisarda, P., & Romeo, F. V. (2022). Prebiotic effects of citrus pectic oligosaccharides. *Natural Product Research*, 36(12), 3173–3176. https://doi.org/10.1080/14786419.2021.1948845
74. Maran, J. P., Sivakumar, V., Thirugnanasambandham, K., & Sridhar, R. (2013). Optimization of microwave assisted extraction of pectin from orange peel. *Carbohydrate Polymers*, 97(2), 703–709. https://doi.org/10.1016/j.carbpol.2013.05.052
75. Su, D. L., Li, P. J., Quek, S. Y., Huang, Z. Q., Yuan, Y. J., Li, G. Y., & Shan, Y. (2019). Efficient extraction and characterization of pectin from orange peel by a combined surfactant and microwave assisted process. *Food Chemistry*, 286, 1–7. https://doi.org/10.1016/j.foodchem.2019.01.200
76. Yousuf, O. (2019). Comprehensive valorization of Citrus peel through sequential microwave-assisted solvent extraction of bioactive compounds and pectin for their application in food. Ph.D. Dissertation. Pantnagar, Uttarakhand, India: G.B. Pant University of Agriculture and Technology.
77. Yang, Y. Y., Ma, S., Wang, X. X., & Zheng, X. L. (2017). Modification and application of dietary fiber in foods. *Journal of Chemistry*, 2017. https://doi.org/10.1155/2017/9340427
78. Wang, L., Xu, H., Yuan, F., Fan, R., & Gao, Y. (2015). Preparation and physicochemical properties of soluble dietary fiber from orange peel assisted by steam explosion and dilute acid soaking. *Food Chemistry*, 185, 90–98. https://doi.org/10.1016/j.foodchem.2015.03.112
79. Dervisoglu, M., & Yazici, F. (2006). Note. The effect of citrus fibre on the physical, chemical and sensory properties of ice cream. *Food Science and Technology International*, 12(2), 159–164. https://doi.org/10.1177/1082013206064005
80. Marín, F. R., Soler-Rivas, C., Benavente-García, O., Castillo, J., & Pérez-Alvarez, J. A. (2007). By-products from different citrus processes as a source of customized functional fibres. *Food Chemistry*, 100(2), 736–741. https://doi.org/10.1016/j.foodchem.2005.04.040

81. Gertenbach, D. D. (2002). Solid-liquid extraction technologies for manufacturing nutraceuticals. In J. Shi, G. Mazza, & M. L. Maguer (Eds.), *Functional Foods: Biochemical and Processing Aspects* (vol. 2, pp. 331–66). CRC Press.
82. Chen, Y., Barzee, T. J., Zhang, R., & Pan, Z. (2019). Citrus. In *Integrated Processing Technologies for Food and Agricultural By-products* (pp. 217–242). Academic Press.
83. Barba, F. J., Rosell_o-Soto, E., Marszałek, K., Kovacevic, D. B., Jambrak, A. R., Lorenzo, J. M., & Putnik, P. (2019). Green food processing: Concepts, strategies, and tools. In *Green Food Processing Techniques* (pp. 1–21). Academic Press.
84. Azmir, J., Zaidul, I. S. M., Rahman, M. M., Sharif, K. M., Mohamed, A., Sahena, F., & Omar, A. K. M. (2013). Techniques for extraction of bioactive compounds from plant materials: A review. *Journal of Food Engineering*, 117(4), 426–436. https://doi.org/10.1016/j.jfoodeng.2013.01.014
85. Feng, W., Li, M., Hao, Z., & Zhang, J. (2019). Analytical methods of isolation and identification. In *Phytochemicals in Human Health*. IntechOpen.
86. Safdar, M. N., Kausar, T., Jabbar, S., Mumtaz, A., Ahad, K., & Saddozai, A. A. (2017). Extraction and quantification of polyphenols from kinnow (*Citrus reticulate* L.) peel using ultrasound and maceration techniques. *Journal of Food and Drug Analysis*, 25(3), 488–500. https://doi.org/10.1016/j.jfda.2016.07.010
87. Wu, C., Wang, F., Liu, J., Zou, Y., & Chen, X. (2015). A comparison of volatile fractions obtained from Lonicera macranthoides via different extraction processes: Ultrasound, microwave, Soxhlet extraction, hydrodistillation, and cold maceration. *Integrative Medicine Research*, 4(3), 171–177. https://doi.org/10.1016/j.imr.2015.06.001
88. Petigny, L., Périno, S., Minuti, M., Visinoni, F., Wajsman, J., & Chemat, F. (2014). Simultaneous microwave extraction and separation of volatile and non-volatile organic compounds of boldo leaves. From lab to industrial scale. *International Journal of Molecular Sciences*, 15(5), 7183–7198. https://doi.org/10.3390/ijms15057183
89. He, J. Z., Shao, P., Liu, J. H., & Ru, Q. M. (2012). Supercritical carbon dioxide extraction of flavonoids from Pomelo (*Citrus grandis* (L.) Osbeck) peel and their antioxidant activity. *International Journal of Molecular Sciences*, *13*(12), 13065–13078. https://doi.org/10.3390/ijms131013065
90. Ameer, K., Shahbaz, H. M., & Kwon, J. H. (2017). Green extraction methods for polyphenols from plant matrices and their byproducts: A review. *Comprehensive Reviews in Food Science and Food Safety*, 16(2), 295–315. https://doi.org/10.1111/1541-4337.12253
91. Tiwari, B. K. (2015). Ultrasound: A clean, green extraction technology. *TrAC Trends in Analytical Chemistry*, 71, 100–109. https://doi.org/10.1016/j.trac.2015.04.013
92. Silva, R. P. F. F., Rocha-Santos, T. A. P., & Duarte, A. C. (2016). Supercritical fluid extraction of bioactive compounds. *Trac Trends in Analytical Chemistry*, 76, 40–51. https://doi.org/10.1016/j.trac.2015.11.013
93. Lenardao, E. J., Freitag, R. A., Dabdoub, M. J., Batista, A. C. F., & Silveira, C. D. C. (2003). Green chemistry: The 12 principles of green chemistry and it insertion in the teach and research activities. *Quimica Nova*, 26(1), 123–129.
94. Rosello-Soto, E., Koubaa, M., Moubarik, A., Lopes, R. P., Saraiva, J. A., Boussetta, N., & Barba, F. J. (2015). Emerging opportunities for the effective valorization of wastes and by-products generated during olive oil production process: Non-conventional methods for the recovery of high-added value compounds. *Trends in Food Science & Technology*, 45(2), 296–310. https://doi.org/10.1016/j.tifs.2015.07.003
95. Rajha, H. N., Boussetta, N., Louka, N., Maroun, R. G., & Vorobiev, E. (2015). Effect of alternative physical pretreatments (pulsed electric field, high voltage electrical discharges and ultrasound) on the dead-end ultrafiltration of vine-shoot extracts. *Separation and Purification Technology*, 146, 243–251. https://doi.org/10.1016/j.seppur.2015.03.058
96. Briones-Labarca, V., Plaza-Morales, M., Giovagnoli-Vicuña, C., & Jamett, F. (2015). High hydrostatic pressure and ultrasound extractions of antioxidant compounds, sulforaphane and fatty acids from Chilean papaya (*Vasconcellea pubescens*) seeds: Effects of extraction conditions and methods. *LWT-Food Science and Technology*, 60(1), 525–534. https://doi.org/10.1016/j.lwt.2014.07.057
97. Kek, S. P., Chin, N. L., & Yusof, Y. A. (2013). Direct and indirect power ultrasound assisted pre-osmotic treatments in convective drying of guava slices. *Food and Bioproducts Processing*, 91(4), 495–506. https://doi.org/10.1016/j.fbp.2013.05.003
98. Riera, E., Golas, Y., Blanco, A., Gallego, J. A., Blasco, M., & Mulet, A. (2004). Mass transfer enhancement in supercritical fluids extraction by means of power ultrasound. *Ultrasonics Sonochemistry*, 11(3–4), 241–244. https://doi.org/10.1016/j.ultsonch.2004.01.019
99. Rifna, E. J., Misra, N. N., & Dwivedi, M. (2021). Recent advances in extraction technologies for recovery of bioactive compounds derived from fruit and vegetable waste peels: A review. *Critical Reviews in Food Science and Nutrition*, 1–34. https://doi.org/10.1080/10408398.2021.1952923
100. Garavand, F., Rahaee, S., Vahedikia, N., & Jafari, S. M. (2019). Different techniques for extraction and micro/nanoencapsulation of saffron bioactive ingredients. *Trends in Food Science and Technology*, 89, 26–44. https://doi.org/10.1016/j.tifs.2019.05.005

101. Adetunji, L. R., Adekunle, A., Orsat, V., & Raghavan, V. (2017). Advances in the pectin production process using novel extraction techniques: A review. *Food Hydrocolloids*, 62, 239–250. https://doi.org/10.1016/j.foodhyd.2016.08.015
102. Mena-Garcia, A., Ruiz-Matute, A. I., Soria, A. C., & Sanz, M. L. (2019). Green techniques for extraction of bioactive carbohydrates. *Trends in Analytical Chemistry*, 119, 115612. https://doi.org/10.1016/j.trac.2019.07.023
103. Li, Y., Fabiano-Tixier, A. S., Vian, M. A., & Chemat, F. (2013). Solvent-free microwave extraction of bioactive compounds provides a tool for green analytical chemistry. *Trends in Analytical Chemistry*, 47, 1–11. https://doi.org/10.1016/j.trac.2013.02.007
104. Inoue, T., Tsubaki, S., Ogawa, K., Onishi, K., & Azuma, J. I. (2010). Isolation of hesperidin from peels of thinned *Citrus unshiu* fruits by microwave-assisted extraction. *Food Chemistry*, *123*(2), 542–547. https://doi.org/10.1016/j.foodchem.2010.04.051
105. Hayat, K., Zhang, X., Chen, H., Xia, S., Jia, C., & Zhong, F. (2010). Liberation and separation of phenolic compounds from citrus mandarin peels by microwave heating and its effect on antioxidant activity. *Separation and Purification Technology*, 73(3), 371–376. https://doi.org/10.1016/j.seppur.2010.04.026.
106. Nayak, B., Dahmoune, F., Moussi, K., Remini, H., Dairi, S., Aoun, O., & Khodir, M. (2015). Comparison of microwave, ultrasound and accelerated-assisted solvent extraction for recovery of polyphenols from *Citrus sinensis* peels. *Food Chemistry*, 187, 507–516. https://doi.org/10.1016/j.foodchem.2015.04.081.
107. Petrotos, K., Giavasis, I., Gerasopoulos, K., Mitsagga, C., Papaioannou, C., & Gkoutsidis, P. (2021). Optimization of vacuum-microwave-assisted extraction of natural polyphenols and flavonoids from raw solid waste of the orange juice producing industry at industrial scale. *Molecules*, 26(1), 246. https://doi.org/10.3390/molecules26010246.
108. Leadbeater, N. E. (2014). Organic synthesis using microwave heating. reference module in chemistry, molecular sciences and chemical engineering. In *Comprehensive Organic Synthesis II: Enabling Technologies for Organic Synthesis* (2nd ed., Vol. 9, pp. 234–286). Elsevier Ltd. https://doi.org/10.1016/B978-0-08-097742-3.00920-4
109. Silva, L. P. S., & Martínez, J. (2014). Mathematical modeling of mass transfer in supercritical fluid extraction of oleoresin from red pepper. *Journal of Food Engineering*, 133, 30–39. https://doi.org/10.1016/j.jfoodeng.2014.02.013
110. Pouliot, Y., Conway, V., & Leclerc, P. L. (Eds.). (2014). Separation and concentration technologies In *Food Processing: Principles and Applications* (2nd ed., pp. 33–60). https://doi.org/10.1002/9781118846315.ch3
111. Ndayishimiye, J., Lim, D. J., & Chun, B. S. (2018). Antioxidant and antimicrobial activity of oils obtained from a mixture of citrus byproducts using a modified supercritical carbon dioxide. *Journal of Industrial and Engineering Chemistry*, 57, 339–348. https://doi.org/10.1016/j.jiec.2017.08.041.
112. Romano, R., De Luca, L., Aiello, A., Rossi, D., Pizzolongo, F., & Masi, P. (2022). Bioactive compounds extracted by liquid and supercritical carbon dioxide from citrus peels. *International Journal of Food Science & Technology*, 57(6), 3826–3837. https://doi.org/10.1111/ijfs.15712
113. Zakaria, S. M., & Kamal, S. M. M. (2016). Subcritical water extraction of bioactive compounds from plants and algae: Applications in pharmaceutical and food ingredients. *Food Engineering Reviews*, 8, 23–34. https://doi.org/10.1007/s12393-015-9119-x/
114. King, J. W., & Grabiel, R. D. (2007). Isolation of polyphenolic compounds from fruits or vegetables utilizing sub-critical water extraction. *U.S. Patent 7,208,181*, 24 April 2007.
115. Lu, J., Feng, X., Han, Y., & Xue, C. (2014). Optimization of subcritical fluid extraction of carotenoids and chlorophyll a from *Laminaria japonica* Aresch by response surface methodology. *Journal of the Science of Food and Agriculture*, 94, 139–145. https://doi.org/10.1002/jsfa.6224
116. Amashukeli, X., Pelletier, C. C., Kirby, J. P., & Grunthaner, F. J. (2007). Subcritical water extraction of amino acids from Atacama Desert soils. *Journal of Geophysical Research Biogeosciences*, 112, G04S16. https://doi.org/10.1029/2006JG000308
117. Marcet, I., Álvarez, C., Paredes, B., & Díaz, M. (2016). The use of sub-critical water hydrolysis for the recovery of peptides and free amino acids from food processing wastes. Review of sources and main parameters. *Waste Management*, 49, 364–371. https://doi.org/10.1016/j.wasman.2016.01.009
118. Ndlela, S., De Moura, J., Olson, N., & Johnson, L. (2012). Aqueous extraction of oil and protein from soybeans with subcritical water. *Journal of the American Oil Chemists Society*, 89, 1145–1153. https://doi.org/10.1007/s11746-011-1993-7/
119. Thani, N. M., Kamal, S. M. M., Taip, F. S., Sulaiman, A., & Omar, R. (2019). Effect of sub-critical water hydrolysis on sugar recovery from bakery leftovers. *Food and Bioproducts Processing*, 117, 105–112. https://doi.org/10.1016/j.fbp.2019.07.002
120. Kim, D.-S., & Lim, S.-B. (2020). Semi-continuous subcritical water extraction of flavonoids from *Citrus unshiu* peel: Their antioxidant and enzyme inhibitory activities. *Antioxidants*, 9(5), 360. https://doi.org/10.3390/antiox9050360

121. Panja, P. (2018). Green extraction methods of food polyphenols from vegetable materials. *Current Opinion in Food Science*, 23, 173–182. https://doi.org/10.1016/j.cofs.2017.11.012
122. Grassino, A. N., Ostojic, J., Miletic, V., Djakovic, S., Bosiljkov, T., Zoric, Z., Jezek, D., Brncic, S. R., & Brncic, M. (2020). Application of high hydrostatic pressure and ultrasound-assisted extractions as a novel approach for pectin and polyphenols recovery from tomato peel waste. *Innovative Food Science & Emerging Technologies*, 64, 102424. https://doi.org/10.1016/j.ifset.2020.102424
123. Lombardelli, C., Liburdi, K., Benucci, I., & Esti, M. (2020). Tailored and synergistic enzyme-assisted extraction of carotenoid-containing chromoplasts from tomatoes. *Food and Bioproducts Processing*, 121, 43–53. https://doi.org/10.1016/j.fbp.2020.01.014
124. Mushtaq, M., Sultana, B., Anwar, F., Adnan, A., & Rizvi, S. S. (2015). Enzyme-assisted supercritical fluid extraction of phenolic antioxidants from pomegranate peel. *Journal of Supercritical Fluids*, 104, 122–131. https://doi.org/10.1016/j.supflu.2015.05.020
125. Soliva-Fortuny, R., Balasa, A., Knorr, D., & Martín-Belloso, O. (2009). Effects of pulsed electric fields on bioactive compounds in foods: A review. *Trends in Food Science & Technology*, 20(11–12), 544–556. https://doi.org/10.1016/j.tifs.2009.07.003
126. Parniakov, O., Barba, F. J., Grimi, N., Lebovka, N., & Vorobiev, E. (2014). Impact of pulsed electric fields and high voltage electrical discharges on extraction of high-added value compounds from papaya peels. *Food Research International*, 65, 337–343. https://doi.org/10.1016/j.foodres.2014.09.015
127. Barba, F. J., Boussetta, N., & Vorobiev, E. (2015). Emerging technologies for the recovery of isothiocyanates, protein and phenolic compounds from rapeseed and rapeseed press-cake: Effect of high voltage electrical discharges. *Innovative Food Science & Emerging Technologies*, 31, 67–72. https://doi.org/10.1016/j.ifset.2015.06.008
128. Oroian, M., & Escriche, I. (2015). Antioxidants: Characterization, natural sources, extraction and analysis. *Food Research International*, 74, 10–36. https://doi.org/10.1016/j.foodres.2015.04.018
129. Luengo, E., Álvarez, I., & Raso, J. (2013). Improving the pressing extraction of polyphenols of orange peel by pulsed electric fields. *Innovative Food Science & Emerging Technologies*, 17, 79–84. https://doi.org/10.1016/j.ifset.2012.10.005
130. Xue, D., & Farid, M. M. (2015). Pulsed electric field extraction of valuable compounds from white button mushroom (*Agaricus bisporus*). *Innovative Food Science & Emerging Technologies*, 29, 178–186. https://doi.org/10.1016/j.ifset.2015.03.012
131. Giacometti, J., Bursac Kovacevic, D., Putnik, P., Gabric, D., Bilusic, T., Kresic, G. Stulic, V., Barba, F. J., Chemat, F., Barbosa-Canovas, G., & Rezek Jambrak, A. (2018). Extraction of bioactive compounds and essential oils from mediterranean herbs by conventional and green innovative techniques: A review. *Food Research International*, 113, 245–262. https://doi.org/10.1016/j.foodres.2018.06.036
132. Mehmood, A., Ishaq, M., Zhao, L., Yaqoob, S., Safdar, B., Nadeem, M., Munir, M., & Wang, C. (2019). Impact of ultrasound and conventional extraction techniques on bioactive compounds and biological activities of blue butterfly pea flower (*Clitoria ternatea* L.). *Ultrasonics Sonochemistry*, 51, 12–19. https://doi.org/10.1016/j.ultsonch.2018.10.013
133. Wu, H., Zhu, J., Diao, W., & Wang, C. (2014). Ultrasound-assisted enzymatic extraction and antioxidant activity of polysaccharides from pumpkin (*Cucurbita moschata*). *Carbohydrate Polymers*, 113, 314–324. https://doi.org/10.1016/j.carbpol.2014.07.025.
134. Chemat, F., Rombaut, N., Sicaire, A.-G., Meullemiestre, A., Fabiano-Tixier, A.-S., & Abert-Vian, M. (2017). Ultrasound assisted extraction of food and natural products. Mechanisms, techniques, combinations, protocols and applications. A review. *Ultrasonics Sonochemistry*, 34, 540–560. https://doi.org/10.1016/j.ultsonch.2016.06.035.
135. Zhang, R., Li, S., Zhu, Z., & He, J. (2019). Recent advances in valorization of Chaenomeles fruit: a review of botanical profile, phytochemistry, advanced extraction technologies and bioactivities. *Trends in Food Science and Technology*, 91, 467–482. https://doi.org/10.1016/j.tifs.2019.07.012
136. Cheng, Z., Song, H., Yang, Y., Liu, Y., Liu, Z., Hu, H., & Zhang, Y. (2015). Optimization of microwave-assisted enzymatic extraction of polysaccharides from the fruit of *Schisandra chinensis* Baill. *International Journal of Biological Macromolecules*, 76, 161–168. https://doi.org/10.1016/j.ijbiomac.2015.01.048
137. Gan, J., Huang, Z., Yu, Q., Peng, G., Chen, Y., Xie, J., Nie, S., & Xie, M. (2020). Microwave assisted extraction with three modifications on structural and functional properties of soluble dietary fibers from grapefruit peel. *Food Hydrocolloids*, 101, 105549. https://doi.org/10.1016/j.foodhyd. 2019.105549.
138. Cravotto, G., Boffa, L., Mantegna, S., Perego, P., Avogadro, M., & Cintas, P. (2008). Improved extraction of vegetable oils under highintensity ultrasound and/or microwaves. *Ultrasonics Sonochemistry*, 15(5), 898–902. https://doi.org/10.1016/j.ultsonch.2007.10.009
139. Milman, B. L., & Zhurkovich, I. K. (2017). The chemical space for non-target analysis. TrAC *Trends in Analytical Chemistry*, 97, 179–187. https://doi.org/10.1016/j.trac.2017.09.013.

140. Maric, M., Grassino, A. N., Zhu, Z., Barba, F. J., Brncic, M., & Brncic. S. R. (2018). An overview of the traditional and innovative approaches for pectin extraction from plant food wastes and by-products: Ultrasound, microwaves, and enzyme-assisted extraction. *Trends in Food Science & Technology*, 76, 28–37. https://doi.org/10.1016/j.tifs.2018.03.022
141. Liew, S. Q., Ngoh, G. C., Yusoff, R., & Teoh, W. H. (2016). Sequential ultrasound-microwave assisted acid extraction (UMAE) of pectin from pomelo peels. *International Journal of Biological Macromolecules*, 93, 426–435. https://doi.org/10.1016/j.ijbiomac.2016.08.065
142. Laldinchhana, Lalrengpuii, J., Ray, S., & Pachuau, L. (2020). Indian tropical fruits and their bioactive compounds against human diseases. In M. K. Swamy (Ed.), *Plant-derived Bioactives* (pp. 455–494). Springer Nature. https://doi.org/10.1007/978-981-15-2361-8_21
143. Mann, S., & Aggarwal, K. S. M. P. (2013). Development of phytochemical rich ice cream incorporating kinnow peel. *Global Journal of Science Frontier Research Agriculture and Veterinary*, 13(4), 1–3.
144. Gómez-Mejía, E., Sacristán, I., Rosales-Conrado, N., León-González, M. E., & Madrid, Y. (2023). Valorization of *Citrus reticulata* blanco peels to produce enriched wheat bread: Phenolic bioaccessibility and antioxidant potential. *Antioxidants*, 12(9), 1742. https://doi.org/10.3390/antiox1209174
145. Kalita, M., Medhi, M., Yumnam, M., Gupta, A. K., & Mishra, P. (2021). Synthesis of carbon dots from *Citrus limon* peel by microwave-assisted process and its application for detection of ferric ion (Fe3+) and development of pH paper. In: L. S. Badwaik, C. N. Aguilar, & A. K. Haghi (Eds.), *Food Loss and Waste Reduction* (pp. 83–103). Apple Academic Press. https://doiorg.uml.idm.oclc.org/10.1201/9781003083900
146. Muzaffar, K., Ahmad Sofi, S., Kamran, F., Teixeira-Costa, B. E., Mir, S. A., & Dar, B. N. (2022). Fruit processing wastes and by-products. In: K. Muzaffar, S. Ahmad Sofi, & S. A. Mir (Eds.), *Handbook of Fruit Wastes and By-Products: Chemistry, Processing Technology, and Utilization* (pp. 1–16). CRC Press. https://doi.org/10.1201/9781003164463-1
147. Widmer, W., Stewart, D., Grohmann, K., & Wilkins, M. (2011). Method of pretreating citrus waste. *United States Patent US 7,879,379 B1*, issued 1 February.
148. Nafisi-Movaghar, K., Druz, L. L., & Victoria, C. P. (2013). Conversion of citrus peels into fiber, juice, naringin, and oil. *United States Patent US 2013/0064947 A1*, issued 14 March.
149. Clark, J. H., Pfaltzgraff, L. A., Budarin, V. L., & De Bruyn, M. (2013). Microwave assisted citrus waste biorefinery. *World Intellectual Property WO 2013/150262 A1*, issued 10 October.
150. Bala, S., Hitchcock, B. W., Hsieh, M. J., Jordan, R. L., Mathews, J. D., Rivera, T., Shin, J-E., & Small II, W. B. (2020). Preparation and incorporation of co-products into beverages to enhance nutrition and sensory attributes. *United States Patent US10,667,546 B2*, issued 2 June.
151. Oganaesjan, E. T., Khochava, M. R., Dorkina, E. G., Reps, V. F., & Andreeva, O. A. (2005). Agent from citrus waste eliciting anti-ulcer activity. *Russian Patent RU 2262944 C2*, issued 10 October.
152. Anonymous. (2023). Available from https://www.indiamart.com/manak-citrus-products/profile.html
153. Doolan, K. (2023). 'Wonky fruit' cosmetics brand FRUU opens its first store. *Cosmetic Design Europe*, 18 September. Available from https://www.cosmeticsdesign-europe.com/

Valorization of Lemon Peel Waste

A Potential Treasure Trove against Neurodegenerative Diseases

9

Jhoti Somanah, Manish Putteeraj, and Veronique François-Newton

9.1 INTRODUCTION

The concept of functional foods has led to the consensual recognition that eventual human health status is associated with the foods we consume. But it is only through understanding the mechanisms by which food modulates our metabolism that its nutritional value can be appreciated, and more mindful eating habits can be adopted. Rincón León (2003) suggested that functional food can be proven experimentally or through well-designed intervention studies to possess health-benefiting properties that positively impact an individual's physiological function and mental well-being, in addition to providing essential nutrients for normal body function. Examples of functional foods originate from natural foods rich in antioxidants and vitamins but can also be extended to legumes, fortified mineral-rich milk products, citrus fruits, herbs, spices, cereals, and fermented products. The steady stabilization of the functional food market has seen a boost in sales recently, despite the disruption caused by the COVID-19 pandemic. A recent global analysis reported this industry was estimated to be worth USD 280.7 billion in 2021 and forecasted to expand by 8.5% by 2030 (IndexBox 2024). The demand for functional foods arises from the increasing awareness that a hectic, unmanaged lifestyle can have consequential impacts on health and the cost of medical treatment, particularly in adults aged 40 or older (Baker et al. 2022). The possibility of preventing or attenuating the progression of debilitating neurodegenerative disorders (i.e., epilepsy, Alzheimer's, Parkinson's, and Huntington's diseases, and multiple and amyotrophic lateral sclerosis) by altering the activities of physiologically critical enzymes or proteins through the diet is an appealing strategy many are willing to try.

DOI: 10.1201/9781003352341-9

This neuroprotective approach focuses on modulating mechanisms such as oxidative stress, neuroinflammation, faulty autophagy, and mitochondrial dysfunction, among others (Chou et al. 2021). Neurological disorders can affect anyone, irrespective of gender, age, household income, educational background, or country of residence. The World Health Organization estimates that 6.8 million people are affected by neurodegenerative pathologies. Despite the availability of various treatment options, the burden of this disorder will be felt more profoundly by the older (≥ 65 years) population. For many people in low- and middle-income households, primary health care and low-technology interventions are the only plausible forms of available treatment. The lack of adequate health delivery systems, regular screening or monitoring facilities, essential drugs, and medically trained personnel poses a great challenge to families and caregivers. This situation has led to a significant increase in the prevalence of innovative medical approaches using traditional medicine, beliefs, and practices to treat the ill (World Health Organization 2007)—a field of medicine known commonly as phytotherapy. Numerous plants have been experimentally shown to improve brain function and can potentially be useful in the treatment of certain neurological diseases (John et al. 2022; Alzobaidi et al. 2021). Many of these reported plants contain one of the most bioactive polyphenolic compounds—flavonoids. Flavonoids are well acknowledged in the literature for their antioxidant, anti-inflammatory, anti-proliferative, and anti-carcinogenic properties (Ullah et al. 2020). Of particular interest to the scientific community are citrus fruits, for their high levels of flavonoids naturally abundant in their outer peels, juice sacs, and *flavedo*, attributing their characteristic color and taste. β-Carotene, lycopene, hesperidin, and eriocitrin are components found in the peel and other solid residues of lemon waste, while chlorophyll *a* and *b* are responsible for the green peel color of mature limes (Mahato et al. 2019). Hence, this chapter will focus on the beneficial effects of natural products, emphasizing lemon peel, whose neuroprotective components can act as a buffer against oxidative and inflammatory responses as a measure to mitigate neurodegenerative diseases.

9.2 INDIAN LEMON

Citrus lemon, an Indian folk medicine, has been used to treat many diseases, including diabetes, cardiovascular and inflammatory diseases, hepatobiliary dysfunction, and neurodegenerative disorders (Falls et al. 2018). The term "Indian lemon" refers to indigenous lemon varieties such as the Nepali Oblong, Nepali Round, Rough lemons, and Assam lemons. Indian lemons are used in traditional Ayurvedic medicine as a digestive aid. Lemons are deeply intertwined with India's spiritual and culinary traditions and have been cultivated in India for many centuries, leading the country to become one of the largest lemon producers worldwide. Apparently, lemons descended from ancient wild species native to an area spanning northeastern India (Specialty Produce 2023). India produces about 17% of the world's lemons (Yadav 2022). Commercial lemon is produced in Andhra Pradesh, Maharashtra, Tamil Nadu, Gujarat, Rajasthan, and Bihar. Gujarat, Andhra Pradesh, Madhya Pradesh, and Karnataka account for most of the production (20%, 18%, 10%, and 10% of total Indian lemon production, respectively). India is considered the largest lemon producer in the world; however, only 20% to 25% of the fresh lemon meets the international market standards, with the rest ending up in local markets. In 2021–2022, India exported US $8.93 million of lemon and lime products (fresh and dried), primarily to UAE, Nepal, and Bhutan. Karnataka produced 511,166 million tons of lemon from 21,660 ha (Government of Karnataka 2023).

Eureka lemon is grown in Jammu and Kashmir on about 2565 million ha, producing 7337 million tons of fruits and is generally used for making pickles, culinary purposes, and small-scale juice extraction. Fresh Eureka lemon peel with 73.5% moisture contains crude fiber, total and reducing sugars, and ash (16.1%, 6.5%, 4.4%, and 6.45%, respectively), as well as ascorbic acid, β-carotene, tannin, and phosphorus (35.2, 7.3, 168, and 18.1 mg/100 g, respectively) (Kour et al. 2020). Oven-dried (70 °C), ground (~1 mm) lemon peel from Nagpur is highly porous (~50% porosity), containing moisture, ash, and volatile matter (6.1%, 5.4%, and 87.2%, respectively). The dried peel consists (% dry basis) of cellulose (23.1), hemicellulose (8.1), acid-detergent lignin (7.6), total sugars (6.5), protein (7), pectin (13), and ash (2.5) (Pathak et al. 2017 and references therein). Chennai lemon peel contains total phenolic content (26.6 and 119.6 μg gallic acid equivalent [GAE]/mg for methanol and acetone extracts, respectively) that is higher than other sources (15 μg GAE/mg), and flavonoids (4.5 and 56.3 μg quercetin equivalent [QE]/mg) (John et al. 2017; Putnik et al. 2017).

Assam lemon (Kaji Nemu) is an important indigenous variety valued for its fragrance and content of acid and natural compounds widely used for culinary, beverage, industrial, and medicinal purposes. Chetia et al. (2023) optimized essential oil extraction from fresh Assam lemon peels using microwave-assisted extraction to develop an alternative green preservative for improving fruit storage. The essential oil of Assam lemon peel consists primarily of limonene, neral, trans-verbenal, decanal, ethyl cinnamate, and ethyl *p*-methoxycinnamate (50.2%, 11.25%, 6.72%, 3.65%, 2.22%, and 2.2% concentration, respectively), along with other compounds (each <2% concentration) (Chetia et al. 2023). Essential oil obtained by hydrodistillation of lemon peel from Assam, Northeast India, contained 43 compounds, with limonene, neral, trans-verbenol, and decanal as major components (55.4%, 10.4%, 6.4%, and 3.3%, respectively) (Paw et al. 2020). The major classes of lemon peel essential oil were monoterpene hydrocarbons (55.4% limonene), oxygenated monoterpenes (neral and *trans*-verbenal), sesquiterpenes, aldehydes, and esters (56.7%, 25.2%, 3.4%, 4.4%, and 4.6% concentration, respectively). Earlier studies report limonene (54.6%–98.4%; generally in the 55%–74% range) in lemon peel essential oil (Paw et al. 2020 and references therein). Jain and Sharma (2017) identified 45 essential oil components in lemon peel from Jaipur, Rajasthan. The predominant essential oil compounds were limonene (43%), β-pinene (12.6%), γ-terpinene (11.5%), α-terpineol (7.2%), α-pinene (3.4%), α-terpinolene (2.4%), and small amounts (1%–2%) of myrcene, geraniol, α-terpinene, linalool, and cis-α-bergamotene.

9.3 MOLECULAR AND CELLULAR CHANGES UNDERLYING NEURODEGENERATIVE DISEASES

Neurodegenerative diseases are on a steep ascent and contribute extensively to morbidity rates on a global scale. Although much progress has been made in terms of delineating the molecular and cellular changes that occur upon initiation and progression of neurodegeneration, much information on the pathologies and concomitant changes reflected in idiosyncrasies is still being added to the existing pool of literature. Loss of neuronal cell population is a normal process that accompanies aging, the latter of which has also been characterized by a regression of the white matter during the healthy aging process rather than a significant loss in gray matter (Peters 2006), further endorsing the disruption of neuronal connectivity through a retraction of the myelin fibers, hence, altering synaptic functionality (Bishop et al. 2010). However, under specific circumstances, this process can accelerate as a result of altered cellular and molecular functions (Table 9.1), leading to a plethora of emerging disorders characterizing those neurological changes. Within those lines, some of the proposed mechanisms that precipitate the neurodegenerative pathways include changes in circadian processes. Also termed chrono-disruption, the latter potentiates neurodegenerative diseases in older patients (≥ 60 years) through alterations in clock gene expression such as the *Period 1* gene (*Per1*), cascading into decreased neuronal activity, neurogenesis, and inducing neuroinflammation (Verma et al. 2023).

Under the concept of aging and cognitive decline, there has been a tendency to link "healthy" neuronal decline, as demarcated by the decrease in synaptogenesis and loss of dendritic arborization (Peters 2006), to internal and external stressors as precursors to the acceleration of neurodegeneration. The production of free radicals, which was previously thought to be one of the major stepping stones toward neurodegenerative diseases, has now been associated with an increase in reactive oxygen species (ROS), with frailty as the most conducive combination for such neurological decline (Robertson et al. 2013). Indeed, studies have shown that the decreased ability to cope with stressors has a major impact on cognitive decline and is further associated with numerous physiological, psychological, and biological changes (Robertson et al. 2013). Interestingly, changes accelerated by aging and frailty also include microinfarcts, along with Lewy bodies and other pathologies marking the onset of neurodegenerative diseases (Buchman et al. 2013). Given the high oxygen consumption

TABLE 9.1 Neurodegenerative-induced biological changes

MOLECULAR AND CELLULAR CHANGES	*DESCRIPTION*
Abnormal protein dynamics	Disruption in protein clearance through the downregulated roles of the Ubiquilin 2 protein and its mechanistic actions through the proteasomal pathways has been associated with many degenerative diseases (Zhang et al. 2014; Kaye and Shows 2000; Massey et al. 2004; Rutherford et al. 2013). The ubiquitin/proteasome mechanism has been extensively characterized as a potent pathway toward neurodegeneration through models replicating proteasomal inhibition (Ekimova et al. 2023; van Leeuwen et al. 1998). Other proteins such as caspases have also been associated with the onset of neurodegenerative diseases through an increase in toxic protein fragments, activation of precursor proteins, and axonal degradation (Wellington and Hayden 2000; Graham et al. 2011).
Oxidative stress and free radical formation	One of the hallmarks of neurodegeneration is oxidative stress associated with an accumulation of reactive oxygen species (ROS) at the neuronal level. Excessive ROS can affect cell signaling pathways (Castellani et al. 2001), cause neuronal respiratory dysfunction (Nascimento et al. 2022), and lead to calcium loading resulting from increased ROS, which precipitates apoptotic processes (Bello-Medina et al. 2022; Uddin et al. 2020). It can also precipitate the formation of intrinsically disordered proteins, which may aggregate, as evidenced in neurological disorders (Uversky 2011); cause oxidation of K+ channels leading to altered neuronal excitability (Sesti et al. 2010); and result in oxidized DNA bases as a precursor to cognitive disorders mediated by neurodegeneration (Jenner 1996).
Altered cellular metabolic efficiency and mitochondrial dysfunction	The imbalance in ROS production within mitochondrial sites and the scavenging capacity of the antioxidant system is the prime determinant leading to ROS-mediated mtDNA damage and culminating in altered functionality (Warraich et al. 2020: Richter 1995). An increase in mtDNA damage is also a precursor to respiratory-chain-deficient cells, leading to reduced cellular density (Trifunovic and Larsson 2008; Gonzalez-Freire et al. 2015). Mitophagy disruption—hence, mitochondrial maintenance—coupled with altered bioenergetic profiles can contribute to mitochondrial-dysfunction-mediated neurodegenerative diseases (Gonzalez-Freire et al. 2015), a process that involves the rapamycin (mTOR) and AMPK pathways (Hirota et al. 2012; Guarente 2014).

at the brain level, the augmented ROS levels—as a result of the billions of neurons organized in a highly complex network and the differential roles adopted by the variants of cells at the brain level with their respective bioenergetic profiles—is expected (Stefanatos and Sanz 2018). Aging and neurodegeneration are accompanied by increased ROS, which typically cannot be compensated by regulatory mechanisms to buffer against toxic levels (Cui et al. 2004). This ROS increase occurs as a result of multiple changes at cellular levels, including mitochondrial dysfunction (López-Otin et al. 2013), deregulation of antioxidant systems (Leutner et al. 2001; Reiter 1995; Zemlan et al. 1989), alterations in membrane lipid composition due to reductions in arachidonic acid levels through peroxidation (Gemma et al. 2007), glutamate- and dopamine-induced release of ROS (Cui et al. 2004), and micronutrient deficiencies, among others (Mohajeri et al. 2015; Douaud et al. 2013).

9.4 BIOFUNCTIONAL CHARACTERIZATION OF LEMON PEEL WASTE

The citrus fruit industry is a booming one. The volume of citrus fruits imported worldwide contracted slightly to 17 million tons, with China, the US, Brazil, Argentina, and Mexico being the leaders in citrus fruit per capita consumption (IndexBox 2024). However, over 30% of the global citrus production primarily focuses on the production of fruit juices and essential oils. Such industries generate substantial amounts of waste, such as after-wash wastewater and semi-solid residues, like pulp, and solid residues—of which peels, membranes, and seeds form a significant part of what is discarded (Satari and Karimi 2018). Even though the composition of citrus wastes and by-products varies depending on the processing industry, every year over 1.3 billion tons go to waste. Generally, lemon peel accounts for over 10% of the whole fruit, whereas grapefruit peel and pomelo peel account for up to 35%, and oranges up to 70% (Wedamulla et al. 2022). While conventional solid waste treatments and home fertilization methods are common disposal methods for citrus fruit peels, advanced disposal management is required due to their low pH values and high moisture content, which predispose them to rot rapidly (Kim and Om 2021). Repurposing and commercializing citrus peel waste is an exciting avenue for consumers and food technologists. Since the properties of citrus peel are largely owed to secondary components with substantial antioxidant activity, preservation of the latter is crucial for maintaining its biological properties.

Phenolic characterization of lemon (*Citrus limon*) has shown that it is rich in flavonoids, especially flavanones and flavone glycosides (Ballistreri et al. 2019; Dugo et al. 2003; Gattuso et al. 2007). Flavonoids are a broad group of polyphenolic compounds that have several biological effects, mainly related to their antioxidant activity through scavenging of free radicals and activation of antioxidant enzymes, and that can reduce the concentration of substances that have an active role in ROS generation (Gattuso et al. 2007). The main flavanones and flavones detected in lemon are eriocitrin, hesperidin, C-glycosides of diosmetin and apigenin, and diosmetin 7-O-rutinoside (diosmin) (Dugo et al. 2003). For lemon peel essential oil, the major classes detected were monoterpene hydrocarbons (55.4% limonene), oxygenated monoterpenes (neral and trans-verbenal), sesquiterpenes, aldehydes, and esters (Paw et al. 2020). Nuzzo et al. (2020) developed pre-industrial scale high-phenolic lemon pectin "IntegroPectin" rich in bioflavonoids and nanoemulsified oil by hydrodynamic cavitation of waste lemon peel in water.

9.4.1 Lemon Pomace

Zappia et al. (2023) extracted lemon (*Citrus limon* (L.) Osbeck) pomace (i.e., peels, pulp, and seeds after commercial lemon juice and essential oils extraction) with a 50% hydroalcoholic mixture using ultrasound extraction [UAE] (25°C, 1 h). The extract (LP_E) contained hesperidin and eriocitrin (47 and 34 mg/100 g dw), total phenolic compounds (6.75 mg GAE/g dw), total flavonoids (2.04 mg catechin equivalent [CE]/g dw), exhibiting antioxidant (ABTS and DPPH) activity (19.42 and 8.25 μmol Trolox equivalent [TE]/g dw). The same group obtained the highest recovery of antioxidant compounds from lemon pomace at different UAE conditions (70°C, 30 min) with a hydroalcoholic (ethanol:water, 50:50) mixture (Imeneo et al. 2022). The extract (LP_E) contained hesperidin and eriocitrin (1695 and 1167 mg/kg dw), total phenolic compounds (6.93 mg GAE/g dw), total flavonoids (2.07 mg CE/g dw), exhibiting antioxidant (ABTS and DPPH) activity (18.36 and 7.43 μmol TE/g dw). The extract also contained the highest amount of gallic acid (104 mg/kg), rutin (77 mg/kg), narirutin (141 mg/kg), naringin (44 mg/kg), and neohesperidin (12 mg/kg) compared to other extraction methods (conventional and microwave-assisted extraction) and conditions. Imeneo et al. (2021) obtained hydroalcoholic (50% ethanol) lemon pomace extract by ultrasonic homogenization (40°C, 40 min) after commercial lemon juice and essential oils production. The extract (LP_E) contained hesperidin and eriocitrin (232 and 190 mg/100 g dw), total phenolic compounds (0.59 mg GAE/g dw), total flavonoids (0.16 mg CE/g dw), exhibiting antioxidant (ABTS and DPPH) activity (36 and 1.86 μmol TE/g dw). Enriched biscuits prepared by substituting skim milk with lemon pomace extract or fresh lemon peel exhibited higher resistance to lipid oxidation compared to control biscuits, thereby preserving the biscuits from oxidation. Lemon peel and lemon pomace extract incorporated in biscuit formulation increased the content of bioactive compounds without affecting their overall appearance, resulting in improved functional and nutraceutical properties.

Lemon juice squeezed residue [LJSR], a by-product similar to commercial lemon pomace, is an underutilized source of beneficial compounds that can be treated as functional food to prevent lifestyle-related diseases. LJSR, a food processing by-product, contains several functional components (i.e., carotenoids, flavonoids, coumarins, and dietary fiber) with therapeutic benefits. Hesperidin, eriocitrin, and auraptene (0.83%, 0.77%, and 0.55%, respectively) were the flavonoids and coumarins present in LJSR that exhibited antioxidant activity (83.7% vs 95.5% DPPH radical scavenging activity for α-tocopherol). Administration of a diet containing 5% LJSR reduced blood pressure in rats due to NADH/NADPH oxidase suppression, NO bioavailability increase, blocking of calcium channels, and alleviation of endothelial dysfunction by LJSR components (hesperitin, eriocitrin, and auraptene) (Kawakami et al. 2021).

9.4.2 Lemon Peel

Lemon peels possess the highest antioxidant potential among citrus fruits (lemon, orange, and grapefruit) and the most fiber (soluble and insoluble) (~14 g/100 g dry matter). Whole lemon peel powder contained 47% dietary fiber (34.8% and 12.3% water-insoluble and -soluble fiber, respectively), 15% (by weight, the sum of protein, lipid, ash, and moisture), polyphenols (hesperidin, eriocitrin, diosmin, and narirutin; 427, 174, 146, and 26.5 mg/100 g powder, respectively). Dietary fiber is primarily responsible for the anti-inflammatory effects of lemon peel (powder) in reducing colitis in male Balb/c mice by suppressing inflammatory cytokine expression, including IL-6 and Cxcl2 (a murine IL-8 homolog) (Tinh et al. 2021). Lemon peel fiber consists primarily of cellulose, pectin, hemicellulose, and lignin (23%, 13%, 8.1%, and 7.6% dry weight, respectively) (Putnik et al. 2017 and references therein). The main lemon peel flavanones (400–600 mg/100 g peel) are neoeriocitrin, naringin, and neohesperidin (Putnik et al. 2017 and references therein). Lemon peel is rich in macronutrients (potassium, calcium, phosphorus, and magnesium; 127, 32, 24, and 13 mg/100 g) and micronutrients (iron, zinc, copper, manganese, and selenium; 0.34, 0.28, 0.04, 0.05 mg/100 g, and 4.13 μg/100 g, respectively) (Czech et al. 2020). Zinc (0.28 mg/100 g) in lemon peel protects the body against oxidative stress and stimulates immune mechanisms.

Russo et al. (2014) investigated the distribution of bioactive compounds in all by-products, including peels (Table 9.2), produced by a local Italian (Barcellona Pozzo di Gotto, Messina) industrial processing of lemon fruits. Flavanones (eriocitrin and hesperidin) and flavones (apigenin 6,8-di-C-glucoside and diosmetin 6,8-di-C-glucoside) were the most abundant compounds in lemon peels and other lemon by-products. The ratio between the most abundant flavanones (eriocitrin and hesperidin) and flavones (apigenin 6,8-di-C-glucoside and diosmetin 6,8-di-C-glucoside) in lemon peel was 1:1. However, the lemon peel flavones ratio was 1:3. Lemon peel had the lowest narirutin content (85 ± 2.9 mg/kg) compared to lemon juice, seeds, or pulp (163, 152, and 124 mg/kg, respectively). Lemon peel was quantitatively the second-richest source of bioactive compounds after lemon pulp (5991 vs 6528 mg/kg compared to 13,349 mg/kg for the juice), primarily due to high diosmetin 6,8-di-C-glucoside, eriocitrin, and hesperidin contents (900, 5012, and 6551 mg/kg, respectively). Manually separated Femminello Santa Teresa lemon peel contained a considerable amount of total phenolic (59.45 vs 19.24 mg GAE/100 g for lemon pulp) and high antioxidant activity (11.77 vs 1.85 mM TEAC/100 g [ABTS assay] for lemon pulp) (Pieracci et al. 2022).

Xi et al. (2017) investigated the phenolic composition and antioxidant capacity of peels among other fruit parts (pulp, juice, whole fruit, and seed) from five Chinese lemon cultivars (Table 9.2). Lemon peel had significantly higher levels of phenolics than other fruit parts. Hesperidin was the predominant flavanone (10.27–3315 μg/g fresh weight [FW]), and nobiletin was the main polymethoxylated flavonoid. The total phenolic and flavonoid contents of lemon peel varied among the cultivars (3.17–4.71 and 5.12–8.3 μg/g FW, respectively). Fourteen individual phenolic compounds were identified by HPLC from lemon fruit parts, including four phenolic compounds (gallic, chlorogenic, caffeic, and ferulic acids), seven flavanones (eriocitrin, hesperidin, rutin, diosmin, eriodictyol, naringenin, and hesperetin), and three polymethoxylated flavanones (sinensetin, nobiletin and tangeretin). Lemon peel flavanones (eriocitrin, hesperidin, rutin, and eriodictyol) varied among cultivars: eriocitrin (7.73–49.61 μg/g FW), hesperidin (1465–3315 μg/g FW), rutin (4.92–60.59 μg/g FW), and eriodictyol (1.57–79.27 μg/g FW); diosmin (1.8–14.66 μg/g FW) was present in only four varieties, while naringenin (3.52–12.35 μg/g FW) was present in only three

TABLE 9.2 Polyphenolic composition of lemon peel

LEMON PEEL	*RUSSO*	*PIERACCI*	*XI*
Compounds	mg/g	mg/100 g FW	μg/g FW
Caffeic acid	31		294–742
p-Hydroxybenzoic acid	54		
Ferulic acid	46	2.08	
Synapic acid	60		
Apigenin 6,8-di-C-glucoside	273		
Diosmetin 6,8-di-C-glucoside	399	1.38	
Eriocitrin (Flavanone)	2670	2.17	7.73–49.61
Apigenin 7-(malonylpiosyl)-glucoside	68	35.4	
Diosmetin 6-C-β-D-glucoside	152		
Narirutin	85		
Rutin	52	0.734	
Hesperidin	2198		1465–3315
Ichangin	48	24.9	
Phenolic acids	111		3.17–4.71
Flavonols	52		
Flavones	892		
Flavanones	4953		
Limonoids	117		
Coumarins	19.4		
Furanocoumarins	9.6		
Total Bioactive	5991		

out of five varieties. Sinensetin, nobiletin, and tangeretin contents in peels varied (5.45–19.27, 7.45–107.3, and 1.23–62.08 μg/g FW, respectively).

Antioxidant flavonoid glycosides are present in lemon peel, including six flavanone glycosides (eriocitrin, neoeriocitrin, narirutin, naringin, hesperidin, and neohesperidin) and three flavone glycosides (diosmin, 6,8-di-C-β-glucosyldiosmin [DGD], and 6-C-β-glucosyldiosmin [GD]). Eriocitrin, neoeriocitrin, and DGD have higher antioxidant activity compared to other biophenols. Lemon flavonoids exert strong antioxidant activity (eriocitrin has higher antioxidant activity than α-tocopherol in a low-density lipoprotein oxidation system). Lemon flavonoids exert anti-inflammatory activity via several mechanisms, from antioxidant and radical scavenging activities through modulation of proinflammatory cytokine production and proinflammatory gene expression (Nuzzo et al. 2020 and references therein).

Lemon peels, after 48 h ethanol infusion (for liquor production), contained low total polyphenols (0.38 mg/g of dry waste), primarily benzoic acid, naringenin, and quercetin rutinoside (0.29, 6.53, 0.92 yield mg/g dry waste, respectively). Lemon peel and its polyphenol extract strongly inhibited acetylcholinesterase (AChE), an enzyme involved in the metabolic hydrolysis of acetylcholine at cholinergic synapses in the central and peripheral nervous system (Di Donato et al. 2018). Lemon peel polyphenol extract (LPE; 0.46 mg/mL GAE total polyphenol content) from lemon waste liquor production strongly inhibits acetylcholinesterase (IC_{50} 36 μg GAE/mL; Ellman colorimetric method), an enzyme involved in Alzheimer's disease (Pagliara et al. 2019).

Ingestion of lemon peel flavanone glycosides (FG) [eriodictyol, homoeriodictyol, and hesperetin] with water was undetected in human plasma after 4 h (Miyake et al. 2006). FG contains 30% eriocitrin and small amounts of 6,8-C-diglucosyldiosmetin (0.1%) and hesperidin (0.05%). Flavanone aglycones (FA) contain 14.7% eriodictyol and other flavonoids (3% hesperetin and 0.3% eriocitrin). FA (2–3.7 g [1.89

mmol eriodictyol] with 500 mL water) was absorbed faster and at higher levels than FG (1–4.5 g [1.35 g eriocitrin] with 500 mL water) in ten male volunteers (mean age 35.9 years; 68 kg mean body weight) because eriodictyol has no sugar moiety and requires no deglycosylation. FG is absorbed slowly and in low amounts up to 4 h in the presence of flavonoid-free lemon juice.

9.4.3 Lemon Peel Essential Oil

Essential oil of dried (to 12% moisture content) Nigerian lemon peel contained limonene (53.07%), β-pinene (9.53%), borneol (5.57%), neral (4.7%), sabinene (4.18%), α-pinene (3.82%), linalool (3.7%), 1,8-cineole (3.38%), myrecene (3.33%), geranial (3.33%), and small amounts of other components (Oboh et al. 2014). The total phenolic and flavonoid contents of the essential oil were 7.42 mg gallic acid equivalent and 3.56 mg quercetin equivalent per 100 g, respectively.

Essential oil components of lemon peel vary depending on origin, variety, and environmental conditions (Table 9.3). Limonene is the main essential oil (EOs) compound present in lemon peels, ranging from 40% to 86%. Most lemon peels contain γ-terpinene, β-myrcene, pinene, ocimene, linalool, α-terpinene, and sabinene. These essential oils exert many beneficial biological effects, such as sedative, analgesic, anesthetic, antimicrobial, and anti-inflammatory activities useful for the pharmaceuticals, cosmetic, and food products (Ben Hsouna et al. 2023).

Paw et al. (2020) extracted Assam (India)-grown lemon peel oil using hydrodistillation (6 h boiling in Clevenger apparatus) and analyzed it by gas chromatography/mass spectroscopy (GC/MS). Limonene and neral were the major compounds among 43 identified compounds (55.4% and 10.39%, respectively), followed by *trans*-verbenol and decanal (6.43% and 3.25%, respectively). Other compounds present were ethyl cinnamate and ethyl *p*-methoxycinnamate (2.21% each), *cis*-*α*-bergamotene (1.60%), geraniol (1.48%), trans-carveol (1.33%), nonanal (1.19%), linalool (1.16%), α-terpineol (1.07%), and other minor (<1%) compounds including α-pinene (0.17%), terpinen-4-ol (0.17%), germacrene-D (0.16%), and *o*-cymene (0.12%). Fresh Assam lemon peels had similar essential oil composition after optimized microwave-assisted extraction (0.43:1 liquid-to-solid ratio [v/w], 17 min extraction time, 850 W power) (Chetia et al. 2023). Limonene and neral were the two most prevalent components (50.2% and 11.25%) among the 43 essential oil compounds, followed by *trans*-verbenol (6.72%), decanal (3.65%), ethyl cinnamate (2.22%), ethyl *p*-methoxycinnamate (2.2%), *cis*-*α*-bergamotene (1.5%), geraniol (1.43%), *trans*-carveol (1.2%), and linalool (1.05%). Gupta et al. (2025) identified 17 essential oil components (97%) in hydrodistilled local market (Palampur, Himachal Pradesh, India) fresh and waste lemon peel. Major essential oil constituents

TABLE 9.3 Essential oil components of lemon peel of different origin

COMPOUND (%)	INDIA-LF	INDIA-LW	INDIA	IRAN	18 ITALIAN	4 ITALIAN	5 SPANISH
D-Limonene	54.2	52.4	55.4	61.4	72.5–76.4	63.7–76.9	61.1–66.9
β-Pinene	9.2	6.7	1.2	13.1	11.6–18.7	7.7–14.7	11.7–15.5
γ-Terpinene	3.6	2.2	0.05	11.3	2.9–8.3	5.5–10.4	10.3–11.9
Sabinene				2.3			1.9–2.6
Myrcene	1.12	0.96		1.6	0.95–1.1		
Linalool	1.09	0.72	1.16				
α-Pinene	1.5	1.2	1.2	2.4	1.4–1.5		2–2.5
Geranial				1.5			
Neral			10.4	1.1			

Adapted from Paw et al. (2020) and Ben Hsouna et al. (2023) for peels from India and Iran, respectively; LF-fresh lemon peel and LW-lemon waste peel. Data was also extracted from Gupta et al. (2025) and Magalhães et al. (2023).

included *d*-limonene (52.4%–54.2%), α-terpineol (16.8%–21.2%), β-pinene (6.7%–9.2%), and γ-terpinene (2.2%–3.6%). The essential oil of lemon waste peel was effective against pulse beetle (*Callosobruchus chinensis*; 2899 μL/L LC_{50}) without food after 96 h and exerted high ovipositional inhibitory potential (50%) after 24 h. The potent insecticidal activity (fumigant toxicity, repellence, and ovipositional deterrence) of lemon peel waste essential oil was attributed to its enzymes (acetylcholinesterase [AChE] and glutathione-S-transferase [GST]) inhibition activities. Magalhães et al. (2023) demonstrate the range in essential oil components of lemon peel depending on Italian and Spanish varieties.

9.5 ANIMAL AND *IN VITRO* STUDIES

Naringenin, abundant in lemon peel, alleviates anxiogenic-like behavior impairment induced by Fe-dextran exposure (50 mg/kg/day intraperitoneally for 28 days) in rats. Naringenin restores iron-induced acetylcholinesterase suppression, attenuates mitochondrial membrane potential, and mitochondrial complexes activities in rat hippocampus (Dike et al. 2021 and references therein). It also restores changes in the activity and expression of ectonucleotidases (i.e., adenosine triphosphate diphosphorylase and 5′-nucleotidase) enzymes that hydrolyze and regulate extracellular ATP and adenosine concentrations in the synaptic cleft and iron-induced cholinergic deficits in the cerebral cortex of rats. Naringenin exerts a protective effect against cerebral ischemic injury, attenuates β-amyloid toxicity, induces MAP kinase activation, modulates glutamate uptake, and prevents neurodegeneration with cognitive impairment caused by the intracerebroventricular-streptozotocin in the diabetic oxidative damage rat model (Dike et al. 2021 and references therein).

Lemon peel flavonoids (LPF) significantly increased the exhaustion swimming time (88 vs 40 min for untreated mice), hepatic glycogen and free fatty acid content, and reduced lactic acid and BUN content (energy supply) in a dose-dependent manner. LPF dose dependently increases antioxidant enzymes (superoxide dismutase [SOD] and catalase [CAT]) and reduces lipid peroxidation (malondialdehyde levels). It also elevates hepatic tissue antioxidant activities (SOD and CAT) and reduces skeletal muscle tissue iNOS and TNF-α levels in mice. At the cellular level, LPF (12.5 mg/mL) reduced hydrogen peroxide-induced oxidative damage of human kidney (293T) cells [decreases chronic ROS release]. LPF contains isomangiferin, rutin, astragalin, naringin, and quercetin (291, 235, 15.5, 26.7, and 10.6 mg/g, respectively). LPF acts as an anti-fatigue agent; it regulates nitric oxide that can stimulate the cerebral cortex and excite the brain (modulate brain function) (Bao et al. 2020).

Lemon peel (70%) ethanol extract [LEE] (200 mg/kg daily for 9 weeks) administered orally improved mice performance without significantly affecting body weight. Oral administration of LEE promoted neurogenesis (increased proliferation of neuronal progenitors in doublecortin-positive cells and tropomyosin receptor kinase B) in aged (12-month-old) male C57BL/6 mice. The study indicates that LEE can improve synaptic plasticity due to the presence of active components (i.e., hesperidin, hesperetin, and other flavonoids, essential oils, vitamin C, citric acid). In fact, hesperidin administered orally (100 and 200 mg/kg body weight daily for 15 days) prevented intracerebroventricular-streptozotocin-induced cognitive impairment in Swiss male albino mice (Javed et al. 2015). The treatment improved memory consolidation (Morris water maze), probably by modulating acetylcholinesterase activity (AChE). Moreover, hesperidin modulates neuronal cell death by suppressing inflammatory markers (NF-κB, inducible NO synthase, and COX-2).

Arcone et al. (2023) analyzed the effect of *Citrus limon* peel polyphenol (80% ethanol) extract (LPE) on key enzymes involved in neurodegeneration, antioxidant defense system, and *in vitro* inhibition of Aβ1–40 aggregation. LPE dose dependently inhibited acetylcholinesterase [AChE] (21 μM IC_{50}), butyrylcholinesterase [BuChE] (73 μM IC_{50}), Aβ1–40 aggregation (176 μM IC_{50}), monoamine oxidases [MAO-A and MAO-B (82 and 78 μM IC_{50})], and superoxide dismutases [SOD1 and SOD2 (22 and 10 μM IC_{50})]. LPE inhibition of neurodegenerative enzymes is attributed to its major compounds naringenin and quercetin rutinoside (6.53 and 0.92 mg/g, respectively) and potentially other abundant compounds (naringin, p-coumaric acid, and hesperetin). The study demonstrated that LPE functions as a multitarget agent against key factors involved in Alzheimer's diseases (AD) and PD.

9.5.1 Animal Studies

The essential oil of dried Nigerian lemon peel dose dependently inhibited acetylcholinesterase and butyrylcholinesterase (EC_{50} values 164 and 169 μL/L vs 45 μg/L for prostigmine). Butyrylcholinesterase increases neurotoxicity of certain plaques, elevating Alzheimer's disease susceptibility; its inhibition is therefore essential to manage neurodegenerative conditions. The anticholinesterase activity of the essential oil is attributed to its phenolic content and presence of volatile compounds (borneol, β-pinene, and 1,8-cineole) and synergistic action of volatile compounds (monoterpenoids and phenylpropanoids). The essential oil inhibited Fe^{2+} and quinolinic acid-induced malondialdehyde (MDA) production in rat brain homogenate (~90 μL/L EC_{50}), attributed to radical (Fe^{2+}, OH, and NO) scavenging activity. The lemon peel essential oil dose dependently scavenged DPPH, OH, NO, and Fe^{2+} (EC_{50} values 206, 192, 340, and 134 μL/L, respectively). Limonene and linalool are known to inhibit (7,12-dimethylbenz[a]anthracene) (DMBA)-induced MDA production in female rats (Oboh et al. 2014).

Falls et al. (2018) investigated the effects of lemon oil on memory changes of stressed and unstressed Swiss young albino mice. Lemon essential oil (100 mg/kg) enhanced memory movement, significantly altered and reduced cerebrum acetylcholinesterase (AChE), and reversed scopolamine-induced amnesia in both stressed and unstressed mice comparable to the cholinesterase inhibitor drug donepezil. Lemon oil significantly increased cellular antioxidant enzyme (catalase, superoxide dismutase, and reduced glutathione) levels and reduced lipid peroxidation (TBARS). This memory-enhancing activity of lemon oil was attributed to reduced AChE (rebound) pertinent to the acetylcholine neurotransmitter and TBARS activity and elevated antioxidant enzymes due to corticosterone regulation (Falls et al. 2018).

Lemon peel essential oil (1.4 g/kg body weight) administered for 14 days improves motor coordination, reduces brain lipid peroxidation (MDA levels), attenuates oxidative stress by increasing antioxidant enzymes (SOD, CAT, and GPx), and functions as an anxiolytic, antidepressant, and antioxidant agent (Yousuf et al. 2023). These neuroprotective effects are attributed to the primary active components of lemon peel oil (limonene, linalool, and citral) that regulate the Nrf2 signaling pathway by increasing brain antioxidant enzyme (SOD, CAT, and GPx) activities. However, higher doses (2.1–3.5 g/kg) of the essential oil are neurotoxic and disrupt the redox balance due to increased plasma corticosterone levels associated with inflammatory responses in rats (Yousuf et al. 2023). Depression in Sprague Dawley rats is influenced by monoamine (serotonin, norepinephrine, and dopamine) levels in the brain. Limonene in lemon peel exerts antidepressant effects by increasing brain monoamine levels in stress-induced (tail suspension) white rats (Setiawan et al. 2021). Rats administered aromatherapy (5% lemon peel inhalation dose) significantly reduced immobility time (depression marker), presumably due to elevated monoamine levels induced by limonene in lemon peel essential oil.

Administration of lemon essential oil (LEO) components significantly enhances monoamines including dopamine (DA) and serotonin (5-HT) release from brain slices. These components can also relieve physiological and psychological stress after oral administration. Dopamine reduction and the functional degradation of dopamine receptors in the brain are recognized as common causes of Parkinson's (PD) and AD. Passive avoidance test (PA) detects associative memory, and open field habituation test (OFP) checks non-associative memory in animal behavior models. These tests were used to evaluate *s*-limonene and *s*-perillyl alcohol dementia attenuation induced by scopolamine. Administration of *s*-limonene significantly attenuated the deficiency in both kinds of memory, whereas its derivative, *s*-perillyl alcohol, only improved associative memory significantly. These results indicate that *s*-limonene and *s*-perillyl alcohol stimulate different neuron networks because of their different effects on non-associative memory. Dopamine is the most potent neurotransmitter influencing the passive avoidance task. Apparently, *s*-limonene and *s*-perillyl alcohol improved scopolamine-induced impaired memory partly by keeping a balance of dopamine concentration in the relevant brain regions. Memory is improved by *s*-limonene and *s*-perillyl alcohol, presumably by inhibiting acetylcholine hydrolysis in the synaptic cleft, thereby enhancing acetylcholine concentration, resulting in enhanced memory. The study demonstrates that dopamine and acetylcholine play important roles in the anti-dementia effects of *s*-limonene and *s*-perillyl alcohol (Zhou et al. 2009).

9.5.2 Clinical Trials

Ueda et al. (2023) investigated the effects of inhaling lemon essential oil obtained from cold-pressed Italian lemon peels on human brain activity and memory function using electroencephalography (EEG) and brain source activity. During EEG, 24 healthy young men (22 years mean age) performed a working memory task (neurobehavioral probe) before and after essential oil (10% in dipropylene glycol [DGP]) inhalation. Lemon essential oil inhalation significantly improved task performance and activated alpha (working memory), delta (associated with the frontal lobe), and theta bands in the prefrontal cortex. The study suggests that lemon essential oil inhalation contributes to memory encoding and retrieval, as well as to the control of attentional selection, eventually resulting in improved performance in the working memory. Inhalation of lemon essential oil enhances positive mood, can facilitate word recall, reduce errors in humans, and reduce scopolamine-induced memory impairment in rats according to previous studies (Ueda et al. 2023 and references therein). These studies indicate that odor molecules are actively transported to task-related brain regions, selectively activating these regions.

Lemon essential oil aromatherapy is an effective non-pharmacological therapy for dementia, since it improves cognitive function, especially in Alzheimer's (AD) patients (Jimbo et al. 2009). The study involved 28 elderly people (86 years mean age) with AD (15F/2M; three women had vascular dementia, and eight had other diagnoses [mixed AD and cerebrovascular lesions]). After a 28-day control period, patients were exposed to aroma (0.04 mL lemon and 0.08 mL rosemary essential oil) in the morning (9–11 AM) and 0.08 mL lavender and 0.04 mL orange essential oils in the evening (7:30–9 PM) for 28 days. The lemon and rosemary mixed essential oils activate the sympathetic nervous system to strengthen concentration and memory. Aromatherapy significantly ($P < 0.05$) improved recognition and cognitive function in dementia patients and can be effective in improving cognitive disorders, particularly for patients with moderate AD.

Lemon essential oil aromatherapy significantly reduced cognitive test anxiety scores among nursing students (35F/4M) (Johnson 2019). In the quantitative randomized pretest/posttest study, lemon essential oil (8–9 drops) obtained from cold-pressed lemon peel (Aura Cacia, Italy) was deposited in a personal handheld nasal inhaler. Students inhaled the aroma when experiencing anxiety symptoms during examinations.

9.6 PHYTOTHERAPEUTICAL BENEFITS OF LEMON PEEL WASTE

Nobiletin, a ubiquitous flavone isolated from citrus peels, is considered an anti-inflammatory and immunomodulatory drug (Lin et al. 2003) that also shows anti-cancer activity. In A549 lung cancer cells, nobiletin attenuates the expression of P-glycoprotein (P-gp), whose enhanced expression has been associated with malignant behavior and the progression of cancer cells via the stimulation of epithelial-to-mesenchymal transition (Ashrafizadeh et al. 2020). Nobiletin can also inhibit tumor angiogenesis by modulating the signaling pathway of the tyrosine kinases Src and FAK in estrogen receptor-positive (ER^+) breast cancer cells (Sp et al. 2017). Numerous flavonoids including eriocitrin, quercetin, rutin, and hesperidin induced apoptosis in cell cultures such as A549 lung cancer cells, leukemia HL-60 cells, hepatocellular carcinoma, and estrogen receptor-positive and -negative breast cancer cells. The inhibitory effect of citrus peel is related to inhibition of key signaling pathways including the PKC, PI3K, and Ras/Raf/ERK pathways (Koolaji et al. 2020; Liao et al. 2015; Kawaii et al. 1999; Kim et al. 2008).

9.6.1 Potential Neuroprotective Effects of Lemon Peel Waste Components

Neurodegenerative disorders are multifactorial diseases that result in the progressive loss of structure and function of neuronal cells in the brain and spinal cord. This further results in neuronal dysfunction and memory impairment, which is frequently associated with major neurodegenerative diseases such as Alzheimer's disease and Parkinson's disease. Oxidative stress and neuroinflammation are the initiators of neurodegenerative disorders. Previous studies have shown that citrus peel flavonoids such as hesperidin may confer neuroprotection by inhibiting the oxidative stress-mediated neuroinflammation, apoptotic cell death, and neurodegeneration in the LPS-induced neurodegeneration mouse model of Alzheimer's disease (Badshah et al. 2019; Batista et al. 2019; Khan et al. 2020). Moreover, hesperidin also shows attenuated neuronal oxidative stress and neuroinflammation through regulation of the transcription factors Nrf2 and NF-$_{\kappa}$B respectively, and can markedly reduce motor dysfunction in the 6-hydroxydopamine (environmental neurotoxin)-induced Parkinson's disease rats (Khan et al. 2020; Kiasalari et al. 2016). The amyloid hypothesis implies that elevated levels of redox-reactive iron reportedly trigger amyloid-β aggregation in the brain (reviewed in Zhao 2019). Iron-chelation therapy is a prospective pathway for maintaining neuronal function and health. Noteworthy studies by Senol et al. (2016) reported butyrylcholinesterase (BChE) inhibitory activities of 17 citrus cultivars, whereas Ramful et al. (2010) observed remarkable radical scavenging and iron-chelating activities in 21 varieties of citrus cultivars.

Hesperidin is the primary flavonoid (flavanones) in Eureka lemons, mostly concentrated in the peel and albedo (Nogata et al. 2006) (Table 9.4). It represents over 70% of the total flavonoids in whole fruit and epidermis, and approximately ¾ of the total in peel and albedo. Hajialyani et al. (2019) reviewed the clinical evidence for the neuroprotective potential of hesperidin to elucidate its mechanism of action in humans.

High flavanone [HF] (305 mg) 100% orange juice and an equicaloric low flavanone [LF] (37 mg) orange-flavored cordial (500 mL) were consumed daily for 8 weeks by healthy adults (24F/13M, 67 years mean age, 26 kg/m^2 median BMI) in a crossover, double-blind, randomized clinical trial separated by a 4-week washout (Kean et al. 2015). The hesperidin and narirutin contents of the drinks per liter were 549 and 60 mg for HF, and 64 and 10 mg for LF, respectively. HF consumption significantly ($P < 0.01$) improved cognitive and executive functions, and episodic memory, compared to the LF group. Moreover, hesperidin-rich juice significantly ($P = 0.006$) enhanced immediate recall after the follow-up period and lowered diastolic blood pressure. This study demonstrated that hesperidin-rich dietary interventions can prevent cognitive decline in neurodegenerative patients, presumably by increasing cerebral blood flow.

TABLE 9.4 Flavonoid composition (mg/100 g fresh weight) of Eureka lemon

	FRUIT	*PEEL*	*FLAVEDO*	*ALBEDO*	*EPIDERMIS*	*JUICE VESICLE*
Hesperidin	356	711	193	918	479	63.8
Eriocitrin	102	132	69.4	156	129	81.5
Isorhoifolin	15.8	35.5	10	45.7	12.5	1.6
Diosmin	20.8	43.2	37.7	45.4	12.7	6.1
Narirutin	11.4	22.5	2.7	30.4	21.2	0.8
Rhoifolin	1.3	2.9	5.8	1.7	1.3	0
Sinensetin	0.3	0.7	2.5	0	0	0

Adapted from: Nogata et al. (2006)

Lamport et al. (2017) investigated the association of flavanone-rich [FR] juice with acute cognitive benefits and increased cerebral blood flow [CBF] in healthy young adults (mean age 22 years, 23 kg/m^2 median BMI). The flavanone-rich juice (Tropicana Ruby Breakfast juice) naturally contained 70.5 mg flavonoids (specifically, 42.15 mg hesperidin, 17.25 mg naringin, 6.75 mg narirutin, and 4.3 mg caffeic acid), and the control was a zero-flavonoid drink. Flavanone-rich juice (500 mL) consumption significantly ($P < 0.001$) improved CBF of the right frontal gyrus and executive function (Digit Symbol Substitution Test, $P < 0.01$). However, improvement in behavioral parameters was not associated with CBF increase. Flavonoids are believed to acutely induce vasodilation and enhance CBF by increased nitric oxide synthesis in the endothelium.

FR juice consumption improved cognitive function and subjective alertness after 6 h relative to an energy-matched placebo in 24 healthy middle-aged male adults (51 years mean age, 28 kg/m^2 median BMI) free of disease or mild cognitive impairment (Alharbi et al. 2016). Participants were selected for behavioral, cognitive, and arterial spin labeling arms of the study (n = 24 and 16, respectively) for the acute single-blind randomized crossover trial with 500 mL HF and control (zero flavonoids) drinks. The FR juice consisted of Tropicana Pure Premium Orange juice with 5.5 g added orange pomace (272 mg flavonoids; 220.5 mg hesperidin, 34.5 mg narirutin, and 17.14 mg other flavonoids). Cognitive benefits occurred within 6 h postprandial FR juice consumption, the anticipated peak in plasma flavanone metabolites. A pilot bioavailability data set (six healthy adults) showed peak hesperetin and naringenin (hesperidin and narirutin metabolites) concentrations (0.25 and 0.12 μM) in the plasma at 6 h.

In another randomized, double-blind, placebo-controlled crossover study, hesperidin (500 mg/d orally for 3 weeks) increased flow-mediated dilation and reduced levels of circulating inflammatory biomarkers (high-sensitivity C-reactive protein, serum amyloid A protein, and soluble E-selectin) in 12 subjects (6M/6F, 50 years average age, 35.4 kg/m^2 median BMI) with metabolic syndrome (Rizza et al. 2011). Hesperidin improved endothelial function and reduced inflammation by stimulating phosphorylation of Src, Akt, AMP kinase, and endothelial NO synthase.

Ávila-Gálvez et al. (2021) compared the bioavailability and metabolism of hesperidin and eriocitrin in healthy humans after consuming hesperidin-rich orange extract (CBCs, Citrus Bioflavonoids Complex, 15% standardized in hesperidin) and an aqueous soluble eriocitrin-rich lemon extract (Wellemon™, Euromed S.A., Mollet del Vallès, Barcelona, Spain). The orange extract contained only hesperidin (~134 mg/g), while the lemon extract was a complex mixture of citrus flavanones (~240 mg/g) for equal amounts (260 mg) of hesperidin and eriocitrin. The volunteers consumed a single dose of 744 and 260 mg citrus flavanones after consuming lemon and orange extract, respectively, with equal administration of 260 mg eriocitrin and hesperidin. Eriocitrin from lemon extract (260 mg from 3.1 g lemon extract) showed higher bioavailability than hesperidin (260 mg from 1.95 g orange extract) in a randomized crossover (2-week washout period) single-dose healthy human (8M/8F, 33 years average age, 23 kg/m^2 median BMI) pharmacokinetic trial. Eriocitrin bioavailability was ascribed to its higher solubility, and therefore, consumption of eriocitrin-rich lemon can provide better health benefits.

Eriomin® is a nutraceutical supplement of citrus flavonoids composed of 70% eriocitrin, 5% hesperidin, 4% naringin, and 1% didymin that exhibits anti-inflammatory, antihyperglycemic, and antioxidant properties (Minato et al. 2003). Eriomin® supplementation (200, 400, and 800 mg daily) significantly ($P < 0.05$) reduced inflammatory markers (12% hsCRP levels, 13% IL-6, 12% TNF-α, and 17% lipid peroxidation levels) and increased serum antioxidant capacity after 12 weeks in prediabetic patients (n = 78, 50 years average age, 34 kg/m^2 median BMI). Reduction of proinflammatory cytokines (IL-6 and TNF-α) modulates neuroinflammation (likely mediated through mitigation of hyperactive immune cells due to NF-κB reduction that governs [regulates] chemokine and inflammatory mediator transcription) (Ribeiro et al. 2019).

Despite hesperidin's various physiological functions, its low water solubility limits its use in beverages or other liquids. The solubility problem has been resolved by adding glucose to hesperidin, producing α-glucosyl hesperidin (GHES). GHES has physiological functions similar to and more effective than hesperidin. GHES at a single dose (100 mg) facilitates the recovery of skin blood flow and temperature following cold stress in a double-blind crossover study of 18 healthy adults (8M/10F, mean age 41 years,

20 kg/m^2 BMI) (Morishita et al. 2020). Two potential mechanisms are proposed for GHES-improved blood flow in this study: through autonomic nerves and via nitric oxide (NO) production. In an earlier human study, GHES inhibited sympathetic activity and enhanced parasympathetic activity compared to placebo, thereby supporting the autonomic nerves mechanism of action (Morishita et al. 2020 and references therein). The physiological effect of this low GHES dose enables and/or extends its use in a wide range of food applications.

The polymethoxylated flavone nobiletin is abundant in citrus peel; however, only one study reports its presence in *Citrus limon*. Xi et al. (2017) investigated the phenolic composition and antioxidant capacity of different fruit parts (whole fruit, peel, pulp, juice, and seed) of five lemon cultivars, including three *Citrus limon* cultivars (Feiminailao [FM], Pangdelusaningmeng [PD], and Beijingningmeng [BJ]). Nobiletin was most abundant in peels (7.45–107.3 μg/g fw), pulp (nil–5.51 μg/g fw), juice (nil–17.13 μg/g fw), and whole fruit (undetected to 72.44 μg/g fw). Generally, nobiletin is largely in peels > whole fruit > juice > pulp. PD peel and whole fruit have the highest nobiletin level. Nobiletin was the most abundant polymethoxylated flavonoid in lemons; lemon whole fruit, pulp, and juice were also good sources of sinensetin, nobiletin, and tangeritin.

Seki et al. (2013) examined the safety and feasibility of nobiletin-rich *Citrus reticulate* peel extract (NChinpi), a kampo medicine cotreatment with donepezil in Alzheimer's disease patients. Mild to moderate AD patients (2M/4F, 75 years average age) were treated with NChinpi (30 g NChinpi, 0.44% nobiletin; 100 mL decoction, thrice daily for 1 year) and donepezil, whereas five patients (2M/3F, 77 years mean age) were pre-administered donepezil treatment (5 mg daily) alone as control in this pilot clinical study. NChinpi intervention significantly (P = 0.02) reduced cognitive impairment (Alzheimer's Disease Assessment Scale-Cognitive Subscale [ADAS-J cog] score from 4.3 to −2.9).

Hashimoto et al. (2022) investigated nobiletin's effects on cognitive functions in elderly Japanese patients in a randomized double-blind parallel clinical trial. Patients received soft gelatin capsules containing perilla seed oil [PO] (0.88 g α-linolenic acid [ALA]) or PO enriched with nobiletin (2.91 mg nobiletin). The nobiletin-enriched PO group (12M/11F, 69 years mean age, 23 kg/m^2 median BMI) showed significantly higher cognitive index scores than the PO group (11M/10F) after 12 months of intervention. The nobiletin-rich intake did not affect the lipid profiles or hematological parameters compared to those in the PO group. The pro-cognitive effects of nobiletin-enriched PO were accompanied by increases in serum brain-derived neurotrophic factor (BDNF) levels and biological antioxidant potential. Nobiletin was sourced from air-dried immature ponkan (*Citrus poonensis*) powder (16 g hesperidin, 1.2 g narirutin, 260 mg nobiletin, and 4.1 mg other polyphenols per 100 g of dry powder).

Nobilex® is a commercially available supplement in Japan to improve the quality of elderly life. It contains a high-purity nobiletin powder (PMF90: 65% nobiletin, 25% tangeretin) extracted from *Citrus depressa* (Sikwasa) by-product (peel), combined with *Kaempferia parviflora* and *Peucedanum japonicum* extracts. Yamada et al. (2021) investigated the cognitive benefits of nobiletin-containing test food—three capsules daily (each capsule contains 10 mg nobiletin, 5.8 mg tangeretin, 126.7 mg *K. parviflora* dried root powder, and 33.3 mg *P. japonicum* dried leaf powder)—for 16 weeks in a randomized double-blind placebo-controlled trial in healthy elderly (≥ 65 years old) subjects. Nobilex® significantly improved memory function based on Wechsler Memory Scale-Revised [WMS-R] scores and "visual memory" of healthy elderly subjects (39F/15M, 73 years mean age, 23 kg/m^2 median BMI) compared to placebo (36F/18M). Nobiletin-containing food can exert beneficial effects in a group with mild memory dysfunction and lapses. The study supports the efficacy of nobiletin against neurodegenerative disorders such as AD. Nobiletin is known to exert neuroprotective effects by attenuating cholinergic deficits, reducing the abnormal accumulation of neurotoxic amyloid-beta peptides, reversing N-methyl-D-aspartate receptor hypofunction, ameliorating ischemic injury, inhibiting tau protein phosphorylation, suppressing β-secretase, increasing neprilysin, and modulating several signaling cascades in several *in vitro* and *in vivo* studies.

Chai et al. (2022) explored the mechanism of nobiletin attenuating Alzheimer's disease by inhibiting neuroinflammation. Nobiletin (40 μM) significantly lowered mouse microglia [MG] viability and reversed lipopolysaccharide [LPS]- and Nigericin-induced pyroptosis by suppressing inflammatory cytokines (interleukin [IL]-1β, IL-18, and tumor necrosis factor-alpha [TNF-α]). Moreover, nobiletin repressed the expression of pyroptosis-related proteins and high mobility group box-1 [HMGB-1], and improved AD in mice by reducing the expression of neuroinflammation-related proteins. Nobiletin also improved the spatial learning and memory ability of mice by repressing pyroptosis-related proteins, HMGB-1, and inflammatory cytokines in the hippocampal tissues of mice. The study suggests that nobiletin can mitigate AD, since AD patients (n = 30, mean age 70 years; Montreal Cognitive Assessment [MoCA] score <26 indicating cognitive impairment) showed significantly ($P < 0.001$) higher serum inflammatory cytokines (IL-1β, IL-8, and TNF-α) and HMGB-1 levels than their healthy cohorts (n = 30, mean age 65 years).

Fiorito et al. (2021) developed a process to obtain auraptene-rich *Citrus limon* peels powder with functional food and nutraceutical potential and antioxidant properties. The process used bentonite to selectively absorb 98% of the auraptene from the parent EtOAc lemon peels extract (0.33 mg/mL) with 13.7-fold enrichment. Moreover, auraptene-rich extract (0.5 mg/mL) had comparable or higher antioxidant activity than ascorbic acid (72.5% vs 71.1% for ascorbic acid by the DPPH assay) or Trolox (76.6% vs 73.4% for Trolox by the ABTS assay). Thus, auraptene-rich lemon peel can reduce ROS and oxidative markers and scavenge free radicals from high metabolic brain activity susceptible to oxidation damage.

Auraptene (AUR) has neuroprotective effects on the central nervous system (Okuyama 2015). AUR ameliorates lipopolysaccharide-induced inflammation in the brain, evidenced by inhibition of microglial activation, cyclooxygenase (COX)-2 suppression, and neuronal cell death in the hippocampus. In a randomized, placebo-controlled, double-blind clinical trial, administration of AUR-enriched juice (6 mg AUR daily for 24 weeks) significantly ($P < 0.05$) improved cognitive function in healthy adults (n = 41, mean age 71 years, median BMI 23 kg/m^2) compared to placebo (0.1 mg AUR daily). AUR-enriched juice has been approved as a "Food with Functional Claims" by the Consumer Affairs Agency of Japan in 2018 because it prevents cognitive decline in humans (Furukawa et al. 2021).

In a similar randomized, placebo-controlled, double-blind study, AUR (6 mg daily)-enriched juice did not improve cognitive function (Mini-Mental State Examination [MMSE]) after 24 weeks compared to baseline in normal adults (55F/27M, mean age 71 years, median BMI 23 kg/m^2) (Igase et al. 2018). However, auraptene-rich juice significantly ($P < 0.05$) increased cognitive function (immediate recall test) compared to placebo orange juice containing 0.1 mg AUR (6.3% vs −2.4% change in immediate recall test), suggesting that auraptene is a safe supplement in healthy humans to prevent cognitive decline (a surrogate marker for dementia) characteristic of age-related dementia. AUR's cognitive decline protective effects may be due to its improvement in spatial learning and amelioration of cognitive impairment evidenced in a vascular dementia rat model study. The anti-inflammatory effects of AUR in the ischemic brain underlie its preventive mechanism for cognitive impairment (Igase et al. 2018).

IntegroPectin is lemon pectin obtained by hydrodynamic cavitation of citrus processing industrial waste. Its antibacterial activity, largely superior to that of commercial citrus pectin, was ascribed to the combined action of lemon oil and abundant water-soluble lemon flavonoids at the surface of the lyophilized IntegroPectin. Lemon IntegroPectin shows exceptionally high *in vitro* neuro- and mitoprotective activity towards neuronal model cells. Eriocitrin predominates in lemon IntegroPectin, followed by significant amounts of hesperidin, kaempferol, kaempferol 7-O-gluconoride, and naringin (3.35, 0.60, 0.26, 0.26, and 0.11 mg/g) (Scurria et al. 2021). The same group previously identified four major volatile components (α-terpineol, terpinen-4-ol, 2-methyl-1-butanol, and 3-methyl-1-butanol [27.42%, 21.85%, 17.11%, and 12.43% area, respectively]) and 11 other components in lemon IntegroPectin (Scurria et al. 2020). The minor components in area % included α-linalool (4.94), limonene (3.39), 1-hexanol (2.52), 2-hexen-1-ol (2.33), eucalyptol (2.32), 6-hepten-1-ol, 2-methyl (1.52), safranal (1.11), isoprenyl alcohol (1.08), 1-butanol, 3-methyl acetate (0.73), α-phellandren-8-ol (0.65), and α-citral (0.58).

Lemon IntegroPectin exerts significant neuroprotective activity *in vitro* against H_2O_2-induced human neuronal cells (SH-SY5Y), presumably due to mitochondria protection (Nuzzo et al. 2021). Lemon IntegroPectin (1 mg/mL) totally prevented H_2O_2-induced increases in neuronal cell debris due to DNA damage and rescued cell body size. It also hindered the H_2O_2-induced ROS increase by lowering and delaying ROS production and significantly counteracted H_2O_2-induced mitochondrial remodeling. Lemon IntegroPectin inhibits H_2O_2-induced mitochondrial membrane damage by scavenging ROS radicals, thereby protecting the cell's central organelle involved in several neurological dysfunctions. These neuroprotective actions of lemon IntegroPectin are attributed to the combined (synergistic) action of α-terpineol, hesperidin, eriocitrin, and the unique lemon pectin (Nuzzo et al. 2021).

Nomilin, the major limonoid and the most abundant aglycone, has also been reported to demonstrate neuroprotective function in an experimental stroke rat model, on cerebral ischemia-reperfusion injury by improving the infarct size, brain edema, and neurological deficits (Zhou et al. 2023). Moreover, nomilin attenuates blood-brain barrier (BBB) disruption and alleviates the loss of tight junction proteins. The neuroprotective effects of nomilin in ischemia-reperfusion rats were associated with the nuclear factor erythroid 2–related factor 2 (Nrf2)/NAD(P)H dehydrogenase (quinone) 1 pathway (Zhou et al. 2023).

As described, citrus flavonoids exert their neuroprotective effects in animal models of various neurological disorders through oral, subcutaneous, and intraperitoneal administration (Dike et al. 2021). Additionally, since these compounds and their metabolites readily cross the blood-brain barrier, they can be tested in randomized clinical trials in the form of citrus peel powder, tablets, capsules, extracts, or juices in either healthy subjects of all age groups or in patients presenting depressive symptoms or diagnosed with schizophrenia or Alzheimer's disease while taking donepezil. The results proved the safety of the different compounds tested and indicated that citrus peel could act beneficially on cognitive function and mental health (Matsuzaki and Ohizumi 2021; Matsuzaki et al. 2022). One of these clinical trials evaluated the anti-dementia effect of nobiletin-rich *Citrus reticulata* peel extract on Alzheimer's disease patients taking donepezil (5 mg) for more than 5 years and presenting a mild to moderate cognitive impairment. Comparing the cognitive scores of the control and intervention groups showed that only subjects in the intervention group presented amelioration in their cognitive function, suggesting that long-term intake of nobiletin-rich citrus peel extract prevents Alzheimer's disease progression (Matsuzaki and Ohizumi 2021).

9.7 CURRENT AND FUTURE PERSPECTIVES

Citrus is one of the popular sources of polyphenols, a class of over 8,000 compounds found in many plant foods. The global polyphenol market, valued at $1.68 billion in 2022, is projected to grow at a compound annual growth rate of 7.4%, according to Grand View Research (Adrien 2024). About 55%–72% of a lemon is directly wasted after squeezing, depending on the variety, according to a recent systematic review (Martínez-Zamora et al. 2023). This amounts to 11.8–15.4 million tons of annual recoverable global food losses with potential high-value addition. Currently, these food losses are used by flavor and extraction companies to obtain essential oils and fiber for their application as flavorings and odorizing or emulsifying agents.

The Spanish botanical company Monteloeder already manufactures a weight management ingredient, Metabolaid, a combination of lemon, verbena, and hibiscus flower extracts used in over 70 finished global food supplement products. The French company Fytexia also uses citrus polyphenols for its weight management ingredient Sinetrol, sold as a dietary supplement in over 50 countries. Intakes of dietary polyphenols are higher in the Japanese population than in European countries (1500 vs 283–1100 mg per day), which partly relates to their obesity rate (Adrien 2024). Lemon peel powder is offered by many Indian manufacturers. For example, MR Ayurveda offers 100% natural lemon peel powder for skin, face, and hair care due to its anti-aging, oil control, cleansing, hydrating, exfoliating, freshening, and moisturizing effects. Mana Ayurvedam 100% pure and natural organic lemon peel powder and Panchkarma Ayurveda lemon peel powder claim antibacterial, antimicrobial, and antioxidant properties for their skin health benefits. ENEA, in association with Navhetec and Agrumaria Corleone (Palermo, Italy), has patented an innovative membrane separation technology to transform lemon processing by-products into food supplements and nutraceuticals to prevent several diseases (obesity, hypercholesterolemia, and cardiovascular disorders). Navhetec studies also demonstrate anti-tumor activity of these products (Agrumaria Corleone 2022).

The Okinawa Research Centre developed a method to manufacture PMF90 (65% nobiletin and 25% tangeretin) from *Citrus depressa* waste products, primarily peels, by ethanol extraction and alkaline treatment. PMF90 is now commercially available as a supplement, Nobilex® in Japan. Nobilex® had beneficial effects in nocturia, considered a circadian rhythm disorder, in a recent randomized placebo-controlled double-blind crossover clinical trial in nocturia patients (≥ 50 years) (Ito et al. 2023). Yamada et al. (2021) investigating the cognitive benefits of nobiletin, showed that Nobilex® significantly improved memory function and "visual memory" of healthy elderly subjects.

A transition towards an aging population is being witnessed at a global scale. While this will certainly trigger a cascading effect on countries' economic stability with the increase in healthcare expenditure to tackle the growing number of epidemic non-communicable diseases and the increased number of irreversible pathologies leading to degenerative diseases, enhancing the use of protective factors has become quintessential. The promotion of phytochemicals through nutraceuticals paves the way forward to slow down the onset of such debilitating diseases, while further research is required in the field of phytomedicine to understand the holistic effects of phytochemicals in a bio-model that would replicate the molecular, cellular, and physiological features observed across cellular senescence and the developmental spectrum. Understanding cellular bioenergetic profiles and the benefits associated with the use of phytochemicals under a range of conditions has become increasingly important.

REFERENCES

Adrien, C. (2024). Brands harness plant polyphenols for weight management. *NutraIngredients USA*, January 30; https://www.nutraingredients-usa.com/Article/2024/01/30/Brands-harness-plant-polyphenols-for-weight-management

Agrumaria Corleone. (2022). February 18; https://www.agrumariacorleone.com/en/lemon-by-products-gives-way-to-supplements-and-nutraceuticals-against-cardiovascular-risk.html

Alharbi, M. H., Lamport, D. J., Dodd, G. F., Saunders, C., Harkness, L., Butler, L. T., & Spencer, J. P. (2016). Flavonoid-rich orange juice is associated with acute improvements in cognitive function in healthy middle-aged males. *European Journal of Nutrition*, *55*, 2021–2029.

Alzobaidi, N., Quasimi, H., Emad, N. A., Alhalmi, A., & Naqvi, M. (2021). Bioactive compounds and traditional herbal medicine: Promising approaches for the treatment of dementia. *Degenerative Neurological and Neuromuscular Disease*, 1–14; https://doi.org/10.2147/DNND.S299589

Arcone, R., D'Errico, A., Nasso, R., Rullo, R., Poli, A., Di Donato, P., & Masullo, M. (2023). Inhibition of enzymes involved in neurodegenerative disorders and Aβ1_40 aggregation by *Citrus limon* peel polyphenol extract. *Molecules (Basel, Switzerland)*, *28*(17), 6332; https://doi.org/10.3390/molecules28176332

Ashrafizadeh, M., Zarrabi, A., Saberifar, S., Hashemi, F., Hushmandi, K., Hashemi, F., . . . & Garg, M. (2020). Nobiletin in cancer therapy: How this plant derived-natural compound targets various oncogene and onco-suppressor pathways. *Biomedicines*, *8*(5), 110; https://doi.org/10.3390/biomedicines8050110

Ávila-Gálvez, M. Á., Giménez-Bastida, J. A., González-Sarrías, A., & Espín, J. C. (2021). New insights into the metabolism of the flavanones eriocitrin and hesperidin: A comparative human pharmacokinetic study. *Antioxidants*, *10*(3), 435; https://doi.org/10.3390/antiox10030435

Badshah, H., Ikram, M., Ali, W., Ahmad, S., Hahm, J. R., & Kim, M. O. (2019). Caffeine may abrogate LPS-induced oxidative stress and neuroinflammation by regulating Nrf2/TLR4 in adult mouse brains. *Biomolecules*, *9*(11), 719; https://doi.org/10.3390/biom9110719

Baker, M. T., Lu, P., Parrella, J. A., & Leggette, H. R. (2022). Consumer acceptance toward functional foods: A scoping review. *International Journal of Environmental Research and Public Health*, *19*(3), 1217; https://doi.org/10.3390/ijerph19031217

Ballistreri, G., Fabroni, S., Romeo, F. V., Timpanaro, N., Amenta, M., & Rapisarda, P. (2019). Anthocyanins and other polyphenols in citrus genus: Biosynthesis, chemical profile, and biological activity. In *Polyphenols in Plants* (pp. 191–215). Academic Press

Bao, G., Zhang, Y., & Yang, X. (2020). Effect of lemon peel flavonoids on anti-fatigue and anti-oxidation capacities of exhaustive exercise mice. *Applied Biological Chemistry*, *63*, 1–11; https://doi.org/10.1186/s13765-020-00573-3

Batista, C. R. A., Gomes, G. F., Candelario-Jalil, E., Fiebich, B. L., & De Oliveira, A. C. P. (2019). Lipopolysaccharide-induced neuroinflammation as a bridge to understand neurodegeneration. *International Journal of Molecular Sciences*, *20*(9), 2293; https://doi.org/10.3390/ijms20092293

Bello-Medina, P. C., Rodríguez-Martínez, E., Prado-Alcalá, R. A., & Rivas-Arancibia, S. (2022). Ozone pollution, oxidative stress, synaptic plasticity, and neurodegeneration. *Neurología (English Edition)*, *37*(4), 277–286.

Ben Hsouna, A., Sadaka, C., Generalić Mekinić, I., Garzoli, S., Švarc-Gajić, J., Rodrigues, F., . . . & Mnif, W. (2023). The chemical variability, nutraceutical value, and food-industry and cosmetic applications of citrus plants: A critical review. *Antioxidants*, *12*(2), 481; https://doi.org/10.3390/antiox12020481

Bishop, N. A., Lu, T., & Yankner, B. A. (2010). Neural mechanisms of ageing and cognitive decline. *Nature*, *464*(7288), 529–535.

Buchman, A. S., Yu, L., Wilson, R. S., Schneider, J. A., & Bennett, D. A. (2013). Association of brain pathology with the progression of frailty in older adults. *Neurology*, *80*(22), 2055–2061.

Castellani, G. C., Quinlan, E. M., Cooper, L. N., & Shouval, H. Z. (2001). A biophysical model of bidirectional synaptic plasticity: Dependence on AMPA and NMDA receptors. *Proceedings of the National Academy of Sciences*, *98*(22), 12772–12777.

Chai, W., Zhang, J., Xiang, Z., Zhang, H., Mei, Z., Nie, H., . . . & Zhang, P. (2022). Potential of nobiletin against Alzheimer's disease through inhibiting neuroinflammation. *Metabolic Brain Disease*, *37*(4), 1145–1154.

Chetia, J., Adhikary, P., Devi, L. M., & Badwaik, L. S. (2023). Extraction of essential oil from Assam lemon peels and its incorporation in chitosan based coating for maintaining grape quality. *Sustainable Chemistry and Pharmacy*, *32*, 101034; https://doi.org/10.1016/j.scp.2025.101034

Chou, S. C., Aggarwal, A., Dawson, V. L., Dawson, T. M., & Kam, T. I. (2021). Recent advances in preventing neurodegenerative diseases. *Faculty Reviews*, *10*. 81; https://doi.org/10.12703/r/10-81

Cui, K., Luo, X., Xu, K., & Murthy, M. V. (2004). Role of oxidative stress in neurodegeneration: recent developments in assay methods for oxidative stress and nutraceutical antioxidants. *Progress in Neuro-Psychopharmacology and Biological Psychiatry*, *28*(5), 771–799.

Czech, A., Zarycka, E., Yanovych, D., Zasadna, Z., Grzegorczyk, I., & Kłys, S. (2020). Mineral content of the pulp and peel of various citrus fruit cultivars. *Biological Trace Element Research*, *193*, 555–563.

Di Donato, P., Taurisano, V., Tommonaro, G., Pasquale, V., Jiménez, J. M. S., de Pascual-Teresa, S., . . . & Nicolaus, B. (2018). Biological properties of polyphenols extracts from agro industry's wastes. *Waste and Biomass Valorization*, *9*, 1567–1578.

Dike, C. S., Orish, C. N., Nwokocha, C. R., Sikoki, F. D., Babatunde, B. B., Frazzoli, C., & Orisakwe, O. E. (2021). Phytowaste as nutraceuticals in boosting public health. *Clinical Phytoscience*, *7*(24), 1–23; https://doi.org/10.1186/s40816-021-00260-w

Douaud, G., Refsum, H., de Jager, C. A., Jacoby, R., E. Nichols, T., Smith, S. M., & Smith, A. D. (2013). Preventing Alzheimer's disease-related gray matter atrophy by B-vitamin treatment. *Proceedings of the National Academy of Sciences*, *110*(23), 9523–9528.

Dugo, P., Mondello, L., Morabito, D., & Dugo, G. (2003). Characterization of the anthocyanin fraction of sicilian blood orange juice by micro-HPLC-ESI/MS. *Journal of Agricultural and Food Chemistry*, *51*(5), 1173–1176.

Ekimova, I. V., Belan, D. V., Lapshina, K. V., & Pastukhov, Y. F. (2023). The use of the proteasome inhibitor lactacystin for modeling Parkinson's disease: Early neurophysiological biomarkers and candidates for intranigral and extranigral neuroprotection. In *Handbook of Animal Models in Neurological Disorders* (pp. 507–523). Academic Press

Falls, N., Singh, D., Anwar, F., Verma, A., & Kumar, V. (2018). Amelioration of neurodegeneration and cognitive impairment by lemon oil in experimental model of stressed mice. *Biomedicine & Pharmacotherapy*, *106*, 575–583.

Fiorito, S., Epifano, F., Palumbo, L., & Genovese, S. (2021). A novel auraptene-enriched citrus peels-based blend with enhanced antioxidant activity. *Plant Foods for Human Nutrition*, *76*, 397–398; https://doi.org.uml.idm.oclc.org/10.1007/s11130-021-00911-w

Furukawa, Y., Okuyama, S., Amakura, Y., Sawamoto, A., Nakajima, M., Yoshimura, M., . . . & Yoshida, T. (2021). Isolation and characterization of neuroprotective components from citrus peel and their application as functional food. *Chemical and Pharmaceutical Bulletin*, *69*(1), 2–10.

Gattuso, G., Barreca, D., Gargiulli, C., Leuzzi, U., & Caristi, C. (2007). Flavonoid composition of citrus juices. *Molecules*, *12*(8), 1641–1673.

Gemma, C., Vila, J., Bachstetter, A., & Bickford, P. C. (2007). Oxidative stress and the aging brain: From theory to prevention. In *Brain Aging: Models, Methods, and Mechanisms*. CRC Press/Taylor & Francis, Boca Raton, FL, PMID: 21204345

Gonzalez-Freire, M., De Cabo, R., Bernier, M., Sollott, S. J., Fabbri, E., Navas, P., & Ferrucci, L. (2015). Reconsidering the role of mitochondria in aging. *Journals of Gerontology Series A: Biomedical Sciences and Medical Sciences*, *70*(11), 1334–1342.

Government of Karnataka. (2023). https://vtpc.karnataka.gov.in/storage/pdf-files/Lemon.pdf; Accessed March 5, 2023.

Graham, R. K., Ehrnhoefer, D. E., & Hayden, M. R. (2011). Caspase-6 and neurodegeneration. *Trends in Neurosciences*, *34*(12), 646–656.

Guarente, L. (2014). Aging research—where do we stand and where are we going? *Cell*, *159*(1), 15–19.

Gupta, H., Singh, P. P., & Reddy, S. E. (2025). Exploring the chemical profiling and insecticidal properties of essential oils from fresh and discarded lemon peels, *Citrus limon* against pulse beetle. *International Biodeterioration & Biodegradation*, *196*, 10592; https://doi.org/10.1016/j.ibiod.2024.105924

Hajialyani, M., Hosein Farzaei, M., Echeverría, J., Nabavi, S. M., Uriarte, E., & Sobarzo-Sánchez, E. (2019). Hesperidin as a neuroprotective agent: A review of animal and clinical evidence. *Molecules (Basel, Switzerland)*, *24*(3), 648; https://doi.org/10.3390/molecules24030648

Hashimoto, M., Matsuzaki, K., Maruyama, K., Hossain, S., Sumiyoshi, E., Wakatsuki, H., Kato, S., Ohno, M., Tanabe, Y., Kuroda, Y., Yamaguchi, S., Kajima, K., Ohizumi, Y., & Shido, O. (2022). *Perilla* seed oil in combination with nobiletin-rich ponkan powder enhances cognitive function in healthy elderly Japanese individuals: a possible supplement for brain health in the elderly. *Food & Function*, *13*(5), 2768–2781; https://doi.org/10.1039/d1fo03508h

Hirota, Y., Kang, D., & Kanki, T. (2012). The physiological role of mitophagy: new insights into phosphorylation events. *International Journal of Cell Biology*; https://doi.org/10.1155/2012/354914

Igase, M., Okada, Y., Ochi, M., Igase, K., Ochi, H., Okuyama, S., . . . & Ohyagi, Y. (2018). Auraptene in the peels of *Citrus Kawachiensis* (Kawachibankan) contributes to the preservation of cognitive function: A randomized, placebo-controlled, double-blind study in healthy volunteers. *The Journal of Prevention of Alzheimer's Disease*, *5*, 197–201.

Imeneo, V., Romeo, R., De Bruno, A., & Piscopo, A. (2022). Green-sustainable extraction techniques for the recovery of antioxidant compounds from "citrus Limon" by-products. *Journal of Environmental Science and Health, Part B*, *57*(3), 220–232.

Imeneo, V., Romeo, R., Gattuso, A., De Bruno, A., & Piscopo, A. (2021). Functionalized biscuits with bioactive ingredients obtained by citrus lemon pomace. *Foods*, *10*(10), 2460; https://doi.org/10.3390/foods10102460

IndexBox. (2024). *World—Citrus Fruit—Market Analysis, Forecast, Size, Trends And Insights*; https://www.indexbox.io/store/world-citrus-fruit-market-analysis-forecast-size-trends-and-insights/

Ito, H., Negoro, H., Kono, J., Hayata, N., Miura, T., Manabe, Y., Miyazaki, Y., Mishina, M., Woo, J. T., Sakane, N., & Okuno, H. (2023). Effectiveness and safety of a mixture of nobiletin and tangeretin in nocturia patients: A Randomized, placebo-controlled, double-blind, crossover study. *Journal of Clinical Medicine*, *12*(8), 2757; https://doi.org/10.3390/jcm12082757

Jain, N., & Sharma, M. (2017). Evaluation of Citrus lemon essential oil for its chemical and biological properties against fungi causing dermatophytic infection in human beings. *Analytical Chemistry Letters*, *7*(3), 402–409.

Javed, H., Vaibhav, K., Ahmed, M. E., Khan, A., Tabassum, R., Islam, F., . . . & Islam, F. (2015). Effect of hesperidin on neurobehavioral, neuroinflammation, oxidative stress and lipid alteration in intracerebroventricular streptozotocin induced cognitive impairment in mice. *Journal of the Neurological Sciences*, *348*(1–2), 51–59.

Jenner, P. (1996). Oxidative stress in Parkinson's disease and other neurodegenerative disorders. *Pathologie-Biologie*, *44*(1), 57–64.

Jimbo, D., Kimura, Y., Taniguchi, M., Inoue, M., & Urakami, K. (2009). Effect of aromatherapy on patients with Alzheimer's disease. *Psychogeriatrics*, *9*(4), 173–179.

John, O. O., Amarachi, I. S., Chinazom, A. P., Adaeze, E., Kale, M. B., Umare, M. D., & Upaganlawar, A. B. (2022). Phytotherapy: A promising approach for the treatment of Alzheimer's disease. *Pharmacological Research-Modern Chinese Medicine*, *2*, 100030; https://doi.org/10.1016/j.prmcm.2021.100030

John, S., Monica, S. J., Priyadarshini, S., Sivaraj, C., & Arumugam, P. (2017). Antioxidant and antimicrobial efficacy of lemon (*Citrus limonum* L.) peel. *International Journal of Pharmaceutical Sciences Review and Research*, *46*(1), 115–118.

Johnson, C. E. (2019). Effect of inhaled lemon essential oil on cognitive test anxiety among nursing students. *Holistic Nursing Practice*, *33*(2), 95–100.

Kawaii, S., Tomono, Y., Katase, E., Ogawa, K., & Yano, M. (1999). Antiproliferative activity of flavonoids on several cancer cell lines. *Bioscience, Biotechnology, and Biochemistry*, *63*(5), 896–899.

Kaye, F. J., & Shows, T. B. (2000). Assignment of ubiquilin2 (UBQLN2) to human chromosome xp11. 23→ p11. 1 by GeneBridge radiation hybrids. *Cytogenetics and Cell Genetics*, *89*(1–2), 116–117.

Kawakami, K., Yamada, K., Takeshita, H., Yamada, T., & Nomura, M. (2021). Antihypertensive effect of lemon juice squeezed residue on spontaneously hypertensive rats. *Food Science and Technology Research*, *27*(3), 521–527.

Kean, R. J., Lamport, D. J., Dodd, G. F., Freeman, J. E., Williams, C. M., Ellis, J. A., . . . & Spencer, J. P. (2015). Chronic consumption of flavanone-rich orange juice is associated with cognitive benefits: an 8-wk, randomized, double-blind, placebo-controlled trial in healthy older adults. *The American Journal of Clinical Nutrition*, *101*(3), 506–514.

Khan, A., Ikram, M., Hahm, J. R., & Kim, M. O. (2020). Antioxidant and anti-inflammatory effects of citrus flavonoid hesperetin: Special focus on neurological disorders. *Antioxidants*, *9*(7), 609; https://doi.org/10.3390/antiox9070609

Kiasalari, Z., Khalili, M., Baluchnejadmojarad, T., & Roghani, M. (2016). Protective effect of oral hesperetin against unilateral striatal 6-hydroxydopamine damage in the rat. *Neurochemical Research*, *41*, 1065–1072.

Kim, A., & Om, J. (2021). Repurposing citrus peel waste and its positive effects on our health and communities. *Journal of Emerging Investigators*, *4*, 1–5. ISSN 2638–0870

Kim, D. I., Lee, S. J., Lee, S. B., Park, K., Kim, W. J., & Moon, S. K. (2008). Requirement for Ras/Raf/ERK pathway in naringin-induced G 1-cell-cycle arrest via p21WAF1 expression. *Carcinogenesis*, *29*(9), 1701–1709.

Koolaji, N., Shammugasamy, B., Schindeler, A., Dong, Q., Dehghani, F., & Valtchev, P. (2020). Citrus peel flavonoids as potential cancer prevention agents. *Current Development in Nutrition*, *4*(5), nzaa025; https://doi.org/10.1093/cdn/nzaa025

Kour, P., Gupta, N., & Singh, J. (2020). Development of osmo-dried peel flakes from Eureka lemon. *Chemical Science Review Letters*, *9*(36), 966–971.

Lamport, D. J., Pal, D., Macready, A. L., Barbosa-Boucas, S., Fletcher, J. M., Williams, C. M., . . . & Butler, L. T. (2017). The effects of flavanone-rich citrus juice on cognitive function and cerebral blood flow: An acute, randomised, placebo-controlled cross-over trial in healthy, young adults. *British Journal of Nutrition*, *116*(12), 2160–2168.

Leutner, S., Eckert, A., & Müller, W. E. (2001). ROS generation, lipid peroxidation and antioxidant enzyme activities in the aging brain. *Journal of Neural Transmission*, *108*, 955–967.

Liao, C. Y., Lee, C. C., Tsai, C. C., Hsueh, C. W., Wang, C. C., Chen, I., . . . & Kuo, W. H. (2015). Novel investigations of flavonoids as chemopreventive agents for hepatocellular carcinoma. *BioMed Research International*, *2015*; https://doi.org/10.1155/2015/840542

Lin, N., Sato, T., Takayama, Y., Mimaki, Y., Sashida, Y., Yano, M., & Ito, A. (2003). Novel anti-inflammatory actions of nobiletin, a citrus polymethoxy flavonoid, on human synovial fibroblasts and mouse macrophages. *Biochemical Pharmacology*, *65*(12), 2065–2071.

López-Otín, C., Blasco, M. A., Partridge, L., Serrano, M., & Kroemer, G. (2013). The hallmarks of aging. *Cell*, *153*(6), 1194–1217.

Magalhães, D., Vilas-Boas, A. A., Teixeira, P., & Pintado, M. (2023). Functional ingredients and additives from lemon by-products and their applications in food preservation: A review. *Foods*, *12*(5), 1095; https://doi.org/10.3390/foods12051095

Mahato, N., Sinha, M., Sharma, K., Koteswararao, R., & Cho, M. H. (2019). Modern extraction and purification techniques for obtaining high purity food-grade bioactive compounds and value-added co-products from citrus wastes. *Foods*, *8*(11), 523; https://doi.org/10.3390/foods8110523

Martínez-Zamora, L., Cano-Lamadrid, M., Artés-Hernández, F., & Castillejo, N. (2023). Flavonoid extracts from lemon by-products as a functional ingredient for new foods: A systematic review. *Foods (Basel, Switzerland)*, *12*(19), 3687; https://doi.org/10.3390/foods12193687

Massey, L. K., Mah, A. L., Ford, D. L., Miller, J., Liang, J., Doong, H., & Monteiro, M. J. (2004). Overexpression of ubiquilin decreases ubiquitination and degradation of presenilin proteins. *Journal of Alzheimer's Disease*, *6*(1), 79–92.

Matsuzaki, K., & Ohizumi, Y. (2021). Beneficial effects of citrus-derived polymethoxylated flavones for central nervous system disorders. *Nutrients*, *13*(1), 145; https://doi.org/10.3390/nu13010145

Matsuzaki, K., Nakajima, A., Guo, Y., & Ohizumi, Y. (2022). A narrative review of the effects of citrus peels and extracts on human brain health and metabolism. *Nutrients*, *14*(9), 1847; https://doi.org/10.3390/nu14091847

Minato, K., Miyake, Y., Fukumoto, S., Yamamoto, K., Kato, Y., Shimomura, Y., & Osawa, T. (2003). Lemon flavonoid, eriocitrin, suppresses exercise-induced oxidative damage in rat liver. *Life Sciences*, *72*(14), 1609–1616; https://doi.org/10.1016/S0024-3205(02)02443-8

Miyake, Y., Sakurai, C., Usuda, M., Fukumoto, S., Hiramitsu, M., Sakaida, K., . . . & Kondo, K. (2006). Difference in plasma metabolite concentration after ingestion of lemon flavonoids and their aglycones in humans. *Journal of Nutritional Science and Vitaminology*, *52*(1), 54–60.

Mohajeri, M. H., Troesch, B., & Weber, P. (2015). Inadequate supply of vitamins and DHA in the elderly: Implications for brain aging and Alzheimer-type dementia. *Nutrition*, *31*(2), 261–275.

Morishita, N., Ogihara, S., Endo, S., Mitsuzumi, H., & Ushio, S. (2020). Effects of glucosyl hesperidin on skin blood flow and temperature: A randomized, double-blind, placebo-controlled, crossover study. *Shinryo-to-shinyaku (Medical Consultation & New Remedies Web)*, *57*(2), 129–134.

Nascimento, A. L., Medeiros, P. O., Pedrão, L. F., Queiroz, V. C., Oliveira, L. M., Novaes, L. S., . . . & Falquetto, B. (2022). Oxidative stress inhibition via apocynin prevents medullary respiratory neurodegeneration and respiratory pattern dysfunction in a 6-hydroxydopamine animal model of Parkinson's disease. *Neuroscience*, *502*, 91–106.

Nogata, Y., Sakamoto, K., Shiratsuchi, H., Ishii, T., Yano, M., & Ohta, H. (2006). Flavonoid composition of fruit tissues of citrus species. *Bioscience, Biotechnology, and Biochemistry*, *70*(1), 178–192.

Nuzzo, D., Cristaldi, L., Sciortino, M., Albanese, L., Scurria, A., Zabini, F., . . . & Ciriminna, R. (2020). Exceptional antioxidant, non-cytotoxic activity of integral lemon pectin from hydrodynamic cavitation. *ChemistrySelect*, *5*(17), 5066–5071.

Nuzzo, D., Picone, P., Giardina, C., Scordino, M., Mudò, G., Pagliaro, M., . . . & Di Liberto, V. (2021). New neuroprotective effect of lemon IntegroPectin on neuronal cellular model. *Antioxidants*, *10*(5), 669; https://doi.org/10.3390/antiox10050669

Oboh, G., Olasehinde, T. A., & Ademosun, A. O. (2014). Essential oil from lemon peels inhibit key enzymes linked to neurodegenerative conditions and pro-oxidant induced lipid peroxidation. *Journal of Oleo Science*, *63*(4), 373–381.

Okuyama, S. (2015). Effects of bioactive substances from citrus on the central nervous system and utilization as food material. *Yakugaku zasshi: Journal of the Pharmaceutical Society of Japan*, *135*(10), 1153–1159.

Pagliara, V., Nasso, R., Di Donato, P., Finore, I., Poli, A., Masullo, M., & Arcone, R. (2019). Lemon peel polyphenol extract reduces interleukin-6-induced cell migration, invasiveness, and matrix metalloproteinase-9/2 expression in human gastric adenocarcinoma MKN-28 and AGS cell lines. *Biomolecules*, *9*(12), 833; https://doi.org/10.3390/biom9120833

Pathak, P. D., Mandavgane, S. A., & Kulkarni, B. D. (2017). Fruit peel waste: Characterization and its potential uses. *Current Science*, 444–454.

Paw, M., Begum, T., Gogoi, R., Pandey, S. K., & Lal, M. (2020). Chemical composition of Citrus limon L. Burmf peel essential oil from North East India. *Journal of Essential Oil Bearing Plants*, *23*(2), 337–344.

Peters, R. (2006). Ageing and the brain: This article is part of a series on ageing edited by Professor Chris Bulpitt. *Postgraduate Medical Journal*, *82*(964), 84–88.

Pieracci, Y., Pistelli, L., Cecchi, M., Pistelli, L., & De Leo, M. (2022). Phytochemical characterization of citrus-based products supporting their antioxidant effect and sensory quality. *Foods*, *11*(11), 1550; https://doi.org/10.3390/foods11111550

Putnik, P., Bursać Kovačević, D., Režek Jambrak, A., Barba, F. J., Cravotto, G., Binello, A., . . . & Shpigelman, A. (2017). Innovative "green" and novel strategies for the extraction of bioactive added value compounds from citrus wastes—A review. *Molecules*, *22*(5), 680; https://doi.org/10.3390/molecules22050680

Ramful, D., Bahorun, T., Bourdon, E., Tarnus, E., & Aruoma, O. I. (2010). Bioactive phenolics and antioxidant propensity of flavedo extracts of Mauritian citrus fruits: Potential prophylactic ingredients for functional foods application. *Toxicology*, *278*(1), 75–87.

Reiter, R. J. (1995). Oxidative processes and antioxidative defense mechanisms in the aging brain 1. *The FASEB Journal*, *9*(7), 526–533.

Ribeiro, C. B., Ramos, F. M., Manthey, J. A., & Cesar, T. B. (2019). Effectiveness of Eriomin® in managing hyperglycemia and reversal of prediabetes condition: A double-blind, randomized, controlled study. *Phytotherapy Research*, *33*(7), 1921–1933.

Richter, C. (1995). Oxidative damage to mitochondrial DNA and its relationship to ageing. *The International Journal of Biochemistry & Cell Biology*, *27*(7), 647–653

Rincón-León, F. (2003). *Functional foods. Encyclopedia of Food Sciences and Nutrition* (Second Edition, pp. 2827–2832). Academic Press; https://doi.org/10.1016/b0-12-227055-x/01328-6

Rizza, S., Muniyappa, R., Iantorno, M., Kim, J. A., Chen, H., Pullikotil, P., . . . & Quon, M. J. (2011). Citrus polyphenol hesperidin stimulates production of nitric oxide in endothelial cells while improving endothelial function and reducing inflammatory markers in patients with metabolic syndrome. *The Journal of Clinical Endocrinology & Metabolism*, *96*(5), E782–E792.

Robertson, D. A., Savva, G. M., & Kenny, R. A. (2013). Frailty and cognitive impairment—a review of the evidence and causal mechanisms. *Ageing Research Reviews*, *12*(4), 840–851.

Russo, M., Bonaccorsi, I., Torre, G., Sarò, M., Dugo, P., & Mondello, L. (2014). Underestimated sources of flavonoids, limonoids and dietary fibre: Availability in lemon's by-products. *Journal of Functional Foods*, *9*, 18–26; https://doi.org/10.1016/j.jff.2014.04.004

Rutherford, N. J., Lewis, J., Clippinger, A. K., Thomas, M. A., Adamson, J., Cruz, P. E., . . . & Giasson, B. I. (2013). Unbiased screen reveals ubiquilin-1 and-2 highly associated with huntingtin inclusions. *Brain Research*, *1524*, 62–73.

Satari, B., & Karimi, K. (2018). Citrus processing wastes: Environmental impacts, recent advances, and future perspectives in total valorization. *Resources, Conservation and Recycling*, *129*, 153–167.

Scurria, A., Sciortino, M., Albanese, L., Nuzzo, D., Zabini, F., Meneguzzo, F., . . . & Ciriminna, R. (2021). Flavonoids in lemon and grapefruit IntegroPectin. *ChemistryOpen*, *10*(10), 1055–1058.

Scurria, A., Sciortino, M., Presentato, A., Lino, C., Piacenza, E., Albanese, L., . . . & Ciriminna, R. (2020). Volatile compounds of lemon and grapefruit IntegroPectin. *Molecules*, *26*(1), 51; https://doi.org/10.3390/molecules26010051

Seki, T., Kamiya, T., Furukawa, K., Azumi, M., Ishizuka, S., Takayama, S., . . . & Yaegashi, N. (2013). Nobiletin-rich Citrus reticulata peels, a kampo medicine for Alzheimer's disease: A case series. *Geriatrics & Gerontology International*, *13*(1), 236–238.

Senol, F. S., Ankli, A., Reich, E., & Orhan, I. E. (2016). HPTLC fingerprinting and cholinesterase inhibitory and metal-chelating capacity of various Citrus cultivars and *Olea europaea*. *Food Technology and Biotechnology*, *54*(3), 275–281.

Sesti, F., Liu, S., & Cai, S. Q. (2010). Oxidation of potassium channels by ROS: A general mechanism of aging and neurodegeneration? *Trends in Cell Biology*, *20*(1), 45–51.

Setiawan, I., & Indrawanto, I. S. (2021). The effect of lemon peel (Citrus limon) aromatherapy inhalation as antidepressant on rats using diffuser method. *Jurnal Kesehatan Islam: Islamic Health Journal*, *10*(2), 57–64.

Sp, N., Kang, D. Y., Joung, Y. H., Park, J. H., Kim, W. S., Lee, H. K., . . . & Yang, Y. M. (2017). Nobiletin inhibits angiogenesis by regulating Src/FAK/STAT3-mediated signaling through PXN in ER+ breast cancer cells. *International Journal of Molecular Sciences*, *18*(5), 935; https://doi.org/10.3390/ijms18050935

Specialty Produce. (2023). *India Lemons Information and Facts*; https://specialtyproduce.com/India_Lemons_15803; Accessed March 5, 2023

Stefanatos, R., & Sanz, A. (2018). The role of mitochondrial ROS in the aging brain. *FEBS Letters*, *592*(5), 743–758.

Tinh, N. T. T., Sitolo, G. C., Yamamoto, Y., & Suzuki, T. (2021). Citrus limon peel powder reduces intestinal barrier defects and inflammation in a colitic murine experimental model. *Foods*, *10*(2), 240; https://doi.org/10.3390/foods10020240

Trifunovic, A., & Larsson, N. G. (2008). Mitochondrial dysfunction as a cause of ageing. *Journal of Internal Medicine*, *263*(2), 167–178.

Uddin, M. S., Al Mamun, A., Kabir, M. T., Ahmad, J., Jeandet, P., Sarwar, M. S., . . . & Aleya, L. (2020). Neuroprotective role of polyphenols against oxidative stress-mediated neurodegeneration. *European Journal of Pharmacology*, *886*, 173412; https://doi.org/10.1016/j.ejphar.2020.173412

Ueda, K., Horita, T., & Suzuki, T. (2023). Effects of inhaling essential oils of *Citrus limonum* L., *Santalum album*, and *Cinnamomum camphora* on human brain activity. *Brain and Behavior*, *13*(2), e2889; https://doi.org/10.1002/brb3.2889

Ullah, A., Munir, S., Badshah, S. L., Khan, N., Ghani, L., Poulson, B. G., . . . & Jaremko, M. (2020). Important flavonoids and their role as a therapeutic agent. *Molecules*, *25*(22), 5243; https://doi.org/10.3390/molecules25225243

Uversky, V. N. (2011). Flexible nets of malleable guardians: Intrinsically disordered chaperones in neurodegenerative diseases. *Chemical Reviews*, *111*(2), 1134–1166.

van Leeuwen, F. W., Hol, E. M., & Burbach, P. H. (1998). Mutations in RNA: A first example of molecular misreading in Alzheimer's disease. *Trends in Neurosciences*, *21*(8), 331–335.

Verma, A. K., Singh, S., & Rizvi, S. I. (2023). Aging, circadian disruption and neurodegeneration: Interesting interplay. *Experimental Gerontology*, *172*, 112076; https://doi.org/10.1016/j.exger.2022.112076

Warraich, U.-e.-A., Hussain, F., & Kayani, H. U. R. (2020). Aging-Oxidative stress, antioxidants and computational modeling. *Heliyon*, *6*(5), e04107; https://doi.org.10.1016/j.heliyon.2020.e04107

Wedamulla, N. E., Fan, M., Choi, Y. J., & Kim, E. K. (2022). Citrus peel as a renewable bioresource: Transforming waste to food additives. *Journal of Functional Foods*, *95*, 105163; https://doi.org/10.1016/j.jff.2022.105163

Wellington, C. L., & Hayden, M. R. (2000). Caspases and neurodegeneration: On the cutting edge of new therapeutic approaches. *Clinical Genetics*, *57*(1), 1–10.

World Health Organization. (2007). Neurological disorders affect millions globally. *International Journal of Health Care Quality Assurance*, *20*(4); https://doi.org/10.1108/ijhcqa.2007.0622dab.001

Xi, W., Lu, J., Qun, J., & Jiao, B. (2017). Characterization of phenolic profile and antioxidant capacity of different fruit part from lemon (*Citrus limon* Burm.) cultivars. *Journal of Food Science and Technology*, *54*, 1108–1118; https://doi.org/10.1007/s13197-017-2544-5

Yadav, P. (2022). Explained: India's lemon crisis and why the prices are very high. *IT News*, April 18; https://www.indiatimes.com/explainers/news/indias-lemon-crisis-and-why-the-prices-are-very-high-567241

Yamada, S., Shirai, M., Ono, K., Teruya, T., Yamano, A., & Tae Woo, J. (2021). Beneficial effects of a nobiletin-rich formulated supplement of Sikwasa (*C. depressa*) peel on cognitive function in elderly Japanese subjects; A multicenter, randomized, double-blind, placebo-controlled study. *Food Science & Nutrition*, *9*(12), 6844–6853.

Yousuf, S., Emad, S., ur Rehman, M. M., Batool, Z., Qadeer, S., Sarfaraz, Y., . . . & Perveen, T. (2023). Dose dependent effects of lemon peel oil on oxidative stress and psychological behaviors in rats. *Pakistan Journal of Zoology*, *55*(4), 1627; https://doi.org/10.17582/journal.pjz/20211231041242

Zappia, A., Spanti, A., Princi, R., Imeneo, V., & Piscopo, A. (2023). Evaluation of the efficacy of antioxidant extract from lemon by-products on preservation of quality attributes of minimally processed radish (*Raphanus sativus* L.). *Antioxidants*, *12*(2), 235; https://doi.org/10.3390/antiox12020235

Zemlan, F. P., Thienhaus, O. J., & Bosmann, H. B. (1989). Superoxide dismutase activity in Alzheimer's disease: Possible mechanism for paired helical filament formation. *Brain Research*, *476*(1), 160–162.

Zhang, K. Y., Yang, S., Warraich, S. T., & Blair, I. P. (2014). Ubiquilin 2: A component of the ubiquitin–proteasome system with an emerging role in neurodegeneration. *The International Journal of Biochemistry & Cell Biology*, *50*, 123–126.

Zhao, Z. (2019). Iron and oxidizing species in oxidative stress and Alzheimer's disease. *Aging Medicine*, *2*(2), 82–87.

Zhou, W., Fukumoto, S., & Yokogoshi, H. (2009). Components of lemon essential oil attenuate dementia induced by scopolamine. *Nutritional Neuroscience*, *12*(2), 57–64.

Zhou, Z., Yan, Y., Li, H., Feng, Y., Huang, C., & Fan, S. (2023). Nomilin and its analogues in citrus fruits: a review of its health promotion effects and potential application in medicine. *Molecules*, *28*(1), 269; https://doi.org/10.3390/molecules28010269

Processed *Moringa oleifera* (Lam)

Source of Bioactive and Functional Compounds

10

Annaelle Hip Kam, Shreynish Joy Mawooa, and Vidushi S. Neergheen

10.1 INTRODUCTION

Moringa oleifera Lam. has a rich and long history in the Indian pharmacopoeia and is frequently consumed throughout Asian countries. The origin of *M. oleifera* is believed to be Agra and Oudh, which is in the northwest sub-Himalayan region of India (Sujatha and Patel, 2017). Historical evidence reports the use of *M. oleifera* for different purposes. The whole plant—leaves, roots, bark, flowers, and seeds—has been identified against more than 300 diseases in Ayurvedic medicine and is also registered in the Chinese herbal medicine dictionary in China (Meireles et al., 2020). *M. oleifera*, known as "Nebedaye" in many African dialects, meaning "never die," is a panacea (Matic et al., 2018). Maurian warriors used the leaf extracts for pain and stress relief during wartime, while the ancient Egyptian kings and queens used the leaves and fruits as part of their diet to improve skin health and mental fitness (Sujatha and Patel, 2017; Senthilkumar et al., 2018).

Despite the numerous traditional uses of *M. oleifera*, its functionality became known only in the 1990s. Scientific reports describe its attractive nutritional value and notable biological potential, including anti-inflammatory, antioxidant, and immune-boosting effects (Meireles et al., 2020). With the growing interest in superfoods in recent years, a new market for *M. oleifera* has emerged. The global market for *M. oleifera* in 2020 was 6.9 billion USD and is expected to increase by 9.5% at a compound annual growth rate (CAGR) by 2028 (Patil et al., 2022). According to Fortune Business Insight (2022), the global *M. oleifera* product market was USD 7.08 billion in 2020 and was slated to more than double to USD 14.80 billion in 2028 (9.63% CAGR). Polaris Market Research (2022) conservatively estimated the global *M. oleifera* product market at USD 3.06 billion in 2021 and expected it to reach USD 6.70 billion by 2030 at 9.5% CAGR during 2022–2030. A more bullish estimate is provided by Straits Research (2022), valuing the global *M. oleifera* products market size at USD 7.71 billion in 2021 and projecting it to reach USD 17.55 billion by 2030 (9.57%

DOI: 10.1201/9781003352341-10

CAGR 2022–2030). All these market reports indicate substantial growth of the global *M. oleifera* products market. The primary drivers for demand are the common inclusion of *M. oleifera* products in dietary supplements and growing awareness of its therapeutic advantages. All parts of this plant (flowers, seeds, pods, leaves, gum, and bark) have the ability to treat vitamin and mineral deficiencies, encourage normal blood sugar levels, support healthy cardiovascular and immune systems, enhance the body's anti-inflammatory mechanisms, enrich anemic blood, and neutralize free radicals.

The leaves category dominated the market (45.2% maximum revenue share) due to their abundant minerals, vitamins, and other important phytochemicals (i.e., tannins, sterols, terpenoids, flavonoids, saponins, alkaloids, and anthraquinones). The seeds accounted for 34% market share in 2020 (Fortune Business Insight, 2022). The leaves are used in various industries, including food and beverage, pharmaceuticals, cosmetics, and personal care. The food and beverage segment accounts for a significant share (53.4%) of the revenue because of its use in ready-to-cook products (*M. oleifera* powder, *M. oleifera* tea, *M. oleifera* chocolates, cakes, juices, and lattes). The organically cultivated segment accounted for the largest market share (50% in 2021) and is estimated to continue holding the largest share during 2021–2030 according to The Brainy Insight (2021).

The *M. oleifera* oil segment is estimated to grow with the fastest CAGR of 8.9% during 2021–2026, owing to its advantages for moisturizing and cleansing the skin, boosting its utilization against acne, and acting as a hair-moistening healing agent. *M. oleifera* seed oil is rich in essential fatty acids, making it a perfect moisturizing agent and a healing and smoothing emollient for dry, rough skin; it can be used in therapeutic massages. *M. oleifera* oil is comparable to olive oil in terms of its fatty acids. *M. oleifera* oil is used in lotions, creams, balms, scrubs, body oils, and hair care formulations. Consequently, the tradition of using *M. oleifera* in cosmetics drives the *M. oleifera* products market expansion.

According to the Agricultural and Processed Food Products Export Development Authority (APEDA), the North American market demand for *M. oleifera* ingredients is expected to surpass 5 billion USD by 2025, while Europe's should exceed 2 billion USD. For Asia-Pacific countries, which are led by India, the market size may increase by around 8%. India dominates the current market, supplying more than 80% of the global demand, as it is the main producer of *M. oleifera*, with an annual production of 2.2 million tonnes of drumsticks using 43,600 ha, leading to a productivity of ~51 tonnes/ha. Andhra Pradesh leads both in area and production (15,665 ha), followed by Tamil Nadu (13,042 ha) and Karnataka (10,280 ha), with other states occupying ~4,613 ha in *M. oleifera* production. The unique selling point of India for *M. oleifera* is its low price and leaf powder standardization (protein, omega-9, and vitamin E contents).

10.2 CHEMICAL COMPOSITION

The exploration of *M. oleifera* involves an intricate understanding of its chemical composition. This section delves into the foundational aspects of this plant, unraveling the nuances of both nutritional and phytochemical components.

10.2.1 Nutritional Composition

The high nutritional value of *M. oleifera* is very much revered; it contains 7 times, or 200 mg/100 g, the amount of vitamin C of an orange; 10 times more vitamin A than a carrot; 15 times the content of potassium than a banana; 36 times more magnesium than an egg; and 25 times the iron in spinach after processing, such as parboiling or in powdered form (Jagadeesan et al., 2020). Pilotos et al. (2020) reported that 5 to 8 grams of the leaf powder can provide a child 40% of the calcium, 23% of the iron, and 14% of the protein of its daily recommended allowance. Such nutritionally superior food can be used to combat malnutrition and assure nutritional security. Consequently, *M. oleifera* has been promoted as an ally against malnutrition by the World Health Organization.

TABLE 10.1 Nutritional composition (per 100 g) of *M. oleifera* leaves

	LEAVES		
NUTRIENTS	*FRESH*	*DRIED*	*POWDER*
Protein (g)	6.7	29.4	27.1
Fat (g)	1.7	5.2	2.3
Carbohydrate (g)	12.5	41.2	38.2
Fiber (g)	0.9	12.5	19.2
Minerals (mg)			
Calcium	440	2185	2003
Potassium	259	1236	1324
Iron	0.85	25.6	28.2
Magnesium	42	448	368
Phosphorous	70	252	204
Copper	0.07	0.49	0.57
Vitamins (mg)			
Vitamin A	1.28	3.63	16.3
Vitamin B1	0.06	2.02	2.64
Vitamin B2	0.05	21.3	20.5
Vitamin B3	0.8	7.6	8.2
Vitamin C	220	15.8	17.3
Vitamin E	448	10.8	113
Chlorophyll	80	45	1268

Adapted from Kashyap et al. (2022)

Dried and powdered leaves generally have higher nutritional content than fresh leaves (Table 10.1). For example, vitamins B2 and B1 contents of the dried and powdered leaves were over 400-fold higher and over 34 times the amount found in fresh leaves. Similarly, iron and magnesium contents of the dried and powdered leaves were over 30 and 8 times those in fresh leaves, respectively. The powdered leaves had a remarkably high amount of fiber (21 times that in fresh leaves). Hence, dried and powdered leaves are far superior in nutritional content to the highly perishable fresh leaves.

The nutritional composition of *M. oleifera* exhibits significant differences between its fresh leaves and the powdered form. These variations can be attributed to several factors, including the drying process, storage conditions, and the inherent biochemical changes that occur during the transformation from fresh leaves to powder. Different drying techniques, such as solar drying, hot-air drying, and shade drying, can lead to variations in moisture content and the preservation of bioactive compounds. The drying process can enhance the concentration of certain nutrients in Moringa powder. For example, the protein, fat, and fiber contents can increase due to the removal of water, which concentrates these macronutrients (Rustagi et al., 2023). In addition, optimal storage practices can help to maintain the nutritional integrity of Moringa powder over time. The drying process can lead to a reduction of heat labile nutrients, which are highly susceptible to degradation (Fejér et al., 2019). Conversely, certain minerals may become more concentrated in the powdered form due to the loss of water content, resulting in a higher mineral density per gram of powder compared to fresh leaves (Fatmawati et al., 2023). The tannins and polyphenols from Moringa have beneficial properties, but they can also inhibit the bioavailability of certain minerals, particularly iron (Idohou-Dossou et al., 2011). The processing of Moringa into powder may alter the levels of these antinutritional factors, potentially enhancing the bioavailability of minerals in the powdered form compared to fresh leaves (Srimiati and Agestika, 2022).

TABLE 10.2 Mineral Composition of five Indian *M. oleifera* varieties

COMPOSITION	*JAFFNA*	*PKM-1*	*PKM-2*	*ODC*	*CONVENTIONAL*
MINERALS (%)					
Potassium	15.14	11.07	8.50	6.77	4.94
Sodium	10.98	9.77	8.90	7.70	6.61
Calcium	5.90	4.96	3.96	3.02	2.32
Carbon	5.70	4.99	4.17	3.79	2.89
Nitrogen	2.32	2.08	1.68	1.34	0.99
Phosphorous	1.11	0.86	0.78	0.63	0.51
Sulfur	1.13	0.84	0.76	0.58	0.43

Adapted from Farooq et al. (2021)

The nutritional content of leaves depends on *M. oleifera* varieties (Table 10.2). Farooq et al. (2021) evaluated the leaf composition of five common Indian *M. oleifera* varieties (conventional, PKM-1 [Periyakulam-1], PKM-2 [Periyakulam-2], ODC [Oddanchatram], and Jaffna). Ethanol (80%) leaf extracts were prepared from harvested fresh leaves of young (3-month) and mature (15-month-old) plants, shade-dried, powdered, and sieved through 297 microns. Mature leaves had higher sugar, protein, and total chlorophyll contents compared to young leaves for all varieties.

Saini et al. (2014) reported the fatty acid profile in vegetative and reproductive parts of eight commercially cultivated Indian *M. oleifera* cultivars. The leaves were high in α-linolenic acid (49%–59% of total fatty acids), palmitic and linoleic acids (16%–18% and 6%–13%, respectively). Total lipid contents ranged from 1.9% to 4.8% for flowers and leaves. Fresh leaves were obtained from 2-year-old plants grown in the same field in Mysore (Central Food Technologies Research Institute) in 2012, except for the PAVM-1 cultivar (from a farmer in Tamil Nadu). Table 10.3 represents the range and average values of the fatty acid composition of the eight cultivars evaluated by Saini et al. (2014).

Linoleic, α-linolenic, and erucic acids showed the highest genetic variability in the fatty acid composition of leaves from different cultivars. For example, the cultivars were easily segregated into three groups based on linoleic acid content (high [>11%], average [~11%] and low [<11%]). The leaves are rich in polyunsaturated fatty acids (~66% of the total fatty acid) due to the high amounts of α-linolenic and linoleic acids. Oleic, palmitoleic, and erucic acids constitute the monounsaturated fatty acids of the leaves. The high linolenic acid content primarily accounts for the Cox value at the upper limit of high safflower oils. Moreover, the thrombogenic index was lower, and the atherogenic index higher, than those reported for sunflower and olive oils (0.28 and 0.32 for thrombogenic index, and 0.07 and 0.14 for atherogenic index, respectively (Oomah, 2023 and references therein). These atherothrombotic indices are important for type 2 diabetes mellitus (T2DM), an inflammatory condition associated with a high prevalence of thrombotic cardiovascular disease. *M. oleifera* leaves are a far superior source of α-linolenic (ALA) oil compared to other ALA-rich oils such as flaxseed, walnut, and canola oils (28.8%, 9.1%, and 9.1% ALA, respectively). The high polyunsaturated fatty acid content confers value-added benefits to *M. oleifera* leaves.

10.2.2 Phytochemical Composition

Moringa oleifera, revered for its diverse phytochemical composition, holds significant promise for the production of functional foods, pharmaceuticals, and cosmeceutical agents (Saucedo-Pompa et al., 2018). Rooted in centuries of folkloric medicine, this plant is a cornerstone in the treatment of various ailments, with recognized pharmacological uses deeply embedded in Ayurvedic, traditional Chinese, and Unani medicine systems.

TABLE 10.3 Fatty acid composition (%) of *M. oleifera* leaves

FATTY ACID	*AVERAGE*	*RANGE*
Caprylic acid (C8:0)	0.25	0.02–0.57
Capric acid (C10:0)	0.02	0.01–0.04
Lauric acid (C12:0)	0.19	0.07–0.39
Tridecyclic acid (C13:0)	0.38	0.32–0.51
Myristic acid (C14:0)	0.95	0.62–1.62
Palmitic acid (C16:0)	17.68	16.9–18.4
Palmitoleic acid (C16:1)	2.77	2.52–3.17
Stearic acid (C18:0)	3.94	2.96–5.24
Oleic acid (C18:1)	3.53	2.96–4.87
Linoleic acid (C18:2)	10.27	6.46–13.63
Arachidic acid (C20:0)	0.66	0.16–2.11
Linolenic acid (C18:3)	55.37	49.6–59.5
Behenic acid (C22:0)	0.59	0.41–0.76
Erucic acid (C22:1)	2.12	0.67–5.39
Lignoceric acid (C24:0)	1.29	1.08–1.57
ΣSaturated fatty acids (SFA)	25.95	24.6–27.95
ΣUnsaturated fatty acids (UFA)	74.05	72.1–75.4
ΣMonounsaturated fatty acids (MUFA)	8.42	6.77–11.28
ΣPolyunsaturated fatty acids (PUFA)	65.64	63.2–67.4
PUFA:SFA	2.54	2.32–2.68
PUFA:MUFA	8.01	5.68–9.71
SFA:UFA	0.35	0.33–0.39
Total lipids	4.65	4.55–4.82
Oxidizability (Cox) value	13.05	12.2–13.6
Atherogenic index (A1)	0.29	0.27–0.32
Thrombogenic index (TI)	0.25	0.23–0.27

Adapted from Saini et al. (2014)

All parts of the plant—leaves, flowers, bark, roots, and seeds—have contributed to therapeutic practices over generations (Milla et al., 2021). Notably, phenolic compounds have emerged as the most reported bioactive compounds, closely followed by glucosinolates (Rani et al., 2018).

The phytochemical composition, akin to its nutritional content, exhibits variations among different varieties, influencing their respective bioactivities. In this context, Jaffna stands out with the highest phytochemical content, including β-sitosterol, quercetin, kaempferol, and moringin, correlating with robust anti-cancer activity against HepG2 cancer cells (Table 10.4). Variability in kaempferol and quercetin levels (13% and 7%, respectively) among different varieties contributes to distinctions in radical scavenging activities (10% FRAP variation) and cytotoxicity against liver cancer cells. Jaffna and conventional extracts (aqueous and ethanol leaf and seed extracts) exhibited the strongest and weakest antibacterial activity against *E. coli*, *B. subtilis*, *P. aeruginosa*, and *S. aureus*, respectively. Further exploration into the subsections on flavonoids, phenolic acids, and glucosinolates will shed light on their specific roles in *M. oleifera*'s multifaceted health benefits.

10.2.2.1 Flavonoids

Flavonoids are the major class of phenolic compounds in *M. oleifera* (Rodríguez-Pérez et al., 2015). These compounds have been reported for their numerous pharmacological properties, favorable to various

TABLE 10.4 Composition and bioactivity of five Indian *M. oleifera* varieties

COMPOSITION	*JAFFNA*	*PKM-1*	*PKM-2*	*ODC*	*CONVENTIONAL*
BIOACTIVES (%)					
β-Sitosterol	0.244	0.236	0.204	0.11	0.056
Quercetin	0.216	0.155	0.127	0.073	0.03
Kaempferol	0.013	0.012	0.004	0.002	0.001
Moringin	0.063	0.057	0.046	0.043	0.03
Activity					
FRAP (30 μg/mL)	29.39	22.33	13.18	4.41	2.86
MTT (IC_{50}) mg/mL	0.15	0.35	0.65	0.9	1.22
Cytotoxicity (%)*	48.37	43.97	36.13	27.53	22.37

Adapted from Farooq et al. (2021); FRAP—ferric reducing antioxidant power; MTT assay on human liver cell line (HepG2); *Cytotoxicity at 0.25 mg/mL.

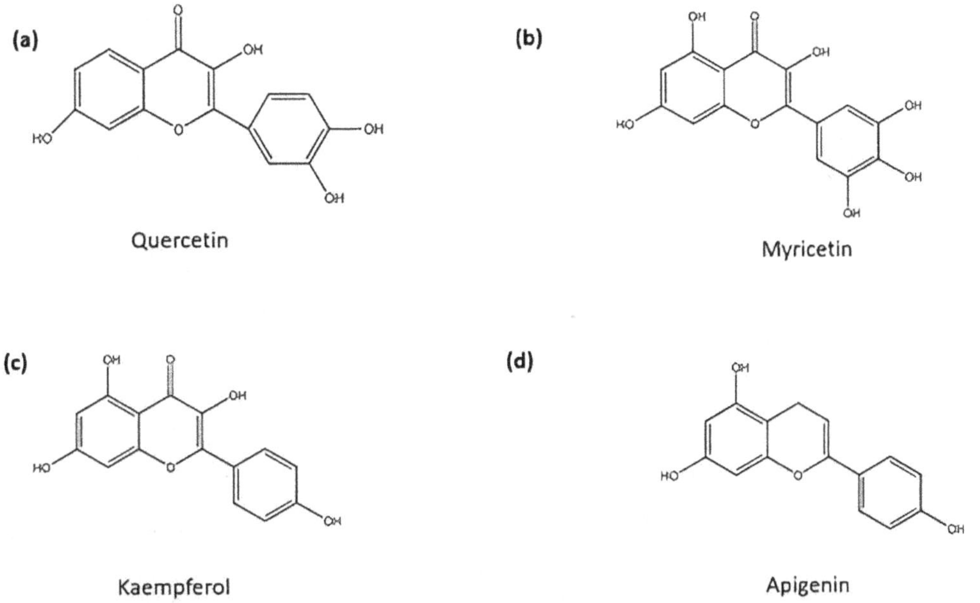

(a) Quercetin

(b) Myricetin

(c) Kaempferol

(d) Apigenin

FIGURE 10.1 Chemical structure of most common flavonoids present in *M. oleifera*. (a) Quercetin (PubChen CID – 5280343), (b) Myricetin (PubChen CID – 5281672), (c) Kaempferol (PubChen CID – 5280863), and (d) Apigenin (PubChen CID – 5280443). (Adapted from Rani, Husain and Kumolosasi, 2018)

diseases, including cancer, neurodegenerative diseases, atherosclerosis, diabetes, and cardiovascular diseases. These benefits have been associated with their broad spectrum of health-promoting effects—antioxidative, anti-inflammatory, antibacterial, anti-mutagenic, and anticarcinogenic properties, coupled with their capacity to modulate key cellular enzyme functions.

Twenty-six flavonoids have been identified from the plant, including quercetin, isoquercetin, rutin, kaempferol, apigenin, astragalin, genistein, daidzein, luteolin, and myricetin (Rani et al., 2018). Figure 10.1 illustrates the main flavonoids in *M. oleifera*. These flavonoids mainly occur, as aglycones, that is, in their free form, or as glycosides bonded to a sugar moiety (Makita et al., 2016). Glycone structures might greatly modulate the bioactivities of flavonoids (Juergenliemk et al., 2003; Soundararajan et al., 2008). Nouman et al. (2016) identified 14 different flavonoids from 7 *M. oleifera* cultivars, where the total flavonoids corresponded to 47%, 30%, and 20% for quercetin, kaempferol, and apigenin derivatives, respectively. These findings corroborate results from Rodríguez-Pérez et al. (2015), who also reported the flavonoids: quercetin, kaempferol, and apigenin derivatives predominantly at 46%, 34%, and 7.7%, respectively. Sultana and Anwar (2008) described the main flavonoids found in *M. oleifera* leaves as myricetin, quercetin, and kaempferol at 5804, 281, and 40 mg/kg dry matter, respectively. Differences in the observed flavonoid contents of *M. oleifera* leaves are most likely due to environmental variations, as well as cultivar and genetic variability (Brunetti et al., 2013). Flavonoids such as quercetin and kaempferol (~47% and 30%, respectively) are widely present in various *M. oleifera* tree plant parts, except the roots and seeds. Quercetin occurs in the glycoside form (quercetin-3-*O*-glucoside) in dried *M. oleifera* leaves. It has high antioxidant activity that has been investigated for its antidiabetic, hypotensive, and hypolipidemic properties in rats. Kaempferol is associated with cancer risk reduction. Most *M. oleifera* isothiocyanates confer anti-inflammatory activity (Mahato et al., 2022).

The flavonoid compositions of *M. oleifera* extracts also depend on the extraction techniques and conditions. Lin et al. (2021) compared the phenolic and flavonoid composition of *M. oleifera* leaves purchased from the New Delhi local market extracted with ethanol (52%) using optimized ultrasonic method with stirring, Soxhlet, and microwave-assisted extraction (Table 10.5).

Optimal ultrasonic-assisted extraction (188 W power, 40 mL/g liquid-to-solid ratio, 30 °C, 20 min) improved analyte (phenolics and antioxidants) extractability, primarily due to cell wall disruption.

Soxhlet extracted the least amount of total flavonoids and antioxidant activities (compounds), due to high temperature (90 °C for 3 h). The major flavonoids in *M. oleifera* extracts were D-(+)-catechin, hyperoside, kaempferol-3-*O*-rutinoside, and astragalin (for the OUAE extract). Pollini et al. (2020) also optimized ultrasound conditions (hydroalcoholic solvent [50% methanol], 60:1 L/S ratio, 45 °C, 35 min)

TABLE 10.5 Phenolics and flavonoid composition (mg/g dry weight) of *M. oleifera* leaves

ANALYTES	*OUAE*	*SAE*	*SOAE*	*MAE*
PHENOLIC ACIDS				
Chlorogenic acid	3.25	1.61	0.65	2.46
Neochlorogenic acid	0.12	4.24	1.44	6.47
Rosmarinic acid	0.77	0.34	0.17	0.53
Flavonoids				
Catechin	10.92	3.69	1.80	6.36
Hyperoside	10.26	4.59	5.27	6.79
Rutin	1.09	0.8	0.84	1.20
Kaemferol-3-*O*-rutinoside	4.85	2.20	0.96	3.21
Astragalin	2.81	1.33	1.67	1.95
Quercetin	0.33	0.11	0.1	0.23
Apigenin	0.12	0.01	ND	0.02
Kaempferol	0.21	0.03	0.01	0.03
Polydatin	0.08	0.01	0.02	0.05
TFC (mg RE/g DW)	43.96	33.25	23.22	36.13
AA (µmol TE/g DW)				
ABTS	492	320	228	414
DPPH	472	282	170	418
OH	86.48	60.36	23.53	79.09

Adapted from Lin et al. (2021); OUAE—optimal ultrasonic-assisted extraction

SAE—stirring assisted extraction, SOAE—Soxhlet extraction, MAE—microwave

Extraction, TFC—total flavonoid content expressed in mg rutin equivalents/g dry weight, AA—antioxidant activity expressed in micromole Trolox equivalents/g dry weight

to extract flavonoids from Italian-origin *M. oleifera* leaves. Eight flavonols were obtained by this method (μg/g DM): quercetin 3-*O*-galactoside (271.8), quercetin 3-*O*-glucoside (272.8), quercetin 3-*O*-(6'-*O*-malonyl)-β-D-glucoside (293.9), quercetin 3-*O*-rhamnoside (216.4), kaempferol 3-*O*-galactoside (217.4), kaempferol 3-*O*-glucoside (218.4), quercetin (65.4), and kaempferol (30.1) (Pollini et al., 2020).

Indian *M. oleifera* leaves from plants grown in Gujarat were evaluated for flavonoid content (Shervington et al., 2018). Acid (0.10 M hydrochloric acid) extracted leaves contained 86.95 and 6.14 mg/kg of quercetin and kaempferol at 3 h extraction. Flavonoid content increased with extended extraction time: 257, 447, and 126 mg/kg at 13 h, and 292, 1099, and 133 mg/kg of myricetin, quercetin, and kaempferol, respectively, at 24 h extraction (Shervington et al., 2018).

Farooq et al. (2021), evaluating the leaf composition of five common Indian *Moringa oleifera* varieties, reported that Jaffna had the highest and the conventional variety the lowest quercetin (0.216% vs 0.03%) and kaempferol (0.013% vs 0.001%) contents. PKM-1 and PKM-2 contained 0.155% and 0.127% quercetin and 0.012% and 0.004% kaempferol. Flavonoids of cultivars PKM-1 and PKM-2 grown in several African countries were higher than those grown in India. Average quercetin and kaempferol values were 0.98% (range 0.51–1.26) and 0.21% (range 0.11–0.30) for PKM-1, and 0.82% (range 0.56–1.03) and 0.47% (range 0.20–0.66) for PKM-2 (Coppin et al. 2013).

M. oleifera seeds also contain significant amounts of phenolic constituents responsible for their antibacterial activities. Dehulled, defatted *M. oleifera* seed flour was separated into free (methanol-soluble) and bound (hydrolyzed methanol-insoluble residue) phenolics (Singh et al., 2013). The total phenolic and flavonoid contents of the bound phenolic extract of dehulled, defatted *M. oleifera* seed flour were significantly higher than those of the free phenolic extract (4173 vs 780 mg gallic acid equivalents [GAE]/100 g DW phenolic [>5-fold], and 234 vs 133 mg catechin equivalents [CAE]/100 g DW), resulting in higher antioxidant capacity (5.62 vs 1.82 g ascorbic acid equivalent/100 g of seed flour extract for free phenolic extract [>3-fold]). Moreover, the bound phenolic extract contained higher amounts of catechin (749 vs 0.41 mg/100 g for free phenolic extract), epicatechin (81.4 vs 8.16 mg/100 g), *p*-coumaric acid (0.74 vs 0.50 mg/100 g), and quercetin (1.88 vs 0.14 mg/100 g). Other phenolic constituents were present only in the free phenolic extracts: ferulic acid, vanillin, caffeic acid, protocatechuic acid, and cinnamic acid (0.24, 0.87, 3.79, 0.34, and 0.25 mg/100 g, respectively). Furthermore, gallic acid content of the free phenolic extract was higher than that of the bound phenolic extract (18.10 vs 1.59 mg/100 g). The bound phenolic fraction was more effective than the free phenolic extract against *Bacillus cereus*, *Staphylococcus aureus*, *Escherichia coli*, and *Yersinia enterocolitica* (Singh et al., 2013).

Methanol leaf extract of *M. oleifera* (grown in Puducherry, South India) contained 22.16 mg quercetin equivalent (QE)/g of total flavonoid and 6.27 mg GAE/100 g of total phenolic compounds (Shanmugavel et al., 2018). Methanol leaf and flower extract of *M. oleifera* (grown in Goa) contained 4.44 and 4.41 mg/mL of total flavonoid and total phenolic (2.28 and 1.08 mg/mL for leaf and flower, respectively) (Sankhalkar and Vernekar, 2016).

10.2.2.2 Phenolic Acids

Phenolic acids have also been identified in *M. oleifera* (Figure 10.2); its leaves include 11 phenolic acids, such as gallic, caffeic, chlorogenic, ellagic, o-coumaric, and p-coumaric acids, and their derivatives, including caffeoylquinic, feruloylquinic, and coumaroylquinic acids and their isomers (Saucedo-Pompa et al., 2018). Nouman et al. (2016) reported 77 to 187 μg/g dry leaves of phenolic acids in seven different *M. oleifera* cultivars. Other studies reported 36.4% coumaroylquinic acid and 45.5% caffeoylquinic acid as the main phenolic acids in *M. oleifera* (Saucedo-Pompa et al., 2018). Hydroxybenzoic acids (gallic acid and p-hydroxybenzoic acid) were the main phenolic acids in *M. oleifera* plants from Sudan (Al Juhaimi et al., 2017). Vergara-Jimenez et al. (2017) reported gallic acid as the most abundant phenolic acid (1.03 mg/g DW), followed by chlorogenic (0.02–0.49 mg/g DW) and caffeic acids (0.41 mg/g DW). These studies emphasize the influence of climate, geographical area, and soil constitution on the secondary metabolite composition of Moringa plants.

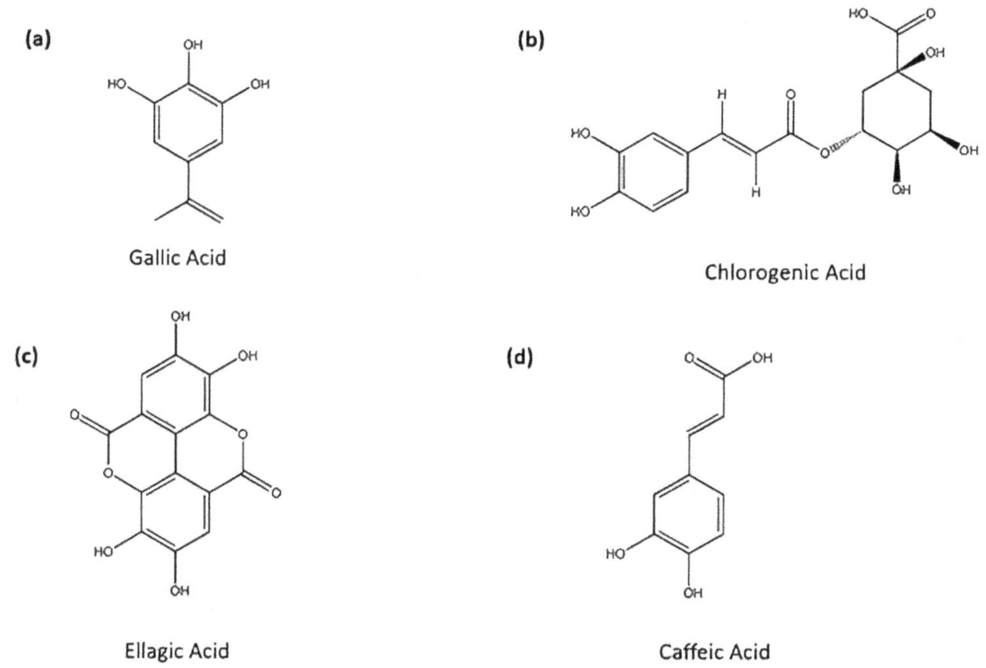

(a) Gallic Acid

(b) Chlorogenic Acid

(c) Ellagic Acid

(d) Caffeic Acid

FIGURE 10.2 Chemical structures of the most common phenolic acids present in *M. oleifera*: (a) gallic acid (PubChem CID – 370), (b) chlorogenic acid (PubChem CID – 1794427), (c) ellagic acid (PubChem CID – 5281855) and (d) caffeic acid (PubChem CID – 689043).

Source: Adapted from Rani et al. (2018).

Fresh leaves were collected from 3-year-old, field-grown *M. oleifera* (cv. PKM-1) plants at the Central Food Technological Research Institute, Mysore, India in 2012 (Saini et al., 2014). Total phenolic content of leaves varied significantly (2.9- to 3.3-fold increase) among the various drying methods (512 mg GAE/100 g for fresh leaves to 1484–1672 mg GAE/100 g). Drying also increased DPPH radical scavenging activity (1230 to 272 μg/mL EC_{50}) and lipid peroxidation activity (1020 to 290 μg/mL EC_{50}).

The therapeutic potential of phenolic acids has been well established. During recent years, gallic acid has received increasing attention for its evident pharmacological effects, including anti-tumor, antibacterial, antidiabetes, anti-obesity, antimicrobial, and protective cardiovascular effects. In *in vitro* and *in vivo* models, gallic acid exhibited considerable antiproliferative activities on cancer cells, via modulation of genes encoding cell cycle, metastasis, angiogenesis, and apoptosis (Kahkeshani et al., 2019). Gallic acid also targets several other beneficial pathways, including reduced expression of proinflammatory mediators, such as tumor necrosis factor (TNF)-α and inducible NO synthase (inos), interferon-γ (INF-γ), interleukins (IL)-1β, IL-6, IL-17, IL-21, IL-23, and cyclooxygenase (Cox)-2.

Chlorogenic acid, reported as one of the main phenolic acids present in Moringa leaves, also has numerous protective effects against metabolic disorders.

This phenolic acid exerts pivotal roles on crucial glucose and lipid biomarkers related to obesity, cardiovascular diseases, cancer, and hepatic diseases (Vergara-Jimenez et al., 2017). *In vivo* studies have shown that chlorogenic acid can regulate several glucose metabolism biomarkers and exert neuroprotective effects by activating the AKT-Bad signaling pathway (Shah et al., 2022).

10.2.2.3 Glucosinolates

The presence of glucosinolates in *M. oleifera* partially explains its amazing medicinal properties. Unique glucosinolates have been documented from the plant; the most abundant is 4-O-(α-L-rhamnopyranosyloxy)-benzyl glucosinolate, also known as glucomoringin. Benzyl glucosinolate (glucotropaeolin) is the most

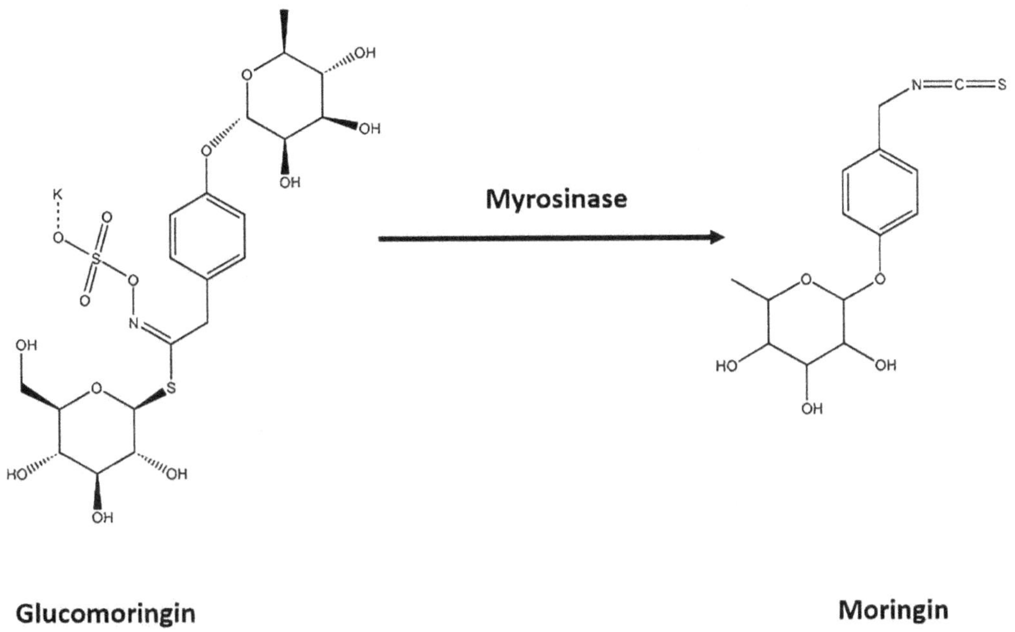

Myrosinase

Glucomoringin **Moringin**

FIGURE 10.3 One of the main glucosinolate identified in *M. oleifera* is glucomoringin, which, under the action of the myrosinase, forms a new compound moringin. (Adapted from Barba et al 2016; Borgonovo et al., 2020)

prominent glucosinolate in the roots. Three isomers of 4-O-(α-L-acetylrhamnopyrosyloxy)-benzyl glucosinolate were also detected in *M. oleifera* leaves, depending on their maturity and physiological properties (Rani et al., 2018).

Disrupting the tissue matrix, usually through cutting or chewing, releases the myrosinase enzyme that catabolizes glucosinolates to isothiocyanates (Figure 10.3), nitriles, and thiocarbamates. Isothiocyanates have attracted much attention recently for their numerous health properties. The most abundant isothiocyanate found in *M. oleifera* is the 4-[(α-L-rhamnosyloxy) benzyl] isothiocyanate, or glucomoringin isothiocyanate (Rani et al., 2018). These isothiocyanate compounds are normally present as volatile oil and usually unstable at room temperature. However, most isothiocyanates from *M. oleifera* have a sugar moiety, which contributes to their stability at room temperature (Waterman et al., 2015). The alkylation of isothiocyanates with proteins and DNA contributes to their biological activity (Nibret and Wink, 2010).

Both glucosinolates and isothiocyanates have attracted great interest for their important antioxidant, anti-inflammatory, anticarcinogenic, antimicrobial, and neuroprotective potentials (Miekus et al., 2020).

These compounds can regulate stress responses via activation of the regulatory protein Nrf2 synthesis, which promotes antioxidants and the induction of phase 2 detoxification enzymes. These sulfur-containing molecules have also been reported to block NF-κB activation through TNF-α, IκBα, and p65 protein inhibition, which also regulates the proinflammatory signaling pathway reported in several disorders.

10.3 HEALTH-ENHANCING PROPERTIES OF *M. OLEIFERA*

M. oleifera is known not only for its remarkable nutritional content but also for its profound health-enhancing properties. The health implications of *M. oleifera* are profound, based on preclinical investigations elucidating molecular targets to human studies exploring antidiabetic potential and nutritional benefits. This section delves into the scientific insights and the multifaceted aspects of *M. oleifera* that make it a compelling subject in the quest for natural health-promoting agents.

10.3.1 Molecular Targets: Insights from Preclinical Studies

Traditionally, *M. oleifera* has been used against several ailments. This plant and its bioactive compounds have demonstrated their pluripharmacological properties, namely antidiabetic, anti-cancerous, antioxidant, anti-inflammatory, and antimicrobial effects (Figure 10.4) (Gopalakrishnan et al., 2016).

10.3.1.1 Antioxidant Mechanisms

M. oleifera and its bioactive molecules have demonstrated potential to reduce oxidative stress via the activation of the nuclear factor erythroid 2-like 2/antioxidant response element (Nrf2/ARE) pathway, which consequently increases antioxidant enzyme expression. *M. oleifera* also exerts anti-inflammatory effects by inhibiting the nuclear factor kappa B (NF-κB) pathway, thus preventing inflammatory cytokine expression.

The antioxidant potential of *M. oleifera* has been well established. Alterations in cellular redox states, often resulting in oxidative stress, are associated with various pathologies, including cancer, diabetes, and cardiovascular and neurological disorders. The two major mechanisms through which reactive oxygen species (ROS) exert their harmful effects are: by oxidizing macromolecules, including lipid membranes, structural proteins, enzymes, and nucleic acids, leading to aberrant cell function and death; and secondly, by acting as secondary messengers, mainly H_2O_2, triggering signaling pathways associated with epigenetic dysregulation, chronic inflammation, endothelial dysfunction, and ultimately apoptotic cell death (Forman and Zhang, 2021). *M. oleifera* has demonstrated several free radical scavenging potentials, metal chelation activities, and *in vitro* reducing abilities (Sandanuwan et al., 2022; Hip Kam et al., 2023). At the cellular level, *M. oleifera* leaf extract significantly reduced lipid peroxidation and increased the antioxidant enzyme activities of glutathione peroxidase (GPx), superoxide dismutase (SOD), and catalase (CAT) induced by oxidative stress in HL60 cells (Dilworth et al., 2020).

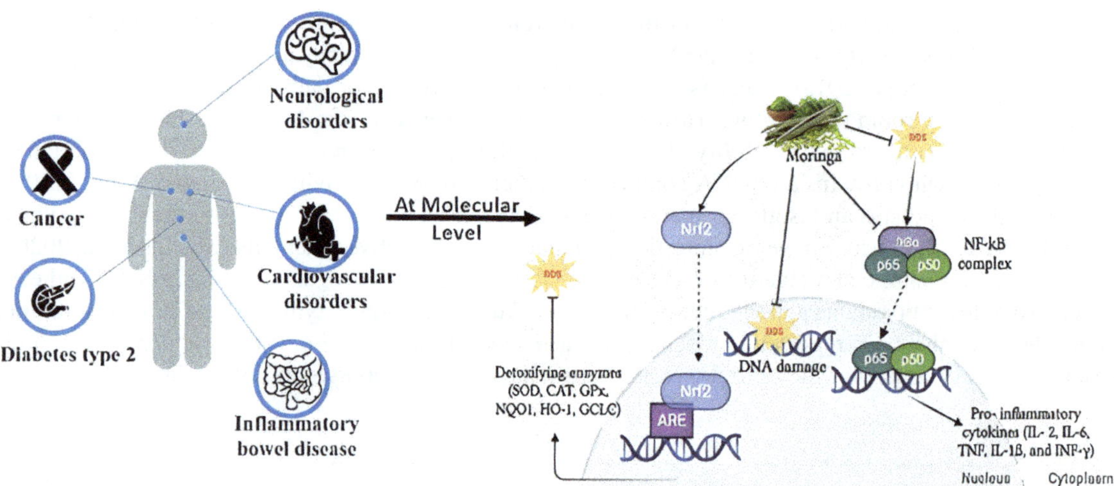

FIGURE 10.4 Oxidative stress and chronic inflammation promote several disorders, such as cancers, cardiovascular diseases, neurodegenerative disorders, diabetes mellitus, and inflammatory bowel diseases. (Adapted from Muangnoi et al., 2012; Berkovich et al., 2013; Fard et al., 2015; Gopalakrishnan et al., 2016; Cheng et al., 2019; Gómez-Martínez et al., 2022; Kamal et al., 2022)

In addition, cells orchestrate a complex network of antioxidants to maintain proper cellular functions to protect macromolecules from indiscriminate damage incurred by free radicals (Chio and Tuveson, 2017). The activity and production of enzymatic antioxidants are tightly regulated by key transcription factors, and one of the main regulators is nuclear factor erythroid 2-like 2 (Nrf2). The activation of Nrf2 expression generates a cascade of responses involved in cytoprotective activities, including xenobiotic metabolism, regulation of proteasomal subunits, and inflammatory responses. Nrf2 dysregulation and its downstream signaling have been observed in many chronic diseases, including diabetes mellitus, obesity, cardiovascular diseases, and atherosclerosis (Cuadrado et al., 2019; Da Costa et al., 2019). Several studies have demonstrated the potential of *M. oleifera* extracts and bioactive compounds to promote Nrf2, thus inducing cellular antioxidant responses. In an H_2O_2-induced oxidative stress kidney epithelial cell model, *M. oleifera* hot water extract reduced intracellular ROS generation in a dose-dependent manner via the activation of the Nrf2/HO-1 signaling pathway, suppressing mitochondrial depolarization and apoptosis, thus protecting the cells against oxidative stress (Sandanuwan et al., 2022). A similar mechanism of action was observed in another model of skeletal muscle cells (C2C12) treated with *M. oleifera* leaf extract. The extract increased Nrf2 and HO-1 protein levels in a dose-dependent manner, thus enhancing glutathione redox homeostasis and the enzyme activities of CAT, SOD, GPx, and glutathione transferase (GST). Isothiocyanate isolated from *M. oleifera* activated Nrf2-ARE, which subsequently increased the downstream gene expression of NQO1, HO-1, and GCLC in human hepatocellular carcinoma cells (Cheng et al., 2019).

In addition to the Nrf2 pathway, *M. oleifera* also modulates the protein kinase C zeta (PKCζ)/NADPH oxidase 4 (Nox4) signaling pathway. Excessive ROS oxidizes low-density lipoprotein (LDL), which, in turn, activates Nox4—dependent on the activation of PKCζ—to promote ROS production, resulting in a vicious cycle and the development of metabolic diseases. Niazirin, a phenolic glycoside isolated from *M. oleifera* seeds, was observed to break this cycle by reducing PKCζ activation and inhibiting Nox4 protein expression (Wang et al., 2019).

10.3.1.2 Anti-inflammatory Properties

Another important mechanism of action of *M. oleifera* is its anti-inflammatory effect. Chronic inflammation is the primary cause of various diseases. It is characterized by the continuous expression of proinflammatory factors and long-lasting tissue damage. *M. oleifera* acts against chronic inflammation and its associated ailments (Xiao et al., 2020). The overactivation of nuclear factor kappa B (NF-κB), a proinflammatory transcription factor, has been described in the etiology of several diseases. *M. oleifera* leaf and pod extracts reduced the expression of p65, p-IkBα, and IkBα proteins, which are essential components for NF-κB signaling pathway activation in several *in vitro* models (Muangnoi et al., 2012; Berkovich et al., 2013; Fard et al., 2015). *M. oleifera* and its isolated compounds also regulate the secretion of nitric oxide (NO), an important proinflammatory signaling messenger. This cytokine provides crucial regulatory control of the cardiovascular system and serves as a biomarker for several diseases, including hypertension, diabetes mellitus, thrombosis, and stroke (Cheenpracha et al., 2010; Louisa et al., 2022).

M. oleifera influences other inflammatory markers, both *in vitro* and *in vivo*. LPS-induced RAW264.7 macrophage cells may be one of the most studied inflammatory models. *M. oleifera* pod, flower, and leaf extracts, and their isolated compounds (mainly isothiocyanate), can modulate the NF-κB signaling pathway by blocking IκBα degradation, thereby preventing NF-κB p65 translocation to the nucleus. *M. oleifera* treatment downregulates NO and proinflammatory cytokines (IL)-6, TNF-α, IL-1β, and prostaglandin E2 in macrophages (Muangnoi et al., 2012; Fard et al., 2015; Park et al., 2011; Lee et al., 2013). In a colitis-associated colon carcinogenesis model, supplementation with 10% *M. oleifera* leaves in infected mice downregulated proinflammatory (IL-2, IL-6, TNF, IL-1β, and IFN-γ) genes in colon tissue, demonstrating chemopreventive properties against inflammatory response and colon carcinogenesis (Cuellar-Núñez et al., 2021). Oral administration of *M. oleifera* leaf powder to heat-stressed rabbits boosted immunological and cecal microecosystem functions by enhancing their antioxidant status and downregulating mucosal tissue inflammatory factors (TNF-α/IL-1α/IL-1β) and myeloperoxidase (MPO) (Yasoob et al., 2021).

10.3.1.3 Antidiabetic Potential

M. oleifera has also been reported for its important antidiabetic potential. Insulin resistance is a key feature of type 2 diabetes, and *M. oleifera* improves insulin sensitivity by activating insulin signaling pathways. Specifically, *M. oleifera* increases the expression of insulin receptors, insulin receptor substrate 1, and glucose transporter 4 in the liver and muscles, thus increasing insulin sensitivity, glucose uptake into cells, and glycogen synthesis in muscle and adipose tissue. *M. oleifera* also alleviates insulin resistance by activating insulin-independent pathways PI3K/AKT and AMPK (Attakpa et al., 2017; Bao et al., 2020; Duranti et al., 2021).

In addition, *M. oleifera* acts as a competitive inhibitor of sodium-glucose linked transporter 1 (SGLT1) in the small intestine, reducing glucose absorption. SGLT1 is a protein responsible for glucose absorption from the small intestine into the bloodstream (Nova et al., 2020). When this process is impaired or inhibited, glucose absorption is reduced, which can help attenuate postprandial hyperglycemia in diabetic subjects. *M. oleifera* can bind to the transporter and prevent glucose absorption. *M. oleifera* leaf extract significantly reduced glucose transport and inhibited SGLT1 activity in a dose-dependent manner in Caco-2 cells in an *in vitro* study. An ethanolic *M. oleifera* leaf extract inhibited SGLT1 activity *in vitro*, with an IC_{50} value of 0.16 mg/mL. In addition, several compounds present in *M. oleifera* have been identified as potential SGLT1 inhibitors. For instance, quercetin and chlorogenic acid inhibited SGLT1 activity *in vitro* (IC_{50} value of 0.86 μM and 14.0 μM, respectively).

M. oleifera also restores the activities of glucose-metabolizing enzymes hexokinase and glucose-6-phosphate dehydrogenase, facilitating glycolysis and glucose utilization by the pentose phosphate pathway. Besides, the bioactive compounds of *M. oleifera* also exhibit antidiabetic activities. Quercetin and kaempferol mono-glycosides have strong binding abilities to pancreatic α-amylase and α-glucosidase enzymes. These glucosides can inhibit the activities of these enzymes, which catalyze the breakdown of polysaccharides and disaccharides into glucose. Additionally, gallic, chlorogenic, and caffeic acids found in *M. oleifera* also have similar abilities to inhibit α-amylase and α-glucosidase activities (Watanabe et al., 2021).

During diabetes mellitus, lipid and protein peroxidation, as well as upregulation of fatty acid synthase (FAS) and 3-hydroxy-3-methylglutaryl-coenzyme A (HMG-CoA) reductase mRNA expression in the liver, contribute to the development of diabetic dyslipidemia. FAS is responsible for the synthesis of palmitate from acetyl-CoA and malonyl-CoA, while HMG-CoA reductase is a rate-limiting enzyme involved in cholesterol biosynthesis. However, the administration of *M. oleifera* has been shown to reduce fatty acid and cholesterol synthesis, leading to the normalization of lipid profiles in diabetic rodent models. Additionally, *M. oleifera* can normalize the mRNA levels of several factors involved in body composition and insulin sensitivity in adipose tissues, including leptin, resistin, adiponectin, melanocortin receptor-4, and peroxisome proliferator-activated receptors α and γ.

10.3.1.4 Cardiovascular Health Protection

M. oleifera seed extract minimizes infarct sizes, alleviates myocardial contractile dysfunction, counteracts myocardial infarct (MI)-induced cardiac remodeling/fibrosis, prevents ventricular failure, and provides post-MI mortality reduction compared to the disease group. The leaf extract significantly improves heart rate and cardiac performance through improved myocardial contraction and relaxation and reduces mortality from congestive heart failure. Isothiocyanate-enriched *M. oleifera* extract (47% isothiocyanates) has anti-obesity and antidiabetic effects because it inhibits the rate-limiting steps in liver gluconeogenesis and increases insulin signaling and sensitivity and lean body mass. In an ischemic stroke model, moringin limits brain damage progression by suppressing the NF-κB pathway, prevents IκB-α degradation, and NF-κBp65 translocation, resulting in suppression of proinflammatory markers and prevention of gait abnormalities (Kamal et al., 2022).

10.3.1.5 Neuroprotective Effects

Various *M. oleifera* plant extracts have demonstrated neuroprotective effects in neurodegenerative disease models. Moringin scavenges free radicals produced by iNOS, NO, and nitrotyrosine and increases the nuclear antioxidant marker Nrf2 in LPS-activated macrophages. Moreover, moringin (1.25 μg/mL) treatment significantly reduces ROS production and gene expressions (cyt-c, p53, Apaf-1, Bax, CASP3, CASP8, and CASP9) in SH-SY5Y cells, while simultaneously increasing Bcl-2 expression. Moringin also suppresses proinflammatory markers (TNF-α, IL-1β, and IL-6) in the brain. Moringin protects neuronal cells from programmed cell death in an animal Parkinson's disease model. Complexation with α-cyclodextrin improves moringin's efficacy and availability by increasing its solubility and stability in aqueous medium. The complex reduces cell growth in a neuroblastoma model by inducing apoptosis and suppresses Alzheimer's disease progression by enhancing neuronal repair and preserving normal neuronal cell viability.

Moringin, the *M. oleifera* isothiocyanate, as a topical cream, can be an innovative, safe, and non-invasive pharmacological agent for many MS patients with conventionally resistant neuropathic pain. Giacoppo et al. (2017) showed that 2% moringin cream treatment effectively relieved neuropathic pain by suppressing the inflammatory pathway and blocking the voltage-gated ion channels in a murine model of multiple sclerosis. Moringin, purified from *M. oleifera* seeds and prepared as a ready-to-use topical treatment (2% moringin cream twice daily from symptom onset), effectively reduced clinical and histological autoimmune encephalomyelitis in a murine model of multiple sclerosis and alleviated neuropathic pain. Moringin reduced the production of proinflammatory cytokines (IL-17 and interferon-γ) and concomitantly increased anti-inflammatory cytokine (IL-10) expression. Moringin treatment increased the expression of genes involved in neuron-critical development of the upper and deep cortical layers (Romeo et al., 2018). The treatment also upregulated genes involved in osteogenesis and adipogenesis. Thus, moringin treatment can improve stem cell therapy against neurological disorders by promoting and accelerating human periodontal ligament stem cells (hPDLSCs) cortex neuronal differentiation.

10.3.1.6 Dermatological Insights

M. oleifera is generally used for skin care preparation in cosmetics, since its leaves are used for facial milks and especially for facial washes (Desam and Al-Rajab, 2021). These cosmetic preparations are based on the leaf's oil, presumed to be the active chemical constituent for its use. Leaves also have significant amounts of vitamins (A, C, and E), polyphenols, and β-carotene with antioxidant and anti-inflammatory properties that provide added cosmetic benefits. The behenic acid-rich seed oil is used for moisturizers, massage oils, and aromatherapy. These active ingredients of leaf and seed oil extracts enable its use in sunscreens, scrubs, body oils, creams, lotions, balms, hair care products, moisturizers, and perfumes preparation (Desam and Al-Rajab, 2021 and references therein).

Albumin extracted from defatted *M. oleifera* seed protein bound to chitin (affinity chromatography) has been proposed as a potential antifungal agent for clinical dermatophytosis treatment because of its potent antidermatophytic activity against *Trichophyton mentagrophytes* (Lopes et al., 2020). Dermatophytosis is considered a serious health problem since it affects up to 25% of the global population according to the World Health Organization. It is contagious and infects keratinized tissues such as skin, hair, and nails of humans and animals. Current antifungal dermatophytosis drugs (griseofulvin, azole derivatives, and allylamines) are inefficient, toxic, with serious side effects, and can develop fungal (*Trichophyton* genus) resistance. This purified *M. oleifera* chitin-binding protein (*Mo*-CBP_4) is a basic protein (pI 10.55, 11.78 kDa) with antinociceptive and anti-inflammatory activities when administered orally and intraperitoneally (Lopes et al., 2020 and references therein). *Mo*-CBP_4 acts antidermatophytically on plasmatic membranes or cell walls, which prevents or alleviates resistance development and lethally damages *T. mentagrophytes* microconidia. *Mo*-CBP_4 (5, 10, and 20 mg/g) reduced the disease time course and lesion severity of Swiss female *T. mentagrophytes*-infected mice, with minor injury severity and complete absence of *T. mentagrophytes* compared to control. *Mo*-CBP_4 showed significant clinical efficacy against dermatophytic infection based on skin lesion scores and histological analysis of *T. mentagrophytes*-infected skin sections (Lopes et al., 2020).

10.3.2 Exploring *M. oleifera's* Impact on Human Health: Clinical Investigations

As clinical trials stand at the forefront of evidence-based medicine, they play a crucial role in assessing the efficacy and safety of interventions, including natural compounds like *M. oleifera*. While preliminary studies have highlighted the bioactive components and pharmacological potential of *M. oleifera* in preclinical settings, it is imperative to bridge the gap between laboratory findings and real-world applications through human-centric trials. These endeavors aim to decipher the impact of *M. oleifera* on human health, addressing questions of dosage, duration, and the potential therapeutic applications in various health conditions.

10.3.2.1 Antidiabetic and Antilipidemic Potential of M. oleifera

The therapeutic health benefits of *M. oleifera* leaf have been the focus of many human studies, particularly related to diabetes. Clinicaltrials.gov lists 38 studies on *M. oleifera*, with 22 completed studies and five clinical trials related to prediabetic/diabetic patients.

In a double-blind, randomized, placebo-controlled, parallel-group intervention study, Gómez-Martínez et al. (2022) investigated glycemia, appetite-controlling hormones, and gut microbiota composition in 31 prediabetic (hemoglobin A1C [HbA1C]: 5.88%) subjects (13F/18M, 56 years mean age, 29.4 kg/m^2 median BMI) consuming six daily *M. oleifera* dry leaf powder capsules (2.4 g/d) over 12 weeks. The capsules consisted of 400 mg organic Spanish-grown *M. oleifera* leaf powder, each containing 27.8% protein, 49.3% carbohydrates, minerals (calcium, potassium, magnesium [4.85, 2.76, and 0.54 g/100 g, respectively]), and 2300 ppm total polyphenols. The intervention acted as a natural antihyperglycemic agent by reducing fasting blood glucose (FBG) and glycated hemoglobin (HbA1C) (2.7% and 1.5%, respectively), with favorable changes in glycemia markers compared to the placebo group. However, *M. oleifera* supplementation showed no significant differences in gut microbiota change rates, gut hormones (ghrelin, glucagon-like peptide, and peptide YY), renal, and hepatic functions (Gómez-Martínez et al., 2022). The same intervention group reported that *M. oleifera* dry leaf powder supplementation did not change secondary outcomes of inflammatory and cardiometabolic markers (blood and fecal inflammatory markers, serum lipid profile, plasma antioxidant capacity, and blood pressure) in prediabetics. However, plasma TNF-α was a significant predictor of HbA1C response for *M. oleifera*-administered subjects (Díaz-Prieto et al., 2022).

Tablets of ground *M. oleifera* leaves were administered to non-insulin-dependent type 2 diabetic patients receiving only oral antidiabetic medications and dietary recommendations. The *M. oleifera* leaves (8 g/day) intake lowered postprandial glucose by 26% after 40 days in 22 diabetic patients (Kumari, 2010). Similarly, an earlier study showed that *M. oleifera* leaf powder (8 g/day) for 14 days reduced fasting plasma glucose and 2 h postprandial glucose level in diabetic subjects (John and Chellappa, 2005). In addition, powdered *M. oleifera* leaves capsule (500 mg × 3/d) supplementation for 12 weeks significantly reduced high-sensitivity C-reactive protein (hsCRP) and HbA1C (3.38 vs 1.69 mg/dL hsCRP and 7% vs 6% HbA1C [pre vs post-treatment]) on 56 (45F/11M, 61 years mean age, 20 kg/m^2 median BMI) diabetics in a prospective quasi-experimental study (Mozo and Caole-Ang, 2015). The authors emphasized the potential of *M. oleifera* antioxidant components such as flavonoids, carotenoids, and vitamins C and E that improve glucose metabolism and reduce insulin resistance. The *M. oleifera* supplementation to the Simvastatin (40 mg/d) treatment ameliorated type 2 diabetes mellitus biomarkers, reorganizing patients from high to moderate risk category based on mean plasma hsCRP (1.5-fold) reduction (Mozo and Caole-Ang, 2015).

In contrast, in a randomized placebo-controlled study, *M. oleifera* leaf (4 g/d, 8 × 500 mg daily capsules for 28 days) had no effect on fasting plasma glucose, HbA1C, mean daily plasma glucose, and mean premeal and post-meal plasma glucose in overweight diabetic patients (7M/9F, 52 years mean age, 28 kg/

m^2 median BMI) compared to placebo (Taweerutchana et al., 2017). Similarly, *M. oleifera* leaf capsules at the same dose (4 g/d) for 6 weeks showed no difference in mean plasma glucose but increased mean insulin secretion by 74% compared to baseline in 10 healthy volunteers (Anthanont et al., 2016).

Higher dose (20 g) *M. oleifera* dried leaf powder as part of a meal lowered postprandial glucose levels compared to control in diabetic patients (Leone et al., 2018). *M. oleifera* leaves (50 g) supplemented to a standardized (isocarbohydrate) meal significantly lowered postprandial blood glucose response (−21%, P < 0.01) compared to 75 g glucose load in a controlled study with untreated diabetic patients (William et al., 1993). Another single-dose study (200 mL) *M. oleifera* tea reduced glucose levels 30 min after glucose overload compared to higher dose (22.8% vs 17.9% for 400 mL) (Fombang et al., 2016). Cookies containing *M. oleifera* leaf powder (5% w/w) significantly reduced blood glucose 30 and 45 min after ingestion and hunger ratings (Ahmad et al., 2018).

M. oleifera powder intake in 50 g pouches used with food regularly for 40 days significantly lowered serum glucose and LDL levels (8.9% and 31%) in 15 (9M/6F) obese non-insulin-dependent diabetic (NIDDM) patients (Kumar and Mandapaka, 2013).

Clinical trials reported the beneficial effects of *M. oleifera* leaf supplementation on circulating lipid profile. *M. oleifera* leaf tablets with high beta carotene, polyphenols, and fiber had a mild positive impact (lower atherogenic) on the lipid profile of overweight hyperlipidemic subjects. Standardized dehydrated *M. oleifera* leaf (8 × 575 mg/day tablets for 50 days) reduced total plasma cholesterol (1.65%) and increased high-density lipoprotein cholesterol (6.25%) in 17 hyperlipidemic subjects (14M/3F, 50 years mean age, ~25 kg/m^2 average BMI, TC > 180 mg/dL) compared to control (Nambiar et al., 2010). In another study, *M. oleifera* leaf powder (8 g/d) for 40 days significantly reduced (~14%) serum total cholesterol and triglycerides in 24 (18M/6F, 41–60 years) adult diabetic patients (Kumari, 2010).

Human studies are limited on administering *M. oleifera* in the form of extracts, in contrast to the most common powdered leaf intake. Ezzat et al. (2020) showed the efficacy of 400 mg leaves ethanol extract administered in capsule to overweight/obese subjects for 8 weeks. The clinical trial resulted in a significant reduction of BMI, total cholesterol, and LDL. Sugunabai et al. (2014) evaluated the antidiabetic and antilipidemic efficiency of aqueous *M. oleifera* leaves extract on 20 male diabetic patients (35–60 years) compared to control group (10 healthy individuals of same age). The extract reduced HbA1C (17.2% and 13.3%; 22.5% and 21.4%; 22.1% and 20.4% after 30-, 60-, and 90-day treatment, respectively). Total cholesterol content decreased (4.7%, 7.3%, and 14.2%) and triglyceride levels decreased by 27%, 36%, and 39% after 30-, 60-, and 90-day treatment, respectively. LDL cholesterol also decreased (9%, 16.3%, and 17.7%) after 30-, 60-, and 90-day treatment, respectively. The hypolipidemic and hypoglycemic effects of the aqueous extract were attributed to its composition—24, 160, 1400, 24, and 18.8 mg/100 g carbohydrates, phenol, cellulose, protein, and carotenoid, respectively (Sugunabai et al., 2014).

Human studies also reported the beneficial effects of *M. oleifera* on blood pressure. A clinical study in Mauritius conducted among young healthy adults (15F/8M, 30–35 years old) consuming 120 g *M. oleifera* leaves (sautéed [80 °C, 10 min] in small amount of vegetable oil) over a weekend showed significant decrease in both postprandial systolic and diastolic blood pressure despite the prior high salt (7 g/d) consumption during the week (Chan Sun et al., 2020). Another human study demonstrated that supplementation of *M. oleifera* leaf powder (30 g/d) significantly (P < 0.05) reduced blood pressure in hypertensive patients (41 years average age), both normal weight (25F/5M, 22 kg/m^2 median BMI) and obese (25F/5M, 32.3 kg/m^2 median BMI), for 6 months (Fombang et al., 2016). Moreover, supplementation benefitted the obese more than the normal group in weight loss parameters (3.2% vs 2.3% body weight, 3.6% vs 2.3% BMI, 2.5% vs 1.3% waist circumference, and 4.4% vs 2.9% hip circumference) (Fombang et al., 2016). *M. oleifera*'s blood pressure-lowering effects were attributed to its hypotensive constituents such as the isocyanates, thiocarbamate glycosides, and high potassium content (Fombang et al., 2016).

In healthy menopausal women, *M. oleifera* leaves powder (7 g/d) reduced fasting glucose (13.5%) and normalized glucose values (<110 mg/dL) after 3 months. Moreover, serum cholesterol and triglyceride levels decreased significantly, with improved levels of antioxidant vitamins (retinol, ascorbic acid) and enzymes (SOD, GSH-Px [glutathione peroxidase], and oxidative stress biomarker MDA [malondialdehyde] (Kushwaha et al., 2014).

The study strongly indicates that *M. oleifera* leaf powder consumption can benefit weight loss, reduce blood pressure, and prevent other diseases with obesity as a risk factor. In therapy-naïve diabetic patients, *M. oleifera* leaf powder (4 g/d for 28 days) lowered blood pressure nonsignificantly, presumably due to low bioactive concentrations (Taweerutchana et al., 2017).

Two randomized double-blind placebo-controlled trials examined the weight management efficacy of *M. oleifera* leaf aqueous extract-containing herbal formulation. The herbal formulation L1858F (also known as Slendacor) contained predominantly *M. oleifera* leaf aqueous extract (6 parts) along with three parts *Murraya koenigii* leaf aqueous extract, and one part *Curcuma longa* extract (standardized to $\nless$ 95% total curcuminoids). In the first study (Sengupta et al., 2012), the formulation (900 mg/day divided in three doses) for 8 weeks significantly reduced body weight and body mass index in obese participants (14F/7M, 42 years mean age, 34 kg/m^2 average BMI). The supplementation improved critical biomarkers associated with type 2 diabetes and obesity (reduced fasting blood glucose, LDL, LDL/HDL ratio, and triglycerides, and increased serum adiponectin [21% compared to placebo group]). In the second trial (Dixit et al., 2018), the formulation (900 mg/day divided in two doses) for 16 weeks significantly reduced body weight and body mass index in healthy overweight adults (44F/26M, 35 years mean age, 29 kg/m^2 average BMI). Moreover, the supplementation significantly improved lipid profiles (reduced serum LDL, VLDL, total cholesterol, triglycerides, and LDL/HDL ratio) compared to placebo, indicating improved fat metabolism status and potential cardiovascular benefits.

10.3.2.2 Nutritional Impact of M. oleifera Consumption

Many studies (170 articles) have focused on the traditional uses, nutraceutical potential, phytochemistry, and pharmacological properties of *M. oleifera*, according to the latest systematic review (Mahato et al., 2022). Several studies demonstrated the nutritional benefits of *M. oleifera* leaf powder, particularly regarding children's health and immunocompromised people. Yadav et al. (2022) evaluated *M. oleifera* leaf powder supplementation in children with severe acute malnutrition (SAM) during facility- and home-based care in a randomized controlled trial. Severely acute malnourished children (n = 50, 7–59 months) received *M. oleifera* leaf powder (15 g × 2/d) in addition to routine SAM supplements for 2 months. *M. oleifera* intervention significantly increased weight gain and reduced severe wasting frequency, but with no significant difference in other anthropometric parameters at the end of the two-month follow-up compared to the control group (routine SAM supplement alone).

M. oleifera biscuits (352 g protein and 4500 kcal) consumed for two months increased children's (17F/14M, 3 years median age) weight and height (Veterini et al., 2023). The average weight gain was 0.50 and 0.20 kg, whereas the average height growth was 0.34 and 0.96 cm for the first and second months, respectively. The difference in growth rate was significant between the boy and girl groups, according to multivariate statistical analysis.

M. oleifera leaf powder (25 g) for 14 days significantly increased body weight in 25 children (11M/14F, 2–5 years old) compared to the control by increasing their appetite (due to high vitamin A content) in Kupang, Indonesia (Allo et al., 2020). Kupang had the highest child malnutrition prevalence (33% in children under 5 years old in 2013). In this experimental pre- and post-test with control group design, *M. oleifera* treatment increased energy intake (31%), resulting in increased average weight gain (0.61 vs 0.03 kg for the control group; 5.7% body weight increase). In the treatment group, only four children consumed all the leaf powder with an average gain of 1.05 kg, whereas 21 children only consumed 5–10 g powder (apparently due to the scent of the leaf). Despite this study, micronutrient deficiencies in Indonesian school-aged children are still a major health problem. A recent cross-sectional study examined the nutritional status of 2,456 Indonesian children aged 5–12 years (Ernawatiet al., 2023). Stunting and thinness prevalence (11.4% and 9.2%) are persistent problems in Indonesian school-aged children and are related to their national micronutrient deficiencies, particularly ferritin, zinc, and vitamin D (13.4%, 19.7%, and 12.7%, respectively). Micronutrient deficiency is still a significant global health challenge affecting an estimated 2 billion people, according to the World Health Organization.

M. oleifera leaf powder (MLF) effect was evaluated on hemoglobin concentration and infants' growth after 4 months of feeding in another randomized controlled trial (Boateng et al., 2019). Infants (n = 237,

8–12 months) were randomly assigned to receive Weanimix (cereal-legume blend formulated with 35 g MLF), 5 g MLF sprinkled on infants' usual complementary foods, and control (Weanimix without MLF). MLF (15% fortification) did not improve their hemoglobin concentration or growth indicators in contrast to the above studies. Moreover, the MLF interventions had lower adherence (62% vs 80%) and higher morbidity outcomes (23% vs 13.8% fever, 26% vs 18% nasal discharge, 19% vs 12% cough, 6.3% vs 2.7% vomiting, and 17% vs 9% diarrhea) than the control.

Shija et al. (2019) investigated the effect of *M. oleifera* leaf powder supplementation (25 g/d mixed in porridge) in alleviating anemia in young (27M/16F, 11.6 months mean age) anemic children in a community-based interventional controlled study. *M. oleifera* leaf powder intervention significantly reduced anemia prevalence (by 54% vs 14% for the control group) and increased hemoglobin (10.9 vs 9.4 g/dL) after 6 months. Moreover, moderate and severe anemia in the intervention communities decreased (by 68% vs 23% for control and 78% vs 57%, respectively). Anthropometric parameters (height, weight, age, illness history) were monitored monthly for 6 months, and the long-term effect was evaluated 3 months post-intervention. The treatment increased hemoglobin (2.6 vs 1.6 g/dL in the control group) after 6 months. Anemia reduction was attributed to *M. oleifera* leaf's high vitamin A content and probably β-carotene, which protects iron availability. The high digestible protein content accounted for the weight gain and improved body immunity for reduced illness.

Gambo et al. (2021) evaluated *M. oleifera* leaf supplementation effects on immune status (CD4 lymphocyte counts and viral load) and anthropometric parameters of HIV-positive adults on antiretroviral therapy. *M. oleifera* supplementation (5 g thrice daily) for 6 months increased CD4 counts of HIV patients (19M/70F) over 10-fold in this double-blind, randomized controlled study. The CD4 increase was attributed to *M. oleifera*'s nutritional constituents [(28, 3.9, and 22 g/100 g protein, fat, carbohydrates) and minerals (calcium, potassium, sodium, zinc, and iron; 1792, 4879, 24, 2.9, and 37.8 mg/100 g, respectively)]. The intervention had no effect on viral load and anthropometric parameters (Gambo et al., 2021). This study contrasts with the results of an earlier single-blind randomized control trial (Tshingani et al., 2017) that reported no significant difference in CD4 lymphocyte counts of 30 HIV patients on antiretroviral therapy after 6 months of *M. oleifera* leaf powder (30 g/d) supplementation.

10.3.2.3 Other Human Studies

The antioxidant and radical scavenging activity of *M. oleifera* leaves were evaluated in a human study (Ngamukote et al., 2016). A single dose of *M. oleifera* leaf aqueous extract (500 mg) increased plasma ferric reducing ability and reduced lipid peroxidation (malondialdehyde [MDA]) 30 min after ingestion in 20 healthy participants.

Finely powdered dried *M. oleifera* seed kernels (3 g/d with water) for three weeks significantly increased hemoglobin values, reduced erythrocyte sedimentation rate, and improved symptom score and severity of asthmatic attacks in 20 patients with mild-to-moderate asthma (Agrawal and Mehta, 2008). The intervention significantly improved lung functions (forced vital capacity, forced expiratory volume in one second, and peak expiratory flow values [by ~33%]) in asthmatic subjects. *M. oleifera* can therefore ameliorate bronchial asthma symptoms with simultaneous improvement in respiratory functions.

M. oleifera stem bark (40 mL decoction obtained by boiling 15 g coarse powder in 100 mL water) is effective against urinary tract infection and helps eradicate urinary pathogens, particularly *E. coli* responsible for UTI (Maurya and Singh, 2014). *coli*, responsible for UTI (Maurya and Singh, 2014). Thirty UTI patients were divided equally into treatment and standard modern medicine (control) groups with three follow-ups at the end of every treatment week. The treatment increased UTI cure and relapse rates (67% vs 47% and 6.7% vs 20%, respectively), presumably due to the antibacterial and anti-inflammatory agents present in the bark (Maurya and Singh, 2014).

Otomycosis is a serious, high-recurrence superficial mycotic outer ear canal infection caused by some pathogenic *Candida* and *Aspergillus* species. *M. oleifera* leaf ethanol extract encapsulated with lecithin–chitosan nanoparticles [LCNPs] (3 drops applied twice daily for 7 days) improved (90% vs 20% for Nystatin) fungal otomycosis in patients (7M/6F, 47 years mean age) within one-week treatment with no recurrence for over 3 months in a randomized controlled (Nystatin-positive control) study (Zahran et al., 2022). Moreover, the nanoparticles inhibited clinical isolates from oomycote patients (*Candida parapsilosis*, *Aspergillus niger*, and *Aspergillus flavus*) (75, 55, and 55 mm disk diffusion inhibition zone compared to Nystatin [35, 25, and 20 mm, respectively]). Subsequent *in silico* investigation indicated that some alkaloids (niazinin A, niazimicin A, niazidin, and niazirinin) may be responsible for the activity, targeting the fungal 14-α-demethylase enzyme (CYP51B) for enhanced antifungal efficacy compared to Nystatin.

Aqueous *M. oleifera* leaf extract (40.5 mg/kg bw/day) for one month significantly suppressed the inflammatory process by reducing IL-6 levels and the simplified disease activity index (SDAI) scores of 15 female rheumatoid arthritis patients compared to the control group (Prabowo et al., 2023). These effects are ascribed to isothiocyanate compounds present in the extract, which strongly affect nitric oxide production in this randomized controlled trial.

M. oleifera leaf petroleum ether extract (two 500 mg leaf extract capsules daily for 28 days) significantly improved Beck Depression Inventory (BDI-II) scores (43–17 vs 48–39 for control) and serum cortisol (44–25 vs 50–40 μg/dL) in rheumatoid arthritis patients (9F/7M) in a quasi-experimental study with a pretest–post-test control group (Ummah et al., 2023). *M. oleifera* leaf extract reduced peripheral cytokine upregulation by inhibiting three pathways (neuronal, direct humoral, and the microglial activation pathways) within brain structures. The extract presumably alleviated joint inflammation by reducing serum prostaglandin E2, CRP, and TNF-α levels, and downregulating the NF-κB and cyclooxygenase 2, thereby restoring the normal joint's histopathological features. The antidepressant effect of *M. oleifera* leaf extract is ascribed to its phytochemicals, particularly flavonoids, which possess potent anti-inflammatory and antioxidant activities that can treat brain neurochemical imbalance-induced depression.

10.4 PROCESSED *M. OLEIFERA* AND ITS RELATED HEALTH BENEFITS

M. oleifera, with its high nutritional value and rich phytochemical content, is a promising functional food. The rapid growth of *M. oleifera* trees in subtropical and tropical areas, even under conditions of prolonged drought, makes this plant a reliable resource for enhancing the nutritional status of local populations. Among the different parts of the plant, the leaves and the young pods are most commonly consumed. In Africa and Asia, the leaves and the immature pods are eaten as greens, in salads, as pickles, or cooked as vegetables (Razzak et al., 2022; Agbogidi and Ilondu, 2012). Recently, with high valorization, numerous commercial *M. oleifera* products have been developed, notably, infusions, powder, and capsules from leaves, oil from seeds, and infusions from flowers, as well as fortifying *M. oleifera* in snacks such as cookies, cream, and crackers (Ramjane et al., 2021).

Processing is essential to minimize microbial growth, thus prolonging shelf life, reducing the risk of health hazards, as well as facilitating the creation of value-added products (Gyamfi et al., 2011). *M. oleifera* can be processed in multiple ways. Dry processing techniques, which include natural sun drying and solar drying, are used due to their easy application and low energy requirement; some wet processing techniques (blanching, boiling, and steaming) are also used to inhibit peroxidase and lipoxygenase enzymes.

10.4.1 Cooking

M. oleifera is traditionally consumed as a common vegetable, in soups, curries, and other different meals. As food, the leaves, flowers, and pods are probably the most utilized. The leaves can be eaten fresh, cooked, or stored as dried powder (Gopalakrishnan et al., 2016). Owing to its high nutritive value, food scientists and nutritionists have been encouraging its cultivation and incorporation into our daily diet.

One of the main nutritional aspects of *M. oleifera* is its high protein content. The leaves possess comparable amounts of protein to those of legumes, such as soybeans and black beans (Alain Mune Mune et al., 2016), in addition to containing 19 of the 20 essential amino acids. Cooking can increase the protein bioavailability of *M. oleifera* leaves. The main factors affecting protein bioavailability depend on amino acid composition and protein digestibility. Fadhilatunnur et al. (2021) showed that thermal treatments—blanching, boiling, and sautéing—enhanced the protein content by 1.98%, 1.72%, and 0.25%, respectively, compared to uncooked leaves.

Plant proteins typically have lower digestibility than animal proteins because of the presence of phytic acid, tannins, and saponins (Sallau et al., 2012). These antinutritional components bind to proteins, as well as minerals like iron, zinc, calcium, and magnesium, inhibiting their absorption (Fadhilatunnur et al., 2021). Nevertheless, a study indicated that boiling significantly reduced the antinutritional compounds cyanide, oxalate, phytate, and trypsin inhibitors in *M. oleifera* leaves. Simmering and blanching also decreased these antinutritional factors to a lesser extent (Sallau et al., 2012). Similarly, Gidamis et al. (2003) found a slight decrease in phytates and a significant decrease in tannins (27% and 45%) after boiling *M. oleifera* leaves and immature pods for 20 min, respectively. They also reported a significant increase in protein digestibility of both leaves and pods after cooking. Mbah et al. (2012), in accordance with previous studies, reported tannin reduction and an increase in protein and saponins after roasting and boiling *M. oleifera* seeds for 20 and 30 min. However, phytates increased after boiling seeds and decreased with increased boiling time. Fadhilatunnur et al. (2021) highlighted the unchanged pepsin digestibility after cooking, the significant decrease in tannin content after blanching and boiling the leaves, while blanching and sautéing increased the saponin content. These studies indicate that cooking can lower the antinutrient content and favor protein and mineral bioavailability.

Thermal treatments are also known to influence phytochemicals (Murador et al., 2018). Since most of the pharmacological properties of *M. oleifera* have been attributed to these bioactive compounds, it is important to consider how cooking affects the benefits of this vegetable. Polyphenols are highly reactive species, which undergo numerous reactions during food processing. These compounds exist in both free and bound forms with macromolecules or are packed in cellular organelles or cell wall components. Cooking can cause complex physical and chemical changes in phenolic compounds, including release from bound forms, degradation, polymerization, and oxidation (Palermo et al., 2013). The phenolic content of both *M. oleifera* leaves and pods was compared before and after cooking. Razzak et al. (2022) recorded loss in the total phenolic and flavonoid content, which was accompanied by a reduction in antioxidant activity after boiling *M. oleifera* pods for 30 min. In contrast, Subramaniam et al. (2017) showed a significant increase in the total phenolic content of *M. oleifera* leaves after both microwaving and boiling for 5 to 20 min. Nevertheless, a 3- to 4-fold lower antioxidant activity was also observed in cooked leaves compared to the raw form. They also analyzed the anti-glycemic potential of the cooked leaves and found that longer microwaving time reduced the α-amylase inhibition activity, while boiling for 5 and 10 min improved the α-glucosidase inhibition activity of the leaves (Subramaniam et al., 2017).

Moreover, a study from our group found that the total phenolic content of the pods was enhanced by 88.6% and 58.8% after boiling and steaming for 20 min, respectively, but the total flavonoid content decreased after the same cooking methods. In general, cooking negatively influenced the antioxidant activities of the pods. The phenolic content of *M. oleifera* leaves was also enhanced after boiling, steaming, and stir-frying. The leaves boiled for 4 min and steamed for 5 min were observed to be the best methods to retain the cytotoxic potential of this vegetable against liver cancer cells (HepG2) (Hip Kam et al., 2023). In addition, new phenolic compounds were identified after cooking *M. oleifera* leaves. Fombang and Romuald (2016) observed improvement in the phenolic content after roasting *M. oleifera* fresh leaves, which was accompanied by the occurrence of two new flavonoids: quercetin-3-O-acetyl glucoside and quercetin-3-O-rhamnoside. Drying and fermentation also positively impact the phenolic content; however, steam blanching had deleterious effects on phenolics, with 31% loss compared to the raw leaves (Fombang and Romuald, 2016). In general, various cooking techniques have been reported to differently influence the phytochemical content of *M. oleifera* leaves and pods.

Glucosinolates have also been reported to be important pharmacological components of *M. oleifera*. During food preparation and cooking, the myrosinase enzyme is released from plant tissues and is in contact with glucosinolates to form isothiocyanates. Mild cooking (60–70 °C) has been reported to retain myrosinase functions and increase isothiocyanate formation (Vanduchova et al., 2019). The most common isothiocyanate found in *M. oleifera* is 4-[(α-L-rhamnosyloxy)benzyl] isothiocyanate, also known as glucomoringin isothiocyanate. This compound has a wide range of biological activities, including anti-inflammatory and antioxidant (Jaafaru et al., 2018a). Glucomoringin isothiocyanate has chemopreventive potential to induce apoptosis and hinder cancer cell proliferation through inhibition of JAK/STAT signaling *in vitro* (Jaafaru et al., 2018b; Michl et al., 2016). This compound has demonstrated its benefits in alleviating ulcerative colitis symptoms in mice (Kim et al., 2017), and neuroprotective potential *in vitro* (Jaafaru et al., 2018b).

Heating has diverse effects on phenolic and glucosinolate compounds, which may affect *M. oleifera*'s bioavailability and bioefficacy. However, further studies on this subject are warranted. *M. oleifera* remains a significant nutritive value food even after cooking, with attractive health benefits.

10.4.2 Fermentation

Fermented foods are produced through the controlled microbial growth and enzymatic conversion of food components. Customarily, foods were fermented as a method of preservation. This technique generated several antimicrobial metabolites, such as organic acids, ethanol, and bacteriocins, reducing the risk of contamination with pathogenic microorganisms. Additionally, fermented foods generate desirable and new organoleptic properties, like taste and texture, that are not present in the starting material (Dimidi et al., 2019).

M. oleifera has also been customarily fermented in some cultures to improve shelf life and develop flavor and nutritional quality (Mehlomakulu and Emmambux, 2022). Zhang et al. (2017) studied solid-state fermentation of *M. oleifera* leaves with *Bacillus pumilus* CICC 10440, a process that occurs in the absence of free water, and found significant nutritional improvement after fermentation. The authors reported a considerable amount of cellulose and pectin, which were broken down by cellulase and pectinase, respectively, promoting nutrient release during the fermentation process. Shi et al. (2020) also reported an improvement in the nutritional quality of solid-state fermented *M. oleifera* leaves with *Aspergillus niger*, *Candida utilis*, and *Bacillus subtilis*. The crude protein, small peptides, and total amino acids were significantly increased after fermentation relative to unfermented leaves and explained this change through the disrupted cellular surface structure and degradation of macromolecular proteins that were observed through electron microscopy and sodium dodecyl sulfate–polyacrylamide gel electrophoresis. Fermentation also improved protein digestibility and significantly reduced the antinutrients tannin and phytic acid. The total free amino acids were also observed to increase from 81.39 ± 2.54 mg/g to 121.95 ± 3.74 mg/g after 48 h *Bacillus subtilis* fermentation (Ali et al., 2020). These results were in accordance with Thierry et al. (2013), who found an increase in protein content, protein digestibility, and reduced phytate content of *M. oleifera* leaves after fermentation.

Fermentation also increased mineral bioavailability of *M. oleifera*. Leaves fermented by *Lactobacillus plantarum* A6 at 30 °C for 5 days increased iron availability from 20.74% to 30.98% (Thierry et al., 2013). Moreover, microbial fermentation significantly enhanced calcium bioavailability and absorption in an *in vitro* cell model. *In vivo*, the supplementation of these fermented leaves promoted the growth and development of calcium-deficient rats, bone calcium deposition, and bone growth, enhanced bone strength, reduced bone resorption, and prevented calcium deficiency (Dai et al., 2020).

The phytochemical content of *M. oleifera* is also influenced by this processing method. Ali et al. (2020) noted an important increase in the total phenolic content associated with higher antioxidant activities of the leaves after fermentation. Furthermore, total isoflavones content increased significantly from 732 µg/g for the unfermented leaves to 881, 1083, 979, and 773 µg/g after 24, 48, 72, and 96 h fermentation, respectively. Both the aglycone isoflavones, that is glycitein, daidzein, and genistein, and the glycoside isoflavones (glycitin, daidzin, and genistin) significantly increased after fermentation. Isoflavones are well documented and a very important subclass of flavonoid compounds, with structural resemblance to estrogen hormones. Due to these similarities, these molecules are recommended to alleviate several pathologies dependent on hormonal disorders, such as breast, endometrial, and prostate cancers, osteoporosis, and cardiovascular diseases, as well as to minimize menopausal symptoms (Gómez-Zorita et al., 2020).

Moreover, supplementation with fermented *M. oleifera* leaves has shown interesting immunomodulatory protective potential. Female Balb/C mice fed fermented *M. oleifera* leaves displayed higher protection against *Salmonella typhi* infection than control mice. The fermented-leaves-fed mice showed better inhibition of TLR3/TLR4, regulatory T cells, and the proinflammatory cytokines followed by naïve Tregs recovery (Wuryandari et al., 2022). Fish infected with *Aeromonas hydrophila* and supplemented with fermented *M. oleifera* leaves for 50 days also showed better immune response than the control group. The authors observed significant regulation of the immune-related genes: toll-like receptors 2 (TLR2), TNF-α, IL-1β, and IL-8; and enhanced antioxidant response of serum SOD, CAT, and GPx activities and decreased serum malondialdehyde and protein carbonyl contents in the fermented *M. oleifera* leaves-fed group. The intervention significantly increased growth, weight, and survival rate, and improved liver function (Zhang et al., 2020).

Besides, fermented *M. oleifera* reduced the risks of metabolic disorders, such as obesity, diabetes, and cardiovascular diseases. Glucose intolerance and non-alcoholic fatty liver disease are often symptoms or consequences of these diseases. The fermented *M. oleifera* leaves attenuated those conditions in C57BL/6 mice previously fed with a high-fat diet for 10 weeks. Consumption of the fermented leaves decreased high-fat-diet-induced endoplasmic reticulum stress, oxidative stress, and lipotoxicity in quadriceps muscles. The high-fat diet induced inflammation in the liver, adipose, and muscle tissues. Following the fermented *M. oleifera* leaf diet, proinflammatory cytokines (TNF-α, IL-6, and IL-12) mRNA expressions markedly decreased in the liver and in the quadriceps, whereas these cytokines and MCP1 expressions were downregulated in the epididymal adipose tissue. Hepatic adiposity and lipid accumulation in skeletal muscles as a consequence of the high-fat diet were reduced through the modulation of lipid metabolism. Moreover, the study showed improved glucose tolerance after the fermented *M. oleifera* leaf diet (Joung et al., 2017). Furthermore, both unfermented and fermented *M. oleifera* seeds showed better potential in reducing hypercholesterolemia markers in rats after an eight-week diet; however, fermented seeds showed better results. The fermented *M. oleifera* seeds-fed rats positively modulated LDL and HDL levels, boosted neutrophil and eosinophil counts, and improved architectural integrity of villi and splenocytes (Adedayo et al., 2018).

10.4.3 *M. oleifera* Oil

Oil is one of the main components of *M. oleifera* seeds, accounting for 36.7% of the seed weight. Solvent (n-hexane) oil extraction is the most efficient extraction method. Cold-press extraction is also used, but with lower yield, and only an average of 69% was extracted using this method (Leone et al., 2016). *M. oleifera* seed oil is high in monounsaturated fatty acids, which aid in the regulation of cardiovascular health. The most abundant monounsaturated fatty acids present are oleic, palmitoleic, and eicosenoic acids (Bhutada et al., 2016; Vaknin and Mishal, 2017), while an average of 1.18% of polyunsaturated fatty acids is present, with an approximate 0.76% and 0.46% of linoleic and linolenic acids, respectively (Leone et al., 2016). Therefore, *M. oleifera* seed oil can represent a healthy alternative to partially hydrogenated vegetable oils.

In addition to its attractive fatty acid composition, *M. oleifera* seed oil has been extensively studied for its pharmacological properties. The seeds have been reported for their interesting composition of phenolic acids, predominantly gallic, ellagic, caffeic, *p*-coumaric, vanillic, protocatechuic, ferulic, and cinnamic acids, as well as flavonoids, represented by catechin, epicatechin, quercetin, and kaempferol, present mainly in the bound form. Alkaloids, glucosinolates, and isothiocyanates have also been identified from the seeds (Leone et al., 2016).

Due to its high oleic content, *M. oleifera* oil is an ideal anti-inflammatory agent, which can help prevent various ailments such as cardiovascular diseases and breast cancer. This oil is rich in vitamins A and E, which have great anti-inflammatory properties and aid in skin hydration (Nadeem and Imran, 2016). In an *in vivo* study, Mahajan et al. (2009) reported the anti-inflammatory effect of *n*-butanol *M. oleifera* extract in guinea pigs with ovalbumin-induced airway inflammation. Cretella et al. (2020) demonstrated that topical *M. oleifera* oil application greatly reduced skin inflammation in Swiss mice as well as epidermal tetradecanoylphorbol-13-acetate-induced hypertrophy.

Ruttarattanamongkol and Petrasch (2016) demonstrated that Soxhlet extraction with light petroleum ether of *M. oleifera* seed oil inhibited the growth of selected bacteria including *Staphylococcus aureus, Bacillus subtilis, Salmonella typhimurium, Enterobactor aerogenes, Pseudomonas aerogonosa* and *Escherichia coli*. Bukar et al. (2010) performed two solvent extractions with chloroform and ethanol; the *M. oleifera* chloroform extract was active against two bacterial isolates, *E. coli* and *S. typhimurium*, as well as the ethanol extract counterpart against three bacterial strains, *S. aureus*, *S. typhimurium*, and *E. coli*.

The hepatoprotective potential of *M. oleifera* seed oil was evaluated by studying the effect of this oil against carbon tetrachloride (CCL_4)—induced hepatotoxicity in Wistar albino rats (Olatosin et al., 2013). The rats were treated with 2 mL of oil alone and in juxtaposition with CCL_4 for 10 days. The collection of blood and liver tissue samples, as well as assays on alanine transaminase (ALT), Transaminase (AST), and alkaline phosphatase (ALP) activity, was carried out. The results showed that *M. oleifera* seed oil has remarkable hepatoprotective activity by anti-necrotic mechanisms observed in histological features of the livers, as well as a decrease and increase in ALT and AST activities in *M. oleifera* seed oil-treated rats and only CCL_4-treated rats, respectively. *M. oleifera* oil was able to reverse lipid peroxidation in liver cells.

The nephroprotective effect of *M. oleifera* seed oil was investigated on gentamicin-induced nephrotoxicity in rats (Edeogu et al., 2019). Gentamicin is an aminoglycoside antibiotic used with great efficacy against gram-negative bacteria; however, gentamicin has one side effect where it induces nephrotoxicity in a non-targeted manner. *M. oleifera* seed oil was used both alone and in juxtaposition with gentamicin; the results showed that the oil was able to attenuate gentamicin-induced nephrotoxicity through antioxidative, anti-inflammatory, and anti-apoptotic action.

The high oleic content of *M. oleifera* oil was helpful in the prevention of cardiovascular diseases due to its anti-inflammatory properties. It was also proposed that this oil could help lower bad cholesterol in non-hypertensive individuals (Ghazali and Mohammed, 2011). *M. oleifera* seed oil helped lower nocturnal heart rate and improved cardiac diastolic function in spontaneous hypertensive rats without changing blood pressure (Randriamboavonjy et al., 2017).

An *in vivo* study investigated the ability of *M. oleifera* oil to abrogate cerebral neurotoxicity (Famurewa et al., 2019). Methotrexate (MTX), an antineoplastic drug, is used for the treatment of various types of cancer. MTX, despite its effectiveness, has some serious side effects associated with various organs, the brain being one of them. This MTX-induced neurotoxicity induces oxidative stress as the leading factor for the etiology of this condition. The study explored the effect of *M. oleifera* oil treatment for 17 days in MTX-induced neurotoxicity rats. Parameters such as acetylcholinesterase, antioxidant enzymes, and lipid peroxidation, as well as reduced glutathione and NO levels, were monitored. *M. oleifera* oil treatment significantly abrogated the MTX-induced neurotoxicity by inhibiting cholinesterase, reducing oxidative stress, and anti-inflammatory actions (Famurewa et al., 2019).

M. oleifera seeds have high oil content (27%–42%) depending on genetic (varieties or cultivars), environmental factors (growing region and condition), and extraction process (Gharsallah et al., 2022). Mechanical (cold) pressing yields the lowest amount of oil (11%–29%) depending on seed treatment prior to pressing and processing conditions. The seed oil health benefits are due to its high oleic acid (>70%) content and the presence of phytosterols and other bioactive compounds such as tocopherols, phenolics, and carotenoids (Gharsallah et al., 2022).

Ogunsina et al. (2014) dehulled *M. oleifera* seeds (Jaffna variety from Tamil Nadu, India) prior to cold hydraulic pressing (CPMSO) or hexane (percolation) extraction (HMSO) at room temperature (25–30 °C) (Table 10.6). The same investigators dehulled (34.6% seed coats removed) dry *M. oleifera* seeds (Jaffna

TABLE 10.6 Fatty acid contents of Indian *M. oleifera* seed oils

	JAFFNA		*PKM-1*		*JANAKI-(2015)*
FATTY ACIDS %	*COLD PRESS*	*HEXANE*	*COLD PRESS*	*HEXANE*	*COLD PRESS*
Myristic (C14:0)	0.24	0.72	0.13	0.13	
Palmitic (C16:0)	5.8	6.1	6.34	6.46	12.97
Palmitoleic (C16:1)	1.2	1.2	1.38	1.45	
Stearic (C18:0)	3.9	4.6	5.7	5.88	2.95
Olec (C18:1)	79.5	78.7	71.6	71.21	77.4
Linoleic (C:18:2)	ND	ND	0.77	0.65	1.4
Linolenic (C18:3)	2.2	1.8	0.2	0.18	1.39
Arachidic (C20:0)	5.1	4.5	3.52	3.62	
Gadeolic (C20:1)	ND	ND	2.24	2.22	
Behenic (C22:0)	5.1	4.5	6.21	6.41	
ΣSFA	17.2	18.3	23.23	23.79	15.92
ΣMUFA	80.7	79.9	75.34	75	77.4
ΣPUFA	2.2	1.8	0.97	0.83	2.79
Oxidizability (Cox) value	1.27	1.18	0.84	0.82	1.22
Thrombogenic Index (TI)	ND	ND	0.31	0.32	0.34

Adapted from Ogunsina et al. (2014) for Jaffna; Lalas and Tsaknis (2002) for PKM-1 and Janaki (2015)

variety) purchased locally in Mysore, Karnataka, India, prior to Soxhlet extraction (1:10 dehulled seeds to hexane ratio, percolated with warm hexane 35–40 °C for 8 h). Jaffna seed kernels contained 39.2% oil with unsaponifiable matter (6.5 g/kg), total tocopherol (88 ppm), total sterols (1700 ppm), total phenolics (119 ppm), total carotenoids (17 ppm), and DPPH radical scavenging activity (35.5 IC_{50}) (Bhatnagar and Krishna, 2013).

Periyakulam 1 (PKM 1) is a new early (fruiting after 6 months from sowing vs 3–4 years for other varieties) variety released from the Horticultural Station of the Tamil Nadu Agricultural University in India. *M. oleifera* (PKM 1 variety grown in Kenya) seeds were cold pressed or solvent (hexane or 1:1 chloroform:methanol mixture) extracted and degummed prior to analysis. However, differences in fatty acid and sterol compositions were not significant for the oils extracted by the three methods. The oil content was 25.1% and 38.3% (or g oil/100 g seed) for cold-pressed and hexane-extracted oils. The extracted oils were not neutralized due to low free fatty acid content, but were degummed to reduce cloudiness; the cold-pressed and hexane-extracted oils contained 2.5% and 0.7% gums, respectively. PKM 1 oil contained other fatty acids (0.04% caprylic [C8:0], 0.08% margaric [C17:0], 0.12% erucic [C22:1], and 1.2% cerotic [C17:0] acids) in addition to those in Table 10.6. β-Sitosterol is the most predominant PKM 1 sterol; PKM 1 oil is relatively high in behenic acid, which can lighten chocolate texture and oily feel, and provide excellent mouthfeel (Lalas and Tsaknis, 2002). Janaki (2015) purchased *M. oleifera* seeds from a private exporter (Coimbatore, Tamil Nadu) without indicating the specific variety.

M. oleifera seed oil is primarily oleic acid-rich (72%–80%), with some saturated fatty acids (mainly palmitic, stearic, arachidic, and behenic acids) (Table 10.6). The high monounsaturated and low polyunsaturated fatty acids confer oxidative stability to the oil, indicated by the low oxidizability (Cox) value. The high oxidative stability of the oil is partly due to low levels of polyunsaturated fatty acids and the presence of the powerful flavone antioxidant myricetin (Gunstone, 2006).

The Cox value of the oil is calculated by the percentage of C18 fatty acids applying an earlier proposed formula (Fatemi and Hammond, 1980);

$$Cox = (1[C18{:}1\%] + 10.3[C18{:}2\%] + 21.6[C18{:}3\%])/100$$

Frying studies also confirm the least deterioration of cold-pressed PKM 1 *M. oleifera* oil compared to organic solvent-extracted oil (Cheikhyoussef et al., 2020 and references therein). The low thrombogenic index was similar to that reported for olive oil (0.32) and indicates potential serum cholesterol reduction and heart disease prevention (Oomah, 2023 and references therein). Arachidic, gadoleic, and behenic acids are in considerably higher amounts compared to some common vegetable oils (Cheikhyoussef et al., 2020).

The presence of behenic acid confers emollient and lubricant properties (in addition to flowability provided by oleic acid) that contribute to skin hydration, enabling the use of the *M. oleifera* seed oil in aromatherapy treatment and as massage oil.

The primary sterols of *M. oleifera* seed oils were β-sitosterol, stigmasterol, campesterol, Δ5-avenasterol, and clerosterol (Table 10.7). These five sterols accounted for over 93% of the total seed oil sterols. The major sterol, β-sitosterol, is the main dietary phytosterol reported to exhibit antiviral, antifungal, anti-inflammatory, anticarcenogenic, and antiartherogenic properties (Cheikhyoussef et al., 2020 and references therein). Nanoparticles (50 nm diameter) of β-sitosterol isolated from *M. oleifera* exert anti-inflammatory activity (Liao et al., 2018). They suppress the secretion of inflammatory cytokines (TNF-α, IL-1β, IL-6, and IL-8) and significantly reduce ROS production from keratinocytes (HaCaT cells) and macrophages (J774A.1 cells). Moreover, β-sitosterol significantly reduced NLRP3 (a key inflammasome component) expression and inhibited caspase-1 activation. Hence, *M. oleifera* β-sitosterol effectively inhibits the generation of inflammatory cytokines and proteins in dermal keratinocytes and immune cells and can therefore enhance innate immunity.

TABLE 10.7 Sterols of cold pressed and hexane extracted *M. oleifera* seed oils

	PKM 1		*LEONE 2016*	
STEROLS (%)	*COLD PRESS*	*HEXANE*	*COLD PRESS*	*HEXANE*
Brassicasterol	0.06	0.05	0.06	0.05
Campesterol	15.81	15.29	17.84	17.26
Campestanol	0.36	0.33	0.28	0.30
Cholesterol	0.18	0.10	0.17	0.09
Clerosterol	2.08	1.22	1.23	1.80
β-Sitosterol	45.58	46.65	47.17	47.07
Ergostadienol	0.30	0.35	0.30	0.30
Δ 5-Avenasterol	8.46	11.61	8.04	9.01
Δ 7-Avenasterol	0.53	ND	0.64	0.79
28, Isoavenasterol	0.27	0.25	0.52	0.74
Stigmasterol	23.1	23.06	19.26	18.59
Stigmastanol	0.76	0.64	0.89	0.79
Δ 7-Stigmastanol	ND	ND	0.58	0.68
Δ 7,14 Stigmastanol	0.35	0.85	0.62	0.52
Δ 5,23 Stigmastadienol	ND	ND	1.23	1.23
Δ 7,14 Stigmastadienol	0.52	0.39	ND	ND
24-Methylene-cholesterol	0.07	0.08	0.61	0.9
Tocopherols (mg/kg)				
α-Tocopherol	5.06	15.38	164.2	121.7
γ-Tocopherol	25.4	4.47	55.5	66.7
δ-Tocopherol	3.55	15.51	146.1	59.6
Total	34.01	35.36	365.8	248
Vitamin E (calculated)	26.1	6.4	73.8	75.1

Adapted from Lalas and Tsaknis (2002) and Leone et al. (2016)

PKM 1 *M. oleifera* seed oil contained 0.52% and 0.56% (w/w) total sterols in cold-pressed and hexane-extracted oil, respectively (Lalas and Tsaknis, 2002). *M. oleifera* (Jaffna variety) hexane-extracted kernel (dehulled) seed oil contained 1700 ppm of total sterols, higher than those of dried coconut kernel, palm, and Niger seed oils (Bhatnagar and Krishna, 2013). The Jaffna kernel oil also had higher tocopherol contents (56.2, 12.6, 19.2, 88, and 57.7 ppm for α-, γ-, δ-, total tocopherols, and calculated vitamin E) than the PKM 1 variety grown in Kenya. Vitamin E was calculated according to the formula of McLaughlin and Weihrauch (1979) where vitamin E = [α-tocopherol + (0.1 × γ-tocopherol) + (0.01 × δ-tocopherol)].

Tocopherol composition of PKM 1 varied (Table 10.7), with the hexane extract containing higher amounts of α- and δ-tocopherols than the cold-pressed oil. Moreover, the calculated vitamin E contrasts the two extractions, with the value for the cold-pressed seed oil over 4-fold that of the hexane extract. Nevertheless, these values were orders of magnitude lower than the average values for cold-pressed and hexane-extracted oils (Leone et al., 2016). Moreover, *M. oleifera* oil contains substantial amounts of vitamins (425, 210, 121, and 0.7 mg/g of vitamins C, E, A, and B1, respectively) and minerals (80, 66, 65, 15, 13, and 6 ppm of potassium, magnesium, calcium, iron, zinc, and selenium, respectively) (Nebolisa et al., 2023). These phytochemicals (sterols, tocopherols, vitamins, minerals, and others) enrich *M. oleifera* seed oil, improving its quality, adding value, and enabling its use in various food and non-food applications.

M. oleifera seed oil and its bioactive β-sitosterol dose-dependently inhibit the central nervous system in male mice (Liu et al., 2020). The seed oil and β-sitosterol dose-dependently reduced sleep latency and increased total sleep time, indicating that these bioactive compounds interact with pentobarbital on the CNS via the GABAergic mechanism in the brain. β-Sitosterol improved GABA levels in brain tissues, and this increased GABA concentration can affect spontaneous locomotor behavior and anticonvulsant actions in mice. Thus, the seed oil can be a potential therapeutic agent to develop functional food to treat insomnia.

10.4.4 *M. oleifera* Tea

After water, tea is the most popular beverage in the world. Recently, "herb tea" has been growing in popularity. Herbal teas are infusions or tisanes of dried fruits, flowers, spices, or herbs in water. Traditionally consumed for their medicinal assets, these drinks have today become fashionable, in particular, because of their special fragrance, new flavor, and caffeine-free content (Pan et al., 2022)

Marketed for its numerous nutritional and health-promoting features, *M. oleifera* tea has garnered a lot of attention and is today well known (Wickramasinghe et al., 2020). *M. oleifera* leaf tea possesses high phytochemical contents with total phenolic and flavonoid content of 1.78 μg gallic acid equivalent/g and 323 μg quercetin equivalent/g, respectively (Rahayu and Timotius, 2022). Five main bioactive compounds were identified from the infusion: three flavonoid glucosides, namely: kaempferol 3-α-L-dirhamnosyl-(1→4)-β-D-glucopyranoside, kaempferol 3-O-neohesperidoside, and quercetin 3-O-glucoside, and the flavonoid undulatoside A, reported for its anti-inflammatory properties (Peng et al., 2016) and the alkaloid gentiatibetine, noted for its anticonvulsant and brain-protective effects (Rahayu and Timotius, 2022).

M. oleifera tea has also been reported for its strong antioxidant potential. A Japanese *M. oleifera* tea, made from 2 g of both dried leaf and stem, infused in water (100 mL) at 95 °C for 20 min, demonstrated higher O_2^- radical scavenging activity than the Trolox standard in the phenazine methosulfate–NADH–nitroblue tetrazolium and xanthine oxidase assay system and in a cellular model of human neutrophils (Sugahara et al., 2018). *M. oleifera* tea also exhibited superior scavenging capacities compared to the standard butylated hydroxytoluene, a well-known antioxidant compound (Rahayu and Timotius, 2022). Sandanuwan et al. (2022) showed that *M. oleifera* hot water extract could exert its antioxidant activity by modulating Nrf2/HO-1 signaling pathway in a kidney epithelial cell model (Vero cells).

Antioxidant compounds have an important role in protecting ROS-induced DNA damage. DNA is initially double-stranded with a supercoiled conformation and high electrophoretic mobility. However, when exposed to ROS, the DNA strand breaks, resulting in an open-loop conformation of low electrophoretic mobility. *M. oleifera* tea demonstrated an antimutagenic and strong DNA protective effect, restoring the conformational condition of pBR322 plasmid exposed to ROS, almost matching the untreated plasmid (Rahayu and Timotius, 2022).

DNA damage has long been recognized as a hallmark of carcinogenesis. Erroneous DNA repair leads to mutations or chromosomal aberrations affecting oncogenes and tumor suppressor genes. Cells undergo malignant transformation, resulting in cancerous growth (Sugahara et al., 2018). Madi et al. (2016) infused *M. oleifera* leaves in hot water, mimicking the people's mode of leaf drink preparation, and tested it on a malignant lung cancer (A549) cell line. The cancerous cells' mitochondrial membrane potential and ATP levels significantly decreased after *M. oleifera* tea treatment, followed by an increase in ROS production and caspase activation, as well as a higher level of proapoptotic expressions of p53, SMAC/Diablo, AIF, and PARP-1 cleavage.

M. oleifera tea is also known for its antiviral potential. In an *in silico* model, five main bioactive compounds identified from the tea (gentiabatine, quercetin, quercetin 3-O-glucoside, kaempferol 3-O-neohesperidoside, and undulatoside A) exhibited strong energy bonds with Mpro, a key enzyme in SARS-CoV-2 virus replication (Rahayu and Timotius, 2022).

M. oleifera tea is also beneficial against stomach ulcers. This chronic disease affects the gastric mucosa and is accompanied by several complications, including bleeding, stenosis, and perforation, as well as a high incidence of recurrence. Pre-treatment with *M. oleifera* leaf infusion protected gastric ulcer-induced rats. Oral administration of the infusion increased the gastric mucosa, malondialdehyde, the antioxidant enzyme activities of SOD, CAT, and GPx, and pH in the stomach, reducing lesion index by 79% compared to non-treated rats (Dalhoumi et al., 2022).

The male reproductive system also benefits from *M. oleifera* tea. Male rats fed with *M. oleifera* leaf tea extract at 0.55, 1.10, or 2.20 mg/kg for 30 days increased their total spermatogenic cells. The infusion stimulated type A spermatogonia proliferation and differentiation to pachytene primary spermatocytes. The male rats receiving 0.55 mg/kg tea were those with the most effective increase in the seminiferous tubule diameter, epithelium area, luminal area, and total spermatogenic cell count. In addition, rats supplemented with 1.10 and 2.20 mg/kg *M. oleifera* tea had a significant increase in the number of Sertoli cells, which consequently stimulated sperm concentration and sperm viability (Laoung-On et al., 2021).

Drinking *M. oleifera* tea may help manage hyperglycemia. In a human study, consumption of *M. oleifera* tea (low dose (200 mL) or high dose (400 mL)) prior to administration of 50 g of glucose suppressed the rise in blood glucose compared to the negative group in a dose-dependent manner. Interestingly, the study also reported that the group receiving the low dose manifested a significantly higher anti-hyperglycemic effect after 30 min of glucose ingestion than the high-dose group. Apparently, *M. oleifera* tea exerted its effect by restricting glucose absorption at the intestinal level, while the high dose worked more in circulation (Fombang and Romuald, 2016).

10.4.5 *M. oleifera*-Supplemented Food Products

M. oleifera as supplements in our daily foods can be beneficial due to its high nutritional value and pharmacological properties. Fortifying foods with *M. oleifera* can provide vitamins, minerals, essential amino acids, and fatty acids, as well as grant prophylactic properties against several diseases. In this regard, many researchers, entrepreneurs, and industrialists have been and are working on the development of fortified foods with *M. oleifera*.

Many *M. oleifera*-fortified foods have been developed, such as bakery products like biscuits (Alam, 2014), bread (Chinma et al., 2014), cake (Kolawole et al., 2013), yoghurt (Hekmat et al., 2015), pasta (Coello et al., 2021), nuggets (Madane et al., 2019), soups (Babayeju et al., 2014), weaning foods (Singla et al., 2019), and noodles (Zula et al., 2021). These fortified foods have garnered much attention, and the results are attractive, as these products have high nutritional value and increased bioactive compounds. Food processing normally reduces the presence of antinutritional components such as phytates; as a result, supplementing *M. oleifera* to these different processed foods maximizes the bioavailability of specific nutrients and beneficial compounds. Enrichment of cereal porridge with *M. oleifera* powder significantly improved vitamin A content by 15-fold, as well as the protein content by 44% to 94% in the different porridge formulations (Milla et al., 2021). In another study, a cereal-based meal composed of maize, soybean, and peanut (60:30:10 ratio) was supplemented with 5%, 10%, and 15% *M. oleifera* leaf powder. This supplementation significantly increased the protein, fiber, and mineral content of the cereal meal. Consuming this cereal meal daily for 28 days significantly improved protein efficiency ratio, with the best performance at 10% *M. oleifera* supplementation in the rat feeding trials (Shiriki et al., 2015). Furthermore, adding 3% *M. oleifera* leaf powder to South African white maize porridge increased the contents of protein, total mineral content, zinc, iron, total provitamin A level, total phenolics, and flavonoids (by 11%, 67%, 60%, 45%, 41%, 40%, and 139%), respectively, as well as the antioxidant activity by 10-fold (Ntila et al., 2020).

These *M. oleifera*-supplemented foods showed great nutritional and functional promise. Nevertheless, addition of *M. oleifera* in our daily common foods affects their organoleptic characteristics, and several studies have shown that high *M. oleifera* concentration in the formulated products is not accepted by the general public due to its astringent and bitter taste and its peculiar odor. Therefore, several trials are required before obtaining the right formulation for *M. oleifera*-fortified food. Different *M. oleifera* flours have been developed and can be used in bakery products, noodles, and pasta. This may represent an advantage since these products are commonly consumed and more easily acceptable. Dehydrated *M. oleifera* seeds and leaf powder have been used in several formulations of various edible products to obtain fortified or functional foods. It was advised that, due to its high oil absorption capacity, untreated leaf flour can be used in bakery food formulations, while alkali-pretreated *M. oleifera* leaf flour (defatted flour) could be more suitable for making low-fat snack products (Saucedo-Pompa et al., 2018). In addition, these flours have a long shelf life without nutrient loss (Conti et al., 2021).

Bread, one of the most common staple foods around the world, is an excellent common food for *M. oleifera* supplementation. The addition of *M. oleifera* leaf flour in the bread dough has shown some promising improvement in nutritional content. *M. oleifera* leaf flour addition (5%) to bread increased its protein and crude fiber (54% and 56%) contents. Some nutrients like magnesium, calcium, and β-carotene also increased from 0.76 to 1.27 mg/100 g, 3.67 to 6.07 mg/100 g, and 0.02 to 3.27 mg/100 g, respectively (Sengev et al., 2013). In another recipe, *M. oleifera*-fortified bread had higher protein and crude fiber contents of approximately 17% and 88%, respectively (Chinma et al., 2014). Though the nutritional composition of the bread improved with supplementation of *M. oleifera* leaf flour, its sensory acceptability decreased with increasing *M. oleifera* concentrations. The texture and the color of the crust showed poor sensory appeal, whereas the loaf volume, weight, and height decreased (Chinma et al., 2014). A more favorable fortification supplement was found in *M. oleifera* seed flour. Bread fortified with *M. oleifera* seed (15%) had increased protein (67%) content without significantly altering the organoleptic properties. Fortifying bread with 5% *M. oleifera* seed flour showed similar physical properties to wheat bread and had a superior sensory appeal compared to the bread supplemented with *M. oleifera* leaves (Oyeyinka and Oyeyinka, 2018).

M. oleifera leaf powder addition (5%) produced the maximum acceptability of whole wheat flour bread (Khan et al., 2023). *M. oleifera* addition significantly increased total phenolic content (0.72–1.99 mg GAE/100 g dw) and improved proximate (mg/100 g) composition (protein 12–19, fat 1.5–2, fiber 16.3–17.8), and mineral profile (Ca 12.7–33.1, Mg 57.2–62.2, K 163–225, Zn 0.53–1.81, and Fe 4.21–4.8 mg/100 g). However, antioxidant content decreased (DPPH 25.3 to 18.6 and FRAP 40 to 28.5 μmol TE/g dw). Acceptability of durum wheat flour noodles decreased with increasing *M. oleifera* leaf powder substitution; *M. oleifera* leaf powder substitution (5%–20%) reduced noodles acceptability by 15% to 32% (Zula et al., 2021). However, 5% *M. oleifera* leaf powder substitution did not affect the proximate composition of durum wheat noodles.

Arif et al. (2022) developed a strawberry-*M. oleifera* extract (90% strawberry juice and 10% blanched aqueous *M. oleifera* leaf extract) juice blend that overcame *M. oleifera*'s limiting bitter taste and green color acceptability. The juice blend positively improved the total phenolic content, antioxidant, and vitamin C from 12 to 40 mg GAE/100 g, 61% to 79%, and 64 to 79 mg/100 mL, respectively. However, cold storage (4 °C for 28 days) reduced the vitamin C (79 to 72 mg/100 mL), antioxidant (79% to 73.7%), and total phenolic contents (40–28 mg GAE/100 g).

Different fortified meat products have also been tested. *M. oleifera* is a rich source of dietary fiber, micronutrients, and phytochemicals, which can complement a meat meal. One example is the attempt to develop chicken functional products by a Spanish team. They formulated a novel chicken nuggets recipe supplemented with 2% *M. oleifera* flowers. These nuggets enhanced the cooking yield, emulsion stability, total mineral content, proteins, phenolics, and dietary fiber. Besides, addition of *M. oleifera* flowers improved the sensory attributes—texture, odor, and color—as well as oxidative stability. The product was stable and acceptable after 15 days of storage (Madane et al., 2019).

These *M. oleifera*-fortified foods show great potential. Some of these foods are already on the market and have been patented. For instance, a nutritious *M. oleifera* biscuit was patented in 2011 by a Chinese company. This biscuit aims at providing a highly nutritious snack and eliminating the bitter taste of *M. oleifera* leaves. The composition of a *M. oleifera* dough for noodle preparation was also patented. The method consists of drying the fresh leaves before the dough-making procedure. Several cakes supplemented with *M. oleifera* leaves, seeds, or flowers have also been licensed.

10.5 PATENTED *M. OLEIFERA* PRODUCTS

Solanki (2004) described a medicinal preparation comprising a mixture of 11 herbs including *M. oleifera* (seeds and leaves, 10% w/w) that is effective to treat cancer, particularly squamous cell carcinomas, tumors, and other metastatic states, including lung cancer. The herbs are ground, extracted by Freon, and the concentrates mixed in appropriate proportions to produce a pharmaceutical or medicinal preparation used to treat lung cancer as a cancer immunomodulatory and anti-inflammatory agent, a radiotherapy adjuvant, and a dietary supplement for cancer patients (Solanki, 2004).

The nitrile glycoside niaziridin was isolated by bioactivity-guided fractionation of *M. oleifera* pods (Khanuja et al., 2005). Niaziridin, its derivatives, and analogues enhance the bioactivity of commonly used antibiotics (rifampicin, tetracycline, and ampicillin) against gram-positive and gram-negative bacteria. Moreover, it improves drugs', vitamins', and nutrients' absorption through the gastrointestinal membrane, thereby increasing their availability. Hence, niaziridin can be used in combination therapy with drugs and nutrients to reduce drug-associated toxicity, cost, and chemotherapy duration (Khanuja et al., 2005).

M. oleifera leaves, seeds, and fruits have been used to develop a nutraceutical beverage to provide biologically metabolized nutritional composition for health, well-being, and treatment of ailments (Andrews, 2006). The leaves and seeds contain potent anti-aging cytokinins (zeatin, dihydrozeatin, and isopentyladenine) that delay aging and protect against human neuronal toxicity by modulating cell division and through their antioxidant properties. Human consumption of the beverage (5.5 fl oz/daily for 30 days) is claimed to alleviate migraine headaches and depression, shorten duration of viral infections, elevate mood of fibromyalgia subjects, reduce allergen sensitivity, blood pressure, and blood glucose level, improve eyesight, moisturize and soothe skin, relieve arthritic pain, improve acid reflux, and increase endurance.

Managoli (2007) describes a medical preparation comprising a combination of two herbal compositions to be administered together to treat a wide range of human physiological and pathological conditions against weakened or deteriorating immune conditions. The first herbal composition consists of 11 herbs, while the second contains *M. oleifera* bark (~15% by weight, >5% glycosides) and six other herbs. The polyherbal gelatin capsule (750 mg) containing *M. oleifera* (115.5 mg) is claimed to beneficially improve human immune response and rejuvenate the immune system.

A process is reported for the preparation of *M. oleifera* gum bio-nanoparticles (<150 nm) that protect DNA and RNA plasmid degradation and induce membrane blebbing, resulting in apoptosis of human epidermoid carcinoma Hep2 cells (Shanmugasundaram et al., 2007).

Hydroalcohol *M. oleifera* leaf extract containing kaempferol-3-O-glucoside, quercetin-3-O-glucoside, and niazinin is presumably useful for hyperlipidemia treatment. The extract (10 μg/mL) inhibited lipid accumulation in differentiated 3T3-L1 human adipocytes with 25.7% anti-obesity (adipogenesis) activity (Gokaraju et al., 2013). The combination (by weight) of *M. oleifera* leaf extract (30%–70%), *Murraya koenigii* leaf extract (15%–45%), and enriched demethylated curcuminoids from *Curcuma longa* (10%) is claimed to ameliorate adipogenesis- and lipolysis-mediated diseases. This synergistic combination effectively reduces adipose tissue mass, body weight, total serum cholesterol level, phospholipids, and triglycerides and inhibits 5-lipoxygenase.

Japanese patent (JP4810078B2, 2004) assigned to Ohta Oil Mill Co. Ltd. discloses a process to increase the oxidative stability of *M. oleifera* seed oil. The process involves roasting the dehulled seed or seed without sheath (130–140 °C) and adding 5%–10% water (by weight) to adjust the bulk specific gravity (0.46 g/cm^3) prior to crushing. The oil is further refined by degumming, deoxidation, decolorization, dewaxing, and deodorization steps similar to those used in conventional vegetable oil refining. The refined oil is claimed to have excellent oxidation stability (AOM value 200–700 h vs 80–100 h for conventional oil). The high oxidative stability of the oil enables its use as an edible oil and fat and for cosmetics. Danoux et al. (2020) disclose a protein extract of defatted non-germinated *M. oleifera* seeds designed to treat sensitive, sensitized, reactive, fragile, and/or weakened skin and/or mucous membrane such as contact urticaria, irritative or allergic contact dermatitis, eczema, psoriasis, seborrheic or atopic dermatitis, and/or treatment or prevention of inflammation, erythema, particularly diaper rash. The extract claims to inhibit the release of proinflammatory cytokines (IL-6 and IL-8) induced by the opportunistic pathogenic *Staphylococcus aureus*, thereby reducing skin inflammation. The protein (7100–11,000 Da) is obtained preferably by aqueous extraction of dehulled, defatted (hexane extraction) *M. oleifera* seeds using conventional protein extraction and pH precipitation methods. It is used in solid form (0.01%–1% dry matter) in cosmetics and dermatological formulation. Purisoft™ LS9726, a BASF product, is a *M. oleifera* seed anti-pollution peptide clinically proven to protect skin from pollution, purify the skin, and enhance skin complexion and a healthy glow.

Pernodet et al. (2017) describe an emulsion comprising an extract (0.01%–5%) obtained from live damaged *M. oleifera* plant parts that release myrosinase enzyme (0.5%–3.5% by weight of total *M. oleifera* extract of *M. oleifera* isothiocyanates) and acetyl hexapeptide-8 (0.0001%–5%), *Laminaria digitata* extract (0.0001%–5%), and whey protein (0.01%–5%). The extract claims to stimulate collagen synthesis in skin cells. High *M. oleifera* isothiocyanate (MICs) concentrations can be prepared by injuring plant parts to increase myrosinase enzyme content that, in turn, promotes *M. oleifera* glucosinolate conversion to MICs. *M. oleifera* seed extract containing 0.5%–3% MICs is available from Nutrasorb LLC under the trade name Nutriga™ (mixture of *M. oleifera* seed extract and isoceteth-20 in 7.5:92.5 ratio) containing about 1% MIC. Nutriga™ is highly anti-inflammatory and protects against aging and oxidative stress according to Nutrasorb LLC. MIC-1 (1 µM), purified from aqueous *M. oleifera* seed extract, possesses strong anti-inflammatory and antioxidant properties. It significantly reduces nitric oxide (NO) and interleukins 1β and 6 (IL-1β and IL-6) production and suppresses LPS-inducible nitric oxide synthase (iNOS). Moreover, it upregulates the nuclear factor (erythroid-derived 2)-like 2 (Nrf2) target genes *NAD(P)H, quinone oxidoreductase 1* (NQO1), *glutathione S-transferase pi 1* (GSTP1), and *heme oxygenase 1* (HO1) (Jaja-Chimedza et al. 2017).

This structurally unique and chemically stable *M. oleifera* isocyanate (MICs) is an effective dietary food to prevent and/or treat obesity and type 2 diabetes. MICs' anti-obesity and antidiabetic activity is due to inhibition of the rate-limiting steps in liver gluconeogenesis that increase insulin signaling and sensitivity. More specifically, MICs supplementation (66 mg/kg/d) improved glucose tolerance and insulin signaling, reduced plasma insulin, leptin, resistin, cholesterol, IL-1β, TNFα, and hepatic glucose-6-phosphate in high-fat diet-fed mice (Waterman et al., 2015). Sugiura and Kulkarni (2018) describe a *M. oleifera* aqueous extract containing benzyl glucosinolate (≥6% mass dry solid content of extract) without alkaloid. The extract claims to prevent diseases such as insulin resistance, hyperinsulinism, type 2 diabetes, hypertension, hyperlipidemia, arterial sclerosis, obesity, fatigue recovery, or endurance improvement by improving basal metabolism due to PPAR activation. The *M. oleifera* glucosinolate, moringin, and its precursor glucomoringin have beneficial physiological functions such as anti-fatigue, antioxidant, nourishment and revitalization, and hormonal regulation. Shimizu and Moriwaki (2020) disclose a composition containing an aqueous *M. oleifera* extract and/or pulverized (ground) *M. oleifera* product with a moringin/glucomoringin ratio from 0.05 to 0.30 by deactivating the myrosinase enzyme. The glucomoringin (preferably 1.5% by mass of dry solid content) enables the use of the claimed product in foods (with high absorption index) and beverages to combat fatigue and in cosmetics as a whitening agent.

PLT Health Solutions' Slendacor Weight Management Complex is a patented formulation of three standardized herbal extracts from *Curcuma longa*, *M. oleifera*, and *Murraya koenigii*. Two clinical trials support the weight management claim due to inhibition of fat accumulation (lipogenesis) and enhanced ability to break down body fat. Slendacor claims to lower blood triglycerides, total and LDL cholesterol, and help support cardiovascular health by reducing total and LDL cholesterol. *M. oleifera* alcohol extract combined with aqueous alcohol *Murraya koenigii* extract and *Curcuma longa* extract (6:3:1 ratio) acts synergistically to inhibit lipogenesis in 3T3-L1 adipocytes (Krishanu et al., 2011). This synergistic combination (L185008F) (900 mg/d in two divided doses) for 16 weeks significantly reduced body weight, BMI, and waist/hip ratio and improved lipid profiles in healthy overweight adults (n = 70, 21–50 years, 27–29.9 kg/m^2 BMI) in a randomized, double-blind, placebo-controlled clinical study (Dixit et al., 2018).

M. oleifera is the predominant (48%–58% by weight) component of a dietary supplement also containing curcumin (37%–47% by weight) and piperine (2%–4% by weight) (Gomez, 2011). The dietary supplement (700 mg tablet or capsule) consisting of *M. oleifera* (250–375 mg), curcumin (200–250 mg), and piperine (10–20 mg) is developed to help fight malnutrition and keep the body in optimum form. *M. oleifera* is used in the dietary supplement for its hypocholesterolemic activity.

10.6 CURRENT AND FUTURE PERSPECTIVES

Over the past decade, the nutraceutical and food supplement market has witnessed significant growth. During the COVID-19 pandemic, the nutraceutical industry experienced an unprecedented surge in demand, driven by a growing interest in supplements to bolster immune function. This intense demand for supplements has continued to drive the industry's global sales and growth. As a result of the increasing demand for natural and functional foods, and the growing awareness of *M. oleifera*'s high nutritional profile and numerous traditional therapeutic properties, the market for this plant is expanding. This resilient plant can withstand long periods of drought and mild frost conditions, in addition to being a fast-growing crop that requires low agricultural input. In the context of the Millennium Sustainable Development Goals (SDG), *M. oleifera* presents an excellent ally in the steps to eradicate malnutrition, especially in areas with harsh climatic conditions. Additionally, the plant's rich reservoir of phytochemicals, especially phenolics, flavonoids, and glucosinolates, confers a range of pharmacological actions. Although there is a considerable body of evidence supporting the health benefits of *M. oleifera*, many of these studies have been conducted *in vitro* or in animal models. Clinical trials are still limited. Therefore, more extensive and well-designed human studies are necessary to establish the safety and efficacy of *M. oleifera* as a therapeutic agent.

M. oleifera is a versatile plant, with almost all parts of the plant, including leaves, pods, bark, roots, and flowers, consumed or used traditionally for therapeutic reasons. In many parts of the world, *M. oleifera* is consumed as a vegetable. The processing of *M. oleifera* is also an inherent component of the food and nutraceutical industry. New trends in developing *M. oleifera*-derived products, such as infusions, oil, and fortified foods, represent an excellent opportunity to increase its intake and offer new scopes to researchers, entrepreneurs, and industries to innovate and develop novel functional foods, nutraceuticals, cosmetics, and pharmaceutical products. While some processing techniques can enhance nutrient and bioactive compound bioavailability, other methods may have the opposite effect and cause the degradation or impede the bioavailability of these beneficial components, ultimately impacting *M. oleifera*'s therapeutic potential. Therefore, it is important to carefully select and optimize processing methods to ensure that the final product retains the desired bioactivities and pharmacological potential.

Despite limited clinical evidence, there is significant market demand for *M. oleifera* products, which has led to a surge in the number of patented products in recent years. This indicates the growing interest in this plant's potential applications in the food, pharmaceutical, and nutraceutical industries. These patents cover a wide range of products and applications, including formulations for dietary supplements, functional foods, cosmetics, and pharmaceuticals. They also cover various extraction and processing techniques aimed at improving the bioavailability and therapeutic potential of *M. oleifera*'s bioactive compounds. While patents do not necessarily indicate the effectiveness of a product, they suggest the commercial potential of *M. oleifera*. They also indicate the need for investing in research and development to create new products and applications. It is essential to conduct further studies to determine the safety and efficacy of these patented products to ensure that they meet regulatory standards and deliver the claimed health benefits.

M. oleifera is a significant player in the functional food and nutraceutical industries as consumers become more health-conscious and seek out natural and plant-based products. However, further human studies are warranted to detail its nutrient bioavailability, safety, and efficacy as a therapeutic agent, toxicity over long periods of consumption, and mode of action to demonstrate the health claims attributed to these products.

REFERENCES

Adedayo, M. R., Akintunde, J. K., Sani, A., & Boligon, A. A. (2018). Effect of dietary supplement from mono-culture fermentation of *Moringa oleifera* seeds by *Rhizopus stolonifer* on hematology and markers linked to hypercholesterolemia in rat model. *Food Science and Nutrition*, *6*(7), 1826–1838. https://doi.org/10.1002/fsn3.729

Agbogidi, O. M., & Ilondu, E. M. (2012). *Moringa oleifera* Lam: Its potentials as a food security and rural medicinal item. *Journal of Biology*, *1*, 156–167.

Agrawal, B., & Mehta, A. (2008). Antiasthmatic activity of *Moringa oleifera* Lam: A clinical study. *Indian Journal of Pharmacology*, *40*(1), 28–31.

Agricultural and Processed Food Products Export Development Authority (APEDA) *Market Intelligence Report: Moringa*. rep. Agricultural and Processed Food Products Export Development Authority (APEDA). Available at: chrome-extension://efaidnbmnnnibpcajpcglclefindmkaj/https://agriexchange.apeda.gov.in/Weekly_eReport/Moringa_Report.pdf (Accessed: 11 November 2023).

Ahmad, J., Khan, I., Johnson, S. K., Alam, I., & Din, Z. U. (2018). Effect of incorporating stevia and moringa in cookies on postprandial glycemia, appetite, palatability, and gastrointestinal well-being. *Journal of the American College of Nutrition*, *37*(2), 133–139.

AL Juhaimi, F., Ghafoor, K., Ahmed, I. A. M., Babiker, E. E., & Özcan, M. M. (2017). Comparative study of mineral and oxidative status of *Sonchus oleraceus*, *Moringa oleifera* and *Moringa peregrina* leaves. *Journal of Food Measurement and Characterization*, *11*(4), 1745–1751. https://doi.org/10.1007/s11694-017-9555-9

Alain Mune Mune, M., Nyobe, E. C., Bakwo Bassogog, C., & Minka, S. R. (2016). A comparison on the nutritional quality of proteins from *Moringa oleifera* leaves and seeds. *Cogent Food and Agriculture*, *2*(1), 4–11. https://doi.org/10.1080/23311932.2016.1213618

Alam, M. A. (2014). Development of fiber enriched herbal biscuits: A preliminary study on sensory evaluation and chemical composition. *International Journal of Nutrition and Food Sciences*, *3*(4), 246–250. https://doi.org/10.11648/j.ijnfs.20140304.13

Ali, M. W., Ilays, M. Z., Saeed, M. T., & Shin, D. H. (2020). Comparative assessment regarding antioxidative and nutrition potential of *Moringa oleifera* leaves by bacterial fermentation. *Journal of Food Science and Technology*, *57*(3), 1110–1118. https://doi.org/10.1007/s13197-019-04146-1

Allo, J., Sagita, S., Woda, R. R., & Lada, C. O. (2020). Effect of *Moringa oleifera* leaf powder supplementation on weight gain of toddler in the working area of Naibonat health center, Kupang regency. *World Nutrition Journal*, *4*(1), 56–62.

Andrews, D. A. (2006). Nutraceutical moringa composition. *U.S. Patent 0222682 A1*, Issued October 5.

Anthanont, P., Lumlerdkij, N., Akarasereenont, P., Vannasaeng, S., & Sriwijitkamol, A. (2016). *Moringa oleifera* leaf increases insulin secretion after single dose administration: A preliminary study in healthy subjects. *The Journal of the Medical Association of Thailand*, *99*(3), 308–313.

Arif, M. A., Inam-Ur-Raheem, M., Khalid, W., Lima, C. M. G., Jha, R. P., Khalid, M. Z., Santana, R. F., Sharma, R., Alhasaniah, A. H., & Emran, T. B. (2022). Effect of antioxidant-rich Moringa leaves on quality and functional properties of strawberry juice. *Evidence-based complementary and alternative medicine: eCAM*, *2022*, 8563982. https://doi.org/10.1155/2022/8563982

Attakpa, E. S., Sangaré, M. M., Béhanzin, G. J., Ategbo, J. M., Seri, B., & Khan, N. A. (2017). *Moringa oleifeira* Lam. Stimulates activation of the insulin-dependent akt pathway. antidiabetic effect in a diet-induced obesity (DIO) mouse model. *Folia Biologica*, *63*(2), 42–51.

Babayeju, A., Gbadebo, C., Obalowu, M., Otunola, G., Nmom, I., Kayode, R., Toye, A. A., & Ojo, F. (2014). Comparison of organoleptic properties of egusi and efo riro soup blends produced with moringa and spinach leaves. *Food Science and Quality Management*, *28*(2012), 15–18. https://core.ac.uk/download/pdf/234683801.pdf

Bao, Y., Xiao, J., Weng, Z., Lu, X., Shen, X., & Wang, F. (2020). A phenolic glycoside from *Moringa oleifera Lam.* improves the carbohydrate and lipid metabolisms through AMPK in Db/db mice. *Food Chemistry*, *311*, 125948. https://doi.org/10.1016/j.foodchem.2019.125948

Barba, F. J., Nikmaram, N., Roohinejad, S., Khelfa, A., Zhu, Z., & Koubaa, M. (2016). Bioavailability of glucosinolates and their breakdown products: Impact of processing. *Frontiers in Nutrition*, *3*, 24. https://doi.org/10.3389/fnut.2016.00024

Berkovich, L., Earon, G., Ron, I., Rimmon, A., Vexler, A., & Lev-Ari, S. (2013). *Moringa oleifera* aqueous leaf extract down-regulates nuclear factor-kappaB and increases cytotoxic effect of chemotherapy in pancreatic cancer cells. *BMC Complementary and Alternative Medicine*, *13*, 212. https://doi.org/10.1186/1472-6882-13-212

Bhatnagar, A. S., & Krishna, A. G. (2013). Natural antioxidants of the Jaffna variety of *Moringa oleifera* seed oil of Indian origin as compared to other vegetable oils. *Grasas Y Aceites*, *64*(5), 537–545. https://doi.org/10.3989/gya.010613

Bhutada, P. R., Jadhav, A. J., Pinjari, D. V., Nemade, P. R., & Jain, R. D. (2016). Solvent assisted extraction of oil from *Moringa oleifera* Lam. seeds. *Industrial Crops and Products*, *82*, 74–80. https://doi.org/10.1016/j.indcrop.2015.12.004

Boateng, L., Quarpong, W., Ohemeng, A., Asante, M., & Steiner-Asiedu, M. (2019). Effect of complementary foods fortified with *Moringa oleifera* leaf powder on hemoglobin concentration and growth of infants in the Eastern Region of Ghana. *Food Science and Nutrition*, *7*(1), 302–311. https://doi.org/10.1002/fsn3.890

Borgonovo, G., De Petrocellis, L., Schiano Moriello, A., Bertoli, S., Leone, A., Battezzati, A., Mazzini, S., & Bassoli, A. (2020). Moringin, a stable isothiocyanate from *Moringa oleifera*, activates the somatosensory and pain receptor TRPA1 channel in vitro. *Molecules (Basel, Switzerland)*, *25*(4), 976. https://doi.org/10.3390/molecules25040976

The Brainy Insights (2021). *Moringa Ingredients Market Size By Origin (Organic), by Application (Conventional), Global Industry Analysis, Share, Growth, Trends, and Forecast 2022 to 2030*. Report TBI-13209. https://www.thebrainyinsights.com/report/moringa-ingredients-market-13209

Brunetti, C., George, R. M., Tattini, M., Field, K., & Davey, M. P. (2013). Metabolomics in plant environmental physiology. *Journal of Experimental Botany*, *64*(13), 4011–4020. https://doi.org/10.1093/jxb/ert244

Bukar, A., Uba, A., & Oyeyi, T. (2010). Antimicrobial profile of *Moringa oleifera* Lam. extracts against some food—borne microorganisms. *Bayero Journal of Pure and Applied Sciences*, *3*(1), 43–48. https://doi.org/10.4314/bajopas.v3i1.58706

Chan Sun, M., Ruhomally, Z. B., Boojhawon, R., & Neergheen-Bhujun, V. S. (2020). Consumption of *Moringa oleifera* Lam leaves lowers postprandial blood pressure. *Journal of the American College of Nutrition*, *39*(1), 54–62. https://doi.org/10.1080/07315724.2019.1608602

Cheenpracha, S., Park, E. J., Rostama, B., Pezzuto, J. M., & Chang, L. C. (2010). Inhibition of nitric oxide (NO) production in lipopolysaccharide (LPS)-activated murine macrophage RAW 264.7 cells by the norsesterterpene peroxide, epimuqubilin A. *Marine Drugs*, *8*(3), 429–437. https://doi.org/10.3390/md8030429

Cheikhyoussef, N., Kandawa-Schulz, M., Böck, R., & Cheikhyoussef, A. (2020). Cold pressed *Moringa oleifera* seed oil. In *Cold Pressed Oils: Green Technology, Bioactive Compounds, Functionality, and Applications*. https://doi.org/10.1016/B978-0-12-818188-1.00042-6

Cheng, D., Gaoa, L., Su, C., Sargsyana, D., Wu, R., Raskinc, I., & Kong, A.-N. (2019). Moringa isothiocyanate activates Nrf2: Potential role in diabetic nephropathy. *AAPS Journal*, *20*(2), 31. https://doi.org/10.1208/s12248-019-0301-6

Chinma, C. E., Abu, J. O., & Akoma, S. N. (2014). Effect of germinated tigernut and moringa flour blends on the quality of wheat-based bread. *Journal of Food Processing and Preservation*, *38*(2), 721–727. https://doi.org/10.1111/jfpp.12023

Chio, I. I. C., & Tuveson, D. A. (2017). ROS in cancer: The burning question. *Trends in Molecular Medicine*, *23*(5), 411–429. https://doi.org/10.1016/j.molmed.2017.03.004

Coello, K. E., Peñas, E., Martinez-Villaluenga, C., Elena Cartea, M., Velasco, P., & Frias, J. (2021). Pasta products enriched with moringa sprout powder as nutritive dense foods with bioactive potential. *Food Chemistry*, *360*(2021), 130032. https://doi.org/10.1016/j.foodchem.2021.130032

Conti, M. V., Kalmpourtzidou, A., Lambiase, S., De Giuseppe, R., & Cena, H. (2021). Novel foods and sustainability as means to counteract malnutrition in Madagascar. *Molecules*, *26*(8), 1–12. https://doi.org/10.3390/molecules26082142

Coppin, J. P., Xu, Y., Chen, H., Pan, M. H., Ho, C. T., Juliani, R., . . . & Wu, Q. (2013). Determination of flavonoids by LC/MS and anti-inflammatory activity in *Moringa oleifera. Journal of Functional Foods*, *5*(4), 1892–1899.

Cretella, A. B. M., Soley, B. D. S., Pawloski, P. L., Ruziska, R. M., Scharf, D. R., Ascari, J., Cabrini, D. A., & Otuki, M. F. (2020). Expanding the anti-inflammatory potential of Moringa oleifera: Topical effect of seed oil on skin inflammation and hyperproliferation. *Journal of ethnopharmacology*, *254*, 112708. https://doi.org/10.1016/j.jep.2020.112708

Cuadrado, A., Rojo, A. I., Wells, G., Hayes, J. D., Cousin, S. P., Rumsey, W. L., Attucks, O. C., Franklin, S., Levonen, A. L., Kensler, T. W., & Dinkova-Kostova, A. T. (2019). Therapeutic targeting of the NRF2 and KEAP1 partnership in chronic diseases. *Nature Reviews Drug Discovery*, *18*, 295–317. https://doi.org/10.1038/s41573-018-0008-x

Cuellar-Núñez, M. L., Gonzalez de Mejia, E., & Loarca-Piña, G. (2021). *Moringa oleifera* leaves alleviated inflammation through downregulation of IL-2, IL-6, and TNF-α in a colitis-associated colorectal cancer model. *Food Research International*, *144*, 110318. https://doi.org/10.1016/j.foodres.2021.110318

Da Costa, R. M., Rodrigues, D., Pereira, C. A., Silva, J. F., Alves, J. V., Lobato, N. S., & Tostes, R. C. (2019). Nrf2 as a potential mediator of cardiovascular risk in metabolic diseases. *Frontiers in Pharmacology*, *10*, 1–12. https://doi.org/10.3389/fphar.2019.00382

Dai, J., Tao, L., Shi, C., Yang, S., Li, D., Sheng, J., & Tian, Y. (2020). Fermentation improves calcium bioavailability in *Moringa oleifera* leaves and prevents bone loss in calcium-deficient rats. *Food Science and Nutrition*, *8*(7), 3692–3703. https://doi.org/10.1002/fsn3.1653

Dalhoumi, W., Guesmi, F., Bouzidi, A., Akermi, S., Hfaiedh, N., & Saidi, I. (2022). Therapeutic strategies of *Moringa oleifera* Lam. (*Moringaceae*) for stomach and forestomach ulceration induced by HCl/EtOH in rat model. *Saudi Journal of Biological Sciences*, *29*(6), 103284. https://doi.org/10.1016/j.sjbs.2022.103284

Danoux, L., Depouilly, P., Leoty-okombi, S., & Vogelgesang, B. (2020). Cosmetic use of a protein extract of *Moringa oleifera* seeds *U.S. Patent 0330366 A1*, Issued October 22.

Desam, N. R., & Al-Rajab, A. J. (2021). The importance of natural products in cosmetics. In D. Pal & A. K. Nayad (eds.), *Bioactive Natural Products for Pharmaceutical Applications*, 643–685, Springer Nature, Switzerland. https://doi.org/10.1007/978-3030-54027-2_19

Díaz-Prieto, L. E., Gómez-Martínez, S., Vicente-Castro, I., Heredia, C., González-Romero, E. A., Martín-Ridaura, M. D. C., Ceinos, M., Picón, M. J., Marcos, A., & Nova, E. (2022). Effects of *Moringa oleifera* Lam. supplementation on inflammatory and cardiometabolic markers in subjects with prediabetes. *Nutrients*, *14*(9), 1937. https://doi.org/10.3390/nu14091937

Dilworth, L. L., Stennett, D., & Omoruyi, F. O. (2020). Effects of *Moringa oleifera* Leaf extract on human promyelocytic leukemia cells subjected to oxidative stress. *Journal of Medicinal Food*, *23*(7), 728–734. https://doi.org/10.1089/jmf.2019.0192

Dimidi, E., Cox, S. R., Rossi, M., & Whelan, K. (2019). Fermented foods: Definitions and characteristics, impact on the gut microbiota and effects on gastrointestinal health and disease. *Nutrients*, *11*, 1806. https://doi.org/10.3390/nu11081806

Dixit, K., Kamath, D. V., Alluri, K. V., & Davis, B. A. (2018). Efficacy of a novel herbal formulation for weight loss demonstrated in a 16-week randomized, double-blind, placebo-controlled clinical trial with healthy overweight adults. *Diabetes, Obesity and Metabolism*, *20*(11), 2633–2641.

Duranti, G., Maldini, M., Crognale, D., Sabatini, S., Corana, F., Horner, K., et al. (2021). *Moringa oleifera* leaf extract influences oxidative metabolism in C2C12 myotubes through SIRT1-Pparα pathway. *Phytomedicine Plus*, *1*, 100014. https://doi.org/10.1016/j.phyplu.2020.100014

Edeogu, C. O., Kalu, M. E., Famurewa, A. C., Asogwa, N. T., Onyeji, G. N., & Ikpemo, K. O. (2020). Nephroprotective effect of *Moringa oleifera* seed oil on gentamicin-induced nephrotoxicity in rats: Biochemical evaluation of antioxidant, anti-inflammatory, and antiapoptotic pathways. *Journal of the American College of Nutrition*, 39(4), 307–315, https://doi.org/10.1080/07315724.2019.1649218

Ernawatiet, F., Nurjanah, N., Aji, G. K., Hapsari Tjandrarini, D., Widodo, Y., Retiaty, F., . . . & Syauqy, A. (2023). Micronutrients and nutrition status of school-aged children in Indonesia. *Journal of Nutrition and Metabolism*, *2023*, 4610038. https://doi.org/10.1155/2023/4610038

Ezzat, S. M., El Bishbishy, M. H., Aborehab, N. M., Salama, M. M., Hasheesh, A., Motaal, A. A., Rashad, H., & Metwally, F. M. (2020). Upregulation of MC4R and PPAR-α expression mediates the anti-obesity activity of *Moringa oleifera* Lam. in high-fat diet-induced obesity in rats. *Journal of Ethnopharmacology*. *251*, 112541. https://doi.org/10.1016/j.jep.2020.112541

Fadhilatunnur, H., Fransisca, F., & Dewi, R. T. K. (2021). Evaluation of in vitro protein digestibility of *Moringa oleifera* leaves with various domestic cooking. *Carpathian Journal of Food Science and Technology*, *13*(1), 214–222. https://doi.org/10.34302/crpjfst/2021.13.1.18

Famurewa, A. C., Aja, P. M., Nwankwo, O. E., Awoke, J. N., Maduagwuna, E. K., & Aloke, C. (2019). *Moringa oleifera* seed oil or virgin coconut oil supplementation abrogates cerebral neurotoxicity induced by antineoplastic agent methotrexate by suppression of oxidative stress and neuro-inflammation in rats. *Journal of Food Biochemistry*, *43*(3), e12748. https://doi.org/10.1111/jfbc.12748

Fard, M., Arulselvan, P., Karthivashan, G., Adam, S., & Fakurazi, S. (2015). Bioactive extract from *Moringa oleifera* inhibits the pro-inflammatory mediators in lipopolysaccharide stimulated macrophages. *Pharmacognosy Magazine*, *11*(44s3), S556–S563. https://doi.org/10.4103/0973-1296.172961

Farooq, B., Koul, B., Mahant, D., & Yadav, D. (2021). Phytochemical analyses, antioxidant and anticancer activities of ethanolic leaf extracts of *Moringa oleifera* Lam. varieties. *Plants (Basel, Switzerland)*, *10*(11), 2348. https://doi.org/10.3390/plants10112348

Fatemi, S. H., & Hammond, E. G. (1980). Analysis of oleate, linoleate and linolenate hydroperoxides in oxidized ester mixtures. *Lipids*, *15*(5), 379–385.

Fatmawati, F., Koro, S., Nadimin, N., Gani, K., Hasan, H., Abadi, E., & Mallongi, A. (2023). Intervention of giving Moringa biscuits (*Moringa oliefera*) mix sori fish flour to increased blood hemoglobin levels in young girls, kendari, indonesia. *Pharmacognosy Journal*, *15*(2), 414–417. https://doi.org/10.5530/pj.2023.15.64

Fejér, J., Kron, I., Pellizzeri, V., Pľuchtová, M., Eliašová, A., Campone, L., Gervasi, T., Bartolomeo, G., Cicero, N., Babejová, A., et al. (2019). First report on evaluation of basic nutritional and antioxidant properties of *Moringa Oleifera* Lam. from Caribbean Island of Saint Lucia. *Plants*, *8*, 537. https://doi.org/10.3390/plants8120537.

Fombang, E. N., Bouba, B., & Ngaroua (2016). Management of hypertension in normal and obese hypertensive patients through supplementation with *Moringa oleifera* Lam leaf powder. *Indian Journal of Nutrition*, *3*(2), 143. https://www.opensciencepublications.com/wp-content/uploads/IJN-2395-2326-3-143.pdf

Fombang, E. N., & Romuald, W. S. (2016). Antihyperglycemic activity of *Moringa oleifera* Lam leaf functional tea in rat models and human subjects. *Food and Nutrition Sciences*, *07*, 1021–1032. https://doi.org/10.4236/fns.2016.711099

Forman, H. J., & Zhang, H. (2021). Targeting oxidative stress in disease: Promise and limitations of antioxidant therapy. *Nature Reviews Drug Discovery*, *20*(9), 689–709. https://doi.org/10.1038/s41573-021-00233-1

Fortune Business Insight. (2022). *Moringa Products Market to Worth USD 14.80 Billion by 2021–2028*, February 28. https://www.fortunebusinessinsights.com/enquiry/queries/moringa-products-market-102280

Gambo, A., Moodley, I., Babashani, M., Babalola, T. K., & Gqaleni, N. (2021). A double-b(Zhao et al., 2021)d, randomized controlled trial to examine the effect of *Moringa oleifera* leaf powder supplementation on the immune status and anthropometric parameters of adult HIV patients on antiretroviral therapy in a resource-limited setting. *PloS ONE*, *16*(12), e0261935. https://doi.org/10.1371/journal.pone.0261935

Gharsallah, K., Rezig, L., & Mahfoudhi, N. (2022). Beneficial effects of *Moringa oleifera* seed oil bioactive compounds. In *Handbook of Research on Advanced Phytochemicals and Plant-Based Drug Discovery*. https://doi.org/10.4018/978-1-6684-5129-8.ch013.

Ghazali, H. M., & Mohammed, A. S. (2011). Moringa (*Moringa oleifera*) seed oil: Composition, nutritional aspects, and health attributes. In *Nuts and Seeds in Health and Disease Prevention*. Elsevier Inc. https://doi.org/10.1016/B978-0-12-375688-6.10093-3

Giacoppo, S., Iori, R., Bramanti, P., & Mazzon, E. (2017). Topical moringin cream relieves neuropathic pain by suppression of inflammatory pathway and voltage-gated ion channels in murine model of multiple sclerosis. *Molecular Pain*, *13*, 1744806917724318.

Gidamis, A. B., Panga, J. T., Sarwatt, S. V., Chove, B. E., & Shayo, N. B. (2003). Nutrient and antinutrient contents in raw and cooked young leaves and immature pods of *Moringa oleifera*, Lam. *Ecology of Food and Nutrition*, *42*(6), 399–411. https://doi.org/10.1080/03670240390268857

Gokaraju, G. R., Gokaraju, R. R., Golakoti, T., Chirravuri, V. R., Somepalli, V., & Bhupathiraju, K. (2013). Synergistic phytochemical composition for the treatment of obesity. *U.S. Patent. 8,541,383*, Issued September 24.

Gomez, R. (2011). *Dietary Supplement Composition*. World intellectual property organization, *WO 098819 A1*, International publication date, August 18.

Gómez-Martínez, S., Díaz-Prieto, L. E., Vicente Castro, I., Jurado, C., Iturmendi, N., Martín-Ridaura, M. C., . . . & Nova, E. (2022). *Moringa oleifera* leaf supplementation as a glycemic control strategy in subjects with prediabetes. *Nutrients*, *14*(1), 57. https://doi.org/10.3390/nu14010057

Gómez-Zorita, S., González-Arceo, M., Fernández-Quintela, A., Eseberri, I., Trepiana, J., & Portillo, M. P. (2020). Scientific evidence supporting the beneficial effects of isoflavones on human health. *Nutrients*, *12*(12), 1–25. https://doi.org/10.3390/nu12123853

Gopalakrishnan, L., Doriya, K., & Kumar, D. S. (2016). *Moringa oleifera*: A review on nutritive importance and its medicinal application. *Food Science and Human Wellness*, *5*(2), 49–56. https://doi.org/10.1016/j.fshw.2016.04.001

Gunstone, F. D. (2006). Minor specialty oils, In F. Shahidi (ed.), *Nutraceutical and Specialty Lipids and their Co-Products*, 91–126, CRC Press, Bocka Raton, FL.

Gyamfi, E. T., Kwarteng, I. K., Ansah, M. O., Anim, A. K., Ackah, M., Kpattah, L., & Bentil, N. O. (2011). Effects of processing on *Moringa oleifera. Proceedings of the International Academy of Ecology and Environmental Sciences*, *1*, 179–185. www.iaees.org

Hekmat, S., Morgan, K., Soltani, M., & Gough, R. (2015). Sensory evaluation of locally-grown fruit purees and inulin fibre on probiotic yogurt in Mwanza, Tanzania and the microbial analysis of probiotic yogurt fortified with *Moringa oleifera. Journal of Health, Population and Nutrition*, *33*(1), 60–67.

Hip Kam, A., Li, W., Bahorun, T., & Neergheen, V. S. (2023). Traditional processing techniques impacted the bioactivities of selected local consumed foods. *Scientific African*, *19*, e01558. https://doi.org/10.1016/j.sciaf.2023.e01558

Idohou-Dossou, N., Diouf, A., Gueye, A., & Guiro, A. (2011). Impact of daily consumption of Moringa (*Moringa oleifera*) dry leaf powder on iron status of Senegalese lactating women. *African Journal of Food Agriculture Nutrition and Development*, *11*(4). https://doi.org/10.4314/ajfand.v11i4.69176

Jaafaru, M. S., Karim, N. A. A., Eliaser, E. M., Waziri, P. M., Ahmed, H., Barau, M. M., Kong, L., & Razis, A. F. A. (2018a). Nontoxic glucomoringin-isothiocyanate (GMG-ITC) rich soluble extract induces apoptosis and inhibits proliferation of human prostate adenocarcinoma cells (PC-3). *Nutrients*, *10*(9), 1174. https://doi.org/10.3390/nu10091174

Jaafaru, M. S., Nordin, N., Shaari, K., Rosli, R., & Abdull Razis, A. F. (2018b). Isothiocyanate from *Moringa oleifera* seeds mitigates hydrogen peroxide-induced cytotoxicity and preserved morphological features of human neuronal cells. *PLoS ONE*, *13*(5), 1–17. https://doi.org/10.1371/journal.pone.0196403

Jagadeesan, S., Sarangharaajan, A., Ravikumar, N., Palani, K., & Ramanathan, R. M. (2020). Development of "ready to use" value added products from Moringa leaves. *International Journal of Food Science and Nutrition*, *5*(2), 60–62. www.foodsciencejournal.com

Jaja-Chimedza, A., Graf, B. L., Simmler, C., Kim, Y., Kuhn, P., Pauli, G. F., & Raskin, I. (2017). Biochemical characterization and anti-inflammatory properties of an isothiocyanate-enriched moringa (*Moringa oleifera*) seed extract. *PloS ONE*, *12*(8), e0182658. https://doi.org/10.1371/journal.pone.0182658

Janaki, S. (2015). Characterization of cold press moringa oil. *International Journal of Science and Research (IJSR)*, *4*(4), 386–389.

John, S., & Chellappa, A. R. (2005). Hypoglycemic effect of *Moringa oleifera* (drumstick) leaf powder on human diabetic subjects and albino rats. *The Indian Journal of Nutrition and Dietetics*, 22–29.

Joung, H., Kim, B., Park, H., Lee, K., Kim, H. H., Sim, H. C., Do, H. J., Hyun, C. K., & Do, M. S. (2017). Fermented *Moringa oleifera* decreases hepatic adiposity and ameliorates glucose intolerance in high-fat diet-induced obese mice. *Journal of Medicinal Food*, *20*(5), 439–447. https://doi.org/10.1089/jmf.2016.3860

Juergenliemk, G., Boje, K., Huewel, S., Lohmann, C., Galla, H. J., & Nahrstedt, A. (2003). In vitro studies indicate that miquelianin (Quercetin 3-O-β-D-Glucuronopyranoside) is able to reach the CNS from the small intestine. *Planta Medica*, *69*(11), 1013–1017. https://doi.org/10.1055/s-2003-45148

Kahkeshani, N., Farzaei, F., Fotouhi, M., Alavi, S. S., Bahramsoltani, R., Naseri, R., Momtaz, S., Abbasabadi, Z., Rahimi, R., Farzaei, M. H., & Bishayee, A. (2019). Pharmacological effects of gallic acid in health and disease: A mechanistic review. *Iranian Journal of Basic Medical Sciences*, *22*(3), 225–237. https://doi.org/10.22038/ijbms.2019.32806.7897

Kamal, R. M., Abdull Razis, A. F., Mohd Sukri, N. S., Perimal, E. K., Ahmad, H., Patrick, R., . . . & Rigaud, S. (2022). Beneficial health effects of glucosinolates-derived isothiocyanates on cardiovascular and neurodegenerative diseases. *Molecules*, *27*(3), 624. https://doi.org/10.3390/molecules27030624

Kashyap, P., Kumar, S., Riar, C. S., Jindal, N., Baniwal, P., Guiné, R. P. F., Correia, P. M. R., Mehra, R., & Kumar, H. (2022). Recent advances in drumstick (*Moringa oleifera*) leaves bioactive compounds: Composition, health benefits, bioaccessibility, and dietary applications. *Antioxidants*, *11*(2), 1–37. https://doi.org/10.3390/antiox11020402

Khan, M. A., Shakoor, S., Ameer, K., Farooqi, M. A., Rohi, M., Saeed, M. S., Asghar, M. T., Irshad, M. B., Waseem, M., Tanweer, S., Ali, U., Mohamed Ahmed, I. A., & Ramzan, Y. (2023). Effects of dehydrated Moringa (*Moringa oleifera*) leaf powder supplementation on physicochemical, antioxidant, mineral, and sensory properties of whole wheat flour leavened bread. *Journal of Food Quality*. *2023*, 4473000. https://doi.org/10.1155/2023/4473000

Khanuja, S. P. S., Arya, J. S., Tiruppadiripuliyur, R. S. K., Saikia, D., Kaur, H., Singh, M., Gupta, S. C., Shasany, A. K., Darokar, M. P., Srivastava, S. K., Gupta, M. M., Verma, S. C., & Pal, A. (2005). Nitrile glycoside useful as a bioenhancer of drugs and nutrients, process of its isolation from *Moringa oleifera*. *U.S. Patent 6,858,588 B2*, Issued February 22.

Kim, Y., Wu, A. G., Jaja-Chimedza, A., Graf, B. L., Waterman, C., Verzi, M. P., & Raskin, I. (2017). Isothiocyanate-enriched moringa seed extract alleviates ulcerative colitis symptoms in mice. *PLoS ONE, 12*(9), 1–20. https://doi.org/10.1371/journal.pone.0184709

Kolawole, F., Balogun, M., Opaleke, D., & Amali, H. (2013). An evaluation of nutritional and sensory qualities of wheat-*Moringa* cake. *Agrosearch, 13*(1), 87. https://doi.org/10.4314/agrosh.v13i1.9

Krishanu, S., Trimurtulu, G., Venkateswara Rao, C., & Ajit Kumar, M. (2011). An herbal formula LI85008F inhibits lipogenesis in 3T3-L1 adipocytes. *Food and Nutrition Sciences, 2*, 809–817.

Kumar, P. K., & Mandapaka, R. T. (2013). Effect of *Moringa oleifera* on blood glucose, LDL levels in types II diabetic obese people. *Innovative Journal of Medical Health Science, 3*(1), 23–25.

Kumari, D. J. (2010). Hypoglycaemic effect of *Moringa oleifera* and *Azadirachta indica* in type 2 diabetes mellitus. *Bioscan, 5*(20), 211–214.

Kushwaha, S., Chawla, P., & Kochhar, A. (2014). Effect of supplementation of drumstick (*Moringa oleifera*) and amaranth (*Amaranthus tricolor*) leaves powder on antioxidant profile and oxidative status among postmenopausal women. *Journal of Food Science and Technology, 51*, 3464–3469.

Lalas, S., & Tsaknis, J. (2002). Characterization of *Moringa oleifera* seed oil variety Periyakulam-1. *Journal of Food Compositions and Analysis, 15*, 65–77.

Laoung-On, J., Saenphet, K., Jaikang, C., & Sudwan, P. (2021). Effect of *Moringa oleifera* Lam. leaf tea on sexual behavior and reproductive function in male rats. *Plants, 10*, 2019. https://doi.org/10.3390/plants10102019

Lee, H. J., Jeong, Y. J., Lee, T. S., Park, Y. Y., Chae, W. G., Chung, I. K., Chang, H. W., Kim, C. H., Choi, Y. H., Kim, W. J., Moon, S. K., & Chang, Y. C. (2013). Moringa fruit inhibits LPS-induced NO/iNOS expression through suppressing the NF-κB activation in RAW264.7 cells. *American Journal of Chinese Medicine*. https://doi.org/10.1142/S0192415X13500754

Leone, A., Bertoli, S., Di Lello, S., Bassoli, A., Ravasenghi, S., Borgonovo, G., . . . & Battezzati, A. (2018). Effect of *Moringa oleifera* leaf powder on postprandial blood glucose response: In vivo study on Saharawi people living in refugee camps. *Nutrients, 10*(10), 1494. https://doi.org/10.3390/nu10101494

Leone, A., Spada, A., Batti, A., Schiraldi, A., Aristil, J., & Bertoli, S. (2016). *Moringa oleifera* seeds and oil: Characteristics and uses for human health. *International Journal of Molecular Sciences, 17*(12), 2141; https://doi.org/10.3390/ijms17122141

Liao, P. C., Lai, M. H., Hsu, K. P., Kuo, Y. H., Chen, J., Tsai, M. C., . . . & Chao, L. K. P. (2018). Identification of β-sitosterol as in vitro anti-inflammatory constituent in *Moringa oleifera*. *Journal of Agricultural and Food Chemistry, 66*(41), 10748–10759.

Lin, X., Wu, L., Wang, X., Yao, L., & Wang, L. (2021). Ultrasonic-assisted extraction for flavonoid compounds content and antioxidant activities of India *Moringa oleifera* L. leaves: Simultaneous optimization, HPLC characterization and comparison with other methods. *Journal of Applied Research on Medicinal and Aromatic Plants, 20*, 100284. https://doi.org/10.1016/j.jarmap.2020.100284

Liu, W. L., Wu, B. F., Shang, J. H., Zhao, Y. L., & Huang, A. X. (2020). *Moringa oleifera* Lam seed oil augments pentobarbital-induced sleeping behaviors in mice via GABAergic systems. *Journal of Agricultural and Food Chemistry, 68*(10), 3149–3162.

Lopes, T. D. P., da Costa, H. P. S., Pereira, M. L., da Silva Neto, J. X., de Paula, P. C., Brilhante, R. S. N., . . . & Sousa, D. O. B. (2020). Mo-CBP4, a purified chitin-binding protein from *Moringa oleifera* seeds, is a potent antidermatophytic protein: In vitro mechanisms of action, in vivo effect against infection, and clinical application as a hydrogel for skin infection. *International Journal of Biological Macromolecules, 149*, 432–442.

Louisa, M., Patintingan, C. G. H., & Wardhani, B. W. K. (2022). *Moringa oleifera* Lam. in cardiometabolic disorders: A systematic review of recent studies and possible mechanism of actions. *Frontiers in Pharmacology, 13*, 792794. https://doi.org/10.3389/fphar.2022.792794

Madane, P., Das, A. K., Pateiro, M., Nanda, P. K., Bandyopadhyay, S., Jagtap, P., Barba, F. J., Shewalkar, A., Maity, B., & Lorenzo, J. M. (2019). Drumstick (*Moringa oleifera*) flower as an antioxidant dietary fibre in chicken meat nuggets. *Foods, 8*, 307. https://doi.org/10.3390/foods8080307

Madi, N., Dany, M., Abdoun, S., & Usta, J. (2016). *Moringa oleifera*'s nutritious aqueous leaf extract has anticancerous effects by compromising mitochondrial viability in an ROS-dependent manner. *Journal of the American College of Nutrition, 35*(7), 604–613. https://doi.org/10.1080/07315724.2015.1080128

Mahajan, S. G., Banerjee, A., Chauhan, B. F., Padh, H., Nivsarkar, M., & Mehta, A. A. (2009). Inhibitory effect of n-butanol fraction of *Moringa oleifera* Lam. seeds on ovalbumin-induced airway inflammation in a guinea pig model of asthma. *International Journal of Toxicology, 28*(6), 519–527. https://doi.org/10.1177/1091581809345165

Mahato, D. K., Kargwal, R., Kamle, M., Sharma, B., Pandhi, S., Mishra, S., . . . & Kumar, P. (2022). Ethnopharmacological properties and nutraceutical potential of *Moringa oleifera*. *Phytomedicine Plus, 2*(1), 100168. https://doi.org/10.1016/j.phyplu.2021.100168

Makita, C., Chimuka, L., Steenkamp, P., Cukrowska, E., & Madala, E. (2016). Comparative analyses of flavonoid content in *Moringa oleifera* and *Moringa ovalifolia* with the aid of UHPLC-qTOF-MS fingerprinting. *South African Journal of Botany*, *105*, 116–122. https://doi.org/10.1016/j.sajb.2015.12.007

Managoli, N. B. (2007). Herbal composition for treatment of immunocompromised conditions. *U.S. Patent 0122496 A1*, Issued May 31.

Matic, I., Guidi, A., Kenzo, M., Mattei, M., & Galgani, A. (2018). Investigation of medicinal plants traditionally used as dietary supplements: A review on *Moringa oleifera* Ivana. *Journal of Public Health in Africa*, *9*(3), 841. https://doi.org/10.4081/jphia.2018.841

Maurya, S. K., & Singh, A. K. (2014). Clinical efficacy of *Moringa oleifera* Lam. stems bark in urinary tract infections. *International Scholarly Research Notices*, *2014*, 906843. http://doi.org/10.1155/2014/906843

Mbah, B. O., Eme, P. E., & Ogbusu, O. F. (2012). Effect of cooking methods (boiling and roasting) on nutrients and anti-nutrients content of *Moringa oleifera* seeds. Pakistan. *Pakistan Journal of Nutrition*, *11*(3), 211–215.

McLaughlin, P. J., & Weihrauch, J. L. (1979). Vitamin E content of foods. *Journal of the American Dietetic Association*, *75*(6), 647–665.

Mehlomakulu, N. N., & Emmambux, M. N. (2022). Nutritional quality of wet and dry processed *Moringa oleifera* Lam. leaves: A review. *Food Reviews International*, 38(8), 1635–1655. https://doi.org/10.1080/87559129.2020.1831527

Meireles, D., Gomes, J., Lopes, L., Hinzmann, M., & Machado, J. (2020). A review of properties, nutritional and pharmaceutical applications of *Moringa oleifera*: Integrative approach on conventional and traditional Asian medicine. *Advances in Traditional Medicine*, *20*(4), 495–515. https://doi.org/10.1007/s13596-020-00468-0

Michl, C., Vivarelli, F., Weigl, J., De Nicola, G. R., Canistro, D., Paolini, M., Iori, R., & Rascle, A. (2016). The chemopreventive phytochemical moringin isolated from *Moringa oleifera* seeds inhibits JAK/STAT signaling. *PLoS ONE*, *11*(6), e0157430. https://doi.org/10.1371/journal.pone.0157430

Miekus, N., Marszałek, K., Podlacha, M., Iqbal, A., Puchalski, C., & Swiergiel, A. H. (2020). Health benefits of plant-derived sulfur compounds, glucosinolates, and organosulfur compounds. *Molecules*, *25*(17). https://doi.org/10.3390/molecules25173804

Milla, P. G., Peñalver, R., & Gema, N. (2021). Health benefits of uses and applications of *Moringa oleifera* in bakery products. *Plants*, *10*(2), 318. https://doi.org/10.3390/plants10020318

Moringa Products Market: Information by product (leaf powder, tea), distribution channel (online, offline), and region-forecast till 2030. Report Code: SRFB2941DR, October 17. https://straitsresearch.com/report/moringa-products-market

Mozo, R. N., & Caole-Ang, I. (2015). The effects of Malunggay (*Moringa oleifera*) leaves capsule supplements on high specificity C-reactive protein and hemoglobin A1c levels of diabetic patients in Ospital ng Maynila Medical Center: A prospective cohort study. *Philippine Journal of Internal Medicine*, *53*(4), 1–10.

Muangnoi, C., Chingsuwanrote, P., Praengamthanachoti, P., Svasti, S., & Tuntipopipat, S. (2012). *Moringa oleifera* pod inhibits inflammatory mediator production by lipopolysaccharide-stimulated RAW 264.7 murine macrophage cell lines. *Inflammation*, *35*(2), 445–455. https://doi.org/10.1007/s10753-011-9334-4

Murador, D., Braga, A. R., Da Cunha, D., & De Rosso, V. (2018). Alterations in phenolic compound levels and antioxidant activity in response to cooking technique effects: A meta-analytic investigation. *Critical Reviews in Food Science and Nutrition*, *58*(2), 169–177. https://doi.org/10.1080/10408398.2016.1140121

Nadeem, M., & Imran, M. (2016). Promising features of *Moringa oleifera* oil: Recent updates and perspectives. *Lipids in Health and Disease*, *15*, 212. https://doi.org/10.1186/s12944-016-0379-0

Nambiar, V. S., Guin, P., Parnami, S., & Daniel, M. (2010). Impact of antioxidants from drumstick leaves on the lipid profile of hyperlipidemics. *Journal of Herbal Medicine &Toxicology*, *4*(1), 165–172.

Nebolisa, N. M., Umeyor, C. E., Ekpunobi, U. E., Umeyor, I. C., & Okoye, F. B. (2023). Profiling the effects of microwave-assisted and soxhlet extraction techniques on the physicochemical attributes of *Moringa oleifera* seed oil and proteins. *Oil Crop Science*, *8*(1), 16–26. https://doi.org/10.1016/j.ocsci.2023.02.003

Ngamukote, S., Khannongpho, T., Siriwatanapaiboon, M., Sirikwanpong, S., Dahlan, W., & Adisakwattana, S. (2016). *Moringa oleifera* leaf extract increases plasma antioxidant status associated with reduced plasma malondialdehyde concentration without hypoglycemia in fasting healthy volunteers. *Chinese Journal of Integrative Medicine*, 1–6. https://doi.org/10.1007/s11655-016-2515-0

Nibret, E., & Wink, M. (2010). Trypanocidal and antileukaemic effects of the essential oils of *Hagenia abyssinica, Leonotis ocymifolia*, *Moringa stenopetala*, and their main individual constituents. *Phytomedicine*, *17*(12), 911–920. https://doi.org/10.1016/j.phymed.2010.02.009

Nouman, W., Anwar, F., Gull, T., Newton, A., Rosa, E., & Domínguez-Perles, R. (2016). Profiling of polyphenolics, nutrients and antioxidant potential of germplasm's leaves from seven cultivars of *Moringa oleifera* Lam. *Industrial Crops and Products*, *83*, 166–176. https://doi.org/10.1016/j.indcrop.2015.12.032

Nova, E., Redondo-Useros, N., Martínez-García, R. M., Gómez-Martínez, S., Díaz-Prieto, L. E., & Marcos, A. (2020). Potential of *Moringa oleifera* to improve glucose control for the prevention of diabetes and related metabolic alterations: A systematic review of animal and human studies. *Nutrients*, *12*(7), 2050.

Ntila, S. L., Ndhlala, A. R., Mashela, P. W., Kolanisi, U., & Siwela, M. (2020). Supplementation of a complementary white maize soft porridge with *Moringa oleifera* powder as a promising strategy to increase nutritional and phytochemical values: A research note. *South African Journal of Botany*, *129*, 238–242. https://doi.org/10.1016/j.sajb.2019.07.021

Ogunsina, B. S., Indira, T. N., Bhatnagar, A. S., Radha, C., Debnath, S., & Gopala Krishna, A. G. (2014). Quality characteristics and stability of *Moringa oleifera* seed oil of Indian origin. *Journal of Food Science and Technology*, *51*(3), 503–510. https://doi.org/10.1007/s13197-011-0519-5

Olatosin, T., Akinduko, D., & Uche, C. (2013). Evaluation of the hepatoprotective efficacy of *Moringa oleifera* seed oil on ccl 4-induced liver damage in Wistar albino rats. *The International Journal Of Engineering And Science (IJES)*, *2*(11), 13–18.

Oomah, B. D. (2023). Hempseed: A functional food source. In R. Campos-Vega & B. D. Oomah (eds.), *Molecular Mechanisms of Functional Food*, 269–356. John Wiley & Sons Ltd., Chichester, West Sussex. https://doi.org/10.1002/9781119804055.ch9

Oyeyinka, A. T., & Oyeyinka, S. A. (2018). *Moringa oleifera* as a food fortificant: Recent trends and prospects. *Journal of the Saudi Society of Agricultural Sciences*, *17*(2), 127–136. https://doi.org/10.1016/j.jssas.2016.02.002

Palermo, M., Pellegrini, N., & Fogliano, V. (2013). The effect of cooking on the phytochemical content of vegetables. *Journal of the Science of Food and Agriculture*, *94*(6), 1057–1070. https://doi.org/10.1002/jsfa.6478

Pan, S. Y., Nie, Q., Tai, H. C., Song, X. L., Tong, Y. F., Zhang, L. J. F., Wu, X. W., Lin, Z. H., Zhang, Y. Y., Ye, D. Y., Zhang, Y., Wang, X. Y., Zhu, P. L., Chu, Z. S., Yu, Z. L., & Liang, C. (2022). Tea and tea drinking: China's outstanding contributions to the mankind. *Chinese Medicine (United Kingdom)*, *17*(1), 1–40. https://doi.org/10.1186/s13020-022-00571-1

Park, E. J., Cheenpracha, S., Chang, L. C., Kondratyuk, T. P., & Pezzuto, J. M. (2011). Inhibition of lipopolysaccharide-induced cyclooxygenase-2 and inducible nitric oxide synthase expression by 4-[(2'-O-acetyl-α-L-rhamnosyloxy) benzyl]isothiocyanate from *Moringa oleifera. Nutrition and Cancer*, *63*(6), 971–982. https://doi.org/10.1080/01635581.2011.589960

Patil, S. V., Mohite, B. V., Marathe, K. R., Salunkhe, N. S., Marathe, V., & Patil, V. S. (2022). Moringa tree, gift of nature: A review on nutritional and industrial potential. *Current Pharmacology Reports*, *8*(4), 262–280. https://doi.org/10.1007/s40495-022-00288-7

Peng, B., Bai, R. F., Li, P., Han, X. Y., Wang, H., Zhu, C. C., Zeng, Z. P., & Chai, X. Y. (2016). Two new glycosides from Dryopteris fragrans with anti-inflammatory activities. *Journal of Asian Natural Products Research*, *18*(1), 59–64. https://doi.org/10.1080/10286020.2015.1121853

Pernodet, N., Corallo, K., Layman, D., & Collins, D. (2017). Methods and compositions for treating aged skin. *U.S. Patent 9,687,439*, Issued June 27.

Pilotos, J., Ibrahim, K. A., Mowa, C. N., & Opata, M. M. (2020). *Moringa oleifera* treatment increases Tbet expression in CD4+ T cells and remediates immune defects of malnutrition in *Plasmodium chabaudi*-infected mice. *Malaria Journal*, *19*(1), 1–16. https://doi.org/10.1186/s12936-020-3129-8

Polaris Market Research. (2022). *Global Moringa Products Market Size Report, 2022–2030. Moringa Products Market Share, Size, Trends, Industry Analysis Report, by Source (Leaves, Seeds, Roots); by Origin (Organic, Conventional); by Application; by Region; Segment Forecast, 2022–2030*. Report ID: PM1889, Published October; accessed 20 October, 2023. https://www.polarismarketresearch.com/industry-analysis/moringa-ingredients-market

Pollini, L., Tringaniello, C., Ianni, F., Blasi, F., Manes, J., & Cossignani, L. (2020). Impact of ultrasound extraction parameters on the antioxidant properties of *Moringa oleifera* leaves. *Antioxidants (Basel, Switzerland)*, *9*(4), 277. https://doi.org/10.3390/antiox9040277

Prabowo, N. A., Nurudhin, A., Werdiningsih, Y., Putra, D. D., Putri, D. P., & Widyastuti, R. (2023). *Moringa oleifera* extract decreases interleukin 6 levels and disease activity in rheumatoid arthritis patients. *Bangladesh Journal of Medical Science*, *22*(2), 416–424.

Rahayu, I., & Timotius, K. H. (2022). Phytochemical analysis, antimutagenic and antiviral activity of *Moringa oleifera* L. leaf infusion: In vitro and in silico studies. *Molecules*, *27*, 4017. https://doi.org/10.3390/molecules27134017

Ramjane, H., Bahorun, T., Ramasawmy, B., Ramful-Baboolall, D., Boodia, N., Aruoma, O. I., & Neergheen, V. S. (2021). Exploration of the potential of terrestrial and marine biodiversity for the development of local nutraceutical products: A case for Mauritius. *American Journal of Biopharmacy and Pharmaceutical Sciences*, *1*(3), 1–36. https://doi.org/10.25259/ajbps_3_2021

Randriamboavonjy, J. I., Rio, M., Pacaud, P., Loirand, G., & Tesse, A. (2017). *Moringa oleifera* seeds attenuate vascular oxidative and nitrosative stresses in spontaneously hypertensive rats. *Oxidative Medicine and Cellular Longevity. 2017*, 4129459. https://doi.org/10.1155/2017/4129459

Rani, N. Z. A., Husain, K., & Kumolosasi, E. (2018). Moringa genus: A review of phytochemistry and pharmacology. *Frontiers in Pharmacology*, *9*, 108. https://doi.org/10.3389/fphar.2018.00108

Razzak, A., Roy, K. R., Sadia, U., & Zzaman, W. (2022). Effect of cooking on physicochemical and antioxidant properties of raw and cooked *Moringa oleifera* pods. *International Journal of Food Science*, *2022*, 1502857. https://doi.org/10.2139/ssrn.4077506

Rodríguez-Pérez, C., Quirantes-Piné, R., Fernández-Gutiérrez, A., & Segura-Carretero, A. (2015). Optimization of extraction method to obtain a phenolic compounds-rich extract from *Moringa oleifera* Lam leaves. *Industrial Crops and Products*, *66*, 246–254. https://doi.org/10.1016/j.indcrop.2015.01.002

Romeo, L., Diomede, F., Gugliandolo, A., Scionti, D., Lo Giudice, F., Lanza Cariccio, V., . . . & Mazzon, E. (2018). Moringin induces neural differentiation in the stem cell of the human periodontal ligament. *Scientific Reports*, *8*(1), 9153. https://doi.org/10.1038/s41598-018-27492-0

Rustagi, S., Khan, S., Jain, T., Singh, R., & Modi, V. (2023). Design optimization and comparative analysis of hypoallergenic muffins to wheat muffins and nutritive improvement using *Moringa* leaves powder. *Nutrition & Food Science*, *54*(1), 71–85. https://doi.org/10.1108/nfs-04-2023-0085

Ruttarattanamongkol, K., & Petrasch, A. (2016). Oxidative susceptibility and thermal properties of *Moringa oleifera* seed oil obtained by pilot-scale subcritical and supercritical carbon dioxide extraction. *Journal of Food Process Engineering*, *39*(30), 226–236. https://doi.org/10.1111/jfpe.12213

Saini, R. K., Shetty, N. P., & Giridhar, P. (2014). GC-FID/MS analysis of fatty acids in Indian cultivars of *Moringa oleifera*: Potential sources of PUFA. *JAOCS, Journal of the American Oil Chemists' Society*, *91*(6), 1029–1034. https://doi.org/10.1007/s11746-014-2439-9

Sallau, A. B., Mada, S. B., Ibrahim, S., & Ibrahim, U. (2012). Effect of boiling, simmering and blanching on the antinutritional content of *Moringa oleifera* leaves. *International Journal of Food Nutrition and Safety*, *2*(1), 1–6.

Sandanuwan Kirindage, K. G. I., Fernando, I. P. S., Jayasinghe, A. M. K., Han, E. J., Madhawa Dias, M. K. H., Kang, K. P., Moon, S. I., Shin, T. S., Ma, A., & Ahn, G. (2022). *Moringa oleifera* hot water extract protects vero cells from hydrogen peroxide-induced oxidative stress by regulating mitochondria-mediated apoptotic pathway and Nrf2/HO-1 igsnaling. *Foods*, *11*, 420. https://doi.org/10.3390/foods11030420

Sankhalkar, S., & Vernekar, V. (2016). Quantitative and qualitative analysis of phenolic and flavonoid content in *Moringa oleifera* Lam and *Ocimum tenuiflorum* L. *Pharmacognosy Research*, *8*(1), 16–21.

Saucedo-Pompa, S., Torres-Castillo, J. A., Castro-López, C., Rojas, R., Sánchez-Alejo, E. J., Ngangyo-Heya, M., & Martínez-Ávila, G. C. G. (2018). Moringa plants: Bioactive compounds and promising applications in food products. *Food Research International*, *111*, 438–450. https://doi.org/10.1016/j.foodres.2018.05.062

Sengev, A. I., Abu, J. O., & Gernah, D. I. (2013). Effect of *Moringa oleifera* leaf powder supplementation on some quality characteristics of wheat bread. *Food and Nutrition Sciences*, *04*(03), 270–275. https://doi.org/10.4236/fns.2013.43036

Sengupta, K., Mishra, A. T., Rao, M. K., Sarma, K. V., Krishnaraju, A. V., & Trimurtulu, G. (2012). Efficacy and tolerability of a novel herbal formulation for weight management in obese subjects: A randomized double blind placebo controlled clinical study. *Lipids in Health and Disease*, *11*(1), 1–10.

Senthilkumar, A., Karuvantevida, N., Rastrelli, L., Kurup, S. S., & Cheruth, A. J. (2018). Traditional uses, pharmacological efficacy, and phytochemistry of *Moringa peregrina* (Forssk.) Fiori: A review. *Frontiers in Pharmacology*, *9*, 465. https://doi.org/10.3389/fphar.2018.00465

Shah, M. A., Kang, J. B., Kim, M. O., & Koh, P. O. (2022). Chlorogenic acid alleviates the reduction of Akt and Bad phosphorylation and of phospho-Bad and 14-3-3 binding in an animal model of stroke. *Journal of Veterinary Science*, *23*(6), e84. https://doi.org/10.4142/JVS.22200

Shanmugasundaram, S., Krupakar, P., & Shanmugasundaram, S. (2007). Preparation of oligosaccharide bionanoparticles from *Moringa oleifera* lam. *IP Patent 375MAS2003 A*, *6*.

Shanmugavel, G., Prabakaran, K., & George, B. (2018). Evaluation of phytochemical constituents of *Moringa oleifera* (Lam.) leaves collected from Puducherry region, South India. *International Journal of Zoology and Applied Biosciences*, *3*(1), 1–8.

Shervington, L. A., Li, B. S., Shervington, A. A., Alpan, N., Patel, R., Muttakin, U., & Mulla, E. (2018). A comparative HPLC analysis of myricetin, quercetin and kaempferol flavonoids isolated from Gambian and Indian *Moringa oleifera* leaves. *International Journal of Chemistry*, *10*(4), 28–37.

Shi, H., Su, B., Chen, X., & Pian, R. (2020). Solid state fermentation of *Moringa oleifera* leaf meal by mixed strains for the protein enrichment and the improvement of nutritional value. *PeerJ*, *8*. https://doi.org/10.7717/peerj.10358

Shija, A. E., Rumisha, S. F., Oriyo, N. M., Kilima, S. P., & Massaga, J. J. (2019). Effect of *Moringa oleifera* leaf powder supplementation on reducing anemia in children below two years in Kisarawe District, Tanzania. *Food Science & Nutrition*, *7*(8), 2584–2594.

Shimizu, K., & Moriwaki, M. (2020). Composition containing moringa extract and/or pulverized product *U.S. Patent No. 10,869,829*, Issued December 22.

Shiriki, D., Igyor, M. A., & Gernah, D. I. (2015). Nutritional evaluation of complementary food formulations from maize, soybean and peanut fortified with *Moringa oleifera* leaf powder. *Food and Nutrition Sciences*, *06*(05), 494–500. https://doi.org/10.4236/fns.2015.65051

Singh, R. S., Negi, P. S., & Radha, C. (2013). Phenolic composition, antioxidant and antimicrobial activities of free and bound phenolic extracts of *Moringa oleifera* seed flour. *Journal of Functional Foods*, *5*, 1883–1891.

Singla, R. K., Dubey, A. K., Garg, A., Sharma, R. K., Fiorino, M., Ameen, S. M., Haddad, M. A., & Al-Hiary, M. (2019). Natural polyphenols: Chemical classification, definition of classes, subcategories, and structures. *Journal of AOAC International*, *102*(5), 1397–1400. https://doi.org/10.5740/jaoacint.19-0133

Solanki, R. (2004). Composition of eleven herbals for treating cancer. *U.S. Patent 6,780,441* B2, Issued August 21.

Soundararajan, R., Wishart, A. D., Rupasinghe, H. P. V., Arcellana-Panlilio, M., Nelson, C. M., Mayne, M., & Robertson, G. S. (2008). Quercetin 3-glucoside protects neuroblastoma (SH-SY5Y) cells in vitro against oxidative damage by inducing sterol regulatory element-binding protein-2-mediated cholesterol biosynthesis. *Journal of Biological Chemistry*, *283*(4), 2231–2245. https://doi.org/10.1074/jbc.M703583200

Srimiati, M., & Agestika, L. (2022). The substitution of fresh Moringa leaves and Moringa leaves powder on organoleptic and proximate characteristics of pudding. *Amerta Nutrition*, *6*(2), 164–172. https://doi.org/10.20473/amnt.v6i2.2022.164-172

Subramaniam, S., Rosdi, M. H. B., & Kuppusamy, U. R. (2017). Customized cooking methods enhance antioxidant, antiglycemic, and insulin-like properties of *Momordica charantia* and *Moringa oleifera*. *Journal of Food Quality*, *2017*, 9561325. https://doi.org/10.1155/2017/9561325

Sugahara, S., Chiyo, A., Fukuoka, K., Ueda, Y., Tokunaga, Y., Nishida, Y., Kinoshita, H., Matsuda, Y., Igoshi, K., Ono, M., & Yasuda, S. (2018). Unique antioxidant effects of herbal leaf tea and stem tea from *Moringa oleifera* L. Especially on superoxide anion radical generation systems. *Bioscience, Biotechnology and Biochemistry*, *82*(11), 1973–1984. https://doi.org/10.1080/09168451.2018.1495552

Sugiura, K., & Kulkarni, A. (2018). Moringa extract. *Japanese Patent 20180318367* (Application No. 15/770,620), Issued November 8.

Sugunabai, J., Jayaraj, M., Karpagam, T., & Varalakshmi, B. (2014). Antidiabetic efficiency of *Moringa oleifera* and *Solanum nigrum*. *International Journal of Pharmacy and Pharmaceutical Sciences*, *6*(Suppl 1), 40–42.

Sujatha, B. K., & Patel, P. (2017). *Moringa oleifera*—Nature ' s Gold. *Imperial Journal of Interdisciplinary Research*, *3*(5), 1175–1179.

Sultana, B., & Anwar, F. (2008). Flavonols (kaempeferol, quercetin, myricetin) contents of selected fruits, vegetables and medicinal plants. *Food Chemistry*, *108*(3), 879–884. https://doi.org/10.1016/j.foodchem.2007.11.053

Taweerutchana, R., Lumlerdkij, N., Vannasaeng, S., Akarasereenont, P., & Sriwijitkamol, A. (2017). Effect of *Moringa oleifera* leaf capsules on glycemic control in therapy-naïve type 2 diabetes patients: A randomized placebo controlled study. *Evidence-Based Complementary and Alternative Medicine*, *2017*, 6581390. https://doi.org/10.1155/2017/6581390

Thierry, N. N., Léopold, T. N., Didier, M., & Moses, F. M. C. (2013). Effect of pure culture fermentation on biochemical composition of *Moringa oleifera* Lam leaves powders. *Food and Nutrition Sciences*, *04*(08), 851–859. https://doi.org/10.4236/fns.2013.48111

Tshingani, K., Donnen, P., Mukumbi, H., Duez, P., & Dramaix-Wilmet, M. (2017). Impact of *Moringa oleifera* Lam. leaf powder supplementation versus nutritional counseling on the body mass index and immune response of HIV patients on antiretroviral therapy: A single-blind randomized control trial. *BMC Complementary and Alternative Medicine*, *17*(1), 1–13. https://doi.org/10.1186/s12906-017-1920-z

Ummah, A. S., Muhammad, F., Rahmawati, Y. E. N., Ridwan, I., Nurudhin, A., Fauzi, E. R., . . . & Werdiningsih, Y. (2023). Effect of *Moringa oleifera* leaf extracts on depression in rheumatoid arthritis patients. *Malaysian Journal of Medicine and Health Sciences*, *19*(4), 166–170.

Vaknin, Y., & Mishal, A. (2017). The potential of the tropical "miracle tree" *Moringa oleifera* and its desert relative Moringa peregrina as edible seed-oil and protein crops under Mediterranean conditions. *Scientia Horticulturae*, *225*, 431–437. https://doi.org/10.1016/j.scienta.2017.07.039

Vanduchova, A., Anzenbacher, P., & Anzenbacherova, E. (2019). Isothiocyanate from broccoli, sulforaphane, and its properties. *Journal of Medicinal Food*, *22*(2), 121–126. https://doi.org/10.1089/jmf.2018.0024

Vergara-Jimenez, M., Almatrafi, M. M., & Fernandez, M. L. (2017). Bioactive components in *Moringa oleifera* leaves protect against chronic disease. *Antioxidants*, *6*, 91. https://doi.org/10.3390/antiox6040091

Veterini, A. S., Susanti, E., Ardiana, M., Adi, A. C., & Rachmawati, H. (2023). Effects of consuming biscuits made from *Moringa oleifera* leaf on body weight and height of children under five in Bangkalan, Madura Island. *National Nutrition Journal/Media Gizi Indonesia*, *18*(2), 150–156.

Wang, F., Bao, Y., Shen, X., Zengin, G., Lyu, Y., Xiao, J., & Weng, Z. (2019). Niazirin from *Moringa oleifera* Lam. attenuates high glucose-induced oxidative stress through PKCζ/Nox4 pathway. *Phytomedicine*, *86*, 153066. https://doi.org/10.1016/j.phymed.2019.153066

Watanabe, S., Okoshi, H., Yamabe, S., & Shimada, M. (2021). *Moringa oleifera* Lam. in diabetes mellitus: A systematic review and meta-analysis. *Molecules (Basel, Switzerland)*, *26*(12), 3513. https://doi.org/10.3390/molecules26123513

Waterman, C., Rojas-Silva, P., Tumer, T. B., Kuhn, P., Richard, A. J., Wicks, S., . . . & Raskin, I. (2015). Isothiocyanate-rich *Moringa oleifera* extract reduces weight gain, insulin resistance, and hepatic gluconeogenesis in mice. *Molecular Nutrition & Food Research*, *59*(6), 1013–1024.

Wickramasinghe, Y. W. H., Wickramasinghe, I., & Wijesekara, I. (2020). Effect of steam blanching, dehydration temperature & time, on the sensory and nutritional properties of a herbal tea developed from *Moringa oleifera* leaves. *International Journal of Food Science*, *2020*, 5376280. https://doi.org/10.1155/2020/5376280

William, F., Lakshminarayanan, S., & Chegu, H. (1993). Effect of some Indian vegetables on the glucose and insulin response in diabetic subjects. *International Journal of Food Science and Nutrition*, *44*(3), 191–196.

Wuryandari, M. R. E., Atho'illah, M. F., Laili, R. D., Fatmawati, S., Widodo, N., Widjajanto, E., & Rifa'i, M. (2022). Lactobacillus plantarum FNCC 0137 fermented red *Moringa oleifera* exhibits protective effects in mice challenged with Salmonella typhi via TLR3/TLR4 inhibition and down-regulation of proinflammatory cytokines. *Journal of Ayurveda and Integrative Medicine*, *13*(2), 100531. https://doi.org/10.1016/j.jaim.2021.10.003

Xiao, X., Wang, J., Meng, C., Liang, W., Wang, T., Zhou, B., Wang, Y., Luo, X., Gao, L., & Zhang, L. (2020). *Moringa oleifera* Lam and its therapeutic effects in immune disorders. *Frontiers in Pharmacology*, *11*, 566783. https://doi.org/10.3389/fphar.2020.566783

Yadav, H., Gaur, A., & Bansal, S. C. (2022). Effect of *Moringa oleifera* leaf powder supplementation in children with severe acute malnutrition in Gwalior District of Central India: A randomised controlled trial. *Journal of Clinical & Diagnostic Research*, *16*(8), SC09–SC14. https://doi.org/10.7860/JCDR/2022/55126.16746

Yasoob, T. B., Yu, D., Khalid, A. R., Zhang, Z., Zhu, X., Saad, H. M., & Hang, S. (2021). Oral administration of *Moringa oleifera* leaf powder relieves oxidative stress, modulates mucosal immune response and cecal microbiota after exposure to heat stress in New Zealand White rabbits. *Journal of Animal Science and Biotechnology*, *12*, 66. https://doi.org/10.1186/s40104-021-00586-y

Zahran, E. M., Mohamad, S. A., Yahia, R., Badawi, A. M., Sayed, A. M., & Abdelmohsen, U. R. (2022). Anti-otomycotic potential of nanoparticles of *Moringa oleifera* leaf extract: An integrated *in vitro*, *in silico* and phase 0 clinical study. *Food & Function*, *13*(21), 11083–11096.

Zhang, M., Zhao, H., Wang, T., Xie, C., Zhang, D., Huang, Y., Wang, X., & Sheng, J. (2017). Solid-state fermentation of *Moringa oleifera* leaf meal using *Bacillus pumilus* CICC 10440. *Journal of Chemical Technology & Biotechnology*, *92*, 2083–2089. https://doi.org/10.1002/jctb.5203

Zhang, X., Sun, Z., Cai, J., Wang, J., Wang, G., Zhu, Z., & Cao, F. (2020). Effects of dietary fish meal replacement by fermented moringa (*Moringa oleifera* Lam.) leaves on growth performance, nonspecific immunity and disease resistance against *Aeromonas hydrophila* in juvenile gibel carp (Carassius auratus gibelio var. CAS III). *Fish and Shellfish Immunology*, *102*, 430–439. https://doi.org/10.1016/j.fsi.2020.04.051

Zula, A. T., Ayele, D. A., & Egigayhu, W. A. (2021). Proximate composition, antinutritional content, microbial load, and sensory acceptability of noodles formulated from moringa (*Moringa oleifera*) leaf powder and wheat flour blend. *International Journal of Food Science*, *2021*, 6689247. https://doi.org/10.1155/2021/6689247

Index

Note: Page numbers in *italics* indicate a figure, and page numbers in **bold** indicate a table on the corresponding page.

N

O

P

R

S

W

Y

For Product Safety Concerns and Information please contact our EU
representative GPSR@taylorandfrancis.com
Taylor & Francis Verlag GmbH, Kaufingerstraße 24, 80331 München, Germany

www.ingramcontent.com/pod-product-compliance
Lightning Source LLC
Chambersburg PA
CBHW080707220326
41598CB00033B/5334

* 9 7 8 1 0 3 2 4 0 2 8 8 8 *